Percutaneous Absorption

DRUGS AND THE PHARMACEUTICAL SCIENCES

DRUGS AND THE PHARMACEUTICAL SCIENCES

A Series of Textbooks and Monographs

ADDITIONAL VOLUMES IN PREPARATION

Bioadhesive Drug Delivery Systems: Fundamentals, Novel Approaches, and Development, *edited by Edith Mathiowitz, Donald E. Chickering, and Claus-Michael Lehr*

Peptide and Protein Drug Analysis, *edited by Ronald E. Reid*

Protein Formulation and Stability, *edited by Eugene McNally*

New Drug Approval Process: Third Edition, Revised and Expanded, *edited by Richard A. Guarino*

Excipient Toxicity and Safety, *edited by Myra Weiner and Lois Kotkoskie*

Percutaneous Absorption
Drugs—Cosmetics—Mechanisms—Methodology

Third Edition, Revised and Expanded

edited by

Robert L. Bronaugh
Food and Drug Administration
Laurel, Maryland

Howard I. Maibach
University of California
School of Medicine
San Francisco, California

MARCEL DEKKER, INC. NEW YORK · BASEL

Library of Congress Cataloging-in-Publication Data

Percutaneous absorption : drugs—cosmetics—mechanisms—methodology /
edited by Robert L. Bronaugh, Howard I. Maibach.—3rd ed., rev.
and expanded.
    p. cm.—(Drugs and the pharmaceutical sciences : 97)
    Includes bibliographical references and index.
    ISBN 0-8247-1966-2 (alk. paper)
    1. Transdermal medication. 2. Skin absorption. 3. Skin—
Permeability. I. Bronaugh, Robert L. II. Maibach,
Howard I. III. Series: Drugs and the pharmaceutical sciences ; v.
97.
    [DNLM: 1. Skin Absorption. W1 DR893B v.97 1999 / WR 102 P4285
1999]
    RM151.P47   1999
    615′.6—dc21
    DNLM/DLC
    for Library of Congress
                                          99-14994
                                          CIP

This book is printed on acid-free paper.

Headquarters
Marcel Dekker, Inc.
270 Madison Avenue, New York, NY 10016
tel: 212-696-9000; fax: 212-685-4540

Eastern Hemisphere Distribution
Marcel Dekker AG
Hutgasse 4, Postfach 812, CH-4001 Basel, Switzerland
tel: 41-61-261-8482; fax: 41-61-261-8896

World Wide Web
http://www.dekker/com

The publisher offers discounts on this book when ordered in bulk quantities. For more
information, write to Special Sales/Professional Marketing at the headquarters address
above.

Current printing (last digit):
10 9 8 7 6 5 4 3 2 1

PRINTED IN THE UNITED STATES OF AMERICA

Preface to the Third Edition

We have revised and updated *Percutaneous Absorption* to include recent advances in the field. The number of chapters has increased from 37 to 56, with new chapters included on barrier function, mathematical models, skin metabolism, drug delivery, and much more. The revised subtitle *Drugs–Cosmetics–Mechanisms–Methodology* reflects the increased emphasis on skin absorption, both local and systemic, of cosmetic ingredients and products. The third edition contains almost twice the number of pages of the first edition, reflecting the many advances that have occurred since.

We would like to thank the authors for preparing outstanding chapters and Sandra Beberman and Barbara Mathieu of Marcel Dekker, Inc., for their expert editorial guidance.

Robert L. Bronaugh
Howard I. Maibach

Preface to the First Edition

The impetus for this book results from the 25 speakers at the "Percutaneous Absorption Symposium" held April 27, 1983 at the Food and Drug Administration in Washington, D.C. The overflow crowd of more than 500 persons at this meeting demonstrates the increasing interest in skin permeation, which is due primarily to the recognition of the skin as an important portal of entry of chemicals into the body. The penetration of substances is sometimes desirable and promoted, as in the case of drug delivery, for either a local or systemic effect. At other times, interest in percutaneous absorption stems from a concern about the dermal exposure to potentially toxic agents in topical drug and cosmetic products or agents encountered in the workplace or environment.

This book is our effort to provide a forum for a more complete exposition of the area than has previously been available. Its multiauthored approach supplements the recent excellent texts by Brian Barry (*Dermatological Formulations: Percutaneous Absorption*, Marcel Dekker) and Hans Schaefer, Achim Zesch and Gunter Stuttgen (*Skin Permeability*, Springer-Verlag).

The chapters are divided into three general areas: Mechanisms of Absorption, Methodology, and Drug Delivery. Of course, some chapters could have been placed in more than one of these areas, so the classification is somewhat arbitrary.

The Mechanisms of Absorption section contains chapters dealing with studies on the various factors that affect percutaneous absorption. Pathways of penetration, mathematical models, metabolism, binding, and pharmacokinetics are examples of topics discussed.

In the Methodology section, the various in vivo and in vitro techniques used to measure skin penetration are covered. Chapters include other factors influencing the methodology such as: animal models, volatility of test compound, multiple dosing, artificial membranes, and blood flow studies.

Procedures are discussed for use in the transdermal delivery of drugs. Topics include the effects of penetration enhancers on absorption, optimizing absorption, and the topical delivery of drugs to muscle tissue.

It is hoped that the book will be useful as a source of ideas and references to those engaged in percutaneous absorption studies. For those just planning to begin, the Methodology section, in particular, should be useful.

Robert L. Bronaugh
Howard I. Maibach

Contents

Contents

Contributors

John I. Ademola, Ph.D.   Department of Dermatology, University of California School of Medicine, San Francisco, California

Yuri G. Anissimov, Ph.D.   Department of Medicine, University of Queensland at Princess Alexandra Hospital, Brisbane, Queensland, Australia

Brian W. Barry, Ph.D., D.Sc., F.R.Pharm.S., C.Chem, F.R.S.C.   School of Pharmacy, University of Bradford, Bradford, West Yorkshire, England

Saqib J. Bashir, B.Sc., M.B., Ch.B.   Department of Dermatology, University of California School of Medicine, San Francisco, California

Bret Berner, Ph.D.   Development, Drug Delivery Systems, Cygnus, Redwood City, California

Kenneth T. Bogen, M.A., M.P.H., Dr.P.H.   Health and Ecological Assessment Division, Lawrence Livermore National Laboratory, University of California, Livermore, California

Keith R. Brain, B.Pharm., Ph.D.   An-eX Analytical Services, Ltd., Cardiff, Wales

Robert L. Bronaugh, Ph.D.   Skin Absorption and Metabolism Section, Office of Cosmetics and Colors, Food and Drug Administration, Laurel, Maryland

Daniel Bucks, Ph.D.   University of California School of Medicine,
San Francisco, and Drug Transport Department, Penederm Inc.,
Foster City, California

Julie Christoffel   Department of Dermatology, University of California
School of Medicine, San Francisco, California

Richard A. Corley, Ph.D.   Molecular Biosciences, Pacific Northwest
National Laboratory, Richland, Washington

Rebecca Cox, M.D.   Department of Dermatology, University of
California School of Medicine, San Francisco, California

Steven M. Dinh, Sc.D.   Research and Development, Lavipharm Inc.,
Hightstown, New Jersey

Piet De Doncker, Ph.D.   International Clinical Research and
Development, Janssen Research Foundation, Beerse, Belgium

William E. Dressler, Ph.D.   Toxicology and Regulatory Affairs, Bristol-
Myers Squibb Worldwide Beauty Care, Stamford, Connecticut

Didier Dupuis, Ph.D.   Department of Biology, Laboratoires de
Recherche Fondamentale, L'Oréal, Aulnay sous Bois, France

Peter M. Elias, M.D.   Dermatology and Medical Services, Veterans
Affairs Medical Center, and Departments of Dermatology and Medicine,
University of California School of Medicine, San Francisco, California

Jerome S. Elkins, B.S.Chem.   Pharmaceuticals Department, Food and
Drug Administration, Dallas, Texas

Kenneth R. Feingold, M.D.   Dermatology and Medical Services,
Veterans Affairs Medical Center, and Professor of Dermatology and
Medicine, University of California School of Medicine,
San Francisco, California

James H. Forsell, Ph.D.   Northern California Transplant Bank, San
Rafael, California

Stephen D. Gettings, Ph.D., J.D.   Environmental and Regulatory Affairs,
Avon Products, Inc., Suffern, New York

Tatiana E. Gogoleva   Department of Dermatology, University of California School of Medicine, San Francisco, California

Richard A. Gonsalvez, M.Sc.   Department of Medicine, University of Queensland at Princess Alexandra Hospital, Brisbane, Queensland, Australia

Richard H. Guy, Ph.D.   Centre Interuniversitaire de Recherche et d'Enseignement, Archamps, France, and Laboratoire de Pharmacie Galénique, University of Geneva, Geneva, Switzerland

John M. Haigh, Ph.D.   School of Pharmaceutical Sciences, Rhodes University, Grahamstown, South Africa

Tracy Hartway   Department of Dermatology, University of California School of Medicine, San Francisco, California

Philip G. Hewitt   Department of Dermatology, University of California School of Medicine, San Francisco, California, and Department of Toxicology, E. Merck, Darmstadt, Germany

Harolyn L. Hood, M.S.*   Skin Absorption and Metabolism Section, Office of Cosmetics and Colors, Food and Drug Administration, Laurel, Maryland

Xiaoying Hui, M.D., M.S.   Department of Dermatology, University of California School of Medicine, San Francisco, California

Yogeshvar N. Kalia, Ph.D.   Centre Interuniversitaire de Recherche et d'Enseignement, Archamps, France, and Laboratoire de Pharmacie Galénique, University of Geneva, Geneva, Switzerland

Garrett A. Keating, Ph.D.   Earth and Environmental Sciences Department, Lawrence Livermore National Laboratory, University of California, Livermore, California

Joseph Kost, Ph.D.   Department of Chemical Engineering, Ben-Gurion University, Beer-Sheva, Israel

---

*Present affiliation: Bristol-Myers Squibb Worldwide Beauty Care, Stamford, Connecticut.

Margaret E. K. Kraeling, M.S.   Skin Absorption and Metabolism Section, Office of Cosmetics and Colors, Food and Drug Administration, Laurel, Maryland

Robert Langer, Sc.D.   Department of Chemical Engineering, Massachusetts Institute of Technology, Cambridge, Massachusetts

Andrea C. Lauer, Ph.D.   Analytical Development, Boehringer Ingelheim–Roxane Laboratories, Columbus, Ohio

Jue-Chen Liu, Ph.D.   Product Development, Johnson & Johnson Consumer Products Worldwide, Skillman, New Jersey

Puchun Liu, Ph.D.   Pharmaceutical Research and Development Department, Novartis Pharmaceuticals Corporation, Suffern, New York

Claire Lotte, Ph.D.   Department of Biology, Laboratoires de Recherche Fondamentale, L'Oréal, Aulnay sous Bois, France

Carolyn H. Lund, R.N., M.S., F.A.A.N.   Intensive Care Unit, Children's Hospital, Oakland, California

Philip S. Magee, Ph.D.   Department of Dermatology, University of California School of Medicine, San Francisco, California

Howard I. Maibach, M.D.   Department of Dermatology, University of California School of Medicine, San Francisco, California

Vivien Mak, Ph.D.   Research and Development, Cellegy Pharmaceuticals Inc., San Carlos, California

Joseph L. Melendres, M.D.   Department of Dermatology, University of California School of Medicine, San Francisco, California

Samir Mitragotri   Department of Chemical Engineering, Massachusetts Institute of Technology, Cambridge, Massachusetts

Hamid R. Moghimi, Pharm.D., Ph.D.   Department of Pharmaceutics, School of Pharmacy, Shaheed Beheshti University of Medical Sciences, Tehran, Iran

Aarti Naik, Ph.D. Centre Interuniversitaire de Recherche et d'Enseignement, Archamps, France, and Laboratoire de Pharmacie Galénique, University of Geneva, Geneva, Switzerland

Avinash Nangia, Ph.D. Oral Product Development Department, ALZA Corporation, Palo Alto, California

Reinhard H. H. Neubert, Prof. Dr. Institute for Pharmaceutics and Biopharmaceutics, Martin Luther University, Halle/Saale, Germany

Lourdes B. Nonato, Ph.D. Department of Biopharmaceutical Sciences, University of California, San Francisco, California

George A. Omura, M.D. Clinical Development, BioCryst Pharmaceuticals, Inc., Birmingham, Alabama

Fabrice Pirot, Pharm.D., Ph.D. Department of Pharmacy, Institut des Sciences Pharmaceutiques et Biologiques, University of Claude Bernard-Lyon, Lyon, France

Nicholas Poblete Department of Dermatology, University of California School of Medicine, San Francisco, California

Torka S. Poet, Ph.D. Molecular Biosciences, Pacific Northwest National Laboratory, Richland, Washington

William John Pugh, B.Pharm., Ph.D., M.R.Pharm.S. Welsh School of Pharmacy, Cardiff University, Cardiff, Wales

Danyi Quan, Ph.D.* Department of Dermatology, University of California School of Medicine, San Francisco, California

Jim E. Riviere, D.V.M, Ph.D. College of Veterinary Medicine, North Carolina State University, Raleigh, North Carolina

Michael S. Roberts, Ph.D. University of Queensland at Princess Alexandra Hospital, Brisbane, Queensland, Australia

André Rougier, Ph.D. Laboratoire Pharmaceutique, La Roche-Posay, Courbevoie, France

---

Present affiliation: Theractech Corporation, Salt Lake City, Utah.

Ulrike Schmalfuß, B.Pharm., Ph.D.   Scientific Development, permamed laboratories ltd., Therwil/Basel, Switzerland

Fabian P. Schwarb, Ph.D.   Department of Dermatology and Institute of Hospital–Pharmacy, University Hospital, Basel, Switzerland

Lena Sedik   Department of Dermatology, University of California School of Medicine, San Francisco, California

Steffany Serranzana, B.A.*   Department of Dermatology, University of California School of Medicine, San Francisco, California

Vinod P. Shah, Ph.D.   Office of Pharmaceutical Science, Center for Drug Evaluation and Research, Food and Drug Administration, Rockville, Maryland

J. Zev Shainhouse, M.D., B.Sc., F.R.C.P.(C)   Dimethaid Research Inc., Markham, Ontario, Canada

Parminder Singh, Ph.D.   Transdermal Pharmaceutical Development, Novartis Pharmaceuticals Corporation, Suffern, New York

Eric W. Smith, Ph.D.   Raabe College of Pharmacy, Ohio Northern University, Ada, Ohio

Ying Sun, Ph.D.   Topical Formulations Technology Resource Center, Johnson & Johnson Consumer Products Worldwide, Skillman, New Jersey

Christian Surber, Ph.D.   Department of Dermatology and Institute of Hospital–Pharmacy, University Hospital, Basel, Switzerland

Hanafi Tanojo, Ph.D.   Department of Dermatology, University of California School of Medicine, San Francisco, California

Carl Thornfeldt, M.D.   Cellegy Pharmaceuticals Inc., Foster City, California

Karla D. Thrall, Ph.D.   Molecular Biosciences, Pacific Northwest National Laboratory, Richland, Washington

*Present affiliation: Penederm Inc., Foster City, California.

Ethel Tur, M.D.   Department of Dermatology, Sourasky Medical Center, Tel Aviv University, Tel Aviv, Israel

Tacey X. Viegas, Ph.D.   Pharmaceutical Development, BioCryst Pharmaceuticals, Inc., Birmingham, Alabama

John S. Vogel, Ph.D.   Center for Accelerator Mass Spectrometry, Lawrence Livermore National Laboratory, University of California, Livermore, California

Gerald M. Walsh   BioCryst Pharmaceuticals Inc., Birmingham, Alabama

Kenneth A. Walters, Ph.D.   An-eX Analytical Services Ltd., Cardiff, Wales

Jonas C. T. Wang, Ph.D.   Research and Technology and Research Development and Engineering Division, Johnson & Johnson Consumer Products Worldwide, Skillman, New Jersey

Ronald C. Wester, Ph.D.   Department of Dermatology, University of California School of Medicine, San Francisco, California

Adrian C. Williams, Ph.D.   School of Pharmacy, University of Bradford, Bradford, West Yorkshire, England

Roger L. Williams, M.D.   Office of Pharmaceutical Science, Center for Drug Evaluation and Research, Food and Drug Administration, Rockville, Maryland

Jeffrey J. Yourick, Ph.D., DABT   Skin Absorption and Metabolism Section, Office of Cosmetics and Colors, Food and Drug Administration, Laurel, Maryland

Hongbo Zhai, M.D.   Department of Dermatology, University of California School of Medicine, San Francisco, California

I
MECHANISMS OF ABSORPTION

1

Mathematical Models in Percutaneous Absorption

Michael S. Roberts, Yuri G. Anissimov and Richard A. Gonsalvez
University of Queensland at Princess Alexandra Hospital, Brisbane, Queensland, Australia

A number of mathematical models have been used to describe percutaneous absorption kinetics. In general, most of these models have used either diffusion-based or compartmental equations. The object of any mathematical model is to a) be able to represent the processes associated with absorption accurately, b) be able to describe/summarize experimental data with parametric equations or moments, and c) predict kinetics under varying conditions. However, in describing the processes involved, some developed models often suffer from being of too complex a form to be practically useful. In this chapter, we attempt to approach the issue of mathematical modeling in percutaneous absorption from four perspectives. These are to a) describe simple practical models, b) provide an overview of the more complex models, c) summarize some of the more important/useful models used to date, and d) examine some practical applications of the models.

The range of processes involved in percutaneous absorption and considered in developing the mathematical models in this chapter is shown in Fig. 1. We initially address in vitro skin diffusion models and consider a) constant donor concentration and receptor conditions, b) the corresponding flux, donor, skin, and receptor amount–time profiles for solutions, and c) amount– and flux–time profiles when the donor phase is removed. More complex issues, such as finite-volume donor phase, finite-volume receptor phase, the presence of an efflux rate constant at the membrane–receptor interphase, and two-layer diffusion, are then considered. We then look at

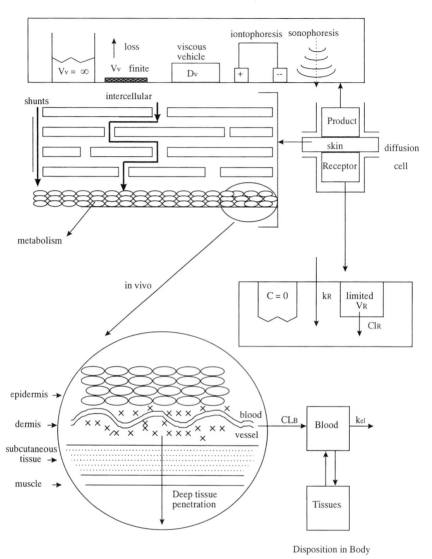

Figure 1 Diagrammatic overview of percutaneous processes associated with mathematical models.

specific models and issues concerned with a) release from topical products, b) use of compartmental models as alternatives to diffusion models, c) concentration-dependent absorption, d) modeling of skin metabolism, e) role of solute–skin–vehicle interactions, f) effects of vehicle loss, g) shunt transport, and h) in vivo diffusion, compartmental, physiological, and deconvolution models. We conclude by examining topics such as a) deep tissue penetration, b) pharmacodynamics, c) iontophoresis, d) sonophoresis, and e) pitfalls in modeling.

Each model is described in diagrammatic and equation form. Given that the analytical solution to most models is in the form of series, often involving solutions to transcendental equations, we have emphasized the Laplace domain and steady-state solutions. Most nonlinear regression programs such as MULTI FILT, MINIM, and SCIENTIST enable analysis of concentration–time data using numerical inversion of Laplace domain solutions and avoid some of the computational difficulties associated with series solutions, especially those involving solving transcendental equations. The steady-state solutions describing the linear portion of a cumulative amount versus time profile for a constant donor concentration are of great practical use, being described by a linear equation with lag time and steady-state flux as the intercept and the slope, respectively. In order to make the equations in this chapter as useable as possible, each equation has been presented in dimensioned form (all variables have their normal dimensions). Simulations and nonlinear regressions presented in this review were undertaken using either SCIENTIST 2.01 or MINIM 3.09.

I. IN VITRO SKIN DIFFUSION MODELS IN PERCUTANEOUS ABSORPTION

We consider first mathematical models associated with solute penetration through excised skin. The simplest of these models is when a well-stirred vehicle of infinite volume is applied to the stratum corneum and the solute passes into a receptor sink (Fig. 2A). Increasing complexity of the models arises when the vehicle volume is finite, when the receptor is no longer a sink (Fig. 2B), and when the vehicle cannot be considered well stirred. We examine each of these models in terms of expressions for amount penetrating, flux, and, where possible, summary parameters such as mean absorption time, normalized variance, peak time for flux, and maximum flux.

A. In Vitro Skin Permeability Studies with a Constant Donor Concentration and Sink Receptor Conditions

Most in vitro skin permeability studies are carried out assuming that both a) the concentration of solute in a vehicle applied to the skin and b) the sink

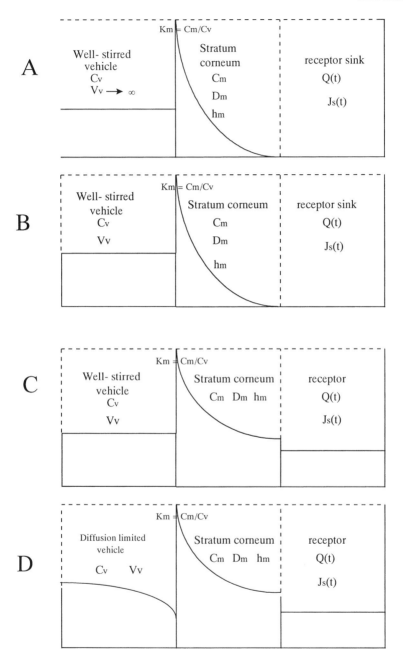

conditions provided by the receptor remain constant over the period of the study. Significant depletion of solute in the donor vehicle or an inadequate receptor sink requires more complex modeling, as discussed later. If transport through the stratum corneum is rate limiting, the steady-state approximation of amount of solute absorbed (Q) when concentration C_v is applied to an area of application (A) for an exposure time (T) is given by Eq. (1) (Roberts & Walters, 1998):

$$Q = k_p A C_v (T - \text{lag}) \tag{1}$$

where k_p is the permeability coefficient (units: distance/time) of the stratum corneum. In reality, absorption does not cease after removal of the vehicle, so that the overall absorption is slightly over $k_p A C_v T$. The permeability coefficient in Eq. (1) is normally defined in terms of the dimensionless partition coefficient between the stratum corneum and vehicle (K_m) and D_m, the diffusivity of a solute in stratum corneum over a diffusion path length h_m:

$$k_p = \frac{K_m D_m}{h_m} \tag{2}$$

K_m is defined as the ratio of solute concentrations in the stratum corneum (C_m) and vehicle (C_v) under equilibrium, that is, $K_m = C_m/C_v$.

In practice, the permeability coefficient k_p is a composite parameter (Scheuplein, 1978). When solute transport occurs via both a lipid pathway of permeability coefficient $k_{p.lipid}$ and a polar pathway of permeability coefficient $k_{p.polar}$ and an aqueous boundary layer of the epidermis provides a rate-limiting permeability coefficient $k_{p.aqueous}$, k_p is more properly expressed as

$$k_p = \left(\frac{1}{k_{p.lipid} + k_{p.polar}} + \frac{1}{k_{p.aqueous}} \right)^{-1} \tag{3}$$

As discussed by Roberts and Walters (1998), for most solutes, $k_p \approx k_{p.lipid}$.

Absorption is more commonly expressed in terms of the steady-state flux J_{ss} or the absorption rate per unit area:

Figure 2 In vitro skin models of transport. (A) Well-stirred vehicle containing solute concentration C_v in volume V_v (where $V_v = \infty$) adjacent to assumed homogeneous stratum corneum with solute concentration C_m at distance x from applied vehicle. Solute moves with a diffusion coefficient D_m over an effective pathlength h_m and penetrates into a receptor sink to give an amount penetrated $Q(t)$ in time t or flux $J(t)$. (B) As for (A) but with V_v finite. (C) As for (B) but the receptor is not a sink. (D) As for (B) but the vehicle is not well stirred.

$$J_{ss} = \frac{Q}{A(T - \text{lag})} = k_p C_v \tag{4}$$

Equations (1) and (4) are the simplified forms of a more complex expression based on the solution of the diffusion equation for transport of solute in the skin:

$$\frac{\partial C_m}{\partial t} = D_m \frac{\partial^2 C_m}{\partial x^2} \tag{5}$$

the initial condition:

$$C_m(x,0) = 0 \tag{6}$$

and boundary conditions:

$$C_m(0,t) = K_m C_v \tag{7}$$

$$C_m(h_m,t) = 0 \tag{8}$$

Traditionally Eq. (5) is solved in terms of the amount of solute $Q(t)$ exiting from the membrane in time t and expressed as a series solution (Crank, 1975):

$$Q(t) = -D_m A \int_0^t \frac{\partial C_m}{\partial x}\bigg|_{x=h_m} dt$$

$$= K_m A C_v h_m \left[\frac{t}{t_d} - \frac{1}{6} - \frac{2}{\pi^2} \sum_{n=1}^{\infty} \frac{(-1)^n}{n^2} \exp(-\frac{t}{t_d} \pi^2 n^2) \right] \tag{9}$$

where the diffusion time is given by:

$$t_d = \frac{h_m^2}{D_m} \tag{10}$$

It should be noted that as the exponent of a very large negative number approaches zero, the summation term in Eq. (9) can be ignored at long times, so that Eq. (9) reduces to the form of Eq. (1):

$$Q(t) = K_m A C_v h_m \left(\frac{D_m t}{h_m^2} - \frac{1}{6} \right) = k_p A C_v \cdot \left(t - \frac{h_m^2}{6D_m} \right)$$

$$= k_p A C_v (t - \text{lag}) \tag{11}$$

where lag is given by

$$\text{lag} = h_m^2/(6D_m) = \frac{t_d}{6} \tag{12}$$

Given the advent of fast inverse Laplace transforms (FILT; Yano et al.,

1989a, 1989b; Purves, 1995) with nonlinear regression modeling, we would normally analyze cumulative amount versus time data numerically inverting from the Laplace domain using Eq. (13) where s is the Laplace variable:

$$\hat{Q}(s) = -D_m A \frac{1}{s} \frac{\partial \hat{C}_m}{\partial x}\bigg|_{x=h_m} = \frac{k_p A C_v}{s^2} \frac{\sqrt{st_d}}{\sinh(\sqrt{st_d})} \tag{13}$$

Figure 3 shows a plot of the cumulative amount penetrated for the diffusion [Eq. (13)] (curve 2) and steady-state [Eq. (11)] (curve 1) models versus time. Equation (9) or (13) can be used to analyze in vitro experimental data by nonlinear regression as shown in Fig. 4.

Figure 3 (curve 3) also shows the amount of solute taken up by the stratum corneum with time. These profiles are of interest for those solutes that may be targeted for retention in this tissue, such as sunscreens, or that

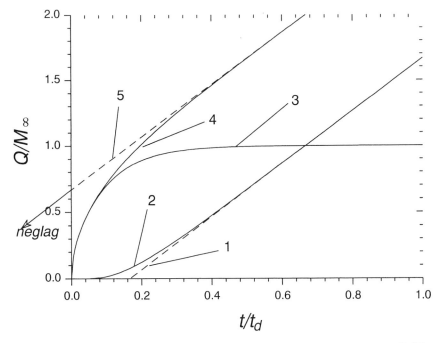

Figure 3 Normalized cumulative amount of solute penetrating Q/M_∞ (curve 2) [Eq. (13)]; taken up by the stratum corneum (curve 3) [Eq. (15)]; and leaving vehicle (curve 4) [Eq. (18)] with normalized time. Curves 1 and 5 represent steady-state approximations of the cumulative amount penetrating the stratum corneum [Eq. (11)] and leaving the vehicle [Eq. (17)] with normalized time.

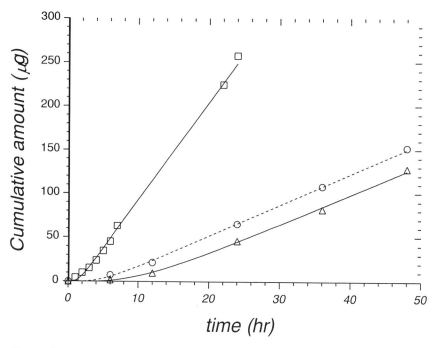

Figure 4 Nonlinear regressions of cumulative amount penetrating human epidermis with time using Eq. (13) and a weighting of $1/y_{obs}$. Data correspond to triethanolamine salicylate ($\square$, $t_d = 9.8$ h, $J_{ss} = 11.1$ µg/h), diclofenac skin 1 ($\bigcirc$, $t_d = 32.7$ h, $J_{ss} = 3.5$ µg/h), and diclofenac skin 2 ($\triangle$, $t_d = 68.0$ h, $J_{ss} = 3.8$ µg/h).

may be sequestered in this tissue, such as steroids. The time domain and Laplace domain solutions for the amount of solute $M(t)$ taken up into an assumed homogeneous stratum corneum with time are:

$$M(t) = M_\infty \left\{ 1 - \frac{8}{\pi^2} \sum_{n=0}^{\infty} \frac{1}{(2n + 1)^2} \exp\left[-\frac{t}{t_d} \pi^2 (2n + 1)^2 \right] \right\} \tag{14}$$

$$\hat{M}(s) = M_\infty \frac{2}{s} \frac{\cosh(\sqrt{st_d}) - 1}{\sqrt{st_d} \sinh(\sqrt{st_d})} \tag{15}$$

where M_∞ is the amount of solute in the skin at steady state and is given by $K_m C_v h_m A/2$ when a linear concentration gradient is assumed.

The summation of $Q(t)$ and $M(t)$ yields the expression for the amount that leaves the vehicle, $Q_{in}(t)$ (the profile shown in Fig. 3, curve 4):

$$Q_{in}(t) = K_m A C_v h_m \left[\frac{t}{t_d} + \frac{1}{3} - \frac{2}{\pi^2} \sum_{n=1}^{\infty} \frac{1}{n^2} \exp\left(-\frac{t}{t_d} \pi^2 n^2\right) \right] \tag{16}$$

When $t \to \infty$, Eq. (16) reduces to:

$$Q_{in}(t) = K_m A C_v h_m \left(\frac{t}{t_d} + \frac{1}{3} \right) = k_p A C_v \left(t + \frac{t_d}{3} \right)$$

$$= k_p A C_v (t + \text{neglag}) \tag{17}$$

Hence, the linear portion of $Q_{in}(t)$ versus t has a slope of $k_p A C_v$ and intercepts on the negative side of the time axis at a point of neglag = $t_d/3$ = $h_m^2/(3D_m)$ (Fig. 3, curve 5).

The corresponding Laplace domain expression for $Q_{in}(t)$ is:

$$\hat{Q}_{in}(s) = \frac{k_p A C_v}{s^2} \sqrt{st_d} \, \text{cotanh}(\sqrt{st_d}) \tag{18}$$

The absorption rate or flux of solutes in the period before steady state is important for many agents topically applied for local effects and in the toxicology of agents applied to the skin. The flux of solutes exiting membrane per unit area of membrane, $J_s(t)$, is defined by $J_s(t) = (1/A) \, \partial Q/\partial t$ or $\hat{J}_s(s) = s\hat{Q}(s)/A$ in the Laplace domain. Using Eqs. (9) and (13) we find therefore

$$J_s(t) = -D_m \left. \frac{\partial C_m}{\partial x} \right|_{x=h_m} = k_p C_v \left[1 + 2 \sum_{n=1}^{\infty} (-1)^n \cdot \exp\left(-\frac{t}{t_d} \pi^2 n^2\right) \right] \tag{19}$$

$$\hat{J}_s(s) = \frac{k_p C_v}{s} \frac{\sqrt{st_d}}{\sinh(\sqrt{st_d})} \tag{20}$$

The corresponding equation for the flux of solute from the vehicle into the membrane, $J_{in}(t)$, is

$$J_{in}(t) = k_p C_v \left[1 + 2 \sum_{n=1}^{\infty} \exp\left(-\frac{t}{t_d} \pi^2 n^2\right) \right] \tag{21}$$

$$\hat{J}_{in}(s) = \frac{k_p C_v}{s} \sqrt{st_d} \, \text{cotanh}(\sqrt{st_d}) \tag{22}$$

Figure 5 shows the flux profiles for solutes leaving the membrane and vehicle, respectively.

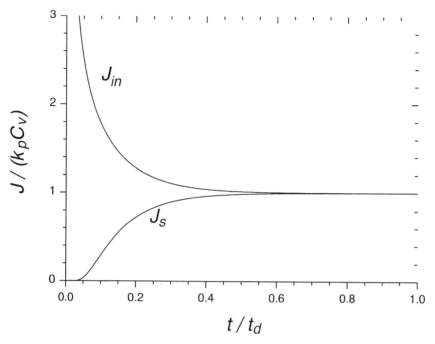

Figure 5 Normalized flux $J/(k_pC_v)$ against normalized time (t/t_d) for flux of solutes penetrating the stratum corneum $[J_s$, Eq. (20)] and entering the stratum corneum $[J_{in}$, Eq. (22)].

B. Amount and Flux–Time Profiles on Removing the Donor Phase After Reaching the Steady State for Conditions Described Section I,A

We now consider the amount– and flux–time profiles for the specific case in which the donor phase has been removed after a steady state has been reached. This equates to a number of practical cases of interest such as patch removal, sunscreen, and other products being washed off and removal of toxins from the skin, when the assumption can be made that there has not been a significant (>10%) depletion in the concentration of solute at the surface. The amount absorbed into a systemic circulation across the skin from the time the dosage form is removed is given by

$$Q(t) = M_\infty \left\{ 1 - \frac{4}{\pi^3} \sum_{n=0}^{\infty} \frac{(-1)^n}{\left(n + \frac{1}{2}\right)^3} \exp\left[-\frac{t}{t_d} \pi^2 \left(n + \frac{1}{2}\right)^2 \right] \right\} \qquad (23)$$

where $M_\infty = K_m C_v A h_m/2 = k_p A C_v t_d/2$ is amount of solute present in the skin before removal of the vehicle. The Laplace domain equivalent of this expression is

$$\hat{Q}(s) = \frac{M_\infty}{s^2} \frac{2}{t_d} \left(1 - \frac{1}{\cosh(\sqrt{st_d})} \right) \tag{24}$$

The corresponding equations for flux are

$$J_s(t) = k_p C_v \frac{2}{\pi} \sum_{n=0}^{\infty} \frac{(-1)^n}{n + \frac{1}{2}} \exp\left[-\frac{t}{t_d} \pi^2 \left(n + \frac{1}{2} \right)^2 \right] \tag{25}$$

$$\hat{J}_s(s) = \frac{k_p C_v}{s} \left(1 - \frac{1}{\cosh(\sqrt{st_d})} \right) \tag{26}$$

Figure 6 shows the amount– and flux–time profiles associated with donor-phase removal.

The mean time for absorption of solute from the skin in this case is given by:

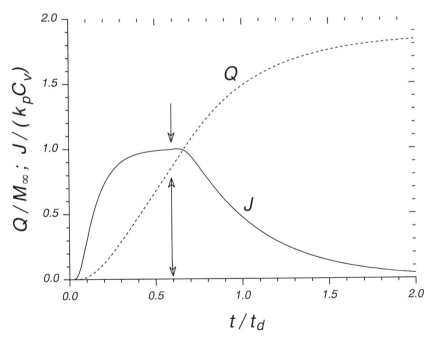

Figure 6 Changes in normalized cumulative amounts penetrating (Q/M_∞) and flux [$J/(k_p C_v)$] when the vehicle is removed (as indicated by arrows) at a specific normalized time (t/t_d) after application.

$$\text{MAT} = \frac{\int_0^\infty J_s(t) \, t \, dt}{\int_0^\infty J_s(t) \, dt} = -\lim_{s \to 0} \frac{d}{ds} \ln(\hat{J}_s) = \frac{5h_m^2}{12D_m} = \frac{5}{12} t_d \tag{27}$$

and the amount absorbed after infinite time is the amount in the skin before the vehicle is removed, which is equal to M_∞.

Given the complexity of solute–distance–time profiles in membranes, the expressions for these profiles are not reproduced here. However, it should be emphasized that these may be important as illustrated in the use of in vivo ATR-FTIR to examine the kinetics of solute uptake into human stratum corneum in vivo (Pirot et al., 1997).

C. In Vitro Permeability Studies with a Constant Donor Concentration and Finite Receptor Volume

In most in vitro studies, it is assumed that sink conditions apply in the receptor phase. However, the receptor phase is a finite volume, and solute accumulation may be possible if there is an inadequate removal rate of the solute penetrating through. Siddiqui et al. (1989) related the steady-state flux of steroids through human epidermis to the differences in concentrations between donor (C_v) and receptor (C_{ss}) concentrations. In the present notation, this equation is:

$$J_{ss} = k_p \left(C_v - C_{ss} \frac{K_R}{K_m} \right) \tag{28}$$

where K_R is partitioning coefficient between membrane and receptor ($K_R = C_m/C_R$) and K_m is partitioning coefficient between the membrane and vehicle ($K_m = C_m/C_v$). Siddiqui et al. (1989) assumed that $K_R/K_m = 1$.

Implicit in the underlying boundary conditions for the receptor phase is a constant clearance of solute Cl_R, being by repeated sampling or use of a flow through cell. If such a clearance was absent, C_{ss} would continually increase and approach $C_v K_m/K_R$. The value of C_{ss} is defined by the relative magnitudes of the clearance (Cl_R) and $k_p A K_R/K_m$:

$$C_{ss} = \frac{k_p A C_v}{Cl_R + k_p A K_R/K_m} \tag{29}$$

Siddiqui et al. (1989) also applied this equation and the dermal clearance of solutes (Cl_R) to predict the steady-state epidermal concentrations of solutes C_{ss}. Roberts (1991) considered the limits of large $k_p A$ as exists for phenols absorption and low $k_p A$ as exists for steroid absorption. He suggested that, as a consequence, the C_{ss} for phenols would eventually approach the donor concentrations used ($C_v K_m/K_R$). In contrast, the steroid C_{ss} approaches $k_p A C_v/Cl_R$.

The derivation of the full equation from which steady-state equations (28) and (29) arise needs to take into account a finite receptor or epidermis volume. The boundary condition at $x = h_m$ in this case is $C_m(h_m, t) = K_R C_R(t)$, together with:

$$V_R \frac{dC_R}{dt} = -AD_m \left. \frac{\partial C_m}{\partial x} \right|_{x=h_m} - Cl_R C_R \tag{30}$$

where Cl_R is the clearance (ml/min) of solution containing solute from the receptor phase, V_R is the volume of the receptor, and C_R is the concentration in the receptor. Using this boundary condition together with boundary condition (7) yields for the amount of solute that penetrated the skin into the receptor (= amount in receptor + amount cleared from receptor) and for the flux of solute into the receptor:

$$\hat{Q}(s) = \frac{k_p A C_v}{s^2} \frac{\sqrt{st_d}}{\sinh(\sqrt{st_d}) + \dfrac{\sqrt{st_d}}{st_d \lambda_2 + \lambda_1} \cosh(\sqrt{st_d})} \tag{31}$$

$$\hat{J}_s(s) = \frac{k_p C_v}{s} \frac{\sqrt{st_d}}{\sinh(\sqrt{st_d}) + \dfrac{\sqrt{st_d}}{st_d \lambda_2 + \lambda_1} \cosh(\sqrt{st_d})} \tag{32}$$

where $\lambda_1 = Cl_R K_m/(k_p A K_R)$ and $\lambda_2 = V_R/(K_R V_m)$.

Figure 7 shows the effect of receptor volume (as defined by λ_2) and clearance of solution from the receptor phase (as defined by λ_1) on $J_s(t)$–time profile.

The steady-state approximation of Eq. (31) is

$$Q(t) \approx A J_{ss}(t - \text{lag}) \tag{33}$$

where

$$J_{ss} = k_p C_v \frac{Cl_R}{Cl_R + k_p A K_R/K_m} = \frac{k_p C_v}{1 + 1/\lambda_1} \tag{34}$$

and

$$\text{lag} = \frac{t_d}{6} \left[1 + \frac{2\lambda_1 - 6\lambda_2}{\lambda_1(\lambda_1 + 1)} \right] \tag{35}$$

We note that if Eq. (29) is substituted into Eq. (28), the expression for J_{ss} is identical to Eq. (34). We also note that when $Cl_R \to \infty$ (infinite sink), J_{ss} and lag reduce to Eqs. (4) and (12), respectively.

The corresponding solution for the receptor/epidermal concentration with the preceding boundary conditions is

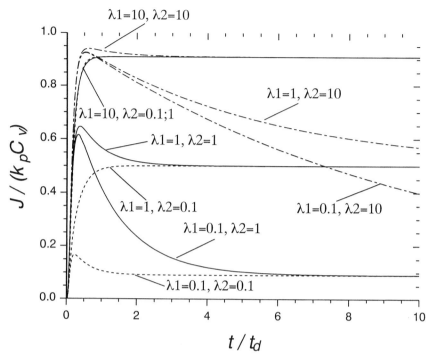

Figure 7 Normalized flux $[J/(k_pC_v)]$ versus normalized time (t/t_d) for a finite receptor volume and limited clearance [Eq. (32)]; $\lambda_1 = Cl_RK_m/(k_pAK_R)$; $\lambda_2 = V_R/(K_RV_m)$.

$$\hat{C}_R(s) = \frac{k_pAC_v}{s} \frac{\sqrt{st_d}}{(V_Rs + Cl_R)\left[\sinh(\sqrt{st_d}) + \dfrac{\sqrt{st_d}}{st_d\lambda_2 + \lambda_1}\cosh(\sqrt{st_d})\right]}$$

(36)

At long times $(t \to \infty)$, C_R is defined by Eq. (29).

Parry et al. (1990) has described a percutaneous absorption model in which both the donor and receptor compartments for an in vitro membrane study were well stirred and finite. Boundary conditions similar in form to that defined by Eqs. (42) and (30), but with $Cl_R = 0$, were used to describe the disappearance of solute from the donor chamber into the membrane and efflux of solute from the membrane into the receptor chamber. The resultant expression included a complex function requiring the solution of transcendental equations. It should be emphasized that this model differs from others described in this section in that it does not have a clearance term to account for sampling.

D. In Vitro Permeability Studies with a Constant Donor Concentration or Defined Input Flux and Finite Clearance of Solute from the Epidermis

The importance of receptor conditions in epidermal transport has been the subject of various studies over the last 30 years. Two models are widely used. In the first model, it is assumed that the viable epidermis or aqueous diffusion layer below the stratum corneum can exert a significant influence on skin penetration (Scheuplein and Blank, 1973; Roberts et al., 1978). The second model is one where there is an effectively desorption-rate-limited step in partitioning from the membrane to the next phase (e.g., epidermis → receptor solution, stratum corneum → epidermis, epidermis → dermis). This rate constant, which we define as k_c, and the interfacial barrier rate constant are identical if the lag time for the interfacial barrier can be assumed to be negligible. In the specific case of an aqueous diffusion layer being a barrier, $k_c = D_{aq}/l_{aq}^2$ where l_{aq} is the thickness of the layer and D_{aq} is the diffusion coefficient in the layer (Roberts et al., 1978).

Guy and Hadgraft (1983) developed a pharmacokinetic model for skin absorption based on the diffusion model with the boundary conditions defined by a) the influx into the membrane being related to an assumed exponential decline in vehicle donor concentration and b) the efflux from the membrane being related to first-order removal at a rate constant k_c. These authors went on to examine short and long time approximations. Kubota and Ishizaki (1985) presented a more generalized diffusion model for drug absorption through excised skin by using the boundary conditions of the fluxes a) into the skin being defined by an arbitrary function $f(t)$ and b) out of the skin being defined by $ClC(x = h_m)$ where Cl is the clearance from the skin and $C(x = h_m)$ is the concentration of solute at the skin-system interface. They considered a boundary condition at the membrane–vehicle interface defined by an input rate into the membrane $f(t)$ together with a first-order rate constant k_c determined efflux from the membrane. Accordingly, the amount of solute absorbed across the skin $Q(t)$ at various times t is defined in the Laplace domain as:

$$\hat{Q}(s) = \frac{A}{s} \frac{k_c t_d \hat{f}(s)}{\sqrt{st_d}\, \sinh(\sqrt{st_d}) + k_c t_d \cosh(\sqrt{st_d})} \tag{37}$$

Of particular interest in this overview is the case of a constant donor concentration (infinite donor) and sink receptor. $\hat{Q}(s)$ is then defined by:

$$\hat{Q}(s) = \frac{k_p C_v A}{s^2} \frac{k_c t_d \sqrt{st_d}}{\sqrt{st_d}\, \cosh(\sqrt{st_d}) + k_c t_d \sinh(\sqrt{st_d})} \tag{38}$$

Figure 8A shows the effect of k_c (as defined by $\alpha = k_c t_d$) on $Q(t)$ versus time profile.

Note that, at long times, the linear portion of $Q(t)$ [defined by Eq. (38)] versus t profile describes a steady-state flux J_{ss} and lag time (lag):

$$Q(t) = J_{ss}A(t - \text{lag}) = C_v k_p A \frac{k_c t_d}{1 + k_c t_d}(t - \text{lag}) \tag{39}$$

$$\text{lag} = \frac{t_d}{6}\left[1 + \frac{2}{1 + k_c t_d}\right] \tag{40}$$

Thus, both the slope and lag of the steady-state portion of a $Q(t)$ versus t plot depend on k_c.

When Cl_R is defined as $k_c K_R A h_m$ to be analogous to the definition in Section I.C, J_{ss} in Eq. (39) can be reexpressed as:

$$J_{ss} = C_v k_p \frac{1}{1 + 1/k_c t_d} = C_v k_p \frac{1}{1 + Ak_p K_R/(Cl_R K_m)} \tag{41}$$

which is equivalent to Eq. (34).

E. In Vitro Skin Permeability Studies with Finite Donor Volume and Receptor Sink Conditions

In practice, the solute concentration applied to the skin does not remain constant but declines owing to the finite volumes of vehicles or pure substances applied to the skin. We therefore need to examine solutions for Eq. (5) in which the boundary condition allows for a depletion in solute concentration. We assume, initially, the simplest boundary conditions applying to the sorption of solutes into a membrane from the well stirred vehicle at $x = 0$ and from membrane into a systemic circulation [Eqs. (7) and (8)] together with a condition of depletion of concentration in the vehicle at $x = h_m$ (Crank, 1975):

$$V_v \frac{dC_v}{dt} = AD_m \left.\frac{\partial C_m}{\partial x}\right|_{x=0} \tag{42}$$

where $V_v = Ah_v$ is the volume of the vehicle applied to the skin. Solution of Eq. (5) with Eqs. (7), (8), and (42) as boundary conditions gives

$$Q(t) = M_0 \left[1 - \sum_{n=1}^{\infty} \frac{2\exp\left(-\frac{t}{t_d}\gamma_n^2\right)}{\cos\gamma_n(1 + \gamma_n^2/\beta + \beta)}\right] \tag{43}$$

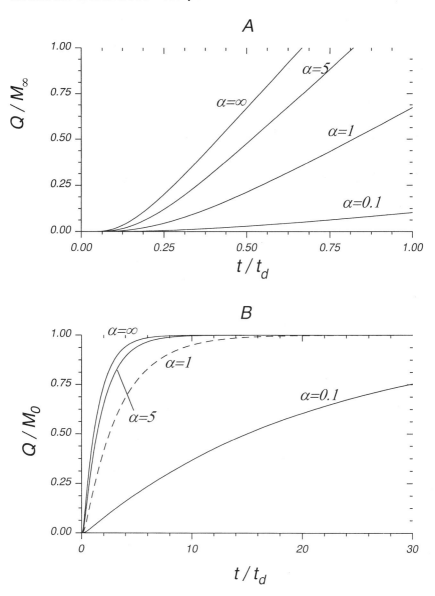

Figure 8   Effect of interfacial barrier rate constant (expressed as $\alpha = k_c t_d$) on exit from stratum corneum on normalized amount penetrating the epidermis with normalized time (t/t_d). (A) Constant donor concentration and (B) a finite dose in well-stirred vehicle where $\beta (= K_m h_m / h_v) = 1$ for time normalized to diffusion time.

where $M_0 = C_{v0}V_v = C_{v0}Ah_v$ is the initial amount of solute in the vehicle, C_{v0} is the initial concentration in the vehicle, h_v is the effective thickness of the vehicle, β is a dimensionless parameter defined by

$$\beta = \frac{K_m h_m}{h_v}$$

and γ_n are positive roots of the transcendental equation

$$\tan\gamma = \frac{\beta}{\gamma}$$

The Laplace transform of $Q(t)$ is given by

$$\hat{Q}(s) = \frac{M_0}{s} \frac{\beta}{\sqrt{st_d}\sinh(\sqrt{st_d}) + \beta\cosh(\sqrt{st_d})} \tag{44}$$

The corresponding expressions for flux are given by Eqs. (45) and (46):

$$J_s(t) = \frac{C_{v0}k_p}{\beta} \sum_{n=1}^{\infty} \frac{2\gamma_n^2 \exp\left(-\frac{t}{t_d}\gamma_n^2\right)}{\cos\gamma_n(1 + \gamma_n^2/\beta + \beta)} \tag{45}$$

$$\hat{J}_s(s) = C_{v0}k_p \frac{t_d}{\sqrt{st_d}\sinh(\sqrt{st_d}) + \beta\cosh(\sqrt{st_d})} \tag{46}$$

Figure 9 shows the predicted profiles for the flux of solute [Eq. (46)] with varying $\beta = K_m h_m/h_v$. It is apparent that both the peak time and area under the curve decrease with the increasing β. The longer peak time with increasing h_v reflects the movement from a finite to a infinite donor source. The larger area under the curve reflects the higher dose associated with an increase in h_v.

Two summary parameters can be derived from Eq. (46):

1. Mean absorption time measuring from systemic side of the skin is

$$\text{MAT}_s = -\frac{d \ln \hat{J}_s(s)}{ds}\bigg|_{s=0} = \frac{1}{2}t_d + \frac{t_d}{\beta} \tag{47}$$

It needs to be emphasized that MAT_s differs from MTT, which is mean transit time through stratum corneum. MTT can be calculated as

$$\text{MTT} = \text{MAT}_s - \text{MAT}_v$$

where MAT_v is the mean absorption time from the vehicle:

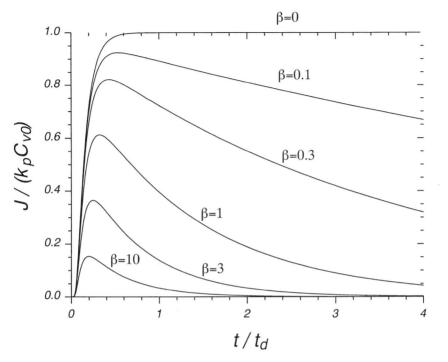

Figure 9 Normalized flux for penetration of a solute from a finite dose in a well-stirred vehicle [$J/(k_pC_{v0})$] against normalized time (t/t_d) for diffusion time with varying $\beta(= K_mh_m/h_v)$.

$$\text{MAT}_v = \frac{d \ln \hat{J}_v(s)}{ds}\bigg|_{s=0}$$

where $\hat{J}_v(s)$ is the Laplace transform of the flux from the vehicle into the skin. It can be found that

$$\hat{J}_v(s) = C_{v0}k_p \frac{t_d}{\sqrt{st_d} \tanh(\sqrt{st_d}) + \beta}$$

and $\text{MAT}_v = t_d/\beta$. We therefore have $\text{MTT} = t_d/2$.

2. CV^2 for absorption:

$$CV^2 = \frac{\dfrac{d^2}{ds^2} \ln \hat{J}_s(s)}{\dfrac{d}{ds} \ln \hat{J}_s(s)}\bigg|_{s=0} = \frac{2}{3} + \frac{4}{3(2 + \beta)^2} \qquad (48)$$

Another two summary parameters can be derived from Eq. (46) when $\beta \gg 1$ ($h_v \ll h_m K_m$). This applies in a specific case of solvent-deposited solids. Scheuplein and Ross (1974) described the application of drugs when 25 µl of acetone is applied to 2.54 cm^2 and allowed to evaporate, leaving a very thin layer of solid material. When $\beta \gg 1$, Eq. (46) reduces to:

$$\hat{J}_s(s) = \frac{C_{v0}k_p}{\beta} \frac{t_d}{\cosh(\sqrt{st_d})}$$

and therefore $J(t)$ can be written as

$$J(t) = \frac{C_{v0}k_p}{\beta} f(t/t_d) = \frac{C_{v0}k_p}{\beta} f(\tau)$$

where $\tau = t/t_d$, and $f(\tau)$ is a function independent of β and whose Laplace transform is $\hat{f}(s) = 1/\cosh \sqrt{s}$. It can be shown by the numerical inversion of $\hat{f}(s)$ that the maximum of the function $f(\tau)$ occurs at $\tau_{max} = 1/6$ with the value $f(\tau_{max}) = 1.850$. The maximum flux, J_{max}, and the time of maximum flux, t_{max}, for finite-dose absorption solvent-deposited solutes are therefore described by the simple equations:

$$J_{max} = 1.85 \frac{C_{v0}k_p}{\beta} = \frac{1.85 C_{v0} D_m h_v}{h_m^2}; \qquad t_{max} = \frac{t_d}{6} = \frac{h_m^2}{6 D_m} \tag{49}$$

Hence the peak time corresponds to the lag time observed after application of a constant donor solution [Eq. (12)]. Scheuplein and Ross (1974) provided experimental data to show: a) J_{max} is proportional to C_{v0} for benzoic acid, b) t_{max} for different solutes is inversely related to their D_m values, and c) penetration was facilitated by hydrating the stratum corneum.

F. In Vitro Permeability Studies with a Finite Donor Volume and a Finite Clearance from the Epidermis into the Receptor

Another case of particular practical interest is when the donor phase is assumed to be well stirred and finite in volume and there is limiting clearance from the epidermis to the receptor phase. Applying the boundary condition defined by Eq. (42), together with boundary condition for $x = h_m$,

$$D_m \frac{\partial C_m}{\partial x}\bigg|_{x=h_m} = h_m k_c C_m(h_m, t) \tag{50}$$

yields for $\hat{Q}(s)$

$$\hat{Q}(s) = \frac{M_0}{s} \frac{\beta}{\sqrt{st_d}\,\sinh(\sqrt{st_d})\left(1 + \dfrac{\beta}{t_d k_c}\right) + \cosh(\sqrt{st_d})\left(\beta + \dfrac{s}{k_c}\right)} \tag{51}$$

The profiles for $Q(t)$ versus t defined by Eq. (51) for different values of k_c ($\alpha = k_c t_d$) and $\beta = 1$ are shown in Fig. 8B.

G. In Vitro Skin Permeability Studies with Finite Donor Volume in Which Solute Diffusion Is Limiting and with Sink Receptor Conditions

One of the first attempts at modeling percutaneous absorption with diffusion-limiting uptake from both the vehicle and the skin was made by Kakemi et al. (1975). Their one-dimensional model is shown in Fig. 2D. Guy and Hadgraft (1980) used a similar model with sink receptor conditions as shown in Fig. 10.

In this model, the solute has a diffusivity D_v in a finite vehicle of volume V_v, which is in contact with stratum corneum in which a solute has a diffusivity D_m down a path length h_m. The Laplace transform for the amount penetrating the epidermis $\hat{Q}(s)$ into an absorbing "sink" is:

$$\hat{Q}(s) = \frac{M_0}{s} \frac{\beta\,\sinh(\sqrt{st_{dv}})}{\sqrt{st_{dv}}[\sqrt{t_d/t_{dv}}\,\sinh(\sqrt{st_{dv}})\,\sinh(\sqrt{st_d}) + \beta\,\cosh(\sqrt{st_{dv}})\,\cosh(\sqrt{st_d})]} \tag{52}$$

where, as usual, $M_0 = C_{v0}V_v$ and t_{dv} is the diffusion time in the vehicle, $t_{dv} = h_v^2/D_v$.

When the transport across the epidermis is also dependent on a first-order rate constant k_c for removal from the epidermis, the Laplace transform becomes:

$$\begin{aligned}
\hat{Q}(s) = \frac{M_0}{s} \frac{\beta\,\sinh(\sqrt{st_{dv}})}{\sqrt{st_{dv}}} & \left[\sqrt{\frac{t_d}{t_{dv}}}\,\sinh(\sqrt{st_{dv}})\,\sinh(\sqrt{st_d}) \right. \\
& + \beta\,\cosh(\sqrt{st_{dv}})\,\cosh(\sqrt{st_d}) \\
& + \frac{\sqrt{st_d}}{t_d k_c}\left. \left(\sqrt{\frac{t_d}{t_{dv}}}\,\cosh(\sqrt{st_d})\,\sinh(\sqrt{st_{dv}}) + \beta\,\cosh(\sqrt{st_{dv}})\,\sinh(\sqrt{st_d}) \right) \right]^{-1}
\end{aligned} \tag{53}$$

Figure 11 shows profiles of $Q(t)$ versus t as defined by Eq. (53) for different vehicle diffusivities ($\gamma = t_{dv}/t_d$) and k_c ($\alpha = k_c t_d$) for $\beta = 1$. In the particular cases when $t_{dv} \gg t_d$ and $t_d \gg t_{dv}$, Eq. (52) reduces to Eqs. (58) and (44), respectively.

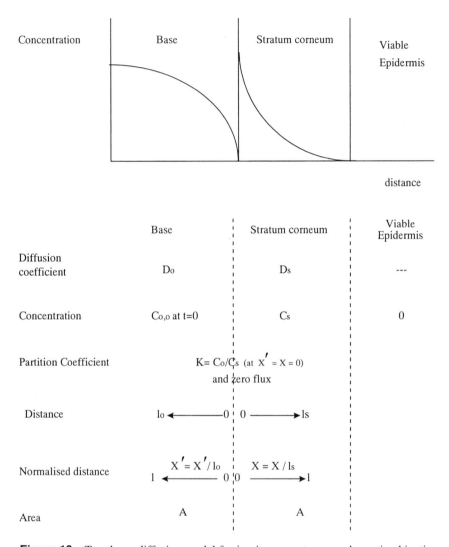

Figure 10   Two-layer diffusion model for in vitro percutaneous absorption kinetics as defined by Guy and Hadgraft (1980).

Kubota and Ishizaki (1986) have presented a time domain solution with simulations for a similar model to that described in Eq. (53). The moments for this model, with a method for estimating parameters, have been described by Kubota and Yamada (1990).

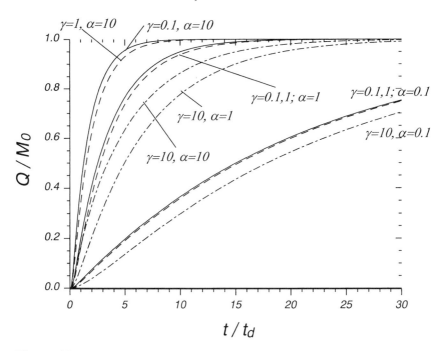

Figure 11 Cumulative amount penetrated Q normalized for amount applied M_0 versus time normalized for diffusion time t_d for the case of both vehicle and stratum corneum limited diffusion. The two parameters varied define the relative diffusion time in vehicle relative to that in stratum corneum $\gamma(= t_{dv}/t_d)$ and the interfacial barrier rate constant effect on the exit of solutes from the stratum $\alpha(= k_c t_d)$ [Eq. (54)].

H. In Vitro Permeability Studies with Two-Layer Diffusion Limitations in Transport

The complex cases of diffusion being a limitation in the transport through both the stratum corneum and epidermis have been considered by Hadgraft (1979b). He considered the case when solute exists as a reservoir in the stratum corneum. In his approach, the solute initially present in the stratum corneum diffuses from it into and through the epidermis. The case of rate-limiting removal from the epidermis (k_c) is considered. The resulting equation for the amount of solute exiting the epidermis $Q(t)$ is:

$$\hat{Q}(s) = \frac{M_0}{s} k_c t_{de} \frac{h_e}{h_{sc}} \left\{ \sqrt{st_{de}} \sinh(\sqrt{st_{de}}) \left[K_e k_c t_{de} + \frac{s}{k_1} + \frac{h_e}{h_{sc}} \sqrt{st_{ds}} \operatorname{cotanh}(\sqrt{st_{ds}}) \right] \right.$$
$$\left. + \cosh(\sqrt{st_{de}}) \left[st_{de} \left(\frac{k_c}{k_1} + K_e \right) + k_c t_{de} \frac{h_e}{h_{sc}} \sqrt{st_{ds}} \operatorname{cotanh}(\sqrt{st_{ds}}) \right] \right\}^{-1}$$
$$(54)$$

where $M_0 = C_{sc0}V_{sc}$, C_{sc0} is initial concentration in the stratum corneum, $t_{de} = h_e^2/D_e$, $t_{ds} = h_{sc}^2/D_{sc}$, and h_e, D_e, and h_{sc}, D_{sc} are the thickness and diffusion coefficients of epidermis and stratum corneum, respectively. Further, K_e is the partition coefficient between stratum corneum and epidermis ($K_e = C_{sc}/C_e$), k_1 is the interfacial rate constant, and k_c is the rate constant for removal from the epidermis.

Cleek and Bunge (1993) presented the mass entering this two-phase in-series model $[Q_{in}(t)]$ as a function of time in terms of both an analytical solution and simulations. This model was then extended to include solute properties as a determinant of uptake (Bunge and Cleek, 1995). They suggested that steady-state permeability will be underestimated if not corrected for the relative permeabilities of the stratum corneum and epidermis. The result of these considerations is a steady-state equation (55), similar to Eqs. (39) and (40):

$$Q_{in}(t) = C_{v0}k_pA \frac{1}{1 + B}\left[t + t_{ds}\frac{G(1 + 3B) + B(1 + 3BG)}{3G(1 + B)}\right] \qquad (55)$$

where $k_p = K_{sv}D_{sc}/h_{sc}$, $G = t_{ds}/t_{de}$, $B = D_{sc}h_eK_e/(D_eh_{sc})$, and K_{sv} is the partition coefficient between stratum corneum and the vehicle ($K_{sv} = C_{sc}/C_v$).

II. RELEASE PROFILES FROM TOPICAL PRODUCTS

A number of transdermal systems are now available for clinical use. Hadgraft (1979a) considered the solutions for release from patches for a range of boundary conditions. When a drug is contained in both the contact adhesive (priming dose) and patch, the release rate R_s approximates to (Chandrasekaran et al., 1978):

$$R_s = R_0 + H \exp(-at) \qquad (56)$$

where R_0 is the zeroth order flux from the patch assuming no depletion, and H and a are constants defining the release kinetics of the priming dose.

A. Diffusion-Controlled Release

Of practical interest is a homogenous phase in which the drug is released by diffusion. The expression for the drug release from slabs is well known to be that of the "burst" effect (Crank, 1975):

$$Q(t) = M_0\left\{1 - \frac{8}{\pi^2}\sum_{n=0}^{\infty}\frac{1}{(2n + 1)^2}\exp\left[-\frac{t}{4t_{dv}}\pi^2(2n + 1)^2\right]\right\} \qquad (57)$$

where again $M_0 = C_{v0}V_v$ and $t_{dv} = h_v^2/D_v$. The Laplace expression for Eq. (57) is:

$$\hat{Q}(s) = \frac{M_0}{s} \frac{\tanh(\sqrt{st_{dv}})}{\sqrt{st_{dv}}} \tag{58}$$

At short times when the amount released is less than 30%, Eq. (57) can be approximated by:

$$Q(t) = 2M_0 \left(\frac{t}{\pi t_{dv}}\right)^{1/2} \tag{59}$$

B. Release of a Suspended Drug by Diffusion

Another special case is that for a vehicle or patch containing a suspended drug. In this case, the amount of solute released into a perfect sink is given by (Higuchi, 1967):

$$Q(t) = A\sqrt{tD_v(2C_{v0} - C_s)C_s} \tag{60}$$

where C_{v0} in this context has the meaning of the total amount of drug (soluble and suspended) in the vehicle per unit volume, and C_s is the saturation concentration of the drug in the vehicle.

III. COMPARTMENTAL MODELS AS AN ALTERNATIVE TO DIFFUSION MODELS IN PERCUTANEOUS ABSORPTION

Riegelman (1974) analyzed a range of in vivo skin absorption data using a unidirectional absorption and simple compartment based pharmacokinetic models. A more complex series of models (Fig. 12) was used by Wallace and Barnett (1978) to describe in vitro methotrexate absorption across the skin. The solution for model 1 is given by:

$$A_3 = \frac{A_1 k_{12} k_{23}}{k_{21} + k_{23}} \left\{ t - \frac{1 - \exp[-t(k_{21} + k_{23})]}{k_{21} + k_{23}} \right\} \tag{61}$$

Since the exponential term approaches zero as $t \to \infty$, Eq. (61) can be reexpressed as:

$$A_3 = \frac{A_1 k_{12} k_{23}}{k_{21} + k_{23}} \left(t - \frac{1}{k_{21} + k_{23}} \right) \tag{62}$$

Hence, at long times there is a steady-state flux of $(A_1 k_{12} k_{23})/(k_{21} + k_{23})$ and lag time of $1/(k_{21} + k_{23})$. It should be noted that $A_1 k_{12} = k_p C_v$ in the present notation.

Two approximations of model 1 appear to be widely used by Guy and Hadgraft (1984d) and Kubota (1991), respectively, without explicitly ac-

A. Model Riegelman (1974)

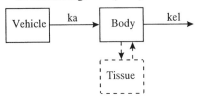

B. Model 1 Wallace and Barnett (1978)

C. Model 2 Wallace and Barnett (1978)

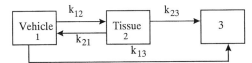

D. Model 3 Wallace and Barnett (1978)

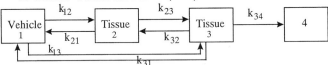

E. Compartment Model of Guy and Hadgraft (1983)

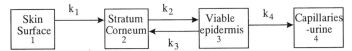

F. Compartment Model of Kubota (1991)

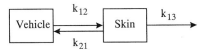

G. Isolated Perfused Model of Williams et al (1990)

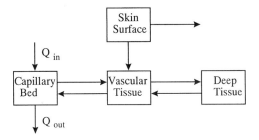

Figure 12 Examples of compartmental models of skin penetration.

knowledging their origins. As a consequence, much of the recent literature appears to have ignored the modeling undertaken by Wallace and Barnett (1978). For instance, Shah (1996) states, "this model (an approximation in which k_{21} is negligible) has not been applied to describe the in vitro percutaneous permeation data of any drug." Wallace and Barnett (1978) applied their full model to describe the pH dependence of methotrexate transport across excised skin. We now consider the two approximations in more detail.

The first approximation of a negligible k_{21} was introduced by Guy et al. (1982). This model a) facilitated in vivo modeling of blood and urine levels of solutes after topical application and b) enabled absorption to be related to the physical processes underlying absorption. An example of the use of this model with physically defined parameters is discussed elsewhere in this volume (Walters et al., 1998) in relation to the prediction of sunscreen percutaneous absorption. Guy et al. (1983) used this model to describe the plasma levels resulting from multiple dosing in percutaneous absorption. When k_3 is large, a low plasma peak is to be expected after an initial dose due to the significant uptake into the skin. With the accumulation of solute in the skin, the extent of uptake from subsequent doses will be lower and the peak higher. Such a phenomenon exists for hydrocortisone but not for malathion. Kubota et al. (1991), apparently unaware of the work of Guy et al. (1983), applied a random walk method using finite differences to show that absorption from patches decreased on repeated dosing. Guy and Hadgraft (1987) also used their model to explore the effect of penetration enhancers on percutaneous absorption kinetics. They suggested that, while both prostaglandin (PG)E$_1$ and PGE$_2$ increased k_1 by 10-fold, PGE$_2$ affected the ratio of k_3/k_2 by the same order of magnitude.

The model of Guy et al. (1983) (Fig. 12E and Fig. 13) would appear to have two advantages: a) Its parameters have some physical relevance, and b) its transfer function $F_{skin}(s)$ is suitable for Laplace convolution in relation to either a defined release-time profile for a vehicle, as defined by a Laplace function $F_v(s)$ or disposition in the body by a function $F_b(s)$:

$$F_{body}(s) = F_v(s)F_{skin}(s)F_b(s) \tag{63}$$

where

$$F_{skin}(s) = \frac{k_2 k_4}{(s + k_2)(s + k_3 + k_4) - k_3 k_2}$$

However, this model also has two limitations. First, the presence of a single absorption step as defined by k_{12} means the only input function is one that has been predefined by release from the vehicle phase. Release kinetics defined, in part, by the changing concentration at the epidermal interface with the vehicle are not directly catered for. Hence, this model is not the

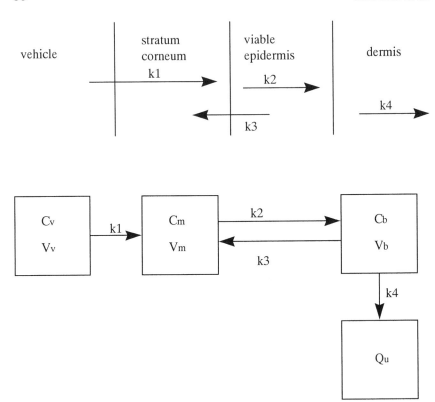

Figure 13 The compartmental model of skin penetration of Guy et al. (1983).

most suitable for analyzing the kinetics of percutaneous absorption for a finite dose of solute. Second, as suggested by Kubota (1991), this model does not represent the diffusion process as well as other compartmental models.

The compartmental model of Kubota (1991) assumes a negligible k_{32} (Fig. 12F). Interestingly, Kubota claims his model is an improvement on its predecessors, although it is a reduced form of the uncited model reported by Wallace and Barnett (1978) (Fig. 12A). Kubota suggests that the rate constant k_{23} (k_{1B} in his notation) is defined (in the present notation) as:

$$k_{23} = \left(\frac{t_d}{2} + \frac{1}{k_c} \right)^{-1} \tag{64}$$

whereas k_{12} and k_{21} can be found solving Eqs. (65) and (66):

$$\frac{k_{12}}{k_{21}} = k_{23}^2 \left[\frac{t_{dv}^2}{45} + \frac{h_v}{h_m K_m} \left(\frac{t_d^2}{3} + \frac{t_d}{k_c} + \frac{1}{k_c^2} \right) \right] \tag{65}$$

$$\frac{1}{k_{21}} + \frac{k_{12}}{k_{21}k_{23}} = \frac{h_v}{h_m K_m} \left(t_d + \frac{1}{k_c} \right) + \frac{t_{dv}}{3} \tag{66}$$

Kubota and Maibach (1992) have argued that only their compartmental model was compatible with the two lag times and steady-state flux of the diffusion model. Under the equivalent conditions defined in Section I.A, namely, a constant donor concentration and $k_c \to \infty$, the equivalent compartment representation of the diffusion equation solutions (9) and (16) are:

$$Q(t)_{compart,K} = K_m A C_v h_m \left[t - \frac{t_d}{6} + \frac{t_d}{6} \exp\left(-\frac{6t}{t_d} \right) \right] \tag{67}$$

and

$$Q_{in}(t)_{compart,K} = K_m A C_v h_m \left[t + \frac{t_d}{3} - \frac{t_d}{3} \exp\left(-\frac{6t}{t_d} \right) \right] \tag{68}$$

Differences are therefore most apparent in the exponent terms of diffusion and compartment models. Hence, as $t \to \infty$ both sets of models approach the steady-state approximations defined by Eqs. (11) and (16), respectively. Accordingly, all equations define a common steady-state flux $k_p A C_v$. A lag of $t_d/6$ is defined for $Q(t)$ versus t and neglag (intercept with negative time axis) of $t_d/3$ is defined for $Q_{in}(t)$ versus time. Kubota and Maibach (1992) go on to show that the mean absorption time for the system MAT_s is defined by the MTT_{skin} and the $MAT_{vehicle}$:

$$MAT_s = MTT_{skin} + MAT_{vehicle} \tag{69}$$

when the input function from the vehicle is defined independently of the skin.

Figure 14 shows a comparison of predictions of the diffusion model, steady-state approximation of the diffusion model, and Kubota's compartmental model. Also shown in Fig. 14 is a specific form of the Guy et al. (1982) model in which the compartmental rate constants have been chosen to yield a steady-state flux and lag time equivalent to the diffusion model $[k_1 k_2 = 2(k_2 + k_3)/t_d; \ 6k_1 k_2 = (k_2 + k_3)^2; \ k_2 = 1/t_d]$. The resulting equation is:

$$Q(t) = K_m A C_v h_m \left[t - \frac{t_d}{6} + \frac{t_d}{6} \exp\left(-\frac{12t}{t_d} \right) \right] \tag{70}$$

It is apparent that both the Kubota model and the one we have derived from Guy et al. (1982) are intermediate between the full diffusion equation profile

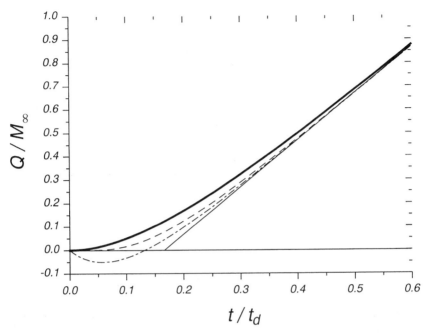

Figure 14 Normalized cumulative amount penetrated (Q/M_∞) versus normalized time (t/t_d) for diffusion model (thick solid line) [Eq. (13)], steady-state approximation of diffusion model (solid line) [Eq. (11)], and Kubota's compartmental approximation (dashed line) [Eq. (67)]. Also shown is the steady-state matched form of a compartmental model of Guy et al. (dash dotted line) [Eq. (70)] (see text for details).

and its steady-state approximation, with the Kubota compartmental model being closer to the diffusion model.

Recently, McCarley and Bunge (1998) examined the matching of various one-compartment model rate expressions with a diffusion model for a single membrane. They examined 11 models and found that the matching depended on the vehicle conditions (solute concentration, nature of vehicle, exposure time) and the outcome prediction (concentration, mass into skin, or flux) being sought. Differences between models are most evident in the transient time period. Interestingly, their favored model 4 appears to be identical to that originally proposed 20 years earlier by Wallace and Barnett (1978). The latter, however, lacked the matching exercise and consideration of transient conditions undertaken by McCarley and Bunge (1998). The key models developed (McCarley and Bunge, 1998) match a) steady-state stratum corneum concentration and penetration rate with an assumption of equilibrium at the two boundaries of the membrane (equilibrium model); b) steady-state stratum corneum concentration, penetration rate, and lag time for all blood and vehicle concentration ratios, and for large blood concentrations (general time lag model); c) as in b) but for low blood concentrations (simplified time lag model); and d) the traditional model. The real contribution of this work has been a systematic derivation of equations for coefficients of compartmental models, expressed in terms of physicochemical parameters, which are documented in the paper in a readily accessible form.

IV. OTHER PROCESSES AFFECTING IN VITRO PERCUTANEOUS ABSORPTION

A. Concentration-Dependent Diffusive Transport Processes

It is well known that the permeability of solutes through the skin may be affected by their concentration-dependent interaction with the skin as shown for the alcohols (Scheuplein and Blank, 1973) and phenols (Roberts et al., 1977). At this time, relatively little work has been published on mathematical models for diffusion in a swelling or denaturing skin environment. Wu (1983) has attempted to relate water diffusivity $D(C)$ as a function of its water concentration (C) in a keratinous membrane. The expression $D(C) = D_0 + AC^B$, where D_0, A, and B are constants, best described the results. Giengler et al. (1986) have described the numerical solution of solute concentrations in a system consisting of a polymer film, microporous membrane, adhesive, skin layer, and capillary sink and in which the diffusion coefficients were time dependent.

Higuchi and Higuchi (1960) summarized the theory associated with diffusion through heterogeneous membranes and vehicles. They suggested that transport through a two-phase system was a function of the volume fraction and permeability of each phase. They also derived an expression for lag time across a membrane when simultaneous diffusion and Langmuir adsorption has occurred. The lag time across the membrane is increased by binding of solute to membrane's binding sites:

$$\text{lag} = \frac{h_m^2}{4D_m} + \frac{h_m^2 A}{2D_m K_1 V_1 C_v} \tag{71}$$

where D_m is effective diffusion constant in heterogeneous membrane, A is amount of solute taken up by filler per unit volume of membrane material, K_1 is partitioning coefficient of the solute in vehicle and phase 1 in the membrane, V_1 is volume fraction of phase 1, and C_v is concentration of solute in the vehicle.

When the solute concentration in the vehicle increases, the second term in Eq. (71) decreases and the overall lag time is shorter.

Chandrasekaran et al. (1976, 1978) assumed that uptake of solutes by skin was described by a dual-sorption model,

$$Z \frac{dC}{dt} = D \frac{d^2C}{dx^2}$$
(72)

where

$$Z = B \left[1 + \frac{C_i b/k_D}{(1 + C_D b/kD)^2} \right]$$

However, these authors then assumed that Z was a constant. This model reduces to the conventional diffusion model in which the effective diffusivity is the diffusion coefficient of free solute modified by the instantaneous partitioning of solute into immobile sites in the diffusion path. Expressions were presented for plasma concentrations and urinary excretion rate profiles after single- and multiple-patch applications of scopolamine to humans. Kubota et al. (1993b) applied the dual sorption model to account for the nonlinear percutaneous absorption of timolol. The model accounted for the prolongation in timolol lag time associated with the decrease in applied concentration.

B. Bioconversion/Metabolism of Solutes in the Skin

Roberts and Walters (1998) related the in vitro (metabolically inactive) skin flux $J_{s,in\ vitro}$ to that in vivo, $J_{s,\ in\ vivo}$, by a first-pass bioavailability F_s and a fraction released from the product into the skin F_R:

$$J_{s,in\ vivo} = F_s F_R J_{s,in\ vitro}$$
(73)

The importance of recognizing F_s is illustrated by methylsalicylate where $F_s < 0.05$ (Roberts and Walters, 1998). Caution must therefore be applied in extrapolating in vitro data into likely in vivo absorption.

The modeling of percutaneous absorption kinetics in the epidermis when diffusion and metabolic processes occur simultaneously leads to relatively complex solutions. Ando et al. (1977) examined the diffusive transport of a solute through a metabolically inactive stratum corneum and hence through the epidermis, where it was assumed that there was homogeneous distribution of metabolizing enzymes. Subsequent work developed this model to examine the bioconversion prodrug → drug → metabolite (Yu et al., 1979a,1979b). The work applied the diffusion equation and derived expressions for the steady state fluxes and cutaneous concentration–distance relationships for each of the species. Yu et al. (1980) then solved this model

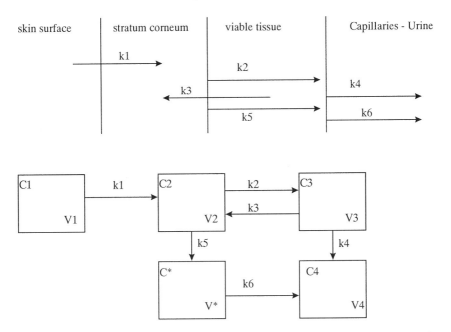

Figure 15   The compartmental model for skin penetration modified to include epidermal metabolism (adapted from Guy and Hadgraft, 1984b).

for nonuniform enzyme distribution in the skin. Fox et al. (1979) considered Michaelis–Menten kinetics in their examination of prodrug, drug, and metabolite concentrations in the epidermis and dermis. Approximations to diffusion-based models were applied by Hadgraft (1980), Guy and Hadgraft (1982), and Kubota et al. (1995) to describe the effect of linear and saturable (Michaelis–Menten) epidermal metabolism on percutaneous absorption.

Guy and Hadgraft (1984b) adapted their pharmacokinetic model defined by k_1, k_2, k_3, and k_4 (Fig. 15) to include a metabolic step k_5 and removal of metabolite k_6. Linear kinetics was assumed to enable solution in the Laplace domain and inversion to give analytical solutions. A number of theoretical plasma concentration profiles were then constructed. In reality, Michaelis–Menten kinetics may be operative for a number of solutes.

C.   Solute–Vehicle, Vehicle–Skin, and Solute–Skin Interactions

The practical application of mathematical models in percutaneous absorption to therapeutics or risk assessment is dependent on an understanding of solute–skin, solute–vehicle, and vehicle–skin interactions. Some aspects of each of these areas have been discussed by Roberts and Walters (1998), Hadgraft and Wolff (1998), and Robinson (1998).

The present analysis has generally been limited to percutaneous absorption kinetics in which the underlying physicochemical parameters are time independent. In practice, the application of a vehicle to the skin will lead to a time-dependent change in permeability due to either a solute–skin interaction or a vehicle–skin interaction. The solutions of the resultant concentration-dependent diffusion processes also lead to a time- and space-dependent change in solute diffusivity, are relatively complex, and are felt to be beyond the scope of this overview.

Of critical importance in both therapeutics and toxicology is the maximum flux of a solute J_{max}. This flux is normally attained at the solubility of the solute in the given vehicle S_v, consistent with the solubility of the solute in the stratum corneum transport pathway S_m:

$$J_{max} = \frac{D_m K_m S_v}{h_m} = k_p S_v = \frac{D_m S_m}{h_m} \tag{74}$$

Hence, J_{max} is the same for all vehicles unless the vehicle affects either D_m or S_m. J_{max} is also solute dependent and may be expressed in terms of their molecular weight (MW) and octanol solubility (S_{oct}) by applying the Potts–Guy equation (Roberts & Walters, 1998):

$$\log J_{max} = 0.71 \log S_{oct} - 0.0061 \text{MW} - 2.72 \tag{75}$$

Roberts and Walters (1998) have related J_{max}/activity for a number of groups of solutes to their octanol–water partition coefficients. The resulting profiles are parabolic, consistent with a maximum in the vicinity of a partition coefficient corresponding to solubility in octanol. More sophisticated analysis of percutaneous absorption-solute structure relationships have been reviewed by Pugh et al. (1998).

D. Effect of Surface Loss Through Processes Such as Evaporation and Adsorption to Skin Surface

There is a potential change in solute concentration as a consequence of surface loss during percutaneous absorption. The loss may result in a) an effective reduction in the volume of the vehicle alone due to evaporation and an increase in solute concentration as a consequence, b) a reduction in both solute and vehicle due to a removal process and c) a loss of solute only due to volatilization or adsorption to skin surface. For instance, Reifenrath and Robinson (1982) have shown that mosquito repellents may be lost due to evaporation at a rate comparable to their percutaneous absorption. The loss of vehicle at a defined rate creates a moving boundary problem and does not appear to have been considered to any great extent in the literature. Guy and Hadgraft (1984a,1984c) examined the first- and zero-order loss of solute from the vehicle surface using diffusion and compartment models, respectively.

E.   Shunt Transport

The importance of shunt transport by appendages has been well recognized. Scheuplein (1967) and Wallace and Barnett (1978) assumed a parallel pathway with a minimal lag time relative to transepidermal transport for diffusion and compartmental models, respectively. In our attempted modeling of epidermal and shunt diffusion, we assumed that the overall amount penetrating was the sum of the amounts penetrating through independent epidermal and shunt pathways (Siddiqui et al., 1989). The amount penetrating through each pathway was assumed to be defined by Eq. (9), in which K_m and D_m were defined in terms of the corresponding constants for the two pathways.

More recently, the presence of polar and nonpolar pathways through the intercellular region of the stratum corneum has been recognized as described in Eq. (3). Mathematical models described include those for steady-state conditions (Ghanem et al., 1987; Hatanaka et al., 1990) and an infinite dosing condition (Tojo et al., 1987). Yamashita et al. (1993) have presented the Laplace solution for a well-stirred finite donor phase in contact with stratum corneum in which solutes can diffuse through both polar and nonpolar routes. The solute can then diffuse through the epidermis into a sink. Numerical inversion of the Laplace transform with FILT was then undertaken to generate real-time profiles. More recently, Edwards and Langer (1994) have derived expressions for a range of conditions and suggested that their theory confirmed the importance of shunt and intercellular transport for small ions and uncharged solutes, respectively.

V.   SIMPLE IN VIVO MODELS IN PERCUTANEOUS ABSORPTION

A.   Compartmental Pharmacokinetic Models

One of the first evaluations of the pharmacokinetics of skin penetration was reported by Riegelman (1974). Absorption of solutes through the skin was generally assumed to follow first-order kinetics with a rate constant k_a (units: time^{-1}). Much of the data analyzed appeared to be characterized by "flip-flop" kinetics where the absorption half-time is much longer than the elimination half-time, as illustrated later in Fig. 17. This modeling approach has been used by a number of authors, including the recent work of Rohatagi et al. (1997). This work described an integrated pharmacokinetic–metabolic model for selegiline after application of a transdermal system for 24 h. A series of differential equations and nonlinear regressions were then used to solve drug and metabolite concentrations in accordance with scheme presented in Fig. 16.

Roberts and Walters (1998) have suggested that four processes are commonly used to describe plasma concentrations C_{plasma} with time after topical application when the body is assumed to be a single compartment, with an elimination rate constant k_{el} and apparent distribution volume V_B.

1. Depletion of the applied dose at a first order rate constant k_a is described by

$$C_{plasma} = k_a F \text{Dose}\{\exp[-k_a(t - \text{lag})] - \exp[-k_{el}(t - \text{lag})]\} \quad t > \text{lag} \qquad (76)$$

where lag is the lag time for absorption through the skin and F is the fraction that would be absorbed if the product were applied for an infinite time.

Figure 17 shows the nonlinear regression of norephedrine urinary excretion rate after topical application using Eq. (76). Also shown in Fig. 17 is the profile for an equimolar dose given orally. It is apparent that the topical application is associated with a lag of 2 h and an absorption half-life (the

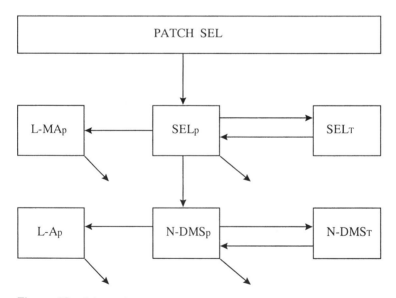

Figure 16 Schematic model for selegiline (SEL) and metabolites (L-MA, L-methamphetamine; L-A, L-amphetamine; N-DMS, N-desmethyl selegiline) in plasma (subscript p) and tissues (subscript T) after application of a transdermal patch (adapted from Rohatagi et al., 1997).

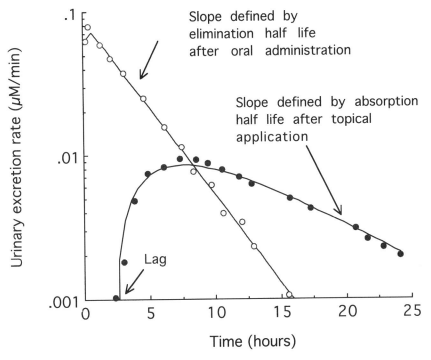

Figure 17 Simultaneous nonlinear regression of urinary excretion rate–time data for norephedrine hydrochloride administered orally (○) and free base applied topically (●) using a weighting of $1/y_{obs}$ and a common elimination half-life for norephedrine. The regression yielded an absorption half-life of 0.09 h for oral administration. There is a lag of 2.2 h and an absorption half-life of 6.0 h for topical application and an elimination half-life of 2.5 h ($r^2 = .999$). Data from Beckett et al. (1972).

terminal phase) of 8 h ($0.693/k_a$). A common elimination half-life ($0.693/k_{el}$) of 3.3 h is found for both routes of administration.

Cross et al. (1994, 1996) related the cumulative amount eluted $Q(t)$–time t profiles, after dermal absorption of solutes into a perfused limb preparation, to

$$Q(t) = M_0 \left[1 + \frac{k_a}{k_{el} - k_a} \exp(-k_{el}t) - \frac{k_{el}}{k_{el} - k_a} \exp(-k_a t) \right] \qquad (77)$$

where M_0 is the initial amount applied to the dermis, k_a is the absorption rate constant for solute absorption from the dermis, defined by fraction remaining in dermis with time [$F_{dermis} = \exp(-k_a t)$], and k_{el} is the elimination rate constant from the preparation. It can be shown that after some rearrangement, Eq. 16 in Williams et al. (1990) for the perfused porcine skin flap model (Fig. 12G) simplifies to Eq. (77).

Recently, Reddy et al. (1998) have described a one-compartment skin pharmacokinetic model to describe in vivo absorption. The model is identical to model III in Wallace and Barnett (1978) when the shunt transport in the latter is assumed to be negligible. They found that the general time lag model derived by McCarley and Bunge (1998), and discussed earlier in Section III, best predicted diffusion in a membrane prediction for most situations.

Each of these models may lead to erroneous estimates for the absorption rate constant if significant tissue distribution occurs in the body (as represented by the dotted lines in Fig. 12A). We have recently reported that estimates for k_a based on a single exponential (one compartmental) disposition are often twice those for the more correct biexponential (two compartmental) disposition model (Roberts et al., 1997). Surprisingly, to date, most analyses of in vivo percutaneous absorption kinetics have assumed monoexponential disposition kinetics and have not considered this potential error.

2. Delivery at a constant rate J_s, for a time period T, is described by

$$
C_b = \begin{cases}
\dfrac{J_{s,in\ vitro}F}{Cl_{body}} (1 - \exp[-k_{el}(t - \text{lag})]) & t < \text{lag} + T \\
\dfrac{J_{s,in\ vitro}F}{Cl_{body}} (1 - \exp[-k_{el}(t - \text{lag})]) \exp[-k_{el}(t - T - \text{lag})] & t \geq \text{lag} + T
\end{cases}
$$

(78)

Singh et al. (1995) used Eq. (78) to describe the in vivo absorption kinetics of a number of solutes iontophoresed transdermally in vivo at a presumed constant flux for a time T. Good fits were obtained in the nonlinear regression of each of the data sets.

Imanidis et al. (1994) used a similar approach to describe a constant total flux of acyclovir from a patch $J_T[= C_s/(1/k_{pm} + 1/k_{pp})]$ into the bloodstream when acyclovir disposition is described by a biexponential elimination after iv administration $[C_b = A \exp(-\alpha t) + B \exp(-\beta t)]$. The resulting blood concentration C_b–time t profile is:

$$
C_b = \frac{J_T}{V_c k_{el}} \left[1 + \frac{\beta - k_{el}}{\alpha - \beta} \exp(-\alpha t) + \frac{k_{el} - \alpha}{\alpha - \beta} \exp(-\beta t) \right]
$$

(79)

where C_s is the acyclovir concentration in the patch, k_{pm} is the permeability coefficient of acyclovir in the skin, k_{pp} is the permeability coefficient of acyclovir in the rate-controlling patch of the membrane, V_c is the apparent volume of distribution of the central compartment, and A, B, α, and β are constants describing the disposition process.

Hadgraft and Wolf (1998) have recently summarized the prediction of in vivo plasma data after topical application. Their modeling appeared to adequately describe the in vivo percutaneous absorption kinetics for a range of drugs.

3. Steady-state conditions are described by

$$C_{bss} = \frac{FJ_{s,in\ vitro}}{Cl_{body}} = \frac{Fk_p C_v A}{Cl_{body}} \tag{80}$$

This equation is a reduced form of Eq. (78) for $t \to \infty$ and $t < \text{lag} + T$. Equation (80) has been used by Roberts and Walters (1998) to define desired patch release rate for a number of drugs in vivo from a knowledge of the drug's clearance and desired plasma concentration.

4. A time-dependent transdermal flux is best analyzed assuming a model deduced from in vitro absorption kinetics (Section I.G) or deconvolution analysis (Section V.D).

B. Diffusion Pharmacokinetic Models

In vivo absorption models usually represent the body as one or more compartments with input into the body via percutaneous absorption. Cooper (1976) derived an expression for the total amount of solute excreted into the urine after topical absorption. Other models of Guy, Hadgraft, Kubota, and Chandresekaran (described earlier) have adopted a similar approach in describing either plasma concentrations or urinary excretion rates from one or two compartment models. Cooper assumed diffusion through the skin according to Eq. (19) into the body, represented as a single compartment. When his model is modified to include a stratum corneum–vehicle partition coefficient K_m, the plasma concentration, $C_p(t)$, and the amount excreted into urine, $M(t)$, are defined by these equations:

$$C_p(t) = \frac{Ak_p C_v}{V_c} \left\{ \frac{1 - \exp(-k_{el}t)}{k_a} + 2 \sum_{n=1}^{\infty} \frac{(-1)^n}{k_{el} - n^2\pi^2/t_d} \right.$$
$$\left. \cdot [\exp\left(-\frac{t}{t_d} n^2\pi^2\right) - \exp(-k_{el}t)] \right\} \tag{81}$$

$$M(t) = Ak_p C_v k_u \left(\frac{tk_{el} - 1 + \exp(-k_a t)}{k_{el}^2} + 2 \sum_{n=1}^{\infty} \frac{(-1)^n}{k_{el} - n^2\pi^2/t_d} \right.$$
$$\left. \cdot \left\{ \frac{td}{n^2\pi^2} \left[1 - \exp\left(-\frac{t}{t_d} n^2\pi^2\right) \right] - \frac{1}{k_{el}} [1 - \exp(-k_{el}t)] \right\} \right) \tag{82}$$

where k_{el} is the total effective elimination rate constant, k_u is the rate constant for excretion in the urine, and V_c is the total effective volume of the compartment.

The steady-state portion of the $M(t)$ versus t plot from Eq. (82) yields a slope of $k_u k_p A / k_{el} = f_e k_p A$ where k_p is the permeability coefficient ($= K_m D_m / h_m$), A is the area of application, k_{el} is the elimination rate constant of the solute from the body, and f_e is the fraction of the solute excreted in the urine. This plot is associated with a lag time t_L of

$$t_L = \frac{1}{k_{el}} + \frac{t_d}{6} \qquad (83)$$

where again $t_d = h_m^2/D$, D is the diffusivity of the solute in stratum corneum, and h_m is the distance of the pathway of diffusion. Even for multicompartmental disposition kinetics, the total lag time for elimination of a solute is uncoupled and is the sum of epidermal diffusion and pharmacokinetic lag times (Cooper, 1979). When only a finite dose of solute is applied to the skin, the urinary excretion rate is also a function of the vehicle thickness (Cooper and Berner, 1985).

In practice, the direct application of Eq. (19) to in vivo absorption may be limited. Equation (32), which takes into account the effectiveness of blood flow in removal of solute from the epidermis and the accumulation of solute in the epidermis in vivo, may be more appropriate. Accordingly, the actual steady-state flux is less than $k_p A C_v$ due to this limitation in blood flow clearance as defined by Eq. (34).

C. Physiological Pharmacokinetic Models (Fig. 18)

A number of authors have advocated the use of physiological rather than compartmental representations of the body. McDougal (1998) recently summarized the modeling in this area. These models utilize the numerical integration of a series of differential equations representing each compartment, to solve for blood concentration–time profiles after topical application. Individual organs or types of tissues are represented as the compartments with blood flow into and out of the organs defining the transport in the body system. Input into the skin, as a perfused organ, is assumed to follow Fick's first law and may allow for evaporation. Jepson and McDougal (1997) recently used this model to estimate the permeability constants for halogenated methanes from an aqueous solution after topical application in a whole-animal study.

D. Deconvolution Analysis in Pharmacokinetic Modeling

Deconvolution analysis is based on the principle that the observed plasma or blood concentrations–time profiles, $C_b(t)$, are defined by the percutaneous

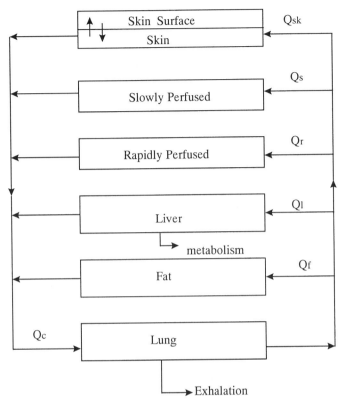

Figure 18 Physiological pharmacokinetic model for percutaneous absorption (adapted from McDougal, 1998, and Jepson and McDougal, 1997).

absorption flux, $J_s(t)$, and the disposition kinetics in the body after an intravenous bolus (impulse) done at $C_{iv}(t)$:

$$\hat{C}_b(s) = \hat{J}_s(s)\hat{C}_{iv}(s) \tag{84}$$

Hence the observed $C_b(t)$ and $C_{iv}(t)$ and inversion of the resulting Laplace domain expression for $\hat{J}_s(s)$ enable $J_s(t)$ to be defined. This technique is especially useful when the mathematical model for the percutaneous absorption process is not known. A comparison of the observed profile with theoretical profiles may define the underlying model for percutaneous absorption kinetics.

Examples of deconvolution analysis applied in this area include the evaluation of the absorption function from nicotine patches (Benowitz et al.,

1991) and the modeling of subcutaneous absorption kinetics (Roberts et al., 1997).

E. Penetration into Tissues Underlying Topical Application Site

Epidermal concentrations in vivo after topical application, assuming D_e is sufficiently large to approximate well stirred (i.e., compartmental representation), is defined by Eq. (36) and at long times ($t \to \infty$) by Eq. (29) via

$$C_{ss} = \frac{k_p A C_v}{Cl_R + k_p A K_R / K_m} \tag{85}$$

where Cl_R is the in vivo epidermal clearance.

A similar expression can be defined for subsequent deeper tissues using a compartment-in-series model in parallel with removal to the systemic circulation and recirculation (Fig. 19) to define deeper tissue concentrations after topical application (Singh et al., 1998b). Transport into deeper tissues could occur either by "convective" blood flow (McNeil et al., 1992) or by diffusion. Nonlinear regressions of experimentally treated and contralateral tissue data with the model used simultaneous numerical integration of a series of differential equations (Singh et al., 1998b; Singh and Roberts, 1993). The analysis showed that although direct deep tissue penetration was apparent at early times, recirculation of drug from the systemic circulation accounted for tissue levels at longer times to define deep tissue penetration of dermally applied solutes (Fig. 19).

F. Pharmacodynamic Modeling

In principle, established pharmacodynamic models used in whole-body pharmacokinetic modeling can be directly used when solutes are delivered by skin. Complexities can exist when the site of drug targeting is the skin itself. Imanidis et al. (1994) showed that the antiviral efficacy to HSV-1 skin infections of acyclovir was directly related to the logarithm of the flux from transdermal patches, consistent with classical log dose-response relationships. However, an equivalent systemic dose was relatively ineffective.

Beastall et al. (1986) examined the onset of erythema (t_E) as a function of solute concentration (C_0). Applying Fick's law of diffusion they obtained the expression:

$$\log \frac{n_E}{h_m} = \log(C_v t_E^{3/2}) + \log \left(\frac{K_m}{1 + K_m / p^{1/2}} \right)$$
$$+ \log \left(\frac{8 D_{sc}^{3/2}}{\pi^{1/2} h_m^3} \right) - \frac{h_m^2}{9.2 D_{sc} t_E} \tag{86}$$

where D_{sc} is the diffusion coefficient of nicotinate in the stratum corneum,

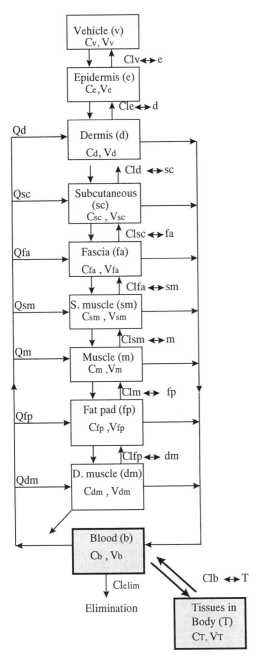

Figure 19 A pharmacokinetic model for local deep tissue penetration after topical application (adapted from Singh et al., 1998b). The symbols are: Q, blood flow; C, concentration; V, volume; Cl, clearance as related to the various tissues and the rest of the body.

K_m is its partitioning coefficient between vehicle and skin, h_m is diffusion path length, p is the ratio of the diffusion coefficients of the nicotinate in the vehicle and the skin, and n_E is the concentration of nicotinate required to trigger erythema. This expression showed a linear relationship should and did exist between $\log(C_v t_E^{2/3})$ and $1/t_E$. The gradient of the relationship D_{sc}/h_m^2 was greatly affected by the coadministration of the enhancer urea.

The human skin blanching assay for evaluating the bioequivalence of topical corticosteroid products should follow standardized guidelines as developed by the U.S. Food and Drug Administration (FDA) in 1995. Demana et al. (1997) evaluated the area under the effect curve (AUEC), also called the effect (E), for both visual and chromameter-derived data. The visual data were best described by a sigmoidal E_{max} model [Eq. (87)], while the chromameter data were described by a simple E_{max} model [Eq. (88)]:

$$E = E_0 \pm \frac{E_{max}D}{D + \text{ED}_{50}} \tag{87}$$

$$E = E_0 \pm \frac{E_{max}D^\gamma}{D^\gamma + \text{ED}_{50}^\gamma} \tag{88}$$

where E_{max} is the maximal AUEC, D is the dose duration, ED_{50} is the dose duration for half maximal E, and γ is a sigmoidicity factor related to the shape of the curve. The parameter E_0, not explicitly stated in the modeling by Demana et al. (1997), should be included in the model fitting to correct for baseline readings (Singh et al., 1998a). Smith et al. (1998) have pointed out that they had corrected for E_0 using unmedicated site values in their earlier work (Demana et al., 1997). A key aspect in this mathematical modeling is varying the dose administered by varying the duration of application. Varying dose duration is then used to relate the vasoconstrictor response to a range of corticosteroid amounts. Demana et al. (1997) used a weighting of 1/AUEC and a number of goodness-of-fit criteria in their analyses.

VI. MODELING WITH FACILITATED TRANSDERMAL DELIVERY

A. Iontophoresis

There are a number of mathematical models used in iontophoresis. As described by Kasting (1992), these are generally defined by the Nernst Planck and Poisson equations. Of particular practical usefulness is the iontophoretic flux of a solute through the epidermis. This flux can be incorporated into various pharmacokinetic models for the body to enable the description of plasma concentration and urinary excretion–time data. Singh et al. (1995) examined in vivo plasma data after iontophoretic transport with simple pharmacokinetic models. In vivo blood concentrations for most solutes delivered by iontophoresis appear to be able to be described by zero-order input into a one-compartment model [Eqs. (78) and (80)] (Singh et al., 1995).

The iontophoretic flux depends on a number of factors, including solute ionization, interaction of solutes with pore walls, solute size, solute shape, solute charge, Debye layer thickness, solute concentration, and presence of extraneous ions. We recently proposed an integrating expression for the flux of the jth solute (Roberts et al., 1998):

$$J_{j,iont} = C_j \left[\frac{2\mu_j fi_j Fz_j I_T \Omega PRT_j}{(k_{s,a} + k_{s,c})(1 + fu_i\theta_{ju} + fi_j\theta_{ji})} \pm (1 - \sigma_j)v_m \right] \tag{89}$$

where C_j is concentration of the jth solute, μ_j is its mobility, fi_j and fu_j are ionized and unionized fractions of the solute, z_j is its charge, PRT_j is a partial restriction term, σ_j is the reflection coefficient term, v_m is the velocity of water flow across the membrane, I_T is the total current across the membrane, Ω is the permselectivity for cations, $k_{s,a}$ and $k_{s,c}$ are conductivities of the anode and cathode solutions, θ_{ju} and θ_{ji} are parameters describing interaction of unionized and ionized fractions of the solute with the pore, and F is Faraday's constant.

Dermal and subcutaneous concentrations of solutes after in vivo iontophoretic application can also be determined in terms of clearance by blood supply to the tissue, clearance to deeper tissues, and influx by iontophoresis (Cross and Roberts, 1995).

B. Sonophoresis

The sonophoretic iontophoretic flux can also be included in pharmacokinetic models in a manner analogous to that described with iontophoresis. Mitragotri et al. (1997) have suggested that sonophoresis induces cavitation. They have suggested that the sonophoretic permeability $k_{p\ sono}$ can be defined in terms of the passive permeability coefficient k_p (units: cm/h) and solute octanol–water partition coefficient (K_{ow}) as

$$k_{p\ sono} = k_p + (2.5 \times 10^{-5})K_{ow}^{3/4} \tag{90}$$

VII. PRACTICAL ISSUES IN APPLYING MATHEMATICAL MODELS TO PERCUTANEOUS ABSORPTION DATA

A major limitation in a number of reported percutaneous absorption studies, including those from our laboratories, has been the assumption of a given mathematical model. Whether that model is strictly the most appropriate one is often difficult to confirm. Most studies appear to have used the simplest model, as defined by Eq. (1), in which the steady-state flux and lag time are defined by the steady-state portion of the curve. There are a number of limitations in using such a model, as discussed by Robinson and other authors.

Robinson (1998) points out that errors can be made if a) the burst influx and lag containing through flux are represented by a steady-state approximation at early times, b) an infinite vehicle is assumed when the concentration is actually declining due to the finite volume used, c) penetration of a solute by passive diffusion also involves modification of the skin barrier properties (solute-skin interactions), d) vehicle effects on solute concentration, e.g., evaporation or skin permeability (vehicle–skin interactions) exists, e) skin reservoir effects exist, as illustrated by the extensive uptake of sunscreens into, but not necessarily through, the skin (Jiang et al., 1998), f) discrepancies exist between in vitro and in vivo absorption due to the role of capillaries in absorption in vivo, and g) the resistance barrier of the skin is compromised.

The expressions for a number of the more complex models contain the necessary correction factors to overcome some of the inherent limitations in the simplest model (Section I.A) representation of data. For instance, the steady-state flux may be affected by the sampling rate from the receptor compartment as defined by Eq. (34). The lag time will be dependent on both this clearance and the volume ratios of the membrane and receptor phases, corrected for partitioning effects, as defined by Eq. (35). A different set of correction factors applies if an interfacial barrier or desorption rate constant exists [Eqs. (39) and (40)]. As Kubota et al. (1993a) point out, although a simple compartmental model may describe percutaneous penetration kinetics, the parameters obtained may not necessarily represent the membrane diffusion and partition coefficient.

Relating data to a specific model using nonlinear regression techniques also requires an appropriate weighting of the data in accordance with the underlying errors associated with the data. In the absence of known error structures, a weighting of $1/y_{obs}$ may be appropriate. This weighting assumes that the coefficient of variation (standard deviation/mean) of the data is relatively constant.

Some of the dilemmas in the mathematical modeling of percutaneous absorption are enunciated in the letters to *Pharmaceutical Research* written by Singh et al. (1998a) and Smith et al. (1998), especially in relation to pharmacodynamic modeling of skin blanching after topical application. Issues raised include a) reliability of visual and chromameter methods, b) analysis of "naive" pool data by nonlinear regression versus mixed-effect modeling, c) baseline correction, d) consistency of parameter values such as sigmoidicity with independent literature estimates, e) precision of critical short and long dose duration data, and f) subject (skin) selection. Smith et al.

(1998) suggest that the current methodology prepared by the FDA requires further evaluation.

Finally, there is probably a greater need for deconvolution techniques to be used with in vivo data. Such techniques do not make any assumption as to the underlying mathematical model of the absorption kinetics. Indeed, such an approach is a powerful way of determining whether assumed models are indeed applicable (Roberts et al., 1997).

VIII. CONCLUSION

This chapter has attempted to overview some of the more important mathematical models used in percutaneous absorption. Given the substantive number of reported models and the complexity in many of the models, the overview is limited in its ability to give each of the models the credit they may deserve. However, it is hoped that the emphasis on the more practical models has enabled this fairly complex area to be presented in a manner useful for ready reference.

Our analysis has considered a number of boundary conditions associated with solute transport across a membrane, including clearance from the receptor solution, clearance from the membrane, and diffusion in an underlying layer (e.g., epidermis below stratum corneum). Each situation is defined by a steady-state flux J_{ss} and lag time of the forms:

$$J_{ss} = \frac{k_p C_v}{1 + M} \qquad (91)$$

$$\text{lag} = \frac{t_d}{6} N \qquad (92)$$

where M and N are functions of the transport processes below the membrane. When the clearance of the solute is very high ($Cl \gg k_p$), J_{ss} approaches the usual $k_p C_v$. Approximations for the lag time are less well defined, so that the use of lag time as an estimate for t_d is much less justified. Consequently, there are dangers of parameter misspecification with obvious consequences when extensions such as structure–transport relationships are based on the uncorrected parameters. Ultimately, therefore, mathematical modeling in this area is a balance between simplicity and an accurate representation of the underlying processes.

ACKNOWLEDGMENTS

We thank the NH&MRC of Australia and the New South Wales and Queensland Lions Medical Research Foundation for the support of this work.

REFERENCES

Ando HY, Ho NFH, Higuchi WI. Skin as an active metabolising barrier. 1. Theoretical analysis of topical bioavailability. J Pharm Sci 66(11):1525–1528, 1977.

Beastall J, Guy RH, Hadgraft J, Wilding I. The influence of urea on percutaneous absorption. Pharm Res 3(5):294–297, 1986.

Beckett AH, Gorrod JW, Taylor DC. Comparison of oral and percutaneous routes in man for systemic administration of ephedrine. J Pharm Pharmacol 24:Suppl. 65–70, 1972.

Benowitz NL, Chan K, Denaro CP, Jacob P. Stable isotope method for studying transdermal drug absorption: nicotine patch. Clin Pharmacol Ther 50(Sep): 286–293, 1991.

Bunge AL, Cleek RL. A new method for estimating dermal absorption from chemical exposure: 2 Effect of molecular weight and octanol-water partitioning. Pharm Res 12(1):88–95, 1995.

Chandrasekaran SK, Michaels AS, Campbell PS, Shaw JE. Scopolamine permeation through human skin in vitro. AIChE J 22(5):828–832, 1976.

Chandrasekaran SK, Bayne W, Shaw JE. Pharmacokinetics of drug permeation through human skin. J Pharm Sci 67(10):1370–1374, 1978.

Cleek RL, Bunge AL. A new method for estimating dermal absorption from chemical exposure. 1. General approach. Pharm Res 10(4):497–506, 1993.

Cooper ER. Pharmacokinetics of skin penetration. J Pharm Sci 65(9):1396–1397, 1976.

Cooper ER. Effect of diffusional lag time on multicompartmental pharmacokinetics for transepidermal infusion. J Pharm Sci 68(11):1469–1470, 1979.

Cooper ER, Berner B. Finite dose pharmacokinetics of skin penetration. J Pharm Sci 74(10):1100–1102, 1985.

Crank J. The Mathematics of Diffusion. 2nd ed. Oxford: Clarendon Press, 1975.

Cross SE, Roberts MS. The importance of dermal blood supply and the epidermis on the transdermal iontophoretic delivery of monovalent cations. J Pharm Sci 84:584–592, 1995.

Cross SE, Wu ZY, Roberts MS. Effect of perfusion flow rate on the tissue uptake of solutes after dermal application using the rat isolated perfused hindlimb preparation. J Pharm Pharmacol 46:844–850, 1994.

Cross SE, Wu ZY, Roberts MS. The effect of protein binding on the deep tissue penetration and elution of transdermally applied water, salicyclic acid, lignocaine and diazepam in the perfused rat. J Pharmacol Exp Ther 277:366–374, 1996.

Demana PH, Smith EW, Walker RB, Haigh JM, Kanfer I. Evaluation of the proposed FDA pilot dose-response methodology for topical corticosteroid bioequivalence testing. Pharm Res 14(3):303–308, 1997.

Edwards DA, Langer R. A linear theory of transdermal transport phenomena. J Pharm Sci 83(9):1315–1334, 1994.

Fox JL, Yu CD, Higuchi WI, Ho NFH. General physical model for simultaneous diffusion and metabolism in biological membranes. Computational approach for the steady state case. Int J Pharm 2:41–57, 1979.

Ghanem AH, Mahmoud H, Higuchi WI, Rohr UD, Borsadia S, Liu P, Fox JL, Good WR. The effects of ethanol on the transport of b estradiol and other permeants in the hairless mouse skin. II. A new quantitative approach. J Control Rel 6: 75–83, 1987.

Giengler G, Knoch A, Merkle HP. Modeling and numerical computation of drug transport in laminates: Model case evaluation of transdermal delivery system. J Pharm Sci 75(1):9–15, 1986.

Guy RH, Hadgraft J. Theoretical description relating skin penetration to the thickness of the applied medicament. Int J Pharm 6:321–332, 1980.

Guy RH, Hadgraft J. Percutaneous metabolism with saturable enzyme kinetics. Int J Pharm 11:187–197, 1982.

Guy RH, Hadgraft J. Physicochemical interpretation of the pharmacokinetics of percutaneous absorption. J Pharmacokinet Biopharm 11(2):189–203, 1983.

Guy RH, Hadgraft J. A theoretical description of the effects of volatility and substantivity on percutaneous absorption. Int J Pharm 18:139–147, 1984a.

Guy RH, Hadgraft J. Pharmacokinetics of percutaneous absorption with concurrent metabolism. Int J Pharm 20:43–51, 1984b.

Guy RH, Hadgraft J. Percutaneous absorption kinetics of topically applied agents liable to surface loss. J Soc Cosmet Chem 35:103–113, 1984c.

Guy RH, Hadgraft J. Prediction of drug disposition kinetics in skin and plasma following topical administration. J Pharm Sci 73(7):883–887, 1984d.

Guy RH, Hadgraft J. The effect of penetration enhancers on the kinetics of percutaneous absorption. J Control Rel 5:43–51, 1987.

Guy RH, Hadgraft J, Maibach HI. A pharmacokinetic model for percutaneous absorption. Int J Pharm 11:119–129, 1982.

Guy RH, Hadgraft J, Maibach HI. Percutaneous absorption: Multidose pharmacokinetics. Int J Pharm 17:23–28, 1983.

Hadgraft J. Calculations of drug release rates from controlled release devices. The slab. Int J Pharm 2:177–194, 1979a.

Hadgraft J. The epidermal reservoir: A theoretical approach. Int J Pharm 2:265–274, 1979b.

Hadgraft J. Theoretical aspects of metabolism in the epidermis. Int J Pharm 4:229–239, 1980.

Hadgraft J, Wolff HM. In vitro/in vivo correlations in transdermal drug delivery. In: Roberts MS, Walters KA, eds. Dermal Absorption and Toxicity Assessment. New York: Marcel Dekker, 1998.

Hatanaka T, Inuma M, Sugibayashi K, Morimoto Y. Prediction of skin permeability of drugs. II. Development of composite membrane as a skin alternative. Int J Pharm 79:3452–3459, 1990.

Higuchi WI. Diffusional models useful in biopharmaceutics. Drug release rate process. J Pharm Sci 56(3):315–324, 1967.

Higuchi WI, Higuchi T. Theoretical analysis of diffusional movements through heterogeneous barriers. J Am Pharm Assoc Sci Ed 49(9):598–606, 1960.

Imanidis G, Song W, Lee PH, Su MH, Kern ER, Higuchi WI. Estimation of skin target site acyclovir concentrations following controlled transdermal drug delivery in topical and systemic treatment of cutaneous HSV-1 infections in hairless mice. Pharm Res 11(7):1035–1041, 1994.

Jepson GW, McDougal JN. Physiologically based modeling of nonsteady state dermal absorption of halogenated methanes from an aqueous solution. Toxicol Appl Pharmacol 144:315–324, 1997.

Jiang R, Roberts MS, Collins DM, Benson HAE. Absorption of sunscreens into human skin: an evaluation of commercial products for children and adults. Br J Clin Pharmacol 1998.

Kakemi K, Kameda H, Kakemi M, Ueda M, Koizumi T. Model studies on percutaneous absorption and transport in the ointment. I. Theoretical aspects. Chem Pharm Bull 23(9):2109–2113, 1975.

Kasting GB. Theoretical models for iontophoretic delivery. Adv Drug Delivery Rev 9:177–199, 1992.

Kubota K. A compartment model for percutaneous drug absorption. J Pharm Sci 80(5):502–504, 1991.

Kubota K, Ishizaki T. A theoretical consideration of percutaneous drug absorption. J Pharmacokinet Biopharm 13(1):55–72, 1985.

Kubota K, Ishizaki T. A diffusion-diffusion model for percutaneous drug absorption. J Pharmacokinet Biopharm 14(4):409–439, 1986.

Kubota K, Maibach HI. Compartment model for percutaneous absorption: compatibility of lag time and steady-state flux with diffusion model. J Pharm Sci 81(9):863–865, 1992.

Kubota K, Yamada T. Finite dose percutaneous drug absorption: Theory and its application to in vitro timolol permeation. J Pharm Sci 79(11):1015–1019, 1990.

Kubota K, Sznitowska M, Maibach HI. Percutaneous absorption: A single-layer model. J Pharm Sci 82(5):450–456, 1993a.

Kubota K, Koyama E, and Twizell EH. Dual sorption model for the nonlinear percutaneous permeation kinetics of timolol. J Pharm Sci 82(12):1205–1208, 1993b.

Kubota K, Koyama E, Yasuda K. Random walk method for percutaneous drug absorption pharmacokinetics: Application to repeated administration of a therapeutic timolol patch. J Pharm Sci 80(8):752–756, 1991.

Kubota K, Ademola J, Maibach HI. Simultaneous diffusion and metabolism of betamethasone 17-valerate in the living skin equivalent. J Pharm Sci 84(12):1478–1481, 1995.

McCarley KD, Bunge AL. Physiologically relevant one-compartment pharmacokinetic models for skin. 1. Development of models. J Pharm Sci 87(4):470–481, 1998.

McDougal JN. Prediction-physiological models. In: Roberts MS, Walters KA, eds. Dermal Absorption and Toxicity Assessment. New York: Marcel Dekker, 1998.

McNeil SC, Potts RO, Francoeur ML. Local enhanced topical delivery (LETD) of drugs: Does it truly exist? Pharm Res 9:1422–1427, 1992.

Mitragotri S, Blankschtein D, Langer R. An explanation for the variation of the sonophoretic transdermal transport enhancement from drug to drug. J Pharm Sci 86(10):1190–1192, 1997.

Parry GE, Bunge AL, Silcox GD, Pershing LK, Pershing DW. Percutaneous absorption of benzoic acid across human skin. In vitro experiments and mathematical modeling. Pharm Res 7(3):230–236, 1990.

Pirot F, Kalia YN, Stinchcomb AL, Keating G, Bunge A, and Guy RH. Characterization of the permeability barrier of human skin in vivo. Proc Natl Acad Sci USA 94(4):1562–1567, 1997.

Pugh WJ, Hadgraft J, Roberts MS. Physicochemical determinants of stratum corneum permeation. In: Roberts MS, Walters KA, eds. Dermal Absorption and Toxicity Assessment. New York: Marcel Dekker, 1998.

Purves RD. Accuracy of numerical inversions of Laplace transforms for pharmacokinetic parameter estimation. J Pharm Sci 84:71–74, 1995.

Reddy MB, McCarley KD, Bunge AL. Physiologically relevant one-compartment pharmacokinetic model for skin. 2. Comparison of models when combined with a systemic pharmacokinetic model. J Pharm Sci 87(4):482–490, 1998.

Reifenrath WG, Robinson PB. In vitro skin evaporation and penetration characteristics of mosquito repellents. J Pharm Sci 71(9):1014–1018, 1982.

Riegelman S. Pharmacokinetics. Pharmacokinetic factors affecting epidermal penetration and percutaneous adsorption. Clin Pharmacol Ther 16(5 Part 2):873–883, 1974.

Roberts MS. Structure-permeability considerations in percutaneous absorption. In: Scott RC, Guy RH, Hadgraft J, eds. Prediction of Percutaneous Penetration. Vol. 2. London: IBC Technical Services, 1991.

Roberts MS, Walters KA. The relationship between structure and barrier function of skin. In: Roberts MS, Walters KA, eds. Dermal Absorption and Toxicity Assessment. New York: Marcel Dekker, 1998.

Roberts MS, Anderson RA, Swarbrick J. Permeability of human epidermis to phenolic compounds. J Pharm Pharmacol 29(11):677–683, 1977.

Roberts MS, Anderson RA, Swarbrick J, Moore DE. The percutaneous absorption of phenolic compounds: The mechanism of diffusion across the stratum corneum. J Pharm Pharmacol 30(8):486–490, 1978.

Roberts MS, Lipschitz S, Campbell AJ, Wanwimolruk S, McQueen EG, McQueen M. Modeling of subcutaneous absorption kinetics of infusion solutions in the elderly using technetium. J Pharmacokinet Biopharm 25(1):1–19, 1997.

Roberts MS, Lai PM, Anissimov YG. Ionic mobility—Pore model for iontophoretic transport. Pharm Res 15(10):1569–1578, 1998.

Robinson PJ. Prediction: Simple risk models and overview of dermal risk assessment. In: Roberts MS, Walters KA, eds. Dermal Absorption and Toxicity Assessment. New York: Marcel Dekker, 1998.

Rohatagi S, Barrett JS, Dewitt KE, Morales RJ. Integrated pharmacokinetics and metabolic modeling of selegiline and metabolites after transdermal administration. Biopharm Drug Dispos 18(7):567–584, 1997.

Scheuplein RJ. Mechanism of percutaneous absorption. II. Transient diffusion and the relative importance of various routes of skin penetration. J Invest Dermatol 48(1):79–88, 1967.

Scheuplein R. Permeability of the skin: A review of major concepts. Curr Probl Dermatol 7:172–186, 1978.

Scheuplein RJ, Blank IH. Mechanism of percutaneous absorption IV. Penetration of nonelectrolytes (alcohols) from aqueous solutions and from pure liquids. J Invest Dermatol 60:286–296, 1973.

Scheuplein RJ, Ross LW. Mechanism of percutaneous absorption. V. Percutaneous absorption of solvent deposited solids. J Invest Dermatol 62:353–360, 1974.

Shah JC. Application of kinetic model to in vitro percutaneous permeation of drugs. Int J Pharm 133:179–189, 1996.

Siddiqui O, Roberts MS, Polack AE. Percutaneous absorption of steroids relative contributions of epidermal penetration and dermal clearance. J Pharmacokinet Biopharm 17(Aug):405–424, 1989.

Singh P, Roberts MS. Blood flow measurements in skin and underlying tissues by microsphere method: Application to dermal pharmacokinetics of polar non-electrolytes. J Pharm Sci 82(9):873–879, 1993.

Singh P, Roberts MS, Maibach HI. Modeling of plasma levels of drugs following transdermal iontophoresis. J Control Rel 33:293–298, 1995.

Singh GJ, Fleischer N, Lesko L, Williams R. Evaluation of the proposed FDA pilot dose-response methodology for topical corticosteroid bioequivalence testing. Pharm Res 15(1):4–7, 1998a.

Singh P, Maibach HI, Roberts MS. Site of effects. In: Roberts MS, Walters KA, eds. Dermal Absorption and Toxicity Assessment. New York: Marcel Dekker, 1998b.

Smith EW, Walker RB, Haigh JM, Kanfer I. Evaluation of the proposed FDA pilot dose-response methodology for topical corticosteroid bioequivalence testing. The authors reply. Pharm Res 15(1):5–7, 1998.

Tojo K, Chiang CC, Chien YW. Drug permeation across the skin: Effect of penetrant hydrophilicity. J Pharm Sci 76(2):123–126, 1987.

Wallace SM, Barnett G. Pharmacokinetic analysis of percutaneous absorption: Evidence of parallel penetration pathways for methotrexate. J Pharmacokinet Biopharm 6(4):315–325, 1978.

Walters KA, Roberts MS, Gettings S. Percutaneous absorption of sunscreens. In: Bronaugh R, Maibach HI, eds. Percutaneous Absorption. 3rd ed. New York: Marcel Dekker, 1998.

Williams PL, Carver MP, Riviere JE. A physiologically relevant pharmacokinetic model of xenobiotic percutaneous absorption using the isolated perfused porcine skin flap. J Pharm Sci 79(4):305–311, 1990.

Wu MS. Determination of concentration-dependent water diffusivity in a keratinous membrane. J Pharm Sci 72(12):1421–1423, 1983.

Yamashita F, Yoshioka T, Koyama Y, Okamoto H, Sezaki H, Hashida M. Analysis of skin penetration enhancement based on a two layer skin diffusion model with polar and nonpolar routes in the stratum corneum: Dose-dependent effect of 1-geranylazacycloheptan-2-one on drugs with different lipophilicities. Biol Pharm Bull 16(7):690–697, 1993.

Yano Y, Yamaoka K, Aoyama Y, Tanaka H. Two-compartment dispersion model for analysis of organ perfusion system of drugs by fast inverse Laplace transform (FILT). J Pharmacokinet Biopharm 17:179–202, 1989a.

Yano Y, Yamaoka K, Tanaka H. A non-linear least squares program, MULTI(FILT), based on fast inverse Laplace transform (FILT) for microcomputers. Chem Pharm Bull 37:1535–1538, 1989b.

Yu CD, Fox JL, Ho NFH, Higuchi WI. Physical model evaluation of topical prodrug delivery—Simultaneous transport and bioconversion of vidarabine 5′-valerate. Part 1. Physical model development. J Pharm Sci 68(11):1341–1346, 1979a.

Yu CD, Fox JL, Ho NFH, Higuchi WI. Physical model evaluation of topical prodrug delivery—Simultaneous transport and bioconversion of vidarabine 5′-valerate. Part 2. Parameter determinations. J Pharm Sci 68(11):1347–1357, 1979b.

Yu CD, Gordon NA, Fox JL, Higuchi WI, Ho NFH. Physical model evaluation of topical prodrug delivery—Simultaneous transport and bioconversion of vidarabine 5′-valerate. Part 5. Mechanistic analysis of influence of nonhomogeneous enzyme distributions in hairless mouse skin. J Pharm Sci 69(7):775–780, 1980.

2
Cutaneous Metabolism During In Vitro Percutaneous Absorption

Robert L. Bronaugh, Margaret E. K. Kraeling, Jeffrey J. Yourick, and Harolyn L. Hood*
Food and Drug Administration, Laurel, Maryland

I. INTRODUCTION

The skin is a portal of entry and the largest organ of the body. It has been shown to contain the major enzymes found in other tissues of the body (1). Topically applied compounds may be metabolized in skin resulting in altered pharmacologic or toxicologic activity.

The metabolism of benzo[a]pyrene applied to mouse skin floating in an organ culture demonstrated the potential importance of metabolism of chemicals during percutaneous absorption (2). A flow-through diffusion cell was subsequently developed to aid in quantitating skin absorption and metabolism (3). Viability of skin in the diffusion cell, which was assessed by light microscopy, was found to be maintained for at least 17 h (3).

The viability of pig skin was maintained in flow-through cells by using tissue culture media; skin-mediated hydrolysis of diethyl malonate was observed (4). The suitability of these conditions to maintain skin viability was assessed in initial studies by the ability to graft the skin to nude mice (5).

II. SKIN VIABILITY

The use of viable skin in percutaneous absorption studies is essential for investigating skin metabolism of absorbed compounds. The viability of skin in flow-through diffusion cells was systematically examined (6). Viability

---

Present affiliation: Bristol-Myers Squibb Worldwide Beauty Care, Stamford, Connecticut.

could be conveniently determined from glucose utilization by skin by measuring lactate appearing in the receptor fluid. Viability could be assessed throughout the course of the experiment. It was observed that a HEPES-buffered Hanks balanced salt solution was equivalent to minimal essential media in maintaining skin viability for at least 24 h. The viability of skin was also confirmed by electron microscopy and by the maintenance of estradiol and testosterone metabolism.

III. SKIN METABOLISM

The following summary of skin absorption/metabolism studies illustrates the types of compounds that are metabolized in skin. In many cases the metabolites formed from the parent compounds have been determined and therefore important metabolic reactions in skin have been identified.

In early studies from our laboratory the penetration and metabolism of estradiol and testosterone (6), acetylethyl tetramethyl tetralin (AETT) and butylated hydroxytoluene (BHT) (7), benzo[a]pyrene and 7-ethoxycoumarin (8), and azo colors (9) were examined by using viable dermatome skin sections from mice, rats, hairless guinea pigs, and humans. These early studies are not discussed here, but may be examined separately by the interested reader.

The percutaneous absorption and metabolism of three structurally related compounds, benzoic acid, p-aminobenzoic acid (PABA), and ethyl aminobenzoate (benzocaine), were determined in vitro with hairless guinea pig and human skin (10). Approximately 7% of the absorbed benzoic acid was conjugated with glycine to form hippuric acid (Figure 1).

Acetylation of primary amines was found to be an important metabolic step in skin. For benzocaine, a molecule susceptible to both N-acetylation and ester hydrolysis, 80% of the absorbed material was acetylated, while less than 10% of the absorbed ester was hydrolyzed (Fig. 1). PABA was much more slowly absorbed than benzocaine and was also less extensively N-acetylated (Fig. 1). Acetyl-PABA was found primarily in the receptor fluid at the end of the experiments, but the receptor fluid contained only 20% of the absorbed dose. Much of the absorbed PABA remained unmetabolized and in the skin, as might be expected for an effective sunscreen agent. The compound in skin would probably not have been exposed to N-acetylating enzymes if it was localized primarily in the stratum corneum. A similar pattern of benzocaine metabolism was observed in human and hairless guinea pig skin; however, there appeared to be less enzyme activity in human skin.

The extent of metabolism of radiotracer doses of benzocaine in the studies just cited was compared to metabolism of much larger doses of

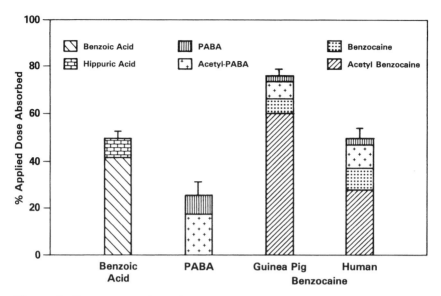

Figure 1 Percutaneous absorption and metabolism of benzoic acid, *p*-aminobenzoic acid (PABA), and benzocaine. All compounds were studied in hairless guinea pig skin. Benzocaine was also studied in human skin.

benzocaine in formulations simulating exposure from use of topical benzocaine anesthetic products (11). The distribution of [^{14}C]benzocaine and metabolites as the percentage of the absorbed dose in receptor fluid and skin was compared by using a radiotracer dose (2 μg/cm^2), an intermediate dose (40 μg/cm^2) and a therapeutic dose (200 μg/cm^2) (Table 1). When a therapeutic dose was applied, the metabolism of benzocaine was reduced, presumably because of the saturation of enzyme activity in skin. However, 34% of the absorbed dose was still converted to acetylbenzocaine. Although the percent applied dose absorbed decreased with increasing dose of benzocaine, total absorption of benzocaine and metabolites increased as the applied dose increased (Table 2). Benzocaine and acetylbenzocaine were found to have similar potencies in blocking nerve conduction in the isolated squid giant axon.

Esterase activity and alcohol dehydrogenase activity were characterized in hairless guinea pig skin with the model compounds methyl salicylate and benzyl alcohol (12). Subsequently, the absorption and metabolism of the cosmetic ingredient retinyl palmitate were determined in human and hairless guinea pig skin.

The metabolism of methyl salicylate was determined in viable and nonviable hairless guinea pig skin (Table 3). In viable skin over 50% of the

Table 1 Percent of Absorbed Dose of Benzocaine and Metabolites Over 24 h in Hairless Guinea Pig Skin at Different Dose Levels

Location and compound[a]	Dose level		
	2 µg/cm²	40 µg/cm²	200 µg/cm²
Receptor fluid			
Benzocaine	9.6 ± 4.2	50.7 ± 6.6	54.0 ± 5.2
AcBenz	83.8 ± 4.4	43.8 ± 5.7	37.9 ± 3.6
PABA	1.0 ± 0.3	0.1 ± 0.1	0.9 ± 0.5
AcPABA	5.1 ± 1.0	5.8 ± 0.9	7.2 ± 1.7
Skin			
Benzocaine	40.0 ± 8.5	4.9 ± 4.9	62.7 ± 12.2
AcBenz	10.3 ± 10.3	66.6 ± 8.2	20.9 ± 11.7
PABA	4.3 ± 4.3	12.6 ± 12.6	1.5 ± 1.3
AcPABA	24.7 ± 14.9	16.0 ± 16.0	14.9 ± 1.2
Total[b]			
Benzocaine	10.7 ± 3.3	49.9 ± 6.5	57.3 ± 3.7
AcBenz	80.5 ± 3.8	43.6 ± 5.6	34.3 ± 3.4
PABA	1.4 ± 0.2	0.1 ± 0.1	0.9 ± 0.5
AcPABA	6.5 ± 1.0	5.9 ± 1.0	7.6 ± 2.0

Note. Values are the mean ± SE for 1–6 determinations in each of 4 animals. From Kraeling et al. (1996).
[a]AcBenz = *N*-acetylbenzocaine; PABA = *p*-aminobenzoic acid; AcPABA = *N*-acetyl-*p*-aminobenzoic acid.
[b]Percentages are similar to receptor fluid values, since most of the absorbed dose was in the receptor fluid.

Table 2 Percent of Applied Dose of Benzocaine Absorbed Over 24 h in Hairless Guinea Pig Skin at Different Dose Levels

Location	Dose level		
	2 µg/cm²	40 µg/cm²	200 µg/cm²
Receptor fluid	64.2 ± 8.6	75.5 ± 3.5	31.5 ± 4.6
Skin	3.1 ± 0.5	0.9 ± 0.3	3.1 ± 1.2
Total absorbed	67.3 ± 8.8	76.4 ± 3.3	34.0 ± 3.6

Note. Values are the mean ± SE for 1–6 determinations in each of 4 animals. From Kraeling et al. (1996).

Table 3 Metabolism of Methyl Salicylate in Hairless Guinea Pig Skin (Percent Absorbed Dose Metabolized)

| Sex | Viable skin | | | Nonviable skin, salicylic acid |
	Salicyluric acid	Salicylic acid	Total	
Male	20.9 ± 5.4	35.6 ± 6.5[a]	56.5 ± 5.1[a,b]	38.3 ± 5.0[a,b]
Female	12.5 ± 3.5	12.3 ± 2.5[a]	24.8 ± 3.0[a,b]	13.4 ± 2.8[a,b]

Note. Values are the mean ± SE of determinations in 3 animals (3–4 repetitions per animal). From Boehnlein et al. (1994).
[a]Significant male vs. female difference by the two-tailed t-test ($p < .01$).
[b]Significant viable vs. nonviable skin difference, same sex ($p < .05$).

absorbed compound was hydrolyzed by esterases in skin to salicylic acid. Twenty-one percent of the absorbed compound was further conjugated with glycine to form salicyluric acid. Greater esterase activity was observed in male skin. Esterase is a stable enzyme, and hydrolysis of methyl salicylate also occurred in nonviable skin. However, no conjugation of salicylic acid was observed in nonviable skin.

Oxidation of benzyl alcohol was also observed in hairless guinea pig skin (Table 4). Approximately 50% of the absorbed benzyl alcohol was oxidized to benzoic acid in viable skin, with a small portion of this compound being further metabolized to the glycine conjugate, hippuric acid. As with the ester, significant activity was also observed in nonviable skin and greater oxidation of the alcohol was obtained with male skin.

Table 4 Metabolism of Benzyl Alcohol in Hairless Guinea Pig Skin (Percent Absorbed Dose Metabolized)

| Sex | Viable skin | | | Nonviable skin, benzoic acid |
	Hippuric acid	Benzoic acid	Total	
Male	8.5 ± 1.9	44.2 ± 8.0[a]	52.7 ± 9.6[a]	44.0 ± 11.2[b]
Female	4.1 ± 1.8	16.0 ± 8.4[a]	20.1 ± 9.6[a]	12.2 ± 6.0[b]

Note. Values are the mean ± SE of determinations in 3 animals (3–4 repetitions per animal). From Boehnlein et al. (1994).
[a]Marginally significant male vs. female difference by the two-tailed t-test ($p < .08$).
[b]Significant male vs. female difference ($p < .05$).

The absorption and metabolism of retinyl palmitate were measured to see if ester hydrolysis and alcohol oxidation occurred with this cosmetic ingredient (Table 5). Skin absorption for this lipophilic material is the sum of the absorbed compound in skin and in the receptor fluid at the end of the 24-h study. Most of the absorbed radiolabel remained in the skin. Substantial amounts of the absorbed compound were hydrolyzed to retinol, but no oxidation of the alcohol to retinoic acid was observed. Any effects of retinyl palmitate on the structure of skin may be due to the formation of retinol during percutaneous absorption.

Absorption values from in vitro studies with viable hairless guinea pig skin have been found to compare closely with in vivo results for phenanthrene (13) and for pyrene, benzo[a]pyrene, and di(2-ethylhexyl) phthalate (14). Also, significant metabolism was observed in vitro during the absorption of all four compounds. Phenanthrene was metabolized in vitro to 9,10-dihydrodiol, 3,4-dihydrodiol, 1,2-dihydrodiol, and traces of hydroxy phenanthrenes (13). After topical administration of phenanthrene, approximately 7% of the percutaneously absorbed material was metabolized to the dihydrodiol metabolites.

Numerous metabolites of benzo[a]pyrene were formed during percutaneous absorption through hairless guinea pig skin (14). Of particular interest was the identification of benzo[a]pyrene 7,8,9,10-tetrahydrotetrol in the diffusion cell receptor fluid. This metabolite is the hydrolysis product of the ultimate carcinogen, 7,8-dihydroxy-9,10-epoxy-7,8,9,10-tetrahydrobenzo[a]pyrene. This study demonstrates that skin metabolism is likely re-

Table 5 Percutaneous Absorption and Metabolism of Retinyl Palmitate

	Skin		Receptor fluid	
	Radiolabel absorbed (%)[b]	Metabolized (%)[c]	Radiolabel absorbed (%)[a,b]	Metabolized (%)[c]
Male guinea pig	29.8 ± 4.5	38.2 ± 13.0	0.5 ± 0.2	100
Female guinea pig	33.4 ± 2.3	30.2 ± 16.3	0.6 ± 0.3	100
Female human	17.9 ± 1.3	43.9 ± 5.0	0.2 ± 0.01	100

Note. Values are the mean ± SE of determinations from 2 human donors (3–4 repetitions per donor) and 3 animals (3 repetitions per animal). From Boehnlein et al. (1994).
[a]0–24 h fractions are combined.
[b]Absorption is expressed as percent of applied dose in skin and receptor fluid.
[c]Metabolism is expressed as percent of the absorbed retinyl palmitate hydrolyzed to retinol.

sponsible for skin tumors formed following topical benzo[a]pyrene administration. In the earlier phenanthrene study (13), no know carcinogenic metabolites were formed during skin permeation. This finding is consistent with the lack of tumorigenicity of phenanthrene in rodents.

The percutaneous absorption and metabolism of trinitrobenzene have recently been examined in human, rat, and hairless guinea pig skin (15). Rapid absorption of trinitrobenzene was observed through human and animal skin. The two major metabolities found were 1,3,5-benzene triacetamide and 3,5-dinitroaniline. It appears that nitro groups on trinitrobenzene can be reduced in skin to amino groups, which are sometimes further metabolized by acetylation to an acetamide derivative.

The effect of skin metabolism on the biological response to topically applied chemicals is only beginning to be investigated. The task is complicated since skin metabolism is difficult to measure in vivo without interference from systemic enzymes. In addition, certain metabolic systems in skin, such as cytochrome P-450, have relatively low activity when compared with liver. In vitro studies indicate that significant metabolism can occur during the percutaneous absorption process.

REFERENCES

1. Pannatier A, Jenner P, Testa B, Etter JC. The skin as a drug metabolizing organ. Drug Metab Rev 1978; 8:319–343.
2. Smith LH, Holland JM. Interaction between benzo[a]pyrene and mouse skin in organ culture. Toxicology 1981; 24:47–57.
3. Holland JM, Kao JY, Whitaker MJ. A multisample apparatus for kinetic evaluation of skin penetration in vitro: The influence of viability and metabolic status of skin. Toxicol Appl Pharmacol 1984; 72:272–280.
4. Chellquist EM, Reifenrath WG. Distribution and fate of diethyl malonate and diisopropyl fluorophosphate on pig skin in vitro. J Pharm Sci. 1988; 77:850–854.
5. Hawkins GS, Reifenrath WG, Influence of skin source, penetration cell fluid, and partition coefficient on in vitro skin penetration. J Pharm Sci. 1986; 75:378–381.
6. Collier SW, Sheikh NM, Sakr A, Lichtin JL, Stewart RF, Bronaugh RL. Maintenance of skin viability during in vitro percutaneous absorption/metabolism studies. Toxicol Appl Pharmacol. 1989; 99:522–533.
7. Bronaugh RL, Stewart RF, Strom JE. Extent of cutaneous metabolism during percutaneous absorption of xenobiotics. Toxicol Appl Pharmacol. 1989; 99:534–543.
8. Storm JE, Collier SW, Stewart RF, Bronaugh RL. Metabolism of xenobiotics during percutaneous penetration: Role of absorption rate and cutaneous enzyme activity. Fundam Appl Toxicol. 1990; 15:132–141.

9. Collier SW, Storm JE, Bronaugh RL. Reduction of azo dyes during in vitro percutaneous absorption. Toxicol Appl Pharmacol. 1993; 118:73–79.

10. Nathan D, Sakr A, Lichtin JL, Bronaugh RL. In vitro skin absorption and metabolism of benzoic acid, p-aminobenzoic acid, and benzocaine in the hairless guinea pig. Pharm Res. 1990; 7:1145–1151.

11. Kraeling MEK, Lipicky RJ, Bronaugh RL. Metabolism of benzocaine during percutaneous absorption in the hairless guinea pig: Acetylbenzocaine formation and activity. Skin Pharmacol. 1996; 9:221–230.

12. Boehnlein J, Sakr A, Lichtin JL, Bronaugh RL. Characterization of esterase and alcohol dehydrogenase activity in skin. Metabolism of retinyl palmitate to retinol (vitamin A) during percutaneous absorption. Pharm Res. 1994: 11: 1155–1159.

13. Ng KME, Chu I, Bronaugh RL, Franklin CA, Somers DA. Percutaneous absorption/metabolism of phenanthrene in the hairless guinea pig: Comparison of in vitro and in vivo results. Fundam Appl Toxicol. 1991; 16:517–524.

14. Ng KME, Chu I, Bronaugh RL, Franklin CA, Somers DA. Percutaneous absorption and metabolism of pyrene, benzo[a]pyrene, and di(2-ethylhexyl) phthalate: Comparison of in vitro and in vivo results in the hairless guinea pig. Toxicol Appl Pharmacol. 1992; 115:216–223.

15. Kraeling MEK, Reddy G, Bronaugh RL. Percutaneous absorption of trinitrobenzene. Animal models for human skin. J Appl Toxicol (in press).

3
Cutaneous Metabolism of Xenobiotics

Saqib J. Bashir and Howard I. Maibach
University of California School of Medicine, San Francisco, California

I. INTRODUCTION

The human skin is exposed to many topical agents, either intentionally or by accident. The variety of these foreign agents (xenobiotics) reflects the variety of their intended uses: Cosmetics are intended, in theory, to decorate the skin rather than penetrate it, while dermatological drugs such as corticosteroids are intended to act locally within the skin, with little or minimal systemic action. Some drugs, such as nitroglycerin, are not intended to act at the skin, but at distant target organs, in this case, the coronary arteries.

Therefore, it is clear that the application of substances to human skin is widespread. One must pause to consider the consequences of this behavior with respect to the skin, and the body as a whole. Although many preparations are placed on the skin on the assumption that the skin is biologically inert, this chapter demonstrates that many exogenous compounds are metabolized in skin (xenobiotic metabolism). We review the existence of enzymes that are capable of metabolizing cutaneous xenobiotics, and some of the factors regulating their activity. Recent work documenting the metabolism of commonly prescribed drugs and also the metabolism of environmental agents on the skin is also reviewed.

The role of cutaneous xenobiotic metabolism in the production of toxic metabolites, irritants, and allergens is discussed, in addition to the implication of cutaneous metabolism on transdermal drug delivery in healthy and damaged skin (Fig. 1).

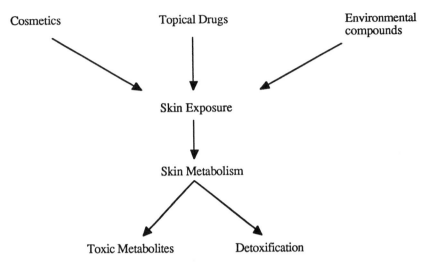

Figure 1   Metabolism of xenobiotics.

II.   XENOBIOTIC-METABOLIZING ENZYMES

These are enzymes participating in the metabolism of foreign compounds. They metabolize substrates that are predominantly lipophilic (and thus penetrate the skin well) into substances that are hydrophilic and less active and can then be excreted in the urine via the kidney.

There are two distinct metabolic steps in this process. The first step is known as the phase I reaction and introduces a polar reactive group into a molecule, which renders the molecule suitable for further metabolism as part of the phase II reaction.

Phase I reactions include metabolism by cytochrome P-450-dependent monooxygenases, which have been demonstrated in skin (Goerz, 1987). These enzymes add a single oxygen atom from a molecule of O_2 to a carbon atom, resulting in the formation of an -OH group on the substrate (hydroxylation) and one molecule of water, H_2O.

Subsequently, these metabolites formed by the phase I reaction by undergo further metabolism, known as phase II reactions. These are conjugation reactions, which render the substrate more hydrophilic, allowing renal excretion. Metabolites can be conjugated with substances such as glucuronic acid, sulfur, or glutathione, resulting in the production of easily excretable products.

III. PHASE I METABOLISM: CYTOCHROME P-450 MONOOXYGENASES

The cytochrome P-450 nonooxygenase enzymes are microsomal enzymes demonstrated in the liver and other organs including skin (Gonzalez, 1989). They play an important role in the phase I metabolism of both exogenous and endogenous compounds such as fatty acids, prostaglandins, leukotrienes, and steroid hormones, and it is has been suggested that many dermatological topical drugs are suitable substrates for this enzyme (Ahmad et al., 1996).

Cytochrome P-450 enzymes are cofactor-dependent enzymes: They require energy from an external source such as NADPH to catalyze the reaction. This is in contrast to cofactor-independent reactions, which require only the enzyme to catalyze the reaction.

Cytochrome P-450 exists in both prokaryotes and eukaryotes. In eukaryotes, the enzyme is mainly located in the membranes of the endoplasmic reticulum and the mitochondria. The structure of cytochrome P-450 is a protoporphyrin ring that contains a centrally placed Fe^{3+} and a polypeptide chain of approximately 45,000 to 55,000 kD (Goeptar et al., 1995).

The substrate to be metabolized binds to the protein moiety of the cytochrome P-450, inducing a conformational change. This triggers the necessary cofactor NADPH–P-450 reductase, which donates an electron to the cytochrome P-450; the Fe^{3+} is reduced to Fe^{2+}.

This reduced cytochrome P-450–substrate complex may now bind to a molecule of oxygen. Another electron is donated from NADPH–P-450 reductase; the oxygen molecule is split into two oxygen atoms, with one binding to the substrate, which is then released from the enzyme as a hydroxylated product. The second oxygen atom is released as water (Fig. 2) (Goeptar et al., 1995).

Evidence for the existence of cutaneous cytochrome P-450 was initially obtained from the study of the carcinogenic effects of polycyclic aromatic hydrocarbons on the skin. The carcinogenic consequences of cutaneous metabolism are discussed later. Other more recent studies have demonstrated that cytochrome P-450 metabolizes topically applied medications in a fashion similar to the metabolism of systemic medications by the liver.

For example, recent work has demonstrated the relevance of cytochrome P-450 in the context of therapeutic agents. Ademola et al. (1992) studied the diffusion and metabolism of theophylline using a flow-through in vitro system. Theophylline is metabolized in the liver by monooxygenases to the metabolites 1,3-dimethyluric acid, 3-methylxanthine, and 1-methyluric acid. In this study of in vitro human skin metabolism of theophylline, these metabolites were also found, suggesting that the cytochrome P-450-dependent enzymes in the skin had metabolized the xenobiotic.

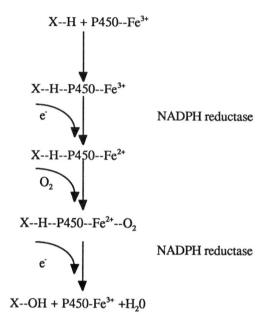

$$X\text{--}H + P450\text{--}Fe^{3+}$$

$$X\text{--}H\text{--}P450\text{--}Fe^{3+}$$

NADPH reductase

$$X\text{--}H\text{--}P450\text{--}Fe^{2+}$$

$$O_2$$

$$X\text{--}H\text{--}P450\text{--}Fe^{2+}\text{--}O_2$$

NADPH reductase

$$X\text{--}OH + P450\text{-}Fe^{3+} + H_2O$$

Key: X =substrate
       P450 = cytochromeP-450 enzyme

Figure 2   Mechanism of action of cytochrome P-450.

A. Isoenzymes of Cytochrome P-450

Many isoenzymes of cytochrome P-450 exist (Table 1), and there are many genes that encode for them. No particular isoenzyme has unique substrate specificity; rather, there is an overlap of substrates. The isoenzymes are categorized by their amino acid similarities into families, named with the root CYP followed by the family number, a capital letter denoting the sub-family, and a number identifying the particular form.

The family CYP1 has been implicated in xenobiotic metabolism and the families CYP2 and CYP3 in the metabolism of both xenobiotics and steroids. The CYP1A1 is a well-studied member of the cytochrome P-450 family, and is expressed in the skin (Bickers et al., 1982). The CYPs, including CYP1A1, are normally expressed at a low level in the skin; however, their activity can be induced by a variety of agents, discussed later in this chapter.

Table 1 Cytochrome P-450 Isomers Determined in
Mammals and Their Functions

Isomer	Function
CYP 1	Metabolism of xenobiotics
CYP 2	Metabolism of xenobiotics and steroids
CYP 3	Metabolism of xenobiotics and steroids
CYP 4	Fatty acid ω and ω-1 hydroxylation
CYP 5	Thromboxane synthase
CYP 7	Cholesterol 7α-hydroxylase
CYP 11	Steroid 11β-hydroxylase
CYP 17	Steroid 17β-hydroxylase
CYP 19	Aromatase
CYP 21	Steroid 21-hydroxylase
CYP 24	Vitamin D-25 hydroxylase
CYP 27	Cholesterol 27-hydroxylase

IV. PHASE II METABOLISM

Much of the literature on cutaneous metabolism of xenobiotics focuses on phase I reactions, especially on the role of cytochrome P-450 enzymes. However, phase I reactions are only part of the metabolic process. Following the phase I reaction, the metabolite must be conjugated to facilitate its elimination.

A. Transferases

Transferase activities in the skin can be as high as 10% of that of liver. In comparison, the relative activity of cytochrome P-450 in skin may be only 1–5% of the liver's (Merk et al., 1996). Raza et al. (1991) have demonstrated the presence of glutathione S-transferase in skin. Higo et al. (1992a) have shown that the cutaneous metabolism of nitroglycerin (GTN) to 1,2-GDN (glyceryl di-nitrate) and 1,3-GDN is heavily dependent on the presence of glutathione, which is a cofactor for the transferase enzyme. They exposed GTN to skin homogenates with and without the glutathione cofactor to determine its role in cutaneous metabolism. In the tissue with the cofactor, 30% of the GTN was metabolized within 2 h, whereas only 5% of the GTN was metabolized in the tissue without glutathione.

Glycine conjugation is another mechanism of metabolism in the skin. Nasseri-Sina et al. (1997) described glycine conjugation in both human and rat keratinocytes. The metabolic pathway involves the activation of the carboxylic acid group with coenzyme A (CoA) in an ATP-dependent reaction.

This is followed by the reaction of the S-CoA derivative with the glycine molecule, catalyzed by a mitochondrial acyltransferase. The resulting glycine conjugation renders the metabolite more polar than the parent compound, and it can then be excreted renally (Fig. 3).

Nasseri-Sina et al. (1997) investigated the metabolism of benzoic acid, used topically for the treatment of tinea infestations. Benzoic acid, when administered systemically, is excreted as hippuric acid in urine. This group

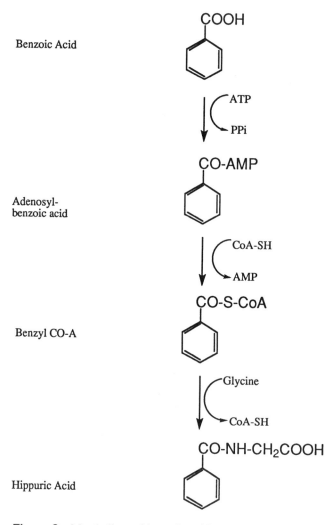

Figure 3   Metabolism of benzoic acid: an example of a conjugation reaction.

found that cultured keratinocytes in both humans and rats also metabolized benzoic acid to hippuric acid, but to a much smaller extent than hepatocytes.

Both of the studies just described demonstrate that transferase activity may play a significant role in the metabolism of topically applied compounds.

V. EXAMPLES OF XENOBIOTIC METABOLISM

A. Corticosteroids

Topical corticosteroids are extensively prescribed for dermatological conditions. Kubota et al. (1993) studied the metabolism of betamethasone 17-valerate (B-17) in the living skin equivalent (LSE) model. The betamethasone 17-valerate was initially isomerised to betamethasone 21-valerate (B-21), before it was hydrolyzed to the more polar betamethasone. The rate of conversion of B-17 to B-21 was the same with or without skin homogenate, suggesting that the initial isomerization step was not enzyme dependent but possibly a passive chemical degradation.

Taking this study further, Kubota et al. (1995) compared the rates of metabolism of the two isomers B-17 and B-21. When the B-17 isomer was applied to the LSE, half of the drug was left unchanged. In contrast, when B-21 was applied to the LSE, almost all of the drug was metabolized. Thus, the esterases that are responsible for this second, enzyme-dependent step demonstrate preference for the B-21 isomer.

Applying this knowledge in the human setting, Ademola and Maibach (1995) demonstrated that B-17 isomer is metabolized to form the B-21 isomer and betamethasone in both human skin in vivo and in the LSE model. This work showed that the B-17 isomer accumulated in the human skin. This was possibly because the B-21 isomer was metabolized faster than the B-17 isomer, which would be consistent with isomeric preference shown in vitro.

These studies have therefore shown that corticosteroids are metabolized in human skin. Further, this metabolism may involve a passive step of chemical degradation as well as active enzyme-dependent metabolism. Importantly, the isomeric structure of the topical agent may influence the rate of metabolism within the skin. Therefore, different isomers of the same compound may be more or less suitable than one another for topical application. This must be considered in the study of any agent to which the skin is exposed.

B. Beta-Adrenoceptor Antagonists

Propanolol is a widely prescribed, highly lipophilic beta-adrenoceptor antagonist. As it is lipophilic, having a partition coefficient of 5.39 at pH 7.0, the topical route of administration may theoretically achieve a steady drug release and plasma concentration. Oral propanolol is subject to first-pass metabolism, leading to variable absorption and low systemic bioavailability.

Ademola et al. (1991), studying percutaneous absorption and metabolism of propanolol in vitro, using intact human skin and microsomal preparations, found that between 10.4% and 36.6% of the drug was absorbed, but only 4.1% to 16.1% of the drug penetrated the skin. Some propanolol was retained in the skin, and metabolites of propanolol were found. Naphthoxyacetic acid, 4-hydroxypropanolol, and *N*-desisopropyl propanolol were formed by intact human skin. The concentration of these metabolites was lower compared with hepatic metabolism, suggesting less enzymatic activity in the skin compared to the liver. These metabolites were also formed by the skin microsomes, in a greater concentration than in intact skin. This may be the result of the greater surface area that the microsomes (everted endoplasmic reticulum) had to react with the drug. Additionally, the microsomes biotransformed propanolol to norpropanolol, which the intact skin did not form.

Ademola and Maibach (1995) took this study further, using human, LSE, and keratinocyte models. Propanolol does indeed accumulate in human skin, which may be responsible for irritant or toxic effects. They suggest that the differences between the metabolism of propanolol in skin and liver may explain the accumulation of the drug in the skin. This difference could not be attributed to the degree of enzyme activity, as the enzyme saturation points in the metabolism of propanolol in liver and skin were similarly high.

Ademola and Maibach postulated that the difference in metabolism may lie in the stereoisomeric structure. Using racemic propanolol, they demonstrated that the *S*-enantiomer was eliminated more efficiently by the skin than the *R*-enantiomer. This is in contrast to hepatocytes, which are more efficient at removing the *R*-enantiomer (Ward et al., 1989). Therefore, the irritation caused by the topical application of propanolol may be the result of accumulation of the *R*-enantiomer (Melendres et al., 1992).

These studies therefore suggest that propanolol is metabolized by human skin, and that its metabolism may be stereoselective. This metabolism and the retention of propanolol in the skin may explain both the low plasma concentration and irritant dermatitis after topical application.

C. Topical Nitrates

Higo et al. (1992a) used intact skin and homogenates from hairless mice to study the metabolism of nitroglycerin. In the homogenate study, GTN was incubated with homogenized tissue. After 2 h of incubation, 30% of the GTN had been metabolized to the breakdown products 1,2- and 1,3-GDN.

This metabolism was shown to be heavily dependent on the presence of glutathione (see earlier). Using the intact skin model, the investigators compared the extent of metabolism using different formulations of the GTN: a 1-mg/ml aqueous solution, a 2% ointment, and a transdermal delivery system. The percentage of metabolites formed was greatest with the aqueous solution (61%), followed by the patch (49%), and least of all with the ointment (35%). This difference is thought to be the explained by the greater transdermal flux with the patch and ointment compared to the solution: The smaller the flux, the greater the relative level of skin metabolism.

D. Theophylline

Theophylline is a xanthine derivative that is used as a bronchodilator. Ademola et al. (1992) studied the effect of cutaneous metabolism on its topical administration. This drug has a narrow therapeutic index at which optimal bronchodilation is maintained with minimal adverse effects occurring. Considering this, topical administration may give theoretical advantage over the oral route, as the latter results in variable plasma concentrations and is subject to altered absorption with the presence or absence of food in the gastrointestinal tract. Using both human skin samples and its microsomes, Ademola et al. determined that theophylline was metabolized with the production of 1,3-dimethyl uric acid, 3-methyl uric acid, and 3-methylxanthine from the skin samples. These metabolites of theophylline are produced via cytochrome P-450-dependent metabolism in the liver, and the authors proposed that a similar mechanism may occur in skin (Fig. 4).

VI. METABOLISM OF ENVIRONMENTAL XENOBIOTICS

An important consideration in this subject is the metabolism by the skin of compounds it is exposed to in the environment. The skin forms a barrier against our environment and is constantly exposed to compounds both natural and manmade. In this section, we address the effects of their metabolism.

A. Polycyclic Aromatic Hydrocarbons

Polycyclic aromatic hydrocarbons (PAHs) are produced by the incomplete combustion of fossil fuels and other organic matter. Their potential role in human carcinogenesis is suggested by their presence in the environment and the carcinogenicity of their metabolites.

Cutaneous metabolism of PAH is capable of forming carcinogenic metabolites (see Kao and Carver, 1991, for review). Studies with model compounds such as benzo[a]pyrene have demonstrated that cutaneous metabo-

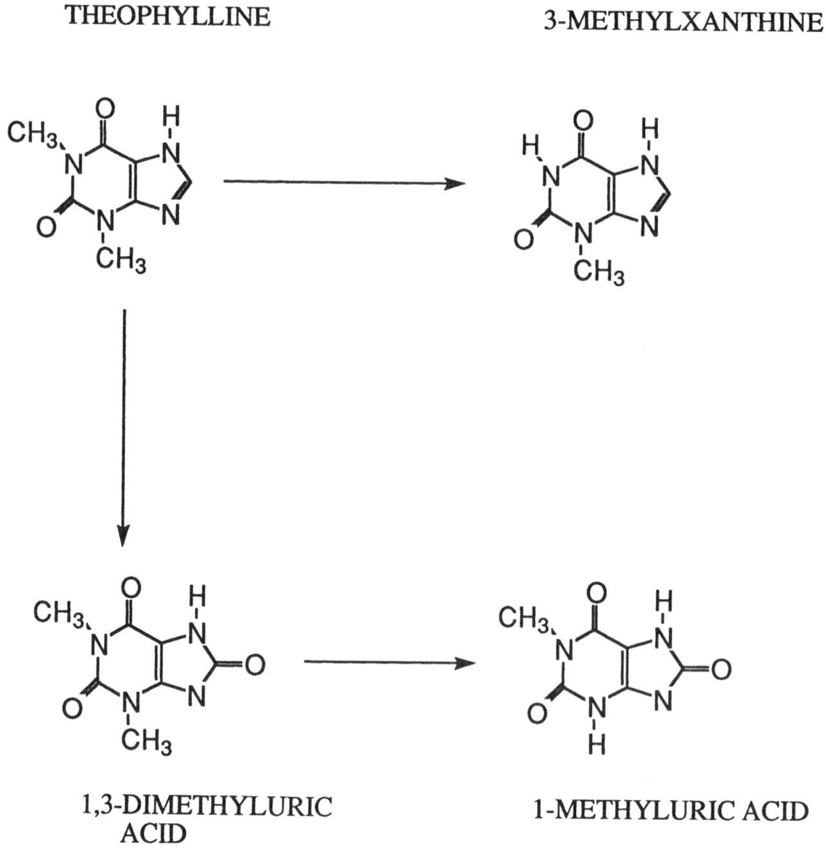
THEOPHYLLINE
3-METHYLXANTHINE
1,3-DIMETHYLURIC ACID
1-METHYLURIC ACID

Figure 4 Pathways of theophylline metabolism.

lism of PAHs can lead to the formation of phenols, quinones, dihydrodiols, and reactive diol epoxides. The diol epoxides are thought responsible for the carcinogenic effect, biding covalently to macromolecules. Covalent binding with DNA correlates well with the tumorigenicity of the metabolites of benzo[a]pyrone (Mukhtar, 1992).

PAHs are present in crude coal tar, which is extensively used in dermatological practice. Merk et al. (1987) demonstrated that exposure of crude coal tar to the human hair follicle results in the induction of aromatic hydrocarbon hydroxylase (AHH), which is a cytochrome P-450-dependent enzyme. This resulted in the production of benzo[a]pyrene derivatives that were shown to bind to DNA.

These studies of PAHs therefore exemplify the potentially hazardous nature of the cutaneous metabolism of environmental xenobiotics. We discuss the metabolism of other environmental agents and their potential for toxicity next.

B. Pesticides

Ademola et al. (1993a) studied the cutaneous metabolism of an environmental pesticide, 2-chloro-2,6-diethyl-*N*-(butoxymethyl) acetanilide (butachlor), on human skin in vitro. This study is significant in our discussion of the role of cutaneous metabolism in our everyday lives and of its potential consequences. Skin is the most important route of exposure to such agents; topical exposure could result in systemic absorption, which may be toxic, and also could result in cutaneous or systemic metabolism, either of which could toxify or detoxify the compound. In this study, the butachlor was metabolized to 4-hydroxybutachlor and was NADPH dependent, implying that the metabolism may be dependent on monooxygenases in the skin. The 4-hydroxybutachlor metabolite was noted to accumulate in skin.

Cysteine- and glutathione-conjugated metabolites were also found. The formation of glutathione conjugates is consistent with the known presence of glutathione in human skin (Raza et al., 1991). Although the significance of these metabolism is not yet known, their formation and accumulation in the skin may be potentially hazardous.

Ademola et al. (1993b) also investigated the metabolism of a widely used herbicide, atrazine, within the skin. The metabolites 2-chloro-4-ethyl-amino-6-amino-*s*-triazine (desisopropylatrazine) and 2-chloro-4,6-diamino-*s*-triazine were found in the receptor fluid and the skin supernates. An additional metabolite (2-chloro-4-amino-6-isopropylamino-*s*-triazine) was found in the skin supernates. This study again showed that metabolites of an environmental agent can be produced in the skin, further reinforcing the need for the detailed study of skin metabolism as a possible source of pathology (Table 2).

VIII. FACTORS AFFECTING CUTANEOUS METABOLISM

The factors that influence the metabolism of cutaneous xenobiotics can be dynamic or static. Dynamic metabolism may vary according to the physiological and pathological condition of the skin. In contrast, static factors may be related to the structure of the skin at a particular site.

A. Dynamic Factors

The dynamic response of enzymes to inductive and inhibitory stimuli could be an important factor in determining the extent of metabolism within the

Table 2 Examples of Some Xenobiotics and Their Metabolites

Compound	Major metabolites	Comment
Betamethasone 17-valerate	Betamethasone 21-valerate	Chemical degradation
	Betamethasone	Active metabolism
Propanolol	Naphthoxyacetic acid	Produced by intact skin
	4-Hydroxypropanolol	Produced by intact skin
	N-Desisopropyl propanolol	Produced by intact skin
	Norpropanolol	Only produced by microsomes
Nitrogylcerin	1,2-GDN	
	1,3-GDN	
Theophylline	1,3-Dimethyl uric acid	
	3-Methyl uric acid	
	3-Methyl xanthene	
Polycyclic aromatic hydrocarbons	Phenols	
	Quinones	
	Dihydrodiols	
	Diol epoxides	Carcinogenic
Butachlor	4-Hydroxybutachlor	
	Cysteine conjugates	
	Glutathione conjugates	
Atrazine	Desisopropylatrazine	
	2-Chloro-4,6-diamino-s-triazine	

skin. Also, in the case of isoenzymes, such as the cytochrome P-450 family, which particular isoenzymes are induced and in what proportions must also be considered.

1. Enzyme Induction

The induction of enzymes that metabolize xenobiotics may increase the rate and/or amount of metabolites produced. Some xenobiotics may induce enzymes for which they themselves are substrates, or may induce enzymes that act on other exogenous or endogenous substrates.

Schlede and Connely (1970) demonstrated a 10-fold increase in aryl hydrocarbon hydroxylase (AHH) activity in skin homogenates from rats pretreated with 3-methychloranthene. AHH is a cytochrome P-450-dependent enzyme associated with the expression of CYP1A1 (Ahmad et al., 1996). Further studies have shown that topically applied polycyclic hydrocarbons, coal tar, and petroleum derivatives are also effective in the induction of AHH in human skin (Kao and Carver, 1991).

Jugert et al. (1993) studied the effect of topically applied dexamethasone on the induction of cutaneous cytochrome P-450 isoenzymes in murine skin. The induction of cytochromes 1A1, 2B1, 2E, and 3A was seen, in addition to induction of the monoxygenase enzymes catalyzed by these CYPs. The group further employed immunohistochemistry to localize the expression of the CYP2B1 isoenzyme within the epidermis. This particular isoenzyme was investigated as it was involved in the greatest enzyme induction. The isoenzyme was localized to the suprabasal layer of the epidermis and the cells of the hair follicle.

That dexamethasone can induce several isoenzymes of cytochrome P-450 is a significant finding because the cytochrome P-450 monooxygenases are not substrate specific. Therefore, if one substrate induces a series of enzymes, other xenobiotics that are applied to the skin, either intentionally or unintentionally, may be metabolized at an increased rate. For example, if one came in contact with benzo[a]pyrone while using topical corticosteroids for atopic dermatitis, the metabolism of carcinogenic metabolites could be increased.

2. Enzyme Activity Inhibition

In contrast to induction, the inhibition of enzymes must also be considered. Inhibition of the cutaneous metabolism of xenobiotics has several theoretical advantages. For example, selectively inhibiting an enzyme may increase the overall percutaneous absorption of a particular medication. Kao and Carver (1991) reviewed work in this field. The imidazole antifungal agents, widely prescribed in dermatological practice, are potent inhibitors of the microsomal P-450-dependent monooxygenases. In skin, they inhibit the activity of AHH and epoxide hydrolase activity (EPOH). Also, imidoazoles induce glutathione-s-transferase activity and inhibit the cutaneous metabolism, marcromolecular binding, and carcinogenicity of topically applied benzopyrene in cultured mouse keratinocytes (Das et al., 1986). Plant phenols also inhibit the monooxygenase metabolism of benzo[a]pyrene in vitro (Das et al., 1987). These studies suggest that inhibitors of xenometabolizing enzymes may be useful in the prevention of polycyclic hydrocarbon skin malignancies.

B. Barrier Disruptions and Cutaneous Xenometabolism

Several studies have attempted to study the metabolism of xenobiotics following disruption of the skin barrier. Higo et al. (1992b) studied the effect of skin condition in vitro on the cutaneous metabolism of nitroglycerin. Full-thickness excised skin from hairless mice was placed in a plastic bag prior to immersion in boiling water for 10 min. Heating the skin disrupted its barrier function, a fact that can be inferred from the increased total nitrate flux across the heated skin compared to controls. The skin did continue to metabolize the GTN; however, compared to control skin, the heated skin showed a preference for the formation of 1,3-GDN rather than 1,2-GTN. The heated tissue continued to metabolize the nitroglycerin at a steady rate during the 10-h experiment, whereas the control specimen's metabolism decreased with time. This suggests that the altered metabolism may be the result of nonezymatic metabolism of the drug. In the same study, the authors also damaged the skin barrier using tape stripping. The greater was the number of strippings, the more was damaged the skin, with greater flux of nitrates and less metabolic activity.

However, Shaikh et al. (1996) demonstrated that freezing human skin did not alter its metabolic capacity. Investigating the metabolism of 8-methoxypsoralen (8-MOP) on human skin in vitro, it was demonstrated that the skin barrier had been perturbed as there was a greater flux of 8-MOP in the frozen specimen compared to the control. However, the metabolic capacity of the skin remained constant.

Different insults to the human skin barrier may alter the metabolism of xenobiotics in different ways. These studies demonstrate that further investigation of skin barrier function in cutaneous metabolism is necessary. As products for topical use become increasingly popular, their use on damaged skin must be investigated.

VIII. CONSEQUENCES OF CUTANEOUS XENOBIOTIC METABOLISM

For any drug metabolized in the skin, the potentially toxic nature of any metabolite must be considered. For example, a metabolite may be irritant, allergenic, or even carcinogenic, either locally or systemically. The precarcinogen benzo[a]pyrene was described earlier as an example of this, as was the metabolism of propanolol.

The ability of the enzymes responsible for xenometabolism to be induced or inhibited mat affect the rate and extent of metabolism of any compound on the skin. This may affect the efficacy of drugs applied topically either for local or systemic administration. Indeed, there is potential for topical formulations to include inhibitors of enzymes to enhance drug delivery.

Metabolism of the drug at the cutaneous level constitutes "first-pass metabolism," which may result in subtherapeutic doses reaching the systemic circulation. Indeed, the metabolic activity of the enzymes may be dynamic rather than static; this implies that the under different physiological and pathological conditions, variable doses of the drug may be delivered through the skin, perhaps resulting in toxicity or decreased effectiveness. Particular regard must therefore be paid to drugs of narrow therapeutic index.

In conclusion, this chapter has demonstrated that cutaneous metabolism is relevant in the application of any topical agent to the skin, whether for superficial use or end-organ effect.

REFERENCES

Ademola JI, Maibach HI. (1995). Cutaneous metabolism and penetration of methoxypsoralen, betamethasone 17-valerate, retinoic acid, nitroglycerin and theophylline. *Curr Problems Dermatol* 22:201–213.

Ademola JI, Wester RC, Maibach HH. (1992). Cutaneous metabolism of theophylline by human skin. *J Invest Dermatol* 98(3):310–314.

Ademola JI, Wester RC, Maibach HI. (1993a). Absorption and metabolism of 2-chloro-2,6-diethyl-*N*-(butoxymethyl)acetanilide (butachlor) in human skin in vitro. *Toxicol Appl Pharmacol* 121(1):78–86.

Ademola JI, Chow CA, Wester RC, Maibach HI. (1991). Metabolism of propanolol during percutaneous absorption in human skin. *J Pharm Sci* 82(8):767–770.

Ademola JI, Sedik LE, Wester RC, Maibach HI. (1993b). In vitro percutaneous absorption and metabolism in man of 2-chloro-4-ethyl amino-6-isopropylamine-*s*-triazine (atrazine). *Arch Toxicol* 67(2):85–91.

Ahmad N, Agarwal R, Mukhtar H. (1996). Cytochrome P-450-dependent drug metabolism in skin. *Clin Dermatol* 14:407–415.

Bickers DR, Dutta-Chaudhury T, Mukhtar H. (1982). Epidermis: A site of drug metabolism in rat skin. Studies on cytochrome P-450 content and mixed function oxidase and epoxide hydrolase activity. Mol Pharmacol 21:239–247.

Das M, Mukhtar H, Del Tito BJ Jr, Marcelo CL, Bickers DR. (1986). Clotrimazole, an inhibitor of benzo(a)pyrene metabolism and its subsequent glucuronidation, sulfation, and macromolecular binding in BALB/c mouse culured keratinocytes. *J Invest Dermatol* 87:4–10.

Das M, Mukhtar H, Bik DP, Bickers DR (1987). Inhibition of epidermal xenobiotic metabolism in SENCAR mice by naturally occurring phenols. *Cancer Res* 47: 760–766.

Goeptar AR, Scheerens H, Vermeulen NP. (1995). Oxygen and xenobiotic reductase activities of cytochrome P450. *Critical Reviews in Toxicology* 25(1):25–65.

Goerz G. (1987). Animal models for cutaneous drug metabolising enzymes. In: Maibach HI, Lowe NJ, eds. Models in Dermatology, Vol III. Basel: Karger, pp. 93–105.

Gonzalez FJ. (1989). The molecular biology of cytochrome P-450s. *Pharmacol Rev* 40:243–388.

Higo N, Hinz RS, Lau DTW, Benet LZ, Guy RH. (1992a). Cutaneous metabolism of nitroglycerin in vitro. *Pharm Res* 9(2):187–191.

Higo N, Hinz RS, Lau DTW, Benet LZ, Buy RH. (1992b). Cutaneous metabolism of nitroglycerin in vitro. II. Effects of skin condition and penetration enhancement. *Pharm Res* 9(3):303–306.

Jugert FK, Agarwal R, Kuhn A, Bickers DR, Merk HF, Mukhtar H. (1994). Multiple cytochrome P450 isozymes in murine skin: induction of P450 1A, 2B, 2E, and 3A by dexamethasone. *J Invest Dermatol* 102(6):970–975.

Kao J, Carver MP. (1991). Skin metabolism. In: Marzulli N, Maibach HI, eds. Dermatotoxicology, 4th ed. Washington, DC: Hemisphere, pp. 143–200.

Kubota K, Ademola JI, Maibach HI. (1993). Metabolism of topical drugs within the skin, in particular glucocorticoids. In: Korting HC, Maibach HI, eds. Topical Glucocorticoids with Increased Benefit/Risk Ratio. Current Problems in Dermatology, Vol. 21. Basel: Karger, pp. 61–66.

Kubota K, Ademola JI, Maibach HI. (1995). Simultaneous diffusion and metabolism of betamethasone 17-valerate in the living skin equivalent. *J Pharm Sci* 84(12):1478–1481.

Melendres JL, Nangia A, Sedik L, Mitsuhiko H, Maibach HI. (1992). Nonane enhances propanolol hydrochloride penetration in human skin. *Int J Pharm* 92: 243–248.

Merk HF, Jergert FK, Frankenberg S. (1996). Biotransformations in the skin. In: Marzulli FN, Maibach HI, eds. Dermatoxicology, 5th ed. Washington, DC: Taylor & Francis, Ch. 6, pp. 61–74.

Merk HF, Mukhtar H, Kaufmann I, Das M, Bickers DR. (1987). Human hair follicle benzo(a)pyrene and benzo(a)pyrene 7,8-diol metabolism: Effect of exposure to a crude coal tar containing shampoo. *J Invest Dermatol* 88:71–76.

Mukhtar H, Agarwal R, Bickers DR. (1992). Cutaneous metabolism of xenobiotics and steroid hormones. In: Mukhtar H, ed. Pharmacology of the Skin. Boca Raton, FL: CRC Press, pp. 89–110.

Nasseri-Sina P, Hotchkiss SA, Caldwell J. (1997). Cutaneous xenobiotic metabolism: glycine conjugation in human and rat keratinocytes. *Food and Chemical Toxicology* 35(3–4):409–416.

Raza H, Awasthi YC, Zaim MT, Eckert RL, Mukhtar H. (1991). Glutathione-*S*-transferase in human and rodent skin; Multiple forms and species specific expression. *J Invest Dermatol* 96:463–467.

Schlede E, Connely AH. (1970). Induction of benzo(a)pyrene hydroxylase activity in rat skin. *Life Sci* 9(II):1295–1303.

Shaikh NA, Ademola JI, Maibach HI. (1996). Effects of freezing and azide treatment of in vitro human skin on the flux and metabolism of 8-methoxypsoralen. *Skin Pharmacol* 9:274–280.

Ward S, Walle T, Walle K, Wilkinson GR, Branch RA. (1989). Propanolol's metabolism is determined by both mephenytoin and debrisoquin hydroxylase. *Clin Pharm Ther* 45:72–78.

4
Occlusion Does Not Uniformly Enhance Penetration In Vivo

Daniel Bucks
University of California School of Medicine, San Francisco and Penederm Inc., Foster City, California

Howard I. Maibach
University of California School of Medicine, San Francisco, California

I. INTRODUCTION

Mammalian skin provides a relatively efficient barrier to the ingress of exogenous materials and the egress of endogenous compounds, particularly water. Loss of this vital function results in death from dehydration. Compromised function is associated with complications seen in several dermatological disorders. Stratum corneum intercellular lipid domains form a major transport pathway for penetration (1–4). Perturbation of these lamellar lipids causes skin permeation resistance to fall and has implicated their crucial role in barrier function. Indeed, epidermal sterologenesis appears to be modulated by the skin's barrier requirements (5). Despite the fact that the skin is perhaps the most impermeable mammalian membrane, it is permeable to a degree, that is, it is semipermeable; as such, the topical application of pharmaceutical agents has been shown to be a viable route of entry into the systemic circulation as well as an obvious choice in the treatment of dermatological ailments. Of the various approaches employed to enhance the percutaneous absorption of drugs, occlusion (defined as the complete impairment of passive transepidermal water loss at the application site) is the simplest and perhaps one of the most common methods in use. In this chapter we have summarized the literature to evaluate the effect of occlusion on the percutaneous absorption of topically applied compounds and to look at how certain compound physicochemical properties (such as volatility, partition coefficient, and aqueous solubility) may predict what effect occlusion may have.

The increased clinical efficacy of topical drugs caused by covering the site of application was first documented by Garb (6). Subsequently, Scholtz (7), using fluocinolone acetonide, and Sulzberger and Witten (8), using hydrocortisone, reported enhanced corticoid activity with occlusion in the treatment of psoriasis. The enhanced pharmacological effect of topical corticosteroids under occlusion was further demonstrated by the vasoconstriction studies of McKenzie (9) and McKenzie and Stoughton (10). Occlusion has also been reported to increase the percutaneous absorption of various other topically applied compounds (11–17). However, as shown later, short-term occlusion does not necessarily increase the percutaneous absorption of all chemicals.

II. PERCUTANEOUS ABSORPTION OF *p*-PHENYLENEDIAMINE (PPDA) IN GUINEA PIGS

The in vivo percutaneous absorption of *p*-phenylenediamine (PPDA) from six occlusive patch test systems was investigated by Kim et al. (18). The extent of absorption was determined using ^{14}C radiotracer methodology. The ^{14}C-PPDA was formulated as 1% PPDA in petrolatum (USP) and applied from each test system at a skin surface dose of 2 mg/cm^2. Thus, the amount of PPDA was normalized with respect to the surface area of each patch test system (and, hence, to the surface area of treated skin). A sixfold difference in the level of skin absorption ($p < .02$) was found between the patches (Table 1). It should be noted that a nonocclusive control was not included in this study.

The rate of ^{14}C excretion following topical application of the radiolabeled PPDA in the various patch test systems is shown in Fig. 1. Clearly, the rate and extent of PPDA absorption were dependent upon the patch test system employed. The mechanism responsible for differences in PPDA percutaneous absorption from these patch test systems is not known. However, magnitude of occlusiveness of each dressing is hypothesized to correspond to enhanced absorption.

III. PERCUTANEOUS ABSORPTION OF VOLATILE COMPOUNDS

The effect of occlusion on the in vivo percutaneous absorption of two fragrances (safrole and cinnamyl anthranilate) and two chemical analogs (cinnamic alcohol and cinnamic acid) in rhesus monkeys was evaluated by Bron-

Table 1 Percutaneous Absorption of PPDA from Patch
Test Systems

Patch test system	Mean % dose absorbed (SD)
Hill Top chamber	53 (21)
Teflon (control)	49 (9)
Small Finn chamber	30 (9)
Large Finn chamber	23 (7)
AL-Test chamber	8 (1)
Small Finn chamber with paper disc insert	34 (20)

Note. The rate of ^{14}C excretion following topical application of the radiolabeled PPDA in the various patch test systems is shown in Fig. 1. The extent of PPDA absorption was dependent upon the occlusive patch test system employed. It should be noted that a nonocclusive control study was not conducted. The test system used 2 mg/mm^2 PPDA for 48 h on the dorsal mid-lumbar region of the guinea pig. Data from Kim et al. (18).

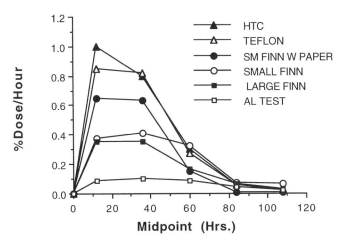

Figure 1 In vivo percutaneous absorption of PPDA (2 mg/mm^2) following a 48-h exposure on the dorsal lumbar region of guinea pigs. HTC, Hill Top chamber; Teflon, sheet of Teflon; sm Finn w paper, small Finn chamber with paper insert included; small Finn, small Finn chamber with paper insert removed; large Finn, large Finn chamber with paper insert removed. Redrawn from Kim et al. (18).

augh et al. (19). Each compound was applied at a topical dose of 4 μg/cm^2 from a small volume of acetone. Occlusion was achieved by covering of the site of application with plastic wrap (Saran Wrap, a chlorinated hydrocarbon polymer) after the acetone had evaporated from the skin surface. The extent of absorption following single-dose administration was determined using ^{14}C radiotracer methodology. The fragrance materials were well absorbed through monkey skin. Plastic-wrap occlusion of the application site resulted in large increases in absorption (Table 2). The authors also presented in vitro data documenting the significant increase in percutaneous absorption of these chemicals under occluded compared to nonprotected conditions, that is, left open to the air.

Investigation of the effect of occlusion on the percutaneous absorption of six additional volatile compounds (benzyl acetate, benzamide, benzoin, benzophenone, benzyl benzoate, and benzyl alcohol) was conducted using the same in vivo methodology. These studies included occlusion of the site of application with a glass cylinder secured to the skin by silicone glue and capped with Parafilm, occlusion with plastic wrap, and nonprotected conditions (20). As shown in Table 3, occlusion, in general, enhances the percutaneous absorption of these compounds. However, differences in percutaneous absorption were observed between plastic wrap and "glass chamber" occlusive conditions. The absorption of benzoin and that of benzyl acetate were lower under plastic wrap compared to the nonprotected condition. This discrepancy might be due to compound sequestration by the plastic wrap. Glass chamber occlusion resulted in greater bioavailability than nonprotected or plastic wrap occlusion except for benzyl benzoate, where plastic wrap conditions resulted in greater absorption, and for benzophenone,

Table 2 In Vivo Percutaneous Absorption of Fragrances in Monkeys

	Percent dose absorbed[a]	
	Nonprotected	Plastic wrap occlusion
Cinnamyl anthranilate	26.1 (4.6)	39.0 (5.6)
Safrole	4.1 (1.6)	13.3 (4.6)
Cinnamic alcohol	25.4 (4.4)	74.6 (14.4)
Cinnamic acid	38.6 (16.6)	83.9 (5.4)

Note. Values were corrected for incomplete renal elimination. Mean ± SD (*N* = 4). Data from Bronaugh et al. (19).
[a]Single 4-μg/cm^2 dose with a 24-h exposure prior to soap and water washing.

Table 3 In Vivo Percutaneous Absorption of Benzyl Derivatives in Monkeys

| | Percent dose absorbed[a] | | | |
	Nonprotected	Plastic wrap occlusion	Glass chamber occlusion	log $K_{o/w}$
Benzamide	47 (14)	85 (8)	73 (20)	0.64
Benzyl alcohol	32 (9)	56 (29)	80 (15)	0.87
Benzoin	49 (6)	43 (12)	77 (4)	1.35
Benzyl acetate	35 (19)	17 (5)	79 (15)	1.96
Benzophenone	44 (15)	69 (12)	69 (10)	3.18
Benzyl benzoate	57 (21)	71 (9)	65 (20)	3.97

Note. Values corrected for incomplete renal elimination. Mean $\pm$ SD ($N = 4$). Data from Bronaugh et al. (20)
[a]Single 4-μg/cm^2 dose with a 24-h exposure prior to soap and water washing.

where glass chamber and plastic wrap conditions resulted in the same magnitude increase over nonprotected test conditions.

An attempt was made to correlate occlusion-enhanced bioavailability with each compound's octanol/water partition coefficient. Unexpectedly, no apparent trends were noted for these volatile fragrance compounds. Absence of a trend is in contradiction to results obtained with steroids and phenol derivatives discussed later, given the range of log $K_{o/w}$ evaluated with these fragrances.

Gummer and Maibach (16) studied the penetration of methanol and ethanol through excised, full-thickness guinea pig skin. Occlusion significantly enhanced the cumulative amount penetrating as well as the profiles of the amount penetrating per hour for both methanol and ethanol. Consistent with results from other investigators reported already, occlusion-induced penetration enhancement was dependent upon the nature of the occlusive material, with the greatest enhancement observed with a plastic Hill Top chamber (21). Intuitively, occlusion-induced enhancement in the penetration of volatile compounds should be one of the items related to the degree to which the occlusive device inhibits evaporative loss of compound from the skin surface.

IV. PERCUTANEOUS ABSORPTION OF STEROIDS IN HUMANS

The earliest attempt to correlate the increased pharmacological effect of hydrocortisone under occlusive conditions with the pharmacokinetics of absorption was reported by Feldmann and Maibach (12). In this study, the rate and extent of ^{14}C-label excretion into the urine following topical application of [^{14}C]hydrocortisone to the ventral forearm of normal human volunteers were measured. Radiolabeled hydrocortisone (75 μg) was applied in acetone solution (1000 μl) as a surface deposit over 13 cm^2 of skin. The authors estimated that this was equivalent to a sparing application of a 0.5% hydrocortisone topical preparation (5.8 μg/cm^2). The site of application was either nonprotected or occluded with plastic wrap (Saran Wrap). When the skin was nonprotected, the dosing site was washed 24 h postapplication. On the other hand, when the skin was occluded, the plastic wrap remained in place for 96 h (4 days) postapplication before the application site was washed. The percent of the applied dose excreted into the urine, corrected for incomplete renal elimination, was (mean ± SD) 0.46 ± 0.20 and 5.9 ± 3.5 under nonprotected and occluded conditions, respectively (Fig. 2). A paired t-test of the results indicates a significant difference ($p = .01$) in cumulative absorption of hydrocortisone between the two exposure conditions. Quantitatively, the occlusive conditions employed increased the cumulative absorption of hydrocortisone (HC) by about an order of magnitude. However, note that the occlusive system retained the drug in contact with the skin for 96 h, compared to the 24-h exposure period under nonprotected conditions, and this could affect absorption as measured by the cumulative measurement of

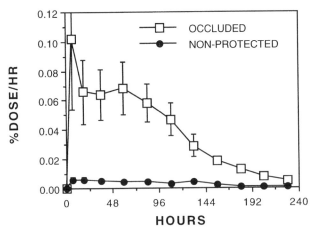

Figure 2   Percutaneous absorption of hydrocortisone in humans. Human 96-h occluded versus 24-h nonprotected exposure of hydrocortisone at 4 μg/cm^2 prior to soap and water washing. Occlusion was with plastic wrap. Data from Feldmann and Maibach (12).

drug excreted into the urine, but the dramatic difference in percent dose absorbed per hour between occluded and nonprotected at 12 and 24 h is not expected to be dependent upon differences in times of washing. This enhancement in HC absorption afforded by occlusion is not consistent with the additional studies reported next.

Guy et al. (13) investigated the effect of occlusion on the percutaneous absorption of steroids in vivo following single and multiple application. The extent of absorption of four steroids (progesterone, testosterone, estradiol, and hydrocortisone), using radiotracer elimination into the urine following topical application to the ventral forearm of male volunteers, was reported. The chemical dose was 4 μg/cm^2 over an application area of 2.5 cm^2. The ^{14}C-labeled chemicals were applied in 20 μl acetone. In the occlusive studies, after evaporation of the vehicle, the site of application was covered with a plastic (polyethylene–vinyl acetate copolymer, Hill Top) chamber. In all cases, after 24 h, the site of application was washed with soap and water using a standardized procedure (22). In the occlusive studies, the administration site was then covered again with a new chamber. An essentially identical protocol was also performed following a multiple dosing regime (23). Daily topical doses of three of the steroids (testosterone, estradiol, and hydrocortisone) were administered over a 14-day period. The first and eighth doses were ^{14}C-labeled and urinary excretion of radiolabel was followed. As described earlier, the 24-h washing procedure was performed daily and a new chamber was applied. Occlusive chambers and washes were collected and assayed for residual surface chemical. The results of this study are summarized in Table 4. Steroid percutaneous absorption as a function of penetrant octanol/water partition coefficient ($K_{o/w}$) is shown in Fig. 3. The studies indicate that:

1. The single-dose measurements of the percutaneous absorption of hydrocortisone, estradiol, and testosterone are predictive of percutaneous absorption following a comparable multiple dose regimen (see Chapter 27 on the effect of repetitive application), under both occluded and nonoccluded conditions.
2. Occlusion significantly ($p < .05$) increased the percutaneous absorption of estradiol, testosterone, and progesterone, but not that of hydrocortisone (the compound with the lowest $K_{o/w}$ value in this series of steroids).
3. Percutaneous absorption increases with increasing $K_{o/w}$ up to testosterone but declines for progesterone, under occluded and nonoccluded conditions.
4. The occlusive procedure generally permits excellent dose accountability (Table 5).

Table 4 Percutaneous Absorption of Steroids in Humans

	Mean % applied dose absorbed (± SD)	
	Nonprotected	Occluded
Hydrocortisone		
Single application	2 ± 2[a]	4 ± 2
Multiple application		
1st Dose	3 ± 1	4 ± 1
8th Dose	3 ± 1	3 ± 1
Estradiol		
Single application	11 ± 5[a]	27 ± 6
Multiple application		
1st Dose	10 ± 2	38 ± 8
8th Dose	11 ± 5	22 ± 7
Testosterone		
Single application	13 ± 3[a]	46 ± 15
Multiple application		
1st Dose	21 ± 6	51 ± 10
8th Dose	20 ± 7	50 ± 9
Progesterone		
Single application	11 ± 6[a]	33 ± 9

[a]Data from Feldmann and Maibach (42). Other data from Bucks et al. (11,23).

The percutaneous absorption of these same four steroids under "protected" (i.e., covered, but nonocclusive) conditions has also been measured in vivo (11,24) using the same methodology. The data obtained from these later experiments permitted the effect of occlusion to be rigorously assessed (since complete mass balance of the applied dose was possible). With the exception of hydrocortisone (Table 6), occlusion significantly increased the percutaneous absorption ($p < .01$) of the steroids. These results were in excellent agreement with the comparable nonprotected studies described earlier. As stated before, excellent dose accountability was reported (Table 7).

To investigate the apparent discrepancy between the effect of plastic wrap occlusion (12) and that of the plastic chamber on hydrocortisone absorption (13), we repeated the measurements of penetration using plastic wrap (Saran Wrap) with the experimental protocol of Guy et al. (13). Under these circumstances, we found no difference between plastic wrap and plastic chamber occlusion on the percutaneous absorption of hydrocortisone (Table 8).

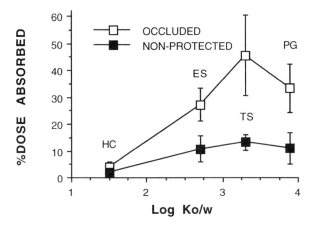

Figure 3   Percutaneous absorption of four steroids (HC, hydrocortisone; ES, estradiol; TS, testosterone; PG, progesterone) in humans as a function of penetrant octanol/water partition coefficient. Exposure period 24 h at 4 μg/cm^2 prior to soap and water washing. Redrawn from Guy et al. (13).

V.   PERCUTANEOUS ABSORPTION OF PHENOLS IN HUMANS

We subsequently investigated the effect of occlusion on the in vivo percutaneous absorption of phenols following single-dose application. The occlusive and protective chamber methodology described by Bucks et al. (24,25) was utilized. Nine ^{14}C-ring-labeled *para*-substituted phenols (4-aminophenol, 4-acetamidophenol, 4-propionylamidophenol, phenol, 4-cyanophenol, 4-nitrophenol, 4-iodophenol, 4-heptyloxyphenol, and 4-pentyloxyphenol) were used. As in the earlier steroid studies, the site of application was the ventral forearm of male volunteers and the area of application 2.5 cm^2. Penetrants were applied in 20 μl ethanol (95%). The chemical dose was 2 to 4 μg/cm^2. After vehicle evaporation, the application site was covered with either an occlusive or protective device. After 24 h, the patch was removed and the site washed with a standardized procedure (22). The application site was then re-covered with a new chamber of the same type. Urine was collected for 7 days. On the seventh day: a) the second chamber was removed, b) the dosing site was washed with the same procedure, and c) the upper layers of stratum corneum from the application site were removed by cellophane tape stripping. Urine, chambers, washes, and skin tape strips were collected and assayed for radiolabel. Percutaneous absorption of each compound under protected and occluded conditions is presented in Figs. 4 through 12. Phenol percutaneous absorption as a function of the penetrant octanol–water par-

Table 5   Accountability of Applied Dose[a] in Occluded Studies

	Absorbed (%)	Removed from skin (%)	Total % dose
Hydrocortisone			
Single dose[b]	4 ± 2	64 ± 5	68 ± 4
1st MD[c]	4 ± 1	82 ± 5	85 ± 4
8th MD[d]	3 ± 1	78 ± 2	81 ± 3
Estradiol			
Single dose[b]	27 ± 6	60 ± 12	87 ± 13
1st MD[c]	38 ± 8	62 ± 6	100 ± 4
8th MD[d]	22 ± 7	59 ± 8	81 ± 6
Testosterone			
Single dose[b]	46 ± 15	44 ± 7	90 ± 8
1st MD[b]	51 ± 10	48 ± 9	99 ± 4
8th MD[d]	50 ± 9	42 ± 9	92 ± 17
Progesterone			
Single dose[b]	33 ± 9	47 ± 10	80 ± 6

Note. Values corrected for incomplete renal elimination, mean ± SD. Occluded with a plastic (Hill Top) chamber. Adapted from Bucks et al. (43).
[a]Single 4-μg/cm^2 dose with a 24-h exposure prior to soap and water washing.
[b]Single dose study.
[c]First dose of a 14-day multiple-dose study.
[d]Eighth dose of a 14-day multiple-dose study.

Table 6   Percutaneous Absorption of Steroids in Humans: Single Dose Application for 24-h at 4 μg/cm^2

	Mean % dose absorbed ($\pm$ SD; $N \geq 5$)	
	Protected[a]	Occluded[b]
Hydrocortisone	4 ± 2	4 ± 2
Estradiol	3 ± 1	27 ± 6
Testosterone	18 ± 9	46 ± 15
Progesterone	13 ± 6	33 ± 9

Note. Data from Guy et al. (13) and Bucks et al. (11,43).
[a]Dose site covered with a ventilated plastic chamber.
[b]Dose site covered with an occlusive plastic chamber.

Table 7 Accountability of Applied Dose[a] in Protected Studies Using Ventilated Plastic Chambers

	Absorbed (%)	Removed from skin (%)	Total % accounted for
Hydrocortisone	4 ± 2	85 ± 6	89 ± 6
Estradiol	3 ± 1	96 ± 1	100 ± 1
Testosterone	18 ± 9	77 ± 8	96 ± 2
Progesterone	13 ± 6	82 ± 7	96 ± 3

Note. Data from Bucks et al. (11,24).
[a]Single 4-μg/cm^2 dose with a 24-h exposure prior to soap and water washing; penetration corrected for incomplete renal elimination.

tition coefficient ($K_{o/w}$) is shown in Fig. 13. Phenol percutaneous absorption is summarized in Table 9. The methodology permitted excellent dose accountability (Tables 10 and 11). The studies indicate that:

1. Occlusion significantly increased (unpaired t-test, $p < .05$) the absorption of phenol, heptyloxyphenol, and pentyloxyphenol.
2. Occlusion did not statistically enhance the absorption of aminophenol, acetaminophen, propionylamidophenol, cyanophenol, nitrophenol, and iodophenol.
3. The methodology employed again permitted excellent dose accountability.
4. In general, the two compounds with the lowest $K_{o/w}$ values of this series of compounds showed the least enhancement in absorption afforded by occlusion.

Table 8 Percutaneous Absorption of Hydrocortisone in Humans

	Percent dose absorbed[a]
Plastic wrap occlusion	4.7 ± 2.1
Plastic chamber occlusion	4.0 ± 2.4
"Protected" condition	4.4 ± 1.7

Note. Adapted from Bucks et al. (43).
[a]Single 4-μg/cm^2 dose with a 24-h exposure prior to soap and water washing; penetration corrected for incomplete renal elimination.

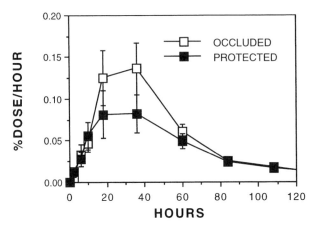

Figure 4 Percutaneous absorption of aminophenol in humans under occluded and protected conditions (mean ± SEM, $N = 6$). Exposure 24 h prior to soap and water washing. Redrawn from Bucks et al. (34,35).

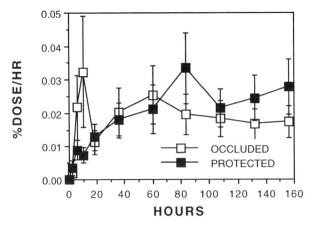

Figure 5 Percutaneous absorption of acetaminophen in man under occluded and protected conditions (mean ± SEM, $N = 6$). Exposure 24 h prior to soap and water washing. Redrawn from Bucks et al. (34,35).

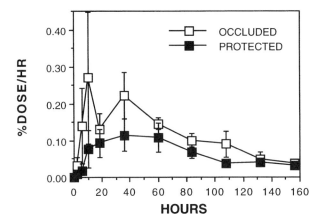

Figure 6 Percutaneous absorption of propionylamidophenol in humans under occluded and protected conditions (mean ± SEM, $N = 6$). Exposure 24 h prior to soap and water washing. Redrawn from Bucks et al. (34,35).

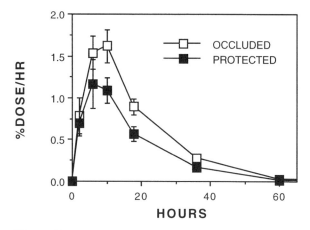

Figure 7 Percutaneous absorption of phenol in humans under occluded and protected conditions (mean ± SEM, $N = 6$). Exposure 24 h prior to soap and water washing. Redrawn from Bucks et al. (34,35).

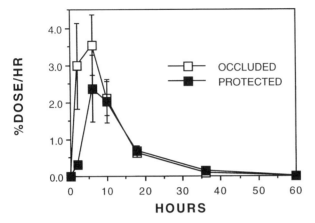

Figure 8   Percutaneous absorption of cyanophenol in humans under occluded and protected conditions (mean ± SEM, $N = 6$). Exposure 24 h prior to soap and water washing. Redrawn from Bucks et al. (34,35).

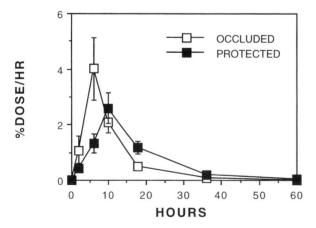

Figure 9   Percutaneous absorption of nitrophenol in humans under occluded and protected conditions (mean ± SEM, $N = 6$). Exposure 24 h prior to soap and water washing. Redrawn from Bucks et al. (34,35).

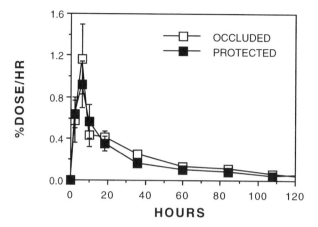

Figure 10 Percutaneous absorption of iodophenol in humans under occluded and protected conditions (mean ± SEM, $N = 6$). Exposure 24 h prior to soap and water washing. Redrawn from Bucks et al. (34,35).

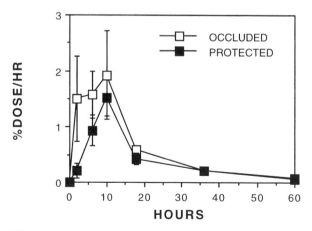

Figure 11 Percutaneous absorption of heptyloxyphenol in humans under occluded and protected conditions (mean ± SEM, $N = 6$). Exposure 24 h prior to soap and water washing. Redrawn from Bucks et al. (34,35).

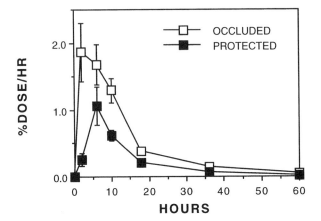

Figure 12 Percutaneous absorption of pentyloxyphenol in humans under occluded and protected conditions (mean ± SEM, $N = 6$). Exposure 24 h prior to soap and water washing. Redrawn from Bucks et al. (34,35).

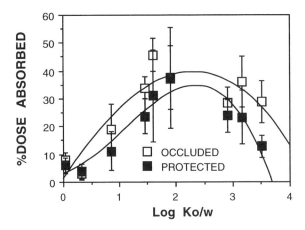

Figure 13 Percutaneous absorption of phenols in humans under occluded and protected conditions as a function of penetrant octanol/water partition coefficient ($K_{o/w}$). Redrawn from Bucks et al. (34,35).

Table 9 Percutaneous Absorption of Phenols[a] in Humans

Compound	log $K_{o/w}$	Mean % dose absorbed (SD)	
		Occluded	Protected
Aminophenol	0.04	8 (3)	6 (3)
Acetaminophen	0.32	3 (2)	4 (3)
Propionylamidophenol	0.86	19 (9)	11 (7)
Phenol[b]	1.46	34 (4)	24 (6)[c]
Cyanophenol	1.60	46 (6)	31 (16)
Nitrophenol	1.91	37 (18)	38 (11)
Iodophenol	2.91	28 (6)	24 (6)
Heptyloxyphenol	3.16	36 (9)	23 (10)[d]
Pentyloxyphenol	3.51	29 (8)	13 (4)[c]

Note. Data from Bucks et al. (34,35).
[a]Single dose application from 95% ETOH ($N = 6$) at 2 to 4 μg/cm^2 to the ventral forearm; 24-h exposure prior to soap and water washing.
[b]Data analysis accounts for 27.2% of applied dose evaporating off skin surface during application.
[c]Significant difference at $p < .01$.
[d]Significant difference at $p < .05$.

Table 10 Accountability of Applied Dose[a] in Occluded Studies; Mean Percent Dose Absorbed (SD)

Compound	Absorbed %	Removed from skin (%)	Total % dose
Aminophenol	8 (3)	55 (18)	63 (17)
Acetaminophen	3 (2)	61 (24)	64 (24)
Propionylamidophenol	19 (9)	77 (9)	96 (2)
Phenol[b]	34 (4)	61 (13)	95 (10)
Cyanophenol	46 (6)	41 (9)	87 (7)
Nitrophenol	37 (18)	50 (11)	87 (13)
Iodophenol	28 (6)	63 (4)	91 (3)
Heptyloxyphenol	36 (9)	59 (7)	95 (3)
Pentyloxyphenol	29 (8)	71 (8)	100 (2)

Note. Data from Bucks et al. (34,35).
[a]Single dose application from 95% ETOH ($N = 6$) at 2 to 4 μg/cm^2 to the ventral forearm; 24-h exposure prior to soap and water washing.
[b]Data analysis accounts for 27.2% of applied dose evaporating off skin surface during application.

Table 11   Accountability of Applied Dose[a] in Protected Studies: Mean Percent
Dose Absorbed (SD)

Compound	Absorbed %	Removed from skin (%)	Total % dose
Aminophenol	6 (3)	85 (4)	91 (2)
Acetaminophen	4 (3)	93 (5)	97 (4)
Propionylamidophenol	11 (7)	84 (7)	95 (2)
Phenol[b]	24 (6)	68 (15)	92 (18)
Cyanophenol	31 (16)	70 (12)	101 (5)
Nitrophenol	38 (11)	65 (12)	103 (4)
Iodophenol	24 (6)	73 (7)	97 (2)
Heptyloxyphenol	23 (10)	71 (10)	95 (4)
Pentyloxyphenol	13 (4)	85 (3)	98 (2)

Note. Data from Bucks et al. (34,35).
[a]Single dose application from 95% ETOH ($N = 6$) at 2 to 4 $\mu g/cm^2$ to the ventral forearm;
24-h exposure prior to soap and water washing.
[b]Data analysis accounts for 27.2% of applied dose evaporating off skin surface during
application.

VI.   DISCUSSION

A predominant effect of occlusion is to increase hydration of the stratum
corneum, thereby swelling the corneocytes, and promoting the uptake of
water into intercellular lipid domains. The magnitude of increased stratum
corneum hydration is related to the degree of occlusion exerted and is dependent upon the physicochemical nature of the dressing (26). The normal
water content of stratum corneum is 5 to 15%, a value that can be increased
up to 50% by occlusion (27,28). Upon removal of a plastic occlusive dressing after 24 h of contact, transepidermal water loss values are increased by
an order of magnitude (24); the elevated rate then returns rapidly (~15 min)
to normal with extraneous water dissipation from the stratum corneum. With
occlusion, skin temperature generally increases from 32°C to as much as
37°C (29). Faergemann et al. (30) showed that occlusion: a) increases the
transepidermal flux of chloride and carbon dioxide, b) increases microbial
counts on skin, and c) increases the surface pH of skin from a preoccluded
value of 5.6 to 6.7. Anhidrosis results from occlusion (31,32). Plastic chamber occlusion can also cause skin irritation (personal observation). Occlusion-induced increases in mitotic rate of skin and epidermal thickening have
been documented by Fisher and Maibach (33).

With respect to percutaneous absorption, occlusion or a protective cover may prevent loss of the surface-deposited chemical by evaporation, friction, and/or exfoliation; bioavailability may, thereby, be increased. However, comparison of the data in Tables 6 and 8 for the percutaneous absorption of steroids under nonprotected and protected conditions shows clearly that the potential increase in bioavailability from protection of the site of application does not explain the increase in steroid absorption under occluded conditions.

Occlusion does not necessarily increase percutaneous absorption. Hydrocortisone absorption under occluded conditions was not enhanced in single dose or multiple dose application studies (Table 12). This lack of penetration enhancement under occluded conditions has also been observed with certain *para*-substituted phenols (34,35) as well as with ddI (2′,3′-dideoxy-inosine, aqueous solubility of 27.3 mg/ml at pH ~6) (36). However, a trend of occlusion-induced absorption enhancement with increasing penetrant lipophilicity is apparent. This trend is also supported by the results of Treffel et al. (37), who have shown that the in vitro permeation of citropten (a lipophilic compound) increased 1.6 times under occlusion whereas that of caffeine (an amphiphilic compound) remained unchanged. However, the degree of lipophilicity (such as measured by octanol/water partition coefficient) exhibited by a penetrant in order for occlusion-induced enhanced skin permeation to be manifested is not clear and may be chemical-class dependent.

The increase in percutaneous absorption of hydrocortisone under occlusive conditions observed by Feldmann and Maibach (12) may be due to an acetone solvent effect. In this early work, the chemical was applied in 1.0 ml acetone over an area of 13 cm. Might the pretreatment of the skin with a large volume of acetone compromise stratum corneum barrier function? It has been reported that acetone can damage the stratum corneum (38,39). It is conceivable that the large volume of acetone used (76.9 μl/cm^2) may be responsible for the observed increase in hydrocortisone penetration under plastic wrap occlusion. In addition, it is reasonable to suggest that the increased duration of exposure (96 h compared to 24 h) may also contribute to the increase in observed hydrocortisone percutaneous absorption. This enhancement in absorption was not observed in the experiments with hydrocortisone under occlusion (24) when the acetone surface concentration was only 8.0 μl/cm^2 (20 μl over 2.5/cm^2) and skin surface exposure was limited to 24 h.

The occlusion-induced enhancement of lipophilic compounds may be understood by considering the steps involved in the percutaneous absorption process. Minimally, after application in a volatile solvent, the penetrant must a) dissolve/partition into the surface lipids of the stratum corneum, b) diffuse through the lamellar lipid domains of the stratum corneum, c) partition from

Table 12 Percutaneous Absorption of Hydrocortisone in Humans
Following Application from Acetone Solution

	Percent dose absorbed[a]	Applied dose (μg/cm^2)
Plastic wrap occlusion; single dose[b]	9 (6)	5.8
Plastic wrap occlusion; single dose[c]	5 (2)	4.0
Plastic chamber occlusion[d]	4 (2)	4.0
Plastic chamber occlusions[e]	4 (1)	4.0
Plastic chamber occlusion[f]	3 (1)	4.0
"Protected" condition; single doses[g]	4 (2)	4.0
Nonprotected; single dose[h]	1 (0.3)	5.8
Nonprotected; single dose[i]	2 (2)	4.0
Nonprotected; multiple dose[j]	3 (1)	4.0
Nonprotected; multiple dose[k]	3 (1)	4.0

[a]Absorption values, mean (SD), corrected for incomplete renal elimination.
[b]Occluded for 4 days, washed 96 h postapplication with soap and water (12).
[c]Occluded for 1 day, washed 24 h postapplication with soap and water (35).
[d]Single dose occluded for 7 days, washed 24 h postapplication with soap and water (13).
[e]Percent of first dose absorbed following daily doses. Occluded for 14 days, washed 24 h postapplication with soap and water (13).
[f]Percent of eighth dose absorbed following daily doses. Occluded for 14 days, washed 24 h postapplication with soap and water (13).
[g]Site covered for 7 days with a ventilated plastic chamber, washed 24 h postapplication with soap and water (11,24).
[h]Site washed 24 h postapplication with soap and water (12).
[i]Site washed 24 h postapplication with soap and water (42).
[j]Percent of first dose absorbed following daily doses for 14-day site washed 24 h postapplication with soap and water (23).
[k]Percent of eighth dose following daily doses (or 14-day site washed 24 h postapplication with soap and water (23).

the stratum corneum into the more hydrophilic viable epidermis, d) diffuse through the epidermis and upper dermis, and e) encounter a capillary of the cutaneous microvasculature and gain access to the systemic circulation (Fig. 14).

As stated earlier, occlusion hydrates the stratum corneum, and if the effect of hydration were simply to decrease the viscosity of the stratum corneum intercellular domain, then the penetration of all chemicals should be equally enhanced by occlusion. In other words, the relative increase in the effective diffusion coefficient of the penetrant across the stratum cor-

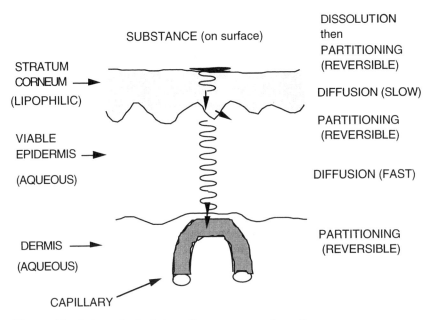

Figure 14 Schematic depiction of percutaneous absorption.

neum would be independent of the nature of the penetrant. But this is not the situation observed; the degree of enhancement is compound specific. To account for this effect, we postulate that stratum corneum hydration alters the stratum corneum–viable epidermis partitioning step. Occlusion hydrates the keratin in corneocytes and increases the water content between adjacent intercellular lipid lamellae. A penetrant diffusing through the intercellular lipid domains will distribute between the hydrophobic bilayer interiors and the aqueous regions separating the head groups of adjacent bilayers. Stratum corneum hydration magnifies the latter environment and increases the "hydrophilic" character of the stratum corneum somewhat. It follows that this leads, in turn, to a reduction in the stratum corneum–viable epidermis partition coefficient of the penetrant (because the two tissue phases now appear more similar). The decrease should facilitate the kinetics of transfer of penetrant through the stratum corneum and from the stratum corneum to the viable epidermis, and the relative effect on this rate should become greater as the lipophilicity of the absorbing molecule increases (40). The limit of this mechanism of enhancement would occur when penetrant is either a) completely insoluble in the aqueous phases of the stratum corneum or b) sterically hindered from penetrating the skin at a measurable rate due to, for example, large molecular size.

The importance of the partitioning step is implied by the dependence of percutaneous absorption with compound lipophilicity, as would be predicted if the skin behaved as a simple lipid membrane. Attenuation in absorption may be explained by a shift in the rate-limiting step from diffusion through the stratum corneum to the transfer across the stratum corneum–viable epidermis interface with increasing compound lipophilicity. The effect should be most apparent when the penetrant's aqueous solubility is extremely low; thus, it follows that this transfer process, or partitioning into the viable epidermis, should become slower as penetrant lipophilicity increases. This suggested mechanism is further supported by results obtained with the *para*-substituted phenols described earlier (25,41) (Fig. 13).

Restricting evaporative loss of volatile compounds using plastic wrap occlusion enhances percutaneous absorption (38). Clearly, this effect may increase the extent of absorption of these lipophilic, volatile compounds, in addition to the possible enhancement afforded by occlusion-induced hydration of the stratum corneum.

As noted earlier, occlusion does not always increase the percutaneous absorption of topically applied agents. Furthermore, the extent of penetration may depend upon the method of occlusion. This finding has important implications in the design of a transdermal drug delivery system (TDS) for which the duration of application exceeds 24 h. We have found that about a third of normal, healthy, male volunteers experience plastic chamber occlusion-induced irritation following contact periods greater than 24 h; however, we have not observed any irritation of the skin using the nonocclusive patch system (made from these occlusive chambers) on the same volunteers following identical contact periods with the same penetrant. In those situations for which occlusion does not significantly increase the percutaneous absorption of a topically applied drug, or an occlusion-induced enhancement in percutaneous absorption is not required, a nonocclusive TDS is an approach worthy of consideration.

Conclusions drawn from the preceding discussion are as follows:

1. Studies from multiple investigators indicate that the extent of percutaneous absorption may depend upon the occlusive system used (16,18–20).
2. Occlusion does not necessarily increase percutaneous absorption. Penetration of hydrophilic compounds (e.g., compounds with low $K_{o/w}$ values), in particular, may not be enhanced by occlusion.
3. Mass balance (dose accountability) has been demonstrated using occlusive and nonocclusive patch systems in vivo in humans. Dose

accountability rigorously quantifies percutaneous absorption measured using radiotracer methodology and allows objective comparison between different treatment modalities.

4. Occlusion, per se, can cause local skin irritation, and the implication of this observation in the design of TDS should be considered.

REFERENCES

1. Elias PM, Brown BE. The mammalian cutaneous permeability barrier. Lab Invest 1978; 39:574–583.
2. Elias PM, Cooper ER, Korc A, Brown B. Percutaneous transport in relation to stratum corneum structure and lipid composition. J Invest Dermatol 1981; 76: 297–301.
3. Elias P. Epidermal lipids, barrier function, and desquamation. J Invest Dermatol 1983; 80:44s–49s.
4. Golden GM, Guzek DB, McKie JE, Potts RO. Role of stratum corneum lipid fluidity in transdermal drug flux. J Pharm Sci 1987; 76:25–31.
5. Menon GK, Feingold KR, Moser AH, Brown BE, Elias PM. De novo sterologenesis in the skin. II. Regulation by cutaneous barrier requirements. J Lipid Res 1985; 26:418–427.
6. Garb J. Nevus verrucosus unilateralis cured with podophyllin ointment. Arch Dermatol 1960; 81:606–609.
7. Scholtz JR. Topical therapy of psoriasis with fluocinolone acetonide. Arch Dermatol 1961; 84:1029–1030.
8. Sulzberger MB, Witten VH. Thin pliable plastic films in topical dermatological therapy. Arch Dermatol 1961; 84:1027–1028.
9. McKenzie AW. Percutaneous absorption of steroids. Arch Dermatol 1962; 86: 91–94.
10. McKenzie AW, Stoughton RB. Method for comparing percutaneous absorption of steroids. Arch Dermatol 1962; 86:88–90.
11. Bucks DAW, McMaster JR, Maibach HI, Guy RH. Bioavailability of topically administered steroids: a "mass balance" technique. J Invest Dermatol. 1988; 90:29–33.
12. Feldmann RJ, Maibach HI. Penetration of ^{14}C hydrocortisone through normal skin. Arch Dermatol 1965; 91:661–666.
13. Guy RH, Bucks DAW, McMaster JR, Villaflor DA, Roskos KV, Hinz RS, Maibach HI. Kinetics of drug absorption across human skin in vivo. In: Shroot B, Schaefer H, eds. Skin Pharmacokinetics. Basel: Karger, 1987:70–76.
14. Wiechers JW. The barrier functions of the skin in relation to percutaneous absorption of drugs. Pharm Weekbl [Sci] 1989; 11:185–198.
15. Qiao GL, Riviere. Significant effects of application site and occlusion on the pharmacokinetics of cutaneous penetration and biotransformation of parathion in vivo in swine. J Pharm Sci 1995; 84:425–432.

16. Gummer CL, Maibach HI. The penetration of [^{14}C]ethanol and [^{14}C]methanol through excised guinea-pig skin in vitro. Food Chem Toxicol 1986; 24:305–306.

17. Riley RT, Kemppainen BW, Norred WP. Penetration of aflatoxins through isolated human epidermis. J Toxicol Environ Health 1985; 15:769–777.

18. Kim HO, Wester RC, McMaster JR, Bucks DAW, Maibach HI. Skin absorption from patch test systems. Contact Dermatitis 1987; 17:178–180.

19. Bronaugh RL, Stewart RF, Wester RC, Bucks DAW, Maibach HI. Comparison of percutaneous absorption of fragrances by humans and monkeys. Food Chem Toxicol 1985; 23:111–114.

20. Bronaugh RL, Wester RC, Bucks DAW, Maibach HI, Sarason R. In vivo percutaneous absorption of fragrance ingredients in rhesus monkeys and humans. Food Chem Toxicol 1990; 28:369–373.

21. Quisno RA, Doyle RL. A new occlusive patch test system with a plastic chamber. J Soc Cosmet Chem 1983; 34:13–19.

22. Bucks DAW, Marty J-PL, Maibach HI. Percutaneous absorption of malathion in the guinea pig: effect of repeated skin application. Food Chem Toxicol 1985; 23:919–922.

23. Bucks DAW, Maibach HI, Guy RH. Percutaneous absorption of steroids: Effect of repeated application. J Pharm Sci 1985; 74:1337–1339.

24. Bucks DAW, Maibach HI, Guy RH. Mass balance and dose accountability in percutaneous absorption studies: Development of non-occlusive application system. Pharm Res 1988; 5:313–315.

25. Bucks DAW, McMaster JR, Maibach HI, Guy RH. Percutaneous absorption of phenols in vivo (abstr). Clin Res 1987; 35–672A.

26. Berardesca E, Vignoli GP, Fideli D, Maibach H. Effect of occlusive dressings on the stratum corneum water holding capacity. Am J Med Sci 1992; 304: 25–28.

27. Blank IH, Scheuplein RJ. The epidermal barrier. In: Rook AJ, Champion RH, eds. Progress in the Biological Sciences in Relation to Dermatology, Vol. 2. Cambridge: Cambridge University Press, 1964:245–261.

28. Potts RO. Stratum corneum hydration: Experimental techniques and interpretation of results. J Soc Cosmet Chem 1986; 37:9–33.

29. Kligman AM. A biological brief on percutaneous absorption. Drug Dev Ind Pharm 1983; 9:521–560.

30. Faergemann J, Aly R, Wilson DR, Maibach HI. Skin occlusion: Effect on Pityrosporum orbiculate, skin permeability of carbon dioxide, pH, transepidermal water loss, and water content. Arch Dermatol Res 1983; 275:383–387.

31. Gordon B, Maibach HI. Studies on the mechanism of aluminum anhidrosis. J Invest Dermatol 1968; 50:411–413.

32. Orentreich N, Berger RA, Auerbach R. Anhidrotic effects of adhesive tapes and occlusive film. Arch Dermatol Res 1966; 94:709–711.

33. Fisher LB, Maibach HI. The effect of occlusive and semipermeable dressings on the mitotic activity of normal and wounded human epidermis. Br J Dermatol 1972; 86:593–600.

34. Bucks DAW. Prediction of Percutaneous Absorption, Ph.D. dissertation. University of California, San Francisco, 1989.

35. Bucks D, Guy R. Maibach H. Effect of occlusion. In: Bronaugh RL, Maibach HI, eds. In Vitro Percutaneous Absorption: Principles, Fundamentals, and Applications. Boston: CRC Press, 1991:85–114.

36. Mukherji E, Millenbaugh HJ, Au JL. Percutaneous absorption of 2′,3′-dideoxyinosine in rats. Pharm Res 1994; 11:809–815.

37. Treffel P, Muret P, Muret-D'Aniello P, Coumes-Marquet S, Agache P. Effect of occlusion on in vitro percutaneous absorption of two compounds with different physicochemical properties. Skin Pharmacol 1992; 5:108–113.

38. Bond JR, Barry BW. Damaging effect of acetone on the permeability barrier of hairless mouse skin compared with that of human skin. Int J Pharmaceut 1988; 41:91–93.

39. Schaefer H, Zesch A, Stuttgen G. Skin Permeability. Berlin: Springer-Verlag, 1982:541–896.

40. Guy RH, Hadgraft J, Bucks DAW. Transdermal drug delivery and cutaneous metabolism. Xenobiotica 1987; 17:325–343.

41. Bucks DAW, McMaster JR, Maibach HI, Guy RH. Prolonged residence of topically applied chemicals in the stratum corneum: Effect of lipophilicity (abstr). Clin Res 1987; 35:672A.

42. Feldmann R, Maibach HI. Percutaneous absorption of steroids in man. J Invest Dermatol 1969; 52:89–94.

43. Bucks DAW, Maibach HI, Guy RH. Occlusion does not uniformly enhance penetration in vivo. In Bronaugh R, Maibach H, eds. Percutaneous Absorption, Vol. 2. New York: Marcel Dekker, 1989: 77–94.

5
Regional Variation in Percutaneous Absorption

Ronald C. Wester and Howard I. Maibach
University of California School of Medicine, San Francisco, California

I. INTRODUCTION

The first occupational disease in recorded history was scrotal cancer in chimney sweeps (Wester, 1995). The historical picture of a male worker holding a chimney brush and covered from head to toe with black soot is vivid. But why the scrotum?

Percutaneous absorption in humans and animals varies depending on the area of the body on which the chemical resides. This is called *regional variation*. When a certain skin area is exposed, any effect of the chemical will be determined by how much is absorbed through the skin. Where systemic drug delivery is desired, such as transdermal delivery, a high-absorbing area may be desirable to deliver sufficient drug. Scopolamine transdermal systems are supposedly placed in the postauricular area (behind the ear) because at this skin site the percutaneous absorption of scopolamine is sufficiently enhanced to deliver effective quantities of the drug. A third example is with estimating human health hazard effects of environmental contaminants. This could be pesticide residue on exposed parts of the skin (head, face, neck, hands) and trying to determine the amount of pesticide that might be absorbed into the body. The estimate for skin absorption is an integral part of the estimate for potential hazard; thus, accuracy of estimate is very relevant.

Therefore, when considering skin absorption in humans, the site of application is important.

II. REGIONAL VARIATION IN HUMANS

Feldmann and Maibach (1967) were the first to systematically explore the potential for regional variation in percutaneous absorption. The first absorption studies were done with the ventral forearm, because this site is convenient to use. However, skin exposure to chemicals exists over the entire body. They first showed regional variation with the absorption of hydrocortisone (Fig. 1). The scrotum was the highest absorbing skin site (scrotal cancer in chimney sweeps is the key). Skin absorption was lowest for the foot area, and highest around the head and face.

Table 1 gives the effect of anatomical region on the percutaneous absorption of pesticides in humans (Maibach et al., 1971). There are two major points in this study. First, regional variation was confirmed with the different chemicals parathion and malathion. Second, those skin areas that would be exposed to the pesticides, the head and face, were among the higher absorbing sites. The body areas most exposed to environmental contaminants are the areas with the higher skin absorption.

Table 2 gives site variability for parathion skin absorption with time. Soap and water wash in the first few minutes after exposure is not a perfect decontaminant. Site variation is apparent early in skin exposure (Wester and Maibach, 1985).

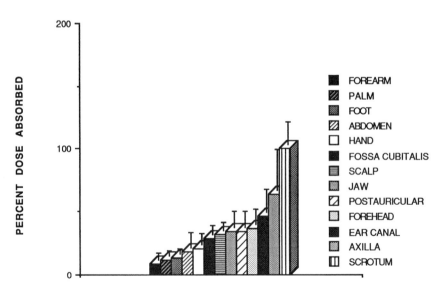

Figure 1   Anatomic regional variation with parathion percutaneous absorption in humans.

Table 1 Effect of Anatomical Region on In Vivo Percutaneous Absorption of Pesticides in Humans

	Dose absorbed (%)		
Anatomical region	Hydrocortisone	Parathion	Malathion
Forearm	1.0	8.6	6.8
Palm	0.8	11.8	5.8
Foot, ball	0.2	13.5	6.8
Abdomen	1.3	18.5	9.4
Hand, dorsum	—	21.0	12.5
Forehead	7.6	36.3	23.2
Axilla	3.1	64.0	28.7
Jaw angle	12.2	33.9	69.9
Fossal cubitalis	—	28.4	
Scalp	4.4	32.1	
Ear canal	—	46.6	
Scrotum	36.2	101.6	

Guy and Maibach (1985) took the hydrocortisone and pesticide data and constructed penetration indices for five anatomical sites (Table 3). These indices should be used with their total surface areas (Table 4) when estimating systemic availability relative to body exposure sites.

Van Rooy et al. (1993) applied coal tar ointment to various skin areas of volunteers and determined absorption of polycyclic aromatic hydrocarbons (PAH) by surface disappearance of PAH and the excretion of urinary

Table 2 Site Variation and Decontamination Time for Parathion

Skin residence time before soap and water wash	Parathion dose absorbed[a] (%)		
	Arm	Forehead	Palm
1 min	2.8	8.4	
5 min			6.2
15 min	6.7	7.1	13.6
30 min		12.2	13.6
1 h	8.4	10.5	11.7
4 h	8.0	27.7	7.7
24 h	8.6	36.3	11.8

[a]Each value is a mean for four volunteers. The fact that there were different volunteers at each time point accounts for some of the variability with time for each skin site.

Table 3 Penetration Indices for Five Anatomical Sites Assessed Using Hydrocortisone Skin Penetration Data and Pesticide (Malathion and Parathion) Absorption Results

Site	Penetration index based on	
	Hydrocortisone data	Pesticide data
Genitals	40	12
Arms	1	1
Legs	0.5	1
Trunk	2.5	3
Head	5	4

1-OH-pyrene. Using PAH disappearance, skin ranking (highest to lowest) was shoulder > forearm > forehead > groin > hand (palmar) > ankle. Using 1-OH-pyrene excretion, skin ranking (highest to lowest) was neck > calf > forearm > trunk hand. Table 5 compares their results with Guy and Maibach (1985).

In another study, Wester et al. (1984) determined the percutaneous absorption of paraquat in humans. Absorption was the same for the leg (0.29 ± 0.02%), hand (0.23 ± 0.1%), and forearm (0.29 ± 0.1%). Here, the chemical nature of the low-absorbing paraquat overcame regional variation.

Rougier et al. (1986) examined the influence of anatomical site on the relationship between total penetration of benzoic acid in humans and the quantity present in the stratum corneum 30 min after application. Figure 2 shows the total penetration of benzoic acid according to anatomical site. Figure 3 shows the correlation between level of penetration of benzoic acid

Table 4 Body Surface Areas Distributed Over Five Anatomical Regions for Adult and Neonate

Anatomical region	Adult		Neonate	
	Body area (%)	Area (cm^2)	Body area (%)	Area (cm^2)
Genitals	1	190	1	19
Arms	18	3420	19	365
Legs	36	6840	30	576
Trunk	36	6840	31	595
Head	9	1710	19	365
Totals		19,000		1920

Table 5 Absorption Indices of Hydrocortisone and Pesticides (Parathion/ Malathion) Calculated by Guy and Maibach (1985) Compared with Absorption Indices of Pyrene and PAH for Different Anatomical Sites by Van Rooy et al. (1993)

Anatomical site	Absorption index			
	Hydrocortisone[a]	Pesticides[b]	Pyrene[c]	PAH[d]
Genitals	40	12	—	—
Arm	1	1	1	1
Hand	1	1	0.8	0.5
Leg/ankle	0.5	1	1.2	0.8/0.5
Trunk/shoulder	2.5	3	1.1	/2.0
Head/neck	5	4	/1.3	1.0

[a]Based on hydrocortisone penetration data (Feldmann & Maibach, 1967).
[b]Based on parathion and malathion absorption data (Maibach et al., 1971).
[c]Based on excreted amount of 1-OH-pyrene in urine after coal-tar ointment application (Van Rooy et al., 1993).
[d]Based on the PAH absorption rate constant (K_a) after coal-tar ointment application (Van Rooy et al., 1993).

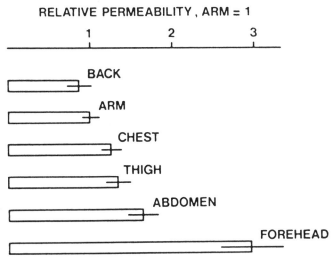

Figure 2 Total penetration of benzoic acid according to anatomic site.

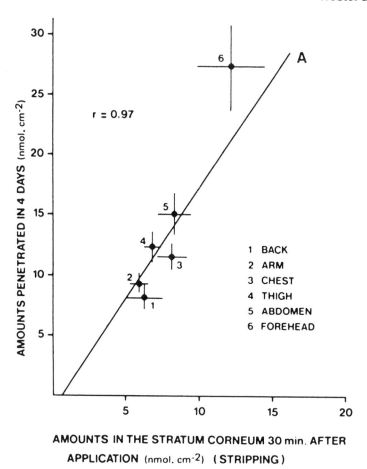

Figure 3 Correlation between the level of penetration of benzoic acid within 4 days and its level in the stratum corneum after 30 min of application according to anatomical site. Curve A: $y = 1.83$, $x = 0.52$.

within 4 days and its level in the stratum corneum after a 30-min application relative to anatomical site.

Wertz et al. (1993) determined regional variation in permeability through human and pig skin and oral mucosa (Table 6). In the oral mucosa of both species, permeability ranked floor of mouth > buccal mucosa > palate. Skin remains a greater barrier; absorption is some 10-fold less than for oral mucosa.

Table 6 Permeability Constants for the Diffusion of ^3H_2O Through Human and Porcine Skin and Oral Mucosa

Site	K_p (10^{-7} cm/min) (mean $\pm$ standard error)	
	Human	Pig
Skin	44 $\pm$ 4	62 $\pm$ 5
Palate	450 $\pm$ 27	364 $\pm$ 18
Buccal	579 $\pm$ 16	634 $\pm$ 19
Floor of mouth	973 $\pm$ 3	808 $\pm$ 23

The barrier properties of skin relative to oral mucosa have been a benefit for longer term transdermal delivery. Nitroglycerin buccal tablets are effective for about 20 min, due to rapid buccal absorption. In contrast, transdermal nitroglycerin is prescribed for 24 h of continuous-dose delivery. The transdermal nitroglycerin patch is placed on the chest more for psychological reasons than those related to scientific regional variation skin absorption. Some transdermal systems take advantage of regional variation skin absorption and some do not (Table 7).

Shriner and Maibach (1996) studied skin contact irritation and showed that areas of significant response were neck > perioral > forehead. The volar forearm was the least sensitive of eight areas tested. This is in contrast to the commonly held belief that the forearm is one of the best locations to test for immediate contact irritation.

Table 7 Site Variation in Transdermal Delivery

Transdermal drug	Body site	Reason
Nitroglycerin	Chest	Psychological: the patch is placed over the heart
Scopolamine	Postauricular	Scientific: behind the ear was found to be the best absorbing area
Estradiol	Trunk	Convenience: easy to place, and out of view
Testosterone	Scrotum	Scientific: highest skin absorbing area
Testosterone	Trunk	Scientific/convenience: removal from trunk skin is easier than scrotal skin

III. REGIONAL VARIATION IN ANIMALS

Percutaneous absorption data obtained in humans is most relevant for human exposure. However, many estimates for humans are made from animal models. Therefore, regional variation in animals may affect prediction for humans. Also, if regional variation exists in an animal, that variation should be relative to humans.

Bronaugh (1985) reported the effect of body site (back vs. abdomen) on male rat skin permeability. Abdominal rat skin was more permeable to water, urea, and cortisone. Skin thickness (stratum corneum, whole epidermis, whole skin) is less for the abdomen than for the back. With the hairless mouse, Behl et al. (1985) showed dorsal skin to be more permeable than abdominal skin (reverse of that of the male rat) (Table 8). Hairless mouse abdominal skin is thicker than dorsal skin (also reverse of that of the male rat) (Table 8).

Skin absorption in the rhesus monkey is considered to be relevant to that of humans. Table 9 shows the percutaneous absorption of testosterone (Wester et al., 1980), fenitrothion, aminocarb, and diethyltoluamide (DEET) (Moody and Franklin, 1987; Moody et al., 1988) in the rhesus monkey compared with the rat. What is interesting is that for the rhesus monkey there is regional variation between forehead (scalp) and forearm. If one determines the ratio of forehead (scalp/forearm for the rhesus monkey and compares the results with humans, the similarities are the same (Table 10). Therefore, the rhesus monkey probably can be a relevant animal model for human skin regional variation.

Table 8 Effect of Body Site on Rat Skin Permeability

Compound and site	Permeability constant (cm/h $\times$ 10^4)
Water	
Back	4.9 ± 0.4
Abdomen	13.1 ± 2.1
Urea	
Back	1.6 ± 0.5
Abdomen	18.8 ± 5.5
Cortisone	
Back	1.7 ± 0.4
Abdomen	12.2 ± 0.6

Table 9 Percutaneous Absorption of Fenitrothion, Aminocarb, DEET, and Testosterone in Rhesus Monkey and Rat

Chemical	Species	Applied dose absorbed (%) skin site		
		Forehead	Forearm	Back
Fenitrothion	Rhesus	49	21	
	Rat			84
Aminocarb	Rhesus	74	37	
	Rat			88
Testosterone	Rhesus	20.4[a]	8.8	
	Rat			47.4
DEET	Rhesus	33	14	
	Rat			36

[a]Scalp.

Table 10 Percutaneous Absorption Ratio for Scalp and Forehead to Forearm in Humans and Rhesus Monkey

Chemical	Species	Percutaneous absorption ratio	
		Scalp/forearm	Forehead/forearm
Hydrocortisone	Human	3.5	6.0
Benzoic acid	Human		2.9
Parathion	Human	3.7	4.2
Malathion	Human		3.4
Testosterone	Rhesus	2.3	
Fenitrothion	Rhesus		2.3
Aminocarb	Rhesus		2.0
DEET	Rhesus		2.4

IV. CONCLUSION

This chapter outlines 30 years of progress. Careful review of the data shows some general trends; however, most parts of various animal skins have not been explored, and many possible special areas in humans remain unstudied, such as finger and toe nails, eyelids, perirectal skin, upper versus lower arm, thigh versus leg, and so on. It is hoped that as more complete maps are available that cover a range of chemical moieties, we should be in a position to further refine those aspects of dermatopharmacology and dermatotoxicology that require knowledge of skin penetration.

REFERENCES

Behl, C.R., Bellantone, N.H., and Flynn, G.L. (1985). Influence of age on percutaneous absorption of drug substances. In *Percutaneous Absorption*. Edited by R. Bronaugh and H. Maibach. Marcel Dekker, New York, pp. 183–212.

Bronaugh, R.L. (1985). Determination of percutaneous absorption by *in vitro* techniques. In *Percutaneous Absorption*. Edited by R. Bronaugh and H. Maibach. Marcel Dekker, New York, pp. 267–279.

Feldmann, R.J., and Maibach, H.I. (1967). Regional variation in percutaneous penetration of [^{14}C] cortisol in man. *J. Invest. Dermatol. 48*:181–183.

Guy, R.H., and Maibach, H.I. (1985). Calculations of body exposure from percutaneous absorption data. In *Percutaneous Absorption*. Edited by R. Bronaugh and H. Maibach. Marcel Dekker, New York, pp. 461–466.

Maibach, H.I., Feldmann, R.J., Milby, T.H., and Sert, W.F. (1971). Regional variation in percutaneous penetration in man. *Arch. Environ. Health 23*:208–211.

Moody, R.P., and Franklin, C.A. (1987). Percutaneous absorption of the insecticides fenitrothion and aminocarb. *J. Toxicol. Environ. Health 20*: 209–219.

Moody, R.P., Benoit, F.M., Reedel, D., Retter, L., and Franklin, C. (1988). Dermal absorption of the insect repellent Deet in rats and monkeys: Effect of anatomic site and multiple exposure. (Personal communication.)

Rougier, A., Dupuis, D., Lotte, C., Roquet, R., Wester, R.C., and Maibach, H.I. (1986). Regional variation in percutaneous absorption in man: Measurement by the stripping method. *Arch. Dermatol. Res. 278*:465–469.

Shriner, D.L., and Maibach, H.I. (1996). Regional variation of nonimmunological contact urticaria. *Skin Pharmacol. 348*:1–11.

Van Rooy, T.G.M., De Roos, J.H.C., Bodelier-Bode, M.M., and Jongeneelen, F.J. (1993). Absorption of polycyclic aromatic hydrocarbons through human skin: Differences between anatomic sites and individuals. *J. Toxicol. Environ. Health 38*:355–368.

Wertz, P.W., Swartzendruber, D.C., and Squier, C.A. (1993). Regional variation in the structure and permeability of oral mucosa and skin. *Adv. Drug Delivery Rev. 12*:1–12.

Wester, R.C. (1995). Twenty absorbing years. In *Exogenous Dermatology*. Edited by C. Surber, P. Elsner, and A.J. Bircher. Karger, Basel, pp. 112–123.

Wester, R.C., and Maibach, H.I. (1985). *In vivo* percutaneous absorption and decontamination of pesticides in humans. *J. Toxicol. Environ. Health 16*:25–37.

Wester, R.C., Noonan, P.K., and Maibach, H.I. (1980). Variation on percutaneous absorption of testosterone in the rhesus monkey due to anatomic site of application and frequency of application. *Arch. Dermatol. Res. 267*:299–235.

Wester, R.C., Maibach, H.I., Bucks, D.A.W., and Aufrere, M.B. (1984). *In vivo* percutaneous absorption of paraquat from hand, leg and forearm of humans. *J. Toxicol. Environ. Health 14*:759–762.

6

In Vivo Relationship Between Percutaneous Absorption and Transepidermal Water Loss

André Rougier
Laboratoire Pharmaceutique, La Roche-Posay, Courbevoie, France

Claire Lotte
Laboratoires de Recherche Fondamentale, L'Oréal, Aulnay sous Bois, France

Howard I. Maibach
University of California School of Medicine, San Francisco, California

In its role as a barrier the skin participates in homeostasis by limiting a) water loss (Blank et al., 1984; Scheuplein and Blank, 1971) and b) percutaneous absorption of environmental agents (Marzulli, 1962; Malkinson, 1958).

The stratum corneum's role of a double barrier is intimately linked to its degree of hydration (Fritsch and Stoughton, 1963; Wurster and Kramer, 1961), transport mechanisms being diffusional (Marzulli, 1962; Scheuplein, 1978). In humans (Guillaume et al., 1981; Smith et al., 1961) and in animals (Lamaud et al., 1984), an increase in water permeability of the skin corresponds to an increase in permeability to topically applied compounds. However, most of the studies dealing with this topic are only quantitative observations, and the relationship linking these two factors is unknown.

At present, transepidermal water loss (TEWL) can be considered a determinant indicative of the functional state of the cutaneous barrier (Maibach et al., 1984; Surinchack et al., 1985; Wilson and Maibach, 1982). Apart from pathological considerations, the functional state of the cutaneous barrier may vary considerably under physiological conditions (Wilson and Maibach, 1982). Thus, in humans, cutaneous permeability to applied compounds varies from one site to another (Feldmann and Maibach, 1967; Maibach et

117

al., 1971; Rougier et al., 1986). The present chapter investigates the influence of anatomical site, age, and sex in humans, on both TEWL and percutaneous absorption, to establish the precise relationship between these two indicators of the functional state of the cutaneous barrier.

I. PERCUTANEOUS ABSORPTION MEASUREMENTS

The penetration of benzoic acid was measured at 10 anatomical sites, the location of which are shown in Figure 1. A group of six to eight informed male volunteers, aged 20–30 years, was used for each anatomical site. The influence of aging on skin absorption of benzoic acid was studied on groups of seven to eight male volunteers, aged 45–55 and 65–80 years. The anatomical site involved was the upper outer arm. The influence of gender on skin absorption was assessed on the upper outer arms and on the foreheads of groups of seven to eight female volunteers, aged 20–30 years.

One thousand nanomoles of benzoic acid (ring-^{14}C) (New England Nuclear), with a specific activity of 10^{-3} μCi/nmol, was applied to an area of 1 cm^2 in 20 μl of a vehicle consisting of ethylene glycol to which 10% Triton X-100 had been added as surfactant. The treated area was demarcated by an open circular cell fixed by silicone glue to prevent chemical loss. After 30 min, excess chemical was quickly removed by two successive

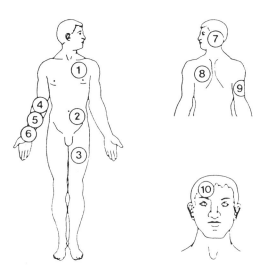

Figure 1   Anatomical sites tested: 1, chest; 2, abdomen; 3, thigh; 4, forearm (ventral-elbow); 5, forearm (ventral-mid); 6, forearm (ventral-wrist); 7, postauricular; 8, back; 9, arm (upper outer); 10, forehead.

washes (2 × 300 μl) with a 95:5 ethanol/water mixture, followed by two rinses (2 × 300 μl) with distilled water and light drying with a cotton swab.

Benzoic acid was selected because of the rapidity and high level of its urinary excretion. Thus, from literature data on the kinetics of urinary excretion of this compound when administered intravenously and orally (Bridges et al., 1970; Bronaugh et al., 1982) or percutaneously (Dupuis et al., 1984; Feldmann and Maibach, 1970; Rougier et al., 1983) in different species, the proportion of the total amount of benzoic acid absorbed that would be excreted in the urine within the first 24 h was 75%. The total quantities absorbed during the 4 days after application could therefore be calculated, after scintillation counting, from the quantities found in the urines during the first 24 h.

II. TRANSEPIDERMAL WATER LOSS MEASUREMENTS

After completion of the benzoic acid treatment, TEWL was measured with an Evaporimeter EPIC (Servo Med, Sweden) from a contralateral site (same anatomical region) in each subject. The handheld probe was fitted with a 1-cm tail chimney, to reduce air turbulence around the hydrosensors, and metallic shield (supplied by Servo Med), minimizing the possibility of sensor contamination. Measurements $(g \cdot m^{-2} \cdot h^{-1})$ stabilized within 30–45 s. As the room environment was comfortable (room temperature 20°C, relative humidity 70%) and the subjects were physically inactive, the TEWL should closely reflect stratum corneum water flux without significant sweating interference.

Table 1 gives the figures for permeability to water (TEWL) and to benzoic acid (percutaneous absorption), according to anatomical site, age, and sex. In male subjects aged from 20 to 30, cutaneous permeability to both water and benzoic acid was: forearm (ventral-elbow) < forearm (ventral-mid) < back < arm (upper outer) ≤ chest ≤ thigh = abdomen < forearm (ventral-wrist) < postauricular < forehead. In the sites studied (upper outer arm, forehead) no differences between sexes were observed.

In relation to age, no alterations in skin permeability appeared to occur before the age of 55. In subjects aged from 65 to 80 (upper outer arm), although there was no change in TEWL, percutaneous absorption of benzoic acid decreased appreciably (factor of 4; $p < .001$). Irrespective of anatomical site and sex, there exists a linear relationship (Fig. 2; $r = .92$, $p < .001$) between total penetration of benzoic acid and TEWL. (Point number 3, corresponding to subjects aged 65–80 measured on the upper outer arm, is a special case that will be discussed, and that was not taken into account in the calculation of the correlation coefficient.)

Table 1 Influence of Anatomical Site, Age, and Sex on Transepidermal Water Loss (TEWL) and Percutaneous Absorption of Benzoic Acid

Group no.	Volunteers per group	Anatomical site	Age (Years)	Sex	Transepidermal water loss (TEWL)[a] $(g \cdot m^{-2} \cdot h^{-1})$	Total penetration of benzoic acid within 4 days[a] $(nmol \cdot cm^{-2})$
1	8	Arm (upper outer)	20–30	M	4.24 (0.35)	9.15 (1.01)
2	8	Arm (upper outer)	45–55	M	5.07 (0.23)	10.02 (1.02)
3	7	Arm (upper outer)	65–80	M	4.73 (0.45)	2.53 (0.82)
4	7	Arm (upper outer)	20–30	F	5.12 (0.35)	11.20 (1.20)
5	7	Abdomen	20–30	M	4.40 (0.51)	12.50 (1.64)
6	8	Postauricular	20–30	M	8.35 (0.41)	22.49 (5.14)
7	7	Forehead	20–30	M	10.34 (0.70)	26.80 (3.19)
8	8	Forehead	20–30	F	9.39 (0.73)	28.99 (1.81)
9	7	Forearm (ventral-elbow)	20–30	M	2.50 (0.30)	3.48 (0.33)
10	8	Forearm (ventral-mid)	20–30	M	4.00 (0.32)	5.51 (1.70)
11	7	Forearm (ventral-wrist)	20–30	M	7.19 (0.39)	12.18 (1.62)
12	8	Back	20–30	M	4.51 (0.57)	8.55 (1.32)
13	8	Chest	20–30	M	4.73 (0.26)	11.70 (1.30)
14	7	Thigh	20–30	M	4.39 (0.32)	12.50 (1.43)

[a]Values are means, with SD in parentheses.

Among factors that might modify skin permeability, both in vitro (Scheuplein, 1979) and in vivo (Feldmann and Maibach, 1967; Maibach et al., 1971; Rougier et al., 1986, 1987; Wester et al., 1984), the anatomical location is of great importance, even if the connection between the differences observed, the structure of the skin, and the physicochemical nature of

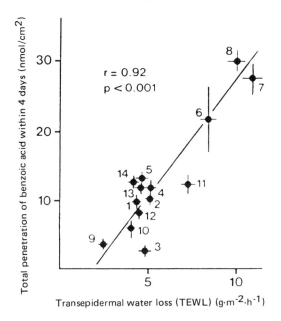

Figure 2   In vivo relationship between transepidermal water loss (TEWL) and per-cutaneous absorption of benzoic acid according to the anatomical site, age, and sex in humans (cf. Table 1).

the penetrant remain obscure. As a matter of fact, general reviews often give contradictory explanations of the differences observed from one site to another (Idson, 1975; Scheuplein, 1979).

The laws describing diffusion through a membrane accord a major role to membrane thickness. These general laws have been applied to several biophysical phenomena, including TEWL and percutaneous absorption. However, examination of the literature shows that it is not unusual that these laws, predicated on pure mathematical logic, are found wanting when applied to a discontinuous membrane of great physicochemical complexity such as the stratum corneum. Thus, the skin permeability of the forehead to water and benzoic acid is, respectively, four and eight times higher than on the forearm (ventral-elbow), despite the fact that the stratum corneum thicknesses of these two sites are comparable (on average 12 μm and 18 cell layers; Holbrook and Odland, 1974; Pathack and Fitzpatrick, 1974). This example, therefore, seems to run counter to the inverse relationship that should exist between permeability and membrane thickness. Hence, simple consideration of the thickness of the stratum corneum cannot, of itself, explain the differences in TEWL and penetration observed between anatomical sites. Other criteria must be considered.

A possible explanation of the higher permeability of the forehead to both benzoic acid and water could partly be because of the great number of sebaceous and sweat glands found in this area. However, even if it is relatively easy to imagine a molecule's penetration by simultaneously adopting the follicular, sweat, and transcorneal routes, it is, unfortunately, difficult to evaluate the relative extent of each route. Because of the density of active sebaceous glands, the forehead is the richest of all the sites tested in terms of sebum. It forms a discontinuous film on the surface of the skin, between 0.4 and 4 μm thick (Kligman, 1983a). It is reasonable to question to what extent the physicochemical nature of benzoic acid interacts with this film and to what extent this initial contact influences its absorption. However, previous studies have shown that the same ratio exists between the permeability levels of areas such as the forehead, which is rich in sebum, and the arm, which has very little sebum, for molecules with totally different lipid/water solubilities (Rougier et al., 1987). Moreover, it has been demonstrated that removing the lipidic film from the skin surface or artificially increasing its thickness has no effect on TEWL (Kligman, 1983b), which is another indicator of skin permeability.

Regional variations in cell size of the desquaming part of the human stratum corneum have been demonstrated by Plewig and Marples (1970) and by Marks et al. (1981). When taking into account that in all the sites tested the individual thickness of the corneocytes does not change (Plewig et al., 1983), the relationship between the flat area of corneocytes and barrier function of the horny layer must be addressed.

In permeability phenomena, the current trend is to assign priority to intercellular, rather than transcellular, penetration. Thus, if we consider two anatomical sites with an equal volume of stratum corneum but that contain corneocytes of unequal volume, such as the abdomen and the forehead, it is obvious that the intercorneal space will be greater in the stratum corneum that has smaller corneocytes. In adults (Marks et al., 1981; Plewig and Marples, 1970), the flat surface of the forehead stratum corneum cells is approximately 30% less in area than that of cells from the arm, abdomen, or thigh. Moreover, as a function of the anatomical site, it has been demonstrated that an inverse relationship exists between the area of the horny layer cells and the value of the TEWL (Marks et al., 1981). What influence the intercorneal volume has on percutaneous absorption presents a question we have undertaken to answer.

As our results show (Table 1), there is no difference in benzoic acid absorption and TEWL between the 20–30 and 45–55 age groups. In subjects aged 65–80, on the other hand, absorption of this molecule is greatly reduced (factor of 4). These findings agree with those of Malkinson and Fergusson (1955), who failed to show any difference in percutaneous absorption of hydrocortisone in adults aged from 41 to 58. The reduced absorption of benzoic acid observed in the elderly (65–80) was similar to that obtained with testosterone by others (Christophers and Kligman, 1964; Roskos et al., 1986). It is a reasonable presumption that this change could be partly a consequence of alterations in keratinization and epidermal cell production, and itself results in altered structure and function of the stratum corneum. It has been established that corneocyte surface area increases with age (Grove, 1979; Plewig, 1970; Grove et al., 1981). Moreover, recent studies have shown that, at the same time, a linear decrease occurred in the size of epidermal cells (Marks, 1981).

Although there is no change in total thickness of the stratum corneum with age (Lavker, 1979), the question of whether or not there is an increase in surface area of corneocytes concomitant with a decrease in corneocytes thickness has not been finally decided. However, preliminary studies do not suggest that this value alters in aging stratum corneum (Marks and Barton, 1983; Marks et al., 1981). So, if we assume that this latter factor does not change with advancing age, then the volume of intercellular spaces must decrease as the surface area of the corneocytes increases. The spaces between the corneocytes probably act as the molecular "reservoir" of the stratum corneum (Dupuis et al., 1984). For a given molecule, the smaller the capacity of this reservoir, the less it is absorbed (Dupuis et al., 1984, 1986; Rougier et al., 1985, 1986. 1987). As the general morphological organization of the stratum corneum does not appear to be affected by aging (Lavker, 1979; McKenzie et al., 1981), it is tempting to conclude that the great differences observed in percutaneous absorption of benzoic acid, according to age, are solely due to the change in corneocyte size. However, this would be too simplistic an approach, and we should take into account other factors that affect the physical and physicochemical properties of the barrier, such as changes in the lipid composition of the intercellular cements, cohesion between corneocytes, which decreases with age (Leveque et al., 1984; Marks et al., 1983), or the hydration level of the horny layer (Blank, 1952; Blank et al., 1984). Moreover, morphological and functional changes in adjacent structures, in particular the dermis, should also be taken into consideration. Thus, in advancing age, the underside of the epidermis becomes flattened, with this flattening being accompanied by diminution of superficial blood vessels (Ellis, 1958; Montagna and Carlisle, 1979). These alterations in the vascular bed and extracellular matrix may lead to decreased clearance of transdermally absorbed materials from the dermis (Küstala, 1972; Marks, 1981).

Although the barrier function of the stratum corneum to the penetration of environmental agents appears to increase with age, we agree with others that TEWL does not statistically vary (Kligman, 1979; Leveque et al., 1984; Marks et al., 1983). This is a strange situation because a particular feature of aged skin is the roughness and apparent dryness of its surface. If it is true that the stratum corneum is the differentiated cellular end of the viable epidermis and, as such, must share the general effect of aging that takes place in all cells, the absolute need to maintain homeostasis suggests that to maintain its functional integrity, functional alterations of the horny structure will be subtle and difficult to detect.

In the areas studied (upper outer arm, forehead), no differences have been found between male and female subjects, either in percutaneous absorption of benzoic acid or in water loss. There has been no systematic study showing the effects of sex on cutaneous permeability in humans. We cannot, therefore, compare our results with the literature.

Although most authors recognize the importance of the anatomical site either on the degree of absorption of molecules or on TEWL, the literature does not include any quantitative data on the relationship that may exist in humans between these two functions. Our results show (Fig. 2) that for the anatomical sites studied, and within the range of TEWL and penetration values determined, a highly significant linear relationship exists ($r = .92$, $p < .001$) between the permeability of skin to water and the percutaneous absorption of a non-water-soluble compound such as benzoic acid. Only those values obtained for aged subjects (65–80 years, upper outer arm, point number 3) do not fit on this correlation curve.

Although it is generally agreed that either of these factors can be considered as a reflection of the functional state and integrity of the cutaneous barrier, our results have demonstrated that they are directly linked. However, before overgeneralizing, we have examined the relationship existing between TEWL and percutaneous absorption of different molecules of varying physicochemical properties. The percutaneous absorption of three radiolabeled compounds (New England Nuclear)—acetylsalicylic acid (carboxyl-^{14}C), caffeine (1-methyl-^{14}C), and benzoic acid sodium salt (ring-^{14}C)—was determined for four anatomical sites, the exact locations of which are shown in Fig. 3. For each molecule and for each site, six to eight male caucasion volunteers aged 28 ± 2 years were studied.

One thousand nanomoles of each compound, with a specific activity adjusted to 10^{-3} μCi/nmol, was applied to an area of 1 cm^2, in 20 μl of the appropriate vehicle. The composition of these vehicles, shown in Table 2, was selected according to the solubility of each compound. Triton X-100 was added as a surfactant to obtain smooth spreading of the vehicle over the treated area, the boundaries of which were circumscribed by an open

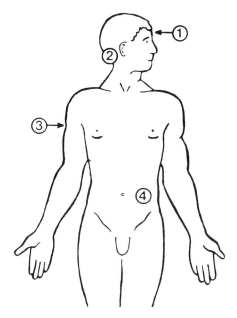

Figure 3 Anatomical sites tested: 1, forehead; 2, postauricular; 3, arm (upper outer); 4, abdomen.

circular cell fixed by silicone glue to prevent any chemical loss. After 30 min, excess substance was rapidly removed by washing, rinsing, and drying the treated area as previously described.

The molecules tested were selected on the basis of the rapidity and high level of their urinary excretion. In view of the literature concerning urinary excretion kinetics for these substances after administration by various routes in different species (Bronaugh et al., 1982; Dupuis et al., 1984; Feldmann and Maibach, 1970; Rougier et al., 1983, 1985), the total amounts that had penetrated during the 4 days following application could be calculated, after scintillation counting, from the quantities found in the urine during the first 24 h. The proportion of the total amounts of benzoic acid sodium salt, caffeine, and acetylsalicylic acid absorbed that were excreted within 24 h were 75, 50, and 31%, respectively.

After topical administration of the tested compound, TEWL was measured from a contralateral site (same anatomical region) in each subject as described earlier. The results (Table 2) show that the percutaneous penetration of the test molecules varied with the anatomical location. The area behind the ear and that on the forehead were the most permeable, regardless

Table 2 Percutaneous Absorption and TEWL Values According to Anatomical Site

n and Anatomical site	Amount in urines after 24 h	Total amount penetrated within 4 days[a]	Transepidermal water loss[b]	Relative permeability to arm	
				Penetration	TEWL
Compound: benzoic acid sodium salt, vehicle A					
6 arm	3.02	4.02	6.06	1	1
(upper outer)	(0.34)	(0.45)	(0.36)		
6 abdomen	5.73	7.65	5.37	1.9	0.9
	(0.54)	(0.72)	(0.46)		
6 postauricular	7.54	10.06	7.72	2.5	1.3
	(0.62)	(0.82)	(0.64)		
8 forehead	9.31	12.32	12.29	3.1	2
	(1.76)	(2.30)	(0.96)		
Compound: caffeine, vehicle B					
7 arm	6.04	12.09	7.04	1	1
(upper outer)	(0.92)	(1.84)	(0.95)		
6 abdomen	3.76	7.53	6.05	0.6	0.9
	(0.67)	(1.34)	(0.43)		
7 postauricular	5.87	11.72	8.74	1	1.2
	(0.52)	(1.05)	(0.62)		
6 forehead	11.17	22.35	12.77	1.9	1.8
	(1.20)	(2.39)	(1.05)		

Compound: acetylsalicylic acid, vehicle A

7 arm	5.27 (0.18)	17.00 (0.37)	5.08 (0.79)	1	1
6 abdomen	5.34 (1.03)	17.20 (3.35)	5.16 (0.43)	1	1
6 postauricular	11.04 (2.50)	29.17 (5.37)	9.04 (0.84)	2.1	1.8
6 forehead	10.89 (1.02)	35.14 (3.29)	11.22 (0.96)	2.1	2.2

Note. Values are expressed in nmol $\cdot$ cm^{-2} with SD in parentheses. Vehicle A: (ethylene glycol/Triton X-100); 90:10. Vehicle B: (ethylene glycol/Triton X-100); 90:10/H$_2$O:50:50.

[a] Calculated from urinary excretion for benzoic acid sodium salt, (amount in urines)/0.75; for caffeine, (amount in urines)/0.5; for acetylsalicylic acid, (amount in urines)/0.31.

[b] Measured just before the application in g $\cdot$ m^{-2} $\cdot$ h^{-1}.

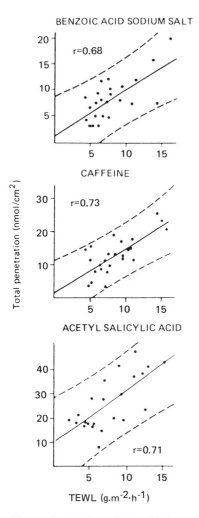

Figure 4 In vivo relationship between transepidermal water loss (TEWL) and percutaneous absorption of some organic compounds according to the anatomical site in humans.

of the physicochemical properties of the compound tested. In general, the order of cutaneous permeability was as follows: arm $\leq$ abdomen < post-auricular < forehead. Although the literature provides few details about the influence of anatomical site on the absorption of molecules, the rank order obtained here agrees with studies performed with other chemicals (Feldmann and Maibach, 1967, 1970; Maibach et al., 1971; Taskovitch and Shaw, 1978).

Cutaneous permeability is generally considered a mirror of the integrity of the horny layer. Even in normal skin, the efficacy of this barrier is not constant. Thus, as shown in Table 2, for a given anatomical site the permeability varies widely in relation to the nature of the molecule administered because this is related to the physicochemical interactions that may occur between the molecule, the vehicle, and the stratum corneum. For the anatomical sites investigated and for the range of TEWL and penetration observed, there exists a linear relationship between the permeability of the skin to the outward movement of water and the inward uptake of molecules. Table 2 shows that this relationship fits with all the compounds tested, with caffeine being an apparent exception (decreased penetration on abdomen and postauricular regions as compared with the arm, whereas TEWL showed a slight decrease on the abdomen and an increase in the postauricular region). It is, however, worth noting that the same relationship, when expressed with individual values (Fig. 4), shows correlation coefficients between .68 and .73 ($p < .05$), the latter (and better one) corresponding to caffeine (the confidence limits represent a risk of 5%).

Thus, it appears that the linear relationship linking TEWL and percutaneous absorption is independent of the physicochemical properties of the compound applied. Moreover, with the three molecules investigated, a mean increase of 2.7 in percutaneous absorption corresponded to an increase of 3 in the TEWL value. This fact supports the hypothesis that the efficiency of the barrier is dependent on the physicochemical properties of the molecule administered, but its functional state is independent of these. Consequently, as with determinations of TEWL, percutaneous absorption measurement provides a good marker of the cutaneous barrier integrity.

REFERENCES

Blank, I.H. (1952). Factors which influence the water content of the stratum corneum. J. Invest. Dermatol. 18:433–437

Blank, I.H., Moloney, J., Emslie, A.G., Simon, I., Apt, C.H. (1984). The diffusion of water across the stratum corneum as a function of its water content. J. Invest. Dermatol. 82:188–194.

Bridges, J.W., French, M.R., Smith, R.L., Williams, R.T. (1970). The fate of benzoic acid in various species. Biochem. J. 188:47–51.

Bronaugh, R.L., Stewart, R.F., Congdon, E.R., Giles, A.L. (1982). Methods for in vitro percutaneous absorption studies. I. Comparison with in vivo results. Toxicol. Appl. Pharmacol. 62:474–480.

Christophers, E., Kligman, A.M. (1964). Percutaneous absorption in aged skin. In Advances in Biology of the Skin. Edited by W. Montagna. Pergamon Press, New York, pp. 163–175.

Dupuis, D., Rougier, A., Roguet, R., Lotte, C., Kalopissis, G. (1984). In vivo relationship between horny layer reservoir effect and percutaneous absorption in human and rat. J. Invest. Dermatol. 82:353–356.

Dupuis, D., Rougier, A., Roguet, R., Lotte, C. (1986). The measurement of the stratum corneum reservoir: A simple method to predict the influence of vehicles on in vivo percutaneous absorption. Br. J. Dermatol. 115:233–238.

Ellis, R.A. (1958). Aging of the human male scalp. In The Biology of Hair Growth. Edited by W. Montagna and R. A. Ellis. Academic Press, New York, pp. 469–485.

Feldman, R.J., Maibach, H.I. (1967). Regional variations in percutaneous penetration of ^{14}C cortisol in man. J. Invest. Dermatol. 48:181–183.

Feldmann, R.J., Maibach, H.I. (1970). Absorption of some organic compounds through the skin in man. J. Invest. Dermatol. 54:399–404.

Fritsch, W.F., Stoughton, R.B. (1963). The effect of temperature and humidity on the penetration of ^{14}C acetylsalicylic acid in excised human skin. J. Invest. Dermatol. 41:307–311.

Grove, G.L. (1979). Exfoliative cytological procedures as a nonintrusive method for dermatogerontological studies. J. Invest. Dermatol. 79:67–69.

Grove, G.L., Lavker, R.M., Holzle, E., Kligman, A.M. (1981). The use of nonintrusive tests to monitor age associated changes in human skin. J. Soc. Cosmet. Chem. 32:15–26.

Guillaume, J.C., de Rigal, J., Leveque, J.L., Galle, P., Touraine, R., Dubertret, L. (1981). Etude comparée de la perte insensible d'eau et de la pénétration cutanée des corticoïdes. Dermatologica 162:380–390.

Holbrook, K.A., Odland, G.F. (1974). Regional differences in the thickness (cell layers) of the human stratum corneum: an ultrastructural analysis. J. Invest. Dermatol. 62:415–422.

Idson, B. (1975). Percutaneous absorption. J. Pharm. Sci. 64:901–924.

Kligman, A.M. (1979). Perspectives and problems in cutaneous gerontology. J. Invest. Dermatol. 73:39–46.

Kligman, A.M. (1983a).The use of sebum. Br. J. Dermatol. 75:307–319.

Kligman, A.M. (1983b). A biological brief on percutaneous absorption. Drug Dev. Ind. Pharm. 9:521–560.

Küstala, G. (1972). Dermal epidermal separation: Influence of age, sex and body region on suction blister formation in human skin. Ann. Clin. Res. 4:10–22.

Lamaud, E., Lambrey, B., Schalla, W., and Schaefer, H. (1984). Correlation between transepidermal water loss and penetration of drugs. J. Invest. Dermatol. 82: 556.

Lavker, R.M. (1979). Structural alterations in exposed and unexposed aged skin. J. Invest. Dermatol. 73:59–66.

Leveque, J.L., Corcuff, P., de Rigal, J., Agache, P. (1984). In vivo studies of the evolution of physical properties of the human skin with age. Int. J. Dermatol. 23:322–329.

Maibach, H.I., Feldmann, R.J., Milby, T., Serat, W. (1971). Regional variation in percutaneous penetration in man. Arch. Environ. Health 23:208–211.

Maibach, H.I., Bronaugh, R., Guy, R., Turr, E., Wilson, D., Jacques, S., Chaing, D. (1984). Noninvasive techniques for determining skin function. In Cutaneous Toxicity. Edited by V.A. Drill and P. Lazar. Raven Press, New York, pp. 63–97.

Malkinson, F. D., and Ferguson, E. H. (1955). Percutaneous absorption of hydrocortisone-4-^{14}C in two human subjects. J. Invest. Dermatol. 25:281–283.

Malkinson, F.D. (1958). Studies on percutaneous absorption of ^{14}C-labelled steroids by use of the Gaz low cell. J. Invest. Dermatol. 31:19–28.

Marks, R. (1981). Measurement of biological aging in human epidermis. Br. J. Dermatol. 104:627–633.

Marks, R., Nicholls, S., King, C.S. (1981). Studies on isolated corneocytes. Int. J. Cosmet. Sci. 3:251–258.

Marks, R., Barton, S.P. (1983). The significance of size and shape of corneocytes. In Stratum Corneum. Edited by R. Marks and G. Plewig. Springer-Verlag, Berlin-Heidelberg, New York, pp. 161–170.

Marks, R., Lawson, A., Nicholls, S. (1983). Age-related changes in stratum corneum structure and function. In Stratum Corneum. Edited by R. Marks and G. Plewig. Springer-Verlag, Berlin-Heidelberg, New York, pp. 175–180.

Marzulli, F.N. (1962). Barriers to skin penetration. J. Invest. Dermatol. 39:387–393.

McKenzye, I.C., Zimmerman, K., Peterson, L. (1981). The pattern of cellular organization of human epidermis. J. Invest. Dermatol. 76:459–461.

Montagna, W., Carlisle, K. (1979). Structural changes in aging human skin. J. Invest. Dermatol. 73:47–53.

Pathiak, M.A., Fitzpatrick, T.B. (1974). The role of natural photoprotective agents in human skin. In Sunlight and Man. Edited by T. B. Fitzpatrick. University Tokyo Press, Tokyo, pp. 725–750.

Plewig, G. (1970). Regional differences of cell sizes in the human stratum corneum. Part II: Effect of sex and age. J. Invest. Dermatol. 54:19–23.

Plewig, G., Marples, R.R. (1970). Regional differences of cell sizes in human stratum corneum. Part I. J. Invest. Dermatol. 54:13–18.

Plewig, G., Scheuber, E., Reuter, B., Waidelich, W. (1983). Thickness of corneocytes. In Stratum Corneum. Edited by R. Marks and G. Plewig. Springer-Verlag, Berlin-Heidelberg, New York, pp. 171–174.

Roskos, K., Guy, R., Maibach, H.I. (1986). Percutaneous penetration in the aged. Dermatol. Clin. 4:455–465.

Rougier, A., Dupuis, D., Lotte, C., Roguet, R. (1985). The measurement of the stratum corneum reservoir. A predictive method for in vivo percutaneous absorption studies: Influence of application time. J. Invest. Dermatol. 84:66–68.

Rougier, A., Dupuis, D., Lotte, C., Roguet, R., Schaefer, H. (1983). In vivo correlation between stratum corneum reservoir function and percutaneous absorption. J. Invest. Dermatol. 81:275–278.

Rougier, A., Dupuis, D., Lotte, C., Roguet, R., Wester, R., Maibach, H.I. (1986). Regional variation in percutaneous absorption in man: Measurement by the stripping method. Arch. Dermatol. Res. 278:465–469.

Rougier, A., Lotte, C., Maibach, H.I. (1987). In vivo percutaneous penetration of some organic compounds related to anatomic site in man: Predictive assessment by the stripping method. J. Pharm. Sci.

Scheuplein, R.J. (1978).The skin as a barrier. In The Physiology and Pathophysiology of the Skin, Vol. 5. Edited by A. Jarrett. Academic Press, New York, pp 1669–1692.

Scheuplein, R.J. (1979). Site variations in diffusion and permeability. In The Physiology and Pathology of the Skin, Vol. 5. Edited by A. Jarrett. Academic Press, New York, pp. 1731–1752.

Scheuplein, R.J., Blank, I.H. (1971). Permeability of the skin. Physiol. Rev. 51:702–747.

Smith, J.G., Fischer, R.W., Blank, I.H. (1961). The epidermal barrier: a comparison between scrotal and abdominal skin. J. Invest. Dermatol. 36:337–343.

Surinchak, J.S., Malinowski, J.A., Wilson, D.R., Maibach, H.I. (1985). Skin wound healing determined by water loss. J. Surg. Res. 38:258–262.

Taskovitch, L., Shaw, J.E. (1978). Regional differences in the morphology of human skin: Correlation with variations in drug permeability. J. Invest. Dermatol. 70: 217.

Wester, R.C., Maibach, H.I., Bucks, D.A., Aufrere, M.B. (1984). In vivo percutaneous absorption of paraquat from hand, leg and forearm of humans. J. Toxicol. Environ. Health 84:759–761.

Wilson, D.R., Maibach, H.I. (1982). A review of transepidermal water loss: Physical aspects and measurements as related to infants and adults. In Neonatal Skin. Edited by H. I. Maibach and E. K. Boisits. Marcel Dekker, New York, p. 83.

Wurter, D.E., Kramer, S.F. (1961). Investigation of some factors influencing percutaneous absorption. J. Pharm. Sci. 50:288–293.

7
Skin Contamination and Absorption of Chemicals from Water and Soil

Ronald C. Wester and Howard I. Maibach
University of California School of Medicine, San Francisco, California

I. INTRODUCTION

Contamination of soil and water (ground and surface water) and the transfer of hazardous chemicals is a major concern. When the large surface area of skin is exposed to contaminated soil and water (work, play, swim, daily bath), skin absorption may be significant. Brown et al. (1984) suggested that skin absorption of contaminants in water has been underestimated and that ingestion may not constitute the sole, or even the primary route of exposure. Soil has become an environmental depository for potentially hazardous chemicals. Exposure through work in pesticide-sprayed areas on chemical dump sites seems obvious. However, there may be hidden dangers in weekend gardening or in the child's play area where the soil has become laden with lead or other hazardous chemicals.

II. WATER

Tables 1 and 2 give concentrations of chemical contaminants in drinking water. These lists are extensive, but not inclusive. And, because the problem of drinking water contamination is just becoming appreciated, we must assume that the lists and concentrations of chemicals are also just beginning.

Methodology to calculate an acceptable level of a chemical in drinking water has been developed (National Academy of Sciences, 1977), but the underlying assumption of this methodology is that ingestion constitutes the

Table 1 Drinking Water Contamination Levels

Compound	Ranges detected in ground water (μg/L)	Ranges detected in surface water (μg/L)
Trichloroethylene	Trace–35,000	Trace–3.2
Tetrachloroethylene	Trace–3000	Trace–21
Carbon tetrachloride	Trace–379	Trace–30
1,1,1-Trichloroethane	Trace–401,300	Trace–3.3
1,2-Dichloroethane	Trace–400	Trace–4.8
Vinyl chloride	Trace–380	Trace–9.8
Methylene chloride	Trace–3600	Trace–13
1,1-Dichloroethylene and cis-1,2-dichloroethylene	Trace–860	Trace–2.2
trans-1,2-Dichlorethylene and ethylbenzene	Trace–2000	
Xylene	Trace–300	
Toluene	Trace–6400	

chief route of exposure to the contaminant. Such an assumption disregards other routes of exposure, such as skin absorption during daily bathing or swimming. Indeed, this assumption is so overlooked that it is not uncommon to see bottled water being brought to a home where the well water has been shown to be contaminated. Certainly, this bottled water will not be connected to the shower or swimming pool.

Brown et al. (1984) showed that the skin absorption rates for solvents are high. They concluded that skin absorption of contaminants in drinking water has been underestimated, and that ingestion may not constitute the sole, or even the primary, route of exposure. Table 3 gives their estimates of the relative contribution for skin absorption versus ingestion. The dermal route usually equals or exceeds the oral.

Wester and Maibach (1985) estimated the body burden of environmental contaminants from short-term exposure to skin absorption while bathing or swimming. They concluded that, for chemicals for which there was published data, the body burden was low for short-term exposure at water concentrations detected in surface water. However, for higher contaminant concentrations, such as those detected in groundwater, skin exposure has a potential for being hazardous to human health.

Wester et al. (1987) further investigated the interactions of chemical contaminants in water and their skin absorption and potential systemic availability. Table 4 shows that when chemicals in water solution come in contact

Table 2   Concentration of Drinking Water Contaminants

Compound	NAS[a] 10^{-6} (μg/L)	CAG[b] 10^{-6} (μg/L)
Acrylonitrile	0.77	0.034
Arsenic		0.004
Benzene		3.0
Benzo[a]pyrene		
Beryllium		0.02
Bis(2-chloroethyl)ether	0.83	
Carbon tetrachloride	9.09	0.086
Chlordane	0.056	0.012
Chloroform	0.59	0.48
DDT	0.083	
1,2-Dichlorethane	1.4	1.46
1,1-Dichlorethylene		.28
Dieldrin	0.004	
Ethylenedibromide	0.11	0.0022
ETU	0.46	
Heptachlor	0.024	2.4
Hexachlorobutadiene		1.4
Hexachlorobenzene	0.034	
N-Nitrosodimethylamine		0.0052
Kepone	0.023	
Lindane	0.108	
PCB	0.32	
PCNB	7.14	
TCDD		5.0×10^{-6}
Tetrachloroethylene	0.71	0.82
Trichloroethylene	9.09	5.8
Vinyl chloride	2.13	106.0

Note. From Office of Technological Assessment (1981).
[a]NAS, National Academy of Science.
[b]CAG, Cancer Assessment Group.

with skin (powered human stratum corneum), the chemicals partition/bind to human skin, depending upon their physiochemical interactions with skin. This partitioning was also shown to be linear for a 10-fold concentration range of nitroaniline.

Table 5 shows that each of these chemicals is able to distribute into the inner layers of skin after binding and that a significant portion of the chemical is absorbed. Other experiments showed that the absorption of ni-

Table 3 Estimated Dose and Contribution per Exposure for Skin Absorption Versus Ingestion

Compound	Concentration (mg/L)	Case 1[a]		Dose (mg/kg) Case 2[b]		Case 3[c]	
		Dermal	Oral	Dermal	Oral	Dermal	Oral
Toluene	0.005	67	33	44	56	91	9
	0.10	63	37	46	54	89	11
	0.5	59	41	45	55	89	11
Ethylbenzene	0.005	75	25	44	56	91	9
	0.10	63	37	46	54	89	11
	0.5	68	31	45	55	89	11
Styrene	0.005	67	33	29	71	83	17
	0.10	50	50	35	65	84	16
	0.5	59	41	29	71	83	17

Note. From Office of Technological Assessment (1981).
[a]For 7-kg adult bathing 15 min, 80% immersed (skin absorption); 2 L water consumed per day (ingestion).
[b]For 10.5-kg infant bathed 15 min, 75% immersed (skin absorption); 1 L water consumed per day (ingestion).
[c]For 21.9-kg child swimming 1 h, 90% immersed (skin absorption); 1 L water consumed per day (ingestion).

troaniline was linear over a 10-fold concentration range. Also, multiple doses of benzene were at least additive and, perhaps, exceeded single-dose predictions. This is due to benzene evaporation into the air during the absorption period. There is a partition of benzene between the air and skin. This is illustrated in Table 6, which gives the human skin absorption of 0.1% benzene in water and in toluene. From water vehicle a total of 5.7% of the dose benzene was absorbed, while only 0.16% of the dose was absorbed from toluene vehicle. This is because the benzene and toluene rapidly evaporate

Table 4 Partition of Contaminants from Water Solution to Powdered Human Stratum Corneum Following 30-min Exposure

Chemical contaminant	Dose partitioned to skin (%)
Benzene (21.7 μg/ml)	16.6 ± 1.4
54% PCB (1.6 μg/ml)	95.7 ± 0.6
Nitroanaline (4.9 μg/ml)	2.5 ± 1.1

Table 5 In Vitro Percutaneous Absorption and Skin Distribution of Contaminants in Water Solution for 30-min Exposure

Parameter	Chemical contaminant (% dose)[a]		
	Benzene	54% PCB	nitroaniline
Percutaneous absorption (systemic)	0.045 ± 0.037	0.03 ± 0.00	5.2 ± 1.6
Surface-bound/stratum corneum	0.036 ± 0.005	6.8 ± 1.0	0.2 ± 0.17
Epidermis and dermis	0.065 ± 0.057	5.5 ± 0.7	0.61 ± 0.25
Total (skin/systemic)	0.15	12.3	6.1
Skin wash/residual	2.51 ± 0.94	71.2 ± 0.2	0.47 ± 0.03
Apparatus wash	0.006 ± 0.005	0.25 ± 0.2	0.47 ± 0.03
Total (accountability)	2.67	83.3	99.0

[a]Percentage of applied dose ($n = 4$ for each parameter): benzene, 21.7 μg/ml; 54% PCB, 1.6 μg/ml; nitroaniline, 4.9 μg/ml.

off the skin, leaving little benzene absorbed. Contrary, water is less volatile than toluene so there is more time for benzene absorption.

The in vivo percutaneous absorption of nitroaniline in the rhesus monkey, an animal model relevant for skin absorption in humans (Wester and Maibach, 1983), was 4.1 ± 2.3% of applied dose. This compared well with the 2.5 ± 1.6% binding to powdered human stratum corneum and the 5.2 ± 1.6% in vitro absorption (Table 5).

Table 7 gives a hypothetical percutaneous absorption of a chemical contamination from water while bathing or swimming for a single 30-min

Table 6 In Vitro Percutaneous Absorption of Benzene in Human Skin

Parameter	Percent dose absorbed[a]	
	Water vehicle	Toluene vehicle
Receptor fluid (RF)	5.0 ± 1.9	0.11 ± 0.08
Skin		
Epidermis	0.4 ± 0.3	0.04 ± 0.02
Dermis	0.3 ± 0.2	0.01 ± 0.01
RF + skin (absorbed)	5.7	0.16
Skin surface wash	1.8 ± 0.9	0.04 ± 0.02
Total	7.5 ± 1.0	0.19 ± 0.07

[a]0.1% Benzene in water or toluene dosed at 5 μl/cm^2 skin area.

Table 7   Hypothetical Percutaneous Absorption of Chemical Contaminant from
Water While Bathing or Swimming for 30 min

Parameters
    Concentration of chemical in water   1.0 ug/ml
    Volume of water on skin surface in study   1.5 ml
    Skin Surface area in study 5.7 cm^2
    Total skin surface area of human adult $17,000 \text{ cm}^2$
    Percentage dose absorbed, 30-min exposure   5%
Calculation of body burden per single bath or swim[a]

$$\frac{1.0 \ \mu g}{ml} \times \frac{1.5 \ ml}{5.7 \ cm^2} \times \frac{17,000 \ cm^2}{1} \times 0.05 = 223.7 \ \mu g$$

[a]The amount will obviously vary owing to concentration of chemical contaminant in water, percutaneous absorption of contaminant, and frequency of bathing, and swimming.

period. Thus, it is possible for milligram amounts to be absorbed for single exposure, and would be at least cumulative in daily life.

A more detailed overview of the complexities of percutaneous flux and more specific implications of penetration from bathing and swimming are outlined in Wester and Maibach (1984). The estimates provided are tentative; few chemical moieties have been examined. Specialized skin sites (ear canal, genitalia, face, etc.) offer substantial opportunity for penetration enhancement. Workers, such as housewives, cooks, mechanics, and the like, may wash their hands and arms many times daily. Damaged skin and occluded skin (diapers), as well as skin in the infant and aged, offer other possibilities for enhanced flux.

A.   Water: Finite and Infinite Doses

Table 8 compares the in vitro and in vivo absorption from water for boric acid, borax, and disodium octaborate tetrahydrate (DOT). The K_p values from the infinite dose are a 1000-fold higher than from the in vivo study, while the finite-dose K_p of boric acid was only 10-fold higher. Clearly, use of a similar dose volume was critical to the comparison. The in vitro system used a phosphate-buffered saline perfusion as well as a water solution dose. The in vivo dose was applied and allowed to dry, and absorption took place in whatever water content the skin of each volunteer contained. The preponderance of water in the infinite dose may have influenced the skin membrane during 24-h continual exposure, and this may have increased the permeability of the borate chemicals. This would make the in vivo data and the in vitro finite-dose data more relevant for the usual exposure conditions.

Table 8 In Vivo and In Vitro Percutaneous Absorption of Boron Administered as Boric Acid, Borax, and Disodium Octaborate Tetrahydrate (DOT) in Human Skin

Dose	In vitro		In vivo	
	Flux (μg/cm^2/h)	K_p (cm/h)	Flux (μg/cm^2/h)	K_p (cm/h)
Boric acid				
5% at 2 μl/cm^2	0.07	1.4×10^{-6}	0.009	1.9×10^{-7}
5% at 1 ml/cm^2	14.6	2.9×10^{-4}		
Borax				
5% at 2 μl/cm^2			0.009	1.8×10^{-7}
5% at 1 ml/cm^2	8.5	1.7×10^{-4}		
DOT				
5% at 2 μl/cm^2			0.010	1.0×10^{-7}
5% at 1 ml/cm^2	7.9	0.8×10^{-4}		

Note. 2 μl/cm^2 is a finite water dose on in vivo skin. More will just run off the body. Thus, 1 ml/cm^2 is an infinite dose only capable of being used with an in vitro system.

The only exception might be where a subject bathed in a boric acid, borax, or DOT solution for an extended period of time. The in vitro infinite dosing conditions may better predict the bathing situation. Note that there was 1000-fold difference between in vivo absorption and in vitro absorption where the in vitro dose was infinite. General risk assessment from in vitro infinite dosing may be greatly overestimated.

III. SOIL

The study design should be such that percutaneous absorption from soil is relative to actual exposure conditions. The gardener, or the child playing in dirt, will have some stationary contact with the soil (sitting on it; dust that settles on the skin). There also will be dirt manipulated with the hands where skin contacts an ever-changing layer of dirt. The exposed worker will certainly have airborne dust settle on skin; manipulated dirt will depend on the job. In all cases, the dirt covering skin is exposed to air (dirt in clothing is noted as an exception). Laboratory studies, to be relevant, should have the soil on skin open to air exchange. This is simple with in vitro studies where the soil is contained. With in vivo studies the soil needs to be contained and the best method is to use a nonocclusive cover that allows free passage of air and moisture. A Gore-Tex film accomplishes this (Wester, et al., 1990a). Use of an occlusive cover such as glass (Turkall et al., 1994) will cause changes in the microenvironment under the cover, which will enhance skin absorption. The in vitro diffusion cell is a truly stationary system where chemical passage from soil to skin is physically undisturbed and will function according to the chemical kinetics of the components. An in vivo study may have some animal or human movement, which could shift/rotate soil confined under the nonocclusive patch if space under the patch permits this (the patch can be sufficiently concave to allow soil movement in the open space).

It is important to know the relevant characteristics of the test soil. This should include percent sand, clay, silt, and organic content. Soil can be passed through mesh sieves to be uniform in size. Mixing of the test chemical added in solvent to the soil should be done open to air to allow dissipation of the solvent. The "dust" fraction of soil can be avoided by the sieving method. This is for safety purposes so that airborne particles containing radioactivity do not contaminate laboratory personnel. An uncontrolled airborne fraction may also affect study results.

Table 9 gives the in vitro (human skin) and in vivo (rhesus monkey) percutaneous absorption of organic chemicals from soil and a comparative vehicle (water or solvent, depending upon vehicle). The soil is the same source for all chemicals. For each chemical the concentration of mass (μg) per unit skin area (cm^2) is the same for each vehicle. Receptor fluid (human plasma) accumulation of DDT was negligible in the in vitro study due to solubility restriction. Chemicals with higher log P values are lipophilic and there are not soluble in biological fluid receptor fluid (plasma, buffered saline) (Wester et al., 1990a, 1190b, 1992a). Human skin content was 18.1% of dose from acetone vehicle. In vivo absorption in the rhesus monkey was

Table 9 In Vitro and In Vivo Percutaneous Absorption of Organic Chemicals

		Percent dose		
		In vitro		
Compound	Vehicle	Skin	Receptor	In Vivo
DDT	Acetone	18.2 ± 13.4	0.08 ± 0.02	18.9 ± 9.4
	Soil	1.0 ± 0.7	0.04 ± 0.01	3.3 ± 0.5
Benzo[a]pyrene	Acetone	23.7 ± 9.7	0.09 ± 0.06	51.0 ± 22.0
	Soil	1.4 ± 0.9	0.01 ± 0.06	13.2 ± 3.4
Chlordane	Acetone	10.8 ± 8.2	0.07 ± 0.06	6.0 ± 2.8
	Soil	0.3 ± 0.3	0.04 ± 0.05	4.2 ± 1.8
Pentachlorophenol	Acetone	3.7 ± 1.7	0.6 ± 0.09	29.2 ± 5.8
	Soil	0.11 ± 0.04	0.01 ± 0.00	24.4 ± 6.4

18.9% of dose from acetone vehicle. These values are comparable to the published 10% of dose absorbed in vivo in humans from acetone vehicle. Percutaneous absorption from soil was predicted to be 1.0% of dose in human skin in vitro and a comparative 3.3% of dose in vivo in rhesus monkey.

In vivo percutaneous absorption of benzo[a]pyrene is high, 51.0% reported here for rhesus monkey and 48.3% (Bronaugh and Stewart, 1986) and 35.3% (Yang et al., 1989) for the rat. Benzo[a]pyrene absorption from soil was approximately one-fourth that of solvent vehicle (Wester et al., 1990a; Shu et al., 1988).

For chlordane, pentachlorophenol, and 2,4-D, the in vivo percutaneous absorption in rhesus monkey from soil was equal to or slightly less than that obtained from solvent vehicle (Table 9). Validation to humans in vivo is available for 2,4-D, where the percutaneous absorption is the same for rhesus monkey and humans (Wester et al., 1990a). In vitro percutaneous absorption is variable, probably due to solubility problems relative to high lipophilicity.

In vivo studies have an advantage over in vitro studies in that in vivo pharmacokinetic data can be obtained and these data applied to better understand the potential toxicokinetics of a chemical (Table 10). The percutaneous absorption of PCP from acetone vehicle was $29.2 \pm 5.8\%$ of total dose applied for a 24-h exposure period. Compared to other compounds, the absorption of PCP would be considered high. In vivo absorption from soil was 3.3% for DDT, 13.2% for benzo[a]pyrene, and 4.2% for chlordane. Additionally, the ^{14}C excretion for PCP in urine was slow, a half-life of 4.5 days for both intravenous and topical application. If biological exposure is considered in terms of dose $\times$ time, then PCP biological exposure can be

Table 10 In Vivo Percutaneous Absorption of Pentachlorophenol in Rhesus Monkey

	Percent dose[a]		
	Topical		
	Soil	Acetone	Intravenous
Percent dose absorbed	24.4 ± 6.4	29.2 ± 5.8	—
Surface recovery[b]	38.0 ± 13.4	59.6 ± 4.1	—
Half-life (days)	4.5	4.5	4.5

[a]Means $\pm$ SD for four animals.
[b]Includes chamber, residue, and surface washes.

considered high. The percutaneous absorption of PCP from soil vehicle was also high (24.4 ± 6.4%) and not statistically different. The study of Reigner et al. (1991) and this study show PCP to have high bioavailability, both topical and oral, and PCP also exhibits an extensive half-life. This suggests that PCP has the potential for extensive biological interactions.

Table 11 gives the in vitro and in vivo percutaneous absorption of PCBs (Wester et al., 1993a). As with the other organic chemicals with high log P, receptor fluid accumulation in vitro was essentially nil. Skin accumulation in vitro did exhibit some PCB accumulation. In vivo, PCB percutaneous absorption for both Aroclor 1242 and 1254 was a) high, ranging from 14 to 21%, and b) generally independent of formulation vehicle. PCBs thus have a strong affinity for skin and are relatively easily absorbed into and through skin.

Selected salts of arsenic, cadmium, and mercury are soluble in water, and thus are amenable to in vitro percutaneous absorption with human skin (Table 12) (Wester et al., 1992b, 1993b). Arsenic absorption in vitro was 2.0% (1.0% plus 0.9% receptor fluid), and the same in vivo in rhesus monkey. Absorption from soil was equal to (in vivo) or approximately one-third (in vitro) that from water. Cadmium and mercury both accumulate in human skin, and are slowly absorbed into the body. (Note that in vivo studies with cadmium and mercury are difficult to perform; cadmium accumulates in the body, and mercury is not excreted via urine.) Note the high skin content with cadmium and mercury.

A. Soil Load

Percutaneous absorption of 2,4-D for 24-h exposure was 8.6 ± 2.1% dose for a dose load of 4.2 µg/cm^2 in acetone vehicle. The same 2,4-D dose was then loaded into 1 mg soil or 40 mg soil/cm^2 skin surface area. Percutaneous absorption for 24 h from the 1-mg soil load was 9.8 ± 4.0% of dose (Table 13) (Wester et al., 1996).

Percutaneous absorption from the 40-mg soil load for 24 h was 15.9 ± 4.7% of dose ($p = .178$, nonsignificant for paired t-test). Thus, the in vivo percutaneous absorption of 2,4-D was not affected by soil load (1 mg vs. 40 mg soil; chemical dose constant). Additionally, the percutaneous absorption from the two soil loads was the same as from acetone vehicle.

2,4-D at a constant chemical dose 2 µg/cm^2 was applied to human skin in vitro in soil loads of 5, 10 and 40 mg. Dose accumulation in buffered saline receptor fluid was low (0.02 ± 0.02%), presumably due to relative low solubility of 2,4-D in water. The 2,4-D human skin content was analyzed after the 24-h exposure period and the results show no difference relative to the three soil loads. The in vitro skin content would predict a 2,4-D dose

Table 11 In Vitro and In Vivo Percutaneous Absorption of PCBs

		Percent dose		
		In vitro		
Compound	Vehicle	Skin	Receptor fluid	In vivo
PCBs (1242)	Acetone	—	—	21.4 ± 8.5
	TCB	—	—	18.0 ± 8.3
	Mineral Oil	6.4 ± 6.3	0.3 ± 0.6	10.8 ± 8.3
	Soil	1.6 ± 1.1	0.04 ± 0.05	14.1 ± 1.0
PCBs (1254)	Acetone	—	—	14.6 ± 3.6
	TCB	—	—	20.8 ± 8.3
	Mineral Oil	10.0 ± 16.5	0.1 ± 0.07	20.4 ± 8.5
	Soil	2.8 ± 2.8	0.04 ± 0.05	13.8 ± 2.7

absorption from soil of approximately 2%, which is approximately one-fifth of that in monkey in vivo.

B. Skin Contact Time

Table 14 provides the effect of skin deposition time on 2,4-D percutaneous absorption. In acetone vehicles the percutaneous absorption of 2,4-D over 8 h was 3.2 ± 1.0% of dose. This is approximately one-third of the absorption seen for 24 h (8.6 ± 2.1%). Thus, with acetone vehicle (where the dose is in immediate contact with the skin) the percutaneous absorption was linear over time. With soil as the vehicle, the absorption was only 0.05 ± 0.04% for 1 mg soil/cm^2 and 0.03 ± 0.02% for 40 mg/cm^2 at 8 h. The 16-h ab-

Table 12 In Vitro and In Vivo Percutaneous Absorption of Metals

		Percent dose		
		In vitro		
Compound	Vehicle	Skin	Receptor fluid	In vivo
Arsenic	Water	1.0 ± 1.0	0.9 ± 1.1	2.0 ± 1.2
	Soil	0.3 ± 0.2	0.4 ± 0.5	3.2 ± 1.9
Cadmium	Water	6.7 ± 4.8	0.4 ± 0.2	—
	Soil	0.09 ± 0.03	0.03 ± 0.02	—
Mercury	Water	28.5 ± 6.3	0.07 ± 0.01	—
	Soil	7.9 ± 2.2	0.06 ± 0.01	—

Table 13   Effect of Soil Load on 2,4-D Percutaneous Absorption

Compound	Soil load[b] (mg/cm)	Percent dose absorbed[c]
In vivo, rhesus monkey[a]	1	9.8 ± 4.0
	40	15.9 ± 4.7
In vitro, human skin[a]	5	2.8 ± 1.7
	10	1.7 ± 1.3
	40	1.4 ± 1.2

[a]Mean ± SD ($n = 4$).
[b]Concentration of 2,4-D chemical/cm^2 skin area was kept constant, while soil load/cm^2 skin area was varied.
[c]In vivo percutaneous absorption measured by urinary ^{14}C accumulation; in vitro absorption determined by ^{14}C skin content.

sorption for 1 mg/cm^2 soil was 2.2 ± 1.2%. This suggests that percutaneous absorption of 2,4-D from soil was not linear over time. There may be a "lag" time where chemical must partition from soil into skin.

C.   Soil and Risk Assessment Assumptions

In the study with 2.4-D and the other compounds mentioned previously, an experimental soil load of 40 mg/cm^2 skin surface area was used. Studies on dermal soil adherence show that less than 1 mg to perhaps 5 mg of soil will adhere to skin (U.S. EPA, 1992). It then becomes the practice in estimating potential health hazard to assume linearity and divide the results of 40 mg by a soil adhesion factor, thus reducing estimated body burden by 1/40 or 5/40, etc. (U.S. EPA, 1992). The data generated here for 2,4-D, where soil

Table 14   Effect of Skin Contact Time on In Vivo Percutaneous Absorption of 2,4-D

Skin contact time	Percent dose absorbed[a]		
	Acetone vehicle	Soil vehicle	
		1 mg/cm^2	40 mg/cm^2
8 h	3.2 ± 1.0	0.05 ± 0.04	0.3 ± 0.02
16 h	—	2.2 ± 1.2	—
24 h	—	8.6 ± 2.1	15.9 ± 4.7

[a]Mean ± SD ($n = 4$).

load (range 1–40 mg) had no effect on percutaneous absorption, suggest that this mathematical practice can severely underestimate absorption.

A second assumption, which transferred to the laboratory, is that chemical and soil are "static." The gardener planting summer flowers will certainly have multiple contacts with soil. Chemicals are also mobile within soil. Over the short term, as seen in the data, chemicals move from soil to skin. Calderbank (1989) reports that with time adsorbed residues in soil will become more stable. Therefore, some dynamics in the system need to be understood.

A third assumption of linearity is also placed on time. That 2,4-D absorption in acetone vehicle was 3.2 ± 1.0% for 8 h and 8.6 ± 2.1% for 24 h suggests that absorption is linear over time and that the compound can be removed with washing (Fig. 1). With acetone vehicle the total dose is deposited on the skin surface, and percutaneous absorption is dependent on skin barrier diffusion. With soil vehicle a lag time is needed for the 2,4-D to "mobilize" within soil and become available for absorption. Sufficient time is needed for 2,4-D to become available; then skin barrier function

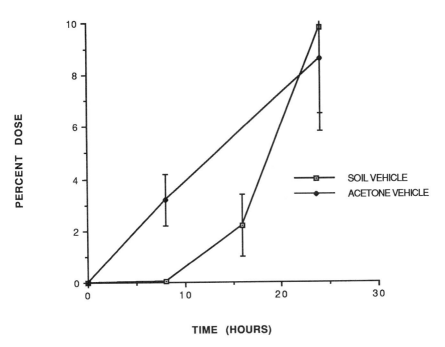

TIME (HOURS)

Figure 1 In vivo percutaneous absorption of 2,4-D from soil and acetone vehicles is linear over time, except when there is an initial lag time from soil vehicle.

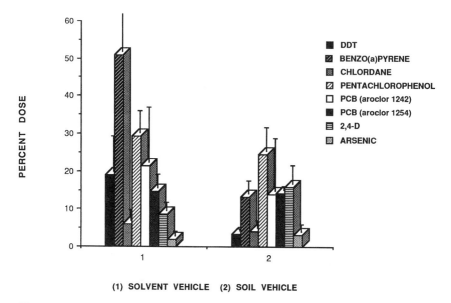

Figure 2 In vivo percutaneous absorption of several hazardous substances from soil and solvent (either acetone or water). Overall, soil reduced absorption to about 60% compared to solvent. There is caution, however, because the absorption of some compounds is the same for soil and solvent.

becomes the rate-limiting step in absorption. Should it turn out that a substantial lag time does exist for transfer from soil to skin, this would favor risk assessment for the worker. However, the data to date show that substantial amounts of hazardous chemicals can be absorbed from soil (Fig. 2).

IV. DISCUSSION

The evolution of skin resulted in a tissue that protects precious body fluids and constituents from excessive uptake of water and contaminants in the external environment. The outermost surface of the skin that emerged for humans is the stratum corneum, which restricts but does not prevent penetration of water and other molecules. This is a complex lipid–protein structure that is exposed to contaminants during bathing, swimming, and exposure to the environment. Industrial growth has resulted in the production of organic chemical and toxic metals whose disposal results in contamination of water supplies. As one settles into a tub or pool, the skin (with a surface area of approximately 18,000 cm^2) acts as a lipid sink (stratum corneum) for the lipid-soluble contaminants. Skin also serves as transfer membrane for water and whatever contaminants may be dissolved in it. Note that a) water transfers through skin and can carry chemicals, and b) the outer layer of skin is lipid in nature. Thus, highly lipophilic chemicals such as DDT, PCBs, and chlordane residing in soil will quickly transfer to skin. Percutaneous absorption can be linear, orderly, and predictive (a measured flux from water). However, evidence exists that chemicals may transfer to skin with short-term exposure.

REFERENCES

Bronaugh, R.L., Stewart, R.F. (1986). Methods for in vitro percutaneous absorption studies. VI. Preparation of the barrier layer. J. Pharm. Sci. 75:487–491.

Brown, H.S., Bishop D.R., Rowan, C.A. (1984). The role of skin absorption as a route of exposure for volatile organic compounds (VOCs) in drinking water. Am. J. Public Health 74:479–484.

Calderbank A. (1989). The occurrence and significance of bound pesticide residues in soil. Rev Environ. Contam. Toxicol. 108:71–103.

National Academy of Sciences. (1977). Drinking Water and Health. NAS/NRC, Washington, DC.

Office of Technological Assessment. (1981). Factors associated with cancer. In: Assessment of Technologies for Determining Cancer Risks from the Environment. Office of Technological Assessment, Washington, DC, Chap. 3.

Reigner, B.G., Gungon, R.A., Hoag, M.K., Tozer, T.N. (1991). Pentachlorophenol toxicokinetics after intravenous and oral administration to rat. Zenobiotica 21: 1547–1558.

Shu, H., Teitelbaum, P., Webb, A.S., Marple, L., Brunck, B., Del Rossi, D., Murray, F.J., Paustenbach, D. (1988). Bioavailability of soil-bound TCDD. Dermal bioavailability in the rat. Fundam. Appl. Toxicol. 10:335–343.

Turkall, R.M., Skowronski, G.A., Kadry, A.M., Abdel-Rahman, M.S. (1994). A comparative study of the kinetics and bioavailability of pure and soil-absorbed naphthalene in dermally exposed male rates. Arch. Environ. Contam. Toxicol. 26:504–509.

U.S. Environmental Protection Agency. (1992). Dermal Exposure Assessment: Principles and Applications. EPA/600/8-91/011B, Office of Research and Development, Washington, DC, Section 5:48–49.

Wester, R.C., Maibach, H.I. (1983). Cutaneous pharmacokinetics: 10 steps to percutaneous absorption. Drug. Metab. Rev. 14:1699–205.

Wester, R.C., Maibach, H.I. (1984). Assessment of dermal absorption of contaminants in drinking water. U.S. EPA Contract No. 68002-3168, 1984.

Wester, R.C., Maibach, H.I. (1985). Body burden of environmental contaminants from acute exposure to percutaneous absorption while bathing or swimming. Toxicologist 5:177.

Wester, R.C., Mobayen, M., Maibach, H.I. (1987). In vivo and in vitro absorption and binding to powdered stratum corneum as methods to evaluate skin absorption of environmental chemical contaminants from ground and surface water. J. Toxicol. Environ. Health 21:367–374.

Wester, R.C., Maibach, H.I., Bucks, D.A.W., Sedik, L., Melendres, J., Liao, C., Di Zio, S. (1990a). Percutaneous absorption of [^{14}C]DDT and [^{14}C]benzo[a]pyrene from soil. Fundam. Appl. Toxicol. 15:510–516.

Wester, R.C., Maibach, H.I., Sedik, L., Melendres, J., Wade, M., Di Zio, S. (1990b). In vitro percutaneous absorption of pentachlorophenol from soil. Fundam. Appl. Toxicol. 19:68–71.

Wester, R.C., Maibach, H.I., Sedik, L., Melendres, J., Liao, C., Di Zio, S. (1992a). Percutaneous absorption of [^{14}C]chlordane from soil. J. Toxicol. Environ. Health 35:269–277.

Wester, R.C., Maibach, H.I., Sedik, L., Melendres, J., Di Zio, S., Wade, M. (1992b). In vitro percutaneous absorption of cadmium from water and soil into human skin. Fundam. Appl. Toxicol. 19:1–5.

Wester, R.C., Maibach, H.I., Sedik, L., Melendres, J., Wade, M. (1993a). Percutaneous absorption of PCBs from soil: In vivo rhesus monkey, in vitro human skin, and binding to powdered human stratum corneum. J. Toxicol. Environ. Health 39:375–382.

Wester, R.C., Maibach, H.I., Sedik, L., Melendres, J., Wade, M. (1993b). In vivo and in vitro percutaneous absorption and skin decontamination of arsenic from water and soil. Fundam. Appl. Toxicol. 20:336–340.

Wester, R.C., Melendres, J., Logan, F., Hui, X., Maibach, H.I. (1996). Percutaneous absorption of 2,4-dichlorophenoxyacetic acid from soil with respect to soil load and skin contact time: In vivo absorption in rhesus monkey and in vitro absorption in human skin. J. Toxicol. Environ. Health 47:335–344.

Yang, J.J., Roy, T.A., Kruger, A.J., Neil, W., Mackerer, C.R. (1989). In vitro and in vivo percutaneous absorption of benzo[a]pyrene from petroleum crude-fortified soil in the rat. Bull. Environ. Contam. Toxicol. 43:207–214.

8
Characterization of Molecular Transport Across Human Stratum Corneum In Vivo

Aarti Naik
*Centre Interuniversitaire de Recherche et d'Enseignement,
Archamps, France, and Laboratoire de Pharmacie Galénique,
University of Geneva, Geneva, Switzerland*

Yogeshvar N. Kalia
*Centre Interuniversitaire de Recherche et d'Enseignement,
Archamps, France, and Laboratoire de Pharmacie Galénique,
University of Geneva, Geneva, Switzerland*

Fabrice Pirot
*Institut des Sciences Pharmaceutiques et Biologiques, University of
Claude Bernard–Lyon, Lyon, France*

Richard H. Guy
*Centre Interuniversitaire de Recherche et d'Enseignement,
Archamps, France, and Laboratoire de Pharmacie Galénique,
University of Geneva, Geneva, Switzerland*

I. INTRODUCTION

A percutaneously delivered therapeutic agent, whether directed at the systemic circulation or the local tissues, must traverse the stratum corneum (SC), which effectively restricts molecular transport between the external environment and the interior of the human body. The composition of the SC, and the highly tortuous nature of the extracellular pathway (1), makes this relatively thin biomembrane perhaps the body's most efficient barrier. This has been a great boon to the transdermal formulation scientist, who can

149

effortlessly evaluate transcutaneous drug transport by relatively uncompli-
cated in vitro diffusion experiments using excised or "simulated" skin.
These experiments have been, and remain, instrumental to the preliminary
screening of transdermal candidates, formulation excipients, and in mechanistic assessments, but, as with all methodology, present a number of obvious limitations, which collectively generate a cogent argument for human
in vivo evaluations in many situations.

First, the SC unfortunately lacks the consistent performance desirable
of most synthetic rate-determining membranes in that its barrier properties
vary both spatially and temporally on any individual, from individual to
individual, and of course between different animal species. Second, although
the diffusivity of a drug across the SC is effectively independent of the
supporting and surrounding tissues, the time taken for a molecule to be
transported across the skin will be determined by its physicochemical properties and interactions with the biological barrier. These not only include
bulk characteristics such as molecular weight and lipophilicity, but also the
drug's ionization profile across the membrane, its susceptibility to cutaneous
metabolism and binding, and its ability to induce pharmacological responses
capable of modifying the molecule's own clearance. These factors (some of
which can only be actuated in vivo) will all affect, to varying degrees, the
rate and extent of the appearance of a drug in the systemic circulation.
Moreover, what of the local bioavailability when the drug is intended for
delivery to the skin or subcutaneous tissues? In this instance, the systemic
bioavailability may be of little consequence (except for the purposes of toxicological evaluation), but measurement within the skin strata, particularly
at the site of action, is substantially more informative. This may not always
be feasible; for instance, how does one accurately, and noninvasively, measure drug concentrations within the viable epidermis, the target for the treatment of numerous dermatological conditions? If the SC is depicted as a rate-
limiting membrane, then quantification of drug within this layer must
provide an estimate of the rate and extent of delivery to the underlying
tissues. This is of particular relevance when the skin is exposed to a chemical
for a short duration, insufficient to attain steady-state diffusion kinetics or
to elicit appreciable (i.e., easily detectable) systemic levels. More importantly, the significance of this determination when performed noninvasively
in vivo in humans is clearly better than its in vitro counterpart.

As stated already, because the efficacy of the barrier severely restricts
percutaneous uptake, plasma concentrations of topically applied drugs are
frequently at, or below, the limits of analytical detection. Therefore, the most
commonly used in vivo techniques to establish the rate of transport have
involved the administration of radioactively labeled tracer molecules with
subsequent analysis of either the excreta or plasma to estimate bioavailability

(2,3). Surface recovery of the residual material following topical application offers an alternative approach for establishing percutaneous uptake and relies on the assumption that the difference between the quantity applied and that recovered corresponds to the amount absorbed. The analogous technique of surface disappearance has also been used to monitor percutaneous absorption. In situations where a compound can elicit a physiological response, this can be used as a marker to report on transport of the compound through the skin (4). The stripping methodology developed by Rougier and coworkers (5) uses short-time application (typically 30 min) of the compound followed by serial tape-stripping and subsequent assay to determine the amount in the SC. This quantity has been shown to be well correlated with that which, under identical experimental conditions but without stripping, becomes systemically bioavailable.

Theoretical algorithms have been devised to predict percutaneous uptake based on molecular physicochemical parameters (6–14). However, SC structural heterogeneity and the complex balance of physicochemical factors that drive drug passage from an aqueous medium into, and across, the lipophilic SC and eventually out into the aqueous viable epidermis ensure that *ab initio* prediction of molecular diffusion across the SC is a considerable challenge.

The development of noninvasive biophysical technologies, such as transepidermal water loss (TEWL) measurements, impedance spectroscopy (IS), and attenuated total reflectance infrared spectroscopy (ATR-FTIR), has provided powerful tools to quantify chemically or electrically induced disruption of SC barrier function in vivo. Since TEWL and IS techniques rely on changes in the transport properties of molecules through the SC to report on alterations in barrier function, the methods also deliver information on molecular mobility within the membrane. ATR-FTIR reports more directly on molecular diffusivity within the SC. The goal of this chapter is to summarize how these noninvasive techniques (TEWL, IS, and ATR-FTIR) can be used to obtain quantitative measures of molecular transport within human SC in vivo as a function of depth from the outermost layers to the SC–viable epidermis interface.

II. WATER DIFFUSIVITY IN VIVO: USING TRANSEPIDERMAL WATER LOSS

One of the principal functions of the SC is to restrict TEWL; as a result, this parameter has been used extensively to report on the efficacy of SC barrier function in diseased skin states or following externally-induced perturbation. However, in addition to assessing barrier integrity, the flux measured is directly related to the diffusivity of water molecules through the

SC. A recent study, using TEWL measurements in conjunction with serial tape-stripping, showed that the SC behaves as a homogeneous (Fickian) barrier to water transport (15). That is, the barrier function properties of the SC were distributed uniformly throughout the membrane, with all the strata contributing equally to the regulation of transcutaneous water loss.

In mathematical terms, using Fick's First Law of Diffusion, progressive removal of the SC by serial tape-stripping increases TEWL as the membrane thickness is reduced, that is,

$$J_x = \frac{(K \cdot \Delta C)\bar{D}_x}{H - x} = K_P^x \cdot \Delta C \tag{1}$$

where J_x is the TEWL value when x μm of SC has been removed by tape-stripping (calculated from the amount of tissue removed using preweighed tape strips of known surface area and assuming an average SC density of 1 g cm^{-3}) (16); K is the SC lipid–viable tissue partition coefficient of water ($K \approx 0.06$) (1); ΔC is the water concentration difference across the membrane (≈ 1 g cm^{-3}); H (μm) is the initial SC thickness; $\bar{D}_x$ is the average apparent diffusivity of water through the remaining $(H - x)$ μm of the SC; and K_P^x is the corresponding permeability coefficient, that is, $K_P^x = K \cdot \bar{D}_x/(H - x)$. Inversion of Eq. (1) gives

$$\frac{1}{J_x} = \frac{H}{\gamma \bar{D}_x} - \frac{x}{\gamma \bar{D}_x} \tag{2}$$

where $\gamma = K \Delta C$ and can be assumed to be constant. Experimental results obtained in vivo, plotted as $1/J_x$ versus x (Fig. 1), were linear, and the slopes and intercepts of the graphs yielded explicit values for $\bar{D}$ (the average apparent diffusivity of water in the SC) ($3.8 \pm 1.3 \times 10^{-9}$ cm^2 s^{-1}) and H (12.7 $\pm$ 3.3 μm), respectively. The corresponding calculated permeability coefficient, K_P^x, was $1.8 \pm 0.5 \times 10^{-7}$ cm s^{-1}. These values agree well with published in vitro measurements (Table 1). It should be noted that the diffusivity is an *apparent* value, since it assumes that the diffusion pathlength of water across the SC is equal to the membrane thickness.

III. IONIC MOBILITIES IN VIVO: USING IMPEDANCE SPECTROSCOPY

Tape-stripping experiments have been used to show that the SC also functions as the primary barrier to current flow through the skin (17,18). The intact integument has an impedance of several hundred kiloohms (kΩ), which drops progressively as successive SC layers are removed, finally attaining a value of ~1 kΩ, which corresponds to removal of the entire barrier.

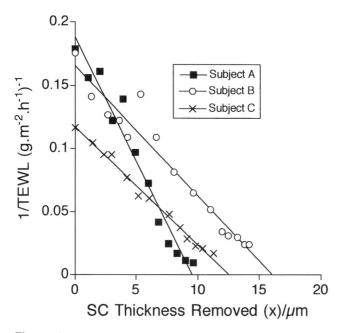

Figure 1 The dependence of 1/TEWL upon stratum corneum thickness. Highly linear fits were obtained for all three subjects. Subject A, $(\text{TEWL})^{-1} = 0.188 - 0.0197x$ $(r^2 = .962)$; Subject B, $(\text{TEWL})^{-1} = 0.165 - 0.0103x$ $(r^2 = .949)$; Subject C, $(\text{TEWL})^{-1} = 0.117 - 0.00932x$ $(r^2 = .991)$. Adapted from Reference 15.

Impedance spectroscopy has been used to measure skin hydration indirectly, and as a method to assess the effects of iontophoresis, chemical enhancers, and ultrasound on barrier function (19–21).

The relationship between skin admittance [admittance (Y) = 1/impedance (Z)] and water permeability is shown in Fig. 2. The correlation between ion flow and water transport is expected, as ion mobility is obviously more facile in an aqueous environment. For intact, fully functional SC, skin impedance is very high; as the barrier is progressively stripped away, the water concentration in the membrane increases, allowing ionic mobility to increase and skin impedance to fall.

Because a specific *low* frequency impedance (1.61 Hz) was used to report on changes in barrier integrity, the measurement is essentially resistive. The apparent ion diffusion pathlength through the SC, the principal source of skin impedance, equals $2(H - x)$, where H is the initial SC thickness and x is the quantity of tissue removed by the tape-stripping procedure. It follows that the SC resistivity as a function of position, ρ_x, can be expressed as

Table 1 Water Permeability Coefficients and Diffusivities Across Human Stratum Corneum

Region	Tissue	T (°C)	$10^7 K_p$ (cm s^{-1})	$10^9 D$ (cm^2 s^{-1})	Reference
Abdomen	Epidermis	25	2.78	0.28	55
Abdomen	Epidermis + dermis (420 μm)	31 ± 1	3.89 ± 0.11		56
Abdomen	Epidermis + dermis (430 μm)	30 ± 1	3.29		57
Abdomen	Epidermis + dermis (400 μm)	30	3.61 ± 1.53		58
Abdomen		30	3.33		59
Leg		32	3.61		60
Forearm (in vivo)		33	1.84 ± 0.47	3.83 ± 1.32	15

Note. Measurements are in vitro unless indicated otherwise.

$$\rho_x = \frac{A^2}{2(H-x)} \cdot \frac{Z_x}{A} = \frac{1}{\kappa_x} \tag{3}$$

where Z_x/A is the impedance per unit area as a function of increasing normalized skin depth (x/H, where x is the absolute depth and H is the overall SC thickness) and κ_x, the conductivity, is the reciprocal of ρ_x. The molar conductivity (or equivalent) conductivity, (Λ_m) is linearly related to the individual ion conductivities (λ_x) and is given by:

$$(\Lambda_m)_x = \kappa_x/(K_x c_x) \tag{4}$$

where the subscript x designates a position dependence. Thus, K_x is the position dependent, membrane-aqueous solution partition coefficient of the ion and c_x is the corresponding concentration. Ion conductivity and ion mobility (u_x) are related as follows:

$$K_{x,i} u_{x,i} = \frac{\lambda_{x,i}}{F} = \frac{\Lambda_x t_{x,i}}{F} = \frac{1}{Fc} \cdot \frac{2(H-x)}{A^2} \cdot \frac{t_{x,i}}{Z_x/A} \tag{5}$$

where the subscript i designates a specific ion, in this case either Na$^+$ or Cl$^-$ and F is Faraday's constant (96,485 C mol^{-1}). The transport number of the ion in the biomembrane as a function of position ($t_{x,i}$) is determined by

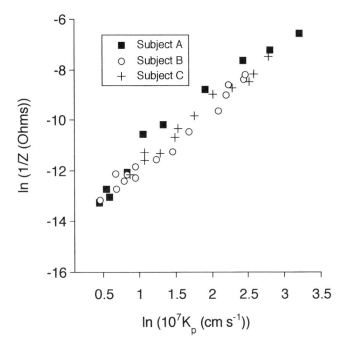

Figure 2 Relationship between skin admittance $(1/Z)$ and water permeability across the stratum corneum. Highly linear correlations were obtained for all subjects: A, $r^2 = .965$; B, $r^2 = .976$; C, $r^2 = .982$. Adapted from Reference 15.

the local membrane permselectivity. For an uncharged membrane, assuming that $t_{Na^+} = t_{Cl^-} = 0.5$ throughout the membrane and that $K_{x,Na^+} = K_{x,Cl^-} = K_{x,i}$, Eq. (4) simplifies to give

$$K_{x,i} u_{x,i} = \frac{1}{Fc} \cdot \frac{(H - x)}{A^2} \cdot \frac{1}{Z_x/A} \tag{6}$$

The data clearly show that ion mobility is position dependent, increasing through two orders of magnitude on passing from the outer SC surface to the SC–viable epidermis interface (Fig. 3).

The results shown in Figs. 1 and 3 indicate the differences between water diffusion, with its constant diffusivity across the SC, and ion transport, which is significantly faster (the mobility increasing by two orders of magnitude) on approaching the SC–viable epidermis interface. This distinction may be due, in part, to the respective underlying driving forces. In the case of water diffusion, it is solely the water concentration gradient between the body's interior and the external environment that is sufficient to create a

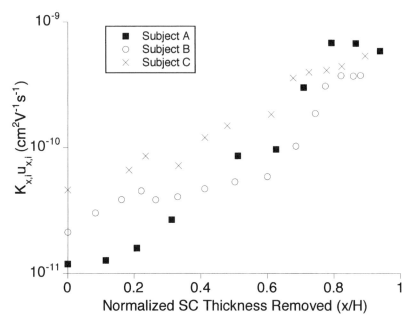

Figure 3 Ionic mobility (u_x) in human stratum corneum (SC) in vivo as a function of position in the membrane. Note that $x/H = 0$ corresponds to the SC surface, and that $x/H = 1$ represents the SC–variable epidermis interface.

water flux and hence give rise to a measurable transepidermal water loss. However, when an alternating current is applied to the skin using a physiologically buffered electrolyte solution (typically containing ~150 mM NaCl), there is no ion concentration gradient across the SC. Instead, it is the applied potential that is required to facilitate ion motion. This mechanism of ion transport is further complicated by the need for ions to have an aqueous milieu in which to move.

IV. TRACKING MOLECULAR TRANSPORT IN VIVO: ATTENUATED TOTAL REFLECTANCE–FOURIER TRANSFORM INFRARED SPECTROSCOPY

An additional technique reporting on the in vivo transport kinetics of permeants within the SC is attenuated total reflectance–Fourier transform infrared (ATR-FTIR) spectroscopy. FTIR spectroscopy has found substantial application in the biophysical examination of skin barrier function (22,23), playing an important role in the elucidation of SC composition–structure–

function relationships, while the reflectance mode (ATR-FTIR) has further extended the use of this technique to in vivo research.

In vivo studies are permitted through an arrangement comprising an IR-transparent crystal (IRE, internal reflection element) in the sample compartment, which transmits the IR beam from the interferometer to the detector. In ATR, consequently, the sample does not transect the IR beam (as in dispersive IR spectrometers), but is placed on an IRE, enabling spectral acquisition from the material at the crystal–sample interface. Hence, for in vivo (or in situ) experimentation, the region of skin under study, or samples removed from this site, may be placed directly in contact with the crystal (Fig. 4a). Alternatively, a remote fiber-optic probe (with IRE head) may be used to convey the IR beam from the source to the sample and ultimately to the detector. A thorough account describing the principles of the ATR phenomenon can be found in the literature (24). Essentially, the ATR device allows one to obtain a spectrum from the surface of a bulk tissue, without tissue separation; importantly, the spectrum is equivalent to one that would be obtained by transmission through this surface layer (Fig. 4b).

With respect to skin, the depth sampled is in the order of 0.3–3 μm, depending on the IRE substrate, the wavelength of interest, and the hydration level of the sample. Thus, when the skin is directly analyzed by this technique, the information obtained from a single ATR-IR spectrum pertains only to the immediate layers in contact with the crystal, that is, the superficial strata of the SC. Information from the deeper regions of the SC may, however, be obtained through sequential tape-stripping, where successive layers of the SC are progressively revealed and spectrally examined. Ultimately, a layer-by-layer spectroscopic profile of the SC is assembled to furnish a depth-dependent diffusion profile.

What information can this technique provide with respect to percutaneous transport? How does one monitor, measure, and characterize, using IR spectroscopy, the movement of molecules across the SC, in vivo? The answer is surprisingly simple. For monitoring, IR spectroscopy records the molecular vibrations of an absorbing species, and consequently can either identify an unknown, or track a known chemical entity. With respect to skin, the only additional requirement is that the exogenous species to be monitored possess a uniquely identifiable IR absorbence distinct from those of the SC components; theoretically, though, even this limitation can be overcome using spectral correction techniques. For measurement, the integrated IR absorbance is directly proportional to, and indicative of, the amount of permeant present at the crystal–sample interface. For characterization, the measured permeant levels as a function of SC depth (assessed by successive tape strippings) can be analyzed using a Fickian diffusion model to obtain pertinent transport parameters. Finally, and most germane to this discussion,

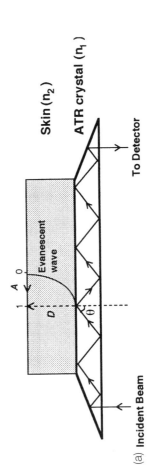

(a) Incident Beam

Figure 4 (a) A schematic representation of attenuated total reflectance. An incident beam propagating through an IR-transparent crystal of refractive index n_1 strikes the skin interface of lower refractive index n_2 at an angle θ, which is greater than the critical angle $[\theta_c = \sin^{-1}(n_2/n_1)]$. As a result, the beam is totally internally reflected and an evanescent beam established at the interface propagates into the skin. The amplitude (A) of the wave decays exponentially with increasing distance (D) from the interface. $A = $ (intensity at distance, D)/(intensity at interface). From Reference 23.

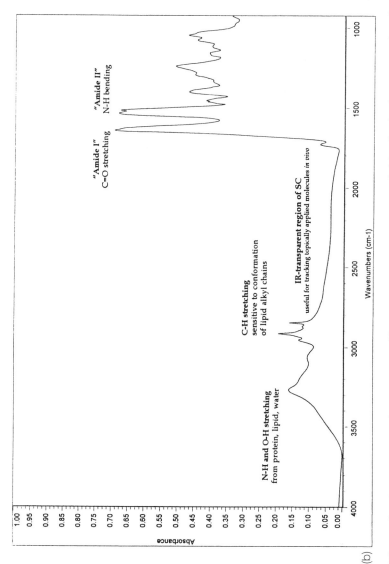

Figure 4 (b) A representative spectrum of human stratum corneum recorded *in vivo* in the attenuated total reflectance mode.

all of these can be accomplished in vivo in humans by virtue of the ATR-IR technique.

The application of ATR-IR spectroscopy to the evaluation of human skin in vivo dates back to the 1960s (25–28), but it is only the relatively recent advent of Fourier transform techniques, with the accompanying improvements in speed, sensitivity, and resolution, that have enabled quantitative measurements. One of the earliest applications of IR spectroscopy to the study of barrier function related to the estimation of water content in the uppermost layers of the stratum corneum. A comprehensive account of these earlier in vitro and in vivo ATR-IR studies, together with those of other noninvasive methods, has been presented (29,30). Several regions of the IR spectrum have been used to determine the concentration of water, including the broad OH stretching absorbance near 3400 cm^{-1} (25), the ratio of amide I (1650 cm^{-1}) to amide II (1550 cm^{-1}) bands (31–33), a combination band near 2100 cm^{-1} (34), and OD oscillations following hydration with D_2O (35). The use of deuterated probe molecules, as in the latter study, can offer a number of advantages. The IR spectrum of water (superimposed on that of the SC) can be difficult to interpret due to the overlapping and pervasive nature of O-H absorbances. The use of deuterated water (D_2O) avoids this problem by providing OD stretching vibrations that can be easily distinguished and monitored.

Reflectance IR spectroscopy in conjunction with the use of a deuterated marker has also proved to be a valuable approach for the noninvasive evaluation of the penetration enhancer oleic acid in human subjects (36). Since the SC lipids comprise esterified and free fatty acids (including oleic acid), the value of using a deuterated analogue here is that it allows the separation of absorbances arising from the exogenously applied perdeuterated oleic acid (^{2}H-OA) and the endogenous SC lipids (Fig. 5). Volunteers were treated with an ethanolic solution of perdeuterated oleic acid (5% v/v) on the inner ventral forearm, and ethanol alone on the control site (on the contralateral arm). At the end of the application period (16 h), the sites were swabbed clean with ethanol and then air-exposed for 2 h. An ATR-IR spectrum of the dosed site was then obtained by positioning the forearm on the horizontal ATR crystal. The application site was then tape-stripped once prior to a second spectral examination. This sequential tape-stripping and spectral acquisition procedure was repeated ~20 times in order to obtain an incremental spectral profile as a function of SC depth. The SC *depth* was defined by the cumulative amount of SC removed; these two parameters are mutually proportional since the mass removed by each tape stripping can be converted to a volume (assuming a SC density of 1 g cm^{-3}), and finally, as the area exposed is fixed, to an effective thickness of SC per tape strip. IR spectra thus collected consequently facilitated an in vivo assessment of ^{2}H-OA up-

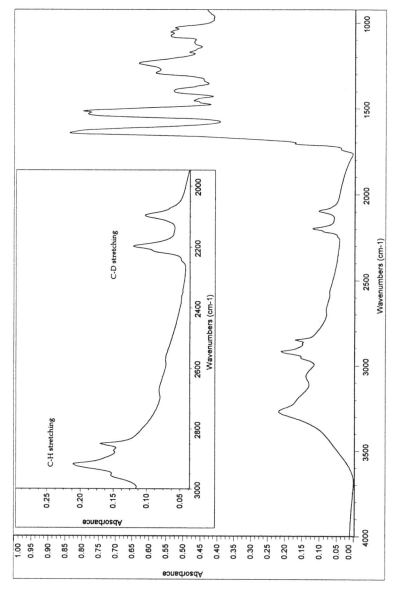

Figure 5 Reflectance infrared spectrum of human stratum corneum, in vivo, treated with perdeuterated oleic acid. The spectral separation of CH_2 and CD_2 stretching bands, originating from the SC lipids and the perdeuterated oleic acid, respectively, is illustrated in the inset.

take into human skin and its subsequent effect on SC lipids, at discrete intervals across the membrane.

The concentration profile of ^{2}H-OA in the SC was evaluated by measuring the area encompassed by the CD_2 stretching absorbances; this signal was detected at all levels of the SC examined and decreased with increasing mass of SC removed, reaching a limiting value near zero in the deepest layers of the SC probed (Fig. 6).

The conformational order of the SC lipids as a function of depth, together with the phase behavior of the topically administered oleic acid within the SC, could also be deduced by examination of the CH_2 and CD_2 stretching absorbance maxima, respectively. The results demonstrated that upon application to human skin in vivo, under conditions that enhance transdermal permeability, ^{2}H-OA did not globally modify the conformational order of SC lipids; rather, it appeared to decrease lipid viscosity only in the

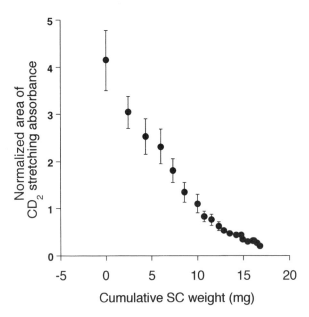

Figure 6   Normalized area of CD_2 stretching absorbance (indicating [^{2}H]oleic acid levels) as a function of SC weight removed, following treatment with a solution of 5% (v/v) [^{2}H]oleic acid in ethanol. Mean ± SE; $n = 7$ or 8. (From Ref. 36. Reprinted from *Journal of Controlled Release*, 37, Naik et al. Mechanism of oleic acid-induced skin penetration enhancement in vivo in humans, pp 299–306, 1995, with permission from Elsevier Science, NL, Sara Burgerhartstraat 25, 1055 KV, Amsterdam, The Netherlands.)

superficial layers. That is, lipid disordering was only apparent at the surface and uppermost layers where the concentration of ^{2}H-OA, and intrinsic fluidity of the SC lipids, is greatest (37); the lipid viscosity in the remainder of the membrane was essentially unaffected by ^{2}H-OA treatment. Furthermore, while the SC lipids existed in a solid state, ^{2}H-OA incorporated into the SC was present in fluid domains, consistent with earlier in vitro studies using porcine SC (38). Lipid phase separation has been shown to result in substantially enhanced permeability in lamellar lipid barriers (39–41). Additionally, studies with simpler lipid systems have shown the propensity of *cis*-unsaturated fatty acids (like oleic acid) to distribute inhomogeneously, or form a phase-separated liquid-crystalline domain, when introduced into a solid, saturated lipid mixture (42), epidermal lipids (43), model SC lipids (44), and DPPC liposomes (45). Based on these collective observations, and the in vivo IR data, it is plausible that OA-induced enhanced transdermal permeability occurs through a dual mechanism involving lipid perturbation via both conformational disordering and phase separation.

Although selective deuteration is an extremely useful technique for the separation of potentially overlapping absorbances, it is not the only means by which an exogenous chemical (such as OA) can be observed within the SC by IR spectroscopy. For example, Mak et al. (46) monitored, in vivo in humans, the concentration of OA within the superficial SC layers using the absorbance at 1710 cm^{-1}, arising from C=O stretching vibrations in the molecule. This absorbance is well separated from that of C=O stretching oscillations occurring in esterified carboxyl residues such as those predominating in SC lipids. The ratio of the OA specific absorbance at 1710 cm^{-1} to that of endogenous SC lipids at 1741 cm^{-1} (to normalize the results for variations in the degree of contact between subjects' arms and IREs) was used as an indicator of the level of OA within the outermost layers of the tissue following treatment with increasing concentrations of the enhancer. These results demonstrated that OA uptake was proportional to the enhancer treatment concentration, as shown by in vitro experiments quantifying ^{14}C-OA uptake into excised SC. By use of three different IREs of varying optical configuration, data were obtained from the skin surface to a maximum of 2 μm into the SC.

In vivo studies of the type just described have been used to evaluate the effect of penetration enhancers on percutaneous transport in humans. These studies have similarly relied on the use of a model permeant that has a distinct IR absorbance, in a region where the SC absorbs little IR radiation, enabling facile spectroscopic detection of the compound upon application to the skin. Incorporation of the "IR active" permeant into a formulation that includes a test penetration enhancer thus allows the effect of this enhancer on the transport kinetics of the permeant to be investigated. An example of such a probe molecule is 4-cyanophenol (CP), which has an intense IR absorbance at 2230 cm^{-1} due to the C$\equiv$N bond stretching vibrations. Mak et al. (47) administered CP topically as a 10% w/v solution either in propylene glycol (PG) or in propylene glycol containing 5% v/v oleic acid to the forearm of human subjects. The absorbance at 2230 cm^{-1}, representing a measure of the CP level within the superficial SC layers, diminished significantly faster when CP was codelivered with OA, indicating the facilitated throughput of the penetrant in the presence of OA. Similarly, the fate of the vehicle, propylene glycol, was followed via measurement of the peak at 1040 cm^{-1} (C—O stretch) and, as for CP, demonstrated the enhanced clearance of PG from the skin in the presence of OA. Subsequent studies have attempted to quantitatively evaluate the effect of OA on the distribution of CP in human SC by simultaneous spectroscopic and radiochemical assays (48). Radiochemical quantification of CP penetration required the incorporation of a known amount of ^{14}C-CP into the above solutions. At the end of the treatment periods (1, 2, or 3 h), SC at the application site was progressively removed by tape-stripping and IR spectra were obtained at each newly exposed skin surface, thus generating a spectroscopic distribution profile as a function of SC depth. Meanwhile, the tape strips were analyzed by liquid scintillation counting to determine the absolute amount of CP in each layer removed. The presence of OA in the applied formulation significantly increased the rate and extent of CP delivery as evaluated by both IR spectroscopy and radiochemical analysis. Furthermore, the ATR-IR and direct ^{14}C analysis of CP as a function of SC position were highly correlated, providing, therefore, initial validation of reflectance IR spectroscopy for quantitative analysis in vivo. The illustration of OA-induced skin penetration enhancement in vivo by IR has been similarly achieved with the model permeant *m*-azidopyrimethamine ethanesulphonate (MZPES), bearing the intensely IR-active azide (-N$_3$) functionality (49). In the studies described here evaluating the OA-modified distribution of CP and MZPES within the SC, penetration of the vehicle, propylene glycol, was also followed (Fig. 7). The acquired profiles closely resembled theoretical curves simulating the effect of OA on the steady-state diffusion profile (50), and suggest that coapplication of an enhancer produces a depth-dependent alteration in diffusion coefficient that modifies the steady-state transport kinetics of the more rapidly permeating solvent molecule.

More recently, this approach has been further extended with a view to developing a general model to predict the rate and extent of chemical absorption for diverse exposure scenarios from simple, and safe, short-duration studies (51). Access to such a model is crucial for the reliable prediction of topical and transdermal bioavailabilities of cutaneously applied drugs and to the accurate estimation of risk following dermal exposure to potentially toxic

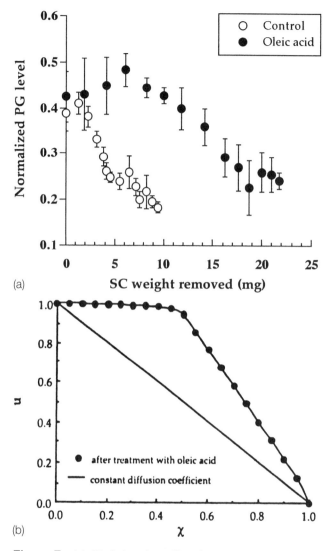

Figure 7 (a) IR-deduced profile of propylene glycol (PG) levels within the SC following exposure of the skin to either a 10% (w/v) solution of 4-cyanophenol in PG (○), or a 10% (w/v) solution of 4-cyanophenol in PG containing 5% (v/v) oleic acid (●). (b) Simulated effect of oleic acid on the diffusion profile at steady state. χ = Normalized SC depth, u = Normalized drug concentration. (Figure 7b from Ref. 50. Reprinted from *International Journal of Pharmaceutics*, 87, Watkinson et al. Computer simulation of penetrant concentration profiles in the stratum corneum, pp 175–182, 1992, with permission from Elsevier Science, NL, Sara Burgershartstraat 25, 1055 KV, Amsterdam, The Netherlands.)

chemicals in the environment. Measurement of the concentration profile of 4-cyanophenol in the SC was achieved by ATR-IR in a similar fashion to that described previously; in this instance, however, tape strips themselves, as opposed to the skin surface, were analyzed directly on the IRE, thus minimizing intersubject/contact pressure variations and preventing chemical loss by transfer to the IRE. The CP absorbance measured on each tape strip was quantified by using a previously developed calibration. This was conducted using tapes treated with varying amounts of stripped SC (6–84 μg cm^{-2}), and CP (8–248 nmol cm^{-2}); the linear relationship between CP absorbance and the CP concentration (in nmol cm^{-2}) on the tape strips was modified slightly by the presence of SC. Following exposure to the formulation for 15 min, the spectroscopically measured concentration profile was analyzed using the unsteady-state solution to Fick's Second Law of Diffusion (52):

$$C = KC_{veh}\left[\left(1 - \frac{x}{L}\right) - \frac{2}{\pi}\sum_{n=1}^{\infty}\frac{1}{n}\sin\left(\frac{n\pi x}{L}\right)\exp\left(\frac{-Dn^2\pi^2 t}{L^2}\right)\right] \quad (7)$$

where C represents the permeant concentration as a function of position x and time t; K is the SC/vehicle partition coefficient of the chemical, the concentration of which in the vehicle is C_{veh}; $C_{x=0}$ is the CP concentration at the skin surface, ($=KC_{veh}$); L is the diffusion pathlength of the permeant across the SC; and D is its diffusivity. The derived parameters D/L^2 (ratio of the diffusivity to the diffusion pathlength squared) and K were then used to predict (and experimentally confirm) concentration profiles, in addition to estimating the flux, cumulative transport, and the permeability coefficient following longer exposure times.

A representative ATR-FTIR-assessed concentration profile of CP across human SC in vivo following a 15-min exposure to a saturated aqueous solution, together with the best fit of the unsteady-state equation to the data, is shown in Fig. 8. The values of D/L^2 and K are presented in Table 2, together with the corresponding parameters from two other subjects. These values were then used to predict the concentration profile subsequent to a 1-h exposure of the identical formulation; these theoretical data along with the measured profile are shown in Fig. 9, and illustrate the excellent agreement of the predicted and measured profiles. Furthermore, the linear concentration profile after 1 h is consistent with the estimated time to reach steady-state diffusion ($\sim 2.3L^2/6D = 1.0$–1.5 h), as deduced from the mean D/L^2 (from 15-min data). Typically, the thickness of the SC (assumed to represent the diffusional pathlength, L) is found to be in the order of 10–20 μm. In these analyses, a value of 15 μm was arbitrarily chosen to calculate the in vivo steady-state flux (J_{ss}) and permeability coefficient (K_p);

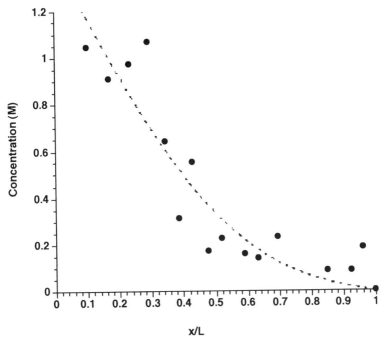

Figure 8 A representative concentration profile of 4-cyanophenol transport across human stratum corneum in vivo following application to the skin of an aqueous solution of the chemical for 15 min. The values on the abscissa (x/L) were calculated from the ratio: (SC mass removed by ith tape strip)/(total SC mass removed by all tape strips combined). Nonlinear regression was used to obtain the best fit of Eq. (7) (dashed line: $r^2 = .84$) to the data. From this analysis it can be deduced that $D/L^2 = 9.90 \times 10^{-5}$ s^{-1} and $K = 5.5$. From Reference 51.

these values were subsequently compared with the parameters obtained from the experimentally determined profiles (Table 2). The similarity of the predicted and measured parameters lends considerable support to the model employed here, and to the predictive nature of the in vivo methodology. Importantly, the concordance of these values with those derived from disparate theoretical models and experimental systems is a further indicator of the robustness of this methodology.

In a separate series of experiments, the applied formulation was spiked with ^{14}C-radiolabeled chemical, and the tape strips were analyzed by accelerator mass spectrometry (AMS), a highly sensitive radioisotope detection technique, and by conventional liquid scintillation, in addition to ATR-FTIR.

Table 2 Measured, Fitted, and Predicted Parameters Characterizing CP Transport Across Human Stratum Corneum In Vivo, Following Application of an Aqueous 4-Cyanophenol (CP) Solution for 15 and 60 min

| | [CP] in SC at 15 min[a] (M) | 15-min fitted parameters[b] | | | | [CP] in SC at 60 min[a] (M) | 60-min fitted parameters[c] | | | | 60-min predicted parameters[e] | |
		K	D/L^2 (10^5 s^{-1})	$C_{x=0}$ (M)	T_{lag} (min)		K	$C_{x=0}$ (M)	J_{ss}[d] (nmol cm^{-2} s^{-1})	$10^7 K_p$ (cm s^{-1})	J_{ss}	$10^7 K_p$
Mean	0.45	8.4	8.4	1.64	32.5	0.63	7.4	1.44	0.18	9.4	0.20	10.4
SD ($n = 3$)	0.15	3.6	1.5	0.70	6.2	0.24	2.6	0.52	0.04	1.9	0.05	2.6

Note. Adapted from Reference 48. K, partition coefficient of CP between SC and vehicle. D/L^2, ratio of the diffusivity to the diffusion pathlength squared. $C_{x=0}$, CP concentration at the skin surface = KC_{veh}. T_{lag}, lag time = $L^2/6D$. J_{ss}, steady-state flux = $(KD/L)C_{veh}$. K_p, permeability coefficient = $KD/L = J_{ss}/C_{veh}$.

[a] Measured experimentally.
[b] From the fit of Eq. (7) to the 15-min in vivo data.
[c] From the fit of the steady-state form of Eq. (7) to the 60-min in vivo data.
[d] Calculated from the gradient of the fitted in vivo data to the steady-state form of Eq. (7), assuming $L = 15$ μm.
[e] Predicted using the values of K and D/L^2 from the 15-min exposure experiments, and assuming $L = 15$ μm.

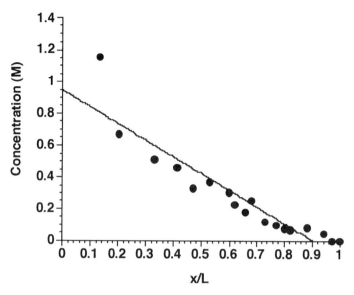

Figure 9 A representative concentration profile of 4-cyanophenol across human stratum corneum in vivo following exposure of the skin to an aqueous solution of the chemical for 1 h. The values on the abscissa (x/L) were calculated as described for Fig. 8. The slope and intercept of the line of linear regression through the data are -1.05 M and 0.95 M, respectively. The values predicted from the 15-min exposure data are -1.43 M and 1.43 M, respectively. From Reference 51.

Table 3 CP Concentrations in 20 Tape Strips of SC Measured by ATR-FTIR Spectroscopy, Liquid Scintillation Counting (LSC), and AMS after a 15-min Exposure to an Aqueous Solution of the Chemical

	Applied concentration		Total amount in SC ($nmol\ cm^{-2}\ mg^{-1}$)		
Subject	mM	nmol cm^{-2}	ATR-FTIR	LSC	AMS
D	196	4410	84	62	129
E	196	4410	40	32	37
F	196	4508	64	51	69
Mean			63	48	78
SD			22	15	47

Note. The average total amounts of CP in the stratum corneum as determined by the three methods were statistically indistinguishable ($p > .05$, Kruskal–Wallis test). From Reference 51.

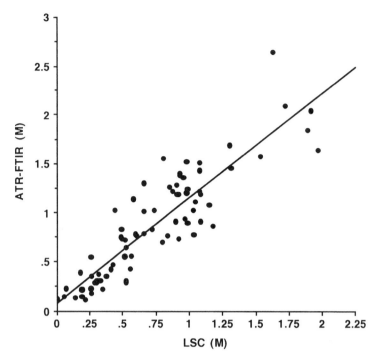

Figure 10 Correlation between the concentrations of CP in the SC, following a 15-min exposure to an aqueous solution of the chemical, as measured by ATR-FTIR spectroscopy and by liquid scintillation counting (LSC). The accumulated data from four different subjects (78 measurements) are shown. The line of linear regression drawn through the data is $y = 1.07 + 0.08$ ($r^2 = .796$, $p < .0001$). The values of the slope and intercept are not significantly different from 1 and 0 ($p < .05$), respectively. From Reference 51.

The total uptake of CP into human SC in vivo, as determined by these different techniques, is shown in Table 3. Correlation between the IR and AMS measurements is illustrated in Fig. 10. The use of 4-nitrophenol as an IR probe in vivo has also been validated using the AMS technique (53,54). Once again, the correspondence of these data obtained by different methodologies emphasizes the value of the reflectance IR technique in acquiring, noninvasively, quantitative data in vivo, in human populations, without recourse to radiochemical methods.

ACKNOWLEDGMENTS

The research summarized in this review was supported by the U.S. National Institutes of Health (HD 23010, HD 27839), the U.S. Air Force Office of Scientific Research, the U.S. Environmental Protection Agency, and Novartis (Switzerland).

REFERENCES

1. Potts, RO, and Francoeur, ML. The influence of stratum corneum morphology on water permeability. Journal of Investigative Dermatology, 1991. 96(4): 495–499.
2. Feldmann, RJ, and Maibach, HI. Percutaneous penetration of steroids in man. Journal of Investigative Dermatology, 1969. 52(1):89–94.
3. Feldmann, RJ, and Maibach, HI. Absorption of some organic compounds through the skin in man. Journal of Investigative Dermatology, 1970. 54(5): 399–404.
4. McKenzie, AW, and Stoughton, RB. Method for comparing percutaneous absorption of steroids. Archives of Dermatology, 1962. 86:608–610.
5. Rougier, A, Dupuis, D, Lotte, C, et al. In vivo correlation between stratum corneum reservoir function and percutaneous absorption. Journal of Investigative Dermatology, 1983. 81(3):275–278.
6. Bunge, A, and Cleek, A. A new method for estimating dermal absorption from chemical exposure. 2. Effect of molecular weight and octanol–water partitioning. Pharmaceutical Research, 1995. 12:88–95.
7. Cleek, RL, and Bunge, AL. A new method for estimating dermal absorption from chemical exposure. 1. General approach. Pharmaceutical Research, 1993. 10(4):497–506.
8. Potts, RO, Guy, RH. Predicting skin permeability. Pharmaceutical Research, 1992. 9(5):663–669.
9. Lien, E, and Gao, H. QSAR analysis of skin permeability of various drugs in man as compared to in vivo and in vitro studies in rodents. Pharmaceutical Research, 1995. 12:583–587.
10. Potts, R, and Guy, R. A predictive algorithm for skin permeability—The effects of molecular size and hydrogen bond activity. Pharmaceutical Research, 1995. 12:1628–1633.
11. Roberts, M, Pugh, W, Hadgraft, J, et al. Epidermal permeability–penetrant structure relationships. 1. An analysis of methods of predicting penetration of monofunctional solutes from aqueous solutions. International Journal of Pharmaceutics, 1995. 126:219–233.
12. Roberts, M, Pugh, W, and Hadgraft, J. Epidermal permeability–penetrant structure relationships. 2. The effect of H-bonding groups in penetrants on their diffusion through the stratum corneum. International Journal of Pharmaceutics, 1996. 132:23–32.

13. Pugh, W, Roberts, M, and Hadgraft, J. Epidermal permeability–penetrant structure relationships. 3. The effect of hydrogen bonding interactions and molecular size on diffusion across the stratum corneum. International Journal of Pharmaceutics, 1996. 138:149–165.

14. Bunge, A, and Cleek, A. A new method for estimating dermal absorption from chemical exposure. 3. Compared with steady-state methods for prediction and data analysis. Pharmaceutical Research, 1995. 12:972–982.

15. Kalia, Y, Pirot, F, and Guy, R. Homogeneous transport in a heterogeneous membrane: Water diffusion across human stratum corneum in vivo. Biophysical Journal, 1996. 71:2692–2700.

16. Anderson, RL, and Cassidy, JM. Variation in physical dimensions and chemical composition of human stratum corneum. Journal of Investigative Dermatology, 1973. 61(1):30–32.

17. Tregear, R. Physical Functions of the Skin. 1966. New York: Academic Press.

18. Yamamoto, T, and Yamamoto, Y. Electrical properties of the epidermal stratum corneum. Medical and Biological Engineering, 1976. March:151–158.

19. Kalia, Y, Nonato, L, and Guy, R. The effect of iontophoresis on skin barrier integrity: Non-invasive evaluation by impedance spectroscopy and transepidermal water loss. Pharmaceutical Research, 1996. 13:957–960.

20. Kalia, Y, and Guy, R. Interaction between penetration enhancers and iontophoresis. Journal of Controlled Release, 1997. 44:33–42.

21. Mitragotri, S, Blankschtein, D, and Langer, R. Transdermal drug delivery using low-frequency sonophoresis. Pharmaceutical Research, 1996. 13(3):411–420.

22. Potts, RO, and Francoeur, ML. Infrared spectroscopy of stratum corneum lipids. In Pharmaceutical Skin Penetration Enhancement, KA Walters and J Hadgraft, Editors. 1993. New York: Marcel Dekker.

23. Naik, A, and Guy, R. Infrared spectroscopic and differential scanning calorimetric investigations of the stratum corneum barrier function. In Mechanisms of Transdermal Drug Delivery, R Potts and R Guy, Editors. 1997. New York: Marcel Dekker.

24. Mirabella, FM, ed. Internal Reflection Spectroscopy: Theory and Practice. 1993. New York: Marcel Dekker.

25. Baier, RE. Noninvasive, rapid characterization of human skin chemistry in situ. Journal of the Society of Cosmetic Chemists, 1978. 29:283–306.

26. Comaish, S. Infra-red studies of human skin by multiple internal reflection. British Journal of Dermatology, 1968. 80:522–528.

27. Puttnam, NA, and Baxter, BH. Spectroscopic studies of skin in situ by attenuated total reflectance. Journal of the Society of Cosmetic Chemists, 1967. 18:469–472.

28. Puttnam, NA. Attenuated total reflectance studies of the skin. Journal of the Society of Cosmetic Chemists, 1972. 23:209–226.

29. Potts, RO. In vivo measurement of water content of the stratum corneum using infrared spectroscopy: A review. Cosmetics and Toiletries, 1985. 100:27–31.

30. Potts, RO. Stratum corneum hydration: Experimental techniques and interpretations of results. Journal of the Society of Cosmetic Chemists, 1986. 37:9–33.

31. Gloor, M, Willebrandt, U, Thomer, G, et al. Water content of the horny layer and skin surface lipids. Archives of Dermatological Research, 1980. 268:221–223.

32. Gloor, M, Hirsh, G, and Willebrandt, U. On the use of infrared spectroscopy for the in vivo measurement of the water content in the horny layer after application of dermatological ointments. Archives of Dermatological Research, 1981. 271:305–314.

33. Triebskorn, A, Gloor, M, and Greiner, F. Comparative investigations on the water content of the stratum corneum using different methods of measurement. Dermatologica, 1983. 167:64–69.

34. Potts, RO, Guzek, DK, Harris, RR, et al. A noninvasive, in vivo technique to quantitatively measure water concentration of the stratum corneum using attenuated total-reflectance infrared spectroscopy. Archives of Dermatological Research, 1985. 277:489–495.

35. Hansen, JR, and Yellin, W. NMR and infrared spectroscopic studies of stratum corneum hydration. In Water Structure at the Water-Polymer Interface, HHG Jellinek, Editor. 1972. New York: Plenum Publishing, pp. 19–28.

36. Naik, A, Pechtold, LARM, Potts, RO, et al. Mechanism of oleic-acid induced skin penetration enhancement in vivo in humans. Journal of Controlled Release, 1995. 37:299–306.

37. Bommannan, D, Potts, RO, and Guy, RH. Examination of stratum corneum barrier function in vivo by infrared spectroscopy. Journal of Investigative Dermatology, 1990. 95(4):403–408.

38. Ongpipattanakul, B, Burnette RR, Potts, RO, et al. Evidence that oleic acid exists in a separate phase within stratum corneum lipids. Pharmaceutical Research, 1991. 8(3):350–354.

39. Blok, MC, van der Neut-Ko, ECM, van Deenan, LLM, et al. The effect of chain length and lipid phase transitions on the selective permeability properties of liposomes. Biochimica et Biophysica Acta, 1975. 406:187–196.

40. Wu, SHW, and McConell, HM. Lateral phase separations and perpendicular transport in membranes. Biochemical and Biophysical Research Communications, 1973. 55:484.

41. Shimshick, EJ, Kleeman, W, Hubbell, WL, et al. Lateral phase separations in membranes. Journal of Supramolecular Structure, 1973. 2:285–295.

42. Ortiz, A, and Gomez-Fernandez, JC. A differential scanning calorimetry study of the interaction of free fatty acids with phospholipid membranes. Chemistry and Physics of Lipids, 1987. 45:75–91.

43. Walker, M, Hollingsbee, DA, Hadgraft, J, et al. Influence of oleic acid on the physical and chemical properties of the human epidermal barrier. In Prediction of Percutaneous Penetration, RC Scott, et al., Editors. 1991. London: IBC, pp. 86–96.

44. Lieckfeldt, R, Villalaín, J, Gómez-Fernández, J-C, et al. Influence of oleic acid on the structure of a mixture of hydrated model stratum corneum fatty acids and their soaps. Colloids and Surfaces, 1994. 90:225–234.

45. Watkinson, AC, Street, PR, Richards, RW, et al. Evidence for phase separation of oleic acid in DPPC liposomes and excised stratum corneum from small angle neutron scattering study. In Prediction of Percutaneous Penetration, RC Scott, et al., Editors. 1991. London: IBC, pp. 380–385.

46. Mak, VHW, Potts, RO, and Guy, RH. Oleic acid concentration and effect in human stratum corneum: Non-invasive determination by attenuated total reflectance infrared spectroscopy in vivo. Journal of Controlled Release, 1990. 12:67–75.

47. Mak, VH, Potts, RO, and Guy, RH. Percutaneous penetration enhancement in vivo measured by attenuated total reflectance infrared spectroscopy. Pharmaceutical Research, 1990. 7(8):835–841.

48. Higo, N, Naik, A, Bommannan, DB, et al. Validation of reflectance infrared spectroscopy as a quantitative method to measure percutaneous absorption in vivo. Pharmaceutical Research, 1993. 10(10):1500–1506.

49. Guy, RH, Higo, N, Naik, A, et al. Mechanism and enhancement of skin penetration in vivo. In Prediction of Percutaneous Penetration, RC Scott, et al., Editors. 1991. London: IBC, pp. 1–12.

50. Watkinson, A, Bunge, A, Hadgraft, J, et al. Computer simulation of penetrant concentration-depth profiles in the stratum corneum. International Journal of Pharmaceutics, 1992. 87:175–182.

51. Pirot, F, Kalia, YN, Stinchcomb, AL, et al. Characterization of the permeability barrier of human skin in vivo. Proceedings of the National Academy of Science, USA, 1996. 94:1562–1567.

52. Crank, J. Mathematics of Diffusion. 1975. Oxford: Oxford University Press.

53. Naik, A, Keating, G, and Guy, RH. Assessment of dermal exposure in humans. In Prediction of Percutaneous Penetration: Methods, Measurements, Modelling. 1995. La Grande Motte: STS Publishing Ltd.

54. Keating, G, McKone, TE, Naik, A, et al. Assessment of dermal exposure to drinking water contaminants—New measurements and models. In Assessing and Managing Health Risks from Drinking Water Contaminants: Approaches and Applications, EG Riechard and GA Zapponi, Editors. 1995, IAHS Press: Wallingford, pp. 235–244.

55. Scheuplein, RJ. Mechanism of percutaneous adsorption I. Routes of penetration and the influence of solubility. Journal of Investigative Dermatology, 1965. 45(5):334–346.

56. Bond, JR, and Barry, BW. Limitations of hairless mouse skin as a model for in vitro permeation studies through human skin: Hydration damage. Journal of Investigative Dermatology, 1988. 90(4):486–489.

57. Akhter, SA, Bennett, SL, Waller IL, et al. An automated diffusion apparatus for studying skin penetration. International Journal of Pharmaceutics, 1984. 21: 17–26.

58. Harrison, SM, Barry, BW, and Dugard, PH. Effects of freezing on human skin permeability. Journal of Pharmacy and Pharmacology, 1984. 36:261–262.

59. Scheuplein, RJ, and Blank, IH. Permeability of the skin. Physiological Reviews, 1971. 51(4):702–747.

60. Astley, JP, and Levine, M. Effect of dimethyl sulfoxide on permeability of human skin in vitro. Journal of Pharmaceutical Sciences, 1976. 65(2):210–215.

9
Relationship Between H-Bonding of Penetrants to Stratum Corneum Lipids and Diffusion

William John Pugh
Welsh School of Pharmacy, Cardiff University, Cardiff, Wales

I. INTRODUCTION

It is generally accepted that the barrier to permeation is effectively the outermost $10-15$ μm of the skin (1–3). To overcome this stratum corneum (SC) obstacle, a molecule must first enter and then cross it.

Fick's law relates the steady-state flux, J_s, to the concentration gradient across the SC. If the viable dermis is regarded as a sink, the gradient determining the flux is C_{sc}/h. Since the partition coefficient K of a solute between the SC and vehicle can be written as C_{sc}/C_v, J_s can be expressed as:

$$J_s = ADKC_v/h$$

The permeability coefficient, k_p, is the steady-state flux per unit area divided by the concentration of solute in solution, so that:

$$k_p = \frac{J_s}{AC_v} = K\frac{D}{h}$$

Most reports on epidermal structure penetration relationships are based on this composite quantity, k_p, for aqueous vehicles. This article concentrates mainly on the diffusion step and examines the molecular features, particularly H-bonding, that determine it. To set this in context, a very brief outline of the overall process is given.

II. THE NATURE OF THE BARRIER

Absorption across the stratum corneum (SC) is a passive process of diffusion. The SC is composed of dead corneocytes in a lipid matrix pierced by hair follicles and sweat glands. In principle there are three routes for diffusion: intercellular diffusion through the lipid matrix, transcellular diffusion through the corneocytes; and pilosebaceous diffusion along the sweat pores and hair follicles.

Absorption via the pores and follicles is considered to be insignificant because the orifices account for only 0.1% of skin area and diffusion along sweat ducts is against an outward aqueous flow (4). Lauer et al. (5) reviewed this pathway, recalling Scheuplein's (6,7) proposal that the initial, transient penetration of steroids is consistent with diffusion through pores. However, after the steady state is established, bulk diffusion through the lipid region accounts for most permeation. Siddiqui et al. (8) compared diffusion of steroids across SC and a pore-free membrane (Silastic) and concluded that the penetration kinetics could be explained without invoking the need for aqueous channels. Although the recent mathematical modeling work of Heisig et al. (9) suggests that permeation through the corneocytes cannot be ignored, the barrier must be essentially lipid in nature since its barrier function is lost after extracting the lipids (10,11), and the equations developed by Edwards and Langer (12) suggest that the intercellular matrix is the significant route for diffusion of uncharged permeants. Roberts et al. (13) propose a mixed model where transport occurs along a continuous pathway but both the lipid and polar portions of the bilayer are used, depending on the polarity of the penetrant.

The lipid composition of SC has been determined by Wertz et al. (14) and the major components are fatty acids, ceramides, and cholesterol. Leickfeldt et al. consider that the precise molecular form of the lipids is unimportant to barrier function. Interaction between cholesterol and fatty acids produces an ordered, impermeable bilayer structure, and the distinctive form of the ceramide molecules enables solubilization of the cholesterol and prevents its separation into a separate phase (15).

III. DETERMINANTS OF THE PERMEABILITY COEFFICIENT k_p

The equation for J_s suggests that partitioning into the SC is an important determinant of permeability, and correlations were sought between k_p and K between various model solvents and water. The partition between octanol and water, $K_{octanol}$, is as good as any other for this purpose (16).

Reasonable correlations between k_p and $K_{octanol}$ were obtained only for families of similar compounds. Thus Roberts et al. (17) found for a set of phenols:

$$\log k_p = -3.3 + 0.69 \log K_{octanol} \qquad N = 19 \qquad r^2 = .72$$

For hydrocortisone esters, El Tayar et al. (18) found

$$\log k_p = -5.3 + 0.80 \log K_{octanol} \qquad N = 11 \qquad r^2 = .88$$

and for alkanols:

$$\log k_p = -2.7 + 0.77 \log K_{octanol} \qquad N = 8 \qquad r^2 = .89$$

Correlations on mixed datasets were unsuccessful. For combined alkanols and phenols, El Tayar et al. found

$$\log k_p = -3.6 + 0.16 K_{octanol} \qquad N = 22 \qquad r^2 = .03$$

When plotted separately the data fall on two distinct lines, and El Tayar's group attributed the difference in the gradients to penetration via intracellular and transcellular routes.

Kasting et al. (19), appreciating that k_p was the product of both partition and diffusion terms, included molecular weight (MW) as a size determinant of diffusion, and following on from this Potts and Guy (20) published their much quoted relationship:

$$\log k_p(cm/h) = -2.74 + 0.71 \log K_{octanol} - 0.0061 MW$$
$$N = 93 \qquad r^2 = .67$$

which encompassed the data set of diverse compounds collected by Flynn.

Roberts et al. (13) analyzed data for 91 of the compounds studied by Potts and Guy and found a similar regression:

$$\log k_p(cm/h) = -2.70 + 0.63 \log K_{octanol} - 0.0054 MW$$
$$N = 91 \qquad r^2 = .65$$

They suggested that interaction might occur between penetrant and both polar and nonpolar components of SC and related k_p to partition into dissimilar (polar and nonpolar) solvents. This enabled them to account for some of the 33% of the variation in k_p not explained by Potts and Guy:

$$\log k_p = -2.29 + 0.24 \log K_{octanol} + 0.40 \log K_{hexane}$$
$$N = 24 \qquad r^2 = .88$$

Taft, Kamlet, Abraham, and co-workers developed the solvatochromic approach to determining permeability. According to this, the essential features of a permeant molecule are its size (V), electronic charge distribution (π^*), and hydrogen bonding donor and receptor capabilities (α and β).

Abraham et al. (21) found for a mixed dataset of alcohols, steroids, and phenols:

$$\log k_p = -1.49 - 0.59\pi^* - 0.63\alpha - 3.48\beta + 0.0179V$$
$$N = 46 \qquad r^2 = .96$$

and were thus able to explain permeation in terms of a single pathway model. The findings of Potts and Guy (20),

$$\log k_p = -1.29 - 1.72\alpha - 3.93\beta + 0.0256V$$
$$N = 37 \qquad r^2 = .94$$

and Roberts et al. (13),

$$\log k_p = -1.35 - 1.37\alpha - 4.53\beta + 0.0205V$$
$$N = 24 \qquad r^2 = .93$$

confirm this observation, and furthermore suggest that π^* is not a significant predictor of k_p.

It seems then that permeation can be explained by a single pathway, and is chiefly determined by H-bonding properties and permeant size.

It has already been noted that k_p is a composite of two factors, K and (D/h). Principal component analysis (PCA) measures how data points in a matrix may be related. In essence, it enables us to see how many processes or mechanisms are involved in relating an outcome ($\log k_p$) and molecular properties (α, β, V) of the permeants (22).

If the data used by Roberts et al. (13) are subjected to PCA (unpublished), the variation in the matrix of four datasets ($\log k_p$, α, β, V) is described in terms of four combinations (principal components) of the variables. The output from the Minitab statistical package (23) is:

Eigenvectors

Variable	PC1	PC2	PC3	PC4
$\log k_p$	0.661	0.197	−0.099	0.718
α	−0.030	−0.880	0.353	0.318
β	−0.561	−0.096	−0.690	0.448
V	0.498	−0.422	−0.625	−0.429
Eigenvalue	2.1352	1.1808	0.6534	0.0306
Proportion explained	0.534	0.295	0.163	0.008
Cumulative proportion	0.534	0.829	0.992	1.000

This output is interpreted as follows: The sum of eigenvalues for four PCs is 4. The eigenvalue for a particular PC is a measure of the amount of variation in the data that can be explained by that PC. Thus variation due to PC1 = 2.1352/5 = 0.534 (53.4%). So we can explain 53.4% of the variation in the data by the interrelationship between the four variables:

$$(0.661 \times \log k_p) - (0.030 \times \alpha) - (0.561 \times \beta) + (0.498 \times V)$$

There is a mechanism/process that accounts for 53.4% of the variation in the data, and the first two PCs mechanisms/processes will together account for 82.9% of the variation. It is reasonable to speculate that these two processes are the partition and diffusion processes that comprise k_p.

The importance of a variable in its PC is given by the square of its eigenvector. Thus in PC1 (process 1) α is unimportant, since 99.9%, that is, $[[1 - (-0.03^2)] \times 100]$, of the variation due to this process can be explained without it. It is, however, very important in process 2.

The importance of H-bonding had been recognized earlier when Roberts (24) showed that k_p was related to the number of H-bonding groups in the penetrant. Recently attempts were made (13) to quantify H-bonding as the difference in partition from water into H-bonding and non-H-bonding solvents. Anderson and Raykar (25) suggested that the SC barrier resembled a hydrogen-bonding organic solvent, and El Tayar et al. (18) concluded that the H-bond donor potential was dominant. However, this differential partitioning approach has been shown to be unreliable by the analyses of Roberts et al. (13) and Pugh et al. (16), and use of α and β values as quantifiers of H-bonding seems to be the way forward. The papers of Abraham (26) and Abraham et al. (21) are valuable data sources for the solvatochromic parameters.

IV. DETERMINANTS OF DIFFUSION ACROSS THE SC

Potential factors reducing diffusion include molecular size, interaction with SC components, and the obstructive effects of the corneocytes. In principle it is possible for penetrants to diffuse both along the lipid pathway and through the corneocytes (12), but Potts and Francoeur (27) argue forcefully against diffusion of water through corneocytes, and it is unlikely that polar organic molecules traverse them if water cannot. Their reasoning may be summarized as follows.

Corneocytes are covered by covalently bound highly nonpolar lipids with exceptionally long (C_{30}–C_{34}) hydrocarbon chains (28), which are significantly more hydrophobic than other SC lamellae (29). The presence of inert "flakes" in a homogenous matrix reduces permeability (30,31) by an obstruction effect, and water permeability falls with increasing corneocyte size (32) as predicted by flake theory. SC is 1000 times less permeable to water vapor than other lipid membranes, and Hadgraft and Ridout (33) found that the passage of drugs through SC was about 1000 times slower than through films of isopropyl myristate. Twenty to 30 layers of corneocytes 0.5 μm thick and 30 μm square spaced 0.1 μm apart (34) would lower permeability by a factor of 1000 by an obstruction effect. They thus seem to act as mechanical barriers and increase the path length of diffusion.

The lipids form the only continuous domain in the SC (35,36), and their removal increases water permeability (37). SC that has been reconstituted from corneocytes and extracted SC lipids behaves similarly to intact SC (38,39), and Friberg et al. (40) showed a variety of lipids could restore SC properties to isolated corneocytes. Rougier et al. (32) showed that permeability of SC to water and benzoic acid are highly correlated for a group of human subjects suggesting a common pathway.

All this evidence points to corneocytes having only a pathlength-increasing effect and not providing a parallel pathway for water—and, by inference, hydrophilic—molecules.

Early methods for finding D (41) involved measurement of the lag time, τ, to establish steady flux across the SC. This is given (42) by

$$\tau = h^2/6D$$

The value of h is uncertain, and estimation of τ involves extrapolation of the "linear" portion of the plot of amount transferred against time. Curve-fitting programs now make it possible to deconvolute the terms in the non-steady-state equation (8)

$$\frac{C}{C_m} = 1 - \frac{4}{\pi} \sum_{n=0}^{\infty} \frac{(-1)^n}{2n+1} \exp\left[\frac{-D(2n+1)^2\pi^2 t}{4h^2}\right]$$

to find D/h^2.

Roberts's group went on to examine the influence of H-bonding on the diffusion (16) using a different approach to find D/h. Since $\log k_p = \log K_{sc} + \log(D/h)$ and

$$\log K_{sc} = -0.024 + 0.59 \log K_{octanol} \qquad N = 45 \qquad r^2 = .84$$

then

$$\log(D/h) = \log k_p - 0.59 \log K_{octanol} + 0.024$$

V. DIFFUSION OF MONOFUNCTIONAL COMPOUNDS

Using experimental values of K_{sc}, Roberts et al. (43) calculated $\log(D/h)$ as ($\log k_p - \log K_{sc}$) and used α and β values as measures of H-bonding potential. A good correlation was found:

$$\log(D/h) = -1.86 - 0.61\alpha - 2.09\beta \qquad N = 37 \qquad r^2 = .90$$

but inclusion of π or MW did not improve the regression. The major determinant of diffusion is H-bonding, implying that each substituent group on the permeant retards diffusion to a characteristic degree. Further, the high coefficient of β shows that SC is predominantly an H-bond donor as proposed earlier by El Tayar et al. (18).

They next set out to quantify the H-bonding powers of various chemical groups (16), first confirming that $\log(D/h)$ as estimated from $\log K_{octanol}$ was related to α and β:

$$\log(D/h) = -1.32 - 1.30\alpha - 2.57\beta \qquad N = 29 \qquad r^2 = .85$$

and going on to regress $\log(D/h)$ against the number (0 or 1) of each functional group present:

$$\log(D/h) = -1.36 - 1.67 \text{ acid} - 1.41 \text{ alcohol} - 1.17 \text{ phenol}$$
$$- 0.986 \text{ carbonyl} - 0.759 \text{ ether} - 0.0502C^*$$

where acid is the number of the acid groups (0 or 1) in permeant and C^* is the number of C atoms not involved in $C{=}O$ bonds. The (negative) coefficients were greater for strong H-bonding functions and were called retardation coefficients (RCs). Thus the presence of an acid would reduce the diffusion across the SC by a factor of about 50 and an ether by about 6. This multiplicative effect explained their earlier observation that introduction of multiple groups caused a dramatic decrease in diffusion.

VI. H-BONDING POTENTIAL OF THE SC

Using the lipid composition of the SC given by Wertz (44) and the α and β values of Abraham (26), Pugh et al. (16) calculated the H-bonding effects of the SC to be in the ratio $\alpha_{sc}{:}\beta_{sc} = 0.4{:}0.6$. This would suggest that SC is predominantly an H-bond acceptor environment and contradicted their earlier conclusion and that of El Tayar et al. The H-bonding for each functional group, g, should be related to the quantity $[(\alpha_g\beta_{sc}) + (\beta_g\alpha_{sc})]$, but a plot of RC against $[(\alpha_g\beta_{sc}) + (\beta_g\alpha_{sc})]$ did not pass through the origin as expected. The $\alpha_{sc}{:}\beta_{sc}$ value of 0.40:0.60 was therefore considered dubious and the H-bonding potential of the SC was reestimated indirectly from RC values as follows.

In the simplest case RC would be directly related to H-bonding between penetrant and SC:

$$RC = X + Y[(\alpha_p\beta_{sc}) + (\beta_p\alpha_{sc})]$$

Since $\alpha_{sc} = (1 - \beta_{sc})$,

$$RC = X + Y\beta_{sc}(\alpha_p - \beta_p) + Y\beta_p$$

The regression is:

$$RC = 0.0024 + 1.36(\alpha_p - \beta_p) + 3.18\beta_p \qquad r^2 = .99$$

The high r^2 value and the insignificant value of the constant 0.0024 imply a satisfactory relationship. Solving $Y\beta_{sc} = 1.36$, $Y = 3.18$ gives $\alpha_{sc} = 0.573$ and $\beta_{sc} = 0.427$, implying that SC is predominantly an H-bond donor, with α and β binding strengths in the approximate ratio 0.6:0.4.

VII. DIFFUSION OF POLYFUNCTIONAL COMPOUNDS

For polyfunctional compounds a plot of D/h against number of H-bonding groups shows the dramatic effect of introducing more than one group (16). The curve (Fig. 1) resembles a Langmuir adsorption plot and shows that maximal retardation is quickly reached. These polyfunctional compounds have a large MW range and a size effect is seen:

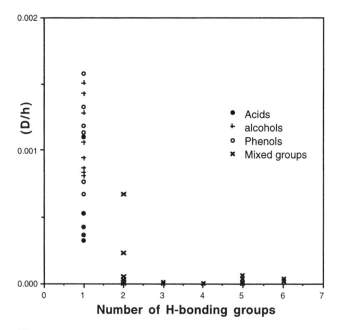

Figure 1   Effect of number of hydrogen-bonding groups on diffusion across stratum corneum.

$$\log(D/h) = -1.50 - 0.91\alpha - 1.58\beta - 0.003MW$$
$$N = 53 \qquad r^2 = .94$$

The intercept ($D_0/h = 0.032$ cm/h; 95% confidence interval 0.02–0.05) represents an intrinsic diffusion term, describing the diffusion of an infinitely small, nonbonding molecule. The regression shown earlier enables the relative importance of α and β to be estimated by comparison of the coefficients -0.91 and -1.58. The relative importance of the H-bonding parameters and size is more difficult to assess since the magnitudes of the predictors are so different. The values of α and β are typically between 0 and 1, while MW ranges from 50 to 500. Thus the low coefficient of the MW term might still result in a large contribution to $\log(D/h)$ when multiplied by a large MW.

Comparison of the importance of such diverse predictors requires that their magnitudes be similar. This standardization of the data can be achieved by subtracting the predictor mean from each value and dividing by the predictor standard deviation. The standardized predictors thus all have means of zero and standard deviations of 1.

Regression of these standardized data (α^* etc.) gives:

$$\log(D/h) = -3.78 - 0.239\alpha^* - 0.752\beta^* - 0.521MV^*$$
$$N = 53 \qquad r^2 = .94$$

indicating that in practice variations in H-bonding and MW have comparable effects on diffusion.

Principal component analysis gives:

Eigenvectors				
Variable	PC1	PC2	PC3	PC4
$\log(D/h)$	0.584	0.070	0.235	0.774
α	−0.099	−0.979	−0.035	0.174
β	−0.569	0.181	−0.552	0.582
MW	−0.570	0.061	0.799	0.182
Eigenvalue	2.8371	1.0130	0.1149	0.0350
Proportion explained	0.709	0.253	0.029	0.009
Cumulative proportion	0.709	0.963	0.991	1.000

The PCA output shows that two processes account for 96.3% of the variation in data relating diffusion, the H-bonding parameters and size; β is probably more important than α.

VIII. MODEL OF THE H-BONDING PROCESS

The plot of (D/h) against number of H-bonding groups (Fig. 1) is a curve resembling an inverted Langmuir adsorption isotherm (43), which describes adsorption as an equilibrium between binding and debinding at an interface. The position of equilibrium is determined by the relative affinities of the adsorbate for the surface and the support phase. The general form of the isotherm is

$$w = \frac{w_{max}}{(K/c) + 1}$$

where w_{max} is the amount needed to saturate the surface, w is the amount adsorbed at concentration c in the support phase, and K is the ratio between the rates of desorption and adsorption, k_d/k_a; c can be considered as the force driving adsorption. In diffusion across the SC the effect analogous to w is the reduction in diffusion $(D_o - D)/h$ and the saturation effect is $(D_o - D_m)/h$, where D_m is the minimum diffusion coefficient relating to an infinitely hindered penetrant. Pugh et al. proposed that the driving force causing binding of the permeant to SC (corresponding to c) is the retardation coefficient, or, more precisely, $(RC - RC_o)$, where RC_o represents the binding of a compound with no H-bonding groups (Fig. 2).

Substituting these values into Langmuir's equation and rearranging gives

$$D/h = D_o/h - [(D_o/h - D_m/h)(RC - RC_o)/(K + RC - RC_o)]$$

and nonlinear curve fitting enables estimation of the parameters:

Parameter	Final value	SD
D_o/h	0.192	0.009
D_m/h	6.6E $-$ 5	12.4E $-$ 5
RC_o	0.222	0.004
K	0.0053	0.0002
$N = 53$ $\quad r^2 = .98$		

The high standard deviation for D_m/h suggests it is indistinguishable from zero as expected, but all the other parameters are statistically valid. The low value for K (the equilibrium constant for debinding) shows that H-bonding is a highly favored process in the SC.

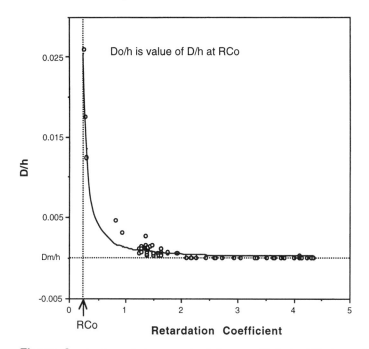

Figure 2 Analogy between retardation coefficient–diffusion relationship and Langmuir's adsorption isotherm.

IX. EFFECT OF PENETRANT SIZE ON DIFFUSION

Diffusion is related to size (42) by

$$D = D_o(MW)^b$$

where D_o refers to diffusion of an infinitely small molecule. Scheuplein and Blank (37) and Roberts (24) used values of b of -0.5 and -0.33, but this assumes the SC is an isotropic fluid medium that does not interact, apart from by physical obstruction, with the diffusant. In fact, the SC is an anisotropic, liquid crystalline structure, and the evidence already described suggests powerful interaction via H-bonding. Diffusion should then be more accurately written as

$$D = D_0 \cdot (binding)^a \cdot (MW)^b$$

If RC is used as a measure of the binding term, then

$$\log(D/h) = 1.62 - 2.6 \log(RC) - 2.2 \log(MW)$$
$$n = 53 \qquad r^2 = .87$$

and the higher size dependency ($b = -2.2$) is consistent with nonfluidity and/or anisotropy.

Regression of the standardized data,

$$\log(D/h) = -3.81 - 0.647 \log(RC)^* - 0.633 \log(MW)^*$$

confirms the equal importance of permeant binding to the SC and molecular size in determining the diffusion process.

Eigenvectors			
Variable	PC1	PC2	PC3
$\log(D/h)$	0.587	−0.166	0.792
RC	−0.576	0.601	0.554
MW	−0.568	−0.782	0.257
Eigenvalue	2.8172	0.1451	0.0377
Proportion explained	0.939	0.048	0.013
Cumulative proportion	0.939	0.987	1.000

Therefore 93.9% of the variation relating diffusion, overall H-bonding measured as RC, and size can be accounted for by a single mechanism. In this mechanism (PC1) the equality of the eigenvectors (0.587, −0.576, −0.568) indicates equal importance of H-bonding and size, and there are negative relationships between these factors and diffusion as expected.

X. SUMMARY

The permeability coefficient, k_p, quantifying the flow of a permeant across the stratum corneum barrier is the product of two terms: $K_{sc/vehicle}$ (transfer from vehicle into the outermost layer), and D/h (diffusion across the SC). The general opinion is that diffusion occurs through the intercellular lipids, with the corneocytes acting as a staggered mechanical barrier giving a high value to the pathlength, h. Both steps are determined by the affinity between the permeant and the SC. The partitioning step from aqueous vehicles can be quantified by $K_{octanol/water}$. The lipid lamellae in the SC form a liquid crystalline, anisotropic barrier and H-bond to functional groups on the permeant. The effects that these groups have on diffusion can be quantified as characteristic retardation coefficients. Diffusion is reduced dramatically if multiple groups are present, with the effect being modeled by an equation analogous to Langmuir's adsorption isotherm. The H-bond acceptor potential (β) of a group has a greater effect on diffusion than its α potential, implying that SC is overall an H-bond donor barrier. Regression of diffusion against standardized H-bonding and size data suggests that in practice both H-bonding interaction and size are equally important in retarding diffusion.

XI. GLOSSARY

A	area (cm^2)
C	concentration in receptor cell at time t
C_m	maximal concentration in receptor cell
C_{sc}	concentration in outermost layer of the stratum corneum
C_v	concentration in vehicle
D	diffusion coefficient (cm^2/h)
D_m	minimum diffusion coefficient attainable by powerfully H-bonding molecule
D_o	diffusion coefficient of infinitely small, non-H-bonding molecule
h	pathlength of diffusion (cm)
J_s	flux $(mol/cm^2/h)$ at the steady state
K	rate of desorption/rate of adsorption at an interface
$K_{a/b}$	partition coefficient in phases a, b
k_p	permeability coefficient (cm/h)
PC	principal component
PCA	principal component analysis
r^2	coefficient of determination adjusted for degrees of freedom
RC_x	retardation coefficient of H-bonding group x
SC	stratum corneum
V	intrinsic molar volume (dm^3/mol)
α	scaled H-bonding donor (acid) potential
β	scaled H-bonding acceptor (base) potential
δ	Hildebrand solubility parameter
π^*	dipole moment/polarizability

REFERENCES

1. Albery, W. J., and Hadgraft, J. Percutaneous absorption: Theoretical description. *J. Pharm. Pharmacol. 31*:129–139 (1979).
2. Albery, W. J., and Hadgraft, J. Percutaneous absorption: In vivo experiments. *J. Pharm. Pharmacol. 31*:140–147 (1978).
3. Bouwstra, J. A., De Vries, M. A., Gooris, G. S., Bras, W., Brussee, J., and Ponec, M. Thermodynamic and structural aspects of the skin barrier. *J. Control. Release 15*:209–219 (1991).

4. Schaefer, H., Watts, J., and Illel, B. Follicular penetration. In *Prediction of Percutaneous Penetration: Methods, Measurements, and Modeling.* Scott, R. C., Guy, R. H. and Hadgraft, J. (eds.). IBC Technical Services, London, pp. 163–173 (1990).

5. Lauer, A., Lieb, L. M., Ramachandran, C., Flynn, G. L., and Weiner, N. D. Transfollicular drug delivery. *Pharm. Res. 12*:179–186 (1995).

6. Scheuplein, R. J., Blank, I. H., Brauner, G. J., and MacFarlane, D. J. Percutaneous absorption of steroids. *J. Invest. Dermatol 52*:63–70 (1969).

7. Scheuplein, R. J. Mechanism of percutaneous adsorption. II. Transient diffusion and the relative importance of various routes of skin penetration. *J. Invest. Dermatol. 48*:79–88 (1967).

8. Siddiqui, O., Roberts, M. S., and Polack, A. E. Percutaneous absorption of steroids: Relative contributions of epidermal penetration and dermal clearance. *J. Pharmacokinetics Biopharmaceutics. 17*:405–424 (1989).

9. Heisig, M., Lieckfeldt, R., Wittum, G., Mazurkevich, G., and Lee, G. Non steady-state descriptions of drug permeation through stratum corneum. I. The biphasic brick-and-mortar model. *Pharm. Res. 13*:421–426 (1996).

10. Scheuplein, R., and Ross, L. *J. Soc. Cosmetic Chem. 21*:853–873 (1970).

11. Anderson, B. D., Higuchi, W. I., and Raykar, P. V. Heterogeneity effects on permeability–partition coefficient relationships in human stratum corneum. *Pharm. Res. 5*:566–573 (1988).

12. Edwards, D. A., and Langer, R. A linear theory of transdermal transport phenomena. *J. Pharm. Sci. 83*:1315–1334 (1994).

13. Roberts, M. S., Pugh, W. J., Hadgraft, J., and Watkinson, A. C. Epidermal permeability–penetrant structure relationships: 1. An analysis of methods of predicting penetration of monofunctional solutes from aqueous solutions. *Int. J. Pharm. 126*:219–233 (1995).

14. Wertz, P. W., Miethke, M. C., Long, S. A., Strauss, J. S., and Downing, D. T. The composition of the ceramides from human stratum corneum and from comedones. *J. Invest. Dermatol. 84*:410–412 (1985).

15. Lieckfeldt, R., Villalain, J., Gomez Fernandez, J. C., and Lee, G. Diffusivity and structural polymorphism in some model stratum corneum lipid systems. *Biochim. Biophys. Acta Biomembranes 1150*:182–188 (1993).

16. Pugh, W. J., Roberts, M. S. R., and Hadgraft, J. Epidermal permeability–penetrant structure relationships: 3. The effect of hydrogen bonding interactions and molecular size on diffusion across the stratum corneum. *Int. J. Pharm. 138*:149–167 (1996).

17. Roberts, M. S., Anderson, R. A., and Swarbrick, J. Permeability of human epidermis to phenolic compounds. *J. Pharm. Pharmacol. 29*:677–683 (1977).

18. El Tayar, N., Tsai, R.-S., Testa, B., Carrupt, P.-A., Hansch, C., and Leo, A. Percutaneous penetration of drugs: A quantitative structure–permeability relationship study. *J. Pharm. Sci. 80*:744–749 (1991).

19. Kasting, G. B., Smith, R. L., and Cooper, E. R. Effect of lipid solubility and molecular size on percutaneous absorption. *Pharmacol. Skin 1*:138–153 (1987).

20. Potts, R. O., and Guy, R. H. Predicting skin permeability. *Pharm. Res. 9*: 663–669 (1992).

21. Abraham, M. H., Chadha, H. S., and Mitchell, R. C. The factors that influence skin penetration of solutes. *J. Pharm. Pharmacol. 47*:8–16 (1995).

22. Armstrong, N. A., and James, K. C. *Pharmaceutical experimental design and interpretation in pharmaceutics.* Taylor & Francis, London (1996).

23. Minitab Release 10Xtra, Minitab Inc. Reading, MA (1995).

24. Roberts, M. S. Percutaneous Absorption of Phenolic Compounds. PhD thesis, University of Sydney (1976).

25. Anderson, B. D., and Raykar, P. V. Solute structure–permeability relationships in human stratum corneum. *J. Invest. Dermatol. 93*:280–286 (1989).

26. Abraham, M. H. Scales of solute hydrogen-bonding: Their construction and application to physicochemical and biochemical processes. *Chem. Soc. Rev. 22*:73–83 (1993).

27. Potts, R. O., and Francoeur, M. L. The influence of stratum corneum morphology on water permeability. *J. Invest. Dermatol. 96*:495–499 (1991).

28. Swartzendruber, D. C., Wertz, P. W., Madison, K. C., and Downing. D. T. Evidence that the corneocyte has a chemically bound lipid envelope. *J. Invest. Dermatol. 88*:709–713 (1987).

29. Rehfeld, S. J., Plachy, W. Z., Hou, S. Y. E., and Elias, P. M. Localization of lipid microdomains and thermal phenomena in murine stratum-corneum and isolated membrane complexes—An electron-spin-resonance study. *J. Invest. Dermatol. 95*:217–223 (1990).

30. Michaels, A. S., Chandrasekaran, S. K., and Shaw, J. E. Drug permeation through human skin. Theory and in vitro experimental measurement. *AIChE J. 21*:985–996 (1975).

31. Cussler, E. L., Hughes, S. E., Ward, W. J., and Aris, R. Barrier membranes. *J. Memb. Sci. 86*:161–174 (1988).

32. Rougier, A., Lotte, C., Corcuff, P., and Maibach, H. I. Relationship between skin permeability and corneocyte size according to anatomic site, age and sex in man. *J. Soc. Cosmet. Chem. 39*:15–26 (1988).

33. Hadgraft, J., and Ridout, G. Development of model membranes for percutaneous absorption measurements. I. Isopropyl myristate. *Int. J. Pharm. 39*: 149–156 (1987).

34. Elias, P. M., and Friend, D. S. The permeability barrier in mammalian epidermis. *J. Cell Biol. 65*:180–191 (1975).

35. Williams, M. L., and Elias, P. M. The extracellular matrix of stratum corneum: Role of lipids in normal and pathological function. *CRC Crit. Rev. Ther. Drug Carrier Syst. 3*:95–112 (1987).

36. Wertz, P. W., Swartzendruber, D. C., Abraham, W., Madison, K., and Downing, D. T. Essential fatty acids and epidermal integrity. *Arch. Dermatol. 123*: 1381–1384 (1987).

37. Scheuplein, R. J., and Blank, I. H. Permeability of the skin. *Physiol. Rev. 51*: 702–747 (1971).

38. Smith, W. P., Christensen, M. S., Nacht, S., and Gans, E. H. Effect of lipids on the aggregation and permeability of human stratum corneum. *J. Invest. Dermatol. 78*:7–11 (1982).

39. Kock, W. R., Berner, B., Burns, J. L., and Bissett, D. L. Preparation and characterisation of a reconstituted stratum corneum membrane film as a model membrane for skin transport. *Arch. Dermatol. Res. 280*:252–256 (1988).

40. Friberg, S. E., Kayali, I., Beckerman, W., Rhein, D. L., and Simion, A. Water permeation of reaggregated stratum corneum with model lipids. *J. Invest. Dermatol. 94*:377–380 (1990).

41. Scheuplein, R. J., Blank, I. H., Brauner, G. J., and MacFarlane, D. J. Percutaneous absorption of steroids. *J. Invest. Dermatol. 52*:63–70 (1969).

42. Crank, J. *The Mathematics of Diffusion.* Clarendon Press, Oxford, chs. 1, 2, 4 (1975).

43. Roberts, M. S., Pugh, W. J., and Hadgraft, J. Epidermal permeability–penetrant structure relationships: 2. The effect of H-bonding groups in penetrants on their diffusion through the stratum corneum. *Int. J. Pharm. 132*:23–32 (1996).

44. Wertz, P. W. Epidermal lipids. *Semin. Dermatol. 11*:106–113 (1992).

10

In Vivo Percutaneous Absorption
A Key Role for Stratum Corneum/Vehicle Partitioning

André Rougier
Laboratoire Pharmaceutique, La Roche-Posay, Courbevoie, France

I. INTRODUCTION

Over the past two decades considerable attention has been paid to understanding the mechanisms and routes by which chemical compounds penetrate the skin. Irrespective of the different theories on mechanisms relating to percutaneous absorption, it is well established that the stratum corneum (SC) constitutes the main barrier (14,16,29,32). Thus, it is to be expected that the overall kinetic process will depend mainly on the pharmacokinetic parameters governing the penetration of compounds through this membrane.

The interaction between the drug, the vehicle, and the SC as a consequence of their physicochemical properties is likely to be an important pharmacokinetic parameter in an early step of the absorption process. In studies in rats (20) and humans (9), we have demonstrated a correlation between the amount of the test substance found in the SC at the end of a 30-min application and the total amount absorbed over 4 days. The predictive aspect of the so-called "stripping method" was found to take into account the main factors influencing percutaneous absorption (10,21,22,23).

It has been suggested that the amount of chemical absorbed within the SC after 30 min of application could reflect its SC vehicle partitioning, and also its rate of entry into the skin (10). Previous studies in hairless rats (21) showed clearly that the amount of various compounds that penetrated in vivo was strictly proportional to the time of application, thus providing indirect evidence that a constant flux of penetration really does exist in vivo.

In the light of these results, the present study was carried out to determine whether the stripping method could also be used to predict the in vivo steady-state flux of a test compound.

II. MATERIALS AND METHODS

Five radiolabeled compounds (Radiochemical Centre, Amersham, UK) with very different physicochemical properties and belonging to different chemical classes were compared: benzoic acid, caffeine, thiourea, hydrocortisone, and inulin (Table 1).

A group of six 12-week-old hairless Sprague-Dawley female rats (IFFA-CREDO, Lyon, France) weighing 200 ± 20 g was used for each compound and each application time.

A. Percutaneous Absorption Measurements

Each compound (1000 nmol) was applied to a 1-cm^2 area of the back of anesthetized animals (ip injection of gammabutyrolactone, 0.5 ml/kg in 20 μl ethylene glycol/Triton W100 mixtures (Table 1). The treated area was delimited by an open circular cell attached with silicone glue in order to prevent spreading. The application times were 1, 2, 3, 4, and 5 h.

At the end of the application time, the treated area was washed twice (2×300 μl) with ethanol/water (95/5), rinsed twice with water, and dried with cotton in order to remove excess product. We considered that the compounds had effectively penetrated when they had crossed the SC and reached the viable epidermis. The SC was then stripped 10 times (3M adhesive tape 810, USA) in order to exclude any compound which had not penetrated during the time of application (previous histological studies in hairless rat have shown that this procedure almost totally removes the SC). The remaining skin (epidermis plus dermis) was sampled and counted by liquid scintillation (Packard 360, Packard Instruments. Downers Grove, IL) after digestion in Soluene 350 (United Technology, Packard).

The carcasses were then lyophilized, homogenized, and samples counted by liquid scintillation after combustion (Oxidizer 306, Packard Instruments). In the case of tritiated molecules (caffeine, hydrocortisone, inulin), the water resulting from the lyophilization of the carcasses was sampled and assayed for radioactivity. The radioactivity found was added to that detected in the carcasses, thus obtaining the overall percutaneous absorption values. The urine excreted during the time of application was collected to take into account any product contained therein in the overall absorption. The total amount of each compound penetrating at each application time was determined by summation of the amount found in the epidermis, dermis, and the carcass.

Table 1 Application Conditions

Compound	Specific activity	Purity (%)	Molecular weight (Da)	Vehicle
Benzoic acid [7–^{14}C]	21.3 mCi/mmol	>98	122	Ethylene glycol/Triton X100 (90/10)
Caffeine [8–^{3}H]	24 Ci/mmol	>98	194	Ethylene glycol/Triton X100/water (45/5/50)
Thiourea [^{14}C]	58 mCi/mmol	>98	76	Ethylene glycol/Triton X100 (90/10)
Hydrocortisone (1.2.6.7) [^{3}H]	98.5 Ci/mmol	>99	362	Ethylene glycol/Triton X100/isopropanol (72/8/20)
Inulin [^{3}H]	3.2 Ci/mmol	>98	5200	Ethylene glycol/Triton X100 (90/10)

Note. Dose applied for all compounds: 1000 nmol/cm^2 in 20 µl vehicle.

B. Stratum Corneum/Vehicle Partitioning Measurement

Six strippings (3M adhesive tape 810) were performed on the treated area of each group of six animals for each compound after a fixed application time of 30 min (the reasons for this choice of time are discussed later). The amount of product within the SC was assessed, after complete digestion of the keratin material in Soluene 350, by liquid scintillation counting.

C. Measurement of the Thickness of the Stratum Corneum

The thickness of the SC was measured in biopsies from the backs of six rats according to the technique described by McKenzie (17). Each biopsy was placed on a strip of acetyl cellulose coated with Tekt issue (Miles Scientific, Naperville, IL), then frozen in dry ice. Transverse sections (8 μm) were then cut with the acid of a cryomicrotome (HRLM Slee. London). The sections obtained were fixed for 10 min in a 70% alcohol bath, then stained for 30 s with a 0.5% aqueous solution of methylene blue. After rinsing with distilled water the sections were mounted between a slide and a coverslip with the aid of aquamount (hydrophilic mounting medium). The thickness of the horny layer was measured at 20 different points on each section by means of a semiautomatic image analyzer (Digipet, Reichert, Wien, Austria) connected to a microscope (Polyvar, Reichert, Austria) and hence a mean thickness could be calculated.

D. Effect of Vehicles Used on the Stratum Corneum Integrity

After anesthesia, 20 μl of each of the vehicles used was applied to a 1-cm^2 area on the backs of groups of five rats. The area was delimited by an open circular cell as described previously. After 5 h of contact, the treated area was washed twice (2 × 300 μl) with water and dried with cotton in order to remove excess vehicle. One hour after completion of the vehicle treatment, transepidermal water loss (TEWL) was measured with an Evaporimeter EP1 (ServoMed, Stockholm, Sweden). The hand-held probe was fitted with a 1-cm tail chimney, to reduce air turbulence around the hydrosensors, and a metallic shield (supplied by ServoMed), to minimize the possibility of sensor contamination. Measurements (G/m^2/h) stabilized within 1 min. The effects of total destruction of the SC have also been studied by measuring TEWL from the backs of a group of five animals, 5 h after a series of 10 strippings (3M adhesive tape 810).

E. Theory and Data Treatment

The mass transfer of compounds from the surface of the skin to the interior of the body through the SC is generally considered to be due to passive diffusion. A classical but oversimplified description of the transport process is represented in Fig. 1. The SC is assumed to be a homogeneous membrane (thickness h); D is the diffusion coefficient of the solute through the membrane. The concentration of solute (C) within the outermost layer of the membrane ($x = 0$) depends on the concentration within the vehicle (c_o) and the partition coefficient (K) between the membrane and the vehicle:

$$C = CK_o \tag{1}$$

For all values of time (t) the concentration of solute within the innermost layer of the membrane ($x = h$) is assumed to be negligible (sink condition). The validity of such an assumption is discussed in the Results section. The change in the cumulative amount of solute (Q) that passes through the membrane per unit area as a function of time is represented in Fig. 2. When a steady state is reached, the curve $Q(t)$ is linear and can be described by the equation:

$$Q = J_s(t - L) \tag{2}$$

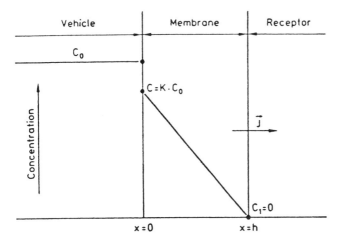

Figure 1 Concentration profile across homogeneous membrane at steady state (zero-order flux case).

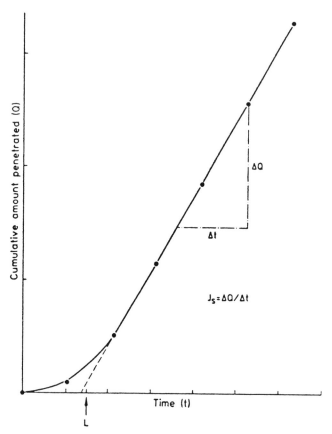

Figure 2 Typical profile of concentration versus time for diffusion through the stratum corneum.

where J_s corresponds to the steady-state flux (the slope of the straight line):

$$J_s = KC_o d/h \tag{3}$$

and L is the lag time (the intercept of the straight line on the time axis):

$$L = \frac{h^2}{6D} \tag{4}$$

In practice, J_s and L were calculated for each compound using a linear regression obtained with the aid of a computer (Vax 11/750, Digital Corporation, Bedford, MA) and standard software (RS/Explore BBN Software Product Corporation, Bedford, MA).

III. RESULTS

Figure 3 shows that, irrespective of the nature of the compound tested, the plot of the cumulative amount of solute that passes through a unit surface area of the SC as a function of the application time appears to be linear ($r = .99$, $p < .001$). As shown in Fig. 3 and Table 2, the steady-state values (J_s) are strongly molecule dependent. Thus, the values for inulin and benzoic acid differ by a factor of 40. The rank order of the J_s values is: inulin < hydrocortisone < thiourea < caffeine < benzoic acid.

In the present case, a constant flux of penetration ought to be attained within a contact time of about 30 min (11×2.7). According to Zatz (34), the attainment of a constant flux would be expected to coincide with the delivery of a constant amount at the SC. In order to test this hypothesis, we measured the amount present in the horny layer after an application time of 30 min (it should be recalled that the time of 30 min corresponds to that used in the stripping method). The results obtained for the five molecules are shown in Table 3. The total amounts of solute accumulated in the first six strippings rank as follows: inulin < hydrocortisone < thiourea < caffeine < benzoic acid.

IV. DISCUSSION

Our results in vivo, like those in vitro, show that the phenomenon of transport across the SC can be considered as a process obeying the general laws governing passive membrane diffusion. Thus, after a time to attain equilibrium, a constant flux of penetration is established. From a theoretical point of view, this can occur only if the solute distribution within the membrane remains constant. This implies that the solute concentration in the outermost layer of the membrane has to remain constant throughout the entire experiment (infinite dose condition), and that the solute concentration in the innermost layer of the membrane has to remain constant and be negligible (sink condition).

As shown in Fig. 4, the amount of the vehicle applied (20 μl/cm^2) changes only during the first hour of administration. Then it remains constant throughout the time of percutaneous absorption measurements (1–5 h). In the case of the most penetrating compound (benzoic acid), the amount that penetrated after 5 h of application (78 nmol) was far below the amount applied (1000 nmol). It can therefore be assumed that the solute concentration in the vehicle remained relatively constant between 1 and 5 h. Experimentally, we can consider that the first condition is met.

It can be assumed that the epidermis and the uppermost part of the papillary layer of the dermis constitute a negligible barrier in comparison

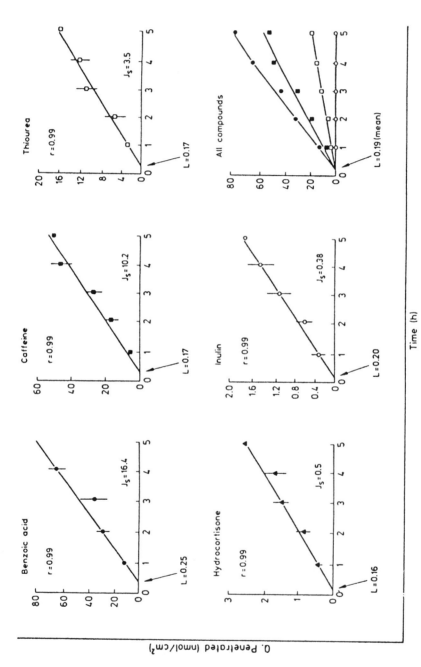

Figure 3 Cumulative amount of solute penetrating through the stratum corneum as a function of application time. (J_s = steady-state flux (nmol/cm^2/h); L = lag time (h).

Table 2 Amount of Chemical Penetrating Through the Stratum Corneum (nmol/cm^2) Measured at the End of the Application Time, and the Steady-State Parameters

Compound	Application time (h)					Steady-state flux (J) (nmol/cm^2/h)	Lag time (h)
	1	2	3	4	5		
Benzoic acid	12.5 (SE 3)	30 (SE 2)	39 (SE 10)	64 (SE 6)	78 (SE 7)	16.4 (SE 0.8)	0.25
Caffeine	7.4 (SE 0.8)	18.6 (SE 2.9)	29.5 (SE 2)	45 (SE 5)	52 (SE 8.5)	10.2 (SE 0.6)	0.17
Thiourea	3 (SE 0.7)	6 (SE 2)	10.7 (SE 2)	13.4 (SE 1.8)	17 (SE 2.2)	3.5 (SE 0.2)	0.17
Hydrocortisone	0.49 (SE 0.05)	0.9 (SE 0.1)	1.5 (SE 0.1)	1.8 (SE 0.3)	2.6 (SE 0.4)	0.5 (SE 0.04)	0.16
Inulin	0.32 (SE 0.04)	0.63 (SE 0.04)	1.1 (SE 0.2)	1.5 (SE 0.2)	1.8 (SE 0.3)	0.38 (SE 0.02)	0.20
						Mean	0.19

Table 3 Percutaneous Absorption Parameters of the Tested Compounds

Compound	Log P octanol/water[a]	K[b]	Q_{sc}[c] calculated (nmol/cm^2)	Q_{sc} measured (nmol/cm^2)	Steady-state flux (J_s) (nmol/cm^2/h)	
					Predicted from Eq. (8)	Measured
Benzoic acid	1.87	0.30	8.77	9.07 (SE 0.66[d])	15.87 (SE 1.15)	16.4 (SE 0.80)
Caffeine	−0.07	0.14	6.46	5.92 (SE 0.46)	10.36 (SE 0.80)	10.2 (SE 0.60)
Thiourea	−1.02	0.066	3.86	3.34 (SE 0.2)	5.85 (SE 0.35)	3.5 (SE 0.20)
Hydrocortisone	1.61	0.077	0.52	2.36 (SE 0.09)	4.1 (SE 0.16)	0.5 (SE 0.04)
Inulin	−3.58	0.078	0.46	0.85 (SE 0.12)	1.49 (SE 0.20)	0.38 (SE 0.02)

[a]From reference 12.
[b]Partition coefficient calculated from Eq. (5) ($h = 13$ µm; $L = 11$ min; C_0 see Note).
[c]Q_{sc} calculated from Eq. (6) ($h = 13$ µm, K = calculated from Eq. (5); C_0 see Note).
[d]Standard error.

Note. C_0 = Solute concentration within the vehicles; (taking into account vehicle evaporation (Fig. 4): benzoic acid, thiourea, inulin = 4.5×10^4 nmol/cm^3; hydrocortisone = 5.2×10^4 nmol/cm^3; caffeine = 7.1×10^4 nmol/cm^3.

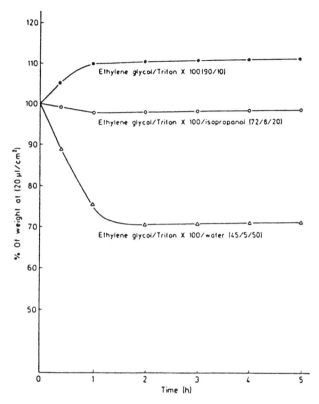

Figure 4 Modification of the vehicles during their administration (20 μl/cm², room temperature 27°C).

with the SC (24), and the microvascularization of the dermal papillae prevents solute accumulation in the region of the capillaries. Thus, the solute concentration in the innermost layer of the SC can be considered to be negligible in comparison with the concentration in the outermost layer. Hence, the sink condition is apparently fulfilled.

Although the existence of a steady-state flux of penetration in vivo was predicted about 20 years ago by Tregear (31) and subsequently by others (2,21,33), the problem is still the subject of debate (11,30). Our results clearly demonstrate (Fig. 3) that a constant flux can be achieved in vivo just as in vitro experiments. Although this seems to be quite logical in our view, this is the first time that it has been demonstrated experimentally. Our results thus fill a gap in the understanding of the mechanisms governing in vivo percutaneous absorption. It should, however, be emphasized that the existence of such a gap is in no way due to negligence on the part of investigators in the field, but rather to the technical difficulties of measuring a steady-state flux of penetration in vivo.

Our results show (Fig. 3 and Table 2) that lag times for the different molecules tested are very close and extremely short. One explanation is that the vehicles used alter the SC and therefore modify the barrier to penetration. However, Table 4 clearly shows that TEWL is not affected by vehicle treatment, whereas removing the SC by 10 successive strippings increases TEWL by a factor of 18.

It is therefore possible to consider that, until the contrary is demonstrated experimentally, such a situation may exist in vivo, even if it upsets some theories that have been built upon in vitro studies. It is important to emphasize that such observations have rarely been made in vitro, perhaps because sink conditions are not necessarily met in vitro. In a dynamic in vitro system, a concentration close to zero in the medium bathing the tissue does not necessarily imply that the concentration in the innermost layer of the membrane is constant with time (24). It is not unusual to observe an increase in the concentration of solute with time in the innermost layers of the epidermis or dermis, depending on the in vitro model adopter (2,11,13,24,30,31,33,35). This increase may be linker either to incomplete resorption by the bathing fluid (6–8,28) or to a possible affinity of the molecule for these structures (18).

On the basis of knowledge concerning the thickness of the SC ($h = 13 + 2$ μm) and the value for the mean lag time ($L = 11$ min), it is possible to deduce a mean value for the apparent diffusion coefficient (D_m) by using Eq. (4): $D_m = 4.3 \times 10^{-10}$ cm^2 s^{-1}. This does not mean that the values of

Table 4 Effect of the Applied Vehicles on the Barrier Function of the Stratum Corneum (Transepidermal Water Loss)

Ethylene glycol/ Triton X100 (90/10) (g/m^2/h)	Ethylene glycol/ Triton X100/ water (45/5/50) (g/m^2/h)	Ethylene glycol/ Triton X100/ isopropanol (72/8/20) (g/m^2/h)	Stratum corneum removed (10 strippings) (g/m^2/h)
5.8 (SE 0.3) Controls:	5.7 (0.4)	5.6 (0.3)	91 (3.0)
5.1 (SE 0.3)	5.3 (0.4)	5.1 (0.3)	5.1 (0.3)

the diffusion coefficients for molecules having physicochemical properties as different as those used in this study are identical. It means that it is impossible to control, with the required degree of precision, all the physical, physicochemical, and biological parameters likely to affect diffusion through a membrane as complex as the SC. From a purely practical point of view, it is thus possible, as a first approximation, to consider different molecules as having the same apparent diffusion coefficient in the case of percutaneous absorption in vivo. On the other hand, it is reasonable to ask whether this coefficient may vary as a function of parameters such as animal species, anatomical site, age, etc.

It follows from Eqs. (3) and (4) that the flux at equilibrium can be written in the form:

$$J_s = \frac{1}{6} K c_o \frac{h}{L} \tag{5}$$

As we have shown earlier, the values of the lag times for the five molecules are similar. This results in the apparent "velocity of diffusion," defined by the ratio h/L, being independent of the nature of the diffusion substance for a given thickness of the horny layer and a given anatomical site. Only the number of molecules in transit ($K c_o$) would be characteristic for a given substance, and would determine the value of its flux at equilibrium. Since, for a given compound, the value of c_o may be considered to be constant within the time of percutaneous absorption measurements (1–5 h), the value of this flux would depend only on the SC/vehicle partition coefficient (K).

Using Eq. (5) and the values of flux (J_s) determined experimentally (Table 3) we have calculated the values of K for each of the five molecules (Table 3) taking into account in the C_o values the evaporation of the vehicles (Fig. 4). The values for the octanol/water partition coefficients (log P) reported in the literature for these five molecules (12) are also shown in Table 3. It appears that no relationship exists between these values and the values for flux at equilibrium. Although many examples appear to support the use of log P for predicting the degree of penetration of a molecule (4,26), there are many others that show the limitations of such a procedure (1,5,19,27). The partition coefficient of a given compound between two solvents can be considered as a constant physical property of that compound. It is now generally accepted that the percutaneous absorption of a compound can vary considerably as a function of the conditions of administration (vehicle, dose, anatomical site, animal species, etc.). This raises the question: How is it possible to predict the value of a variable parameter only from a constant? Thus, in agreement with Scheuplein (25), we consider that, at present, no solvent system is capable of simulating the extreme complexity of the SC.

Only the measurement of the partition coefficient between the SC and the vehicle can be representative of reality.

The amount of substance present in the SC at equilibrium (Q_{sc}) can be measured. According to the model adopted, this quantity is related to the partition coefficient by the equation:

$$Q_{sc} = \frac{1}{2} Kc_o \tag{6}$$

As shown in Table 3 and Fig. 5, there exists a very good agreement between the values of Q_{sc} measured by stripping the treated area after 30 min and the values of Q_{sc} calculated from Eq. (6) (c_o values take into account vehicle evaporation).

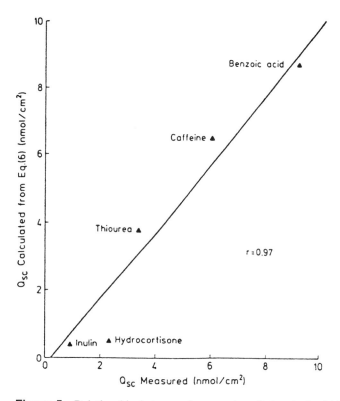

Figure 5 Relationship between the quantity of chemical within the stratum corneum measured after 30 min of contact and predicted using Eq. (6).

In the light of Eqs. (5) and (6), the flux at equilibrium can be written:

$$J_s = \frac{Q_{sc}}{3L} \tag{7}$$

Since the lag times of the molecules under study are similar, the fluxes at equilibrium would be expected to depend only on the amount present in the SC.

According to the theoretical model adopted, using Eq. (7) and a mean lag time of 0.19 h, the theoretical relationship between J_s and Q_{sc} should be:

$$J_s = 1.75 Q_{sc} \tag{8}$$

(J_s being expressed in nmol/cm^2/h). As shown in Fig. 6, the curve derived from Eq. (8) is contained within the 5% confidence limits of the experimental values.

In view of the approximations made in the theoretical model and the inevitable errors arising from the inaccuracies of the measurements and biological variation, we can consider that there exists a very satisfactory agreement between experimental values and theory as in Eq. (8). Only hydrocortisone does not appear to fit well with the theoretical linear relationship linking steady-state flux of penetration (J_s) and amount in the SC (Q_{sc}). This is not really surprising, since steroids are known to form a depot or reservoir within the SC (3,15). A fraction of the available molecules may bind to the keratin or other tissue components, while the remainder diffuses slowly downward.

Six years after the development of the stripping method (9,10,20–23), the results obtained provide a better understanding of why it is possible to predict the total penetration during 4 days of a substance administered for 30 min with satisfactory precision. As shown in Table 3, from a purely practical point of view, the flux of penetration at equilibrium of a substance administered in vivo in a given vehicle can be predicted using Eq. (8) from the simple measurement of the amount present in the SC (Q_{sc}) after a contact time of 30 min. Since the validity of the stripping method has been verified for many molecules administered under different conditions in different species, it is reasonable to think that it would also hold for the predictive assessment of the in vivo steady-state flux of penetration.

Using an original experimental approach we have obtained data leading to a better understanding of the mechanisms implicated in molecular transport across the SC in vivo. Thus, it appears that the SC/vehicle partitioning plays a determining role in the percutaneous absorption of chemicals in vivo.

We can easily conceive that our results, especially those related to lag times and diffusion coefficients, may not be readily accepted. The strength

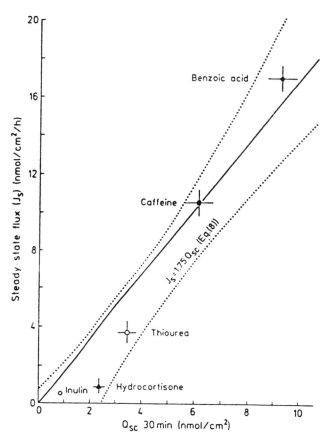

Figure 6 In vivo relationship between steady-state flux of penetration (J_s) and quantity of solute within the stratum corneum after 30 min of contact.

of the data presented lies in the fact that they are experimental. To reason only in terms of in vitro data would be to admit from the outset that there are no differences between the in vitro and in vivo processes of percutaneous absorption. However, considering the theoretical importance of these results, it would be important to see them verified using other chemicals of widely different physicochemical properties. It would also be interesting to ascertain that the theory we have developed concerning the in vivo mechanism of percutaneous absorption is verified when the same chemical is dissolved in different vehicles.

ACKNOWLEDGMENTS

The author thanks Dr. C. Berrebi for her expertise in the histometric measurements, and A. M. Cabaillot, C. Patouillet, and M. Zanini for their excellent technical assistance.

REFERENCES

1. Bronaugh RL (1985). Determination of percutaneous absorption by in vivo techniques. In: Bronaugh RL, Ackerman C, Flynn GL (1987). Ether-water partitioning and permeability through nude mouse skin in vitro. I. Urea, thiourea, glycerol and glucose. Int J Pharmacol 36:61–66.

2. Arita T, Hori R, Anmo T, Washitake M, Akatsu M, Yasima T (1970). Studies on percutaneous absorption of drugs. Chem Pharm Bull 18:1045–1049.

3. Barry BW (1983). Dermatological formulations percutaneous absorption. In: Swarbrick J (ed). Drug and pharmaceutical sciences, vol 18. Marcel Dekker, New York, pp. 49–54.

4. Blank HI, Scheuplein RJ (1969). Transport into within the skin. Br J Dermatol 81(Suppl 4):4–10.

5. Blank HI, Scheuplein RJ, McFarlane DJ (1967). Mechanism of percutaneous absorption. II. The effect of temperature on the transport of non-electrolytes across the skin. J Invest Dermatol 49:582–589.

6. Bronaugh RL (1985). Determination of percutaneous absorption by in vivo techniques. In: Bronaugh RL, Maibach HI (eds). Percutaneous absorption. Marcel Dekker, New York, pp. 267–279.

7. Bronaugh RL, Stewart RF (1984). Methods for in vitro percutaneous absorption studies. III. Hydrophobic compounds. J Pharm Sci 73:1255–1258.

8. Bronaugh RL, Stewart RF (1986). Methods for in vitro percutaneous absorption studies. VI. Preparation the barrier layer. J Pharm Sci 75:487–491.

9. Dupuis D, Rougier A, Roguet R, Lotte C, Kalopissis G (1984). In vivo relationship between horny layer reservoir effect and percutaneous absorption in human and rat. J Invest Dermatol 82:353–356.

10. Dupuis D, Rougier A, Roguet R, Lotte C (1986). The measurement of the stratum corneum reservoir: A simple method to predict the influence of vehicles on in vivo percutaneous absorption. Br J Dermatol 115:233–238.

11. Guy R, Hadgraft J (1985). Mathematical models of percutaneous absorption. In: Bronaugh RL, Maibach HI (eds). Percutaneous absorption. Marcel Dekker, New York, pp. 3–15.

12. Hansch C, Leo AJ (1984). Log P data base. Pomona College Medical Chemistry Project, Pomona, CA.

13. Loden M (1986). The in vivo permeability of human skin to benzene, ethylene glycol, formaldehyde and n-hexane. Acta Pharmacol Toxicol 58:382–389.

14. Malkinson FD (1958). Studies on percutaneous absorption of ^{14}C labeled steroids by use of the gas-flow cell. J Invest Dermatol 31:19–28.

15. Malkinson FD, Ferguson EH (1955). Preliminary and short report. Percutaneous absorption of hydrocortisone 4 C in two human subjects. J Invest Dermatol 25:281–283.

16. Marzulli FN (1962). Barrier to skin penetration. J Invest Dermatol 39:387–393.

17. McKenzie IC (1975). A simple method of orientation and storage of specimens for cryomicrotomy. J Periodont Res 10:49–50.

18. Miselnicky SR, Lichtin JL, Sakr A, Bronaugh RL (1988). The influence of solubility, protein binding and percutaneous absorption on reservoir formation in skin. J Soc Cosmet Chem 39:169–177.

19. Poulsen BJ, Flynn GL (1985). In vitro methods used to study dermal delivery and percutaneous absorption. In: Bronaugh RL, Maibach HI (eds). Percutaneous absorption. Marcel Dekker, New York, pp. 431–459.

20. Rougier A, Dupuis D, Lotte C, Roguet R, Schaefer H (1983). Correlation between stratum corneum reservoir function and percutaneous absorption. J Invest Dermatol 81:275–278.

21. Rougier A, Dupuis D, Lotte C (1985). The measurement of the stratum corneum reservoir. A predictive method for in vivo percutaneous absorption studies: Influence of the application time. J Invest Dermatol 84:66–68.

22. Rougier A, Dupuis D, Lotte C, Roguet R, Wester RE, Maibach HI (1986). Regional variation in percutaneous absorption in man: Measurement by the stripping method. Arch Dermatol Res 278:465–469.

23. Rougier A, Lotte C, Maibach HI (1987). In vivo percutaneous absorption of some organic compounds related to anatomic site in man. J Pharmacol Sci 76: 451–454.

24. Schaefer H, Zesch A, Stuttgen G (1982). Skin permeability, Springer, New York, pp. 607–616.

25. Scheuplein RJ (1965). Mechanism of percutaneous absorption. I Routes of penetration and the influence of solubility. J Invest Dermatol 45:334–346.

26. Scheuplein RJ, Blank HI (1973). Mechanism of percutaneous absorption. IV. Penetration of non-electrolytes (alcohols) from aqueous solution and from pure liquids. J Invest Dermatol 60:286–296.

27. Scheuplein RJ, Blank HI, Branner GJ, McFarlane DJ (1969). Percutaneous absorption of steroids. J Invest Dermatol 52:63–70.

28. Scott RC (1987). Percutaneous absorption in vivo-in vitro comparisons. In: Shroot B, Schaefer H (eds). Pharmacology of the skin. Karger, Basel, pp. 103–110.

29. Stoughton RB (1965). Penetration absorption. Toxicol Appl Pharmacol 7(Suppl 2):1–6.

30. Tojo K, Lee AE-RI (1989). A method for predicting steady-state rate of skin penetration in vivo. J Invest Dermatol 92:105–108.

31. Tregear RT (1966). The permeability of mammalian skin to ions. J Invest Dermatol 46:16–22.

32. Vinson LJ, Singer EJ, Koehler WR, Lehmann MD, Masurat T (1965). The nature of the epidermal barrier and some factors influencing skin permeability. Toxicol Appl Pharmacol 7:7–19.

33.   Wepierre J, Corroler M, Didry JR (1986). Distribution and dissociation of benzoyl peroxide in cutaneous tissues after application on skin in the hairless rat. Int J Cosmet Sci 8:97–104.

34.   Zatz JL (1985). Influence of depletion on percutaneous absorption characteristics. J Soc Cosmet Chem 36:237–249.

35.   Zesch A, Schaefer H (1975). Penetration of radioactive hydrocortisone in human skin for various ointment bases. II. In vivo experiments. Arch Dermatol Forsch 252:245–256.

II
METHODOLOGY

11

In Vivo Methods for Percutaneous Absorption Measurements

Ronald C. Wester and Howard I. Maibach
University of California School of Medicine, San Francisco, California

I. IMPORTANCE OF IN VIVO PERCUTANEOUS ABSORPTION

There is a persistent belief that skin viability has little importance in percutaneous absorption. This concept of skin as a passive membrane has led to the domination of the study of percutaneous absorption by laws of mass action and physical diffusion. This concept has also led investigators to use skin excised from cadavers (human and animal) and then to physically (e.g., by freezing or heat separation) and chemically isolate skin sheets or sections and determine chemical diffusion across these treated tissues. A recent study shows that these methods destroy skin viability (Wester et al., 1998a). Human skin viability currently can be maintained for up to a week under the proper conditions.

A consequence of this earlier concept was the designation of the stratum corneum as the barrier to percutaneous absorption. Many compounds such as low-molecular-weight alcohols were studied, and the barrier properties of the isolated stratum corneum were demonstrated for these chemicals. It has then been assumed that the stratum corneum is the primary barrier for all compounds.

The need to study percutaneous absorption has its reality in dermatotoxicity, by which compounds pose a threat to human health, and in dermatopharmacology, for which drugs need to be delivered into and through the skin to treat disease both locally (skin disease) and systemically (transdermal delivery). Most compounds and defined drugs that are of interest and concern in dermatotoxicology and dermatopharmacology are lipophilic. The stratum corneum, the supposed barrier to percutaneous absorption, is a lipid-saturated tissue that is like a sink to topically applied lipophilic materials; its function more closely approximates that of a sponge capable of absorbing a quantity of material, limited only by the solubility of the chemical in sebaceous and epidermal lipids.

The chemical and physical properties of the topical vehicle and the barrier/sink properties of the living stratum corneum determine the initial absorption of compounds into the skin. The vitality of the living skin will, in part, determine the metabolism, distribution, and excretion of the compounds through the skin and the body (Wester et al., 1998a).

In this chapter, we cannot define and discuss all the factors that determine the nature of in vivo percutaneous absorption. That is the mission of this book, and many of the important aspects of in vivo percutaneous absorption have their own chapters. We discuss only some of the methodology and its limitations, pointing out the steps of in vivo percutaneous absorption as we understand them.

II. IN VIVO METHODS

A. Radioactivity in Excreta

Percutaneous absorption in vivo is usually determined by the indirect method of measuring radioactivity in excreta after topical application of the labeled compound. In human studies plasma levels of compound are extremely low after topical application, often below assay detection level, so it is necessary to use tracer methodology. The compound, usually labeled with ^{14}C or tritium, is applied and the total amount of radioactivity excreted in urine or urine plus feces determined. The amount of radioactivity retained in the body or excreted by some route not assayed (CO_2, sweat) is corrected for by determining the amount of radioactivity excreted after parenteral administration. This final amount of radioactivity is then expressed as the percentage of applied dose that was absorbed (Feldmann and Maibach, 1969, 1970, 1974).

The equation used to determine percutaneous absorption is:

$$\text{Absorption } (\%) = \frac{\text{total radioactivity after topical administration}}{\text{total radioactivity after parenteral administration}} \times 100$$

The limitation on determining percutaneous absorption from urinary or fecal radioactivity, or both, is that the methodology does not account for metabolism by skin. The radioactivity in urine is a mixture of parent compound and metabolites. This type of information is useful in defining the total disposition of the applied topical dose. Because the nature of the radioactivity is undefined, kinetic interpretation should be limited.

B. Radioactivity in Blood

Plasma radioactivity can be measured and the percutaneous absorption determined by the ratio of the areas under the plasma concentration versus time curves (AUC) following topical and intravenous administration (Wester and Noonan, 1978). Note that radioactivity in blood can include both the applied compound and metabolites; thus, the same limitations discussed for excreta also apply here. This method has given results similar to those obtained from urinary excretion (Wester et al., 1983).

    Figure 1 shows plasma radioactivity ipsi- and contralateral to the dosing site (ventral forearm) following [^{14}C]hydroquinone topical dosing in humans. Note that at the first blood draw (0.5 h) a significant amount of radioactivity was in the systemic circulation. Plasma ^{14}C levels peaked at 4 h and continued until 8 h, when the study was stopped. The time course of hydroquinone in humans can also be seen in Figure 2, which shows applied dose recovered by soap-and-water wash and by stratum corneum tape stripping. The visualized picture using these three techniques (blood draw, skin

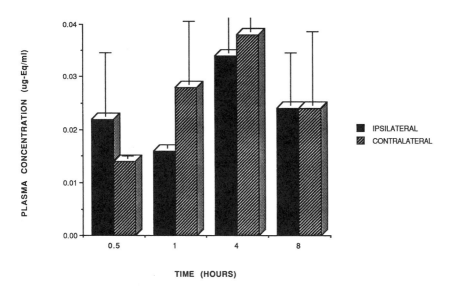

Figure 1 Plasma ^{14}C levels ipsilateral (dosed forearm) and contralateral (opposite arm) following [^{14}C]hydroquinone topical dose in human volunteers. Note that significant plasma levels are achieved at the first 0.5-h blood sampling period, indicating a rapid hydroquinone absorption and systemic circulation.

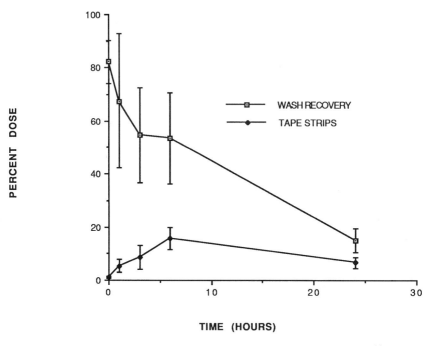

Figure 2 Radioactivity recovered from human forearm dosed with [^{14}C]hydro-
quinone followed by timed skin soap-and-water wash and subsequent skin tape stripping. In conjunction with Figure 1, the visual picture is of hydroquinone leaving the
skin, entering the corneum and the systemic circulation.

wash recovery, skin tape strip) shows [^{14}C]hydroquinone disappearing from
the skin surface, building up in the stratum corneum, and appearing in the
systemic circulation. There is a minimum, if any, lag time and a continuous
absorption through the 24-h dosing period (Wester et al, 1998b).

C. Surface Recovery

This approach to finding in vivo percutaneous absorption is to determine the
loss of material from the surface as it penetrates into the skin. Skin recovery
from an ointment or solution application is difficult because total recovery
of compound from the skin is never assured. With topical application of a
transdermal delivery device, the total unit can be removed from the skin and
the residual amount of drug in the device determined. It is assumed that the
difference between applied dose and residual dose is the amount of drug
absorbed.

D. Surface Disappearance

Related to surface recovery, it is possible to monitor the disappearance of ^{14}C from the surface of skin using appropriate instrumentation. The limitation on this methodology is that the disappearance is due both to movement of ^{14}C-labeled chemical into the skin and to the quenching effect of the skin on the β rays bouncing back to the instrument. This can be simply demonstrated by applying a quantity of radiolabeled chemical on a surface, placing a sheet of stratum corneum over the radiolabeled chemical, and then measuring the radioactivity with an external device. The device will record some radioactivity (β rays penetrating total stratum corneum), yet the radiolabeled chemical is on the other side of the stratum corneum. The degree of quench of chemical in the various cell layers of the skin has not been defined.

E. Biological/Pharmacological Response

Another in vivo method of estimating absorption is to use a biological/pharmacological response (McKenzie and Stoughton, 1962). Biological assay is substituted for a chemical assay, and absorption is estimated. An obvious disadvantage is that biological responses are limited to compounds that elicit responses that can be measured easily and accurately. An example of a biological response would be the vasoconstrictor assay when the blanching effect of one compound is compared with a known compound. This method is more qualitative than quantitative.

Other qualitative methods of estimating in vivo percutaneous absorption include whole-body autoradiography and fluorescence. Whole-body autoradiography provides an overall picture of the dermal absorption followed by the involvement of other body tissues.

F. Stripping Method

The stripping method determines the concentration of chemical in the stratum corneum at the end of a short application period (30 min) and by linear extrapolation predicts the percutaneous absorption of that chemical for longer application periods. The chemical is applied to skin of animals or humans and after the 30-min skin application time the stratum corneum is removed by successive tape application and removal. The tape strippings are assayed for chemical content. Rougier, Lotte, and coworkers have established a linear relationship between this stratum corneum reservoir content and percutaneous absorption using the standard urinary excretion method. The major advantages of this method are a) the elimination of urinary (and fecal) excretion to determine absorption, and b) the applicability to nonradiolabeled determination of percutaneous absorption because the skin strippings contain adequate chemical concentrations for nonlabeled assay methodology. This is an exciting new system for which more research is needed to establish limitations (Rougier et al., 1983, 1986; Dupuis et al., 1986).

G. Absolute Topical Bioavailability

The only way to determine the absolute bioavailability of a topically applied compound is to measure the compound by specific assay in blood or urine after topical and intravenous administration. This is extremely difficult to do in plasma because concentrations after topical administration are often low. However, as advances in analytical methodology bring forth more sensitive assays, estimates of absolute topical bioavailability will become more available.

A comparative example of three methods (Wester et al., 1983) was done using [^{14}C]nitroglycerin in the rhesus monkey (Table 1). Topical bioavailability estimated from urinary excretion was 72.7 ± 5.8%. This was similar to the 77.2 ± 6.7% estimated from plasma total radioactivity AUC. The absolute bioavailability estimated from plasma nitroglycerin unchanged compound AUCs was 56.6 ± 2.5%. The estimates from plasma ^{14}C and urinary ^{14}C were in good agreement. Also, the difference in estimate between that of the absolute bioavailability (56.6%) and that of ^{14}C (72.7–77.2%) is the percentage of compound metabolized in the skin as the compound was being absorbed. For nitroglycerin this is about 20% (Wester and Maibach, 1983).

Table 1 Absolute Bioavailability of Topical Nitroglycerin

Method	Mean bioavailability (%)
Plasma nitroglycerin AUC[a]	56.6 ± 2.5
Plasma total radioactivity AUC[a]	77.2 ± 6.7
Urinary total radioactivity[b]	72.7 ± 5.8

Note: Adapted from Wester et al. (1983).
[a]Absolute bioavailability of nitroglycerin and ^{14}C:

$$\% = \frac{AUC(ng/h \cdot ml^{-1})}{\text{topical dose}} \bigg/ \frac{AUC(ng/h \cdot ml^{-1})}{\text{intravenous dose}} \times 100$$

[b]Percent dose absorbed:

$$\% = \frac{\text{total } ^{14}C \text{ excretion after topical administration}}{\text{total } ^{14}C \text{ excretion after intravenous administration}} \times 100$$

H. Real-Time In Vivo Bioavailability: The Ultimate Method?

Suppose you wanted to know the in vivo topical bioavailability of a chemical ASAP and there was an instrument/system you could use with a subject or animal and get drug or chemical readings every 4 s real-time. That future has started. Figure 3 shows real-time assay of 1,1,1-trichlorethane (methyl chloroform) for a human volunteer with his hand in a bucket of water containing 0.1% methyl chloroform, and a rat exposed to 0.3% methyl chloroform in soil. Each dot represents an average of 4-s readings over a 5-min period.

This method determines the bioavailability of organic solvents following dermal exposures. Breath analysis is used to obtain real-time measurements of volatile organics in expired air following exposure. Human volunteers and animals breathe fresh air via a new breath-inlet system that allows for continuous real-time analysis of undiluted exhaled air. The air supply system is self-contained and separated from the exposure solvent-laden environment. The system uses a Teledyne 3DQ Discovery ion-trap mass spectrometer (MS/MS) equipped with an atmospheric sampling glow discharge ionization source (ASGDI). The MS/MS system provides an appraisal of individual chemical components in the breath stream in the single-digit parts per billion (ppb) detectable range for each of the compounds proposed for study, while maintaining linearity of response over a wide dynamic range (Wester et al., new information in preparation).

I. Skin Decontamination: Reevaluation of Traditional Methods?

Traditional safety in the home and workplace says to wash with soap and water to remove chemical contamination. Many workplaces also have an emergency water shower. It has always been assumed that soap-and-water washing will remove most of the contamination. MDI is an isocyanate chemical in industrial use, for which decontamination potential was determined in vivo in the rhesus monkey. The method is to draw a grid of 1-cm^2 areas on the abdomen of the monkey (the same can be done with humans in vivo). All areas are dosed with the same amount of chemical. Then at set time periods individual grid areas are washed/decontaminated, followed by skin tape stripping for residual chemical not decontaminated. Figure 4 shows MDI in vivo skin decontamination over time with water only, 5% soap, 50% soap, polypropylene glycol, DTAM (a polyglycol-based skin cleaner), and corn oil (Mazola). It is clear that with time water only and soap and water lose their ability to decontaminate skin. On the other hand, polypropylene glycol, DTAM, and corn oil all were more effective ($p < .05$) at removing

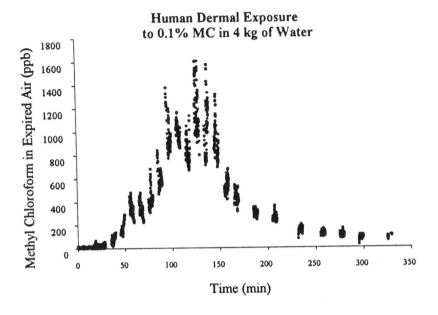

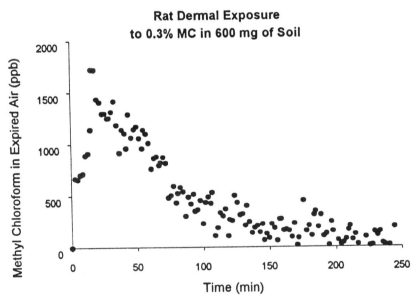

Figure 3   Real-time measurement of methyl chloroform (MC) in human and rat breath after topical dosing. The volunteer or animal breathes into a collection system directly linked to a mass spectrometer, which gives a real-time assay every 4 s.

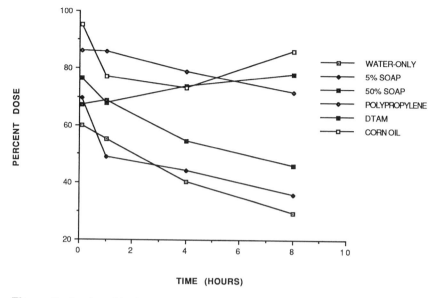

Figure 4 In vivo skin decontamination of MDI (isocyanate) with traditional water or soap and water and with nontraditional polypropylene glycol, DTAM (polyglycol-based skin cleanser), or corn oil. The nontraditional methods are more effective at skin chemical cleansing than traditional soap-and-water wash.

MDI from skin. Figure 5 shows the residual chemical in skin following decontamination. That which was not removed by washing was found in the following skin stripping. Clearly, water only and soap and water were not very effective, especially for the 8-h (workday) period. In the scheme of workplace safety and hygiene, reliability on traditional soap-and-water washing may need to be augmented with more effective systems (Wester et al., new information in preparation).

J. In Vitro Relevance to In Vivo: Infinite and Finite Dosing

Table 2 summarizes the in vivo and in vitro percutaneous absorption of borate compounds in human skin (Wester et al., 1998c). Our focus is on the dosing volume. In vivo dosing was at 2 $\mu l/cm^2$ (if dosed at a greater volume, the dose would flow off the skin) and all K_p values were 10^{-7} cm/h. In vitro, all but one dosing was done using the infinite dose volume of 1.0 ml and all K_p values were 10^{-4}. This is a thousandfold difference in skin absorption. One in vitro dosing was also done at the same finite dose of 2 $\mu l/cm^2$ as was done in vivo. The K_p value for the finite dose was 10^{-6}, certainly more relevant to in vivo than the infinite dosing.

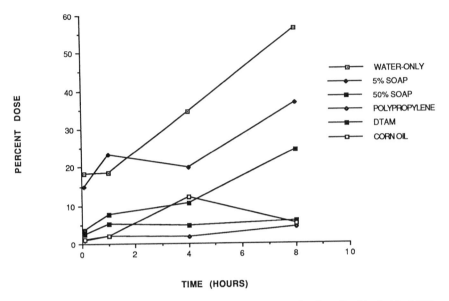

Figure 5 Skin tape stripping after MDI skin decontamination. Residual skin MDI not removed by traditional and nontraditional washing (Figure 4) was recovered in the tape stripping.

Table 2 In Vivo and In Vitro Percutaneous Absorption of Boron Administered as Boric Acid, Borax, and Disodium Octaborate Tetrahydrate (DOT) in Human Skin

	In vitro		In Vivo	
	Flux (μg/cm^2/h)	K_p (cm/h)	Flux (μg/cm^2/h)	K_p (cm/h)
Boric acid				
5% at 2 μl/cm^2	0.07	1.4×10^{-6}	0.009	1.9×10^{-7}
5% at 1 ml/cm^2	14.6	2.9×10^{-4}		
Borax				
5% at 2 μl/cm^2			0.009	1.8×10^{-7}
5% at 1 ml/cm^2	8.5	1.7×10^{-4}		
DOT				
5% at 2 μl/cm^2			0.010	1.0×10^{-7}
5% at 1 ml/cm^2	7.9	0.8×10^{-4}		

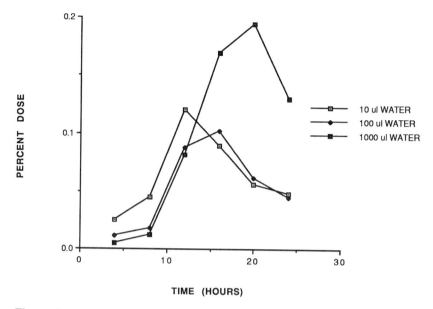

Figure 6 Receptor fluid accumulation of mercury following dosing of human skin in vitro at increasing volumes. The result for finite dose of 10 μl/cm² was less than that of the infinite dose of 1000 μl/cm², showing that dosing volume influences absorption. An infinite dose would run off the skin in vivo.

Table 3 10 Steps to Percutaneous Absorption

1. Vehicle release
2. Absorption kinetics
   a. Skin site of application
   b. Individual variation
   c. Skin condition
   d. Occlusion
   e. Drug concentration and surface area
   f. Multiple dose application
3. Excretion kinetics
4. Effective cellular and tissue distribution
5. Substantivity (nonpenetrating surface adsorption)
6. Wash and rub resistance/decontamination
7. Volatility
8. Binding
9. Anatomical pathways
10. Cutaneous metabolism

Every concept should have confirmation, and Figure 6 does this. Mercury was dosed on human skin in vitro in volumes of 10, 100, and 1000 $\mu l/cm^2$. Absorption was 10- and 100-μl doses are similar, but the infinite dose of 1000 $\mu l/cm^2$ was enhanced. This confirms that the infinite volume dosing of 1000 $\mu l/cm^2$ (1 ml/cm^2) will influence/enhance in vitro percutaneous absorption in human skin.

K.   Factors and Steps in Percutaneous Absorption

Wester and Maibach (1983) defined the steps (Table 3) that require consideration when determining in vivo percutaneous absorption. Many of these individual steps have been expanded into chapters of this book. Each step is important, and in vivo percutaneous absorption is a summation of all these steps. The reader is referred to Wester and Maibach (1997).

REFERENCES

Dupuis, D., Rougier, A., Roguet, R., and Lotte, C. (1986). The measurement of the stratum corneum reservoir: A simple method to predict the influence of vehicles on in vivo percutaneous absorption. Br. J. Dermatol. 115:233–238.

Feldmann, R. J., and Maibach, H. I. (1969). Percutaneous penetration of steroids in man. J. Invest. Dermatol. 542:89–94.

Feldmann, R. J., and Maibach, H. I. (1970). Absorption of some organic compounds through the skin in man. J. Invest. Dermatol. 54:399–404.

Feldmann, R. J., and Maibach, H. I. (1974). Percutaneous penetration of some pesticides and herbicides in man. Toxicol. Appl. Pharmacol. 28:126–132.

McKenzie, A. W., and Stoughton, R. B. (1962). Method for comparing percutaneous absorption of steroids. Arch. Dermatol. 86:608–610.

Rougier, A., Dupuis, D., Lotte, C., Rouguet, R., and Schaefer, H. (1983). In vivo correlation between stratum corneum reservoir function and percutaneous absorption. J. Invest. Dermatol. 81:275–278.

Rougier, A., Dupuis, D., Lotte, C., Roguet, R., Wester, R. C., and Maibach, H. I. (1986). Regional variation in percutaneous absorption in man: Measurement by the stripping method. Arch. Dermatol. Res. 278:465–469.

Wester, R. C., and Noonan, P. K. (1978). Topical bioavailability of a potential antiacne agent (SC-23110) as determined by cumulative excretion and areas under plasma concentration-time curves. J. Invest. Dermatol. 70:92–94.

Wester, R. C., and Maibach, H. I. (1983). Cutaneous pharmacokinetics: 10 Steps to percutaneous absorption. Drug Metab. Rev. 14:169–205.

Wester, R. C., Noonan, P. K., Smeach, S., and Kosobud, L. (1983). Pharmacokinetics and bioavailability of intravenous and topical nitroglycerin in the rhesus monkey. Estimate of percutaneous first-pass metabolism. J. Pharm Sci. 72:745–748.

Wester, R. C., and Maibach, H. I. (1997). Toxicokinetics: dermal exposure and absorption of toxicants. In: Comparative Toxicology, Volume 1, *General Principles*, pp. 99–114. J. Bond, Volume Editor, Elsevier Sciences, New York.

Wester, R. C. Christoffel, J., Hartway, T., Poblete, N., Maibach, H. I., and Forsell, J. (1998a). Human cadaver skin viability for in vitro percutaneous absorption: storage and detrimental effects of heat-separation and freezing. Pharm. Res. 15:82–84.

Wester, R. C. Melendres, J., Hui, X., Cox, R., Serranzana, S., Zhai, H., Quan, D., and Maibach, H. I. (1998b). Human in vivo and in vitro hydroquinone topical bioavailability, metabolism and disposition. J. Toxicol. Environ. Health A 54: 101–117.

Wester, R. C., Hui, X., Hartway, T., Maibach, H., Bell, K., Schell, M. J., Northington, D. J., Strong, P., and Culver, B. D. (1998c). In vivo percutaneous absorption of boric acid, borax and disoderm octaborate tetrahydrate in humans compared to in vitro absorption in human skin from infinite and finite doses. Toxicol. Sci. 45:42–51.

12
Determination of Percutaneous Absorption by In Vitro Techniques

Robert L. Bronaugh, Harolyn L. Hood,*
Margaret E. K. Kraeling, and Jeffrey J. Yourick
Food and Drug Administration, Laurel, Maryland

I. INTRODUCTION

In vitro percutaneous absorption methods have become widely used for measuring the absorption of compounds that come in contact with skin. Safety evaluations of toxic chemicals frequently rely on in vitro studies for human permeation data. Animal data must be used cautiously for estimating human absorption due to differences in barrier properties of animal and human skin (1).

In vitro absorption studies can also be used to measure skin metabolism if viable skin is obtained for the study and if the viability is maintained in the diffusion cells (2). The in vitro system allows for the isolation of skin so that metabolism of the organ can be distinguished from systemic metabolism.

Important considerations in conducting in vitro absorption studies are discussed in the following sections.

II. DEFINITION OF PERCUTANEOUS ABSORPTION

Large differences in percutaneous absorption values can be obtained between studies, depending on whether or not they include the compound remaining in the skin at the end of a study. We believe this material should be included

---

Present affiliation: Bristol-Myers Squibb Worldwide Beauty Care, Stamford, Connecticut.

as percutaneously absorbed unless studies are conducted to demonstrate that the compound does not subsequently diffuse through the skin membrane in a few days.

Recent in vivo and in vitro studies have demonstrated that lipophilic chemicals found in skin at the end of a 24-h study diffuse out of the skin within days and therefore should be included as part of the total dose absorbed. Three radiolabeled chemicals (phenanthrene, benzo[a]pyrene, and di(2-ethylhexyl) phthalate) were applied topically to hairless guinea pigs, and the amount absorbed systemically and locally in skin was determined at 24 h and then again in other guinea pigs after an additional 1 to 2 weeks (3). Absorbed compound found in skin at 24 h was observed to diffuse from the skin and appear in the urine and feces at later times. In 24-h in vitro studies with human and hairless guinea pig skin, substantial amounts of the absorbed fragrance musk xylol were found in skin (4). When another study with human skin was conducted for 7 days (with skin washed at 24 h), only a small amount of the applied dose remained in skin.

Some protocols define the amount of skin absorption solely on material found in the receptor fluid at the end of the experiment. These protocols often require the use of solubilizing agents, such as surfactants or organic solvents, in the receptor fluid to facilitate partitioning of absorbed lipophilic material from skin into the receptor fluid (5). Even if no barrier damage results from the use of solubilizing agents, the efficiency of extraction will vary with the solubility properties of the lipophilic compound. It is not always clear that the extraction of absorbed material is sufficient to simulate the in vivo situation. Furthermore, viability of the skin will likely not be maintained when using solubilizing agents.

III. DIFFUSION CELL

In vitro protocols generally allow the use of either the flow-through or static diffusion cell, but only the flow-through cell allows the maintenance of skin viability. The flow-through cell provides the continual replacement of a nutrient medium necessary to maintain physiological conditions. Diffusion cells should be prepared from a material that is resistant to binding of test material, such as glass or Teflon.

IV. SOURCE OF SKIN

Human skin is generally recommended over animal skin for the most relevant data. When human skin cannot be obtained, animal skin is used for absorption and metabolism studies. Animal skin is more permeable than human skin; thus its use results in a conservative estimate of skin penetration for safety assessments. The use of hairless animals is preferable, since a section can be prepared with a determatome. Metabolism of topically applied compounds by animal skin may be different than human skin (see Chapter 2).

V.   VIABILITY OF SKIN

Viable skin more closely simulates in vivo conditions. Skin metabolism can also be measured when viable skin is used in absorption studies. The skin is capable of biotransformations that can activate or deactivate an absorbed compound. The hydrolysis product of the ultimate carcinogen formed from benzo[a]pyrene was identified in the diffusion cell receptor fluid following topical application of benzo[a]pyrene to viable hairless guinea pig skin (6). The skin has also been shown to have significant capabilities in conjugating percutaneously absorbed compounds. The glycine conjugates of benzoic acid (7) and salicylic acid (8) were observed in hairless guinea pig skin after absorption of the parent compound. Substantial amounts of absorbed benzocaine were found to be acetylated at different doses in human and hairless guinea pig skin (7,9). Viability of skin can be assessed by using glucose utilization techniques or the MTT assay. Cadaver skin is also acceptable for use in a skin absorption study. The cadaver skin must have satisfactorily passed an examination for barrier integrity. Since enzymatic activity is reduced or absent in cadaver skin, metabolism of the test compound must be unimportant or examined in another way (i.e., skin homogenate).

VI.   PREPARATION OF SKIN

A split-thickness preparation of skin should be used in diffusion cells unless full-thickness skin can be justified. A dermatome section containing the epidermis and upper papillary dermal layer (200 μm) most closely simulates the barrier layer of skin (10). Full-thickness skin can artificially retain absorbed compounds that bind or diffuse poorly through it (most lipophilic chemicals) (5). Preparation of an epidermal layer by separation of the epidermis and dermis using heat is effective for nonhairy skin (11), but viability of skin is destroyed.

VII.   RECEPTOR FLUID

A physiological buffer such as a balanced salt solution or tissue culture medium is needed to maintain viability of the skin for at least 24 h (2). Bovine serum albumin is sometimes added to increase the solubility of lipophilic compounds. It is preferable to use a physiological buffer even when metabolism is not measured to simulate in vivo conditions.

Some protocols use solubilizing agents in the receptor fluid so that skin absorption can be more easily determined by simply sampling the receptor fluid (5). Great care must be used with these surfactants and organic solvents, as the skin barrier can be damaged, particularly when split-thickness skin preparations are used. Further investigations are needed to assess the ability of solubilizing agents to adequately remove a wide variety of chemicals from skin without damage to the barrier properties of skin.

VIII. RECOVERY

Determination of total recovery of test compound lends credibility to experimental results. Normally recovery of radiolabel in experiments with labeled compounds exceeds 80%. High recovery values cannot be obtained for volatile compounds unless the evaporating material is trapped.

IX. EXPRESSION OF RESULTS

Studies are usually conducted by applying the test compound under conditions that simulate topical exposure. Sometimes an infinite dose is applied to the skin and absorption is expressed as a permeability constant (steady-state rate divided by applied concentration). Frequently exposure conditions require finite dosing and therefore a steady-state absorption rate is not achieved. Absorption is usually expressed as a percent of the applied dose or as an absorption rate. These absorption values are for the specific dose applied and vehicle used and often cannot be easily extrapolated to other conditions of use.

X. CONCLUSIONS

For greatest acceptability of in vitro data, results from an in vitro method should come from a procedure that simulates in vivo conditions as closely as reasonably possible. We believe that viable skin is preferable, but cadaver skin is acceptable for use in an in vitro replacement test (replaces the need for in vivo data) since adequate supplies of viable human skin are sometimes difficult to obtain. Absorption values for compounds from animal skin would likely exceed the permeability of these same compounds in human skin.

The use of solubilizing agents in the receptor fluid and/or the failure to include skin levels at the end of the study as percutaneously absorbed can lead to errors in absorption measurements. Protocols following these procedures should be restricted to use in screening applications.

REFERENCES

1. Bronaugh RL, Stewart RF, Congdon ER. Methods for in vitro percutaneous absorption studies II. Comparison of human and animal skin. Toxicol Appl Pharmacol 1982; 62:481–488.
2. Collier SW, Sheikh NM, Sakr A, Lichtin JL, Stewart RF, Bronaugh RL. Maintenance of skin viability during in vitro percutaneous absorption/metabolism studies. Tox Appl Pharmacol 1989; 99:522–533.
3. Chu I, Dick R, Bronaugh RL, Tryphonas L. Skin reservoir formation and bioavailability of dermally administered chemicals in hairless guinea pigs. Food Chem Toxicol 1996; 34:267–276.
4. Hood HL, Wickett RR, Bronaugh RL. The in vitro percutaneous absorption of the fragrance ingredient musk xylol. Food Chem Toxicol 1996; 34:483–488.
5. Bronaugh RL, Stewart RF. Methods for in vitro percutaneous absorption studies III. Hydrophobic compounds. J Pharm Sci 1984; 73:1255–1258.
6. Ng KME, Chu I, Bronaugh RL, Franklin CA, Somers DA. Percutaneous absorption and metabolism of pyrene, benzo(a)pyrene and di-(2-ethylhexyl) phthalate: Comparison of in vitro and in vivo results in the hairless guinea pig. Toxicol Appl Pharmacol 1992; 155:216–223.
7. Nathan D, Sakr A, Lichtin JL, Bronaugh RL. In vitro skin absorption and metabolism of benzoic acid, p-aminobenzoic acid, and benzocaine in the hairless guinea pig. Pharm Res 1990; 7:1147–1151.
8. Boehnlein J, Sakr A, Lichtin JL, Bronaugh RL. Metabolism of retinyl palmitate to retinol (vitamin A) in skin during percutaneous absorption. Pharm Res 1994; 11:1155–1159.
9. Kraeling MEK, Lipicky RJ, Bronaugh RL. Metabolism of benzocaine during percutaneous absorption in the hairless guinea pig: Acetylbenzocaine formation and activity. Skin Pharmacol 1996; 9:221–230.
10. Bronaugh RL, Stewart RF. Methods for in vitro percutaneous absorption studies VI. Preparation of the barrier layer. J Pharm Sci 1986; 76:487–491.
11. Bronaugh RL, Congdon ER, Scheuplein RJ. The effect of cosmetic vehicles on the penetration of N-nitrosodiethanolamine through excised human skin. J Invest Dermatol 1981; 76:94–96.

13
Will Cutaneous Levels of Absorbed Material Be Systemically Absorbed?

Robert L. Bronaugh and Harolyn L. Hood*
Food and Drug Administration, Laurel, Maryland

I. INTRODUCTION

For some time it has been recognized that lipophilic compounds can present problems in the in vitro measurement of percutaneous absorption. Compounds that are insoluble in water may not partition freely from excised skin into an aqueous receptor fluid. The problem was alluded to by Franz (1), who in selecting compounds for study omitted highly water-insoluble compounds to avoid results that were "artificially limited due to insolubility in the dermal bathing solution."

II. MODIFICATION OF RECEPTOR FLUID

Attempts have been made to overcome this problem by modifying the receptor fluid composition. Brown and Ulsamer (2) found that the skin permeation of the hydrophobic compound hexachlorophene increased twofold when normal saline was replaced with 3% bovine serum albumin (BSA) (in a physiological buffer) in the diffusion cell receptor fluid.

Other nonphysiological methods have been developed with surfactants and organic solvents. Bronaugh and Stewart (3) achieved more efficient receptor fluid partitioning of two fragrance ingredients by using the nonionic surfactant polyethylene glycol oleyl ether (PEG-20 oleyl ether, Volpo 20). The studies were conducted with 6% PEG-20 oleyl ether and full-thickness

---

*Present affiliation: Bristol-Myers Squibb Worldwide Beauty Care, Stamford, Connecticut.

rat skin. The permeation of a water-soluble control compound was unaffected by the addition of the surfactant to the receptor fluid. The authors concluded, however, that one receptor fluid would not be equally beneficial for all lipophilic compounds. Furthermore, they stated that for extremely lipophilic compounds insoluble in water, less than 50% of in vivo absorption values would probably be obtained by measuring receptor fluid contents. In subsequent studies with split-thickness skin, only 0.5% PEG-20 oleyl ether could be used in the receptor fluid without damage to the skin barrier (4).

III. NEW DEFINITION OF ABSORPTION

The time has come for a revision of the standard definition of ''systemic'' absorption estimated by in vitro diffusion cell experiments. Data can no longer simply be obtained by measuring only receptor fluid levels for most experiments. Addition of solubilizing agents can improve partitioning of compounds from skin, but effectiveness is variable. Surfactants and organic solvents destroy the viability of skin. Absorption should be defined at the end of an experiment as the sum of the amount of compound diffusing into the receptor fluid plus the absorbed material remaining in skin. Skin levels are determined after removal of unabsorbed material with a surface wash using soap and water or other solvents.

A review of the current status of in vivo and in vitro skin absorption methods and recommended protocols was prepared following a 1994 workshop (5). The report mentions that when material remaining in the skin was considered to be percutaneously absorbed, in vivo and in vitro correlations were usually improved.

The viability of skin can be maintained by using a balanced salt solution or a tissue culture medium. Addition of 4% bovine serum albumin may be sufficient in some cases to enhance receptor fluid levels of applied compound without alteration of skin viability.

IV. SKIN RESERVOIR FORMATION

The methodological problem that a skin reservoir creates is not limited to lipophilic compounds. Recent studies with alpha-hydroxy acids, such as glycolic acid and lactic acid, have shown us that very polar compounds can still be substantially retained in skin at the end of a 24-h absorption study (6).

V. FATE OF ABSORBED MATERIAL IN SKIN

The fate of test compounds that are extensively retained in skin after a 24-h exposure has been examined in the following two studies by in vitro and in vivo techniques.

The percutaneous absorption of the fragrance ingredient musk xylol was evaluated in hairless guinea pig and human skin (7). The compound was applied to skin in an oil-in-water emulsion and in a volatile solvent, methanol. After 24 h, the unabsorbed compound was removed from the surface of skin with a soap-and-water wash. Total absorption of musk xylol in hairless guinea pig skin was 55% from the emulsion vehicle and 45% from the methanol vehicle (Table 1). With human skin, permeation of the fragrance ingredient from both vehicles decreased to 22% of the applied dose. A substantial percentage of the absorbed material remained in the skin at the end of the 24-h absorption studies, particularly with the human skin.

An additional study was conducted to assess the fate of the musk xylol remaining in human skin. Diffusion cells were assembled with human skin and dosed with musk xylol in an emulsion vehicle. At 24 h, the skin in all cells was washed, and then half the cells were terminated and the receptor fluid and skin levels were measured (Figure 1). Receptor fluid samples from the rest of the cells were collected for an additional 6 days, and the skin was then removed for analysis. Only about 25% of the musk xylol measured in skin at 24 h remained in skin after 7 days. Most of this material had diffused from skin into the receptor fluid during the 24–72 h period following application. This data suggest that most of the absorbed musk xylol in skin at 24 h would eventually be systemically absorbed in an in vivo study.

The fate of the skin reservoir formed during in vivo skin absorption studies was determined in hairless guinea pigs (8). ^{14}C-Labeled phenanthrene, benzo[a]pyrene, or di(2-ethylhexyl) phthalate was administered dermally in acetone to groups of four animals for each measurement. The absorption values for benzo[a]pyrene are presented in Table 2. At the end

Table 1 Percent Applied Dose Absorbed of Musk Xylol from O/W Emulsion and Methanol Vehicles in 24 h

	Hairless guinea pig skin[a]		Human skin[b]	
	O/W emulsion	Methanol	O/W emulsion	Methanol
Receptor fluid	32.1 ± 1.3	25.8 ± 1.2	4.1 ± 0.7	1.0 ± 0.2
Skin	22.9 ± 2.7	18.8 ± 2.2	17.3 ± 2.3	21.3 ± 0.8
Total absorbed	55.0 ± 2.1	44.6 ± 2.4	21.5 ± 3.0	22.3 ± 0.8
24 h Skin wash	24.9 ± 1.4	5.5 ± 0.3	46.9 ± 3.2	25.9 ± 1.3
Total recovery	80.5 ± 1.9	50.4 ± 2.5	68.5 ± 2.9	45.7 ± 3.3

[a]Values are the mean ± SE of four determinations in each of three animals.
[b]Values are the mean ± SE of four or five determinations from two human subjects.

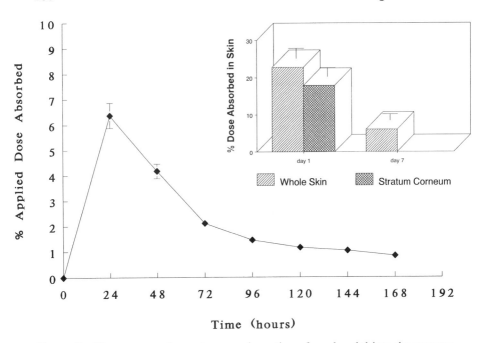

Figure 1 Time course of percutaneous absorption of musk xylol into the receptor fluid. Inset: Skin levels of musk xylol (▨, whole skin; ▩, stratum corneum). Values are means ± SE of determinations in two human subjects.

of 24 h, unabsorbed test compound was removed from all animals with a soap-and-water wash. Some animals were sacrificed at 24 h, and the benzo[a]pyrene absorbed was determined in dosed skin, excreta, and carcass to give a value for total percutaneous absorption. The study was continued in other animals for a total of 48 h and 7 days after application of test compound. The amount of benzo[a]pyrene in skin decreased from 8.9% of the applied dose at 24 h to only 1.4% when the study was continued for an additional 24 h (for a total of 48 h). After 7 days, only 0.2% of the applied dose remained in skin. The results of the studies with all three compounds indicated that the amounts of chemical left in skin after surface washing eventually entered the systemic circulation and should be considered part of the total dose absorbed.

Further studies are needed with additional chemicals to more fully understand the fate of chemicals that form a reservoir in skin during percutaneous absorption studies. This reservoir is likely formed in skin not only by lipophilic compounds like those described in the preceding two studies,

Table 2 Bioavailability of Dermally Applied Benzo[a]pyrene

Recovery site	Percent applied dose		
	24 h	48 h	7 days
Dosed skin	8.9 ± 3.9	1.4 ± 0.4	0.2 ± 0.2
Excreta	6.1 ± 2.0	11.2 ± 1.8	23.9 ± 6.0
Carcass	12.3 ± 3.8	11.4 ± 2.4	0.3 ± 0.1
Total	24.4 ± 6.2	23.9 ± 2.9	24.4 ± 5.9
Wash + cover	58.5 ± 10.9	53.0 ± 2.1	65.3 ± 10.4

Note. Values are the mean ± SD from four hairless guinea pigs.

but also by polar and nonpolar compounds that bind to skin during the absorption process.

REFERENCES

1. Franz TJ. Percutaneous absorption. On the relevance of in vitro data. J Invest Dermatol 1975; 64:190–195.
2. Brown DWC, Ulsamer AG. Percutaneous penetration of hexachlorophene as related to receptor solutions. Food Chem Toxicol 1975; 13:81–86.
3. Bronaugh RL, Stewart RF. Methods for in vitro percutaneous absorption studies III. Hydrophobic compounds. J Pharm Sci 1984; 73:1255–1258.
4. Bronaugh RL, Stewart RF. Methods for in vitro percutaneous absorption studies VI. Preparation of the barrier layer. J Pharm Sci 1986; 75:487–491.
5. Howes D, Guy R, Hadgraft J, Heylings JHU, Kemper F, Maibach H, Marty J, Merk H, Parra J, Rekkas D, Rondelli I, Schaefer H, Tauber U, Verbiese N. Methods for assessing percutaneous absorption: The report and recommendations of the ECVAM workshop. ATLA 1996; 24:81–106.
6. Kraeling MEK, Bronaugh RL. In vitro percutaneous absorption of alpha hydroxy acids in human skin. J Soc Cosmet Chem 1997; 48:187–197.
7. Hood HL, Wickett RR, Bronaugh RL. The in vitro percutaneous absorption of the fragrance ingredient musk xylol. Food Chem Toxicol 1996; 34:483–488.
8. Chu I, Dick R, Bronaugh RL, Tryphonas L. Skin reservoir formation and bioavailability of dermally administered chemicals in hairless guinea pigs. Food Chem Toxicol 1996; 34:267–276.

14

Dermal Decontamination and Percutaneous Absorption

Ronald C. Wester and Howard I. Maibach
University of California School of Medicine, San Francisco, California

Although decontamination of a chemical from the skin is commonly done by washing with soap and water, as it has been assumed that washing will remove the chemical, recent evidence suggests that many times the skin and the body are unknowingly subjected to enhanced penetration and systemic absorption/toxicity because the decontamination procedure does not work or may actually enhance absorption. This chapter reviews some of the recent (although sparse) literature, then offers new in vitro and in vivo techniques to determine skin decontamination.

I. IN VIVO DECONTAMINATION MODEL

Figure 1 shows a skin decontamination model (Wester et al., 1991) where the time course of decontamination for several solvent systems can be tested. The illustration is for the abdomen of a rhesus monkey, but any large skin area can be used, including human. A grid of 1-cm^2 areas is marked on the skin. The illustration shows 24 separate blocks. As an example of use, the four blocks across the grid can represent four different decontaminating systems. The blocks down the abdomen can represent six time periods with which the skin is decontaminated. In our system we use a cotton applicator laden with washing solvent to wash the skin block area. This is illustrated with the following data.

Figure 2 shows glyphosate, a water-soluble chemical, removed from rhesus monkey skin with three successive soap-and-water or water-only washes. Approximately 90% of the glyphosate is removed with the washes,

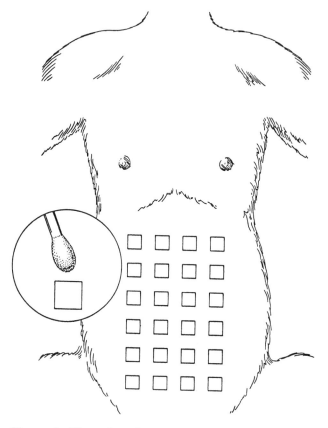

Figure 1 Illustration of grid method where multiple single doses are marked on rhesus monkey abdomen for skin decontamination with time using a cotton-tip applicator laden with appropriate solvent.

most in the first wash. There is no difference between soap-and-water and water-only washing. Figure 3 shows alachlor, a lipid-soluble chemical, also decontaminated with soap and water, and water only. In contrast to glyphosate, more alachlor is removed with soap and water than with water only. Although the first alachlor washing removed a majority of chemical, successive washings contributed to the overall decontamination.

Figures 4 (glyphosate) and 5 (alachlor) illustrate skin decontamination with soap and water or water only over a 24-h dosing period, using the grid methodology from Fig. 1. The three successive washes were pooled for each data time point. Certain observations are made. First, the amount recovered decreases over time. This is because this is an in vivo system and percuta-

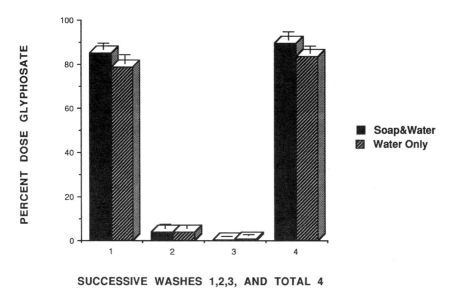

Figure 2 Glyphosate removal from rhesus monkey skin in vivo with successive washes. Note that water only and the addition of soap are equally effective. Glyphosate is water-soluble.

neous absorption is taking place, decreasing the amount of chemical on the skin surface. There also may be some loss due to skin desquamation. Note that alachlor is more readily absorbed across skin than is glyphosate. The second observation is, again, that soap and water and water only remove equal amounts of glyphosate, but alachlor is more readily removed with soap-and-water wash then with water only. The reason is that glyphosate is water-soluble; thus water is a good solvent for it. Alachlor is lipid soluble and needs the surfactant system for more successful decontamination (Wester et al., 1991, 1992).

Figure 6 shows alachlor skin decontamination at four dose concentrations washed with multiple successive soap and water applications. Most of the dose is removed with the first washing, and three successive washes are adequate to remove the dose.

Figure 7 shows 42% PCBs applied in vivo in trichlorobenzene (TCB) or mineral oil (MO) to rhesus monkey skin and washed over a 24-h period with soap and water. With time, the wash recovery of PCBs decreases due to the ongoing skin absorption. PCBs can be removed from skin with soap and water if the decontamination is done soon enough after exposure (Wester et al., 1984).

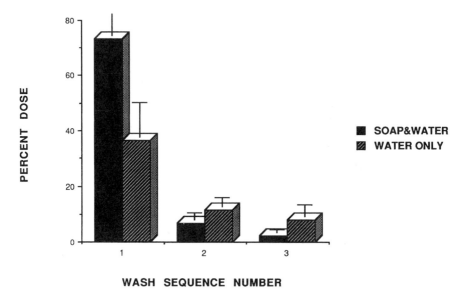

Figure 3 Alachlor removal from rhesus monkey skin in vivo with successive washes. Note that soap and water removes more alachlor than water only. Alachlor is lipid soluble.

The preceding discussion shows (Table 1) that there are two factors with in vivo skin decontamination. The first is the "rubbing effect" that removed loose surface stratum corneum due to natural skin desquamation. The second is the "solvent effect," which is related to chemical lipophilicity, and was illustrated for glyphosate and alachlor.

II. IN VITRO DECONTAMINATION MODEL

In vitro skin mounted in diffusion cells can be decontaminated with solvents. A problem exists in that the mounted skin is fragile and cannot be rubbed, as naturally occurs with people washing their skin. Another in vitro technique that is easy to use for solvent efficiency in decontamination is with powered human stratum corneum (PHSC). Figure 8 shows results when alachlor was added to PHSC and then partitioned against water only or 10% and 50% soap. Alachlor stayed with the PHSC when partitioned against water only; however, it readily partitioned to the solvents 10% and 50% soap. These are the same confirming results shown in Figs. 3, 5, and 6.

The powdered human stratum corneum is made according to Wester et al. (1987). It is then mixed with [^{14}C]alachlor in Lasso formulation, let

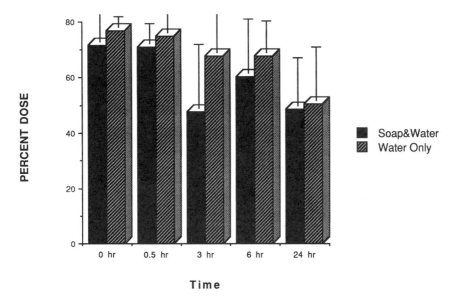

Figure 4 Time course, 0–24 h, for in vivo glyphosate removal with skin washing. Over time the ability to remove glyphosate decreases due to ongoing skin absorption. Soap and water and water only were equally effective; glyphosate is water soluble.

set for 30 min, then centrifuged. The alachlor, a lipid-soluble chemical, partitions (90.3 ± 1.2%) into the powdered human stratum corneum; 5.1 ± 1.2% remains in the Lasso formulation. Water-only wash (and subsequent centrifugation) removes only 4.6 ± 1.3% of the "bound" alachlor. However, when the bound alachlor–stratum corneum is washed with 10% and 50% soap and water (v/v), 77.2 ± 5.7% and 90.0 ± 0.5% of the alachlor is removed from the stratum corneum. Such a model would predict that alachlor in Lasso cannot be removed from the skin with water-only washing, but that the use of soap will decontaminate the skin. The "lipid" constituents of soap probably offer a more favorable partitioning environment for the alachlor.

III. EFFECTS OF OCCLUSION AND EARLY WASHING

Table 2 shows the effect of duration of occlusion on the rate of absorption of malathion (Feldmann and Maibach, 1974). What is important from this table is that 9.6% of the applied malathion was absorbed during a zero time duration. There was an immediate wash of the site of application with soap and water. However, almost 10% of the applied dose was not washed off

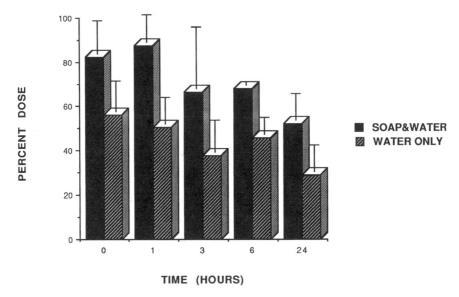

Figure 5 Time course, 0–24 h for in vivo alachlor removal with skin washing. Over time the ability to remove alachlor decreases due to ongoing skin absorption. Soap and water was more effective than water only; alachlor is lipid soluble.

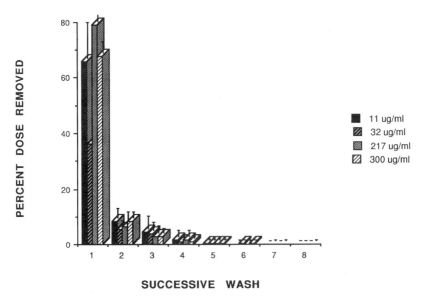

Figure 6 Alachlor soap-and-water skin decontamination with multiple successive washes. Three successive washes seem adequate for decontamination.

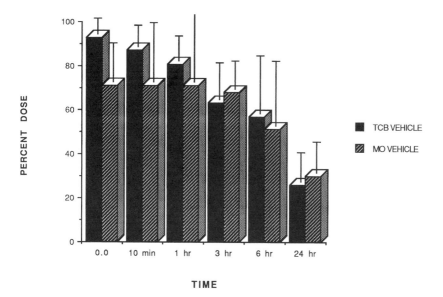

Figure 7 PCBs can be removed with soap and water; however, skin absorption of PCBs is high, and the chance to remove them from skin decreases with time. PCBs were applied to skin in either trichlorobenzene (TCB) or mineral oil (MO) vehicle.

but, in fact, persisted on the skin through the wash procedure and was later absorbed into the body. Table 3 shows the short-term wash recovery for benzo[a]pyrene and DDT. Even when skin washing was initiated within 25 min of dosing, some benzo[a]pyrene and all of the DDT had absorbed sufficiently into skin. Therefore, early washing after exposure is critical, but it may not provide complete decontamination (Wester et al., 1990).

Table 4 shows the effect of washing on the percutaneous absorption of hydrocortisone in a rhesus monkey. With a nonwashing sequence between

Table 1 In Vivo Skin Decontamination

1. Rubbing effect
   Removes "loose" surface stratum corneum
   "Loose" = natural desquamation
2. Solvent effect
   Water only
   Soap and water
   Related to chemical lipophilicity

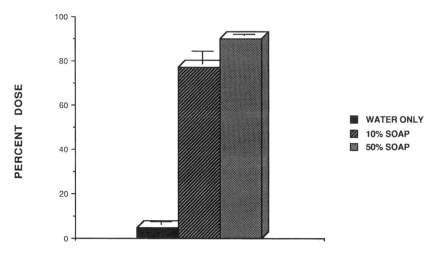

Figure 8 Alachlor was added to powdered human stratum corneum and then partitioned against water only or against water with 10% or 50% soap. Soap is required to remove alachlor, which is lipid soluble. This is a good in vitro model to screen potential decontamination vehicles.

dose applications, the percentage dose absorbed was $0.55 \pm 0.06\%$ of applied dose. When a post-24-h wash procedure was introduced, the percentage of dose absorbed statistically ($p < .05$) increased to $0.72 \pm 0.06\%$. The post-24-h wash was supposed to removed the excess materials and, thus, decrease

Table 2 Effect of Duration of Occlusion on Percutaneous Absorption of Malathion in Humans

Duration (h)	Absorption (%)
0[a]	9.6
0.5	7.3
1	12.7
2	16.6
4	24.2
8	38.8
24	62.8

[a]Immediate wash with soap and water.

Table 3 Short-Term Wash Recovery for Benzo[a]pyrene and DDT: 25-min
Exposure vs. 24-h Exposure

| | Percent dose | | |
| | --- | --- | --- |
Chemical	Short exposure (25 min), in vitro receptor fluid	Skin	Long exposure (24 h), in vivo
Benzo[a]pyrene	0.00 ± 0.00	5.1 ± 2.1	51.0 ± 22.0
DDT	0.00 ± 0.00	16.7 ± 13.2	18.9 ± 9.4

Note. In vitro the chemical in acetone vehicle was dosed on human skin, then washed with soap and water after a 25-min period. The in vivo studies were 24-h exposure with acetone vehicle dosing.

absorption. However, the soap-and-water wash hydrated the skin and the rate of absorption of hydrocortisone increased.

Table 5 shows dermal washing efficiency for polychlorinated biphenyls (PCBs) in the guinea pig. When 42% PCB was applied to guinea pig skin and immediately washed, only 58.9% of the applied dose could be removed. The rest of the material was available for subsequent percutaneous absorption. When 42% PCB was applied to guinea pig skin and 24 h later the site of application was washed, only 0.9% of the applied dose was removed. Thus, all the applied PCB was available for absorption or had already been absorbed into the body. When 54% PCB was applied to guinea pig skin and washed 24 h postapplication, 19.7% of the applied dose was removed. Thus, with 54% PCB, 80% of the applied dose could not be removed or had already been absorbed into the body. Subsequent examination of the rate of absorption of PCB showed that most was absorbed into the body. This study illustrates that the hypothesis that washing or bathing and any other appli-

Table 4 Effect of Washing on Percutaneous
Absorption of Hydrocortisone in
Rhesus Monkey

Treatment	Absorption (% ± SD)
No wash	0.55 ± 0.06
Post-24-h wash[a]	0.72 ± 0.06
	(p < .05)

[a]Soap-and-water wash.

Table 5 Dermal Wash Efficiency for
Polychlorinated Biphenyls (PCBs) in
Guinea Pig

PCB (%)	Wash time[a]	Dose removed (% ± SD)
42	Immediate	58.9 ± 7.5
42	Post-24-h	0.9 ± 0.2
54	Post-24-h	19.7 ± 5.5

[a]Wash procedure: twice with water, twice with acetone, twice with water.

cations of water will remove all material from skin is wrong. Substantivity (the nonspecific absorption of material to skin) can be a strong force.

Figure 9 shows the percent dose absorbed per hour of a herbicide in the rhesus monkey (Wester and Maibach, 1983). At 24 h postapplication, the site of absorption was washed with water and acetone sequentially. The time curve shows a "washing-in effect" following the 24-h postapplication wash. Thus, as we saw previously with hydrocortisone, there is definitely a washing-in effect. The application of water and acetone changed the barrier properties of skin and caused an increase in the rate of absorption of this herbicide. Moody and Nadeau (1997) also recently showed the washing-in effect for 2,4-dichlorophenoxyacetic acid amine during in vitro percutaneous absorption studies in rat, guinea pig, and human skin.

It is generally assumed that pesticides and other chemicals can be easily removed from the skin. Another series of experiments by Feldmann and Maibach (1974) was designed to determine just how effective the removal is in terms of decreasing percutaneous absorption. The experimental variable was the removal of applied pesticide in different groups of subjects at varying times. Decontamination was attempted by a 2-min wash with soap and hot water. Data are presented in Table 6. Absorption of azodrin at a concentration of 4 μg/cm^2 on the human forearm was 14.7% if the site was not washed for 24 h. Washing after 4 h decreased absorption to 8% of the dose. Washing after only 15 min decreased penetration, but still 2.3% of the dose was absorbed. With similar experimental variables, ethion washing in 15 min decreased the penetration from the 24-h time period from only 3.3 to 1.6%. Malathion decrease from the 24-h to 15-min wash was from only 6.8 to 4.3%.

A similar relationship was maintained when the applied concentration was greatly increased. For instance, at a concentration of 4 μg/cm^2 of par-

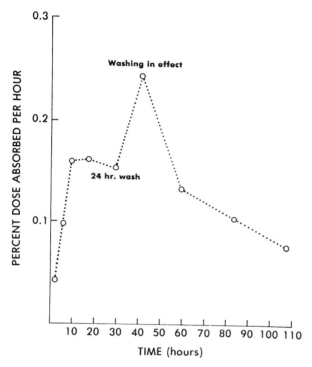

Figure 9 Effect of washing on absorption. Washing at 24 h enhanced absorption, as shown by the peak effect at 48 h.

athion, the penetration after washing at 24 h was 8.6%; this was only decreased to 6.7% by washing in 15 min.

The effect of the anatomical site was also studied. On the palm, the penetration of parathion was 11.8% with washing at 24 h. There was no significant decrease in penetration washing in 15 min; in fact, there was a slight increase. Thus, the very potent pesticide parathion absorbs to skin despite washing and is absorbed into the body.

Experiments with rubbing alcohol had similar results. When used with malathion, alcohol washing at 4 h allowed 16.8% penetration and at 15 min, 17.7%. Penetration with parathion at 4 h was 10.3%, and at 15 min, 8.2%. These data suggest that a careful examination be made of recommendations given to consumers and workers about when they can remove these substances from their skin and what materials should be used. It is obvious that protection from washing and bathing is not what had been predicted.

It was questioned whether a whole-body exposure to a solvent such as water might not be more effective at removing pesticides. For this reason

Table 6 Effect of Washing on Percent Penetration

Compound, dose (μg/cm²), and site	Minutes				Hours				
	1	5	15	30	1	2	4	8	24
Soap and water									
Azodrin, 4, arm			2.3				8.6		14.7
Ethion, 4, arm			1.6				2.9		3.3
Guthion, 4, arm									
Malathion									
4, arm	1.2		4.3	4.5	6.1	8.3	12.1		6.8
40, arm			4.7				6.8		
400, arm			1.4		2.0		4.7		
Parathion									
4, arm	2.8		6.7		8.4		8.0	15.8	8.6
40, arm			3.1				6.9		9.5
400, arm			2.2		2.3		4.2		4.8
4, forehead	8.4		7.1	12.2	10.5	20.1	27.7		36.3
4, palm		6.2	13.6	13.3	11.7	9.4	7.7		11.8
Lindane, 4, arm		1.7	1.8	4.2	3.9	6.7	5.1		9.3
Baygon, 4, arm	1.2		4.7	4.5	4.7	11.8	15.5	11.3	19.6
2,4-D									
4, arm	0.5		0.7	1.8	1.2	3.7	3.7		5.8
40, arm			0.7				2.8		
Rubbing alcohol									
Malathion, 4, arm			17.7		5.8		16.8		
Parathion, 4, arm			8.2		7.0		10.3		

a group of subjects was showered 4 h after application instead of being washed locally with soap and water. The data for malathion, parathion, and Baygon are in Table 7. The shower was no more effective and perhaps less effective than the local application of soap and water. Showering does not appear to be a solution to the problem.

IV. CONCLUSION: SUBSTANTIVITY

Substantivity has been defined as the nonspecific absorption of material from skin. It is obvious that the standard washing procedures do not always readily remove materials from skin. How important is this in terms of occupational exposure? Kazen and coworkers (1974) did hexane hand rinsings on occupationally exposed people. They analyzed the rinsings by electron capture and flame photometric/gas-liquid chromatography for pesticide residues

Table 7 Effect of Shower on Different Types of Removal at 4 h

Compound	Absorption (%) after shower[a]		
	Arm	Forehead	Palm
Malathion	8.8	32.7	7.2
	(12.1)		
Parathion	16.5	41.9	13.4
	(9.0)	(27.7)	(7.7)
Baygon	9.9	20.5	8.7
	(15.5)		

[a]Values in parentheses: after washing.

to determine whether or not these chemicals persisted on the skin long after exposure. Chlordane and dieldrin apparently persisted on the hands of a former pest control operator for at least 2 years. Methoxychlor, captan, and malathion persisted for at least 7 days on the hands of a fruit and vegetable grower. Parathion was found on the hands of one man 2 months after his last known contact with this pesticide. Endosulfan, DDD, kelthane, decthal, trithijon, imidan, and guthion have persisted on the hands of some exposed workers from less than a day to 112 days after exposure.

In conclusion, washing is generally good, but it does not prevent all penetration of some chemicals. Surely, the understanding of the mechanisms and the development of more efficient removal systems must be a high priority for research. The reader is referred to Geno et al. (1996), Hewitt et al. (1995), Kintz et al. (1992) and Merrick et al. (1982) for further information related to skin decontamination.

REFERENCES

Feldmann, R. J., and Maibach, H. I. (1974). Systemic absorption of pesticides through the skin of man. In: *Occupational Exposure to Pesticides*: Report to the Federal Working Group on Pest Management from the Task Group on Occupational Exposure to Pesticides. Appendix B, pp. 120–127.

Geno, P. W., Camann, D. E., Harding, H. J., Villalobos, K., and Lewis, R. G. (1996). Handwipe sampling and analysis procedure for the measurement of dermal contact with pesticides. *Arch. Environ. Contam. Toxicol. 30*:132–138.

Hewitt, P. G., Hotchkiss, P. G., and Caldwell, J. (1995). Decontamination procedures after *in vitro* topical exposure of human and rat skin to 4,4[1]-methylenebis[2-chloroaniline] and 4,4[1]-methylenedianiline. *Fundam. Appl. Toxicol. 26*:91–98.

Kazen, C., Bloomer, A., Welch, R., Oudbier, A., and Price, H. (1974). Persistence of pesticides on the hands of some occupationally-exposed people. *Arch. Environ. Health 29*:315–318.

Kintz, P., Tracqui, A., and Morgin, P. (1992). Accidental death caused by the absorption of 2,4-dichlorophenol through the skin. *Arch. Toxicol. 66*:298–299.

Merrick, M. V., Simpson, J. D., and Liddell, S. (1982). Skin decontamination—A comparison of four methods. *Br. J. Radiol. 55*:317–318.

Moody, R. P., and Nadeau, B. (1997). *In vitro* dermal absorption of two commercial formulations of 2,4-dichlorophenoxyacetic acid dimethylamide (2,4-D amine) in rat, guinea pig and human skin. *Toxicology In Vitro 11*:251–262.

Wester, R. C., and Maibach, H. I. (1983). Advances in percutaneous absorption. In *Cutaneous Toxicity*. Edited by V. A. Drill and P. Lazar. Raven Press, New York, pp. 29–40.

Wester, R. C., Noonan, P. K., and Maibach, H. I. (1977). Frequency of application of percutaneous absorption of hydrocortisone. *Arch. Dermatol. 113*:620–622.

Wester, R. C., Bucks, D. A. W., and Maibach, H. I. (1984). Polychlorinated biphenyls (PCBs): Dermal absorption, systemic elimination, and dermal wash efficiency. *J. Toxicol. Environ. Health 12*:511–519.

Wester, R. C., Mobayen, M., and Maibach, H. I. (1987). *In vivo* and *in vitro* absorption and binding to powdered stratum corneum as methods to evaluate skin absorption of environmental chemical contaminants from ground and surface water. *J. Toxicol. Environ. Health 21*:367–374.

Wester, R. C., Maibach, H. I., Bucks, D. A. W., Sedik, L., Melendres, J., Liao, C., and Di Zio, S. (1990). Percutaneous absorption of [14C]-DDT and [14C]-benzo(a)pyrene from soil. *Fundam. Appl. Toxicol. 15*:510–516.

Wester, R. C., Melendres, J., Sarason, R., McMaster, J., and Maibach, H. I. (1991). Glyphosate skin binding, absorption, residual tissue distribution, and skin decontamination. *Fundam. Appl. Toxicol. 16*:725–732.

Wester, R. C., Melendres, J., and Maibach, H. I. (1992). In vivo percutaneous absorption of alachlor in rhesus monkey. *J. Toxicol. Environ. Health 36*:1–12.

15
Partitioning of Chemicals from Water into Powdered Human Stratum Corneum (Callus)—A Model Study

Xiaoying Hui, Ronald C. Wester, Philip S. Magee, and Howard I. Maibach
University of California School of Medicine, San Francisco, California

I. INTRODUCTION

Chemical delivery/absorption into and through skin is important for both dermatopharmacology and dermatotoxicology. A major endeavor is underway to enable predictions of percutaneous absorption from molecular properties. This is especially important to the U.S. Environmental Protection Agency (EPA) where chemicals produced by industry outdistance the ability to experimentally document absorption properties (U.S. EPA, 1992). Early correlations of percutaneous absorption were restricted to octanol/water partition coefficients, PC(o/w), as a descriptor. These correlations were not completely successful in reducing the experimental variance. Other parameters such as molecular weight have been added to the predictive equations (U.S. EPA, 1992; Potts and Guy, 1992). The substitution of powdered human stratum corneum (PHSC) prepared from callus (sole) for the intact membranous stratum corneum is expected to bring equation predictions closer to actual absorption values. To validate the PHSC system, correlation of PHSC/water partitioning with octanol/water partitioning is necessary.

The human stratum corneum (SC) includes the horny pads of the palms and soles (callus) and the membranous stratum corneum covering the remainder of the body (Barry, 1983). It represents a rate-limiting barrier in the transport of most chemicals across the skin (Blank, 1965; Scheuplein and Blank, 1973; Elias, 1981; Yardley, 1987). Chemicals must first partition into the SC before entering deeper layers of the SC, epidermis, and dermis to reach the vascular system. This partitioning process occurs much faster than complete diffusion through the whole SC, and the process quickly reaches equilibrium (Scheuplein and Bronaugh, 1985). In addition to binding within the SC, a chemical can also be retained within the SC as a reservoir (Zatz, 1993). Thus, understanding the process of chemical partitioning into the SC becomes important in developing an insight into the barrier properties and transport mechanisms. Many experiments have been conducted to predict chemical partitioning into the SC in vitro. However most were based on qualitative structure–activity relationships (QSARs) of related chemicals to determine the partitioning process, and few studies focus on structurally unrelated chemicals (Guy and Hadgraft, 1991). Since the range of molecular structure and physicochemical properties is very broad, any predictive model must address a broad scope of partitioning behavior.

This study assesses the relationship of a number of chemicals with a broad scope of physicochemical properties in the partitioning mechanism between the PHSC and water. Uniqueness and experimental accuracy are added by using the PHSC (Wester et al., 1987). The experimental approach is designed to determine how the PC PHSC/w is affected by a) chemical concentration, b) incubation time, and c) chemical lipophilicity (or hydrophilicity) and other factors. These parameters are used to develop an in vitro model that will aid in predicting chemical dermal exposure for hazardous chemicals.

II. MATERIALS AND METHODS

A. Materials

[Ring UL-^{14}C]atrazine (purity >98%, specific activity 11.5 mCi/mmol), [ring UL-^{14}C]-2,4-dichlorophenoxyacetic acid (2,4-D, >98%, 20.2 mCi/mmol), [8-^{14}C]dopamine hydrochloride (>98%, 15.6 mCi/mmol), and [methoxy-^{14}C]malathion (95%, 8.8 mCi/mmol) were obtained from Sigma Chemical Company (St. Louis, MO). [Dimethylamine-^{14}C]aminopyrine (98%, 107 mCi/mmol), [^{14}C]glyphosate (>98%, 51 mCi/mmol), [8-^{14}C]theophylline (>98%, 51 mCi/mmol), and [U-^{14}C] polychlorinated biphenyls (PCB, Aroclor 1254, 54% chlorine, 32 mCi/mmol) were obtained from Amersham. [U-^{14}C]glycine (98%, 63 mCi/mmol), and [^{14}C]urea (98%, 56 mCi/mmol) were obtained from ICN Biomedical, Inc. (Irvine, CA). [4-^{14}C]Hydrocortisone

(97.6%, 52 mCi/mmol) was obtained from Du Pont Company (Wilmington, DE). [Ring ^{14}C(U)]alachlor (21 mCi/mmol) was a gift from Monsanto Company (St. Louis, MO). The pH values of these chemical solutions in distilled deionized water were determined by a Corning pH/ion analyzer (model 250, Corning Science Products, Corning, NY).

B. Preparation of Powdered Human Stratum Corneum (Callus)

The procedure was based on the method of Wester et al. (1987). Adult foot callus was ground in a Micro-Mill grinder (Bel-Art Products, Pequannock, NJ) in the presence of liquid nitrogen to form a powder. Particles of the PHSC that passed through a 50-mesh sieve but not an 80-mesh were used. Particle sizes within this range were shown to favor the experimental conditions.

C. Incubation Procedure

The experiment was performed by modifying the method of Wester et al. (1987). A given dosage of radiolabeled chemical in 4.0 ml vehicle (distilled deionized water) was mixed with 10.0 mg powdered callus in a glass vial and incubated in a water bath with moderate shaking at 35°C for a given time period. The mixture was then separated by centrifugation at 1500 × g for 15 min and the supernatant carefully removed. The PHSC pellet was resuspended three times in the same volume of deionized distilled water to remove any excess material clinging to the surface. Experiments indicated that further washing (four or five times) did not significantly change the PC value. The radioactivity of the chemical bound to the PHSC, and that remaining in the vehicles was determined by scintillation counting. Five samples were used for each test.

D. Scintillation Counting

The scintillation cocktail was Universal-ES (ICN, Costa Mesa, CA). Background controls and test samples were counted in a Packard model 4640 counter (Packard Instruments). The data was transferred to a computer program (Appleworks/Apple IIE computer; Apple Computer Co., Mountain View, CA) that subtracted background control samples and generated a spreadsheet for statistical analysis. The counting process and the computer program have been verified to be accurate by a quality assurance officer at the University of California at San Francisco.

E. Powdered Human Stratum Corneum/Water Partition Coefficient

The value is determined by the following equation:

$$PC(PHSC/w) = C(PHSC)/C(water) \tag{1}$$

where $C(PHSC)$ is the amount (μg) of the chemical absorbed in 10 mg PHSC and $C(water)$ is the amount (μg) remaining in 4000 mg water after removing the PHSC pellet. Since the number of micrograms of the chemical in the PHSC or water is proportional to the counts (dpm) determined by the scintillation counter, and partitioning of chemical into the PHSC exactly equaled the decrease of the chemical concentration in the vehicle (water), an alternative calculation method was also used:

$$PC(PHSC/w) = \frac{dpm(PHSC)/mg(PHSC)}{[dpm(i) - dpm(PHSC)]/mg(water)} \tag{2}$$

where dpm(i) is the initial chemical concentration in the vehicle (water). These two calculation methods were shown to give similar results, and the reported experiments are calculated by the second method.

F. Statistical Analyses

Statistical analysis (Student's t-test, linear and multiple regressions) were performed in version 6.1 of MINITAB (Minitab, Inc., State College, PA) on an IBM PC-compatible computer.

III. RESULTS

Table 1 shows the effect of varying initial chemical concentrations on the PC(PHSC/w) of 12 test compounds. Under the fixed experimental conditions (2 h incubation time and 35°C incubation temperature), the concentration required to reach a peak value of the partition coefficient varied from chemical to chemical. After reaching the maximum, increasing the chemical concentration in the vehicle did not elevate the PC value; instead, it slightly decreased or was held at approximately the same level. The PC value of each chemical correlated with its lipophilicity or hydrophilicity as described by log PC(o/w). For example, the lipophilicity of malathion [log PC(o/w) = 2.36] is approximately the same as that of atrazine [log PC(o/w) = 2.75]. Therefore their PC(PHSC/w) values are close to each other. However, the concentration of malathion to achieve a peak value was approximately 10-fold higher than that of atrazine.

Figure 1 shows that the maximum amount of chemical partitioning into the SC is limited when the experimental condition is fixed (30 min

Table 1 Effect of Initial Aqueous Phase Chemical Concentration on Partition Coefficient (PHSC/water)

Chemical	Concentration[a] (%, w/v)	Partition coefficient[b] Mean	SD
Dopamine	0.23	5.42	0.22
	0.46	6.04	0.28
	0.92	5.74	0.28
Glycine	0.05	0.36	0.01
	0.10	0.40	0.02
Urea	0.03	0.26	0.02
	0.06	0.15	0.02
	0.12	0.17	0.02
Glyphosate	0.02	0.79	0.04
	0.04	0.68	0.04
	0.08	0.70	0.01
Theophylline	0.18	0.37	0.02
	0.36	0.43	0.03
	0.54	0.42	0.02
Aminopyrine	0.07	0.44	0.09
	0.14	0.46	0.03
Hydrocortisone	0.09	0.37	0.01
	0.18	0.34	0.01
	0.36	0.29	0.02
Malathion	0.47	0.50	0.09
	0.94	0.40	0.03
	1.88	0.53	0.04
Atrazine	0.09	0.53	0.06
	0.14	0.59	0.07
	0.19	0.58	0.03
2,4-D	0.27	7.52	0.81
	0.54	7.53	1.01
	0.82	8.39	1.67
Alachlor	0.32	1.11	0.05
	0.64	1.08	0.04
	1.28	1.96	0.15
PCB	0.04	1237.61	145.52
	0.08	1325.44	167.03
	0.16	1442.72	181.40

[a]Concentration expressed as gm of chemical in 100 ml water, 2 h incubation time at 35°C.
[b]Stratum corneum/water partition coefficient represented the mean of each test ($n = 5$) ± SD.

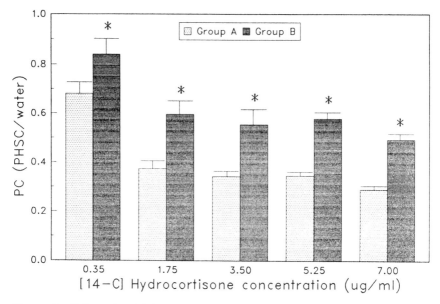

Figure 1 Effect of multiple doses on hydrocortisone partitioning from water into the powdered human stratum corneum. Group A was preincubated with 2.0 ml of saturated unlabeled hydrocortisone for 30 min. Group B was preincubated with the same vehicle without hydrocortisone. Then both groups were incubated with 2.0 ml of various concentration of ^{14}C-labeled hydrocortisone for 2 h. The values for group B were significantly higher than for group A (*$p < .05$).

preincubation followed by 2 h incubation time). Preliminary experiments demonstrated no statistical differences of the PC values of ^{14}C-labeled hydrocortisone between 30 min and 2 h of incubation. Group A was preincubated with saturated unlabeled hydrocortisone while gorup B was preincubated with the same vehicle without hydrocortisone. Both groups were then challenged by different concentrations of ^{14}C-labeled hydrocortisone. The peak value of group A was reduced by approximately 40% when compared with group B.

Figures 2 and 3 show the changes in PC(PHSC/w) of 12 test compounds as a function of varying incubation time. Theophylline and atrazine reached their maximum values as early as 6 h of incubation. For most chemicals, however, increasing incubation time resulted in elevation of PC(PHSC/w). Higher lipophilic compounds such as 2,4-D, alachlor, and PCB, with log PC(o/w) = 2.8–6.4, and very hydrophilic compounds such as dopamine with log PC(o/w) = −3.4, reach their maximum partitioning into the SC around 12 h of incubation. For those compounds of relatively

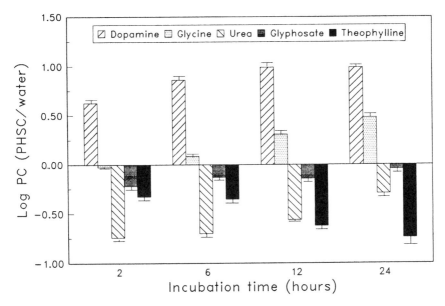

Figure 2 Log PC(PHSC/w) of hydrophilic compounds [log PC(o/w) < 0] determined as a function of incubation time. The PC of each chemical was measured after 2, 6, 12, and 24 h of incubation time. Each bar expressed as the mean ± SD, $n = 5$.

lower lipophilicities or hydrophilicities, PC(PHSC/w) continually increased with incubation time up to 24 h.

Figure 4 describes a smooth, partially curvilinear relationship between the log PC(PHSC/w) and the log PC(o/w) of these chemicals. In this study, lipophilicities and hydrophilicities of compounds were defined as log PC(o/w) larger or smaller than 0, respectively. For lipophilic chemicals, such as aminopyrine, hydrocortisone, malathion, atrazine, 2,4-D, alachlor, and PCB, the logarithms of PHSC/water partition coefficients are proportional to the logarithms of the octanol/water partition coefficients:

$$\log \text{PC(PHSC/w)} = 0.59 \log \text{PC(o/w)} - 0.72 \qquad (3)$$

Student's t-value: 9.93

$n = 7 \qquad r^2 = .95 \qquad s = 0.26 \qquad F = 98.61$

For hydrophilic chemicals, such as theophylline, glyphosate, urea, glycine, and dopamine, the log PC(PHSC/w) values are approximately and reversely proportional to log PC(o/w):

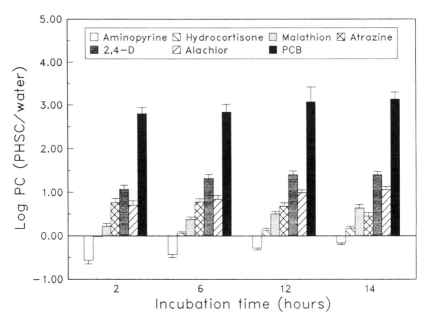

Figure 3 Log PC(PHSC/w) of lipophilic compounds [log PC(o/w) > 0] determined as a function of incubation time. The PC of each chemical was measured after 2, 6, 12, and 24 h of incubation time. Each bar expressed as the mean ± SD, $n = 5$.

$$\log \text{PC(PHSC/w)} = -0.60 \log \text{PC(o/w)} - 0.27 \qquad (4)$$

Student's t-value: -4.86

$n = 5 \qquad r^2 = .88 \qquad s = 0.26 \qquad F = 23.61$

However, the overall relationship of the PC(PHSC/w) of these chemicals and their PC(o/w) is nonlinear. This nonlinear relationship is adequately described by the following equation:

$$\log \text{PC(PHSC/w)} = 0.078 \log \text{PC(o/w)}^2 + 0.868 \log \text{MW} - 2.04 \quad (5)$$

Student's t-values: 8.29, 2.40

$n = 12 \qquad r^2 = .90 \qquad s = 0.33 \qquad F = 42.59$

The logarithm of molecular weight (MW) gave a stronger correlation in this regression than MW ($t = 1.55$). In Figure 4, the calculated values of log PC(PHSC/w) (Y estimate) are compared to the corresponding observed values for these chemicals. As shown, the calculated values are acceptably close to the observed values. The correspondence with minimal scatter suggests that this equation would be useful in predicting in vitro partitioning in the PHSC for important environmental chemicals.

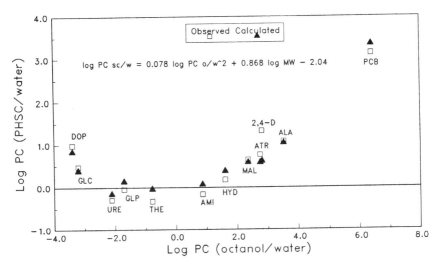

Figure 4 Correlation of the log PC(PHSC/w) and the log PC(o/w) of the 12 test chemicals. Data for PC(PHSC/w) of each chemical were the equilibrium or the maximum values with 24 h of incubation cited from Figs. 2 or 3. Open symbols are expressed as observed values, and each represented the mean of a test chemical ± SD ($n = 5$). Closed symbols express values calculated by Eq. (6). DOP, dopamine; GLC, glycine; URE, urea; GLP, glyphosate; THE, theophylline; AMI, aminopyrine; HYD, hydrocortisone; MAL, malathion; ATR, atrazine; 2,4-D, 2,4-dichlorophenoxy-acetic acid; ALA, alachlor; PCB, polychlorinated biphenyls.

DISCUSSION

Human stratum corneum has been used as an in vitro model to explore percutaneous absorption and risk of dermal exposure (Surber et al., 1990a; Potts and Guy, 1992). The traditional method of preparation is via physicochemical and enzymological processes to separate the membranous layers of the stratum corneum from whole skin (Juhlin and Shelly, 1977; Knutson et al., 1985). However, it is time-consuming and in some cases difficult to control the size and thickness of a sheet of stratum corneum.

In this study, the human callus (stratum corneum of the horny sole pads) was employed as an in vitro model. It is easier to obtain and prepare in powdered form (PHSC), and the particle sizes between 180 and 300 μm were selected. Since a corneocyte is only about 0.5 μm thick and about 30 to 40 μm long, the PHSC is considered to contain intact corneocytes plus intercellular medium structures and to keep its physical–biochemical properties. Moreover, the exterior surface of the PHSC has been greatly extended and therefore favors solutes penetrating the PHSC from any direction. A disadvantage in using the human callus is that it may display some differences in water and chemical permeation when compared with membranous stratum corneum (Barry, 1983).

The effect of varying initial chemical concentration in the vehicle shows a parallel increase in the PC(PHSC/w) until reaching maximum, and then remains approximately the same or slightly decreased (Table 1 and Fig. 1). This is consistent with results of Surber et al. (1990a, 1990b) on whole stratum corneum. Chemical partitioning from the vehicle into the SC includes processes in which molecular binding occurs at certain sites of the SC as well as simple partitioning. Equilibration of partitioning is largely dependent on the saturation of the chemical binding sites of the SC (Surber et al., 1990a; Rieger, 1993). The results from Table 1 and Fig. 1 also indicate that under a given experimental condition, the maximum degree of partitioning was compound specific. As the SC contains protein, lipids, and various lower molecular weight substances with widely differing properties, the many available binding sites will display different selective affinities with each chemical. Thus, the degree of maximum binding or of equilibration varies naturally with molecular structure (Rieger, 1993).

PC(PHSC/w) values of chemicals are sensitive to their lipophilicity or hydrophilicity (Fig. 4). Thus, two types were classified: those for lipophilic compounds where PHSC/water partitioning is proportional to their liposolubilities [as determined by log PC(o/w), Eq. (3)], and those for hydrophilic compounds where partitioning appears to relate to water solubility or reverse lipophilicities [Table 2 and Eq. (4)]. This result demonstrated that the solubility limit of a compound in the SC was important in determining the degree of partitioning, as suggested by Potts and Guy (1992). On the basis of the solubility limit of a chemical, the uptake process of water-soluble or lipid-soluble substances was controlled by the protein domain or the lipid domain, respectively, or a combination of the two (Hansch and Leo, 1979; Raykar et al., 1988). The protein and lipid domains in the SC are histologically revealed as a mosaic of cornified cells containing cross-linked keratin filaments and intercellular lipid-containing regions (Elias, 1981). Since the lipophilicity of the lipid domain in the SC is much higher than that of water, a lipophilic compound would partition into the SC in preference to water. Thus, when water is employed as the vehicle, the PC(PHSC/w) will increase with increasing lipophilicity of solute (Scheuplein and Bronaugh, 1983). Conversely, the protein domain of the SC is significantly more polar than octanol and governs the uptake process of hydrophilic chemicals (Hansch and Leo, 1979; Raykar et al., 1988). For very lipophilic compounds, low solubility in water rather than increased solubility in the SC can be an important factor (Scheuplein and Bronaugh, 1983). Moreover, in addition to

Table 2 Molecular Weight (MW), log PC(o/w), pK_a, and pH Values of the 12 Test Chemicals

Chemical	MW[a]	log PC(o/w)[b]	pK_a[a]	pH (in H$_2$O)	Ionization[f] (in H$_2$O)
Dopamine	153	−3.40	2[c]	4.0	I
Glycine	75	−3.20	2.34, 9.6	4.5	PI
Urea	60	−2.11	0.18[c]	2.1	PI or I
Glyphosate	169	−1.70	2.27, 5.58, 10.25[d]	7.3	N or PI
Theophylline	180	−0.76	8.77	7.6	I (mono-ion)
Aminopyrine	231	0.84	5	4.8	N
Hydrocortisone	362	1.61	7[c]	8.1	N
Malathion	330	2.36	7.5[c]	5.2	N
Atrazine	216	2.75	8.15[c]	6.2	N
2,4-D	221	2.81	4[c]	5.9	P
Alachlor	270	3.52	?[e]	6.6	N
PCB	326	6.40	7[c]	6.5	N

[a]Molecular weight and water solubility of chemical was cited from the *Merck Index*, 11th ed. (1989).
[b]Logarithm partition coefficient was cited from Hansch and Leo (1979).
[c]Data from Sigma Technical Service.
[d]Data from Monsanto Company (St. Louis, MO).
[e]+, Fully dissolved in water; −, almost insoluble in water; ?, unknown.
[f]I, ionized in water; PI, partially ionized in water; N, nonionized in water.

partitioning into these two domains, some amount of chemicals may be taken into the SC as the result of water hydration. This is the so-called "sponge domain" of Raykar et al. (1988). They assume that this water, having the properties of bulk water, carries an amount of solute into the SC equal to the amount of solute in the same volume of bathing solution. Therefore, for hydrophilic compounds and some lower lipophilic compounds, the partitioning process may include both the protein domain and sponge domain.

The data of Figs. 2 and 3, which display the PC(PHSC/w) as a function of incubation time, suggest that the time required to reach PC equilibrium varies chemically. This is consistent with the reports of Surber (1990a, 1990b). Two chemicals containing a similar nitrogen ring structure, theophylline and atrazine, both reach equilibration at around 6 h of incubation. Most higher lipophilic compounds such as 2,4-D, atrazine, and PCB [log PC(o/w) = 2.75–6.40] and higher hydrophilic compounds such as dopamine [log PC(o/w) = −3.40] reached equilibrium at around 12 h. For the other compounds with relatively lower lipophilicities and hydrophilicities, partition coefficients continued to increase up to 24 h, which suggested that the time to reach equilibration may be beyond 24 h. However, in comparing the ratio of the PC value at 12 h of incubation to 2 or 6 h and the PC value at 24 h to 12 h, the latter was much smaller than the former (Figs. 2 and 3). This suggests that the equilibrium may not be far from 24-h incubation times. Moreover, for those compounds with log PC(o/w) in the range of -2 to $+1.5$, the log PC(PHSC/w) values were approximately the same (Table 1 and Fig. 4). These results suggested that for higher lipophilic and hydrophilic compounds, the uptake process might be controlled by the lipid and protein domain, respectively, while for intermediate lipophilic and hydrophilic compounds, partitioning into the SC may partially depend upon the amount of water of hydration in addition to participation of both domains. Another possibility relates to the swelling of whole SC from prolonged hydration (Scheuplein and Morgan, 1967; Robbins and Fermee, 1983; Anderson et al., 1988). To the extent that progressive hydration occurs within the SC, the number of sites accessible for solute sorption is increased (Rieger, 1993), which may result in increasing PC(PHSC/w) for some chemicals. Theophylline (Fig. 2) and atrazine (Fig. 3) reached their equilibrium in around 6 h. These compounds contain a similar nitrogen ring structure with hydrogen bonding capacity that may facilitate binding to the protein domain. Full hydration may disrupt this type of binding and lead to some reversal of adsorption.

In this study, powdered stratum corneum was immersed in the vehicle (water) containing a known concentration of a solute for certain times. Ideally, the model should mimic the physiological pH that exists across human skin in vivo. Experimentally, however, distilled deionized water was employed as the vehicle for chemical partitioning into the SC and natural pH levels were allowed to prevail. Table 2 shows the logarithm of octanol/water partition coefficients, the pK_a values, and the pH values (in distilled deionized water) of the 12 test chemicals. Seven of the chemicals are neutral or nearly neutral in water, while three are partially ionized and two are fully ionized. There are reports suggesting that the pH value of the vehicle can influence the PC(SC/w) of some chemicals (Pardo et al., 1992; Downing et al., 1993).

In conclusion, the powdered human stratum corneum (callus) offered an experimentally easy in vitro model for determination of chemical partitioning from water into the SC. Due to the heterogeneous nature of the SC, the number and affinity of the SC binding sites may vary from chemical to chemical, depending on molecular structure. For the most lipophilic compounds, the PC(PHSC/w) values were governed by the lipid domain, whereas PC values of the more hydrophilic compounds are determined by the protein domain and possibly by the sponge domain. These relationships can be expressed by the log PC(PHSC/w) of these chemicals as a function of the corresponding square of log PC(o/w) and log MW [Eq. (5)], which is useful in predicting various chemical partitioning into the SC in vitro.

The major route for pesticide and herbicide exposure to humans is via the skin; therefore, risk assessment requires an estimate of percutaneous absorption. The U.S. EPA (1992) has adopted the octanol:water partition coefficient method for the many situations where true absorption data are not available. Obtaining true absorption data or partitioning data with intact stratum corneum is time-consuming and expensive. While log P(octanol/water) has been used to model transdermal transport (Potts and Guy, 1992), there are major differences between this medium and PHSC. n-Octanol is homogeneous and exerts its solution forces uniformly and independent of the solute molecular structure and polarity. PHSC is nonhomogeneous and responds differently to variations in polarity and structure by providing a range of physical structures that model the lipid, protein, and sponge domains of intact SC. The PHSC method is easy to prepare and inexpensive to use. Second, the PHSC is derived from human tissue and substitutes for octanol in the U.S. EPA methodology. This should better predict the interactions of hazardous chemicals and human skin. The binding and absorption of hazardous chemicals in water and soil to skin have recently been defined for cadmium (Wester et al., 1992), PCBs (Wester et al., 1993a), and arsenic (Wester et al., 1993b) using PHSC. This chapter defines the in vitro methodology and correlations using PHSC for such studies for toxicology involving human skin.

REFERENCES

Anderson, B. D., Higuchi, W. I., and Raykar, P. V. (1988). Heterogeneity effects on permeability-partition coefficient relationships in human stratum corneum. Pharm. Res. 5(9):566–573.

Barry, B. W. (1983). Structure, function, diseases, and topical treatment of human skin. In Dermatological Formulations: Percutaneous Absorption, ed. B. W. Barry, pp. 1–48. Marcel Dekker, New York.

Blank J. H. (1965). Cutaneous barriers. J. Invest. Dermatol. 45:249–256.

Downing, D. T., Abraham, W., Wegner, B. K., Willma, K. W., and Marchall, J. L. (1993). Partition of sodium dodecyl sulfate into stratum corneum lipid liposomes. Arch. Dermatol. Res. 285:151–157.

Elias, P. M. (1981). Lipids and epidermal permeability barrier. Arch. Dermatol. Res. 270:95–117.

Guy, R. H., and Hadgraft, J. (1991). Principles of skin permeability relevant to chemical exposure. In Dermal and Ocular Toxicology, ed. D. W. Hobson, pp. 221–246, CPC Press, Boston.

Hansch, C., and Leo, A., eds. (1979). Substituent Constants for Correlation Analysis in Chemistry and Biology. John Wiley and Sons, New York.

Juhlin, L., and Shelly, W. B. (1977). New staining techniques for the Langerhans cell. Acta Dermatol. (Stockh.) 57:289–296.

Knutson, K., Potts, R. O., Guzek, D. B., Golden, G. M., Mckie, J. E., Lambert, W. J., and Higuchi, W. I. (1985). Macro and molecular physical-chemical considerations in understanding drug transport in the stratum corneum. J. Contr. Rel. 2:67–87.

Pardo, A., Shiri, Y., and Cohen, S. (1992). Kinetics of transdermal penetration of an organic ion pair: Physostigmine salicylate. J. Pharm. Sci. 81(10):990–995.

Potts, P. O., and Guy, R. H. (1992). Predicting skin permeability. Pharm. Res. 9(5): 663–669.

Raykar, P. V., Fung, M. C., and Anderson, B. D. (1988). The role of protein and lipid domains in the uptake of solutes by human stratum corneum. Pharm. Res. 5(3):140–150.

Rieger, M. (1993). Factors affecting sorption of topically applied substances. In Skin Permeation Fundamentals and Application, ed. J. L. Zatz, pp. 33–72, Allured, Wheaton, IL.

Robbins, C. R., and Fermee, K. M. (1983). Some observations on the swelling of human epidermal membrane. J. Soc. Cosmet. Chem. 34:21–34.

Scheuplein, R. J., and Blank, I. H. (1973). Mechanisms of percutaneous absorption. IV. Penetration of nonelectrolytes (alcohols) from aqueous solutions and from pure liquids. J. Invest. Dermatol. 60:286–296.

Scheuplein, R. J., and Bronaugh, R. L. (1983). Percutaneous absorption. In Biochemistry and Physiology of the Skin, vol. 1, ed. L. A. Goldsmith, pp. 1255–1294, Oxford University Press, Oxford.

Scheuplein, R. J., and Morgan, L. J. (1967). Bound water in keratin membranes measured by a microbalance technique. Nature 214:456.

Surber, C., Wilhelm, K. P., Hori, M., Maibach, H. I., and Guy, R. H. (1990a). Optimization of topical therapy: Partitioning of drugs into stratum corneum. Pharm. Res. 7(12):1320–1324.

Surber, C., Wilhelm, K. P., Maibach, H. I., Hall, H., and Guy, R. H. (1990b). Partitioning of chemicals into human stratum corneum: Implications for risk assessment following dermal exposure. Fundam. Appl. Toxicol. 15:99–107.

U.S. Enviromental Protection Agency. (1992). Interim Report: Dermal Exposure Assessment: Principles and Applications. Washington, DC: U.S. EPA.

Wester, R. C., Mobayen, M., and Maibach, H. I. (1987). In vivo and in vitro absorption and binding to powdered stratum corneum as methods to evaluate skin absorption of environmental chemical contaminants from ground and surface water. J. Toxicol. Environ. Health 21:367–374.

Wester, R. C., Maibach, H. I., Sedik, L., Melenders, J., Dizio, S., and Wade, M. (1992). In vitro percutaneous absorption of cadmium from water and soil into human skin. Fundam. Appl. Toxicol. 19:1–5.

Wester, R. C., Maibach, H. I., Sedik, L., and Melenders, J. (1993a). Percutaneous absorption of PCBs from soil: In vivo rhesus monkey, in vitro human skin, and binding to powdered human stratum corneum. J. Toxicol. Environ. Health 39:375–382.

Wester, R. C., Maibach, H. I., Sedik, L., Melenders, J., and Wade, M. (1993b). In vivo and in vitro percutaneous absorption and skin decontamination of arsenic from water and soil. Fundam. Appl. Toxicol. 20:336–340.

Yardley, A. J. (1987). Epidermal lipids. Int. J. Cosmet. 59:13–19.

Zatz, J. L. (1993). Scratching the surface: Rational approaches to skin permeation. In Skin Permeation Fundamentals and Application, ed. J. L. Zatz, pp. 11–32, Allured, Wheaton, IL.

16
In Vitro Percutaneous Absorption of Model Compounds Glyphosate and Malathion from Cotton Fabric into and Through Human Skin

Ronald C. Wester, Danyi Quan,* and Howard I. Maibach
University of California School of Medicine, San Francisco, California

I. INTRODUCTION

The treated surface that is in most contact with skin is fabric. Consider the clothes we wear day and night, the sheets and blankets, and the fabric in rugs and upholstery. The fabric environment may have been assumed safe in the following exposure problems. Brown (1970) and Armstrong et al. (1969) reported separate cases of phenolic disinfectants in the hospital laundry causing death of infants and sickening others. By having the toxic compounds in clothing such as diapers, the compounds were spread over a large surface area on a skin site with potential absorption. Both parameters (large surface area and application to the urorectal area) enhance absorption (Wester and Maibach, 1982). If the diaper was covered with rubber pants or more clothing, this could enhance absorption. Dermatitis is reported for chemical finish in textiles (possible pesticides in the raw cotton, chemicals in the manufacturing process, chemicals added for the correct color and sheen). These must involve chemical transfer from fabric to skin. The clothing of field workers is filled with pesticides from spraying; the work of Snodgrass (1992) suggests it may not launder out, but remain bioavailable. The pesticide "bomb" in the house settles on rugs, fabric, chairs, etc. and the baby crawls over it. Finally, it should be noted that insecticide sprayed

---
*Present affiliation: Theractech Corporation, Salt Lake City, Utah.

into uniforms of Desert Storm personnel may have transferred from the uniform fabric into and through the soldiers' skin.

Glyphosate and malathion were chosen as model chemicals for their differing chemical–physical properties. Glyphosate is a water-soluble herbicide and malathion is a water-insoluble insecticide. Previous in vivo studies in the rhesus monkey and humans showed that the percutaneous absorption of glyphosate and malathion was 2.2% and 4.48% of the applied doses, respectively (Wester et al., 1983, 1991). In the current study, an in vitro percutaneous absorption study in human skin was performed to examine the extent of absorption across cotton sheets into human skin and residue in the skin under different treatments. The cotton sheets treated with these two chemicals were compared to absorption from solution.

II. MATERIALS AND METHOD

A. Study Design

Wester et al. (1994) showed that a time response exists for chemicals in cloth sheets. Chemicals (including glyphosate and malathion) when applied to cotton sheets are readily extractable within the first few hours. However, after allowing the chemicals to remain in cloth for intervals to 48 h, the extractability of chemicals decreases greatly. Sonication dislodged the chemicals, which suggested that the chemicals were not chemically bonded within the fabric, but were probably sequestered within the fabric away from the solvent. Thus, in this study glyphosate and malathion were added to fabric and allowed to "dry" for up to 2 days. Water was added to "dry" sheets to simulated skin wet from sweating or other wet conditions.

B. Materials

Glyphosate and malathion were of analytical grade from commercial sources. [^{14}C]Glyphosate was purchased from Amersham Co., and [^{14}C]malathion was obtained from Sigma. All other chemicals and reagents purchased from commercial sources were of analytical grade.

C. Pretreatment of Cotton Sheets

An area of 0.8 cm^2 of cotton sheet (100% cotton, J. E. Morgan Working Johns) was treated with 50 μl of 1% glyphosate aqueous solution containing [^{14}C]glyphosate (0.2 μC/μl) or 1% malathion ethanol aqueous solution containing [^{14}C]malathion (0.2 μC/μl), respectively. The treated sheets were dried for 0, 1, or 2 days at room temperature, and then the dried sheets were directly applied to the donor side of human skin in a diffusion cell.

D. In Vitro Human Skin Absorption Experiment

The in vitro assembly consisted of flow-through design glass skin diffusion cells (LG-1084-LCP, Laboratory Glass Apparatus, Inc., Berkeley, CA). Human cadaver skin was dermatomed to a thickness of 1 mm. The dermatomed skin was mounted in the diffusion cells, with an available skin area of 0.8 cm^2. The receiver side consisted of 3.5 ml of phosphate-buffered saline (PBS, pH 7.4), and the donor side consisted of a) 300 μl of 1% glyphosate aqueous solution containing [^{14}C]glyphosate (0.2 μC/μl) or 300 μl of 1% malathion in 50% ethanol aqueous solution containing [^{14}C]malathion (0.2 μC/μl); or b) an area of 0.8 cm^2 of cotton sheet pretreated with these two chemicals. The receiver cells were placed over magnetic stirrers, and all the absorption experiments were carried out at 37°C using a recirculating constant-temperature water bath. Flow rate was 3 ml/h. At designed intervals, samples were automatically collected into scintillation vials.

After each experiment, the residual receiver solution, donor solution or cotton, washing solution from the skin surface and the skin itself were collected for further analysis. Details of the in vitro methodology are found in a recent text (Bronaugh and Maibach, 1991).

E. Scintillation Counting

For the liquid sample, background control samples and liquid test samples were mixed with 10 ml of liquid scintillation solution (Universal ES, Costa Mesa, CA) and counted in a Packard Tri-Carb 4640 counter (Packard Instruments).

For the skin sample, each skin sample was first mixed with 5 ml of tissue solubilizer solution (Soluene 350, Packard) and stored at room temperature for 12 h; then 300 μl of acetic acid was added to neutralize the mixture. Finally, 5 ml of the Ultimate in LSC-Cocktail (Ultima Gold, Packard) was added to the neutralized solution and kept for scintillation counting.

F. Data Calculation

The steady-state flux is calculated from the slope of plots of the amount of chemical absorbed versus time:

$$\text{Flux (J)} = (dc/dt)/A \quad \text{(unit: μg/h-cm}^2\text{)}$$

where dc/dt is the steady-state slope and A is the effective diffusion area. The lag time (unit:h) can be obtained by extrapolating the linear portion to the time axis.

III. RESULTS

A. In Vitro Percutaneous Absorption of Glyphosate Across Cotton Sheets into Human Skin

Figure 1 shows the absorption profiles of 1% glyphosate through human skin (Fig. 1a) and of glyphosate across cotton sheets into human skin (Fig. 1b). The 0-day treatment means that cotton sheets were treated with glyphosate and immediately added to donor side while wet; the 1-day and 2-day treatments mean that treated cotton sheets were dried for 1 or 2 days, and "add water" means that the donor side included the treated cotton sheets and 300 μl distilled water or ethanol–aqueous solution. Clearly, when the glyphosate-treated cotton sheets were used as a donor side, the absorbed amount of glyphosate was less than that of glyphosate solution (control). Comparison of the three treatments showed that 0-day treatment caused two times higher skin absorption than did 1-day or 2-day treatments. Table 1 gives the in vitro absorption parameters of glyphosate. The absorption of glyphosate under 0-day treatment showed about 10 times less than the control sample, and 100 times less than 1-day or 2-day treatments; the absorption (cm/h) of control sample or 0-day, 1-day, and 2-day treatments was 4.59×10^{-4}, 4.90×10^{-5}, 5.21×10^{-6}, and 5.21×10^{-6}, respectively. Adding water to treated cotton sheets in a donor side resulted in an increase in the absorption compared to the 1-day or 2-day treatment. This result indicated that cotton sheets played a protective role. Decreases in lag time of glyphosate were observed (Table 1) with the change of the donor-side conditions.

The residual amount of glyphosate in the skin or on the skin (after the cotton sheet is removed) after 24-h application is shown in Table 2. Note that 0-day treatment showed the highest residual amounts in the skin as well as on the skin (6.71% and 4.32%, respectively). A possible reason is that when the cotton sheets were treated with glyphosate solution and immediately applied to the skin, the glyphosate was not yet bound to the cotton sheet and continued to separate onto the skin surface.

B. In Vitro Percutaneous Absorption of Malathion Across Cotton Sheets into Human Skin

The absorption profiles of malathion under different treatments are shown in Fig. 2. The 1-day and 2-day treatments caused the lowest percutaneous absorption due to the cotton absorption of the compound. However, 0-day treatment and adding solution to treated cotton sheets caused higher absorption. All the in vitro absorption parameters are listed in Table 3. Note that there were no great differences in the skin absorption of malathion among

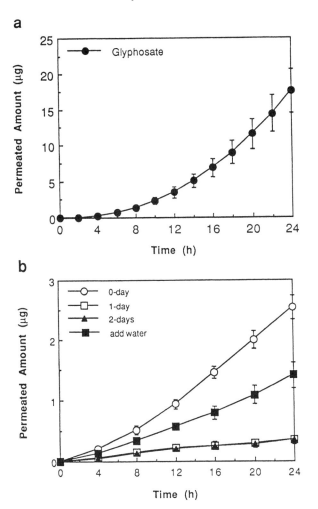

Figure 1 (a) Plot of skin-absorbed amount of glyphosate against time for the in vitro absorption of 1% glyphosate aqueous solution into human skin (control). (b) Plot of skin-absorbed amount of glyphosate against time for the in vitro absorption of 1% glyphosate aqueous solution across the cotton sheets into human skin under different treatments: 0-day treatment; 1-day treatment; 2-day treatment; adding water to cotton sheets. Mean $\pm$ SEM ($n = 6$).

Table 1 Percutaneous Absorption Parameters of Glyphosate Across Cotton Sheets into Human Skin Under Different Donor Conditions

Donor conditions	Treatment[a]	Flux (J) (μg/hr)	Permeability constant (P) (cm/hr)	Lag time (t) (hr)
1% Glyphosate solution	None	4.12 ± 1.35	$4.59 \pm 1.56 \times 10^{-4}$	10.48 ± 1.23
Cotton—1% glyphosate solution	0 day	0.47 ± 0.08	$4.90 \pm 1.41 \times 10^{-5}$	5.00 ± 0.87
Cotton—1% glyphosate solution	1 day	0.05 ± 0.01	$5.21 \pm 0.69 \times 10^{-6}$	2.52 ± 0.53
Cotton—1% glyphosate solution	2 days	0.05 ± 0.01	$5.21 \pm 0.74 \times 10^{-6}$	1.31 ± 0.68
Cotton—1% glyphosate solution	Add water	0.23 ± 0.07	$2.40 \pm 0.54 \times 10^{-5}$	3.11 ± 1.02

[a]Certain amount of 1% glyphosate aqueous solution was applied to cotton sheets, and the sheets were dried at room temperature for 0–2 days. "Add water" means that the donor side included the treated cotton sheet and a certain amount of distilled water. Mean ± SEM (n = 6).

Table 2 Comparison of Percutaneous Absorption Profiles of Glyphosate Across Cotton Sheets into Human Skin Under Different Donor Conditions

Donor conditions	Treatment[a]	Absorbed amount across the skin (%)	Residual amount on the skin (%)	Residual amount in the skin (%)
1% Glyphosate solution	None	1.42 ± 0.25	—	0.56 ± 0.13
Cotton—1% glyphosate solution	0 day	0.74 ± 0.26	6.71 ± 2.80	4.32 ± 1.95
Cotton—1% glyphosate solution	1 day	0.08 ± 0.02	0.69 ± 0.07	0.23 ± 0.13
Cotton—1% glyphosate solution	2 days	0.08 ± 0.01	0.42 ± 0.18	0.06 ± 0.01
Cotton—1% glyphosate solution	Add water	0.36 ± 0.07	—	0.23 ± 0.13

[a]Certain amount of 1% glyphosate aqueous solution was applied to cotton sheets, and the sheets were dried at room temperature for 0–2 days. "Add water" means that the donor side included the treated cotton sheet and a certain amount of distilled water. Mean ± SEM (n = 6).

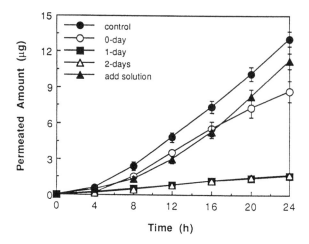

Figure 2   Plot of skin-absorbed amount of malathion against time for the in vitro absorption of 1% malathion in 50% ethanol–aqueous solution across cotton sheets into human skin under different treatments: control (no cotton sheet); 0-day treatment; 1-day treatment; 2-day treatment; adding solution to cotton sheets. Mean ± SEM ($n = 6$).

the control, 0-day treatment, and adding solution to cotton sheets (0.20, 0.14, and 0.20 cm/h, respectively). The skin absorption of malathion under 1-day and 2-day treatments was approximately six times less than that under 0-day treatment. It was found that the lag time of malathion was shorter than that of glyphosate except for the "add soln." samples.

From Table 4, it can be seen that the residual amount of malathion in the skin increased with the following order (listed as the donor conditions): adding solution to treated cotton sheets (9.40%), no cotton (control, 7.63%) > 0-day treatment (4.43%) > 1-day treatment (2.68%) > 2-day treatment (0.85%). The residual amounts of malathion in the skin were higher than the corresponding amounts of glyphosate.

Figure 3 shows the percent dose in residual cotton sheets after treatment with glyphosate and malathion and skin application. Amounts range from 71.18 to 99.5%, thus accounting for the residual cotton sheet doses.

IV. DISCUSSION

For the control solution doses, the absorbed amount through human skin was 1.42% for glyphosate and 8.77% for malathion. These results are in a good agreement with previous in vivo studies in the rhesus monkey and

Table 3 Percutaneous Absorption Parameters of Malathion Across Cotton Sheets into Human Skin Under Different Donor Conditions

Donor conditions	Treatment[a]	Flux (J) (µg/hr)	Permeability constant (P) (cm/hr)	Lag time (t) (hr)
1% Malathion solution	None	2.40 ± 0.34	$2.03 \pm 0.65 \times 10^{-1}$	5.10 ± 1.24
Cotton—1% malathion solution	0 day	1.60 ± 0.18	$1.36 \pm 0.68 \times 10^{-1}$	4.26 ± 1.10
Cotton—1% malathion solution	1 day	0.26 ± 0.04	$2.20 \pm 0.71 \times 10^{-2}$	1.08 ± 0.28
Cotton—1% malathion solution	2 days	0.27 ± 0.01	$2.29 \pm 1.04 \times 10^{-2}$	1.12 ± 0.31
Cotton—1% malathion solution	Add solution	2.34 ± 0.54	$1.98 \pm 0.82 \times 10^{-1}$	5.60 ± 1.04

[a]Certain amount of 1% malathion–aqueous ethanol was applied to cotton sheets, and the sheets were dried at room temperature for 0–2 days. "Add solution" means that the donor side included the treated cotton sheet and a certain amount of aqueous ethanol. Mean ± SEM (n = 6).

Table 4 Comparison of Percutaneous Absorption Profiles of Malathion Across Cotton Sheets into Human Skin Under Different Donor Conditions

Donor conditions	Treatment[a]	Absorbed amount across the skin (%)	Residual amount of the skin (%)	Residual amount in the skin (%)
1% Malathion solution	None	8.77 ± 1.43	—	7.63 ± 1.42
Cotton—1% malathion solution	0 day	3.92 ± 0.49	4.71 ± 1.27	4.43 ± 1.59
Cotton—1% malathion solution	1 day	0.62 ± 0.11	1.67 ± 0.77	2.68 ± 0.19
Cotton—1% malathion solution	2 days	0.60 ± 0.14	0.99 ± 0.02	0.85 ± 0.34
Cotton—1% malathion solution	Add solution	7.34 ± 0.61	—	9.40 ± 2.8

[a]Certain amount of 1% malathion–aqueous ethanol was applied to cotton sheets, and the sheets were dried at room temperature for 0–2 days. "Add solution" means that the donor side included the treated cotton sheet and a certain amount of aqueous ethanol. Mean $\pm$ SEM (n = 6).

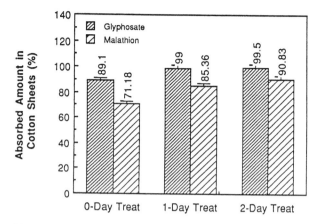

Figure 3 Residual glyphosate and malathion in cotton sheets following application to human skin.

humans (Wester et al., 1983, 1991). Thus, the in vitro data in this study are validated by previous in vivo results. The addition of chemical to cotton sheets then applied to human skin exhibited a time response, in that a "drying" time decreased the availability of chemical from fabric. This corresponds to the previous work of Wester et al. (1994) with extraction of chemicals from fabric. The addition of water or aqueous solvent to simulate sweating or wet conditions caused an increase in chemicals from cotton cloth fabric. Sweat contains water, salt, traces of albumin and urea, organic material, and other compounds. The outer layer of skin is lipophilic (oily). All of these substances could solubilize chemicals in fabric.

Chemicals are introduced to fabric at many steps during manufacture and use. Cotton plants are treated with pesticides. Many chemical dyes and finishes are added to fabric during manufacture, and these chemicals can cause human disease (Hatch and Maibach, 1986). The addition of chemicals (phenolic disinfectants) to fabric has caused human infant sickness and death (Brown, 1970; Armstrong et al., 1969). Now there are implications that the "Gulf War Syndrome" toxicity exhibited by soldiers may have been caused by interactions of insecticides and anti-nerve-gas pills. The insecticides were impregnated into the uniform fabric to ward off insects. This study and that of Snodgrass (1992) show that chemicals in fabric will be absorbed from fabric into skin and then into the systemic circulation. Clothing and other fabric media (rugs, upholstery) must be considered repositories for hazardous chemicals, and these hazardous chemicals within fabric can be transferred to human skin.

REFERENCES

Armstrong RW, Eichner ER, Klein DE, Barthel WF, Bennett JV, Johnsson V, Bruce H, Loveless LE (1969). Pentachlorophenol poisoning in a nursery for newborn infants. II. Epidemiologic and toxicologic studies. J Pediatr 75:317–325.

Bronaugh RL, Maibach HI (1991). In Vitro Percutaneous Absorption: Principles, Fundamentals, and Applications. Boca Raton, FL: CRC Press.

Brown BW (1970). Fatal phenol poisoning from improperly laundered diapers. Am J Public Health 60:901–902.

Hatch KL, Maibach HI (1986). Textile chemical finish dermatitis. Contact Dermatitis 14:1–13.

Snodgrass HL (1992). Permethrin transfer from treated cloth to the skin surface: Potential for exposure in humans. J Toxicol Environ Health 35:91–105.

Wester RC, Maibach HI (1982). Comparative percutaneous absorption. In: Neonatal Skin, Maibach H and Boisits E, eds. Marcel Dekker, New York, pp 137–147.

Wester RC, Maibach HI, Bucks DAW, Guy RH (1983). Malathion percutaneous absorption after repeated administration to man. Toxicol Appl Pharmacol 68: 116–119.

Wester RC, Melendres J, Sarason R, McMaster J, Maibach HI (1991). Glyphosate skin binding, residual tissue distribution, and skin decontamination. Rundam Appl Toxicol 16:725–732.

Wester RC, Melendres J, Maibach HI, Serranzana S, Wade M (1994). Time-response necessary in validation for extraction of herbicides and pesticides from cloth patches used in field exposure studies. Arch Environ Contam Toxicol 27: 276–280.

17

Vaginal Absorption of Prostaglandin E$_1$ in Rhesus Monkey

Ronald C. Wester, Joseph L. Melendres, Steffany Serranzana,* and Howard I. Maibach
University of California School of Medicine, San Francisco, California

I. INTRODUCTION

Pharmacological treatment of human male erectile dysfunction includes many therapeutic modalities. Many patients respond to intracavernous injection of vasoactive drugs; oral treatment has not produced satisfactory results (1). Topical administration of vasoactive drugs in the form of gels, liquid solutions, or plasters is another attractive treatment alternative. Potential topical vasoactive drugs include nitroglycerin (2,3), minoxidil (4), papaverine, and prostaglandin E$_1$ (5). Their vasoactive activity potential in human skin has been demonstrated (6), and these drugs will be absorbed across skin (7,8). Local male treatment followed by coitus could deliver the vasoactive drug into the female vagina; vaginal absorption would make the drug systemically available in the female. This study determined the vaginal absorption potential of the vasoactive drug prostaglandin E$_1$ in female rhesus monkeys.

II. MATERIALS AND METHODS

[^{3}H]Prostaglandin E$_1$ was obtained from NEN Research Products (Boston). Specific activity was 56.0 Ci/mmol, purity was 98.9%, and the radioactivity location on the molecule was [5,6-^{3}H(N)].

The intended human dose for males is projected to be 300 µl gel containing 0.2% PGE$_1$. The assumption of this study is that the maximum

---

Present affiliation: Penederm Inc., Foster City, California.

possible dose will be transferred from male to female during coitus. The rhesus monkey is assumed to be approximately (by weight) one-sixth the size of a human. Therefore a 50-μl (one-sixth dose) was administered to the upper quadrant of the vagina at least 6 cm proximal to where a urinary catheter was inserted in the urethra.

Four rhesus monkeys were administered 120 mg Septra P.O. antibiotic, to protect the animals from possible infection due to catheter insertion. The monkeys were treated twice daily during the 7-day course of vaginal PGE_1 study.

Each monkey was sedated with 50 mg Ketamine HCl (im injection) and then weighed. The monkey was placed in the prone position, and the vaginal labia was retracted using clean retractors, gauze, and hemostats. KY lubricant was aseptically applied to the tip of a Foley catheter. The catheter was inserted into the urethra and the level of vulval opening was noted on the catheter with a marking pen. The catheter was attached to a urine bag, and the access to the bag was opened.

The dosing syringe containing the prostaglandin E_1 was inserted into the upper portion of the vagina and the dose was delivered to the vaginal mucosa. The application was carefully removed and the applicator tip was wiped clean with methanol.

The dosing period was 30 min. Ketamine was administered as needed during this time. After 30 min, the application site was rinsed five times with 1-ml rinses of douching solution, which were captured with cotton balls. The cotton balls were placed into designated vials. The douching solution was 0.75% acetic acid/isotonic water and was autoclaved before use (the same composition as the formulation available commercially).

The catheters were withdrawn 40 min after dosing. The monkeys were transferred to metabolic cages for continued urine collection.

For intravenous administration, [^{3}H]prostaglandin E_1 (5 μCi) in 500 μl propylene glycol was administered as a bolus injection in the anticubital vein of the arm of the same four rhesus monkeys. The animals were housed in individual metabolic cages and complete urine excretion was collected for 7 days. The urine was assayed for ^{3}H content.

Total urine excretion was collected by metabolic cages at room temperature throughout the study. Total urine volume was recorded and a 15-ml aliquot was saved for analysis. All urine samples were well mixed and aliquoted in duplicate for direct counting in scintillation cocktail. Five-milliliter aliquots of urine were counted.

Radioactivity was measured by liquid scintillation counting in glass vials. The scintillation cocktail was Universal ES (ICN Biomedicals, Costa Mesa, CA). Background control samples and test samples were counted in a Packard model 4640 or model 1500 counter (Packard Instruments). Control and test sample counts were transferred to a computer program that subtracted $1\times$ background control samples and generated a spreadsheet with the data.

III. RESULTS

Vaginal absorption was determined by the ratio of urinary ^{3}H excretion following vaginal and intravenous administration of [^{3}H]prostaglandin E₁. Percent dose absorbed was calculated using the equation:

$$\text{Percent dose absorbed} = \frac{^3\text{H urinary excretion from iv dose}}{^3\text{H urinary excretion from vaginal dose}} \times 100$$

After intravenous administration, a total of $47.6 \pm 6.2\%$ of dose was excreted over 7 days. The majority of the dose ($38.4 \pm 1.8\%$) was excreted the first day (Fig. 1).

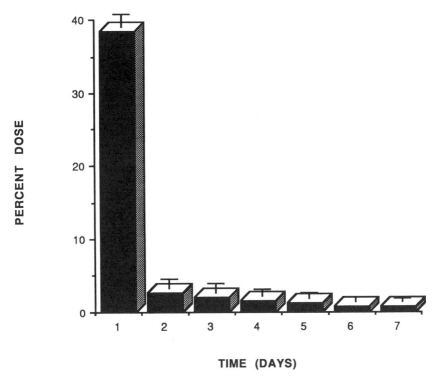

Figure 1 Urinary ^{3}H excretion after intravenous administration of [^{3}H]prostaglandin E₁.

For the 30-min vaginal dosing period, a total of 12.0 ± 5.0% of dose
was excreted over 7 days. No radioactivity (0.06 ± 0.1%) was detected in
urine collected from the catheter during dosing period. Most of the absorbed
dose was excreted over the first 3 days; excretion was less than 1% for days
4–7 (Figure 2). After the 30-min dosing period, the vagina was douched
with 5 consecutive washes and then the vulva was washed once. A total of
76.6 ± 11.1% of dose was recovered in the douche washes. Table 1 gives
the vaginal absorption of [^{3}H]prostaglandin E_1 in Rhesus monkey. Listed are
the individual urinary ^{3}H excretions following vaginal and intravenous administration. From the ratio of these excretions a prostaglandin E_1 vaginal
absorption of 25.2 ± 9.1% of dose (range 17.2–38.2%) is calculated for the
30-min dosing period. Vaginal douche recovery was 76.6 ± 11.1% of dose,
giving a total dose accountability of 101.8 ± 4.9% of dose.

No vaginal bleeding or other toxicological signs were observed during
the course of the study.

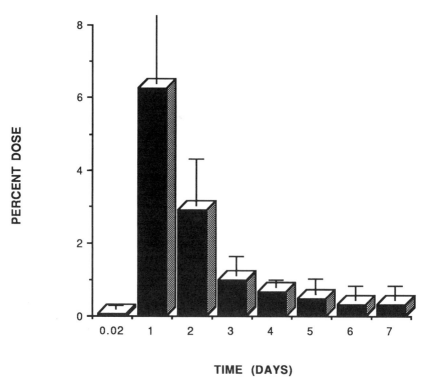

Figure 2 Urinary ^{3}H excretion following vaginal administration of [^{3}H]pros-
taglandin E_1.

Table 1 Vaginal Absorption of [³H]Prostaglandin E₁ in Rhesus Monkey

Animal number	Urinary excretion (vaginal)	Urinary excretion (intravenous)	Vaginal absorption[a]	Recovery (vaginal douche)	Total[b]
18671	9.4	54.7	17.2	86.6	103.9
16082	9.1	42.9	21.2	85.1	106.3
17066	19.5	51.0	38.2	63.9	102.1
16609	10.2	41.9	24.3	70.6	94.9
Mean	12.0	47.6	25.2	76.6	101.8
SD	5.0	6.2	9.1	11.1	4.9

[a]Calculated by ratio:

$$\text{Absorption} = \frac{\text{urinary } ^3\text{H excretion vaginal dose}}{\text{urinary } ^3\text{H excretion intravenous dose}} \times 100$$

[b]Calculated: % vaginal dose absorbed plus % recovery vaginal douche.

IV. DISCUSSION

Vaginal absorption of prostaglandin E₁ in rhesus monkey following a 30-min dose exposure was 25.2 ± 9.1% of dose, suggesting a high-absorbing body site. An absorption study in the rhesus monkey was testosterone showed vaginal absorption to be the highest (63.3 ± 2.6% of dose), compared to scalp (20.4 ± 2.7%), cheek (9.6 ± 0.2%), chest (5.3 ± 0.6%), and ventral forearm (8.8 ± 2.5%) for a 24-h dosing period (9). Note that with prostaglandin E₁ the remainder of the dose was present in the vagina (douche wash recovered 76.6 ± 11.1%). Thus, it must be assumed that without the douche, potentially most of the prostaglandin E₁ could have been absorbed. This information probably projects to other drugs and treatments involving vaginal administration. For prostaglandin E₁ and other potential male treatments, a protective condom would be a necessity. However, for other treatments involving direct vaginal dosing (the currently planned anti-AIDS vaginal creams would be an example), the potentially high vaginal absorption should be part of the risk/benefit evaluation.

REFERENCES

1. Montorsi F, Guazzoni G, Rigatti P, Pozza G. Pharmacological management of erectile dysfunction. Drugs 1995; 50:465–479.
2. Owen JA, Saunders F, Harris C, Fenemore J, Reid K, Surridge D, Condra M, Morales A. Topical nitroglycerin: A potential treatment for impotence. J Urol 1989; 141:546–548.

3. Nunez BD, Anderson DC Jr. Nitroglycerin ointment in the treatment of impotence. J Urol 1993; 150:1241–1243.

4. Cavallini G. Minoxidil versus nitroglycerin: A prospective double-blind controlled trial in transcutaneous erection facilitation for organic impotence. J Urol 1991; 146:50–53.

5. Mahmoud KZ, El Dakhli MR, Fahmi IM, Abdel-Aziz ABA. Comparative value of prostaglandin E1 and papaverine in treatment of erectile failure: Double-blind crossover study among Egyptian patients. J Urol 1992; 147:623–626.

6. Wester RC, Maibach HI, Guy RH, Novak E. Minoxidil stimulates cutaneous blood flow in human balding scalps: Pharmacodynamics measured by laser Doppler velocimetry and photopulse plethysmography. J Invest Dermatol 1984; 82:515–518.

7. Wester RC, Noonan PK, Smeach S, Kosobud L. Pharmacokinetics and bioavailability of intravenous and topical nitroglycerin in the rhesus monkey: Estimate of percutaneous first-pass metabolism. J Pharm Sci 1983; 72:745–748.

8. Adachi H, Irie T, Uekama K, Manako T, Yano T, Saita ME. Combination effects of O-carboxymethyl-O-ethyl-β-cyclodextrin and penetration enhancer HPE-101 on transdermal delivery of prostaglandin E_1 in hairless mice. J Pharm Sci 1993; 1:117–123.

9. Wester RC, Noonan PK, Maibach HI. Variations in percutaneous absorption of testosterone in the rhesus monkey due to anatomic site of application and frequency of application. Arch Derm Res 1980; 267:229–235.

18

Human Cadaver Skin Viability for In Vitro Percutaneous Absorption
Storage and Detrimental Effects of Heat Separation and Freezing

Ronald C. Wester, Julie Christoffel, Tracy Hartway, Nicholas Poblete, and Howard I. Maibach
University of California School of Medicine, San Francisco, California

James H. Forsell
Northern California Transplant Bank, San Rafael, California

I. INTRODUCTION

Human cadaver skin is utilized in hospitals and research laboratories for various reasons. For nearly four decades, hospitals have banked skin for use as an effective temporary covering for burn wounds (1). Research laboratories also use cadaver skin to study the percutaneous absorption of drugs (2) and hazardous chemicals of environmental concern (3). Procedures such as heat treatment to separate epidermis from dermis are performed as part of the skin preparation for these studies. Human cadaver skin is not an easily obtained commodity, and storage for use becomes a necessity. Refrigeration and freezing of skin are commonly done. With treatment and storage, skin viability has become a concern. This study determined human cadaver skin viability from point of death through time of storage, and the effect of heat and freezing treatment.

II. MATERIALS AND METHODS

Human cadaver skin was obtained from the Northern California Transplant Bank (San Rafael, CA). All donors were caucasian, aged 21–53 years, and both genders were represented. The time of a subject's death was recorded; skin was taken from the subject's thighs by use of a dermatome targeted to 500 μm. The skin was immediately placed in Eagle's minimum essential media (MEM) with Earle's balanced salt solution (BSS) (In Vitro Scientific Products Corp., St. Louis, MO) and refrigerated at 4°C. The skin was then transported on ice to the laboratory and stored refrigerated at 4°C in Eagle's MEM-BSS with 50 μg/ml gentamicin until used.

Dermatomed skin samples were mounted in an in vitro assembly consisting of flow-through design glass diffusion cells (Laboratory Glass Apparatus, Inc., Berkeley, CA). Eagle's MEM-BSS gentamicin served as receptor fluid and flow rate was 1.5 ml/h. The receptor fluid was at 37°C; skin surface temperature was 32°C. Eagle's MEM contains glucose, and glucose metabolism to lactate in anaerobic energy metabolism was used as the measure of skin viability. Lactate production was determined using Sigma diagnostic kit 826-UV (St. Louis, MO) and a Hitachi spectrophotometer (San Jose, CA).

Dermatomed skin was used as stored in the refrigerator, frozen at −22°C for storage, or heat separated (60°C water for 1 min) into epidermis and dermis.

III. RESULTS

Figure 1 shows lactate production from four human skin sources mounted in the diffusion system. Each data point represents 4-h receptor fluid collection intervals over the 24-h diffusion period. No chemical was dosed on the skin, just receptor fluid perfusing the skin. Lactate was produced by the skin sources over the full 24-h period. The lactate curves rise in the early part of the period, where glucose is diffusing into the skin and lactate is diffusing out of the skin. Two skin sources reached steady state at about 12 h; lactate from the other two skin sources continued to rise until the process was stopped at 24 h.

Table 1 and Fig. 2 give the cumulative lactate produced (mmol/L) for the 24-h perfusion period. Human skin was either dermatomed skin or heat-separated epidermis used within the time period of 0.75 to 13 days after donor death. The numbers of skin samples and the numbers of skin donors for each time period are also listed. With dermatomed skin, refrigerated for only 0.75 days, the 24-h cumulative lactate had a high of 19.8 ± 8.0 mmol/L. Lactate production decreased by day 2 ($p < .000$) and remained steady through day 8. Lactate production decreased further by approximately one-half between day 8 and day 13. Heat-separated epidermis lactate production was less than dermatomed skin ($p < .01$) at 2.0 ± 1.1 mmol/L. This level

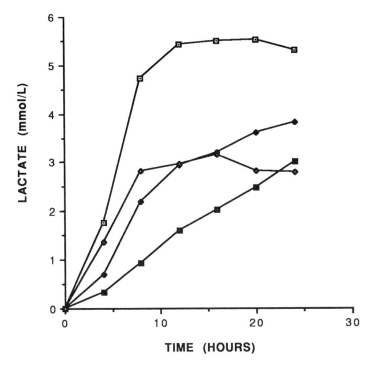

Figure 1 Lactate production from glucose (anaerobic energy metabolism) of four human skin sources perfused for 24 h with Eagle's MEM-BSS with 50 μg/ml gentamicin. An initial time delay is noted where glucose absorbs into skin and lactose perfuses out of skin.

was maintained to the 2-day period, then decreased ($p < .007$) at day 3 and remained less than 1 mmol/L through the 13-day test period.

Dermatomed skin was heat-treated at 60°C for 1 min to simulate the heat-separation procedure to produce epidermis separated from dermis (but no separation was performed) (Table 2). Lactate production decreased significantly ($p < .000$; $p < .04$) for both heat-treated skin samples. Therefore, heating to separate epidermis from dermis damages viability. In another study (Table 3), lactate production was determined in heat-separated epidermis and dermis. The cumulative lactate production was much less than intact dermatomed skin, again showing the detrimental effect of heat separation on skin viability.

Table 4 shows replicates from 6 dermatomed skin and heat-separated epidermis samples that were frozen at $-22°C$. The process of freezing was detrimental to skin viability of dermatomed skin ($p < .04$). Separated epi-

Table 1 In Vitro Human Skin Viability: Glucose Energy Metabolism to Lactate

Time after death (days)[a]	Dermatomed skin		Heat-separated epidermis	
	Lactate[b] (mmol/L/24 h)	Number[c]	Lactate[b] (mmol/L/24 h)	Number[c]
0.75	19.8 ± 8.9[d,e]	6/2	2.0 ± 1.1[e,f]	3/1
2	5.9 ± 4.1[d]	13/4	1.8 ± 0.8[f]	8/3
3	8.0 ± 4.8	8/3	0.6 ± 0.5	6/2
4	6.5 ± 1.7	9/3	0.7 ± 0.4	6/2
6	6.8 ± 3.0	11/3	0.2 ± 0.1	5/2
8	4.6 ± 2.3	6/2	0.2 ± 0.1	3/1
13	2.0 ± 0.6	3/1	0.9 ± 0.4	3/1

[a]Stored refrigerated in Eagle's MEM-BSS with 50 μm/ml gentamicin.
[b]Mean ± SD.
[c]Number of skin samples/number of human skin donors.
[d]Significant at $p < .000$.
[e]Significant at $p < .01$.
[f]Significant at $p < 0.007$.

dermis was not significantly different between refrigerated and frozen ($p > .05$) because the heat-separation process to get the epidermal layer had already been detrimental to skin viability.

IV. DISCUSSION

It is logical that prolonged life and improved quality for stored skin are desirable for any transplant situation (4). During in vivo percutaneous absorption and transdermal delivery, the skin is viable and does metabolize glucose for energy, and metabolism does extend to other enzymes and other chemicals (5,6). Understanding and maintaining human cadaver skin viability places the skin use closer to the in vivo situation. This study shows that, in a sustaining media, skin can be energy viable for up to 8 days. Harvesting the skin and using it within a day of donor death gives the highest viability. Our system gets the skin quickly into sustaining media and refrigeration; it is not known if a delay from harvest to storage will affect viability. Common practices of freezing skin for storage, or heat treatment to separate epidermis from dermis, can destroy skin viability. The effect of enzymatically separating skin is not known. Bhatt et al. (7) also showed that heat treatment of hairless mouse skin for separation purposes eliminates viability.

Cadaver skin viability can be maintained and monitored. Glucose utilization can be measured by conversion of [^{14}C]glucose to $^{14}CO_2$ (8) or by

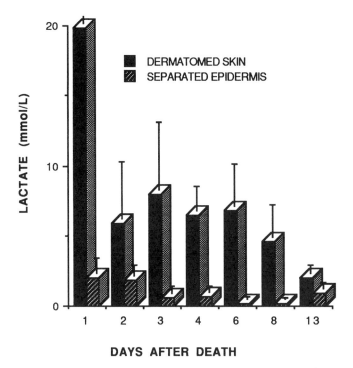

Figure 2   Viability of human skin stored refrigerated at 4°C in Eagle's MEM-BSS with 50 μg/ml gentamicin. Time indicated is that after donor's death.

Table 2   Heat Effect on Human Skin Simulating Epidermis
Heat-Separation Procedure

| | Lactate (mmol/L) produced in 24 h | | |
Source	Control dermatomed skin	Heat-treated skin[c]	Statistics
1[a]	13.6 ± 1.5	1.5 ± 0.6	$p < .000$
2[b]	7.4 ± 4.0	0.7 ± 0.14	$p < .04$

[a]Male, 21, thigh skin, used 69 h after death ($n = 3$).
[b]Female, 30, thigh skin, used 39 h after death ($n = 3$).
[c]Heated for 1 min in 60°C water.

Table 3 Viability of Dermatomed Human Skin, Heat-Separated Epidermis and Dermis

Time (h)	Lactate (mmol/L/24 h) production		
	Dermatomed skin[a]	Epidermis[a]	Dermis[a]
29	9.7 ± 2.3^{b}	1.5 ± 0.4^{b}	0.9 ± 0.4^{b}

[a]Mean $\pm$ SD; male age 25, thigh skin. Epidermis and dermis separated after heat treatment for 1 min in 60°C water. Note that epidermis plus dermis does not equal intact skin.
[b]Significant at $p < .004$.

Table 4 Freezing Effect of Human Skin on Energy Metabolism

Skin sample	Lactate production (mmol/L/24 h)			
	Dermatomed skin		Heat-separated epidermis	
	Refrigerated[b]	Frozen[a,b]	Refrigerated	Frozen[a]
1	12.2 ± 2.1	0.1 ± 0.1	1.0 ± 0.08	0.19 ± 0.13
2	2.4 ± 0.7	0.4 ± 0.3	1.5 ± 0.6	0.18 ± 0.13
3	7.4 ± 4.0	2.6 ± 0.4	—	—
4	9.7 ± 2.3	2.4 ± 0.5	1.5 ± 0.4	2.1 ± 0.2
5	9.5 ± 0.4	1.2 ± 0.04	0.2 ± 0.1	0.1 ± 0.05
6	27.5 ± 4.1	0.0	0.2 ± 0.1	0.0

[a]Frozen 24 h or longer.
[b]Significant at $p < .04$.

lactate production (9) as shown here. The lactate production methodology does not require radioactivity use equipment.

ACKNOWLEDGMENT

We thank the Northern California Transplant Bank, which supplied the skin and made this study possible. Special acknowledgment to George Kositzin for the extra effort to obtain the skin samples.

REFERENCES

1. S.R. May and F.A. DeClement. Skin banking. Part III. Cadaveric allograft skin viability. J. Burn Care Rehab. 2:128–141 (1981).

2. R.C. Wester and H.I. Maibach. Percutaneous absorption of drugs. Clin. Pharmacokinet. 3:253–266 (1992).

3. R.C. Wester, H.I. Maibach, L. Sedik, J. Melendres, and M. Wade. Percutaneous absorption of PCBs from soil: In vivo rhesus monkey, in vitro human skin, and binding to powdered human stratum corneum. J. Toxicol. Environ. Health 39: 375–382 (1993).

4. L.N. Hurst, D.H. Brown, and K.A. Murray. Prolonged life and improved quality of stored skin grafts. Plast. Reconstr. Surg. 73:105–109 (1984).

5. R.C. Wester, P.K. Noonan, S. Smeach, and L. Kosobud. Pharmacokinetics and bioavailability of intravenous and topical nitroglycerin in the rhesus monkey: Estimate of percutaneous first pass metabolism. J. Pharm. Sci. 72:745–748 (1983).

6. R.L. Bronaugh, R.F. Stewart, and J.E. Storm. Extent of cutaneous metabolism during percutaneous absorption of xenobiotics. Toxicol. Appl. Pharmacol. 99: 534–543 (1989).

7. R.H. Bhatt, G. Macali, J. Galinkin, P. Palicharla, R. Koch, D.P. West, and L.M. Solomon. Determination and correlation of in vitro viability for hairless mouse and human neonatal whole skin and stratum corneum/epidermis. Arch. Dermatol. Res. 289:170–173 (1997).

8. S.W. Collier, N.M. Sheikh, A. Sakr, J.L. Lichtin, R.F. Stewart, and R.L. Bronaugh. Maintenance of skin viability during in vitro percutaneous absorption/ metabolism studies. Toxicol. Appl. Pharmacol. 99:522–533 (1989).

9. M.E.K. Kraeling, R.J. Lipicky, and R.L. Bronaugh. Metabolism of benzocaine during percutaneous absorption in the hairless guinea pig: acetylbenzocaine formation and activity. Skin Pharmacol. 9:221–230 (1996).

19
Interrelationships in the Dose Response of Percutaneous Absorption

Ronald C. Wester and Howard I. Maibach
University of California School of Medicine, San Francisco, California

In most medical and toxicological specialties the administered dose absorbed is defined precisely. This has not always been so in dermatoxicology and dermatopharmacology. The absorbed dose is usually defined as percent applied dose absorbed, flux rate, and/or permeability constant. This may suffice for the person creating the data, but it is incomplete for the person judging the worthiness of the data. Chemical absorbed through skin is usually a low percentage of the applied dose. If 5% is absorbed, it is more than curiosity to question where the other 95% resides. Most critical is whether the remaining dose was in place during the course of the study and there is dose accountability. A second critical question is where the clinical or toxicological response lies in relationship to the topically applied dose, the standard safety and efficacy issue that a dose response will provide. The third question is whether absorption is linear to administered dose, that is, the dose response.

This chapter defines our current, albeit far from perfect, understanding of the relation of applied dose to percutaneous absorption. The dose response is a sound scientific principle, and studies on percutaneous absorption need to apply this principle in some portion of a study.

I. DOSE RESPONSE IN REAL TIME

Breath analysis is being used to obtain real-time measurements of volatile organics in expired air following exposure in rats and humans. The exhaled breath data is analyzed using physiologically based pharmacokinetic (PBPK) models to determine the dermal bioavailability of organic solvents. Human volunteers and animals breathe fresh air via a new breath-inlet system that allows for continuous real-time analysis of undiluted exhaled air. The air supply system is self-contained and separated from the exposure solvent-laden environment. The system uses a Teledyne 3DQ Discovery ion trap mass spectrometer (MS/MS) equipped with an atmospheric sampling glow discharge ionization source (ASGDI). The MS/MS system provides an appraisal of individual chemical components in the breath stream in the single-digit parts per billion (ppb) detectable range for compounds under study, while maintaining linearity of response over a wide dynamic range.

1,1,1-Trichlorethane (methyl chloroform, MC) was placed in 4 L of water at 1, 0.1, and 0.01% concentration (MC is soluble at the 0.1 and 0.01% concentrations) and volunteers placed a hand in the bucket of water containing the MC. Figure 1 gives the in vivo human dose response. The analytical system measures MC in exhaled breath every 4 s and displays the amount in real time. For investigator "data analysis sanity," 75 4-s data points are averaged at each 5 min of elapsed time as presented in Fig. 1. At 1% dose MC is immediately and rapidly absorbed through the skin, increasing to over 2500 ppb at 2 h when the hand is removed from the solvent/water bucket. Breath MC levels then decline. The 0.1% dose (MC which is completely soluble in water) peaks at approximately 1200 ppb at 2 h when the hand is removed from the solvent/water exposure. There is an initial delay of 30–60 min in solvent absorption from the water-soluble dose, not seen with the higher dose that exceeded solubility. The lowest dose, 0.01%, was near analytical limitation. These data are then analyzed for human body rate distribution using a physiologically based pharmacokinetic model, and appropriate absorption and excretion rates are produced (Wester et al., unpublished observations).

II. PHARMACODYNAMIC DOSE RESPONSE

Pharmacodynamics is a biological response to the presence of a chemical. The chemical needs to be bioavailable for the response to happen, and a dose response within the limits of the biological response can be measured. This can happen with in vivo human skin, and an example is bloodflow changes measured by laser Doppler velocimetry (LDV). Minoxidil is a direct-acting peripheral vasodilator originally developed for hypertension, but now available to promote hair growth. In a double-blind study, balding volunteers were dosed with 0, 1, 3, and 5% minoxidil solutions once each day on 2 consecutive days and bloodflow was measured by LDV (Fig. 2). On day 1 an LDV dose response showed the highest bloodflow from the 5%

Dose-Response: Methyl Chloroform in 4 L Water

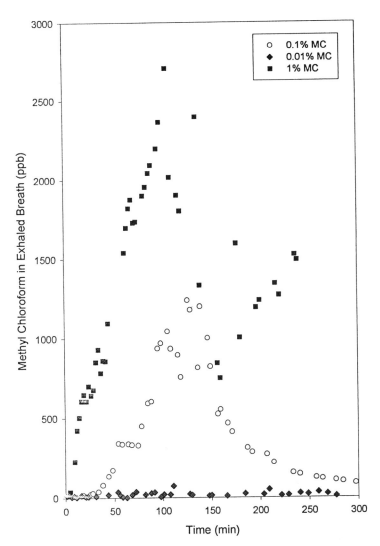

Figure 1 In vivo real-time absorption of the solvent methyl chloroform from human volunteers with a hand in 4 L water containing 1, 0.1, or 0.01% solvent. Methyl chloroform is soluble in water at 0.1% (note lag time) but exceeds solubility at 1% (note no lag time). The hand was removed from the solvent/water at 120 min.

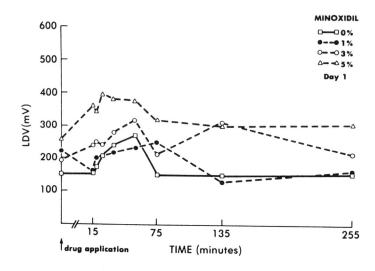

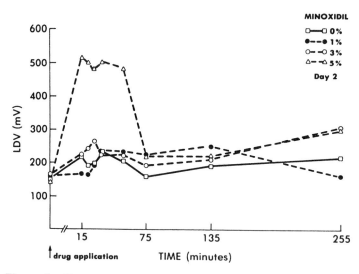

Figure 2 Pharmacodynamic laser Doppler velocimetry dose response of minoxidil stimulating skin blood flow in the balding scalp of male volunteers.

dose, followed by the 3% dose, with the 1% dose with giving no response above control. Day 2 results clearly showed a pharmacodynamic response for the 5% solutions ($p < .0001$) but the other doses were near that of control (Wester et al. 1984, 1985).

This effect of concentration on percutaneous absorption also extends to the penetration of corticoids as measured by the pharmacodynamic vasoconstrictor assay. In this type of assay Maibach and Stoughton (1973) showed that, in general, there is a dose-response relationship, with increasing efficacy closely following increased dose. A severalfold difference in dose can override differences in potency between the halogenated analogues. If this applies to corticoids, it could also apply for other chemicals.

Biological response differs from straight chemical analyses (only limited by analytical sensitivity) in that a threshold is probably needed to initiate the response and there are probably limits to the extent that the biology can respond. However, within these parameters some dose response should exist.

III. DOSE RESPONSE INTERRELATIONSHIPS

The interrelationships of dose response in dermal absorption are defined in terms of accountability, concentration, surface area, frequency of application, and time of exposure. *Accountability* is an accounting of the mass balance for each dose applied to skin. *Concentration* is the amount of applied chemical per unit skin surface area. *Surface area* is usually defined in square centimeter of skin application or exposure. *Frequency* is either intermittent or chronic exposure. "Intermittent" can be one, two, and so on exposures per day. Chronic application is usually repetitive and on a continuing daily basis. *Time of exposure* is the duration of the period during which the skin is in contact with the chemical before washing. Such factors define skin exposure to a chemical and subsequent percutaneous absorption.

A. Accountability (Mass Balance)

Table 1 gives in vivo percutaneous absorption of dinoseb in the rhesus monkey and rat. The absorption in the rat over a dosage range of 52–644 μg/cm^2 is approximately 90% for all of the doses (Shah et al., 1987). Conversely, absorption in the rhesus monkey for the dinoseb dose range of 44–3620 μg/cm^2 is approximately only 5%. There is an obvious difference because of species (rat and rhesus monkey). The question then becomes one of mass balance to determine dose accountability. (If dinoseb was not absorbed through skin, then what happened to the chemical?) Table 1 shows that at least 80% of the applied doses can be accounted for (rat and rhesus monkey) and Table 2 shows that in the rhesus monkey the dinoseb remained

Table 1 In Vivo Percutaneous Absorption of Dinoseb in Rhesus Monkey and Rat

Applied dose (μg/cm^2)	Skin penetration (%)	Dose accountability (%)
	Rat	
51.5	86.4 ± 1.1	87.9 ± 1.8
128.8	90.5 ± 1.1	91.5 ± 0.6
643.5	93.2 ± 0.6	90.4 ± 0.7
	Rhesus monkey	
43.6	5.4 ± 2.9	86.0 ± 4.0
200.0	7.2 ± 6.4	81.2 ± 18.1
3620.0	4.9 ± 3.4	80.3 ± 5.2

Note. Rat treated with acetone vehicle; 72–h application. Monkey treated with Premerge–3 vehicle; 24–h application.

on the skin (skin wash recovery 73.8 ± 6.8%) and was not absorbed over the 24-h application period.

B. Effects of Concentration on Percutaneous Absorption

Maibach and Feldmann (1969) applied increased concentrations of testosterone, hydrocortisone, and benzoic acid from 4 μg/cm^2 in 3 steps to 2000 μg/cm^2 (4 μg/cm^2 is approximately equivalent to the amount applied in a 0.25% topical application; 2000 μg/cm^2 leaves a grossly visible deposit of chemical). Increasing the concentration of the chemical always increased total absorption. These data suggest that as much as gram amounts of some

Table 2 Percutaneous Absorption and Accountability of Dinoseb In Vivo Study in the Rhesus Monkey

Disposition parameter	Applied dose accountability (%) at applied dose		
	43.6 μg/cm^2	200.00 μg/cm^2	3620.0 μg/cm^2
Urine	3.3 ± 1.8	4.4 ± 23.9	3.0 ± 2.1
Feces	0.8 ± 0.5	1.0 ± 0.6	3.0 ± 1.7
Contaminated solids	0.03 ± 0.02	0.02 ± 0.02	0.07 ± 0.08
Pan wash	0.04 ± 0.03	0.8 ± 1.1	0.4 ± 0.3
Skin wash	81.1 ± 4.0	75.0 ± 22.9	73.8 ± 6.8
Total accountability	86.0 ± 4.0	81.2 ± 18.1	80.3 ± 5.2

compounds can be absorbed through normal skin under therapeutic and environmental conditions.

Wester and Maibach (1976) further defined the relationship of topical dosing. Increasing concentration of testosterone, hydrocortisone, and benzoic acid decreased the efficiency of percutaneous absorption (percent dose absorbed) in both the rhesus monkey and human (Table 3), but the total amount of material absorbed through the skin always increased with increased concentration. Scheuplein and Ross (1974) also showed in vitro that the mass of material absorbed across skin increased when the applied dose was increased. The same relationship between dose applied and dose absorbed is also seen with the pesticides parathion and lindane in Table 4.

Wedig and coworkers (1977) compared the percutaneous penetration of different anatomical sites. A single dose of a ^{14}C-labeled magnesium sulfate adduct of dipyrithione at concentrations of 4, 12, or 40 $\mu g/cm^2$ per site was applied for an 8-h contact time to the forearm, forehead, and scalp of human volunteers. The results again indicated that, as the concentration increased, more was absorbed. Skin permeability for equivalent doses on different sites assumed the following order: forehead was equal to scalp, which

Table 3 Percutaneous Absorption of Increased Topical Dose of Several Compounds in Rhesus Monkey and Humans

Compound ($\mu g/cm^2$)	Totals for rhesus		Totals for humans	
	Percent	Micrograms	Percent	Micrograms
Testosterone				
34	18.4	0.7	11.8	0.4
30	—	—	8.8	2.6
40	6.7	2.7	—	—
250	2.9	7.2	—	—
400	2.2	8.8	2.8	11.2
1600	2.9	46.4	—	—
4000	1.4	56.0	—	—
Hydrocortisone				
4	2.9	0.1	1.6	0.1
40	2.1	0.8	0.6	0.2
Benzoic acid				
3	–	—	37.0	1.1
4	59.2	2.4	—	—
40	33.6	134.4	25.7	102.8
2000	17.4	348.0	14.4	288.0

Note. From Maibach and Feldmann (1969) and Wester and Maibach (1976).

Table 4 Effect of Applied Topical Concentration on
Human Percutaneous Absorption

	Totals	
Compound ($\mu g/cm^2$)	Percent	Micrograms
Parathion		
4	8.6	0.3
40	9.5	3.8
400	4.8	19.2
2000	9.0	180.0
Lindane		
4	9.3	0.4
40	8.3	3.3
400	5.7	22.8
1000	3.4	34.0
2000	4.4	88.0

Note. From Feldmann and Maibach (1974).

was greater than forearm. The total amounts absorbed increased even when the percentage of dose excreted at two doses remained approximately the same. On the forehead, proportionately more penetrated from the 40 $\mu g/cm^2$ than from the 4- and 12-$\mu g/cm^2$ doses. On the scalp the difference was even more striking, with almost twice as much proportionately penetrating from 40 $\mu g/cm^2$ than from 4 and 12 $\mu g/cm^2$. Thus, as the concentration of applied dose increased, the total amount of chemical penetrating the skin (and thus becoming systemically available) also increased for all the anatomical sites studied. Therefore, for exposure of many parts of the body, absorption can take place from all of the sites. As the concentration of applied chemical and the total body exposure increase, the subsequent systemic availability will also increase.

Although the penetration at these doses varied between anatomical sites, the percentage of the dose penetrating was similar at the three doses on the forearm and the forehead. However, on the occiput (of the scalp), there was an increasing percentage of penetration with increasing dosage. In other words, at the highest dose, the efficiency of penetration was the greatest.

Reifenrath et al. (1981) determined the percutaneous penetration of mosquito repellents in hairless dogs. As the topical dose increased in concentration, the penetration in terms of percentage of applied dose was about the same (Table 5). However, the mean total amount of material absorbed

Table 5 Percutaneous Penetration and Total Absorption of Repellents in Relation to Dose of Chemical Applied to Hairless Dog

Compound	Topical dose ($\mu g/cm^2$)	Penetration (% of applied dose)	Mean total absorbed ($\mu g/cm^2$)
Ethyl hexanediol	4	8.8	0.35
	320	10.3	33.0
N,N–Diethyl–m–toluamide	4	12.8	0.51
	320	9.4	30.1
Sulfonamide[a]	100	9.1	9.1
	320	7.5	24.0
	1000	5.4	54.0

Note. From Reifenrath et al. (1981).
[a]n–Butane sulfonamide cyclohexamethylene.

increased dramatically. An application of 4 $\mu g/cm^2$ of N,N-diethyl-m-tolu-amide gave a 12.8% absorption, resulting in a total absorption of 0.5 $\mu g/cm^2$. An increase in the dose to 320 $\mu g/cm^2$ decreased the percent absorbed to 9.4; however, the total amount of material absorbed was now up to 30.1 $\mu g/cm^2$, an increase of 60 times!

Roberts and Horlock (1978) examined the effects of concentration and repeated skin application of percutaneous absorption. Following single-treatment application with 1, 5, and 10% ointments, the penetration fluxes for salicylic acid in hydrophilic ointment increased as the concentration increased (Table 6). With extended application (on a daily basis) a change in flux was also observed, the skin underwent a change, and subsequently the penetration flux changed.

Table 6 Mean Penetration Fluxes of Salicylic Acid in Hydrophilic Ointment Base Through Excised Rat Skin After Single Treatment

Salicyclic acid concentration (w/w)	Penetration flux of salicylic acid ($mg/cm^2 \cdot h$, $\pm$ SE)
1	0.014 ± 0.002
5	0.061 ± 0.003
10	0.078 ± 0.003

Note. From Roberts and Horlock (1978).

Table 7 Percutaneous Absorption of Nitroglycerin: Topical Concentration Versus Absorption for Neat Liquid Application

Topical nitroglycerin concentration (mg/cm^2)	Absorption	
	Percent	Total mass (μg)
0.01	41.8	0.004
0.1	43.5	0.04
1.0	36.6	0.4
7.0	26.6	1.9
10.0	7.8	0.8

Note. From R. C. Wester, unpublished observations.

Wester (unpublished observations) looked at percutaneous absorption of nitroglycerin. The topical concentration of nitroglycerin was increased stepwise from 0.01 to 10 mg/cm^2. The percentage dose absorbed remained basically the same between 0.01 and 1 mg/cm^2 (Table 7). But as this dose increased 10 times, the amount of material becoming systemically available increased 10 times. At 10 mg/cm^2 the percentage dose absorbed had markedly decreased. This suggests that the percentage of absorption could become saturated at a high concentration.

Howes and Black (1975) determined the comparative percutaneous absorption of sodium and zinc pyrithione in shampoo, through rat skin. As the concentrations of material in shampoo increased from 0.1 to 2%, the penetration also increased from 0.7 to 1 $\mu g/cm^2$ (Table 8).

C. Concentration and Newborns

Wester and coworkers (1977a) compared the percutaneous absorption in newborn versus adult rhesus monkeys. The total amount absorbed per square

Table 8 Effect of Concentration of Sodium Pyrithione in Shampoo on Absorption Through Rat Skin

Concentration in shampoo (% w/v)	Total absorption (%)
0.1	0.07
0.5	0.27
1.0	0.62
2.0	1.02

Note. From Howes and Black (1975).

centimeter of skin again increased with increased applied dose and was further increased when the site of application was occluded. In the newborn the question of concentration may have special significance because surface area/body mass ratio is greater than in the adult. Therefore, the systemic availability per kilogram of body weight can be increased by as much as threefold.

D. Concentration and Water Temperature

Cummings (1969) determined the effect of temperature on rate of penetration on n-octylamine through human skin. Increasing the temperature increased the rate of penetration, as evidenced by octylamine-induced wheal formation and erythema. The increase in cutaneous blood flow mainly involved areas of the wheal. The increase in cutaneous blood flow mainly involved areas of the epidermal factors. Therefore, increased temperature along with increased concentration will increase the percutaneous absorption.

E. Concentration and Duration of Contact

Howes and Black (1976) studied percutaneous absorption of trichlorcarbon in rat and humans. As the duration of contact increased, penetration increased.

Nakaue and Buhler (1976) examined the percutaneous absorption of hexachlorophene in adult and weanling male rats at exposure times from 1.5 to 24 h and determined the plasma concentrations of hexachlorophene. The plasma concentrations of hexachlorophene increased with time from a low of just a few nanograms per milliliter of plasma up to 80+ ng/ml.

Duration of occlusion enhances percutaneous absorption. The significance of time in occlusion was shown by Feldmann and Maibach (1974), who concluded that the longer clothing occludes a pesticide, the greater the contamination potential becomes (Table 9).

F. Concentration, Duration of Contact, and Multiple-Dose Application

Black and Howes (1979) studied the skin penetration of chemically related detergents (anionic surfactants) through rat skin and determined the absorption effects for multiple variables, mainly concentration of applied dose, duration of contact, and the effect of multiple-dose applications. With alcohol sulfate and alcohol ether sulfate, as the concentration of applied dose increased and the duration of contact increased, penetration increased. With multiple applications there was also an increase in penetration (Table 10). Therefore, again the systemic availability and potential toxicity of a chemical

Table 9   Effect of Duration of
Occlusion on Percutaneous Absorption
of Malathion in Humans

Duration (h)	Absorption (%)
0[a]	9.6
0.5	7.3
1	12.7
2	16.6
4	24.2
8	38.8
24	62.8

Note. From Feldmann and Maibach (1974).
[a]Immediate wash with soap and water.

Table 10   Concentration, Duration of Contact, and
Multiple Application as Variables in Penetration of
Anionic Surfactant Through Rat Skin

Variable	Penetration ($\mu g/cm^2$)
Concentration (% w/v)	
0.2	0.02
0.5	0.11
1.0	0.23
2.0	0.84
Duration of contact (min)	
1	0.25
5	0.47
10	0.69
20	0.97
Multiple application ($\times 5$ min)	
1	0.14
2	0.25
4	0.36

Note. From Black and Howes (1979).

depend on many variables. One of these, the concentration, was discussed in the preceding paragraphs. Other variables, such as duration of contact and multiple application, are also important.

G. Concentration and Surface Area

Sved and coworkers (1981) determined the role of surface area on percutaneous absorption of nitroglycerin. As the surface area of applied dose increased, the total amount of material absorbed and systemic availability of nitroglycerin increased. This was confirmed by the percutaneous absorption studies of Noonan and Wester (1980), but there was no linear relationship between the size of the surface area and increase in absorption. However, the same information held. The surface area of applied dose determined systemic availability of the chemical.

H. Effect of Application Frequency

Wester and coworkers (1977b) studied the effect of application frequency on the percutaneous absorption of hydrocortisone. Material applied once or three times a day showed a statistical difference ($p < .05$) in the percutaneous absorption. One application at each 24 h of exposure gave a higher absorption than material applied at a lower concentration but more frequently, namely, 3 times a day. This study also showed that washing (effect of hydration) enhanced the percutaneous absorption of hydrocortisone. This relationship between frequency of application and percutaneous absorption is also seen with testosterone (Wester et al., 1980a).

The aforenoted studies used intermittent application per single day of application. Another consideration is extended versus short-term administration and the subsequent effect on percutaneous absorption. Wester et al. (1980b) examined the percutaneous absorption of hydrocortisone with long-term administration. The work suggests that extended exposure had some effect on the permeability characteristics of skin and markedly increased percutaneous absorption. With malathion, which apparently has no pharmacological effect on skin, the dermal absorption from day 1 was equivalent to day 8 application (Wester et al., 1983). Therefore, for malathion, the single-dose application data are relevant for predicting the toxic potential for longer-term exposure.

I. Application Frequency and Toxicity

There is a correlation between frequency of application, percutaneous absorption, and toxicity of applied chemical. Wilson and Holland (1982) determined the effect of application frequency in epidermal carcinogenic assays. Application of a single large dose of a highly complex mixture of petroleum or synthetic fuels to a skin site increased the carcinogenic potential of the chemical compared with smaller or more frequent applications (Table 11). This carcinogenic toxicity correlated well with the results of Wester et al. (1977b, 1980a), in which a single applied dose increased the percutaneous absorption of the material compared with smaller or intermittent applications.

IV. DISCUSSION

Many variables affect percutaneous absorption and subsequent dermal toxicity. Increased concentration of an applied chemical on skin increases the body burden, as does increasing the surface area and the time of exposure. The opposite also holds true, namely, that dilution of a chemical will decrease the effects of the applied concentration, provided other factors do not change (such as diluting the chemical but spreading the same total dose over a larger surface area). The body burden is also dependent on the frequency of daily application and on possible effects resulting from long-term topical exposure. Dose accountability (mass balance) completes a dose-response study.

The current data provide a skeleton of knowledge to use in the design and interpretation of toxicological and pharmacological studies, to increase their relevance to the most typical exposures for humans. In essence, we have just begun to define the complexity of the interrelationships between percutaneous absorption and dermatoxicology (Table 12; Wester and Maibach, 1982, 1983). Until an appropriate theoretical basis that has been experimentally verified becomes available, quantitating the various variables listed herein will greatly improve the usefulness of biologically oriented

Table 11 Incidence of Tumors After Application of Shale Oils

Shale oil	Dose and frequency	Total dose per week (mg)	Number animals with tumors
OCSO No. 6	10 mg, 4 times	40	2
	20 mg, 2 times	40	4
	40 mg, once	40	13
PCSO II	10 mg, 4 times	40	11
	20 mg, 2 times	40	17
	40 mg, once	40	19

Note. From Wilson and Holland (1983).

Table 12 Factors in the Dose–Response
Interrelationships of Percutaneous Absorption

Concentration of applied dose ($\mu g/cm^2$)
Surface area of applied dose (cm^2)
Total dose
Application frequency
Duration of contact
Site of application
Temperature
Vehicle
Substantivity (nonpenetrating surface adsorption)
Wash-and-rub resistance
Volatility
Binding
Individual and species variations
Skin condition
Occlusion

Note. From Wester and Maibach (1983).

procedures such as laser Doppler velocimetry, transepidermal water loss, and real-time solvent assay in pulmonary breath exhalation. These will expand our knowledge of not only skin dose-response but the use of skin absorption and dynamics to better the human race.

REFERENCES

Black, J. G., and Howes, D. (1979). Skin penetration of chemically related detergents. J. Soc. Cosmet. Chem. 30:157–165.

Cummings, E. G. (1969). Temperature and concentration effects on penetration of N-octylamine through human skin in situ. J. Invest. Dermatol. 53:64–79.

Feldmann, R. J., and Maibach, H. I. (1974). Systemic absorption of pesticides through the skin of man. In Occupational Exposure to Pesticides: Report to the Federal Working Group on Pest Management from the Task Group on Occupation Exposure to Pesticides, Appendix B, pp. 120–127.

Howes, D., and Black, J. G. (1975). Comparative percutaneous absorption of pyrithiones. Toxicology 5:209–220.

Howes, D., and Black J. G. (1976). Percutaneous absorption of trichlorocarban in rat and man. Toxicology 6:67–76.

Maibach, H.I., and Feldmann, R. J. (1969). Effect of applied concentration on percutaneous absorption in man. J. Invest. Dermatol. 52:382.

Maibach, H. I., and Stoughton, R. B. (1973). Topical corticosteroids. Med. Clin. North Am. 57:1253–1264.

Nakaue, H. S., and Buhler, D. R. (1976). Percutaneous absorption of hexachlorophene in the rat. Toxicol. Appl. Pharmacol. 35:381–391.

Noonan, P. K., and Wester, R. C. (1980). Percutaneous absorption of nitroglycerin. J. Pharm. Sci. 69:385.

Reifenrath, W. G., Robinson, P. B., Bolton, V. D., and Aliff, R. E. (1981). Percutaneous penetration of mosquito repellents in the hairless dog: Effect of dose on percentage penetration. Food Cosmet. Toxicol. 19:195–199.

Roberts, M. S., and Horlock, E. (1978). Effect of repeated skin application on percutaneous absorption of salicylic acid. J. Pharm. Sci. 67:1685–1687.

Scheuplein, R. J., and Ross, L. W. (1974). Mechanism of percutaneous absorption. V. Percutaneous absorption of solvent-deposited solids. J. Invest. Dermatol. 62:353–360.

Shah, P. V., Fisher, H. L., Sumler, M. R., Monroe, R. J., Chernoff, N., and Hall, L. L. (1987). A comparison of the penetration of fourteen pesticides through the skin of young and adult rats. J. Toxicol. Environ. Health 21:353–366.

Sved, S., McLean, W. M., and McGilveray, I. J. (1981). Influence of the method of application on pharmacokinetics of nitroglycerin from ointment in humans. J. Pharm. Sci. 70:1368–1369.

Wedig, J. H., Feldmann, R. J., and Maibach, H. I. (1977). Percutaneous penetration of the magnesium sulfate adduct of dipyrithione in man. Toxicol. Appl. Pharmacol. 41:1–6.

Wester, R. C., and Maibach, H. I. (1976). Relationship of topical dose and percutaneous absorption of rhesus monkey and man. J. Invest. Dermatol. 67:518–520.

Wester, R. C., and Maibach, H. I. (1982). In vivo percutaneous absorption. In Dermatotoxicology, 2nd ed. Edited by F. Marzulli and H. I. Maibach. Hemisphere, Washington, DC, pp. 131–146.

Wester, R. C., and Maibach, H. I. (1983). Cutaneous pharmacokinetics: 10 steps to percutaneous absorption. Drug Metab. Rev. 14:169–205.

Wester, R. C., Noonan, P. K., Cole, M. P., and Maibach, H. I. (1977a). Percutaneous absorption of testosterone in the newborn rhesus monkey: Comparison to the adult. Pediatr. Res. 11:737–739.

Wester, R. C., Noonan, P. K., and Maibach, H. I. (1977b). Frequency of application on percutaneous absorption of hydrocortisone. Arch. Dermatol. Res. 113:620–622.

Wester, R. C., Noonan, P. K., and Maibach, H. I. (1980a). Variations in percutaneous absorption of testosterone in the rhesus monkey due to anatomic site of application and frequency of application. Arch. Dermatol. Res. 267:229–235.

Wester, R. C., Noonan, P. K., and Maibach, H. I. (1980b). Percutaneous absorption of hydrocortisone increases with long-term administration. Arch. Dermatol. Res. 116:186–188.

Wester, R. C., Maibach, H. I., Bucks, D. A. W., and Guy, R. H. (1983). Malathion percutaneous absorption following repeated administration to man. Toxicol. Appl. Pharmacol. 68:116–119.

Wester, R. C., Maibach, H. I., Guy, R. H., and Novak, E. (1984). Minoxidil stimulates cutaneous blood flow in human balding scalps: pharmacodynamics measured by laser Doppler velocimetry and photopulse plethysmography. J. Invest. Dermatol. 82:515–517.

Wester, R. C., Maibach, H. I., Guy, R. H., and Novak, E. (1985). Pharmacodynamics and percutaneous absorption. In Percutaneous Absorption. Edited by R. Bronaugh and H. Maibach. Marcel Dekker, New York, pp. 547–560.

Wilson, J. S., and Holland, L. M. (1982). The effect of application frequency on epidermal carcinogenesis assays. Toxicology 24:45–54.

20

Blood Flow as a Technology in Percutaneous Absorption

The Assessment of the Cutaneous Microcirculation by Laser Doppler and Photoplethysmographic Techniques

Ethel Tur
Sourasky Medical Center, Tel Aviv University, Tel Aviv, Israel

I. INTRODUCTION

Optical techniques for blood flow measurement were first introduced more than 60 years ago with the innovation of photoplethysmography (1), substantiated and expanded by Hertzman (2). Laser Doppler techniques came forth 40 years later (3), followed by the manufacture of commercial devices (4,5), and consequently photoplethysmography was put aside. These optical methodologies enable tracing of the movement of red blood cells in the skin. This is useful in following percutaneous penetration, when the penetrant has an effect on blood vessels or on blood flow. In addition, physiology and anatomy of the skin can be studied, as well as pathology. Moreover, laser Doppler flowmetry (LDF) measurements are applicable in the evaluation of internal diseases and conditions that affect the skin microvasculature.

The diverse application areas of the technique include tissues other than the skin, like the buccal, nasal, or rectal mucosa, as well as the intestine through an endoscope, and kidney, liver, or lung intraoperatively. This chapter exclusively deals with cutaneous laser Doppler flowmetry (LDF) and reviews only investigations where this method was used to measure skin blood flow. In each field of LDF investigation, knowledge has broadened in the last few years. In view of the large number of studies conducted in this area, it is impossible to review each and every one, and we only attempt to demonstrate the possibilities of the technique, concentrating on recent publications.

After a review of some of the relevant basic considerations in experimental designs involving LDF, we discuss several conditions where LDF can be utilized. These include investigations of normal and diseased skin and the influence of the nervous system, environmental temperature, smoking, and pregnancy on skin blood flow. Studies of disease processes, severity, and treatment evaluation are then discussed, including hypertension, peripheral vascular disease, diabetes mellitus, and Raynaud's phenomenon. We conclude with future possibilities and expectations.

II. THE METHOD OF LASER DOPPLER FLOWMETRY

A. Basic Principles

Laser Doppler flowmetry (LDF) is a noninvasive method that continuously follows the flow of red blood cells (6). It operates on the Doppler principle, employing a low-power helium–neon laser emitting red light at 632.8 nm. The radiation is transmitted via an optical fiber to the skin. The radiation is diffusely scattered and its optical frequency is shifted by the moving red blood cells. The reflected light, being coherently mixed with another portion of the light scattered from static tissues, generates a Doppler beat signal in the photodetector output current. A quantitative estimation of the cutaneous blood flow derives from spectral analysis of the beat signal.

B. Advantages

As an objective, moninvasive, and real-time measurement technique, LDF is an attractive practical tool for estimating the cutaneous blood flow. Besides, LDF is relatively simple, fast, and inexpensive, and can provide information that supplements the results of various other techniques.

C. Disadvantages

LDF is inferior in quantitating blood flow as compared to other techniques, such as the ^{133}Xe washout technique. However, it is important to note that different models measure different sections of the microvasculature. It is likely that the flux signal shown by LDF represents the large volume of red blood cells moving within the larger blood vessels, particularly the subpapillary plexus, rather than the much smaller volume of red blood cells residing within the nutritive capillaries. The depth of laser penetration in the wavelengths used is approximately 1 mm. Therefore, in normal skin it is likely to include the subpapillary plexus, as well as the capillaries in the subpapillary dermis. In diseased skin it may measure a different body of vessels; for instance, in psoriasis the epidermal ridges are elongated, and this may alter the relative contribution of superficial and deep blood flow to the laser Doppler signal. Another disadvantage of the LDF is that results obtained by various instruments or by the same instrument in different people cannot be compared.

D. Recent Developments

Improvements of the laser Doppler technique offer new possibilities and present new findings. Progressions are illustrated by refinements such as computerization, and the design of a probe holder that allows repeated measurements over the same site before and after manipulations to the skin, through a multichannel LDF instrument that allows simultaneous measurements of several sites (7). But the most substantial development is scanning LDF (8–11), which records the tissue perfusion in several thousand measurement points. A map of the spatial distribution of the blood flow is obtained in a short period of time. Unlike LDF, which continuously records the blood flow over a single point, scanning LDF maps the blood flow distribution over a specific area. Thus, the two methods do not compete with each other, but are rather complementary. The new scanning LDF has the advantage of operating without any contact with the skin surface, avoiding the influence of the pressure of the probe on the tissue perfusion. It can rapidly measure large areas of skin and allows simultaneous measurement of the extent of blood flow changes in abnormal areas of skin and estimates the area of these changes. Furthermore, objective evaluation of various interventions and therapeutic response may be obtained by serial scans.

III. CONCEPTS AND DESIGN OF EXPERIMENTAL STUDIES USING LDF

A. Planning and Performing LDF Blood Flow Measurements

Both static and dynamic LDF measurements can be used (12–14). Static studies, like baseline blood flow measurements, record only the steady-state blood flow, neglecting all transients. On the contrary, dynamic investigations can examine the competence of the blood vessels by following their response to triggers (12–14). Reactive hyperemia is an example of a dynamic test (12–14), recording the postocclusion time course of the blood flow. Other provocative methods examining the response to external triggers include the cognitive test (15), isotonic (16) and isometric tests (12), vasomotor reflexes

(17), intracutaneous needle stimulation (18), topical vasodilators (19), and the thermal test (12).

Some of these tests are vasoconstrictive (the isometric and cognitive tests and vasomotor reflexes), and some are vasodilative (the arterial post-occlusive reactive hyperemia, intracutaneous needle stimulation, the thermal test, and the isotonic test).

To optimize skin blood flow response to the different tests, vasoconstrictive stimuli should be performed on a high blood flow site, whereas vasodilative stimuli should be performed on a low blood flow site. Consequently, the magnitudes of the changes induced are maximized, providing a significant improvement in the quality of the data obtained. Moreover, vasoconstriction mediated by sympathetic stimulation, as in the cognitive test, should be provoked in the hands and feet, where the local blood supply is under a sympathetic vasconstrictor control, and not in the face, which has a poor sympathetic vasoconstrictor supply (15).

Thus, to enhance the sensitivity of the measurements, the site tested should be carefully selected. For example, the fingers, being rich with microvascular arteriovenous anastomoses, are useful for vasoconstrictive tests, while the forearms are suitable for vasodilative tests. The forearms have several advantages as a preferred site for vasodilative tests: a) abundance of arterioles, capable of reactive hyperemia; b) a local effect, rather than thermoregulatory reflexes (20), is responsible for thermally induced vasodilatation; and c) little inconvenience is experienced by the subjects tested.

The appropriate test should also be carefully selected in order to adequately probe the relevant topic. Before reaching a conclusion, one should bear in mind that differences in experimental settings might lead to different results. For instance, in order to study differences between young and old patients, different tests were used when addressing different questions. When aiming to study the thermoregulatory responses to cold stress, vasoconstrictor responses to inspiratory gasp, contralateral arm cold challenge, and body cooling were measured (20), and differences were indeed found: Elderly subjects had diminished sympathetic vasoconstrictor response. In contrast, in order to evaluate the penetration of drugs through aged skin, the erythema that results from topical application of methyl nicotinate was measured (21), and no differences were found between young and old subjects, indicating that microvascular reactivity to the applied stimulus was comparable.

1. Vasoconstrictive Tests

a. Isometric Test The site of blood flow measurement is the left middle fingertip. After establishing the baseline blood flow, it is continuously recorded when the subject squeezes a hard rubber ball in his right hand for 30 s. He or she then releases the grip and the maximum decrease in blood flow is recorded. The hemodynamic response to isometric hand grip exercise involves activation of the sympathetic nervous system, eliciting an increase in blood pressure that is mainly dependent upon cardiac output (12).

b. Cognitive Test The site of blood flow measurement is again the fingertip. The subject is requested to subtract 7 sequentially from 1000 for a 2-min period. Blood flow is monitored continuously. There is a rapid fall in the blood flow to the finger at the beginning of the mental arithmetic activity, and a rapid recovery at the end. The maximum decrease in blood flow is registered. This decrease is a manifestation of a sudden increase in sympathetic activity (15).

c. Venoarteriolar Response The venoarteriolar reflex measures the ability to decrease flow during venous stasis (normally seen in the feet on dependency), and is assumed to be dependent on an intact sympathetic nerve function (22). The reflex occurs following increased venous pressure, which induces a constriction of the arterioles followed by a decrease in skin blood flow. An increase of venous pressure can be achieved by occlusion with a cuff (for instance at the base of the investigated finger) (23), or by lowering the leg below heart level (24–26). Usually, resting blood flow to the dorsum of the foot is measured with the patient resting in the supine position. Then standing flow is measured (or the flow after venous pressure increase created by a cuff), and the lowest reading over 5 min of standing is registered. The venoarteriolar reflex can be expressed as the percent decrease in skin blood flow on standing. The reaction is mediated by a sympathetic axon reflex, comprised of receptors in small veins and resulting in an increase in pre-capillary resistance.

d. Inspiratory Gasp The subject is instructed to breathe in as deeply and quickly as possible and to hold his or her breath for 10 s. The percentage reduction from the resting flow is calculated. This procedure records the sympathetic vasoconstrictor reflex (27).

2. Vasodilative Tests

a. Cutaneous Postischemic Reactive Hyperemia Test Blood flow in the middle part of the flexor aspect of the forearms (or sometimes the proximal part of the finger) is recorded. The arm is then clamped in a pneumatic cuff and inflated to greater than 40 mm Hg above systolic pressure for a period of 1–5 min, during which blood flow measurements are continuously recorded. The cuff is then deflated, resulting in an increase in blood flow, which is recorded continuously, usually until the blood flow returns to base-line values. Any of the following parameters can be measured: a) baseline flow, b) peak flow above baseline flow, c) the time required to reach the peak, d) the ratio between the peak flow and the time required to reach it, expressing the ability of the tissue to respond to fast external triggers, e) the time required to return to the blood flow at rest, and f) the area under the response-time curve (12–14).

b. Intracutaneous Needle Stimulation (Injection Trauma) Resting blood flow is recorded; then a needle is inserted, usually in the center of the probe holder to a depth previously set by a needle guard. The blood flow reaches a peak within 15 min of injection, and then gradually returns to normal over several hours, depending on the degree of trauma (18).

c. Thermal Test Resting blood flow is measured in the middle part of the flexor aspect of the left forearm, while the probe is mounted through a thermostated probe holder. The temperature setting of the thermistor is adjusted to 26°C. The temperature is maintained at 26°C for 2 min, before turning the setting to 28°C for the next 2-min period. This 2°C step sequence is repeated every 2 min until the temperature reaches 44°C. Blood flow is recorded at the end of each 2-min interval (12).

d. Axon Reflex Vasodilator Response Vasoactive substances such as substance P, capsaicin, or histamine are administered topically or intradermally (28–30). Alternatively, acetylcholine is administered with the aid of electrophoresis (27). The extent of the response is measured at several distances from the site of administration. The same procedure is followed for measuring the response to direct stimulation with a firm mechanical stroke with a dermograph (Lewis triple response) (27).

e. Isotonic Test The subject squeezes a partially inflated blood pressure cuff with maximum effort, at which the pressure is recorded, and one-third of it is calculated. The subject is then instructed to grip the cuff at this value of one-third of the maximum pressure. This isotonic exercise causes vasodilatation, and an increase in skin blood flow results (16).

B. Choosing Subjects

When comparing subjects or various groups of subjects, variations in population regarding sex (31,32), age (20), and race (33) should be taken into account. Assuring that subjects match for these variables will decrease the variance within the results.

IV. APPLICATIONS

LDF may be used to study the time course of circulatory changes caused by physiological or pathological processes, including changes caused by pharmacological substances. Internal and external factors, skin conditions, and general conditions that affect the skin are all candidates for LDF investigations.

A. Skin Physiology, Pharmacology, and Pathology

1. Percutaneous Penetration

LDF was applied for tracing the percutaneous penetration of vasoactive agents such as methyl nicotinate (6) or prostacycline (34), and for studying variations in normal skin (6,19). For instance, LDF assisted in evaluating the enhancement effect of ultrasound on skin penetration (35). Thus spatial variations (19), percutaneous penetration enhancers (36), vehicle effect on percutaneous absorption (37), the appendageal contribution to penetration (38), and age and racial differences (39) were all studied with LDF. A decreased percutaneous penetration was recorded in black subjects at various skin sites (40). Circadian differences in penetration kinetics of methyl and hexyl nicotinate were also demonstrated by LDF (41). On the other hand, LDF is not suitable for studying the percutaneous penetration of vasoconstrictive agents such as glucocorticoids (42).

The effect of ultraviolet treatment may also be evaluated by LDF, but the technique was not adequate for individuals with dark skin (43).

Transdermal delivery utilizing iontophoresis was studied at different sites of the forearms and hands, and vasodilatation was site dependent (44). It was shown that cutaneous vascular responses to iontophoresis of vasoactive agents comprise nonspecific, current-induced hyperemia and specific effects of the administered drug (45).

2. Inflammation and Contact Dermatitis

Inflammation is well suited for LDF studies, because of its marked vasoactive component. An increase in blood flow indicated the induction of erythema by topical application of *Staphylococcus aureus* superantigen on intact skin (46). This occurred with both healthy subjects and patients with atopic dermatitis, suggesting that the superantigen may exacerbate and sustain inflammation. UV-induced inflammation was increased following topical application of estrogen (47), while hypnotic suggestion attenuated UV inflammation (48).

The effect of various topical steroid formulations on UV-induced inflammation was measured by LDF, and it enabled grading of the potency of these topical corticosteroids (49). LDF was also used to assess the effect of systemic antiinflammatory drugs, and it enabled grading of the effect of these drugs (50). Prick tests with allergens and histamine may also be evaluated by LDF (51). Regional variations in response to histamine should be taken into consideration (52).

LDF is widely used for measuring the response to known irritants. Increased duration of exposure resulted in an increased response, and comparison between the back and forearm indicated a greater sensitivity on the back (53). Cumulative effect of subthreshold concentrations of irritants was indicated in studies with LDF (54). The vehicle effect on irritation was also studied by LDF, and the irritant effect depended on the vehicle (55). The damage to the skin by repetitive washing (56) and the protective effect of barrier creams were also assessed by LDF (57), as well as the effect of treatment like topical application of nonsteroidal anti-inflammatory drugs in various vehicles (thus studying the vehicle effect as well) (58).

Nonimmunologic contact urticaria induced by benzoic acid was followed and regional variations mapped by LDF (59), as was the suppressive effect of PUVA treatment (60) and topical nonsteroidal anti-inflammatory drugs (61). Regional variations as well as age-related regional variations in the response to histamine were found (62).

Patch tests for allergic dermatitis may be objectively evaluated with LDF, as were patch tests with calcipotriol ointment in various patients, including psoriasis patients (63).

Scanning LDF was more suitable for the quantification of allergic contact dermatitis than the regular LDF, as readings with the latter are time-consuming, and scanning LDF is valuable for measuring the area of response (64). Both allergic contact dermatitis and irritant contact dermatitis were studied with scanning LDF (65). The technique allowed quantification of subclinical pattern of the allergic inflammation (66).

3. Psoriasis

Psoriasis, with its increased blood flow near the skin surface, is a natural candidate for LDF studies. Several investigators aimed at studying the disease process, whereas others were interested in the effect of several therapeutic modalities. A recent publication concentrated on the question of whether cutaneous blood vessels in psoriasis possess a generalized inherently abnormal response to neuropeptides (67). Calcitonin gene-related peptide (CGRP) was intradermally injected in three concentrations into uninvolved skin of psoriatic patients and in healthy controls. This resulted in an increase in blood flow that did not differ between the two groups, indicating that in uninvolved psoriatic skin the vasculature is not different than normal in its response to CGRP. Effects of treatment were assessed by LDF and compared to clinical evaluation methods (68).

Scanning laser Doppler velocimetry allows rapid measurement of the area and the level of increase of blood flow in psoriatic plaques (9). Plaque severity can be assessed in terms of mean blood flow and area of increased blood flow simultaneously. The obtained scan image reveals the distribution and intensity of the rim of increased blood flow around the psoriatic plaque. This could be used in the study of early biochemical or immunological changes in the skin, before lesions become clinically observable. As an aid in evaluating phototherapy, a reduced sensitivity to both UVA and UVB was demonstrated in psoriasis plaques as compared to uninvolved skin, using the same instrument (10). The method also showed an improved response to PUVA treatment when calcipotriol was topically applied (69).

The application of LDF for the study of psoriasis, including evaluation of therapy, was extensively reviewed (70).

4. Atopic Dermatitis

In a study of dermographism, a significant reduction in the intensity of hyperemia was found in atopic dermatitis patients following pressure on the skin (71). The role of acetylcholine in the etiology of pruritus in atopic dermatitis was studied by injecting acetylcholine and monitoring the vascular reaction. The reaction in the patients started earlier and was longer than in the control group, suggesting an etiological role (72).

5. Age, Chronic Venous Insufficiency, and Cutaneous Ulcers

Skin blood flow to the dorsum of the foot reduces with age, which might contribute to the attenuated cutaneous vasoreactivity to heat and ischemia in elderly people (73). Moreover, the response to heat is attenuated in the aged (74).

The ability of the skin blood vessels to dilate in response to pilocarpin electrophoresis was assessed in patients with chronic venous insufficiency, but the microvasculature showed a normal capacity to vasodilate (75). The effect of external compression was also studied: Compression increased the microcirculatory flow, which might be its mode of action in treating venous insufficiency (76).

In patients with venous ulcers, erythematous ulcer edges exhibited higher blood flow values than nonerythematous edges (77). Postural vaso-regulation caused relative ischemia and reperfusion in venous leg ulcers, but the known mediators of reperfusion injury were not released, and therefore were not associated with the process (78). LDF was also used to monitor the effect of prostanoids on ischemic ulcers (79).

For further examples of the use of LFD in peripheral vascular disease, refer to Section IV.B.7.

6. Burns and Flaps

Flaps can be monitored (80) and their success predicted (81). Similarly, burn severity and treatment may be evaluated (82).

B. General Conditions and Diseases

External factors as well as certain general conditions and diseases may affect the skin or have an effect on its blood flow even when the skin itself is healthy.

1. Nervous System

Skin blood flow is centrally controlled by the autonomic system, and takes part in the general regulatory mechanism. Autonomous activity was reflected in skin blood flow, which fluctuated in response to sympathetic modulation (83).

Higher mental activity alters skin blood flow; during cognitive activity, skin blood flow to the finger decreases, but not to the malar region (15). The reason is that vasoconstriction in the finger is mediated by sympathetic stimulation, which controls the blood supply to the hands and feet, whereas the face has a poor sympathetic vasoconstrictor supply.

Sound stimuli also affect the skin microcirculation in sites rich with sympathetic innervation, as was demonstrated by LDF (84).

The effect of neural blockade at various spinal levels and general anesthesia was followed with LDF, and functions like respiratory movements were correlated (85).

LDF measurements in acral regions ranked the role of alpha-1 and alpha-2 adrenoceptors in mediating sympathetic responses (86). Alpha-2 adrenoceptor was more potent in increasing the cutaneous microvascular resistance and reducing the perfusion.

To study the mechanism of pain relief by vasodilator agents, LDF measurements were conducted following intradermal injection of various vasodilators and compared to pain threshold (87). Pain threshold did not correlate with blood flow, indicating that the effect of vasodilators on primary afferent nociceptors is not related to the vasodilatory effect.

For further examples of the utilization of LDF for the study of the nervous system, please refer to Sections III (concepts and design of experimental studies using LDF), IV.B.8 (diabetes), and IV.B.2 (environmental temperature).

2. Environmental Temperature

Emotional stress induced vasoconstriction after prolonged heating to 34°C, whereas after prolonged cooling to 22°C vasodilatation was induced (88).

The stress-induced decreased flow in warm subjects is probably neurally mediated, since it is preceded by increased skin sympathetic activity. But the increase of flow in cold subjects is more obscure. Different neural commands at different temperatures might be a possibility. Another possibility is that arteriovenous shunts receive a vasoconstrictor sympathetic input, whereas resistant vessels receive a vasodilator input. In cold subjects with high baseline vasoconstriction, the arteriovenous shunts are closed and unable to constrict any further, so only the vasodilatation in the resistance vessels is detected. In warm subjects, thermoregulatory activity is low, the basal flow is high, and the arteriovenous shunts are open. When the emotional stress occurs, the constrictor effect on the open arteriovenous shunts will be more pronounced and will mask the vasodilatation in the resistance vessels.

Both LDF and scanning LDF were utilized to correlate skin blood flow to temperature changes. The two did not correlate at sites with arteriovenous shunts (89).

Central thermoregulatory mechanisms affect the postural vasoconstrictor response. Following heating of the trunk with an electrical blanket, the postural fall in blood flow diminished in skin areas with relatively numerous arteriovenous shunts (plantar surface of the big toe) (90). In contrast, areas with few or no arteriovenous shunts (dorsum of the foot) displayed similar postural flows before and after heating. Therefore, partial release of sympathetic vasoconstrictor tone associated with indirect heating appears to override the local postural control of cutaneous vascular tone in areas where arteriovenous anastomoses are relatively numerous. When measured at heart level, the indirect heating was accompanied by a significant increase in foot blood flow. Many experiments, utilizing both LDF and other techniques, support the view that this reflex thermoregulatory vasodilatation is mainly due to the release of sympathetic vasoconstrictor tone, induced by the elevated core temperature. The normal postural fall in foot skin blood flow was preserved within the skin temperature range of 26–36°C, but at higher temperatures it was markedly attenuated or even abolished (91). This might contribute to some of the problems of cardiovascular adaptations seen in hot environment.

The influence of room temperature on peripheral flow in healthy subjects and patients with peripheral vascular disease was followed by LDF (92). The flow was very little affected when the room temperature was increased from 24 to 30°C, therefore, this range is suitable for skin blood flow studies. For temperatures higher than 30°C the peripheral circulation increased. The relationship between skin flow and room temperature was linear in room temperatures between 23 and 30°C, whereas between 30 and 35°C it was curvilinear.

3. Flushing and Facial Pallor

Flushing, a transient reddening of the face and other body sites, is caused by vasodilatation, which may be provoked by many pharmacologic and physiologic reactions. Emotionally provoked blushing was recorded in the forehead by LDF. There was a sudden increase of blood flow, which returned to baseline value after approximately 3 min (93). Other provoking factors are alcohol and conditions such as menopause, carcinoid, mastocytosis, or drugs like nicotinic acid (94). Flushing occurs after orchidectomy for carcinoma of the prostate, and LDF measurements correspond closely to the intensity of the attacks as experienced by the patients and also to the measurement of sweating by evaporimetry (95).

LDF technique is appropriate for quantitative assessment of alcohol-provoked flushing. Comparing LDF to the change in malar thermal circulation index, a linear correlation was found between the two methods. Moreover, the LDF method was more specific and move sensitive (96). The effects of various therapeutic modalities for facial flushing and rosacea were thus studied.

Systemic administration of nicotinic acids produces a generalized cutaneous erythema, partially mediated by prostaglandin biosynthesis. To further examine the mechanism, local cutaneous vasodilatation was studied following topical application of methyl nicotinate (97). Pretreatment with prostaglandin inhibitors (indomethacin, ibuprofen, and aspirin) significantly suppressed the erythemal response, while doxepin had no effect on this response. Arginine vasopressin was infused in high levels, comparable to those attained during physical stress, and a marked facial pallor in healthy men resulted (98). The pallor was objectively verified by LDF measurements, consistent with a fall in nutritional blood flow to the skin. In contrast, blood flow to the finger rose, indicating an increased blood flow through arteriovenous shunts. Thus, LDF assisted in determining that arginine vasopressin has a selective vasoactive effect in the skin.

4. Smoking

Studies suggest an abnormality in capillary blood flow and its regulation in the skin as an immediate result of cigarette smoking, and a chronic effect as well (13).

Cigarette smoking triggers the release of vasopressin: A significant correlation was found between the skin blood flow response to cigarette smoking and the plasma vasopressin levels after smoking (99). The vasopressin released may mediate some of the vasoconstriction, since pretreatment with a vasopressin antagonist reduced the nicotine-induced vasoconstriction in the skin while vasopressin antagonist alone had no effect.

Calcium ions may play a role in the pathogenesis of the vasoconstriction caused by acute smoking. Elderly habitual smokers were given the calcium channel blocker nifedipine, and calcitonin, a hypocalcemizing hormone that has a vasoactive action. Both drugs prevented the LDF-measured vasoconstriction induced by cigarette smoking, indicating that the process is calcium mediated (100).

These and other studies illuminate some of the mechanisms involved in the changes of the microvascular bed in the skin of habitual smokers, and reveal both acute and chronic changes. Chronic changes occur relatively early in a person's smoking history, but later they become more severe. It would be interesting to assess the correlation between these skin blood flow changes and the future development of peripheral vascular disease in a long-term study.

5. Pregnancy and Gestational Hypertension

Both normal and hypertensive pregnancy manifest microvascular changes. In gestational hypertension baseline blood flow on the fingertip was lower than in normal pregnancy, but this measurement could not discriminate between these subjects and nonpregnant healthy controls (12). Provocative tests were then used (isometric test, cognitive test, reactive hyperemia test, and thermal test), and the isometric test gave the most discriminative results. The control group showed a larger decrease than both the normal pregnancy group and the gestational hypertension group. Gestational hypertension showed a larger decrease than the normal pregnancy group. The cognitive test and the postischemic reactive hyperemia allowed some degree of discrimination, whereas the thermal test did not show any abnormality in the pregnant groups. Thus LDF recording of the response to vasoactive stimuli may differentiate between groups of subjects with normal or hypertensive pregnancy and nonpregnant subjects. Normal pregnancy modifies the response of the skin microvasculature to some vasoactive stimuli, whereas gestational hypertension pushes that response back toward the nonpregnancy state. However, the method cannot yet be applied as a diagnostic tool for the individual patient.

6. Hypertension

Hypertension is associated with, or even originates from, and increased total peripheral resistance. The skin microvasculature plays a role in this peripheral resistance, hence the importance of its investigation in hypertension.

The decrease of skin blood flow that follows smoking of two cigarettes was measured in hypertensive habitual smokers (101). The same measurements were done before any treatment and following intravenous administration of alpha-1 inhibition with doxazosin and beta-1 blockade with atenolol. Skin blood flow decreased under all these conditions, but the decrease was attenuated by doxazosin compared to atenolol. These findings indicate that selective alpha-1 adrenoceptors have a major effect on smoking-induced cutaneous vasoconstriction. In this respect doxazosin is preferable to atenolol for the antihypertensive treatment of patients who smoke.

In hypertensive patients both the resting flow and the standing flow to the feet were significantly lower, but increased after nifedipine treatment (24). The venoarteriolar reflex was lower in hypertensive patients, improving after nifedipine treatment, but still below normal.

Side effects of treatments and their mechanisms may be studies by LDF. Following nifedipine treatment, a weaker venoarteriolar reflex was observed in patients who developed ankle edema, whereas before treatment their response did not differ from those who did not develop edema (102). In another study, calcium channel blockers of different chemical origins antagonized postural vasoconstriction in the skin of the dorsum of the foot, indicating altered postural capillary blood flow regulation (103). Fluid filtration to the extravascular compartment may then eventuate, which may explain ankle edema during treatment with calcium channel blockers.

The cutaneous postischemic reactive hyperemia response does not seem to differentiate between hypertensive patients and normotensive controls (14,104). Thus, a few provocative tests were able to detect differences between hypertensive patients and controls, while other tests could not.

7. Peripheral Vascular Disease

The adequacy of skin blood flow in the ischemic extremity is an important determinant in the assessment of the severity of peripheral vascular disease. Old and young healthy volunteers were compared to patients with lower limb atherosclerosis and intermittent claudication and patients with lower limb atherosclerosis and critical ischemia (105). Elderly controls had higher flux values in the toe compared with claudicators, while claudicators had higher perfusion values than patients with critical ischemia.

Using the ratio between toe and finger flows narrows the range of the results, eliminating differences in cardiac output that occur from patient to patient; this method was able to distinguish between a group of healthy controls and a group of patients with peripheral vascular disease (106). The same technique was useful for evaluating patients with peripheral arterial disease and for distinguishing different etiologies of the disease (25). Patients with intermittent claudication, patients with rest pain, and those with critical foot ischemia had a significantly lower resting flow than normal. Healthy controls showed a reduction of skin blood flow on standing, which was smaller in patients with severe claudication. Patients with rest pain had higher skin blood flow values when standing, indicating a loss of the vasomotor tone and an increase of flow determined by gravity. The inverse effect of an increase in blood flow with standing was associated with the clinical improvement when the patients lowered their limbs (25).

Microcirculatory alterations in limbs with claudication were found, with early occurrence of microcirculatory compensation to atherosclerotic disease of increasing severity (107). The skin perfusion increased with dependency, which explains why patients obtain relief of pain with dependency.

Skin blood flow is an important determinant of healing of ulcers or an inevitable amputation. The baseline skin blood flow in patients with peripheral vascular disease was significantly lower than normal, and the pulse waves attenuated or absent (108). In indeterminate cases, the accuracy of the LDF method could be enhanced by the use of reactive hyperemia.

Using provocative tests, there was a correlation between the impairment of the LDF results and the gravity of the clinical picture, with significant differences being recorded between limbs with no sign of necrosis and limbs affected by slight necrosis (109). In patients with pain at rest, postischemic hyperemia was absent. Following therapy with intra-arterial administration of naftidrofuryl, a statistically significant improvement in several LDF assessed parameters was achieved, and treatment of four limbs with percutaneous transluminal angioplasty resulted in a completely normal test, matching the disappearance of all the symptoms caused by the peripheral occlusive arterial disease.

A reduction in the postischemic reactive hyperemia was found in patients with leg ulcers, with a sensitivity comparable to the measurements of distal systolic blood pressure (110). The healing effect of peripheral sympathectomy and pain relief were also monitored by LDF (111).

These studies again demonstrate that appropriate physiologic tests like hydrostatic pressure loading and the postischemic reactive hyperemia test are more indicative of abnormalities than are static parameters like resting flow. LDF can also provide explanations to therapeutic effects, like that of CO_2 baths in occlusive arterial disease, where an increased skin blood flow and increased oxygen utilization were observed (112).

8. Diabetes Mellitus

Diabetes mellitus is the most illustrative disease for the application of skin blood flow measurements, with its microangiopathic changes serving as a natural target for LDF application to probe the disease's various aspects. Skin blood flow is affected in diabetes, either directly or via the nervous system, as exemplified by deficient responses in the fingertip that resembles premature aging (113). LDF was widely used in studying the process of diabetes mellitus and the etiology of its various complications, in grading disease severity, predicting its outcome, and following treatment. Damage to the microcirculation in diabetes mellitus is responsible for a large number of its grave complications. The neuropathy that further affects the microcirculation adds to the diversity and complexity of the disease, making the exposure of its yet unsolved aspects even more intriguing and rewarding. An understanding of the mechanisms responsible for microvascular complications may help in developing new treatment modalities. Furthermore, detection of early functional changes in the microcirculation might identify patients at risk at a reversible stage. The possibility of using the skin as a model for diabetic microangiopathy, and its correlation to retinopathy and nephropathy, is still under investigation.

a. Insulin-Dependent Diabetes Mellitus (IDDM) Using a thermal probe for provocation, blood flow values were lower than normal at 35°C, and this difference was even more pronounced at 44°C (114). Retinopathy and nephropathy were also associated with such a decrease.

Sympathetic neural dysfunction was indicated by reduction of vasoconstrictor responses to contralateral arm cold challenge and body cooling, and also to inspiratory gasp (20). Vascular disturbances measured on the dorsum of the hand seemed to be exaggerated by parasympathetic neuropathy (115).

Skin microvascular vasodilator response to both injection trauma and local thermal injury was impaired in IDDM, unrelated to diabetic control (18). This impairment in response to injury may be an important factor in the development of foot ulceration that often follows minor trauma. In order to investigate the mechanisms underlying the impaired hyperemia to local injury, substance P was intradermally injected and a reduced peak response was achieved in IDDM patients as compared to controls, whereas the response to capsaicin was the same (116). Following histamine blockade with chlorpheniramine, the response to capsaicin remained unaltered, while the response to substance P was reduced in both groups. Therefore, impaired skin hyperemia may represent decreased vascular reactivity to locally released substance P from peripheral nerve fibers.

Structural pathology in early diabetic neuropathy is best correlated to peroneal motor conduction velocity (117). Transcutaneous oxygen and LDF measurements were correlated with peroneal motor conduction velocity in IDDM patients, suggesting that hypoxia generates diabetic peripheral neuropathy and that in early neuropathy, therapeutic measures to improve blood flow might arrest its progress. Indeed, pentoxifylline treatment (400 mg three times daily) increased the skin blood flow in the lower extremity, as measured after 3 and 6 months of therapy (118).

Nociceptive C fibers were evaluated by measuring the axon reflex vasodilatation of the foot in response to electrophoresis of acetylcholine, to direct mechanical stroke with a dermograph, and to a deep inspiratory gasp (27). The flare was reduced only in patients with foot complications, and it correlated with the clinical diminution of pain sensation, both are components of nociceptive C fiber function. Reduction of the flare indicated an impairment of neurogenic inflammatory response, which, along with an impairment of the protective pain sensation, may be a contributory factor in the poor and slow healing of foot lesions of diabetic patients. Since sympathetic vasoconstrictor reflexes were present in most diabetic patients with foot ulceration, the role of autonomic neuropathy in ulcer development is questioned.

Similar results were obtained by measuring the flare induced by acetylcholine iontophoresis at various current strengths (119). Maximum flare response was reduced in neuropathic patients, especially those with a previous history of foot ulceration, suggesting that small-fiber neuropathy affects ulcer development. The flare was also reduced in some patients with retinopathy without neuropathy, suggesting an early loss of small nociceptor C fibers, preceding large-fiber neuropathy. The curve of the hyperemic response plotted against current strength did not show a rightward shift, indicating that the abnormal response was due to axonal loss rather than to dysfunction. The flare did not correlate with the cardiac autonomic function.

b. Non-Insulin-Dependent Diabetes Mellitus (NIDDM) Subjects at risk of developing NIDDM who have fasting hyperglycemia exhibited decreased microvascular hyperemia, positively correlated with insulin sensitivity and negatively correlated with fasting plasma insulin concentrations (120). Thus, hyperinsulinemia as a result of insulin resistance may affect the microvascular function in the prediabetic state.

Noncritically ischemic feet have a higher hyperemic response in comparison with critically ischemic feet (121).

The venoarteriolar response was lower in diabetic patients than in healthy controls, and lowest in patients with neuropathy (26). Both the increased skin blood flow and the impaired venoarteriolar reflex are causes of edema and may contribute to the thickening of capillary basement membranes detected in diabetes.

A hyperthermal laser Doppler flowmeter was developed to quantify autonomic dysfunction in the skin (122). The technique measures the time to induction of an increase of microcirculation following hyperthermia. Autonomic dysfunction was found in diabetic patients even with a short disease duration, suggesting that it was an early complication of NIDDM, even when the disease was well controlled.

Investigations of the sympathetic nervous function in NIDDM showed that deep breathing induced a decrease in skin blood flow (123). Many other neurovascular functional tests in experimental and human diabetes were studied (124).

Diabetes induces functional microvascular disturbances in the forearms. The cutaneous postischemic reactive hyperemia response was significantly lower than in nondiabetic controls (14). Retinopathy was a further factor in achieving more abnormal results. Evaluation of beraprost sodium treatment revealed a decreased effect when the severity of retinopathy and nephropathy increased (125).

   c. Evaluation of Treatment   Various treatment modalities were assessed by LDF, starting with simple physical measures like elastic stockings, through old and new medications, to evaluation of a combined kidney and pancreas transplantation. Elastic stockings repressed the deterioration of the microcirculation, as was shown by LDF studies (126). The effect of an angiotensin-converting-enzyme inhibitor, captopril, was studied on IDDM (127). It improved skin blood flow, as measured with LDF using postocclusive reactive hyperemia response, and this was independent of its hypotensive effect.

In NIDDM, during insulin infusion foot blood flow was redistributed with an increase in capillary flow relative to arteriovenous shunt flow (128). Since blood that passes through arteriovenous shunts does not enter the capillary bed and plays no role in skin nutrition, this represents an improvement in the skin nutrition.

Hypoglycemia, but not hyperinsulinemia, caused a regional skin vasodilatation in healthy control subjects. Following a hyperinsulinemic euglycemic clamp, hypoglycemia was induced by a stepwise reduction in the intravenous glucose infusion. The increase in blood flow that was observed in healthy controls was absent in patients with non-insulin-dependent diabetes mellitus (129).

Defibrotide, an oral profibrinolytic drug, was given to non-insulin-dependent diabetic patients with microangiopathy. The microcirculation was evaluated by the venoarteriolar response and by the rise in skin blood flow following local heating, both decreased in diabetic patients. Following 6 months of treatment, patients in the new drug treatment group improved their microcirculatory parameters (venoarteriolar reflex) in association with an improvement in signs and symptoms (130). Calcium channel blockers were also evaluated with LDF (131).

A combined kidney and pancreas transplantation improved the skin microvascular reactivity therefore, the possibility that transplantation could reverse or halt diabetic complications was studied (23). Skin blood flow at rest, postischemic reactive hyperemia, and venoarteriolar reflex at 2 and 38 months following transplantation were measured. The rest blood flow was higher 38 months following transplantation as compared to 2 months, but the delayed time to peak during hyperemia was even more impaired at 38 months. This is probably due to a disturbed function of the smooth muscle cells of the vessel wall, and might indicate a progress of structural changes in spite of an improved metabolism. The venoarteriolar reflex was also impaired at both time points, probably due to neuropathy, since this reflex depends on an intact sympathetic nerve function. A trend for improvement in four out of five patients with the most impaired reflex was observed, but it did not reach statistical significance; this may indicate that diabetic neuropathy can be improved after transplantation (23).

These and other studies indicate that LDF can be useful in the investigation of some diabetes pathophysiological mechanisms, disease severity, and the efficacy of its control.

9. Raynaud's Phenomenon

The intermittent blanching that occurs in Raynaud's phenomenon is believed to result from an active microvascular vasoconstriction and emptying. Therefore, LDF is useful for the investigation of the pathophysiological mechanisms underlying Raynaud's phenomenon and for evaluating treatments.

To clarify the etiology of Raynaud's phenomenon, three vasodilators were intravenously administered, and the response of calcitonin gene-related peptide was compared with that of endothelium-dependent adenosine triphosphate and endothelium-independent prostacyclin (132). The first vasodilator induced an increase in blood flow in the hands of patients but not in healthy controls, which may reflect a deficiency of endogenous calcitonin-gene-related peptide release in Raynaud's phenomenon.

The role of the histaminergic and peptidergic axes in primary Raynaud's phenomenon was also studied (29,30). Digital blood flow response to intradermal injections of saline, histamine, histamine-releasing agent (compound 48/80), substance P, and calcitonin gene-related peptide was measured. No evidence of local deficiency in histamine release or in the response to histamine was found (30), even at low temperatures (29), and the patients reacted normally to the neuropeptides substance P (29,30) and calcitonin gene-related peptide (30), providing a rationale for treating Raynaud's phenomenon with vasoactive peptides.

Digital skin blood flow of both hands was measured during local heating of only one hand. Patients with Raynaud's phenomenon showed a decreased digital blood flow during stepwise cooling in both hands, but the reaction in the cooled hand was more pronounced and more consistent (133).

Patients with Raynaud's phenomenon had an abnormal vascular response to temperature change (32). In terms of the hyperemic response to local skin warming, the patients showed vasodilatation at lower skin temperatures than normal, independent of central sympathetic control. A knowledge of skin temperature is therefore important for the interpretation of blood flow studies in Raynaud's phenomenon. Provocative testing (occlusion) under different degrees of finger and body cooling detected an increase in the number of fingers of patients exhibiting vasospasm as the severity of cooling increased (134).

Occupational exposure to vibrations causes Raynaud's phenomenon. For prevention and treatment of vibration syndrome, an objective test was developed, combining LDF with finger cooling (135). It enabled the demonstration of significant differences among four groups: subjects without vibration exposure, subjects with exposure but with no signs of white fingers, subjects with few attacks, and subjects with frequent attacks.

In patients with Raynaud's phenomenon suffering from scleroderma, blood flow decrease during cooling was similar to healthy controls, but the patients had a longer rewarming period (136). In such patients, when resting blood flow was very low, the postischemic reactive hyperemia response was absent (137). After warming the hand in warm water, the hyperemic response was restored and its magnitude corrected, but its time course was longer, with a delay to achieving maximum flow as compared to controls. This may be related to changes in the vessels themselves, or to connective-tissue sclerosis limiting the rate of response.

Provocative tests that evoke sympathetic tone, like the isometric test, were studied in addition to temperature measurements in groups of patients with primary Raynaud's phenomenon, systemic sclerosis, and undifferentiated connective tissue disease as compared to a control group (138). Considerable differences were found in both the level of vessels involved and the relative importance of local finger temperature and discrimination between various etiologies of Raynaud's phenomenon was possible.

Cutaneous postocclusive reactive hyperemia enabled grading obstructive vascular disease in groups of patients with Raynaud's phenomenon, but could not discriminate among individuals in the subgroups (139). Another provocative test, the cognitive test, detected two subgroups within the patients with Raynaud's phenomenon (140). The first subgroup showed a reduction in blood flow similar to healthy controls, whereas the second showed a paradoxical increase, suggesting an organic etiology.

LDF was used to evaluate various treatment modalities in Raynaud's phenomenon. Thus, a single topical application of minoxidil 5% solution to the fingers was ineffective (141). Ketanserin, an antagonist of the serotonin-2 (5-HT-2) receptor, was given to nine patients with generalized scleroderma (142). Finger systolic pressure and LDF after cooling and rewarming of the finger did not improve. Thus, ketanserin in the doses used (20 mg 3 times a day in the first week and 40 mg 3 times a day for 4 weeks) was not effective in the treatment of Raynaud's phenomenon in generalized scleroderma. When given to patients with primary Raynaud's phenomenon, ketanserin normalized digital blood flow (143). Pretreatment with alpha-adrenoceptor antagonists did not abolish this effect, suggesting that in contrast to the effects on the systemic circulation, the mechanism underlying digital vasodilatation after ketanserin does not involved alpha-adrenoceptor antagonism.

Application of the vasodilator hexyl nicotinate at various sites on the upper limb resulted in an increase in blood flow both in patients with Raynaud's phenomenon (13 with primary Raynaud's disease and 12 with systemic sclerosis and Raynaud's phenomenon) and in normal subjects (144). Moreover, increasing the drug concentration increased flow rate.

The effect of nifedipine on perniosis was also studied. An increase in blood flow to a finger or a toe adjacent to a diseased one could be demonstrated after the drug intake (145).

10. Other Diseases

Several diseases, like leprosy, have distinct skin lesions that may be followed by LDF. Blood flow in lesions of leprosy paralleled the clinical appearance and histopathology of the lesions during treatment (146). The amount of hyperemia was useful in monitoring the early changes of reversal reaction during chemotherapy. LDF was also valuable to evaluate peripheral autonomic function in leprosy (147). The skin overlying Kaposi's sarcoma was also studied by LDF (148).

LDF was also used to investigate microvascular physiology in patients with sickle-cell disease (149). Large local oscillations in skin blood flow to the arm was demonstrated, occurring simultaneously at sites separated by 1 cm, suggesting a synchronization of rhythmic flow in large domains of microvessels. The periodic flow may be a compensatory mechanism to offset the deleterious altered rheology of erythrocytes in sickle-cell disease.

Changes in skin microvascular reactivity were demonstrated in hypertriglyceridemia (150) and even in Alzheimer's disease (151).

LDF has potential for use in oncology, as it is able to monitor the vascularization in tumors and adjacent skin. For instance, it may serve as an additional tool in the diagnosis of pigmentary skin lesions. Melanomas showed higher laser Doppler blood flow readings than basal-cell carcinomas, and both showed higher readings than benign lesions (152). Scanning LDF was able to monitor blood flow following photodynamic therapy of non-melanoma skin tumors (153).

V. CONCLUSION AND FUTURE PROSPECTS

The LDF method allows real-time analysis of a wide range of physiological and pathological processes, as well as pharmacological processes and percutaneous penetration. It provides objective numerical data at various time points or at various cutaneous areas. However, results obtained by LDF should not be interpreted as absolute values, but should rather serve as relative estimates. Another disadvantage of the LDF is that results obtained by different instruments or by the same instrument in different subjects cannot be compared. Furthermore, biological variations within the disease state can also produce discord between different studies. There is a need, therefore, for technical improvements, for calibration standards, for better probe designs, and for more reproducible measuring procedures before LDF turns into a useful clinical tool. Improvements may lead to a wider use in the area of skin pharmacology, which naturally requires accuracy and repeatability.

An important development was scanning LDF (28), which quickly and sequentially remotely scans the tissue perfusion in several thousand measurement points. A map of the spatial distribution of the blood flow is thus obtained. Unlike regular LDF, which continuously records the blood flow over a single point, scanning LDF maps the blood flow distribution over a specific area. Thus, the two methods do not compete with each other, but are rather complementary. The scanning LDF has the advantage of operating without any contact with the skin surface, and therefore it does not disturb the local blood flow. A software module was implemented in the scanning LDF, resulting in a duplex mode for recording both spatial and temporal blood perfusion components (154). Another new device combines laser Doppler perfusion imaging and digital photography, and may have many clinical applications like evaluating grafts or treatment of leg ulcers (155).

Other improvements in laser Doppler are the introduction of advanced computerization and the development of new probes, such as multisite probes allowing simultaneous measurements at several sites, and multisubject probes allowing simultaneous measurements of several subjects. These developments are accompanied by new multichannel LDF instruments capable of simultaneous collection of data from many independent probes. Finally, a new probe holder design (7) permits repeated measurements over the same site before and after manipulations to the skin.

Technical innovations may finally improve the accuracy and repeatability of laser Doppler flowmetry to a level that will justify its use as a clinical tool. Then readings will be meaningful on an individual basis rather than in terms of the statistical nature of all the reviewed studies. The several hundred published investigational reports of the last 4 years clearly indicate the increased interest in laser Doppler flowmetry and will undoubtedly eventually lead to successful clinical utilization.

REFERENCES

1. Molitor H, Kniazuk M. A new bloodless method for continuous recording of peripheral circulatory changes. J Pharmacol Exp Ther 1936; 57:6–18.
2. Hertzman AB. Photoelectric plethysmography of the fingers and toes of man. Proc Soc Exp Biol Med 1937; 37:529–534.
3. Stern MD. In vivo evaluation of microcirculation by coherent light scattering. Nature 1975; 254:56–58.
4. Holloway GA, Watkins DW. Laser Doppler measurement of cutaneous blood flow. J Invest Dermatol 1977; 69:306–309.
5. Nilsson GE, Tenland T, Oberg PA. A new instrument for continuous measurement of tissue blood flow by light beating spectroscopy. IEEE Trans Biomed Eng 1980; 27:12–19.
6. Tur E. Cutaneous blood flow: laser Doppler velocimetry. Int J Dermatol 1991; 30:471–476.
7. Braverman IM, Schechner JS. Contour mapping of the cutaneous microvasculature by computerized laser Doppler velocimetry. J Invest Dermatol 1991; 97:1013–1018.
8. Wardell K, Naver HK, Nilsson GE, Wallin BG. The cutaneous vascular axon reflex characterized by laser Doppler perfusion imaging. J Physiol 1993; 460: 185–199.
9. Speight EL, Essex TJH, Farr PM. The study of plaques of psoriasis using a scanning laser-Doppler velocimeter. Br J Dermatol 1993; 128:519–524.
10. Speight EL, Farr PM. Erythemal and therapeutic response of psoriasis to PUVA using high-dose UVA. Br J Dermatol 1994; 131:667–672.
11. Liu D, Svsanber K, Wang I, Andersson-Engels S, Svanberg S. Laser Doppler perfusion imaging: New technique for determication of perfusion of splanchnic organs and tumor tissue. Lasers Surg Med 1997; 20:473–479.
12. Tur E, Tamir A, Guy RH. Cutaneous blood flow in gestational hypertension and normal pregnancy. J Invest Dermatol 1992; 99:310–314.
13. Tur E, Yosipovitch G, Oren-Vulfs S. Chronic and acute effects of cigarette smoking on skin blood flow. Angiology 1992; 43:328–335.
14. Tur E, Yosipovitch G, Bar-On Y. Skin reactive hyperemia in diabetic patients. Diabetes Care 1991; 14:958–962.
15. Wilkin JK, Trotter K. Cognitive activity and cutaneous blood flow. Arch Dermatol 1987; 123:1503–1506.
16. Weinstein L, Janjan N, Droegemueller W, Katz MA. Forearm plethysmography in normotensive and hypertensive pregnant women. South Med J 1981; 74:1230–1232.
17. Low PA, Neumann C, Dyck PF, Fealey RD, Tuck RR. Evaluation of skin vasomotor reflexes by using laser-Doppler velocimetry. Mayo Clin Proc 1983; 58:583–592.
18. Rayman G, Williams SA, Spencer PD, Smaje LH, Wise PH, Tooke JE. Impaired microvascular hyperaemic response to minor skin trauma in type I diabetes. Br Med J 1986; 292:1295–1298.

19. Mayrovitz NH, Smith J, Delgado M. Variability in skin microvascular vasodilatory responses assessed by laser Doppler imaging. Ostomy Wound Manage 1997; 43:66–70.

20. Khan F, Spence VA, Belch JJF. Cutaneous vascular responses and thermoregulation in relation to age. Clin Sci 1992; 82:521–528.

21. Roskos KV, Bircher AJ, Maibach HI, Guy RH. Pharmacodynamic measurements of methyl nicotinate percutaneous absorption: The effect of aging on microcirculation. Br J Dermatol 1990; 122:165–171.

22. Shami SK, Chittenden SJ. Microangiopathy in diabetes mellitus: II Features, complications and investigation. Diabetes Res 1991; 17:157–168.

23. Jorneskog G, Tyden G, Bolinder J, Fagrell B. Skin microvascular reactivity in fingers of diabetic patients after combined kidney and pancreas transplantation. Diabetologia 1991; 34:S135–S137.

24. Cesarone MR, Laurora G, Belcaro GV. Microcirculation in systemic hypertension. Angiology 1992; 43:899–903.

25. Belacro G, Nicolaides AN. Microvascular evaluation of the effects of nifedipine in vascular patients by laser-Doppler flowmetry. Angiology 1989; 40:689–694.

26. Belacro G, Nicolaides AN. The venoarteriolar response in diabetics. Angiology 1991; 42:827–835.

27. Parkhouse N, Le Quesne PM. Impaired neurogenic vascular response in patients with diabetes and neuropathic foot lesions. N Eng J Med 1988; 318:1306–1309.

28. Wardell K, Naver HK, Nilsson GE, Wallin BG. The cutaneous vascular axon reflex characterized by laser Doppler perfusion imaging. J Physiol 1993; 460:185–199.

29. Bunker CB, Foreman JC, Dowd PM. Vascular responses to histamine at low temperatures in normal digital skin and Raynaud's phenomenon. Agents Actions 1991; 33:197–199.

30. Bunker CB, Foreman JC, Dowd PM. Digital cutaneous vascular responses to histamine and neuropeptides in Raynaud's phenomenon. J Invest Dermatol 1991; 96:314–317.

31. Maurel A, Hamon P, Macquin-Mavier I, Largue G. Cutaneous microvascular flow studied by laser-Doppler. Presse Med 1991; 20:1205–1209.

32. Walmsley D, Goodfield MJD. Evidence for an abnormal peripherally mediated vascular response to temperature in Raynaud's phenomenon. Br J Rheumatol 1990; 29:181–184.

33. Gean CJ, Tur E, Maibach HI, Guy RH. Cutaneous responses to topical methyl nicotinate in black, oriental and caucasian subjects. Arch Dermatol Res 1989; 281:95–98.

34. Belch JJ, Mclaren M, Lau CS, Mackay IR, Bancroft A, McEwen J, Thompson JM. Cicaprost, an orally active prostacycline analogue: Its effects on platelet aggregation and skin blood flow in normal volunteers. Br J Clin Pharmacol 1993; 35:643–647.

35. McElnay JC, Benson HA, Harland R, Hadgraft J. Phonopheresis of methyl

nicotinate: A preliminary study to elucidate the mechanism of action. Pharm Res 1993; 10:1726–1731.

36. Ryatt KS, Stevenson JM, Maibach HI, Guy RH. Pharmacodynamic measurement of percutaneous penetration enhancement in vivo. J Pharmacol Sci 1986; 75:374–377.

37. Guy RH, Tur E, Schall LM, Elamir S, Maibach HI. Determination of vehicle effects on percutaneous absorption by laser Doppler velocimetry. Arch Dermatol Res 1986; 278:500–502.

38. Tur E, Maibach HI, Guy RH. Percutaneous penetration of methyl nicotinate at three anatomical sites: Evidence for an appendageal contribution to transport? Skin Pharmacol 1991; 4:230–234.

39. Guy RH, Tur E. Bjerke S, Maibach HI. Are there age and racial differences to methyl nicotinate-induced vasodilatation in human skin? J Am Acad Dermatol 1985; 12:1001–1006.

40. Berardesca E, Maibach HI. Racial differences in pharmacodynamic response to nicotinates in vivo in human skin: Black and white. Acta Derm Venereol 1990; 70:63–66.

41. Reinberg AE, Soudant E, Koulbanis C, Bazin R, Nicolai A, Mechkouri M, Touitou Y. Circadian dosing time dependency in the forearm skin penetration of methyl and hexyl nicotinate. Life Sci 1995; 57:1507–1513.

42. Noon JP, Evans CE, Haynes WG, Webb DJ, Walker BR. A comparison of techniques to assess skin blanching following the topical application of glucocorticoids. Br J Dermatol 1996; 134:837–842.

43. Kollias N, Bager A, Sadiq I. Minimum erythema dose determination in individuals of skin type V and VI with diffuse reflectance spectroscopy. Photodermatol Photoimmunol Photomed 1994; 10:249–254.

44. Gardner-Medwin JM, Taylor JY, Macdonald IA, Powell RJ. An investigation into variability in microvascular skin blood flow and the responses to transdermal delivery of acetylcholine at different sites in the forearm and hand. Br J Clin Pharmacol 1997; 43:391–397.

45. Grossmann M, Jamieson MJ, Kellogg DL Jr, Kosiba WA, Pergola PE, Crandall CG, Shepherd AM. The effect of iontophoresis on the cutaneous vasculature: evidence for current-induced hyperemia. Microvasc Res 1995; 50:444–452.

46. Strange P, Skov L. Lisby S, Nielsen PL, Baadsgaard O. Staphylococcal enterotoxin B applied on intact normal and intact atopic skin induces dermatitis. Arch Dermatol 1996; 132:27–33.

47. Jemec GB, Heideheim M. The influence of sex hormones on UVB induced erythema in man. J Dermatol Sci 1995; 9:221–224.

48. Zacharie R, Oster H, Bjerring P. Effects of hypnotic suggestions on ultraviolet B radiation-induced erythema and skin blood flow. Photodermatol Photoimmunol Photomed 1994; 10:154–160.

49. Bjerring P. Comparison of the bioactivity of monetasone furoate 0.1% fatty cream, betamethasone dipropionate 0.05% cream and betamethasone valerate 0.1% cream in humans. Inhibition of UV-B-induced inflammation monitored by laser Doppler blood flowmetry. Skin Pharmacol 1993; 6:187–192.

50. Duteil L, Queille C, Poncet M, Czernielewski J. Processing and statistical analysis of laser Doppler data applied to the assessment of systemic anti-inflammatory drugs. J Dermatol Sci 1991; 2:376–382.

51. Nyren M, Ollmar S, Nicander I, Emtestam L. An electric impedance technique for assessment of wheals. Allergy 1996; 51:923–926.

52. Tur E, Aviram G, Zeltser D, Brenner S, Maibach HI. Regional variations of human skin blood flow responses to hitamine. Curr Probl Dermatol 1995; 22: 50–66.

53. Dykes PJ, Black DR, York M, Dickens AD, Marks R. A stepwise procedure for evaluating irritant materials in normal volunteer subjects. Hum Exp Toxicol 1995; 14:204–211.

54. Tur E, Eshkol Z, Brenner S, Maibach HI. Cumulative effect of subthreshold concentrations of irritants in humans. Am J Contact Dermatitis 1995; 6: 216–220.

55. Agner T. An experimental study of irritant effects of urea in different vehicles. Acta Derma Venereol (Stockh) 1992; 177:Suppl 44–46.

56. Grunewald AM, Gloor M, Gehring W, Kleesz P. Damage to the skin by repetitive washing. Contact Dermatitis 1995; 32:225–232.

57. Marks R, Dykes PJ, Hamami I. Two novel techniques for the evaluation of barrier creams. Br J Dermatol 1989; 120:655–660.

58. Poelman MC, Piot B, Guyon F, Deroni M, Leveque JL. Assessment of topical non-steroidal anti-inflammatory drugs. J Pharm Pharmacol 1989; 41:720–722.

59. Shriner DL, Maibach HI. Regional variations of nonimmunologic contact urticaria. Functional map of the human face. Skin Pharmacol 1996; 9:312–321.

60. Larmi E. PUVA treatment inhibits nonimmunologic immediate contact reactions to benzoic acid and methyl nicotinate. Int J Dermatol 1989; 28:609–611.

61. Johansson J, Lahti A. Topical non-steroidal anti-inflammatory drugs inhibit non-immunologic immediate contact reactions. Contact Dermatitis 1988; 19: 161–165.

62. Tur E. Age-related regional variations of human skin blood flow response to histamine. Acta Derm Venereol 1995; 75:451–454.

63. Fullerton A, Avnstorp C, Agner T, Dahl JC, Olsen LO, Serup J. Patch test study with calcipotriol ointment in different patient groups, including psoriatic patients with and without adverse dermatitis. Acta Derm Venereol 1996; 76: 194–202.

64. Quinn AG, McLelland J, Esex T, Farr PM. Quantification of contact allergic inflammation: A comparison of existing methods with a scanning laser Doppler velocimeter. Acta Derm Venereol 1993; 73:21–25.

65. Quinn AG, McLelland J, Esex T, Farr PM. Measurement of cutaneous inflammatory reactions using a scanning laser Doppler velocimeter. Br J Dermatol 1991; 125:30–37.

66. Stucker M, Auer T, Hoffman K, Altmeyer P. Two-dimensional blood flow determinations in allergic reactions using laser Doppler scanning. Contact Dermatitis 1995; 33:299–303.

67. Artemi P, Seale P, Satchell P, Ware S. Cutaneous vascular response to calcitonin gene-related peptide in psoriasis and normal subjects. Australas J Dermatol 1997; 38:73–76.

68. Zachariae R, Oster H, Bjering P, Kragballe K. Effects of psychological intervention on psoriasis: A preliminary report. J Am Acad Dermatol 1996; 34: 1008–1015.

69. Speight EL, Farr PM. Calcipotriol improves the response of psoriasis to PUVA. Br J Dermatol 1994; 130:79–82.

70. Tur E, Tamir A, Tamir G, Brenner S. Psoriasis. In: Bioengineering of the Skin: Cutaneous Blood Flow and Erythema. Part 2: Laser Doppler Velocimetry and Photoplethysmography. Berardesca E, Elsner P, Maibach HI, Eds. CRC Press, Boca Raton, FL, pp 123–131, 1995.

71. Hornstein OP, Boissevain F, Wittmann H. Non-invasive measurement of the vascular dynamics of dermographism—Comparative study in atopic and non-atopic subjects. J Dermatol 1991; 18:79–85.

72. Vogelsang M, Heyer G, Hornstein OP. Acetylcholine induces different cutaneous sensations in atopic and non-atopic subjects. Acta Derm Venereol 1995; 75:434–436.

73. Van den Brande P, von Kemp K, De Coninck A, Debing E. Laser Doppler flux characteristics at the skin of the dorsum of the foot in young and in elderly healthy human subjects. Microvasc Res 1997; 53:156–162.

74. Kenney WL, Morgan AL, Farquhar WB, Brooks EM, Pierzga JM, Derr JA. Decreased active vasodilator sensitivity in aged skin. Am J Physiol 1997; 272:H1609–1614.

75. Shami SK, Cheatle TR, Chittenden SJ, Scurr JH, Coleridge-Smith PD. Hyperaemic response in the skin microcirculation of patients with chronic venous insufficiency. Br J Surg 1993; 80:433–435.

76. Abu-Own A, Shami SK, Chittenden SJ, Farrah J, Scurr JH, Smith PD. Microangiopathy of the skin and the effect of leg compression in patients with chronic venous insufficiency. J Vasc Surg 1994; 19:1074–1083.

77. Schemeller W, Roszinski S, Huesmann M. Tissue oxygenation and microcirculation in dermatoliposclerosis with different degrees of erythema at the margins of venous ulcers. A contribution to hypodermitis symptoms. Vasa 1997; 26:18–24.

78. He CF, Cherry GW, Arnold F. Postural vasoregulation and mediators of reperfusion injury in venous ulceration. J Vasc Surg 1997; 25:647–653.

79. Gschwandtner ME, Koppensteiner R, Maca T, Minar E, Schneider B, Schnurer G, Ehringer H. Spontaneous laser Doppler flux distribution in ischemic ulcers and the effect of prostanoids: A crossover study comparing the acute action of prostaglandin E1 and iloprost vs saline. Microvasc Res 1996; 51:29–38.

80. Ribuffo D, Muratori L, Antoniadou K, Fanini F, Mertelli E, Marini M, Messineo D, Trinci M, Scuderi N. A hemodynamic approach to clinical results in the TRAM flap after selective delay. Plast Reconstr Surg 1997; 99:1706–1714.

81. Place MJ, Witt P, Hendricks D. Cutaneous blood flow patterns in free flaps determined by laser Doppler flowmetry. J Reconstr Microsurg 1996; 12: 355–358.

82. Niezgoda JA, Cianci P, Folden BW, Ortega RL, Slade JB, Storrow AB. The effect of hyperbaric oxygen therapy on a burn wound model in human volunteers. Plast Reconstr Surg 1997; 99:1620–1625.

83. Bernardi L, Hayoz D, Wenzel R, Passino C, Calciati A, Weber R, Noll G. Synchronous and baroceptor-sensitive oscillations in skin microcirculation: Evidence for central autonomic control. Am J Physiol 1997; 273:H1867–1878.

84. Kolev OI, Nilsson G, Tibbling L. Influence of intense sound stimuli on skin microcirculation. Clin Auton Res 1995; 5:187–190.

85. Kano T, Shimoda O, Higashi K, Sadanaga M. Effects of neural blockade and general anesthesia on the laser Doppler skin blood flow waves recorded from the finger or toe. J Auton Nerv Syst 1994; 48:257–266.

86. Willette RN, Hieble JP, Sauermelch CF. The role of alpha adrenoceptor subtypes in sympathetic control of the acral-cutaneous microcirculation. J Pharmacol Exp Therap 1991; 256:599–605.

87. Mashimo T, Pak M, Choe H, Inagaki Y, Yamamoto M, Yoshiya I. Effects of vasodilators guanethidine, nicardipine, nitroglycerine, and prostaglandin E1 on primary afferent nociceptors in humans. J Clin Pharmacol 1997; 37:330–335.

88. Elam M, Wallin BG. Skin blood flow responses to mental stress in man depend on body temperature. Acta Physiol Scan 1987; 129:429–431.

89. Bornmyr S, Svensson H, Lilja B, Sundkvist G. Skin temperature changes and changes in skin blood flow monitored with laser Doppler flowmetry and imaging; A methodological study on normal humans. Clin Physiol 1997; 17: 71–81.

90. Hassan AA, Rayman G, Tooke JE. Effect of indirect heating on the postural control of skin blood flow in the human foot. Clin Sci 1986; 70:577–582.

91. Hassan AA, Tooke JE. Effect of changes in local skin temperature on postural vasoconstriction in man. Clin Sci 1988; 74:201–206.

92. Winsor T, Haumschild DJ, Winsor D, Makail A. Influence of local and environmental temperatures on cutaneous circulation with use of laser Doppler flowmetry. Angiology 1989; 40:421–428.

93. Steurer J, Hoffman U, Bollinger A. Skin hyperemia in a habitual blusher. Vasa 1997; 26:135–136.

94. Tur E, Ryatt KS, Maibach HI. Idiopathic recalcitrant facial flushing syndrome. Dermatologica 1990; 181:5–7.

95. Frodin T, Alund G, Varenhorst E. Measurement of skin blood flow and water evaporation as a means of objectively assessing hot flushes after orchidectomy in patients with prostate cancer. Prostate 1985; 7:203–208.

96. Wilkin JK. Quantitative assessment of alcohol-provoked flushing. Arch Dermatol 1986; 122:63–65.

97. Wilkin JK, Fortner G, Reinhardt LA, Flowers OV, Kilpatrick SJ, Streeter WC.

Prostaglandins and nicotinate-provoked increase in cutaneous blood flow. Clin Pharmacol Ther 1985; 38:273–277.

98. Wiles PG, Grant PJ, Davies JA. The differential effect of arginine vasopressin on skin blood flow in man. Clin Sci 1986; 71:633–638.

99. Waeber B, Schaller MD, Nussberger J, Bussien JP, Hofbauer KG, Brunner HR. Skin blood flow reduction induced by cigarette smoking. Role of vasopressin. Am J Physiol 1984; 247:H895–901.

100. Nicita-Mauro V. Smoking, calcium, calcium antagonists, and aging. Exp Gerontol 1990; 25:393–399.

101. Lecerof H, Bornmyr S, Lilja B, De Pedis G, Hulthen UL. Acute effects of doxazosin and atenolol on smoking-induced peripheral vasoconstriction in hypertensive habitual smokers. J Hypertens Suppl 1990; 8:S29–33.

102. Salmasi AM, Belcaro G, Nicolaides AN. Impaired venoarteriolar reflex as a possible cause for nifedipine-induced ankle oedema. Int J Cardiol 1991; 30:303–307.

103. Iabichella ML, Dell'Omo G, Melillo E, Pedrinelli R. Calcium channel blockers blunt postural cutaneous vasoconstriction in hypertensive patients. Hypertension 1997; 29:751–756.

104. Orlandi C, Rossi M, Finardi G. Evaluation of the dilator capacity of skin blood vessels of hypertensive patients by laser Doppler flowmetry. Microvasc Res 1988; 35:21–26.

105. Kvernebo K. Slagsvold CE, Stranden E, Kroese A, Larsen S. Laser Doppler flowmetry in evaluation of lower limb resting skin circulation. A study in health controls and atherosclerotic patients. Scand J Clin Lab Invest 1988; 48:621–626.

106. Winsor T, Haumschild DJ, Winsor D, Mikail A. Influence of local and environmental temperatures on cutaneous circulation with use of laser Doppler flowmetry. Angiology 1989; 40:421–428.

107. Cisek PL, Eze AR, Comerota AJ, Kerr R, Brake B, Kelly P. Microcirculatory compensation to progressive atherosclerotic disease. Ann Vasc Surg 1997; 11:49–53.

108. Karanfilian RG, Lynch TG, Lee BC, Long JB, Hobson RW. The assessment of skin blood flow in peripheral vascular disease by laser Doppler velocimetry. Am Surg 1984; 50:641–644.

109. Leonardo G, Arpaia MR, Del Guercio R. A new method for the quantitative assessment of arterial insufficiency of the limbs: Cutaneous postischemic hyperemia test by laser Doppler. Angiology 1987; 38:378–385.

110. Kristensen JK, Karlsmark T, Bisgaard H, Sondergaard J. New parameters for evaluation of blood flow in patients with leg ulcers. Acta Derm Venereol 1986; 66:62–65.

111. Koman LA, Smith BP, Pollock FE Jr, Smith TL, Pollock D, Russel GB. The microcirculatory effects of peripheral sympathectomy. J Hand Surg Am 1995; 20:709–717.

112. Hartmann BR, Bassenge E, Pittler M. Effect of carbon dioxide-enriched water and fresh water on the cutaneous microcirculation and oxygen tension in the skin of the foot. Angiology 1997; 48:337–343.

113. Stansberry KB, Hill MA, Shapiro SA, McNitt PM, Bhatt BA, Vinik AI. Impairment of peripheral blood flow responses in diabetes resembles an enhanced aging effect. Diabetes Care 1997; 20:1711–1716.

114. Rendell M, Bergman T, O'Donnell G, Dorbny E, Borgos J, Bonner F. Microvascular blood flow, volume and velocity measured by laser Doppler techniques in IDDM. Diabetes 1989; 38:819–824.

115. Bornmyr S, Svensson H, Lilja B, Sundkvist G. Cutaneous vasomotor responses in young type I diabetic patients. J Diabetes Complications 1997; 11: 21–26.

116. Boolell M, Tooke JE. The skin hyperaemic response to local injection of substance P and capsaicin in diabetes mellitus. Diabetic Med 1990; 7:898–901.

117. Young MJ, Veves A, Walker MG, Boulton AJM. Correlations between nerve function and tissue oxygenation in diabetic patients: further clues to the aetiology of diabetic neuropathy. Diabetologia 1992; 35:1146–1150.

118. Rendell M, Bamisedun O. Skin blood flow and current perception in pentoxifylline-treated diabetic neuropathy. Angiology 1992; 243:843–851.

119. Walmsley D, Wiles PG. Early loss of neurogenic inflammation in the human diabetic foot. Clin Sci Colch 1991; 80:605–610.

120. Jaap AJ, Shore AC, Tooke JE. Relationship of insulin resistance to microvascular dysfunction in subjects with fasting hyperglycemia. Diabetologia 1997; 40:238–243.

121. Wall B. Assessment of ischemic feet in diabetes. J Wound Care 1997; 6:32–38.

122. Koltringer P, Longsteger W, Lind P, Klima G, Wakonig P, Eber O, Reisecker F. Autonomic neuropathies in skin and its incidence in non-insulin-dependent diabetes mellitus. Horm Metabol Res 1992; 26:S87–S89.

123. Valensi P, Smagghue O, Paries J, Velayoudon P, Nguyen TN, Attali JR. Peripheral vasoconstrictor responses to sympathetic activation in diabetic patients: Relationship with rheological disorders. Metabolism 1997; 46:235–241.

124. Westerman RA, Low AM, Widdop RE, Neild TO, Delaney CA. Non-invasive tests of neurovascular function in human and experimental diabetes mellitus. In: Endothelial Cell Function in Diabetic Microangiopathy. Problems in Methodology and Clinical Aspects, Vol 9, Molinatti GM, Bar RS, Belfiore F, Porta M, Eds., Karger, Basel, p. 127, 1990.

125. Aso Y, Inukai T, Takemura Y. Evaluation of microangiopathy of the skin in patients with non-insulin-dependent diabetes mellitus by laser Doppler flowmetry; Microvasculatory responses to beraprost sodium. Diabetes Res Clin Pract 1997; 36:19–26.

126. Belcaro G, Laurora G, Cesarone MR, Pomante P. Elastic stockings in diabetic microangiopathy. Long-term clinical and microcirculatory evaluation. Vasa 1992; 21:193–197.

127. Yosipovitch G, Schneiderman J, Erman A, Chetrit A, Milo G, Boner G, van Dijk DJ. The effect of an angiotensin converting enzyme inhibitor on skin

microvascular hyperemia in microalbuminuric insulin-dependent diabetes mellitus. Diabet Med 1997; 14:235–241.

128. Flynn MD, Boolell M, Tooke JE, Watkins PJ. The effect of insulin infusion on capillary blood flow in the diabetic neuropathic foot. Diabet Med 1992; 9:630–634.

129. Aman J, Berne C, Ewald U, Tuvemo T. Cutaneous blood flow during a hypoglycaemic clamp in insulin-dependent diabetic patients and healthy subjects. Clin Sci Colch 1992; 82:615–618.

130. Belcaro G, Marelli C, Pomante P, Laurora G, Cesarone MR, Ricci A, Girardello R, Barsotti A. Fibrinolytic enhancement in diabetic microangiopathy with defibrotide. Angiology 1992; 43:793–800.

131. Rendell MS, Shehan MA, Kahler K, Bailey KL, Eckermann AJ. Effect of calcium channel blockade on skin blood flow in diabetic hypertension: a comparison of isradipine and antenolol. Angiology 1997; 48:203–213.

132. Shawket S, Dickerson C, Hazleman B, Brown MJ. Selective suprasensitivity to calcitonin-gene-related peptide in the hands in Raynaud's phenomenon. Lancet 1989; 2:1354–1357.

133. Suichies HE, Aarnoudse JG, Wouda AA, Jentink HW, de Mul FF, Greve J. Digital blood flow in cooled and contralateral finger in patients with Raynaud's phenomenon. Comparative measurements between photoelectrical plethysmography and laser Doppler flowmetry. Angiology 1992; 43:134–141.

134. Allen JA, Devlin MA, McGrann S, Doherty CC. An objective test for the diagnosis and grading of vasospasm in patients with Raynaud's syndrome. Clin Sci 1992; 82:529–534.

135. Kurozawa Y, Nasu Y, Oshiro H. Finger systolic blood pressure measurements after finger cooling. Using the laser-Doppler method for assessing vibration-induced white finger. J Occup Med 1992; 34:683–686.

136. Kristensen JK, Englhart M, Nielsen T. Laser Doppler measurement of digital blood flow regulation in normals and in patients with Raynaud's phenomenon. Acta Derm Venereol 1983; 63:43–47.

137. Goodfiled M, Hume A, Rowell N. Reactive hyperemic responses in systemic sclerosis patients and healthy controls. J Invest Dermatol 1989; 93:368–371.

138. Engelhart M, Seibold JR. The effect of local temperature versus sympathetic tone on digital perfusion in Raynaud's phenomenon. Angiology 1990; 41: 715–723.

139. Wollersheim H, Reyenga J, Thien T. Postocclusive reactive hyperemia of fingertips, monitored by laser Doppler velocimetry in the diagnosis of Raynaud's phenomenon. Microvasc Res 1989; 38:286–295.

140. Martinez RM, Saponaro A, Dragagna G, Santoro L, Leopardi N, Russo R, Tassone G. Cutaneous circulation in Raynaud's phenomenon during emotional stress. A morphological and functional study using capillaroscopy and laser Doppler. Int Angiol 1992; 11:316–320.

141. Whitmore SE, Wigley FM, Wise RA. Acute effect of topical minoxildil on digital blood flow in patients with Raynaud's phenomenon. J Rheumatol 1995; 22:50–54.

142. Engelhart M. Ketanserin in the treatment of Raynaud's phenomenon associated with generalized scleroderma. Br J Dermatol 1988; 119:751–754.

143. Brouwer RM, Wenting GJ, Schalekamp MA. Acute effects and mechanism of action of ketanserin in patients with primary Raynaud's phenomenon. J Cardiovasc Pharmacol 1990; 15:868–876.

144. Bunker CB, Lanigan S, Rustin MHA, Dowd PM. The effects of topically applied hexyl nicotinate lotion on the cutaneous blood flow in patients with Raynaud's phenomenon. Br J Dermatol 1988; 119:771–776.

145. Rustin MHA, Newton JA, Smith NP, Dowd PM. The treatment of chilblains with nifedipine: The results of a pilot study, a double-blind placebo-controlled randomized study and a long-term open trial. Br J Dermatol 1989; 120: 267–275.

146. Agusni I, Beck JS, Potts RC, Cree IA, Ilias MI. Blood flow velocity in cutaneous lesions of leprosy. Int J Lepr Other Mycobact Dis 1988; 56:394–400.

147. Wilder Smith A, Wilder Smith E. Electrophysiological evaluation of peripheral autonomic function in leprosy patients, leprosy contacts and controls. Int J Lepr Other Mycobact Dis 1996; 64:433–440.

148. Leu AJ, Yanar A, Jost J, Hoffman U, Franzeck UK, Bollinger A. Microvascular dynamics in normal skin versus skin overlying Kaposi's sarcoma. Microvasc Res 1994; 47:140–144.

149. Rodgers GP, Schechter AN, Noguchi CT, Klein HG, Nienhuis AW, Bonner RF. Periodic microcirculatory flow in patients with sickle-cell disease. N Engl J Med 1984; 311:1534–1538.

150. Tur E, Politi Y, Rubinstein A. Cutaneous blood flow abnormalities in hypertriglyceridemia. J Invest Dermatol 1994; 103:597–600.

151. Algotsson A, Nordberg A, Almkvist O, Winblad B. Skin vessel reactivity is impaired in Alzheimer's disease. Neurobiol Aging 1995; 16:577–582.

152. Tur E, Brenner S. Cutaneous blood flow measurements for the detection of malignancy in pigmented skin lesions. Dermatology 1992; 184:8–11.

153. Wang I, Andersson Engels S, Nilsson GE, Wardell K, Svanberg K. Superficial blood flow following photodynamic therapy of malignant non-melanoma skin tumours measured by laser Doppler perfusion imaging. Br J Dermatol 1997; 136:184–189.

154. Wardell K, Nilsson GE. Duplex laser Doppler perfusion imaging. Microvasc Res 1996; 52:171–182.

155. Bornmyr S, Martensson A, Svensson H, Nilsson KG, Wollmer P. A new device combining laser Doppler perfusion imaging and digital photography. Clin Physiol 1996; 16:535–541.

21
Drug Concentration in the Skin

Christian Surber
University Hospital, Basel, Switzerland

Eric W. Smith
Ohio Northern University, Ada, Ohio

Fabian P. Schwarb
University Hospital, Basel, Switzerland

Howard I. Maibach
University of California School of Medicine, San Francisco, California

I. INTRODUCTION

An understanding of the pharmacodynamics and pharmacokinetics of a drug is essential for its safe use. A therapeutic effect of a drug and its appropriate dosage regimens are based upon the relationships of time and drug concentration at the active site (target organ or biophase) and the resultant drug response. For reasons of feasibility, correlation of drug effects and drug concentration in blood or urine are most often monitored for systemically acting agents. The three factors that are generally assessed are a) the area under the serum concentration versus time curve, b) the peak serum concentration, and c) the time to peak serum concentration. Analytical difficulties sometimes preclude the measurement of drug or metabolite levels in the body fluids, in which case drug effects are assessed by observing an appropriate pharmacologic response (1,2). Pharmacokinetics are the current mainstay parameters of bioequivalence assessments because blood and urine assays are available for nearly all drugs, and because pharmacokinetic methods are generally well understood. The key assumption is that equivalent pharmacokinetic parameters indicate equivalent therapeutics—or, more precisely, that observed differences in pharmacokinetic parameters between formulations are predictive of differences in clinical performance of the two formulations. However, little attempt to verify this key assumption is made (3–5).

For topical drug products, such as ointments, creams, and gels (as distinguished from transdermal delivery systems), bioavailability studies paralleling those of orally administered drugs are often difficult to carry out and may provide an inappropriate measure of topical bioequivalence. Furthermore, topical doses tend to be so small that drug concentrations in blood and/or urine are often undetectable using current assay techniques. Moreover, systemic availability may not properly reflect cutaneous bioavailability for medications intended to treat local skin disorders since systemically detected drug is no longer at its intended site of action and is, additionally, subjected to complex central compartment pharmacokinetics.

Assessment of bioavailability or bioequivalence of topical formulations may be made by taking one of several possible approaches. Bioavailability or bioequivalence may be established through a well-controlled clinical trial or through an easily monitored pharmacological endpoint. Establishing that the concentration of drug within the treated tissue correlates with clinical performance or to some relevant pharmacological endpoint is another approach to bioavailability or bioequivalence that is currently being debated. Similarly, some in vitro methods and animal studies that correlate with clinical performance are being investigated.

At present a well-controlled clinical trial is the only uncontested, generally acceptable procedure for demonstrating bioavailabilty and bioequivalence of topical drug products. It is highly desirable to develop alternative in vivo (and in vitro) procedures because of the high costs of conducting clinical trials. The purposes of this chapter are a) to discuss some methods for the direct determination of drug concentration in the skin compartment, b) to review the relationships between the drug concentration in the skin and a clinical or pharmacological effect, and c) to point out some difficulties in the interpretation of the results from these models.

II. SAMPLING TECHNIQUES

A. Suction Blister Technique

The development of a defined suction blister that separates the skin strata subepidermally was first described by Kiistala et al. (6–8). Kiistala's method uses a special dome-shaped Dermovac cap (Instrumentarium Corp., Espoo, Finland) with several small holes of 6 mm diameter. Suction blisters can also be developed using a plane block with holes of similar diameter (8 mm) (9). In a typical suction blister induction with a Dermovac cap, a consistent suction of about 200 mm Hg (2.66 Pa) below the atmospheric pressure is employed for a 2–3 h period, after which 50–150 µl suction blister fluid and small stratum corneum–epidermal sheets can be harvested. The blister fluid corresponds roughly to the interstitial fluid. In contrast to the alternative cantharidin-induced blister fluid (10), the suction blister fluid is free of white cells for the first 5 h, after which leukocytes begin to appear (11). The protein content of suction blisters is 60–70% of the corresponding serum value (9). Secretion of fluid into the capsule of the Dermovac with and without bursting of the blister roof is often observed. In diseased skin, blisters cannot be raised uniformly, and standardization of the methodology in diseased skin has not been possible.

For pharmacokinetic studies up to 20 small blisters are formed; drug is administered systemically or topically (either before or after blister raising), and the blister fluid and blister roof are harvested at various time intervals for subsequent analysis.

Only limited reports exist where the suction blister technique has been used to study topical drug delivery (12–17). Treffel and coworkers (12) studied transepidermal absorption of citropten and bergapten from two different cosmetic tanning products (oil/water emulsion, oil) into suction blister fluid. The drug concentrations determined by high-pressure liquid chromatography were 37 and 51 ng/ml (water/oil emulsion) and 26 and 23 ng/ml (oil), respectively, for citropten and bergapten.

A number of studies have been published concerning 8-methoxypsoralen (8-MOP) absorption. Using the suction blister technique, Huuskonen et al. (13) studied absorption of 8-MOP during PUVA bath therapy. More recently, Averbeck and coworkers (14) compared the photobiological activity (18) and the concentration of 5-MOP (bergapten) after topical application. They observed that the photobiological activity of 5-MOP in the suction blister fluid was more effective when it was delivered from an ethanolic solution than from a bergamot oil with the same 5-MOP content. From these experiments it can be concluded that the suction blister technique is useful not only for pharmacokinetic studies but also for pharmacodynamic evaluations, making it possible to study the biological activity of a drug and its metabolites at the level of the target organ.

Surber and Laugier (16,17) compared acitretin (retinoid) concentration in the skin after topical and systemic administration using the suction blister technique, and two other skin biopsy techniques. These data are discussed in the section on skin biopsy.

B. Cantharidin Blister Technique

Large sheets of stratum corneum and blister fluid can be obtained with some discomfort to the subjects using the cantharidin blister technique (10). Many samples can be obtained from different regions of the same person with little effort and with the advantage of not leaving a permanent scar, although abnormal skin pigmentation at the site of application may persist for several months after harvest. Use is made of the capacity of cantharidin, a substance extracted from "Spanish fly" to produce an intraepidermal blister (in humans only). For most body areas a 0.2% solution of cantharidin in acetone is satisfactory, and for sites where the stratum corneum is thick and dense the concentration is increased to 0.5%. The cantharidin–acetone solution is placed into a glass or metal cylinder and the acetone is evenly evaporated to dryness under a stream of air. A layer of wet cloth is placed over and beyond the application site and fastened occlusively to the skin by impervious plastic tape. Eight to 10 h later, the turgid clear blister can be punctured using an insulin syringe in order to collect the blister fluid. Subsequently the blister itself can be excised with scissors. Remnants of epidermis clinging to the underface of the horny layer are removed by firm rubbing with a cotton-tipped applicator. Circular sheets of stratum corneum as large as 4 cm in diameter can be prepared in this manner, and the volume of blister fluid is dependent on the size of the blister raised. Cantharidin blisters can also be obtained by the various cantharidin plasters described in old pharmacopoeias. The cantharidin blister fluid is an inflammatory exudate, containing 650–12,700 white cells/μl (19,20) and its albumin concentration significantly exceeds that of suction blister fluid (70–80% of the respective serum level) (21).

The blisters (suction blister and cantharidin blister) contain a fluid reservoir that is continuous with the intravascular fluid and the skin surface such that an immediate representation of the tissue concentration of a drug can be assumed. However alterations of the composition of the blister fluid and the morphology of the blister base or blister roof may influence drug penetration into the blister fluid. The induction of an inflammatory response, for example, may result in a slower or enhanced passage of drug from the capillaries into the blister. The preservation of the barrier function of the stratum corneum following blister formation has also been questioned since the stratum corneum is then in an abnormally stretched state (15).

C.   Related Techniques

If the intention of the experimentation is to gain better access to the interstitial fluid as an investigative or analytical medium, then the surface layers of the skin may be removed by suction blistering and subsequent excision or by dermabrasion, in order to obtain a skin window exposing the epidermal tissue. A chamber filled with saline solution or prewetted disks may subsequently be applied to the exposed skin window. At various time intervals, the fluid or the disks are removed from the skin window and are refilled with isotonic saline or replaced by new disks (22,23). The method of dermabrasion leads to an acute inflammatory reaction, and large numbers of granulocytes and albumin are found in the skin chamber fluid. Similarly, access to other peripheral compartments may be achieved by using different experimental "devices" that allow the collection of specific tissue fluids. Pertinent methods include the implantation of tissue cages, fibrin clots, and nylon or cotton threads (24–28).

D. Surface Recovery

Percutaneous absorption may be determined by the loss of material from the application surface as the drug penetrates into skin. Drug recovery from an ointment or a solution application is difficult because total recovery of compound from the skin can never be assured. The technique is usually applicable only in cases where a large amount of drug is absorbed from the applied formulation. In contrast, with the topical application of a transdermal delivery device, the entire unit can be removed from the skin and the residual amount of drug in the device can be determined. It is assumed that the difference between applied dose and residual dose is the amount of drug absorbed. It is also possible to monitor the disappearance of radioactivity from the surface of skin using an external Geiger–Müller counter (29). The limitation of this methodology is that the disappearance is due both to the movement of ^{14}C-labeled chemical into the skin and to the quenching effect of the skin on the β rays incident on the instrument detector. The degree of quench radiolabeled chemicals in the various cell layers of the skin has not been defined. Lee and coworkers (30) recently developed a new methodology using ^{19}F magnetic resonance spectroscopy (MRS) to measure the in vivo percutaneous absorption of flurbiprofen through hairless rat skin. A 2% w/v flurbiprofen gel containing isopropyl alcohol, water, and propylene glycol was used. Gel was applied within a rubber O-ring to the skin of the lower back of an anesthetized hairless rat and occluded with a plastic cover slip. The animal was placed on a magnetic resonance surface coil, and measurements were taken continuously over approximately 3 h at 10-min intervals with a 2 T GE CSI nuclear magnetic resonance (NMR) spectrometer in order to measure the disappearance of magnetic resonance signal intensity per time interval. This assessment relates directly to the percent of drug disappearance over time and can be converted to a flux value.

Related to the surface recovery technique, Phillips et al. (31–34) and Peck et al. (35) developed transcutaneous drug collection device. The device consists of a salt-impregnated absorbent pad (e.g., 0.3 g agarose, 0.5 g activated carbon, 9.2 g 10% saline per 10 g (36), which is covered by water-impermeable tape. Drug is assessed in the pad at continuous time intervals. The technique has been proposed as a method for studying drug disposition in subjects in a fashion that is less intrusive than currently available techniques and that may be suited to the surveillance of drug exposure in ambulatory individuals. In particular, continuous transcutaneous drug collection (CTDC) has been proposed for the estimation of ethanol intake in drinking subjects (32) and for the monitoring of patient compliance with a drug regimens (36,37).

Peck et al. (35) have extensively analyzed the theoretical basis for the use of the CTDC device in assessing various aspects of drug pharmacokinetics. The time course of drug accumulation in the CTDC device was simulated using standard pharmacokinetic and mathematical techniques. The influences and implications of single and polyexponential drug disposition kinetics and backtransfer of drug from the collection device were investigated. Their interpretation of the results suggest two principal factors that affect the utility of this technique: first, the degree of back transfer of drug from the CTDC device into the body and second, the duration of transcutaneous drug collection relative to the elimination rate of the drug from the body. In their theoretical treatment they excluded the notion that the rate of drug transfer into the CTDC device is somehow dependent upon the sweat secretion rate. They were aware of the fact that an occlusive dressing such as a CTDC device surely alters the skin permeability relative to unoccluded skin. However, they assume that the temperature and the hydration of the skin under the dressing become stabilized. Furthermore, their analyses suggested that the utility of a CTDC device is severely restricted when back transfer from the collection device is substantial. The ability to minimize or even exclude back transfer by binding the effused target chemical irreversible would promote the use of this promising tool for assessing the systemic amount of drug exposure.

E. Skin Stripping and Related Techniques

The skin stripping and related techniques are discussed in Chapter 23.

F. Sebum Collection

Sebaceous glands are largest and most numerous on the forehead, face, in the ear, and on the midline of the back. Sebum is a complex mixture of lipids. Its principal components are glycerides, free fatty acids, wax esters, squalene, cholesterol, and cholesterol esters. Sebum is miscible with water and allows water and polar and nonpolar materials to penetrate. The length of time required for sebum to be produced and move from the base of the sebaceous gland to the skin surface is less than 10 days (38–41). Several functions have been ascribed to sebum, such as controlling water loss and protecting the skin from fungal and bacterial infection; however, these claims have been questioned. Sebum collection has been used extensively to study the changes of the lipid composition during therapies with retinoids.

Sebum is normally collected by swabbing the forehead, nose, cheeks, and midback with a sponge soaked in an organic solvent (e.g., isopropylal-cohol or hexane). The sebum is extracted from the sponge for a subsequent analysis. With other techniques a cigarette paper, a special tape, or a ground-glass plate is pressed onto the skin site of interest to adsorb the surface sebum (42–44).

Attempts have also been made to detect different drugs in the sampled lipid mixture. Tetracycline and minocycline are well referenced and are clin-ically considered to be effective drugs in the treatment of acne and other infectious dermatosis. However, documentation of skin pharmacokinetic data for these antibiotics is sparse and controversial. Rashleigh et al. (45) showed that tetracycline detected by a fluorometric method accumulated in sebum after oral administration. Significant quantities of tetracycline (3–18 μg/g sebum lipid) were detected in the surface film taken from the forehead after 4–8 days. However, they did not measure tetracycline levels within the pilosebaceous lumens, the focal point for acneform lesions. There was no direct relationship found between the dose applied, blood levels, and con-centration of the drug in the sebum. They stated that from their experimental design and data a firm conclusion on the route of tetracycline penetration into the skin cannot be drawn. Using a microbiologic method that required approximately 1 g of skin from each subject, Gould et al. (46) showed indirectly that 70% of dosed terramycin was present in the skin when com-pared with the serum terramycin level. Aubin et al. (47) dosed 10 volunteers with minocycline (200 mg/day) for 4 weeks but were not able to detect minocycline in the skin by using high-performance liquid chromatography (HPLC) (C_{18}) at a detection limit of 0.3 ng/ml. However, Luderschmidt et al. (48) reported that they successfully detected approximately 40 μg min-ocycline per 6 cm^2 area by HPLC on day 6 of treatment in patients admin-istered 100 mg minocycline/d. With the exception of Rashleigh et al. (45), none of the investigators discussed the pathway of the drug penetration into the skin, and from the experiments carried out it is purely hypothetical whether tetracyclines accumulate in the sebaceous glands.

The knowledge of the drug concentration in sebum is mainly derived from the examination of surface samples collected from anatomical sites rich in sebaceous glands. An alteration of drug (and sebum) between the site of biosynthesis and the skin surface is highly probable since the correspond-ing lag time is about 8–9 days (40,41). During that delay a number of factors could contribute to an alteration of the initial sample, such as a) biotransformation by microorganisms or by enzymes released from the microorganisms; b) dilution and/or reaction with other endogenous or exogenous species such as epidermal lipids; c) adsorption or binding to other media such as hair; or d) alteration through environmental aggressions such as temperature, light, oxygen, and humidity. Isolation of sebaceous glands (49) from humans or animals during drug treatment is possible and may therefore be an alternative sebum collection procedure.

G. Hair and Nail Collection

The use of hair or nails is advantageous in that xenobiotics are maintained unchanged for long periods of time in these easily sampled tissues. It is thus possible to monitor the intake of drugs and poisons or to follow the chronology of an intoxication by analysis of hair or nail content. The detection of arsenic in hair is probably the most well known of these techniques for forensic purposes. As a result of better analytical equipment, the detection of a variety of drugs such as phencyclidine, phenobarbital, opiates, amphetamines, tricyclic antidepressants, methadone, and nicotine in this matrix has been made possible (50–55). These advances have mainly been made in forensic medicine, while little effort has been made in dermatology to detect drugs (e.g., antimycotics) in the hair after systemic dosing (56).

In a series of papers, Walters and Flynn (57–61) developed the basic concepts of the permeability of the human nail plate and thus created a better understanding of the toxicity potentials and therapeutic possibilities of substances applied to the nail. The collective physicochemical and clinical observations point to one general conclusion, namely, that the nail is permeable to a variety of chemicals ranging from small polar and nonpolar nonelectrolytes and salts to large complex drug molecules. Collectively analyzed, these studies indicate that the nail plate is intrinsically up to one thousand times more permeable to water than is stratum corneum, given that the ratio of nail to stratum corneum thickness is about 100. Information from systematic investigations on penetration of various antifungals into the nail from topical formulations and the influence of the corresponding vehicles are still infrequent and incomplete (62–64).

It is generally understood that in the planning of a drug regimen with, for example, an oral antibiotic for the treatment of an infection, it is necessary to know the pattern of serum and tissue levels following drug administration. When considered in parallel with information concerning the bactericidal levels of an antibiotic, it is possible to plan an appropriate oral dosage, application frequency, and duration of therapy. Although in the past there has been a lack of similar relevant information concerning drugs used to treat disease of the nail plate, the same principles should apply. In a recent study with humans (65), plasma terbinafine levels directly correlated with oral dose, and distal nail levels correlated with plasma levels. Terbinafine was generally detected earlier in proximal nail plate than in distal nail clippings, and the persistence of the drug in the nail plate paralleled plasma levels. Based on this data it should be possible to design a therapeutic regimen based on an optimal oral dose and duration of therapy.

H. Skin Biopsy

The most invasive—but still practicable—method to access skin compartments is the excision of skin tissue. In contrast to the other methods described, the punch and shave biopsies allow direct ingress into the compartment of interest. After removal (optional) of the stratum corneum from skin with an appropriate technique (tape stripping or adhesive), the punch biopsy will contain parts of the subcutaneous tissues, dermis, and epidermis, and the shave biopsy will mainly contain epidermis and some dermis. Parts of the stratum corneum may remain on the epidermis depending on the method used for stratum corneum removal. Subcutaneous tissue can mechanically be divided from the dermis, and the latter can be separated from the epidermis by heating techniques (67). Human skin samples larger than 100 mg are difficult to obtain, and the usual amount harvested is less than 50 mg.

Surber and Laugier (16,17) compared the acitretin concentration (synthetic retinoid) in human skin after oral and topical dosing using three different skin sampling techniques: punch biopsy, shave biopsy, and suction blister. All three techniques have been used by various investigators to quantitate drugs and xenobiotics in the skin. Each technique gives access to distinctive skin compartments (Fig. 1). The acitretin concentration in the skin after systemic application in a steady-state situation was comparable with the drug concentration reached after a single 24-h topical application of a saturated acitretin/isopropylmyristate formulation (Table 1). However, no beneficial effects have so far been observed in psoriasis and other disorders of keratinization by the topical administration of acitretin. Similar observations have been made with methotrexate and cyclosporin (67,68). One may postulate that drug concentration at a particular site within the skin following the two routes of administration could be different due to the direction of the drug concentration gradient. This hypothesis has also been postulated and illustrated by Parry et al. (69) when comparing the clinical efficacy of topical and oral acyclovir. Model predictions and in vivo data agree that topical administration of acyclovir results in a much greater total epidermal drug concentration than that after oral administration. However, mathematical modeling of the acyclovir concentration gradient through the

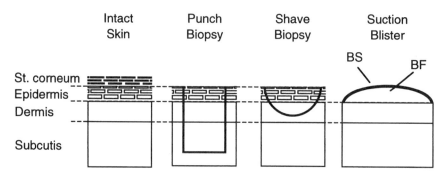

Figure 1 Schematic diagram of intact skin (left) and the skin specimens taken using three sampling techniques. BS, blister skin; BF, blister fluid. For details see text.

epidermis revealed that the drug concentration in the target site of the herpes simplex infection, the basal epidermis, was two to three times less after topical administration than after oral administration. Furthermore, one may postulate that drug metabolism could be different depending on the route of administration. Data supporting this hypothesis are still incomplete. Despite skilled experimentators, sophisticated sampling techniques, and instrumentation, the formation gained from these tissue samples is probably only an estimate of the chemical distribution within the skin. Accurate and specific information on drug localization within a particular skin compartment following both routes of administration is not obtained by these methods, possibly because of interlaminate drug contamination.

The following two techniques have been used to obtain skin sections parallel to the outer skin surface to a depth of about 300–600 μm.

Table 1 Total Drug Concentration in Skin after Oral and Topical Administration of Acitretin

	Total drug concentration, ng/g wet tissue	
Sample	Oral application	Topical application
Punch biopsy	275	160
Shave biopsy	160	360
Suction blister fluid	200	90
Suction blister skin	460	3800

Note. Data from refs. 16 and 17.

1. Semiautomated Skin Sectioning Technique

With this approach the skin tissue from a punch biopsy is placed onto a cryomicrotome table and 10–40 μm sections are cut parallel to the skin surface. This technique is summarized in the form of an excellent standard operating procedure by Schaefer et al. (70,71). When the cylinder-shaped punch biopsy is placed, dermis side down, on the microtome chuck maintained at −17°C it is essential that the stratum corneum surface is parallel to the cutting plane. This can be accomplished by placing several metal rings of internal diameter and composite thickness slightly greater than that of the skin sample around the tissue and filling the rings with embedding medium. Subsequently a flat, precooled surface (glass slide) is pressed against the stratum corneum, resulting in a flat outer surface, and the whole assembly is then rapidly frozen and can be stored. The technique does not take into account the dermal papillae, and a clear separation of histologically distinct compartments is not possible. The method has predominately been used to characterize the pharmacokinetic and metabolic behavior of topically applied drugs (72,73).

Using excised human skin and tissue grafted to athymic mice, the in vitro and in vivo concentration profiles of salicylic acid and salicylate esters were obtained for the outer several hundred microns of the skin (73). The results show significant differences in the extent of enzymatic cleavage and distribution of metabolites between in vitro and in vivo studies. The data also suggest that in vitro results may overestimate metabolism because of increased enzymatic activity and/or decreased capillary removal of biotransformed products (Fig. 2).

2. Manual Skin Sectioning Technique

With the development of the skin sandwich flap model (74), Pershing and coworkers (75) proposed a new manual skin sectioning technique by which the disposition of test compounds may be examined after topical administration. Their technique requires the use of radiolabeled compound and fresh skin that has not been frozen. Briefly (Fig. 3), a 2-mm skin punch biopsy is sectioned into 115-μm-thick sections by dipping the stratum corneum end of the biopsy into cyanoacrylate adhesive (step 1) and fixing this end of the biopsy cylinder to a microscope slide. Two 115-μm-thick microscope cover slips, which act as cutting guides, are positioned on either side of the biopsy and are held in place on the microscope slide with tape (step 2). A single-edged razor blade is used to shave the adhering stratum corneum side of the punch from the remainder of the biopsy. The outermost stratum of the removed biopsy segment is then dipped in the cyanoacrylate adhesive again and fixed in a new location on the microscope slide (b) prior to sectioning

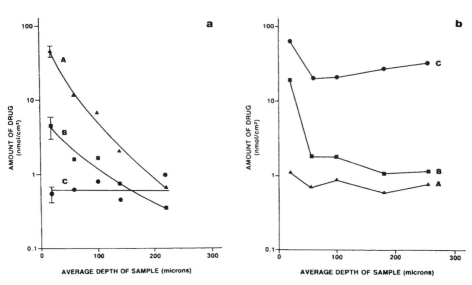

Figure 2   The concentration profile of the diester (A), monoester (B), and salicylate (C), as a function of the depth within human skin either grafted to athymic mice (a) or in vitro (b). Results were obtained after 24 h of topical drug application. From ref. 73, with permission.

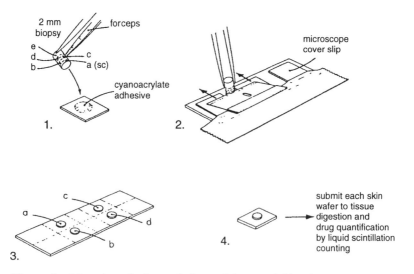

Figure 3   Manual sectioning technique of the punch biopsies. For details see text. From ref. 79, with permission.

in an identical manner using the cover slips as cutting guides. The remainder of the skin biopsy is sectioned similarly, producing multiple 115-μm sections (c–d) on an individual microscope slide (step 3). After complete sectioning of the skin biopsy, the various skin wafers on the microscope slide are individually digested with tissue solublizer and submitted to liquid scintillation counting for drug quantification. Subsequently, a concentration gradient and disposition profile of the test compound in the skin is constructed. The authors also reported that freezing the 2-mm biopsy results in redistribution of the drug from the stratum corneum to the middle sections (400–600 μm depth) of the skin biopsy, thereby portraying an inaccurate distribution profile in the skin, unfortunately, this statement is not documented further, since this redistribution phenomenon may also be relevant for a 4–6 mm biopsy, even though it was only described for a 2-mm biopsy. This observation makes it necessary to carefully reevaluate the skin preparation procedure.

I. Microdialysis

Microdialysis is a sampling technique that allows continuous, real-time drug monitoring and causes minimal tissue damage or physiological alterations in the test subject. Further advantages of microdialysis sampling include the fact that once a molecule crosses the dialysis membrane, enzymatic degradation or protein binding is eliminated and that the tissue can be directly sampled without causing significant tissue fluid loss. Additionally, thermal and hydrolytic degradation of the compounds of interest can be minimized by optimizing the microdialysis sampling rate and the nature of the dialysis medium (76,77). A significant reduction in the number of animals required to perform dermal and transdermal drug delivery research may be possible. A single animal can be used for the continuous real-time study of drug concentration in the skin with microdialysis, compared to numerous animals required for each time point by other currently accepted methods.

The initial questions regarding the feasibility of this technique and the impact of the probe implantation on the flux of drug through the skin were addressed by studies using a commercially available probe implanted into full-thickness rat skin in vitro (78). Results of the initial feasibility studies necessitated the development of a novel probe system based on minimal probe dimensions. While a coated probe of linear geometry has been used previously, a biocompatible silicone coating on a cellulose dialysis fiber for implantation into skin has been designed and validated by the same group (79). Following topical application of 5-fluorouracil cream (Efudex), the implanted dermal microdialysis probe successfully monitored the drug concentration in the skin in vivo. Edema, inflammation, and tissue scarring around the site of implantation assessed by histological examination appeared to be minimal during the first 24–36 h after implantation. The histological studies revealed lymphocyte infiltration into the area of implantation after 6 h and scar-tissue formation around the probe after 72 h. While the problem of tissue damage due to probe implantation is a disadvantage of any invasive procedure, the extent of tissue damage determined by histological examination and drug perfusion studies was minimal.

In a recent report, Matsuyama et al. (80) successfully demonstrated that the intradermal implantation of a microdialysis probe in vivo was practical in order to study percutaneous absorption of methotrexate and to assess changes in drug concentrations following coapplication of an enhancer.

This simple method for determining drug concentration in the skin in vivo provides previously inaccessible information about drug transport in the skin and may, in the future, serve as a reliable technique for studying topical bioavailability and skin pharmacokinetics in order to guide drug development and patient therapy.

III. ANALYTICAL TECHNIQUES

A. Autoradiography

Autoradiography is a photographic technique used to detect the localization of radioactive materials in specimens. The technique itself is over 100 years old, but its application to skin research has increased greatly during the last 30 years. The methodology has improved rapidly, and application of the technique has progressed not only from a tissue level to cell level, but also to a subcellular level (81). In the process of drug delivery to the skin, the localization of the penetrating/permeating compounds in the skin layers and the identification of transport pathways is essential. Only a few approaches to quantify the compound in the different skin layers and appendages have been reported. An important contribution to the quantitative evaluation was made by Schaefer and Schalla (71,82), who introduced the skin sectioning method discussed earlier. To overcome certain disadvantages of their technique, they supplemented their findings with qualitative histoautoradiography. Thus the combination of histoautoradiography and skin sections may indicate predominating permeation routes.

In a recent investigation (83) the development of an intracutaneous depot for drugs was studied using a new pharmaceutical ingredient, diethylene glycol monoethyl ether (Transcutol). A qualitative assessment revealed that the distribution pattern of hydrocortisone (model drug studied) was dependent on the vehicle used. In the control group (treated with hydrocortisone without Transcutol) the distribution was not uniform, whereas in the treatment group (treated with hydrocortisone with Trancutol) the hydrocortisone was uniformly distributed in the skin and the amount of radioactivity present in the skin was higher in the treatment group relative to the control group.

In an autoradiographic study on percutaneous absorption of five ^{14}C-labeled oils in the guinea pig, a decreasing "absorbability" in the following order was found: isopropyl myristate, glyceryl trioleate, n-octadecane, decanoxydecane, 2-hexyldecanoxyoctane (84). The skin irritation potentials (macroscopic observation of erythema) of these oils were in agreement with the absorption observed by microautoradiography, although isopropyl myristate produced much more severe irritation than 2-hexyldecanoxyoctane.

Estradiol retention in the skin was investigated by Bidmon et al. (85) using dry-mount autoradiography. [^{3}H]Estradiol-17β in dimethyl sulfoxide, ethylene glycol, or sesame oil was applied to shaved rat skin in the dorsal neck region. The results of the experimental demonstrate that dry-mount autoradiography provides both regional and cellular resolution for specific tissues and indicates the penetration routes of the diffusible compound. Since embedding and melting of tissue are avoided and permanent contact with photographic emulsion is obtained, translocation of labeled compound during the preparation of the autoradiograms is excluded and the cellular and subcellular resolution is high. After 2 h of topical treatment with [^{3}H]Estradiol-17β dissolved in dimethyl sulfoxide a distinct cellular distribution was apparent. Accumulation of radioactivity was found in the epidermis, sebaceous glands, dermal papillae of hair, and fibroblasts. The stratum corneum accumulated and retained radioactivity, apparently forming a depot for the hormone. Marked concentration and retention of hormone were observed in the sebaceous glands for more than 24 h, suggesting that sebaceous glands serve as second storage sites for the hormone. In all autoradiograms, two penetration pathways to the dermis were visible: one through the stratum corneum and epidermis, and the other through the hair follicles. The same permeation routes and sites of deposition and retention are recognizable for female and male rats, independent of whether dimethyl sulfoxide, ethylene glycol, or sesame oil was used as the vehicle. It is important to note that low topical doses of drug reveal information on specific drug binding in target cells with cellular and subcellular resolution. Higher doses provide the advantage of visualizing permeation routes, gradients and vehicle effects in the various skin compartments (Fig. 4).

By means of a microcomputer-based image-analyzing autoradiographic method, it was possible to measure and visualize the effects of carriers and application time on the localization in the rate skin of two lipophilic compounds, tetrahydrocannabinol and oleic acid (86). It was found that both compounds, dissolved in Transcutol, show similar penetration profiles after

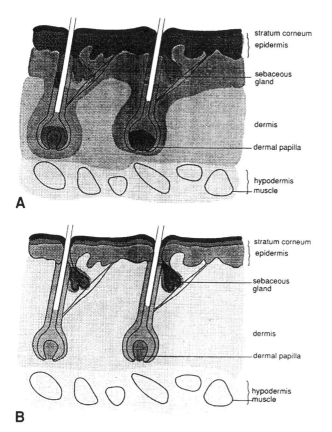

Figure 4 Scheme of distribution of radiolabeled estradiol (A) 2 h and (B) 24 h after topical application of 301 pmol/cm^2. The penetration gradients through the epidermis and the hair follicles are visible, as well as the enhanced uptake of radioactivity in sebaceous glands and dermal papillae. The highest amount of radioactivity is on top of the stratum corneum and in the stratum corneum (B). The different grey levels reflect the different silver grain densities. From ref. 85, with permission.

2 h. No significant difference was found between the concentration in the epidermis and that in the appendages, either for tetrahydrocannabinol or for oleic acid. However, a dramatic effect of formulation composition on the localization of tetrahydrocannabinol was observed (Fig. 5) after 24 h when polyethylene glycol 400, diethylene glycol monoethyl ether, or propylene glycol/ethanol was used.

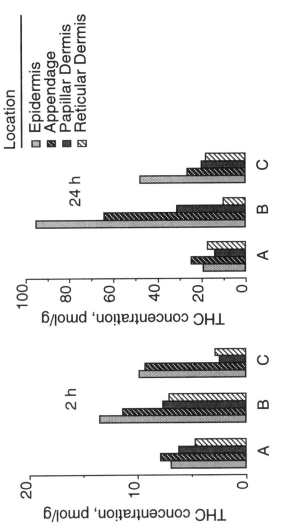

Figure 5 In vivo effect of vehicle and vehicle composition on the localization of tetrahydrocannabinol (THC) in the epidermis, appendages, and dermis of the hairless rat skin after 2 h and 24 h application. Vehicles: (A) polyethylene glycol 400, (B) diethylene glycol monoethyl ether, (Transcutol), (C) propylene glycol:ethanol (7:3). Redrawn from ref. 86.

These data document a first effort in quantitative autoradiography, and the information is well conformed with previously reported results of skin permeation behavior of tetrahydrocannabinol (87).

Autoradiographic approaches are also used to study the effect of topically applied compounds (e.g., retinoic acid, Azone, tetradecane) on epidermal growth using [^{3}H]thymidine (88–90). In a recent study (91), autoradiographic and immunohistochemical techniques were combined to analyze minoxidil localization in cultured vibrissa follicles, revealing incongruent findings. The data suggest that minoxidil is not covalently bound to a cellular receptor in the hair follicle. Therefore, minoxidil's site of activity for stimulation of the hair follicle still remains to be elucidated.

B. Fluorescence

Fluorescent or fluorescence-labeled substances permit the use of the fluorescence microscope for qualitative localization within the skin strata. In 1957 Borelli and Metzger (92) studied the in vitro percutaneous absorption of the fluorescent dye "acridinorange" incorporated into various topical preparations commonly used at that time, using domestic pig and human skin. Using fluorescence microscopy they observed qualitative different penetration characteristics of the dye into the skin depending on the preparations used. The addition of a nicotinic acid ester enhanced penetration of the dye into deeper skin structures. Follicles contributed significantly to the penetration of the dye. Similarly, Meyer and coworkers (93) observed percutaneous absorption of tetracycline derivatives incorporated in various vehicles. Using semisolid formulations, antibiotic penetration into deeper regions of the skin was minimal, while with fluid formulations containing enhancers (dodecan, tetradecan), the penetration into the skin was greater. Fricker et al. (94) investigated the absorption of an intact oligopeptide in rat and dog small intestine using a stable somatostatin analogue. The octapetide, also a potential candidate for iontophoretic topical deliver, was coupled to 4-nitrobenzo-2-oxa-1,3-diazol to have a fluorescent label for direct vizualization. The labeled peptide was successfully used to investigate the existence of preferential absorption sites in the small intestine. However, even though the biological qualities of the labeled and nonlabeled peptide may be comparable, the derivatization of a molecule does change its physiochemical qualities. Therefore, such molecules are only suitable for studying selected aspects of drug transport through the skin.

C. Sample Collection and Drug Analysis:
Critical Parameters

With a few exceptions, tissue collection procedures do involve a high degree of invasiveness and inconvenience and are therefore often impractical. Furthermore, ethical considerations (95) in conducting human studies render it necessary to carefully select the site for sampling, sample size, and frequency of sampling. Hence analytical assays and drug extraction procedure have to meet high demands.

Most of the already mentioned methods yield samples that reflect an instantaneous state of the drug in the tissue. However, changes of drug concentration in a specific compartment are necessary for a pharmacokinetic characterization of a drug. Analytical assays should be done according standardized guidelines, such as HPLC Assays for Bioavailability, Bioequivalence and Pharmacokinetic Studies, a consensus of a symposium held in Washington, DC, in December 1990 (96). There is an important and unsolved problem with the determination of the completeness of drug recovery by extraction from the skin. In order to determine the extraction recovery from the skin tissue, an in vitro experiment is usually performed where excised tissue is spiked with the drug. However, the in vivo drug distribution into skin is different from the in vitro process and the laboratory information is probably only of limited value. The homogenization of certain tissues has been recognized as a major problem in drug analysis, and several methods are available for this purpose (potter, dissolution with strong acids or bases or enzymatic degradation). Skin tissue is one of the most difficult tissue structures to homogenize. We have evaluated several procedures and found that a homoginizer that combines mechanical and shock cavitation forces successfully breaks up the skin tissue structures without adding to the sample treatment procedure other problems, such as isomerization or metabolizing of the compounds of interest (17,97,98).

IV. C^* CONCEPT: RELATIONSHIP OF SKIN TARGET SITE FREE DRUG CONCENTRATION (C^*) TO THE IN VIVO EFFICACY

A novel method based on the measurement of free drug concentration (C^*), or thermodynamic activity of drug, at the skin target site has been developed by Imanidis, Higuchi, Lee, and coworkers (99,100) to assess bioavailability and to predict efficacy for antivirals and other dermatological formulations. To measure C^* following the administration of a topical formulation initially entailed the establishment of a correlation between the steady-state dermal drug flux and an elicited efficacy. This was accomplished by a novel animal model in which hairless mice were infected at a small site at the lumbar skin area with cutaneous herpes simplex virus type 1. Three days postinoculation, this induced a narrow band of skin lesion development along the peripheral neural path toward the spinal cord. Taking advantage of this unique pattern of lesion development, an antiviral agent, such as acyclovir, was applied to an Azone-pretreated skin area, dorsal to the virus inoculation site and in the predicted path of lesion development. Five days after virus inoculation, the lesion development was scored for each mouse and two different antiviral efficacy parameters were separately assessed: a) "topical" (local) efficacy measured the antiviral activity of acyclovir delivery topically to the skin area directly under the drug application site, and b) "systemic" efficacy measured the antiviral activity of acyclovir delivery via systemic circulation to the target site, presumably the epidermal basal layer (101). To quantify drug flux, a transdermal delivery system was developed in this animal model, and the amount of acyclovir delivered to each infected animal could be controlled during the time period of drug therapy through a rate-controlling membrane. The actual (experimental) flux was determined at the end of each in vivo experiment by carrying out an extraction of the residual acyclovir in the transdermal delivery system. This extraction assay served to validate the expected (theoretical) flux or, alternatively, provided the bounds of uncertainty to the drug flux in the particular experiment. The results clearly showed a quantitative relationship between the antiviral efficacy and the experimental flux of acyclovir obtained from in vivo experiments. Topical efficacy increased with increasing acyclovir flux in the range of $10-100$ $\mu g/cm^2/day$. Based on the relatively high precision of topical efficacy results, it is believed that the quantitative nature of this animal model should be valuable in the screening of new antiviral agents for topical treatment of cutaneous herpes virus infections and the optimization of topical formulations. Two factors may limit the applicability of this elegant approach to other classes of dermatological formulations. Firstly, the drug was delivered from a transdermal delivery system at a constant rate over several days. With semisolid formulations, formulation application and formulation changes are difficult to control. Recently Mehta et al. (102) confirmed these results with a series of semisolid formulations using acyclovir as a model drug. However, the same group also showed that while predictions based on the C^* concept may correlate well with the in vivo efficacy for acyclovir, that may not be the case for other drugs such as bromovinyldeoxyuridine (103). Bolger et al. (104) recently reported that the concentration of acyclovir in plasma attained following topical administration was effective in reducing mortality associated with herpes infection of hairless mice. These findings have important ramifications for the use of not only the hairless mouse, but also other small-animal models for investigating the topical therapy of cutaneous herpes infections in which the area is large enough to produce significant antiviral drug concentrations in the blood. Therefore, effective systemic delivery of a topically applied drug might falsely indicate the analysis of its therapeutic potential in small animal models of herpes simplex infection. The possibility of both local and systemic modes of drug action should thus be taken into account in the evaluation of topical drugs when using small-animal models. Second, unfortunately the target sites for many dermatological agents (including acyclovir when treating cutaneous herpes simplex virus type 1) are still unknown.

V. CONCLUSION

In terms of the availability of topical drugs, the measurement of primary interest is the concentration of therapeutic agent within the skin or within a specific tissue layer of the skin. While some of the techniques described here show considerable promise and recent evidence suggests that drug concentration in the skin correlates with measures of drug effect in skin, the techniques still require substantial further development and validation (105,106). For this reason, current regulatory bioequivalence judgements have not been based on dermatopharmacokinetic studies. While it might be argued that acceptance of the method should require documentation of a good correlation to an observed clinical effect, this documentation has never been required for oral dosage forms to assess bioequivalence.

REFERENCES

1. Smolen VF, Williams EJ, Kuehn PB. Bioavailability and pharmacokinetic analysis of chlorpromazine in humans and animals using pharmacological data. Can J Pharm 1975; 10:95–106.
2. Queille-Roussel C, Poncet M, Schaefer H. Quantification of skin-colour changes induced by topical corticosteroid preparations using the Minolta Chroma Meter. Br J Dermatol 1991; 124(264–270).
3. Somberg JC. Bioequivalence or therapeutic equivalence. J Clin Pharmacol 1986; 26:1.
4. Dettelbach HR. A time to speak out on bioequivalence and therapeutic equivalence. J Clin Pharmacol 1986; 26:307–308.
5. Lamy PP. Generic equivalents: Issues and concerns. J Clin Pharmacol 1986; 26:309–316.
6. Kiistala U. Suction blister device for separation of viable epidermis from dermis. J Invest Dermatol 1968; 50(2):129–137.
7. Kiistala U. Dermal-epidermal separation. II. External factors in suction blister formation with special reference to the effect of temperature. Ann Clin Res 1972; 4:236–246.
8. Kiistala U. Dermal-epidermal separation. I. The influence of age, sex and body region on suction blister formation in human skin. Ann Clin Res 1972; 4:10–22.

9. Schreiner A, Hellum KB, Digranes A, Bergman I. Transfer of penicillin G and ampicillin into human skin blisters induced by suction. Scand J Infect Dis 1978; Suppl 14:233–237.

10. Miescher G. Beiträge zur Ekzemfrage. I. Zur Frage der Spezifität der ekzematösen Hautreaktion. Arch Derm Syph 1936; 173:119–154.

11. Kiistala U, Mustakallio KK, Rorsman H. Suction blister in the study of cellular dynamics of inflammation. Acta Derm Venereol (Stockh) 1967; 47: 150–153.

12. Treffel P, Makki S, Faivre B, Humbert P, Blanc D, Agache P. Citropten and bergapten suction blister fluid concentrations after solar product application in man. Skin Pharmacol 1991; 4:100–108.

13. Huuskonen H, Koulu L, Wilen G. Quantitative determination of methoxalen in human serum, suction blister fluid and epidermis by gas chromatography mass spectrometry. Photodermatology 1984; 1:137–140.

14. Averbeck D, Averbeck S, Blais J, Moysan A, Huppe G, Morliere B, Prognon P, Vigny P, Dubertret L. Suction blister fluid: Its use for pharmacodynamic and toxicological studies of drugs and metabolites in vivo in human skin after topic or systemic administration. In: Maibach HI, Lowe NJ, eds. Models in Dermatology. New York: S. Karger AG, 1989:5–11.

15. Agren MS. Percutaneous absorption of zinc from zinc oxide applied topically to intact skin in man. Dermatologica 1990; 180:36–39.

16. Surber C, Wilhelm KP, Berman D, Maibach HI. In vivo skin penetration of acitretin in volunteers using three different sampling techniques. Pharm Res 1993; 10(9):1291–1294.

17. Laugier J-P, Surber C, Bun H, Geiger J-M, Wilhelm K-P, Durand A, Maibach HI. Determination of acitretin in the skin, in the suction blister, and in plasma of human volunteers after multiple oral dosing. J Pharm Sci 1994; 83(5):623–628.

18. Dubertret L, Averbeck D, Prognon P, Blais J, Vigny P. Photobiological activity of the suction blister fluid from patients treated with 8-methoxypsoralen. Br J Dermatol 1983; 109:421–427.

19. Allison JH, Bettley FR. Investigations into cantharidin blisters raised on apparently normal skin in normal and abnormal subjects. Br J Dermatol 1958; 58:330–339.

20. Findlay CD, Wise R, Allcock JE, Durham SR. The tissue penetration as measured by a blister technique and pharmacokinetics of cefsulodin compared with carbenicillin and ticarcillin. J Antimicrob Chemother 1981; 7:637–642.

21. Simon C, Malerczyk V, Brahmstaedt E, Toeller W. Cefazolin, ein neues Breitbandspektrum-Antibiotikum. Dtsch Med Wochenschr 1973; 98:2448–2450.

22. Frongillo RF, Galuppo L, Moretti S. Suction skin blister, skin window, and skin chamber techniques to determine extravascular passage of cefotaxime in humans. Antimicrob Agents Chemother 1981; 19(1):22–28.

23. Shyu WC, Quintiliani R, Nightingale CH. An improved method to determine interstitial fluid pharmacokinetics. J Infect Dis 1985; 152(6):1328–1331.

24. Raeburn JA. A review of experimental models for studying the tissue penetration of antibiotics in man. Scand J Infect Dis 1978; Suppl 14:225–227.

25. Ryan DM, Hodges B, Spencer GR, Harding SM. Simultaneous comparison of three methods for assessing ceftazidime penetration into extravascular fluid. Antimicrob Agents Chemother 1982; 22(6):995–998.

26. Hoffstedt B, Walder M. Penetration of ampicillin, doxycycline and gentamycin into interstitial fluid in rabbits and of penicillin V and pivampicillin in humans measured with subcutaneously implanted cotton threads. Infection 1981; 9:7–11.

27. Holm, SE. Experimental models for studies on transportation of antibiotics to extravasal compartments. Scand J Infect Dis 1978; 13(Suppl):47–51.

28. Wise R. Methods of evaluating the penetration of beta lactam antibiotics into tissues. Rev Infect Dis 1986; 8(Suppl 3):325–332.

29. Marty JP, Bucks DAW. Maibach HI. Noninvasive radioisotope counting on skin: Surface or external counting? In: Bronaugh RL, Maibach HI, eds. Percutaneous absorption: Mechanisms—Methodology—Drug delivery. New York: Marcel Dekker; 1989:435–442.

30. Lee DJ, Burt CT, Koch RL. Percutaneous absorption of flurbiprofen in the hairless rat measured in vivo using ^{19}F magnetic resonance spectroscopy. J Invest Dermatol 1992; 99:431–434.

31. Phillips M. Sweat-patch test for alcohol consumption: Rapid assay with an electrochemical detector. Alcohol Clin Exp Res 1982; 6(4):532–534.

32. Phillips M, McAloon MH. A sweat-patch test for alcohol consumption: Evaluation in continuous and episodic drinkers. Alcohol Clin Exp Res 1980; 4(4): 391–395.

33. Phillips ELR, Little RE, Hillman RS, Labbe RF, Campbell C. A field test of the sweat patch. Alcohol Clin Exp Res 1984; 8(2):233–237.

34. Phillips M. An improved adhesive patch for long-term collection of sweat. Biomater Med Dev Art Org 1980; 8(1):13–21.

35. Peck CC, Lee K, Becker CE. Continuous transepidermal drug collection: Basis for use in assessing drug intake and pharmacokinetics. J Pharmacokinet Biopharmacol 1981; 9(1):41–58.

36. Peck CC, Conner DP, Bolden BJ, Almirez RG, Kingsley TE, Mell LD, Murphy GM, Hill VE, Rowland LM, Ezra D, Kwiatkowski TE, Bradley CR, Abdel-Rahim M. Outward transcutaneous chemical migration: Implications for diagnostics and dosimetry. Skin Pharmacol 1988: 1:14–23.

37. Conner DP, Jacques SL, Almirez RG, McAuliffe DJ, Bolden BJ, Zamani K, Peck CC. Laser-enhanced transcutaneous collection of theophylline. Clin Pharmacol Ther 43(2):137.

38. Downing DT, Stewart ME, Wertz PW, Colton SW, Abraham W, Strauss JS. Skin lipids: An update. J Invest Dermatol 1987; 88:2s–6s.

39. Rieger MD. Skin lipids and their importance to cosmetic science. Cosmet Toiletries. 1987; 102:36–49.

40. Epstein EH, Epstein WL. New cell formation in human sebaceous glands. J Invest Dermatol 1966; 46:453–458.

41. Downing DT, Strauss JS, Ramasastry P, Abel M, Lees CW, Pochi PE. Measurements of the time between synthesis and the surface excretion of sebaceous lipids in sheep and man. J Invest Dermatol 1975; 64:215–219.

42. Faergemann J, Zehender H, Jones T, Maibach HI. Terbinafine levels in serum, stratum corneum, dermis-epidermis (without stratum corneum), hair, sebum and eccrine sweat. Acta Derm Venereol (Stockh) 1991; 71(4):322–326.

43. Nordstrom KM, Schmus HG, McGinley KJ, Leyden JJ. Measurement of sebum output using a lipid absorbent tape. J Invest Dermatol 1986; 87:260–263.

44. Saint-Leger D, Bague A. A simple and accurate routine procedure for qualitative analysis of skin surface lipids in man. Arch Dermatol Res 1981; 271: 215–222.

45. Rashleigh PL, Rife E, Goltz RW. Tetracycline levels in skin surface film after oral administration of tetracycline to normal adults and to patients with acne vulgaris. J Invest Dermatol 1967; 49(6):611–615.

46. Gould JC, Richtie HD. The terramycin content of human skin during therapy: A comparison with serum and urine levels. Br J Plast Surg 1952; 5:208–212.

47. Aubin F, Blanc D, Guinchard C, Agache P. Absence de passage de la minocycline dans le sebum? Ann Dermatol Venereol 1988; 115:977.

48. Luderschmidt C, Nissen, H-P, Neubert U, Knuechel M. Newer methods for measuring therapeutic response in acne. In: Cullen SI, ed. Focus on Acne Vulgaris; March; Athens. London: Royal Society of Medicine Services Ltd. 1983; Congress and Symposim Series no. 95:131–140.

49. Kellum RE. Isolation of human sebaceous glands. Arch Dermatol 1966; 93: 610–612.

50. Suzuki O, Hattori H, Asano M. Nails as useful materials for detection of methamphetamine or amphetamine abuse. Forens Sci Int 1984; 24:9–16.

51. Ochsendorf FR, Runne U, Schöfer H, Schmidt K, Raudonat HW. Sequential chloroquine concentrations in human hair represent ingested dose and duration of therapy—Human hair as a pharmacologic and toxicologic tachograph. Arch Dermatol 1989; 281:121.

52. Balabanova S, Wolf HU. Determination of methadone in human hair by radioimmunoassay. Z Rechtsmed 1989; 102:1–4.

53. Ishiyama I, Nagai T, Toshida S. Detection of basic drugs (methamphetamine, antidepressants and nicotine) from human hair. J Forens Sci 1983; 28(2):380–385.

54. Smith FP, Liu RH. Detection of cocaine metabolite in perspiration stain, menstrual bloodstrain, and hair. J Forens Sci 1986; 31(4):1269–1273.

55. Balabanova S, Brunner H, Nowak R. Radioimmunological determination of cocaine in human hair. Z Rechtsmed 1987; 98:229–234.

56. Van Cutsem J, Van der Flaes M, Thienpont D, Dony J. Hörig C. Quantitative Bestimmung von Ketoconazol in den Haaren oral behandelter Ratten und Meerschweinchen. Mykosen. 1980; 23(8):418–425.

57. Walters KA, Flynn GL, Marvel JR. Physicochemical characterization of the human nail: Permeation pattern for water and the homologous alcohols and differences with respect to the stratum corneum. J Pharm Pharmacol 1983; 35:28–33.

58. Walters KA, Flynn GL, Marvel JR. Physiochemical characterization of the human nail: Solvent effects on the permeation of homologous alcohols. J Pharm Pharmacol 1985; 37:771–775.

59. Walters KA, Flynn GL, Marvel JR. Penetration of the human nail plate: The effects of the vehicle pH on the permeation of miconazole. J Pharm Pharmacol 1985; 37:498–499.

60. Walters KA, Flynn GL. Permeability characteristics of the human nail plate. Int J Cosmetic Sci 1983; 5:231–246.

61. Walters KA, Flynn GL, Marvel JR. Physiochemical characterization of the human nail: I. Pressure sealed apparatus for measuring nail plate permeability. J Invest Dermatol 1981; 76(2):76–79.

62. Stüttgen G, Bauer E. Bioavailability, skin and nail penetration of topically applied antimycotics. Mykosen 1982; 25(2):74–80.

63. Pittrof F, Gerhards J, Erni W, Klecak G. Loceryl® nail lacquer—realization of a new galenical approach to onychomycosis therapy. Clin Exp Dermatol 1992; 17(Suppl 1):26–28.

64. Roncari G, Ponelle C, Zumbrunnen R, Guenzi A, Dingemanse J, Jonkman JHG. Percutaneous absorption of amorolfine following a single topical application of an amorolfine cream formulation. Clin Exp Dermatol 1992; 17(Suppl 1):33–36.

65. Dykes PJ, Thomas R, Finlay AY. Determination of terbinafine in nail samples during systemic treatment for onychomycoses. Br J Dermatol 1990; 123:481–486.

66. Surber C, Wilhelm K-P, Hori M, Maibach, HI, Guy RH. Optimization of topical therapy: Partitioning of drugs into stratum corneum. Pharm Res 1990; 7(12):1320–1324.

67. Fry L, McMinn RMH. Topical methotrexate in psoriasis. Arch Dermatol 1967; 96:483–488.

68. Surber C, Itin P, Büchner S. Clinical controversy on the effect of topical cyclosporin A: What is the target organ. Dermatology 1992; 185(4):242–245.

69. Parry, GE, Dunn P, Shah VP, Pershing LK. Acyclovir bioavailability in human skin. J Invest Dermatol 1992; 98:856–863.

70. Schaefer H, Lamaud E. Standardization of experimental models. In: Shroot B, Schaefer H, eds. Skin Phamacokinetics. New York: Karger; 1987:77–80. (Shroot B, Schaefer H., series editors. Pharmacology and the Skin; vol. 1.)

71. Schaefer H, Stüttgen G, Zesch A, Schalla W, Gazith J. Quantitative determination of percutaneous absorption of radiolabled drugs in vitro and in vivo by human skin. Curr Probl Dermatol 1978: 7:80–94.

72. Zesch A, Schäfer H. Penetrationskinetik von radiomarkierten Hydrocortisone aus verschiedenartigen Salbengrundlagen in die menschliche Haut. Arch Derm Forsch 1975; 252:245–256.

73. Guzek, DB, Kennedy AH, McNeill SC, Wakshull E, Potts, RO. Transdermal drug transport and metabolism. I. Comparison of in vitro and in vivo results. Pharm Res 1989; 6(1)33–39.

74. Wojciechowski Z, Pershing LK, Huether S, Leonard L, Burton SA, Higuchi WI, Krueger GG. An experimental skin sandwich flap on an independent vascular supply for the study of percutaneous absorption. J Invest Dermatol 1987; 88:439–446.

75. Pershing LK, Krueger GG. Human skin sandwich flap model for percutaneous absorption. In: Bronaugh RL, Maibach HI, eds. Percutaneous Absorption. New York: Marcel Dekker, 1989:397–414.

76. Scott DO, Bell MA, Lunte CE. Microdialysis-perfusion sampling for the investigation of phenol metabolism. J Pharm Biomed Analysis 1989; 7(11): 1249–1259.

77. Lunte CE, Scott DO, Kissinger PT. Sampling living systems using microdialysis. Anal Chem 1991; 63:773A–780A.

78. Ault, JM, Lunte CE, Meltzer NM, Riley CM. Microdialysis sampling for investigation of dermal drug transport. Pharm Res 1992; 9(10):1256–1261.

79. Ault, JM, Lunte CE, Meltzer NM, Riley CM. Microdialysis as a dermal sampling technique for percutaneous absorption. In: Brain KR, James VJ, Walters KA, eds. Prediction of Percutaneous Penetration. Cardiff: STS Publishing, 1993:44–48.

80. Matsuyama K, Nakashima M, Nakaboh Y, Ichikawa M, Yano T, Satoh S. Application of in vivo microdialysis to transdermal absorption of methotrexate in rats. Pharm Res 1994; 11(5):684–686.

81. Fukuyama K. Autoradiography. In: Skerrow D, Skerrow CJ, eds. Methods in Skin Research. London: John Wiley & Sons, 1985:71–89.

82. Schalla W, Jamoulle J-C, Schaefer H. Localization of compounds in different skin layers and its use as an indicator of percutaneous absorption. In: Bronaugh RL, Maibach HI, eds. Percutaneous Absorption: Mechanism—Methodology—Drug delivery. 2nd ed. New York: Marcel Dekker, 1989:283–312.

83. Ritschel, WA, Panchagnula R, Stemmer K, Ashraf M. Development of an intracutaneous depot for drugs. Skin Pharmacol 1991; 4:235–245.

84. Suzuki M, Asaba K, Komatsu H, Mochizuka M. Autoradiographic study on percutaneous absorption of several oils useful for cosmetics. J Soc Cosmet Chem 1978; 29:265–282.

85. Bidmon H-J, Pitts JD, Solomon HF, Bondi JV, Stumpf WE. Estradiol distribution and penetration in rat skin after topical application, studies by high resolution autoradiography. Histochemistry 1990; 95:43–54.

86. Fabin B, Touitou E. Localization of lipophilic molecules penetrating rat skin in vivo by quantitative autoradiography. Int J Pharm 1991; 74:59–65.

87. Touitou E, Fabin B, Dany S, Almog S. Transdermal delivery of tetrahydrocannabinol. Int J Pharm 1988; 43:9–15.

88. Plewig G, Fulton JE. Autoradiographische Untersuchungen an Epidermis und Adnexen nach Vitamin A-Säure-Behandlung. Hautarzt 1974; 23(3):128–136.

89. Klein JA, McCullough JL, Weistein GD. Topical tritiated thymidine for epidermal growth fraction determination. J Invest Dermatol 1986; 86(4):406–409.

90. Motoyoshi K, Ito M, Sakamoto F, Sato Y. A time course study of the proliferation of sebaceous glands induced by topically applied tetradecane in rabbit pinna skin. Autoradiography and electron microscopy. J Dermatol 1987; 14: 9–14.

91. Zelei BV, Walker CJ, Sawada GA, Kawabe TT, Knight KA, Buhl AE, Johnson GA, Diani AR. Immunohistochemical and autoradiographic findings suggest that minoxidil is not localized in specific cells of vibrissa, pelage, or scalp follicles. Cell Tissue Res 1990; 262:407–413.

92. Borelli S, Metzger M. Fluorescenzmikroskopische Untersuchungen über die perkutane Penetration fluorescierender Stoffe. Hautarzt 1957; 8(6):261–266.

93. Meyer F. Örtliche Andwendung von Tetracyclinen. Arch Pharmakol Exp Pathol 1966; 225:47–48.

94. Fricker G, Bruns C, Munzer J, Briner U, Albert R, Kissel T, Vonderscher J. Intestinal absorption of octapeptide SMS 201-995 visualized by fluorescence derivatization. Gastroenterology. 1991; 100:1044–1552.

95. Svensson CK. Ethical considerations in the conduct of clinical pharmacokinetic studies. Clin Pharmacokinet 1989; 17(4):217–222.

96. Shah VP, Midha KK, Dighe S, McGilveray IJ, Skelly JP, Yacobi A, Layloff T, Viswanathan CT, Cook CE, McDowall RD, Pittman KA. Spector S. Analytical methods validation: Bioavailability, bioequivalence and pharmacokinetic studies. Pharm Res 1992; 9(4):588–592.

97. Wagener HH, Voegtle-Junkert U. Zur Auswertung von Wirkstoffkonznetrationen in Geweben nach perkutaner Andwendung von nicht-steroidalen Antirheumatika. Arzneim.-Forsch./Drug Res. 1996; 46(1)(3):299–301.

98. Panus, PC, Tober-Meyer B, Ferslew KE. Tissue extraction and high-performance liquid chromatographic determination of ketoprofen. J Chromatogr 1998; 705:295–302.

99. Imanidis G, Song W, Lee PH, Su MH, Kern ER, Higuchi WI. Estimation of skin target site acyclovir concentrations following controlled (trans)dermal drug delivery in topical and systemic treatment of cutaneous HSV-1 infections in hairless mice. Pharm Res 1994; 11(7):1035–1041.

100. Lee PH, Su MH, Kern ER, Higuchi WI. Novel animal model for evaluating topical efficacy of antiviral agents: Flux verus efficacy correlations in acyclovir treatment of cutaneous herpes simplex virus type 1 (HSV-1) infections in hairless mice. Pharm Res 1992; 9(8):979–989.

101. Price RW. Neurobiology of human herpes virus infections. Crit Rev Clin Neurobiol 1986: 2:61–123.

102. Mehta SC, Afouna MI, Ghanem A-H, Higuchi WI, Kern ER. Relationship of skin target site free drug concentration (C^*) to the in vivo efficacy: An extensive evaluation of the predictive value of the C^* concept using acyclovir as a model drug. J Pharm Sci 1997; 86(7):797–801.

103. Afouna MI, Mehta SC, Ghanem A-H, Higuchi WI, Kern ER. Absence of correlation between C^* predictions and in vivo topical antiviral efficacy in hairless mice for bromovinyldeoxyuridine (BVDU). Pharm Res 1997; 14(11): S-35 (abstr. 1105).

104. Bolger GT, Allen T, Garneau M, Lapeyre N, Lird F, Jaramillo J. Cutaneouly applied acyclovir acts systemically in the treatment of herpetic infection in the hairless mouse. Antivir Res 1997; 35:157–165.

105. Caron D, Queille-Roussel C, Shah VP, Schaefer H. Correlation between the drug penetration and the blanching effect of topically applied hydrocortisone creams in human beings. J Am Acad Dermatol 1990; 23:458–462.

106. Pershing LK, Silver BS, Krueger GG, Shah VP, Skelly JP. Feasibility of measuring the bioavailability of topical betamethasone dipropionate in commercial formulations using drug in skin and a skin blanching bioassay. Pharm Res 1992; 9(1):45–51.

22
Stripping Method for Measuring Percutaneous Absorption In Vivo

André Rougier
Laboratoire Pharmaceutique, La Roche-Posay, Courbevoie, France

Didier Dupuis and Claire Lotte
Laboratoires de Recherche Fondamentale, L'Oréal, Aulnay sous Bois, France

Howard I. Maibach
University of California School of Medicine, San Francisco, California

From a practical viewpoint, it remains difficult to draw valid conclusions from the literature concerning the absorption level of a given compound. This is essentially due to the diversity of techniques used, animal species (Tregear, 1964; Winkelman, 1969), anatomical location (Marzulli, 1962; Lindsey, 1963), duration of application (Tregear, 1964), dose applied (Maibach and Feldman, 1969; Wester and Maibach, 1976), and vehicle used (Poulsen, 1973; Tregear, 1964). Furthermore, because this kind of research has interested scientists from widely differing disciplines, each worker has chosen or adapted the methodology to elucidate a particular problem.

From a theoretical viewpoint, over the two past decades, considerable attention has been paid to the understanding of the mechanisms and routes by which chemical compounds may penetrate the skin. Without considering the different interpretation about mechanisms acting on percutaneous absorption, it is well established that the main barrier is the stratum corneum (Malkinson and Rothman, 1963; Marzulli, 1962; Stoughton, 1965), which also acts as a reservoir for topically applied molecules (Stoughton and Fritsch, 1964; Vickers, 1963). Moreover, it is likely that at the early step of the absorption process the interaction between the physicochemical properties of the drug, the vehicle, and the horny layer plays an important role in total absorption.

In the first part of this chapter, we hypothesize that the amount of chemical present in the stratum corneum at the end of application may represent the stratum corneum vehicle partitioning and could also reflect the rate of penetration of the chemical.

In the second part, we ascertain that this hypothesis is independent of the main factors likely to modify the absorption level of a compound, that is, contact time, dose applied, vehicle used, anatomical site involved, and animal species chosen.

I. IN VIVO RELATIONSHIP BETWEEN STRATUM CORNEUM CONCENTRATION AND PERCUTANEOUS ABSORPTION

We chose to test, on the hairless rat, 10 radiolabeled molecules having very different physicochemical properties and belonging to different chemical classes.

For each compound, a group of 12 female hairless Sprague-Dawley rats, aged 12 weeks, weighing 200 ± 20 g was used. The molecules, dissolved in ethanol/water mixtures (chosen according to the solubility of the chemical), were applied on 1 cm^2 of dorsal skin during 30 min. The standard dose applied was 200 nmol $\cdot$ cm^{-2}. At the end of application, the excess product on the treated area was rapidly removed by two washings with ethanol/water (95:5), followed by two rinsings with distilled water, and light drying with cotton wool.

The twelve animals were then divided into two groups (Fig. 1). The animals of group 1, wearing collars to prevent licking, were placed individually in metabolism cages for 4 days. Urinary excretion was established by daily sampling of the urine and liquid scintillation counting (Packard Instruments 460 C). The feces were collected daily, pooled and counted by liquid scintillation after lyophilization, homogenization, and combustion of the samples with an Oxidizer 306 (Packard Instruments).

After 4 days, the animals were sacrificed and series of six strippings were carried out on the treated area to determine the amount of product not penetrated within 96 h. The remaining skin of the treated area (epidermis and dermis) was sampled and counted by liquid scintillation after digestion in Soluene 350 (United Technology Packard). The carcasses were lyophilized, homogenized, and samples were counted by liquid scintillation after combustion. The total amount of chemical penetrating within 4 days was then determined by adding the amounts found in the excreta (urine plus feces), in the epidermis and dermis of the application area, and in the whole animal body.

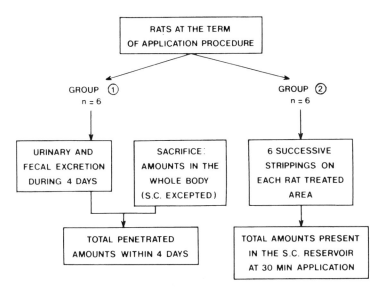

Determination of the four days
penetration and the stratum corneum
reservoir on thirty minutes application,
for each molecule.

RATS AT THE TERM
OF APPLICATION PROCEDURE

GROUP ① n = 6

GROUP ② n = 6

URINARY AND FECAL EXCRETION DURING 4 DAYS

SACRIFICE: AMOUNTS IN THE WHOLE BODY (S.C. EXCEPTED)

6 SUCCESSIVE STRIPPINGS ON EACH RAT TREATED AREA

TOTAL PENETRATED AMOUNTS WITHIN 4 DAYS

TOTAL AMOUNTS PRESENT IN THE S.C. RESERVOIR AT 30 MIN APPLICATION

Figure 1 Procedures for determining total percutaneous absorption, within 4 days, and the stratum corneum reservoir at the end of application. From Rougier et al. (1983).

At the end of application and washing, the stratum corneum of the treated area of the animals from the second group was removed by six strippings, using 3M adhesive tape. The radioactivity on each strip was measured after complete digestion of the keratinic material in Soluene 350 (United Technology Packard), addition of Dimilume 30 (United Technology Packard), and liquid scintillation counting.

In our experimental conditions, the capacity of the stratum corneum reservoir for each compound has been defined as the sum of the amounts found in the first six strippings.

The percutaneous absorption results show (Fig. 2) that after 96 h there are large differences in the amounts of substances that have penetrated through the skin. Thus, one can observe that the most penetrating molecule, benzoic acid, penetrates 50 times more than dexamethasone.

When the compounds benzoic acid, acetylsalicylic acid, dehydroepiandrosterone, sodium salicylate, testosterone, hydrocortisone, and dexametha-

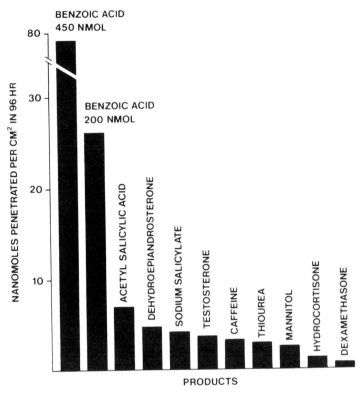

Figure 2 Percutaneous absorption levels of the tested compounds, 4 days after their topical administration in the hairless rat. From Rougier et al. (1983).

sone are classified according to a decreasing order of penetration rate, we observe that this order is similar to that found in the literature concerning studies in humans (Feldmann and Maibach, 1969, 1970). Likewise, it is established that acetylsalicylic acid and salicylic acid have similar penetrating properties (Feldmann and Maibach, 1970), whereas their sodium salts exhibit diminished penetration (Malkinson and Rothman, 1963) and, indeed, we observed that sodium salicylate penetrates less than acetylsalicylic acid.

Figure 3 shows the concentrations of compounds present on each stripping of the dosed area of the animals of group 2, at the end of application procedures. It is noteworthy that, as in the in vitro results (Schaefer et al., 1978), in vivo, the substance concentration decreases inside the stratum corneum, following an exponential relation. Considering the diversity of the compounds tested, it seems that this observation can be outlined as one of the factors governing percutaneous absorption.

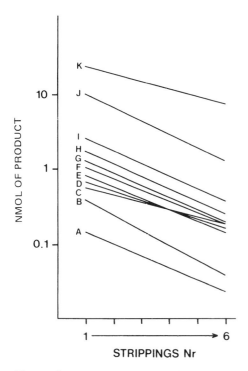

Figure 3 Distribution of the tested molecules within the horny layer of the dosed area, 30 min after their administration in the hairless rat: (A) dexamethasone, (B) hydrocortisone, (C) dehydroepiandrosterone, (D) testosterone, (E) mannitol, (F) thiourea, (G) caffeine, (H) sodium salicylate, (I) acetylsalicylic acid, (J) benzoic acid (200 nmol), (K) benzoic acid (450 nmol). From Rougier, Dupuis, Lotte, and Roguet, unpublished data (1983).

As shown in Figure 4, independent of the physicochemical nature of the tested agent, there exists a highly significant linear correlation between the total amount of chemical penetration within 4 days (y) and the amount present in the stratum corneum at the end of application time (x, 30 min; $r = .98$, $p < .001$).

From a theoretical viewpoint, this correlation sheds some light on a possible explanation of the stratum corneum barrier effect. A weak reservoir capacity would correspond to a weak penetration and therefore a strong barrier. Inversely, a high reservoir capacity corresponds to a high penetration and therefore a weak barrier effect. As a consequence, it is possible that barrier and reservoir functions of the horny layer may reflect the same physiological reality. From a practical viewpoint, the simple measurement of the

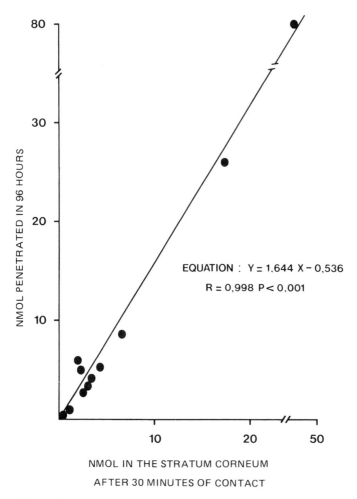

Figure 4   Relationship between the penetration level of the tested compounds after 4 days and their concentrations in the stratum corneum at the end of application (30 min). From Rougier et al. (1983).

amount of a chemical within the stratum corneum at the end of a 30-min application gives a good predictive assessment of the total amount penetrating within 4 days.

As previously mentioned, the absorption level of molecules has been proved to be dependent on their application conditions. It was therefore important to ascertain that the "stripping method" was independent of the main factors able to modify the penetration level of a chemical, that is, application time, dose applied, vehicle used, and anatomical site involved.

II. INFLUENCE OF APPLICATION CONDITIONS ON THE RELATIONSHIP BETWEEN STRATUM CORNEUM CONCENTRATION AND PERCUTANEOUS ABSORPTION

A. Influence of Application Time

The duration of application of a compound may considerably influence the total amount absorbed. Moreover, the time of application of a substance may be closely related to its field of use.

Percutaneous absorption of four radiolabeled compounds—theophylline, nicotinic acid, acetylsalicylic acid, and benzoic acid—was studied in the hairless rat. One thousand nanomoles of each compound was applied onto 1 cm^2 of dorsal skin during 0.5, 2, 4, and 6 h, thus covering most of the usage conditions of the compounds topically applied. Total percutaneous absorption within 4 days for each compound and each application time was carried out as described in the foregoing section. The stratum corneum reservoir was assessed for each compound, after an application time fixed at 30 min by stripping the treated area.

As shown in Fig. 5, the penetration rate of the tested compounds is strictly proportional to the duration of application ($r = .98$, $p < .001$). From a theoretical viewpoint, this relationship provides evidence that, as in the in vitro studies (Scheuplein, 1967; Treherne, 1956), a constant flux of penetration really does exist in vivo. This type of correlation having been found for four compounds with widely different physicochemical properties, it is reasonable to assume that it may constitute one of the laws of the in vivo percutaneous absorption phenomena. From a practical viewpoint, this linear relationship implies that the knowledge of the 4-day penetration of a compound applied for only 30 min has a predictive value for penetration resulting from longer times of application.

As discussed in the preceding section, with a 30-min application, the total amount of compound recovered within the horny layer is strictly correlated with the amount that penetrated in a 4-day period. Figure 6 shows that this correlation is confirmed ($r = .99$, $p < .001$) for the four agents tested in this experiment. The total percutaneous absorption of a compound being directly linked to the duration of application (see Fig. 5), the simple knowledge of the reservoir effect of the stratum corneum for a chemical applied for 30 min allows the predictive assessment of its penetration resulting from longer times of application. Thus, this only mildly invasive method offers

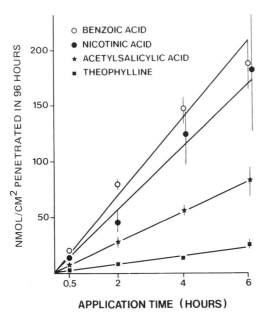

Figure 5 Relationship between the penetration levels of the tested compounds and the application time in the hairless rat. From Rougier et al. (1985).

the advantages, when applied to humans, of reducing skin exposure and of immobilizing subjects only for a short period.

B. Influence of Dose Applied

In most medical and toxicological specialties, the administered dose is defined precisely. This has not always been the case in dermatotoxicology and dermatopharmacology. It is, however, well known that an increased concentration of an applied chemical on the skin increases percutaneous penetration (Maibach and Feldmann, 1969; Scheuplein and Ross, 1974; Wester and Maibach, 1976), as does increasing the surface area treated or the application time. This question of concentration may have special significance in infants because the surface/body weight ratio is greater than in adults.

Percutaneous absorption of four radiolabeled compounds—theophylline, nicotinic acid, acetylsalicylic acid, and benzoic acid, dissolved in ethylene glycol/Triton X-100 (90:10)—was studied in the hairless rat. For each compound, increasing doses from 125 to 1000 nmol were applied onto 1 cm^2 of dorsal skin for 30 min. For each compound and each dose, total percutaneous absorption within 4 days and the stratum corneum reservoir at

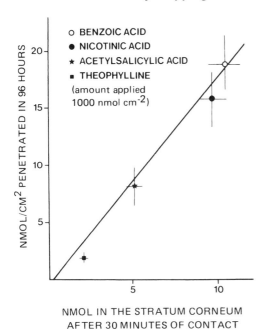

Figure 6 Correlation between the amount of chemical in the stratum corneum at the end of application (30 min) and its overall penetration within 4 days in the hairless rat. (From Rougier et al. (1985).

the end of application time were assessed as described in the preceding section.

As shown in Fig. 7, within the limits of the concentrations used, there exists a linear dose-penetration relationship ($r = .98$, $p < .001$). However, it has been shown by Skog and coworkers (Skog and Wahlberg, 1964) that when the applied concentration was increased, penetration was increased up to a certain point, at which a plateau was reached. Within the range of concentration used in the present study, this phenomenon does not appear. This tends to indicate that the horny layer is unaffected by the concentrated solutions, with the permeability constant being unaltered over the entire concentration range. When one considers the differences in physicochemical properties of the tested compounds, it seems that, at least for a range of concentration, the linear relationship existing between dose applied and percutaneous absorption level might be taken as a general law.

Independently of the physicochemical nature of the chemical and whatever dose was administered, there is (Fig. 8) a highly significant correlation between the total amounts that penetrated over a 4-day period and the

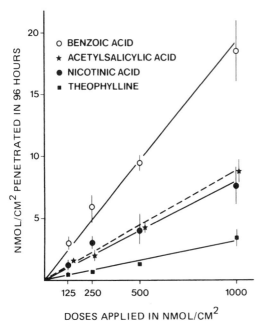

Figure 7   Relationship between dose applied and penetration rate of the tested compounds in the hairless rat. (Rougier et al., unpublished data).

amounts recovered in the stratum corneum at the end of application time ($r = .98$, $p < .001$). From a toxicological viewpoint, the influence of applied concentration on the overall penetration of a drug can therefore be easily predicted using the stripping method. From a pharmacological viewpoint, Sheth and coworkers (1986) have shown that the therapeutic efficacy of increasing doses of an antiviral (iododeoxyuridine) on herpes simplex infection can be predicted by the use of the stripping method.

C.   Influence of Vehicle

In recent years, increasing attention has been devoted to the influence that the components of a vehicle may have on enhancing or hindering skin absorption of drugs. The effects of vehicles have been reviewed in detail by several authors (Barr, 1962; Idson, 1971, 1975; Rothmann, 1954). It is now well established that substances added to formulations as excipients and other factors, such as the physical form of the drug, affect not only its release and absorption, but also its action. Unfortunately, few techniques can be used routinely to elucidate rapidly the role that a vehicle or a component in a vehicle may have on the overall absorption of a drug in vivo.

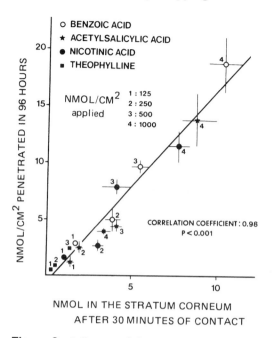

Figure 8 Influence of dose applied on the relationship between the level of penetration of the tested compounds after 4 days and their concentration in the stratum corneum at the end of application (30 min). From Rougier, Lotte, and Dupuis (unpublished data).

The influence of nine vehicles on the in vivo percutaneous absorption of [^{14}C]benzoic acid was studied in the hairless rat using the stripping method. Twenty microliters of each vehicle, containing 200 nmol of benzoic acid, was applied to 1 cm^2 of dorsal skin during 30 min. After this time, total percutaneous absorption and stratum corneum reservoir were assessed as previously described.

As shown in Fig. 9, although the vehicles used were simple in composition, the total amount of benzoic acid that penetrated over 4 days varied by a factor of 50, once more demonstrating the importance of vehicle in skin absorption. Figure 9 also shows the maximum solubility values (mg/ml) of benzoic acid in each vehicle. It is generally admitted that the release of a compound can be favored by the selection of vehicles having low affinity for that compound or by one in which it is least soluble (Blank and Scheuplein, 1964; Schutz, 1957). As may be seen, the solubility of benzoic acid differed by a factor 30 between the least and the most efficient solvent medium (vehicles 4 and 6). However, we can see that there is a weak re-

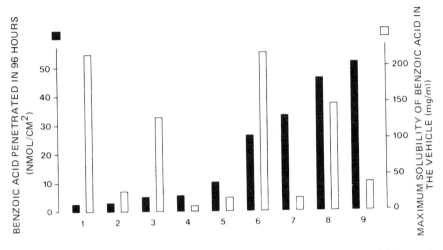

Figure 9 Comparative values of solubility of benzoic acid in the vehicles and corresponding percutaneous absorption levels: (1) propylene glycol/Triton X-100 (90:10), (2) glycerol/Triton X-100 (90:10), (3) ethylene glycol/Triton X-100 (90:10), (4) ethylene glycol/Triton X-100; 90:10)/water (40:60), (5) (propylene glycol/Triton X-100; 90:10)/water (40:60), (6) ethanol/water (95:5), (7) methanol/water (40:60), (8) ethanol/water (60:40), (9) ethanol/water (40:60). From Dupuis et al. (1986).

lationship between penetration level and maximum solubility of benzoic acid. For instance, the greatest penetration is not obtained with the vehicle in which benzoic acid is least soluble, and vice versa.

Applied vehicles have the potential to either increase or decrease the quantity of water in the horny layer and, thereby, to increase or decrease penetration (Shelley and Melton, 1949). It is interesting that the penetration of benzoic acid is enhanced by increasing the water content of the vehicles, whatever the organic phase (vehicles 1 and 5, and 6, 8, and 9).

As shown in Fig. 10, independently of the vehicles composition, the amount of benzoic acid found within the horny layer at the end of application and the amount penetrating in 4 days are linearly correlated ($r = .99$, $p < .001$). The influence of the vehicle composition on the in vivo penetration level of a chemical can therefore be easily predicted by simply stripping the treated area and measuring the amount engaged in the stratum corneum at the end of application.

D. Influence of Anatomical Site

Although all authors agree on the importance of anatomical location in percutaneous absorption, the literature contains relatively little information on

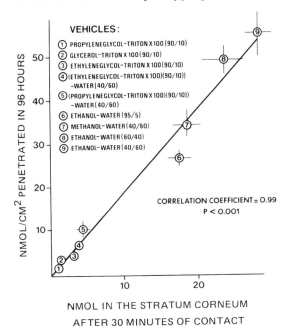

VEHICLES:
① PROPYLENEGLYCOL-TRITON X 100 (90/10)
② GLYCEROL-TRITON X 100 (90/10)
③ ETHYLENEGLYCOL-TRITON X 100 (90/10)
④ (ETHYLENEGLYCOL-TRITON X 100)(90/10))
  -WATER (40/60)
⑤ (PROPYLENEGLYCOL-TRITON X 100)(90/10))
  -WATER (40/60)
⑥ ETHANOL-WATER (95/5)
⑦ METHANOL-WATER (40/60)
⑧ ETHANOL-WATER (60/40)
⑨ ETHANOL-WATER (40/60)

CORRELATION COEFFICIENT = 0.99
P < 0.001

NMOL/CM^2 PENETRATED IN 96 HOURS

NMOL IN THE STRATUM CORNEUM
AFTER 30 MINUTES OF CONTACT

Figure 10 Influence of the tested vehicles on the relationship between the penetration level of benzoic acid within 4 days and its concentration in the stratum corneum at the end of application (30 min). From Dupuis et al. (1986).

the subject. Furthermore, general reviews dealing with this topic, among others (Barry, 1983; Idson, 1975; Scheuplein, 1979), often give contradictory explanations of the differences in permeability observed from one site to another. Moreover, if it is clear that both in vitro (Scheuplein, 1979) and in vivo (Feldmann and Maibach, 1967; Maibach et al., 1971; Wester et al., 1984) the anatomical location is of great importance, the connection between the differences observed, the structure of the skin, and the physicochemical nature of the penetrant, remain obscure.

Percutaneous absorption of four radiolabeled compounds—acetylsalicylic acid, benzoic acid, caffeine, and benzoic acid sodium salt—was measured in humans on four body sites, using the stripping method. For each substance and each location, a group of six to eight male Caucasian informed volunteers, aged 28 ± 2 years, was used. One thousand nanomoles of each compound was applied to an area of 1 cm^2, in 20 μl of ethylene glycol/water/Triton X-100 mixtures, the composition of which was chosen according to the solubility of each compound. After 30 min of contact, the excess substance in the treated area was rapidly removed as described earlier for the rat. On each patient, two strictly identical applications were performed in an interval of 48 h. The first application, designed to measure the total penetration of the chemical involved, was made on the right-hand side of the body. For technical convenience, the compounds chosen for the test were ones that are quickly eliminated in the urine. By using data from the literature on the kinetics of urinary excretion of these substances (Bridges et al., 1970; Bronaugh et al., 1982; Feldmann and Maibach, 1970), administered by different routes in various species, the total amounts penetrating within the following 4 days were deduced, after liquid scintillation counting, from the amounts excreted in the first 24 h urine. These were, respectively, 75, 50, 75, and 31% of the total quantities of benzoic acid sodium salt, caffeine, benzoic acid, and acetylsalicylic acid absorbed.

At the end of the second application, performed on the left-hand side of the body (contralateral site), the stratum corneum of the treated area was removed by 15 successive strippings (3M adhesive tape), and the radioactivity present in the horny layer was measured as previously described for the rat.

To make comparisons easier, Fig. 11 expresses the permeability of each site to the various compounds in relation to that of the arm to benzoic acid sodium salt. This representation offers the advantage of simultaneously showing differences in permeability due to both the physicochemical properties of the penetrants and to the structural peculiarities of the areas where they were applied.

Skin permeability appears to be as follows: arm ≤ abdomen < postauricular < forehead. It is noteworthy that whatever the compound applied, the forehead is about twice as permeable as the arm or the abdomen. It may be pointed out that this average ratio agrees well with those reported for the same areas with other compounds (Feldmann and Maibach, 1967; Maibach et al., 1971).

A possible explanation of the higher penetration in areas where there are more sebaceous glands, such as the forehead, could be that absorption occurs through the follicles, rather than through the epidermis. In our opinion, it is difficult to reconcile the great disproportion, a factor of 50 to 100, existing between the number of sebaceous glands of the arm and the forehead (Benfenati and Brillanti, 1939; Szabo, 1958) and the relatively weak difference, a factor of 2 to 3, observed in skin permeability between these two sites. As a consequence, if it is reasonable to assume that the follicular pathway plays a role in percutaneous absorption, it has to be reevaluated.

In the past, the sebum was believed to reduce absorption of hydrophilic compounds (Calvery et al., 1946). This theory has since been disproved (Blank and Gould, 1961). As our results show (see Fig. 11), the same ratio, a factor of 2, exists between permeability levels of areas such as the fore-

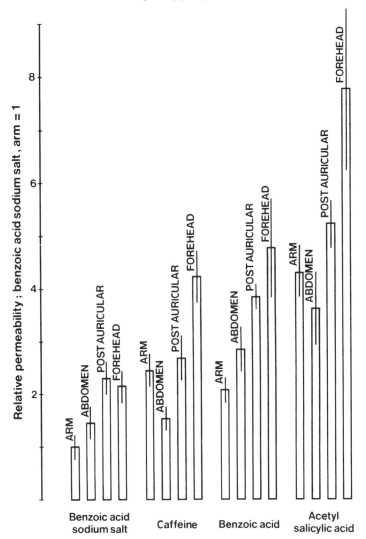

Figure 11 Influence of anatomical site on the total percutaneous absorption of the tested compounds (values expressed relatively to that of benzoic acid sodium salt applied on the arm). From Rougier et al. (1987).

head, which is rich in sebum, and the arm, which has very little, to compounds with totally different lipid/water solubility, such as benzoic acid and its sodium salt.

Among the numerous applications of studies on the relationship existing between skin permeability and anatomical site, considerable attention has been given in recent years to finding favorable "windows" for transdermal treatment of systemic diseases. For various reasons, the postauricular area has been studied most often for scopolamine transdermal drug delivery (Shaw et al., 1975, 1977). According to Taskovitch and Shaw (1978), in this area the closeness of the capillaries to the surface of the skin may promote resorption of substances and give the postauricular skin its good permeability. As our results show (see Fig. 11), whatever the compound applied, this area has a high level of permeability. Apart from caffeine, it is statistically higher than that of the arm or abdomen and often similar to that of the forehead.

As Fig. 12 shows, the correlation between the amount of substance present in the stratum corneum at the end of a 30-min application and the total amount absorbed within 4 days is confirmed in humans, whatever the factors involved in differences of permeability between sites ($r = .97$, $p < .001$). As a consequence, by simply measuring the quantity (x) of a chemical present in the stratum corneum at the end of application, it is now possible to predict the total quantity (y) absorbed over a 4-day period. It should also be mentioned that this correlation curve is similar to that established in the rat (see preceding section), thus demonstrating that the relationship between reservoir function of the horny layer and percutaneous absorption is independent of the animal species.

The consequences of such findings are obvious and far-reaching. They would make it easier to screen new drugs in animals and, thus, predict their toxicological or pharmacological implications. They would also circumvent some ethical difficulties of human experiments, particularly those using potentially toxic agents. It is self-evident that in vivo investigations with animals, and particularly humans, are preferable to in vitro methods. It is also obvious that the experimentor has a higher degree of responsibility when performing in vivo percutaneous absorption studies in humans. Moreover, the blood and urine analyses generally used in the in vivo methods involve severe technical problems because of the low concentrations that must be assayed. Radiolabeled compounds are detected with high sensitivity but imply ethical problems when applied to humans. For technical convenience, labeled compounds were used in our experiments. However, because of the relatively large amount of substance present in the stratum corneum at the end of application, it should be possible, with the stripping method, to measure percutaneous absorption in both animals and humans by appropriate

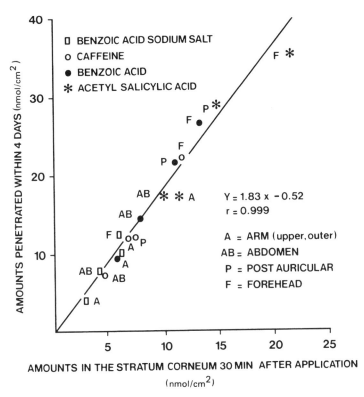

Figure 12 Correlation between total percutaneous absorption within 4 days and the amounts present in the stratum corneum at the end of application time (30 min), for each compound and each anatomical site. From Rougier et al. (1987).

nonradioactive analytical techniques. When it is, nevertheless, essential to use labeled substances, this method makes it possible both to substantially reduce the level of radioactivity administered and to limit contact time.

REFERENCES

Barr, M. (1962). Percutaneous absorption. *J. Pharm. Sci. 51*:395–408.

Barry, B. W. (1983). Dermatological formulations: Percutaneous absorption. In *Drugs and Pharmaceutical Sciences*, Vol. 18. Edited by J. Swarbrick. Marcel Dekker, New York.

Benfenati, A. and Brillanti, F. (1939). Sulla distribuzione delle ghiandole sebacee nella cute del corpo umano. *Arch. Ital. Dermatol. 15*:33–42.

Blank, H. I. and Gould, E. (1961). Penetration of anionic surfactants into the skin. II Study of mechanisms which impede the penetration of synthetic anionic surfactants into skin. *J. Invest. Dermatol. 37*:311–315.

Blank, H. I. and Scheuplein, R. J. (1964). The epidermal barrier. In *Progress in Biological Sciences in Relation to Dermatology*, 2nd ed. Edited by A. Rook and R. H. Champion. Cambridge University Press, London, pp. 245–261.

Bridges, J. W., French, W. R., Smith, R. L., and Williams, R. T. (1970). The fate of benzoic acid in various species. *Biochem. J. 118*:47–51.

Bronaugh, R. L., Stewart, R. F., Congdon, E. R., and Giles, A. L. (1982). Methods for in vitro percutaneous absorption studies. I. Comparison with in vivo results. *Toxicol. Appl. Pharmacol. 62*:474–480.

Calvery, H. O., Draize, J. H., and Lang, E. P. (1946). Metabolism and permeability of normal skin. *Physiol. Rev. 26*:495–540.

Dupuis, D., Rougier, A., Roguet, R., and Lotte, C. (1986). The measurement of the stratum corneum reservoir: A simple method to predict the influence of vehicles on in vivo percutaneous absorption. *Br. J. Dermatol. 115*:233–238.

Feldmann, R. J., and Maibach, H. I. (1967). Regional variations in percutaneous penetration of ^{14}C cortisol in man. *J. Invest. Dermatol. 48*:181–183.

Feldmann, R. J., and Maibach, H. I. (1969). Percutaneous penetration of steroids in man. *J. Invest. Dermatol. 52*:89–94.

Feldmann, R. J. and Maibach, H. I. (1970). Absorption of some organic compounds through the skin in man. *J. Invest. Dermatol. 54*:399–404.

Idson, B. (1971). Biophysical factors in skin penetration. *J. Soc. Cosmet. Chem. 22*: 615–620.

Idson, B. (1975). Percutaneous absorption. *J. Pharm. Sci. 64*:901–924.

Lindsey, D. (1963). Percutaneous penetration. In *Proceedings of the 12th International Congress of Dermatology*. Edited by D. M. Pillsbury and C. S. Livingood. Exerpta Medica, Amsterdam, pp. 407–415.

Maibach, H. I. and Feldmann, R. J. (1969). Effect of applied concentration on percutaneous absorption in man. *J. Invest. Dermatol. 52*:382.

Maibach, H. I., Feldmann, R. J., Milby, T. H., and Serat, W. F. (1971). Regional variations in percutaneous penetration in man. *Arch. Environ. Health 23*: 208–211.

Malkinson, F. D. and Rothman, S. (1963). Percutaneous absorption. In *Handbuch der Haut und Geschlecht Skrauberten, Normale und Pathologische de Haut*, Vol. 1, Part 1. Edited by A. Marchionini and H. W. Spier. Springer-Verlag, Berlin-Heidelberg, pp. 90–156.

Marzulli, F. N. (1962). Barrier to skin penetration. *J. Invest. Dermatol. 39*:387–393.

Poulsen, B. J. (1973). Design of topical drug products: Biopharmaceutics. In *Drug Design*, Vol IV. Edited by E. J. Ariens. Academic Press, New York, pp, 149–190.

Rothman, S. (1954). *Physiology and Biochemistry of the Skin*. University of Chicago Press, Chicago.

Rougier, A., Dupuis, D., Lotte, C., Roguet, R., and Schaefer, H. (1983). In vivo correlation between stratum corneum reservoir function and percutaneous absorption. *J. Invest. Dermatol. 81*:275–278.

Rougier, A., Dupuis, D., Lotte, C., and Roguet, R. (1985). The measurement of the stratum corneum reservoir. A predictive method for in vivo percutaneous absorption studies: Influence of application time. *J. Invest. Dermatol.* *84*:66–68.

Rougier, A., Lotte, C., and Maibach, H. I. (1987). In vivo percutaneous penetration of some organic compounds related to anatomic site in man: Predictive assessment by the stripping method. *J. Pharm. Sci.*

Schaefer, H., Stuttgen, G., Zesch, A., Schalla, W., and Gazith, J. (1978). Quantitative determination of percutaneous absorption of radiolabeled drugs in vitro and in vivo in human skin. In *Current Problems in Dermatology.* Edited by J. W. H. Mali. S. Karger, Basel, pp. 80–94.

Scheuplein, R. J. (1967). Mechanism of percutaneous absorption: II. Transient diffusion and relative importance of various routes of skin absorption. *J. Invest. Dermatol.* *48*:79–88.

Scheuplein, R. J. (1979). Site variation in diffusion and permeability. In *Physiology and Pathophysiology of the Skin.* Edited by A. Jarrett. Academic Press, New York, pp. 1731–1752.

Scheuplein, R. J. and Ross, L. W. (1974). Mechanism of percutaneous absorption of solvents deposited solids. *J. Invest. Dermatol.* *62*:353–360.

Schutz, E. (1957). Der Einflub von polyäthylenglycol 400 auf die percutan resorption von wirkstoffen. *Archiv. Exp. Pathol. Pharmakol.* *232*:237–240.

Shaw, J. E., Chandrasekaran, S. K., Michaels, A. S., and Taskovitch, L. (1975). Controlled transdermal delivery, in vitro and in vivo. In *Animal Modes in Human Dermatology.* Edited by H. I. Maibach, Churchill-Livingstone, Edinburgh, London, pp. 136–146.

Shaw, J. E. Chandrasekaran, S. K., Campbell, P. S., and Schmitt, L. G. (1977). New procedures for evaluating cutaneous absorption. In *Cutaneous Toxicity.* Edited by V. A. Drill and P. Lazar. Academic Press, New York, pp. 83–94.

Shelley, W. B. and Melton, F. M. (1949). Factors accelerating the penetration of histamine through normal intact skin. *J. Invest. Dermatol.* *13*:61–64.

Sheth, N. V., Keough, M. B., and Spruance, S. L. (1986). Measurement of the stratum corneum drug reservoir to predict the therapeutic efficacy of topical iododeoxyuridine for herpes simplex virus infection. Annual Meeting of the American Federation of Clinical Research, Washington, DC, May 1986.

Skog, E. and Wahlberg, J. E. (1964). A comparative investigation of the percutaneous absorption of lethal compounds in the guinea pig by means of the radioactive isotopes ^{51}Cr, ^{58}Co, ^{65}Zn, ^{110m}Ag, ^{115m}Cd, ^{203}Hg. *J. Invest. Dermatol.* *43*:187–192.

Stoughton, R. B. and Fritsh, W. F. (1964). Influence of dimethyl sulfoxide (DMSO) on human percutaneous absorption. *Arch. Dermatol.* *90*:512–517.

Stoughton, R. B. (1965). Percutaneous absorption. *Toxicol. Appl. Pharmacol.* *7*(Suppl. 2):1–6.

Szabo, G. (1958). The regional frequency and distribution of hair follicles in human skin. In *The Biology of Hair Growth.* Edited by W. Montagna and R. A. Ellis. Academic Press, New York, pp. 33–38.

Taskovitch, L. and Shaw, J. E. (1978). Regional differences in morphology of human skin. Correlation with variations in drug permeability. *J. Invest. Dermatol. 70*: 217.

Tregear, R. T. (1964). The permeability of skin to molecules of widely differing properties. In *Progress in Biological Sciences in Relation to Dermatology*, 2nd ed. Edited by A. Rook and R. H. Champion. Cambridge University Press, London, pp. 275–281.

Treherne, J. E. (1956). Permeability of skin to some electrolytes. *J. Physiol. 13*:171–180.

Vickers, C. F. H. (1963). Existence of a reservoir in the stratum corneum. *Arch. Dermatol. 88*:20–23.

Wester, R. C. and Maibach, H. I. (1976). Relationship of topical dose and percutaneous absorption in rhesus monkey and man. *J. Invest. Dermatol. 67*:518–520.

Wester, R. C., Maibach, H. I., Bucks, D. A., and Aufrere, M. B. (1984). In vivo percutaneous absorption of paraquat from hand, leg and forearm of humans. *J. Toxicol. Environ. Health 14*:759–762.

Winkelmann, R. K. (1969). The relationship of structure of the epidermis to percutaneous absorption. *Br. J. Dermatol. 81*(Suppl. 4):11–22.

23
Tape-Stripping Technique

Christian Surber
University Hospital, Basel, Switzerland

Fabian P. Schwarb
University Hospital, Basel, Switzerland

Eric W. Smith
Ohio Northern University, Ada, Ohio

I. INTRODUCTION

Tape stripping is a technique that has been found useful in dermatopharmacological research for selectively and, at times, exhaustively removing the skin's outermost layer, the stratum corneum. Typically an adhesive tape is pressed onto the test site of the skin and is subsequently abruptly removed. The application and removal procedure may be repeated 10 to more than 100 times (1,2).

Skin that has been stripped in this manner has been used as standardized injury in wound healing research. The technique has been adapted for studying epidermal growth kinetics (3–7), and it may also be useful as a diagnostic tool in occupational dermatology to assess the quality of the stratum corneum (8,9).

The observation that the skin may serve as a reservoir for chemicals was originally reported by Malkinson and Ferguson in 1955 (10). The localization of this reservoir within the stratum corneum was later demonstrated for corticosteroids by Vickers in 1963 (11) and has been confirmed by others (12–15). The introduction of the tape stripping method to further investigate the reservoir and barrier function of the skin gave a significant expansion to experimental tools in skin research (16–18).

Differences in the permeability of intact and fully stripped skin have provided information about the diffusional resistance of the various dermal strata (19). It has been recognized that complete removal of the stratum corneum was not possible even after 30–40 strippings (20), and a certain barrier function in the tissue so treated remains (21,22). Öhman et al. showed that after 100 tape strippings the entire stratum corneum could be removed (2) (Fig. 1); however, no permeation data through such completely stripped skin exist.

II. APPLICATION OF THE TAPE-STRIPPING TECHNIQUE IN DERMATOPHARMACOLOGY

The presumption that factors that improve percutaneous absorption also result in an increase in the stratum corneum reservoir (11,15) made the tape-stripping methodology a promising tool for selecting or comparing vehicles for topical drugs. Data from tape-stripping experiments were therefore related to a) chemical penetration into skin, b) chemical permeation through skin, (c) chemical elimination from the skin, d) pharmacodynamic parameters, and e) clinical parameters.

In a series of in vitro and in vivo investigations using the tape-stripping technique, it could be shown that drug penetration into stratum corneum, determined by quantification of radiolabeled chemical on the removed tape, was clearly vehicle dependent. In vitro drug penetration into the stratum corneum was usually higher than in vivo, whereas penetration into deeper tissues was higher in vivo. In vivo drug permeation through the skin, determined as drug excreted in the urine, was found to be vehicle independent. It was noted that different vehicles may bring about different therapeutic drug concentrations into the skin, but similar or different systemic burdens were produced (e.g., corticosteroids, salicylic acid) (16,23–26).

Dupuis, Lotte, Rougier, and coworkers (27–31) standardized the methodology of the tape-stripping technique. Their stripping method determines the concentration of chemical in the stratum corneum at the end of a short application period (30 min). They found a linear relationship between the stratum corneum reservoir content and in vivo percutaneous absorption (total amount of drug permeated in 4 days) using the standard urinary excretion method (32–34). They could also show, for a variety of simple pharmaceutical vehicles, that percutaneous absorption of benzoic acid is vehicle dependent and can be predicted from the amount of drug within the stratum corneum at 30 min after application. They stated that the major advantages of their validated tape-stripping protocol are the subsequent elimination of urinary and fecal excretion to determine absorption, and the applicability to nonradiolabeled determination of percutaneous absorption because the skin

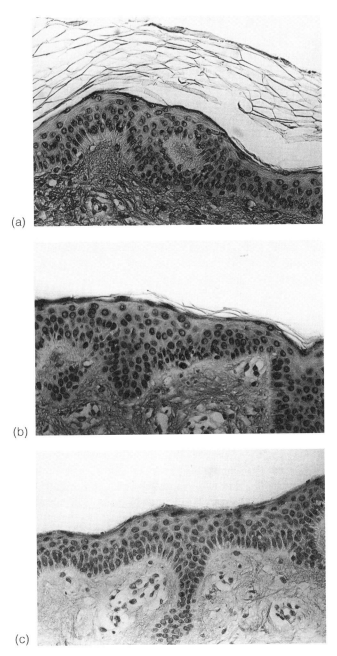

Figure 1 Skin that is (a) intact, (b) 50 times stripped, and (c) 100 times stripped. From ref. 2, with permission.

strippings contain adequate chemical concentrations for non-labeled assay methodologies (see also Table 1). Despite the fact that the assay provides reliable prediction of total absorption for a group of selected compounds, mechanistic interpretations are still rare. Based on the data of Rougier and coworkers, Auton (35) presented an initial first mathematical approach that may help to explain some of the above observations.

The tape-stripping technique has also been used to analyze biological activity, thus taking into account binding, decomposition, and metabolism of a given drug. From a skin area treated with various griseofulvin formulations, stratum corneum was stripped onto tapes, sterilized, and then inoculated with *Trichophyton mentagrohytes* spores. Under controlled, in vitro conditions the degree of growth of the spores in the stratum corneum was graded. Various formulations of topical griseofulvin were tested for activity and duration of effect with this method. The antifungal effect of griseofulvin in the cream base was weak and of short duration, whereas the antifungal formulated into a dimethylacetamide vehicle was effective for 24 h and could be detected 96 h after one application (36).

A direct study evaluating whether differential drug uptake of topical 2% miconazole and 2% ketokonazole from cream formulations into human stratum corneum correlated with differential pharmacological activity (100-fold) against *Candida albicans* was recently done (37). After removal of the residue from a single 24-h topical dose of the two antifungals, the stratum corneum of the test sites was removed by tape stripping. The tapes were extracted for drug quantification (high-performance liquid chromatography, HPLC) and bioactivity against *Candida albicans* growth in vitro. Topical ketaconazole produced significantly higher drug concentrations in stratum corneum than miconazole at 1, 4, and 8 h after a single topical dose. However, after 24 h the concentrations were similar. Tape disk extracts from the ketaconazol-treated skin sites demonstrated significantly greater bioactivity in the bioassay than miconazole, due to the fact that the stratum corneum can easily be sampled, drug activity can be evaluated directly in the target tissue. Data from such experiments are more relevant than the standard in vitro minimal inhibition concentration (MIC) methods. The combination of a pharmacokinetic and a pharmacodynamic method offers a more comprehensive approach with which to identify optimized topical vehicle formulations for antifungal delivery to the skin and to determine bioequivalence between topical antifungal products.

Sheth and coworkers (1) compared the concentration of iododeoxyuridine in the stratum corneum of the guinea pig skin by tape stripping at different points, after single and multiple topical doses of the drug, in various simple formulations. These results were correlated with the efficacy of topical iododeoxyuridine against an experimental cutaneous herpes simplex virus. The results showed an excellent correlation between the quantity of iododeoxyuridine in the stratum corneum and the reduction in lesion severity. Both formulation composition and dosing frequencies had an effect on the iododeoxyuridine concentration in the stratum corneum and on the corresponding clinical efficacy.

Data from in vitro and in vivo human skin model systems indicated that significantly higher acyclovir concentrations were achieved in stratum corneum (tape-stripping/cyanoacrylate method) (38) after topical application as compared to oral administration. However, mathematical modeling of in vitro and in vivo skin biopsy data and sectioning experiments demonstrated that the drug concentration at the target site, the basal epidermis, is two to three times less lower after topical administration than after oral administration (39).

Penetration of two chemicals *dl*-alpha-tocopherol nicotinate and L440, an anti-inflammatory substance (L440 → 2-(*t*-butyl)-4-cyclohexylphenylnicotinate *N*-oxide), into human stratum corneum (by a tape-stripping method) (40) from two liposomal gels was significantly higher than from conventional formulations (oil/water, water/oil bases). Only a slight dependence of the extent of penetration into stratum corneum on lipsome diameter was observed. The anti-inflammatory effect of L440 was determined by the arachidonic acid-induced ear edema in the male NMR mouse. A significant clinical effect of L440 was observed with the liposomal formulation and with the oil-in-water (o/w) base. An insignificant effect of L440 in the water-in-oil (w/o) base was detected (41).

Attenuated total reflectance infrared spectroscopy (ATR-IR) can quantitate the appearance of a drug or vehicle in the skin from human volunteers. However, it will only work for the outer regions of the epidermis, and it requires compounds with an unique infrared spectrum. ATR-IR has been used to study the effect of oleic acid on the percutaneous absorption of *p*-cyanophenol in vivo (42). Higo and coworkers (43) recently showed that the amount of *p*-cyanophenol in the stratum corneum was highly correlated with depth ($r^2 = .84 - .93$) depending on application time. In this study drug determination was carried by ATR-IR and by direct ^{14}C analysis of *p*-cyanophenol following progressive removal of tissue by tape-stripping.

Pershing and coworkers (44–47) simultaneously compared a skin blanching bioassay with drug content in human stratum corneum following topical application of commercial 0.05% betamethasone dipropionate formulations. The rank order of the betamethasone dipropionate formulation potency was found to be is similar between by the visual skin blanching assay and the tape-stripping assessments and by monitoring with the a-scale of the Minolta chromameter. The rank correlation between the tape-stripping method and the skin blanching response was moderate to good.

Usually drug uptake is assessed by applying test and reference products simultaneously to multiple skin sites in a single study subject. Stratum corneum samples are obtained at sequentially increasing times intervals from the time of application. In a similar manner, to assess drug elimination, test and reference products are applied for a specific period of time at multiple sites and removed. The stratum corneum samples are collected at sequentially increasing times after drug formulation removal. Additionally, drug elimination studies after the drug concentration has reached a plateau in the stratum corneum have been proposed (48,49). These data support the application of dermatopharmacokinetic principles to bioavailability and bioequivalence determinations of topical dermatological products (48–50).

III. THE POTENTIAL OF THE TAPE-STRIPPING METHODS

Recently V. P. Shah and colleagues (51) summarized the essential of an international expert panel that presented a series of invited contributions at the AAPS/FDA workshop on Bioequivalence of Topical Dermatological Dosage Forms—Methods for Evaluating Bioequivalence, held in September 1996. The workshop explored the possibility that dermatopharmacokinetic characterization might provide an alternative approach to clinical trials for the determination of bioequivalence of topical dermatological products in a parallel manner, analogous to the use of concentration–time curves for systemically administered drugs. If accepted, this approach might allow dermatopharmacokinetic studies to replace clinical trials as a means of documenting bioequivalence of selected topical drug products. Among a variety of models presented (e.g., microdialysis, confocal laser scanning microscopy or transepidermal water loss), the tape-stripping method seems to have the greatest potential for dermatopharmacokinetic characterization of selected topical drug products.

IV. PROTOCOL OUTLINE FOR A TAPE-STRIPPING EXPERIMENT

The following outlines an example of procedural steps involved in a tape-stripping experiment.

Apply the test and/or the reference drug products concurrently at multiple sites.

After an appropriate interval, remove the excess drug by wiping with tissue or cotton swab. Appropriate time duration should be determined in a pilot study.

Apply the adhesive tape with uniform pressure; remove and discard
the first stripping, as this is believed to represent unabsorbed drug
on the skin surface. Repeat this procedure if one tape strip is not
sufficient to remove all excess/unabsorbed drug from the skin
surface.

Apply (at the same site), remove, and collect 9 to 20 successive tape
strippings from each application site. Repeat the procedure for each
site at other designated time points.

Extract the drug from single or combined tape strippings and determine
the concentration in the extract solution using an appropriate analytical methods (cave extraction recovery).

Express the results as amount of drug per square centimeter area of
the adhesive tape (e.g., ng/cm^2) or by another adequate means (e.g.,
ng/protein content).

This procedure will provide information about the drug uptake in the stratum
corneum.

To determine a drug elimination phase from the stratum corneum, apply the drug product concurrently at multiple sites, allow sufficient exposure
periods until an apparent steady-state level is achieved, and remove excess
drug from the skin surface as described above. After predetermined time
intervals (e.g., 1, 3, 5, and 21 h after drug removal), collect skin samples
using 9 to 20 successive tape strips, and analyze them for drug content.

V. UNANSWERED QUESTIONS AND CONCERNS

Despite the fact that the tape-stripping method has been used in dermatopharmacological research for several decades, several experimental details
have not been addressed and the technique still awaits rigorous validation.

First, in bioequivalence evaluations the vehicle components of a test
and a reference product may be different and may variably influence both
the adhesive properties of the tape as well as the cohesion of the corneocytes.
Hence dermatopharmacokinetic characterization may become extremely
complex and susceptible to error. As described previously by Surber et al.
(52), the vehicle may significantly change the cohesion of the corneozytes.
It was shown in this chapter that the removal of corneocytes can be highly
dissimilar, and dermatopharmacokinetic characterization based on a concentration profile within the stratum corneum is not possible (Fig. 2). A similar
observation was made by van der Molen et al. (53). They showed that normal tape stripping of human stratum corneum yields cell layers that originate
from various depths because of furrows in the skin.

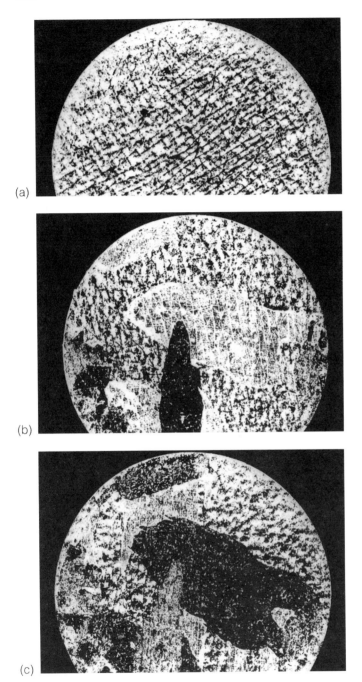

Figure 2 Strip (b) 7 and (c) 8 taken from a skin site 3 h after the application of a water-in-oil (w/o) cream base. (a) Strip 7 taken from untreated skin.

Second, the material and methods sections of most research papers that used a tape-stripping technique rarely describe the use of a template to assure consistent removal of stratum corneum from the exact treatment area. Without such a template, consistency in terms of the stratum corneum removal area is not assured and, hence, dermatopharmacokinetic characterization may become questionable. The recent AAPS/FDA workshop report (51) addresses this issue and recommends also the use of such demarcating templates (52).

Third, the tape-stripping method has primarily been applied in experimental in vivo settings. However the technique has also been used to remove stratum corneum from skin that has been used for in vitro experiments. Depending on the skin origin (human, animal) and source (surgery, morgue), skin storing conditions, and the duration of the in vitro experiment, the integrity of the skin may vary and will markedly influence the amount of stratum corneum that is being removed with each tape strip. As seen in Fig. 3, freshly prepared hairless rat skin used in an in vitro experiment will separate at the basal membrane approximately 3 h after the start of the experiment. Subsequent tape stripping removes the entire epidermis, including the stratum corneum.

Fourth, as previously stated by several experts, the major advantage of the tape-stripping technique is the applicability to nonradiolabeled determination of percutaneous absorption because the skin strippings contain adequate chemical concentrations for nonlabeled assay methodologies. However, using realistic doses of topical dermatological dosage forms on the

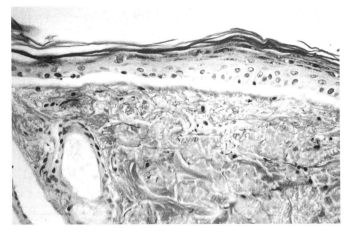

Figure 3 Separation of epidermis from dermis 3 h after the start of an in vitro percutaneous absorption experiment with rat skin.

Table 1 Comparison of Drug Strength, Amount of Drug on the Skin (A: μg/cm^2), Amount of Drug Absorbed When 2 mg of a Topical Dermatological Dosage Form at Various Strengths Is Spread over an Area of 1 cm^2 Assuming a 5% Absorption of the Dose Applied (B), and Amounts When Further Assuming That 90% of the Dose Absorbed Will Be Retained in the Total Skin (C) and That from That Amount 80% Will Be Retained in the Stratum Corneum (D)

Strength (%)	A, Dose applied (μg/cm^2) on the skin surface	B, Dose absorbed (ng/cm^2), 5% of dose applied	C, Dose absorbed (ng/cm^2) as 90% of B in total skin	D, Dose absorbed (ng/cm^2) as 80% of C in the stratum corneum
1	20	1000	900	720
0.1	2	100	90	72
0.01	0.2	10	9	7.2

Note. From these columns, it is obvious that quantification of a drug may be difficult (e.g., corticosteroids/HPLC).

skin (usually less than 2 mg/cm^2); the quantification of the drug in the skin strippings may become a major analytical challenge (low concentration in the strippings and difficult extraction from the tapes). Table 1 may illustrate this issue.

Lastly, even though tape-stripping is considered to be essentially non-invasive, stripped sites in certain dark-skinned individuals may remain pigmented for several months after healing. This effect must be communicated to the volunteers before entering a study.

VI. RELATED TECHNIQUES

The surface biopsy and skin scraping techniques have also been used to remove stratum corneum from the skin.

The skin surface biopsy technique using cyanoacrylate contact cement adhesive was first used to remove the stratum corneum for diagnostic and investigative purposes in superficial mycosis infections. The glue is applied to a glass slide and pressed onto the skin surface. The slide is subsequently removed after the glue has polymerized with the stratum corneum (54,55). This technique allows also the quantification of drug found in the stratum corneum after various application times (56). However, a distinction between the transepidermal and the transfollicular routes of absorption is not possible because the hair follicle content is also removed with this method.

Large amounts of corneocytes can be harvested by scrapping the stratum corneum from the surface with an open dermal steel curette. Usually superficial (<20 μm depth) and, with more vigorous scraping, midlevel (20–40 μm depth) stratum corneum can be collected in this manner (57,58). This technique has also been successfully employed by Faergemann et al. to determine drug concentration (terbinafine, fluconazole) in the stratum corneum (59–61). However, the degree of superficial sebum and sweat contamination of the stratum corneum may influence the amount of drug found in this tissue.

VII. CONCLUSIONS

The skin stripping method has potential for being a specific dermatopharmacokinetic method that assesses drug concentration in stratum corneum as a function of time. Both drug uptake and drug elimination profiles may be evaluated to determine traditional pharmacokinetic metrics, such as AUC, C_{max}, and T_{max}.

Currently two general views on the potential applicability of the skin stripping technique are expressed. Some researchers are of the opinion that only diseases in which stratum corneum is the site of action are amenable to this method of analysis, because only the drug concentration in the stratum corneum is measured (i.e., antifungal or corticosteroid classes of topical dermatological drugs). Others contend that regardless of how far through the skin layers the drug needs to permeate (stratum corneum–epidermis–dermis), the active moiety needs to pass through the stratum corneum first before reaching the deeper skin layers. Since the stratum corneum is the rate-limiting barrier for drug permeation, drug concentrations in this layer may also provide meaningful information for comparative evaluation of topical dosage forms intended for dermal, subdermal, or systemic action.

Despite the application of the tape-stripping methods in dermatopharmacological research for several decades, the technique still awaits rigorous validation in order to become a viable, robust, and reproducible method for bioequivalence evaluation of topical dermatological drug products.

REFERENCES

1. Sheth, N. V., McKeough, M. B., Spruance, S. L. Measurement of stratum corneum drug reservoir to predict the therapeutic efficacy of topical iododeoxyuridine for herpes simplex. J Invest Dermatol. 1987; 89:598–602.
2. Öhman, H., Vahlquist, A. *In vivo* studies concerning a pH gradient in human stratum corneum and upper epidermis. Acta Derm Venereol (Stockh). 1994; 74:375–379.

3. Eriksen, G., Lamke, L. Regeneration of human epidermal surface and water barrier function after stripping. Acta Derm Venereol (Stockh). 1971; 51:169–178.

4. Wilhelm, D., Elsner, P., Maibach, H. I. Standardized trauma (tape-stripping) in human vulvar and forearm skin. Acta Derm Venereol (Stockh). 1991; 71: 123–126.

5. Downes, A. M., Matoltsy, A. G., Sweeney, T. M. Rate of turnover of the stratum corneum in hairless mice. J Invest Dermatol. 1967; 49(4):400–405.

6. Pinkus, H. Examination of the epidermis by the strip method of removing horny layers. I. Observations on thickness of the horny layer, and on mitotic activity after stripping. J Invest Dermatol. 1951; 16:383–386.

7. Pinkus, H. Examination of the epidermis by the strip method of removing horny layers. II. Biometric data on regeneration of human epidermis. J Invest Dermatol. 1951; 16:431–447.

8. Piérard, G. E., Piérard-Franchimont, C., Saint-Léger D., Kligman, A. M. Squamometry: The assessment of xerosis by colorimetry of D-Squame® adhesive discs. J Soc Cosmet Chem. 1992; 47:297–305.

9. Schatz, H., Kligman, A. M., Manning, S., Stoudemayer, T. Quantification of dry (xerotic) skin by image analysis of scales removed by adhesive discs (D-Squame®). J Soc Cosmet Chem. 1993; 44:53–63.

10. Malkinson, F. D., Ferguson, E. H. Percutaneous absorption of hydrocortisone-4-^{14}C in two human subjects. J Invest Dermatol. 1955; 25:281–285.

11. Vickers, C. F. H. Existence of reservoir in the stratum corneum. Arch Dermatol. 1963; 88:20–23.

12. Stoughton, R. B. Dimethyl sulfoxide (DMSO) induction of a steroid reservoir in human skin. Arch Dermatol. 1965; 91:657–660.

13. Carr, R. D., Wieland, R. G. Corticosteroid reservoir in the stratum corneum. Arch Dermatol. 1966; 94:81–84.

14. Carr, R. D., Tarnowski, W. M. The corticosteroid reservoir. Arch Dermatol. 1966; 94:639–642.

15. Munro, D. D. The relationship between percutaneous absorption and stratum corneum retension. Br J Dermatol. 1969; 81(Suppl 4):92–97.

16. Lücker, P., Nowak, H., Stüttgen, G., Werner, G. Penetrationskinetik eines Tritium-markierten 9 alpha-Fluor-16 methylen-prednisolonesters nach epicutaner Applikation beim Menschen. Arzneim.-Forsch./Drug Res. 1968; 18:27–29.

17. Tsai, J.-C., Cappel, M. J., Flynn, G. L., Weiner, N. D., Kreuter, J., Ferry, J. Drug and vehicle deposition from topical applications: Use of *in vitro* mass balance technique with minoxidil solutions. J Pharm Sci. 1992; 81(8):736–743.

18. Tojo, K., Lee, A. C. A method for predicting steady-state rate of skin penetration *in vivo*. J Invest Dermatol. 1989; 92(1):105–108.

19. Moon, K. C., Wester, R. C., Maibach, H. I. Diseased skin models in the hairless guinea pig: *In vivo* percutaneous absorption. Dermatologica. 1990; 180:8–12.

20. Holoyo-Tomoka, M. T., Kligman, A. M. Does cellophane tape stripping remove the horny layer? Arch Dermatol. 1972; 106:767–768.

21. Malkinson, F. D. Studies on the percutaneous absorption of ^{14}C labelled steroids by use of glass-flow cell. J Invest Dermatol. 1958; 31:19–28.
22. Feldman, R. J., Maibach, H. I. Penetration of ^{14}C hydrocortisone through normal skin: Effect of stripping and occlusion. Arch Dermatol. 1965; 91:661–666.
23. Zesch, A., Schäfer, H. Penetrationskinetik von radiomarkierten Hydrocortisone aus verschiedenartigen Salbengrundlagen in die menschliche Haut. Arch Derm Forsch. 1975; 252:245–256.
24. Zesch, A., Schaefer, H., Hoffmann, W. Barriere- und Reservoirfunktion der einzelnen Hornschichtlagen der menschlichen Haut für lokal aufgetragene Arzneimittel. Arch Derm Forsch. 1973; 246:103–107.
25. Zesch, A. Reservoirfunktion der Hornschicht. In: Klaschka, F., ed. Stratum corneum: Struktur und Funktion. Berlin: Grosse Verlag; 1981:63–76.
26. Schwarb, F. P., Gabard, B., Rufli, T., Surber, C. Percutaneous absorption of salicylic acid in man after topical administration of three different formulations. Dermatology. 1999; in press.
27. Dupuis, D., Rougier, A., Roguet, R., Lotte, C., Kalopissis, G. *In vivo* relationship between horny layer reservoir effect and percutaneous absorption in human and rat. J Invest Dermatol. 1984; 82(4):353–356.
28. Dupuis, D., Rougier, A., Roguet, R., Lotte, C. The measurement of the stratum corneum reservoir: A simple method to predict the influence of vehicles *in vivo* percutaneous absorption. Br J Dermatol. 1986; 115:233–238.
29. Rougier, A., Lotte, C., Maibach, H. I. The hairless rat: A relevant animal model to predict *in vivo* percutaneous absorption in humans? J Invest Dermatol. 1987; 88(5):577–581.
30. Rougier, A., Lotte, C., Dupuis, D. An original predictive method for *in vivo* percutaneous absorption studies. J Soc Cosmet Chem. 1987; 38:397–417.
31. Rougier, A., Rallis, M. Krien, P., Lotte, C. *In vivo* percutaneous absorption: a key role for stratum corneum/vehicle partitioning. Arch Dermatol Res. 1990; 282:498–505.
32. Feldmann, R. J., Maibach, H. I. Percutaneous penetration of steroids in man. J Invest Dermatol. 1969; 52(1):89–94.
33. Feldmann, R. J., Maibach, H. I. Absorption of some organic compounds through the skin in man. J Invest Dermatol. 1970; 54:399–404.
34. Feldmann, R. J., Maibach, H. I. Percutaneous penetration of some pesticides and herbicides in man. Toxicol Appl Pharmacol. 1974; 28:126–132.
35. Auton, T. R. Skin stripping and science: a mechanistic interpretation using mathematical modelling of skin deposition as a predictor of total absorption. In: Scott, R. C., Guy, R. H., Hadgraft, J., Boddé, H. E., eds. Prediction of Percutaneous Penetration. London: IBC Technical Services Ltd; 1990:558–576.
36. Knight, A. G. The activity of various topical griseofulvin preparations and the appearance of oral griseofulvin in the stratum corneum. Br J Dermatol. 1974; 91:49–55.
37. Pershing, L. K.; Corlett, J., Jorgensen, C. *In vivo* pharmacokinetics and pharmacodynamics of topical ketoconazole and miconazole in human stratum corneum. Antimicrob Agents Chemother. 1994; 38(1):90–95.

38. Pershing, L. K., Krueger, G. G. Human skin sandwich flap model for percutaneous absorption. In: Bronaugh, R. L., Maibach, H. I., eds. Percutaneous Absorption. New York: Marcel Dekker; 1989:397–414.

39. Parry, G. E., Dunn, P., Shah, V. P., Pershing, L. K. Acyclovir bioavailability in human skin. J Invest Dermatol. 1992; 98:856–863.

40. Bredthauer, D. Vergleichende Untersuchung zur Hornschicht-Penetration von Lichtschutzmitteln: Abrissmethode versus photoakustische Spektroskopie. Göttingen: Georg-August Universität, 1990.

41. Michel, C., Purmann, T., Mentrup, E., Seiller, E., Kreuter, J. Effect of liposomes on percutaneous penetration of lipophilic materials. Int J Pharm. 1992; 84:93–105.

42. Mak, V. H., W., Potts, R. O., Guy, R. H. Percutaneous penetration enhancement in vivo measured by attenuated total reflectance infrared spectroscopy. Pharm Res. 1990; 7:835–841.

43. Higo, N; Naik, A., Bommannan, B. D., Potts, R. O., Guy, R. H. Validation of reflectance infrared spectroscopy as a quantitative method to measure percutaneous absorption *in vivo*. Pharm Res. 1993; 10(10):1500–1506.

44. Pershing, L. K., Silver, B. S., Krueger, G. G., Shah, V. P., Skelly, J. P. Feasibility of measuring the bioavailability of topical betamethasone dipropionate in commercial formulations using drug in skin and a skin blanching bioassay. Pharm Res. 1992; 9(1):45–51.

45. Pershing, L. K., Lambert, L. D., Shah, V. P., Lam S. Y. Variability and correlation of chromameter and tape-stripping methods with the visual skin blanching assay in quantitative assessment of topical 0.05% betamethasone dipropionate bioavailability in humans. Int J Pharm. 1992; 86:201–210.

46. Pershing, L. K., Corlett, J. L., Lambert, L. D., Poncelet, C. E. Circadian activity of topical 0.05% betamethasone dipropionate in human skin *in vivo*. J Invest Dermatol. 1994; 102:734–739.

47. Pershing, K. L., Lambert L., Wright, E. D., Shah, V. P., Williams, R. L. Topical 0.05% betamethasone dipropionate: Pharmacokinetic and pharmacodynamic dose-response studies in humans. Arch Dermatol. 1994; 130:740–747.

48. Lücker, P.W., Beubler, E., Kukovetz, W. R., Ritter, W. Retention time and concentration in human skin bifonazole and clotrimazole. Dermatologica. 1984; 169(Suppl 1):51–56.

49. Shah, V. P., Pershing, L. K. The stripping technique to assess bioequivalence of topical applied formulations. In: Brain, K. R., James, V. J., Walters, K. A., eds. Prediction of Percutaneous Penetration. London: IBC Technical Services Ltd.; 1990: 473–476.

50. von Hattingberg, H. M., Brockmeier, D. Drug concentration control and pharmacokinetic analysis during long term therapy with desk top computers. In: Gladtke, E., Heimann, G., eds. Pharmacokinetics: A 25 year old discipline; 1978 Nov; Cologne. New York: Fischer; 1980:165–179.

51. Shah, V. P., Flynn, G. L., Yacobi, A., Maibach, H. I., Bon, C., Fleischer, N. M., Franz, T. J., Kaplan, S. A., Kawamoto, J.; Lesko, L. J.; Marty, J.-P., Pershing, L. K., Schaefer, H., Sequeira, J. A., Shrivastava, W. J., Williams, R. L.

Bioequivalence of topical dermatological dosage forms—Methods of evaluation of bioequivalence. Pharm Res. 1998; 15(2):167–170.

52. Surber, C., Henn, U., Bieli, E., Schwarb, F. P., Gabard, B., Rufli, T. Skin tape-stripping: Is this technique adequate to explore principles of dermatopharmacokinetics? Dermatology. 1999, in preparation.

53. van der Molen, R. G., Spies, F., van't Noordende, J. M., Boelsma E., Mommaas, A. M., Koerten, H. K. Tape stripping of human stratum corneum yields cell layers that originate from various depths because of furrows in the skin. Arch Dermatol Res. 1997; 289:514–518.

54. Whiting, D. A.; Bisset, E. A. The investigation of superficial fungal infections by skin surface biopsy. Br J Dermatol. 1974; 91:57–65.

55. Marks, R., Dawber, R. P. R. Skin surface biopsy: an improved technique for examination of the horny layer. Br J Dermatol. 1971; 84:117–123.

56. Finlay, A, Marks, R. Determination of corticosteroid concentration profiles in stratum corneum using the skin surface biopsy technique. Br J Dermatol 1982; 107(Suppl 22):33.

57. Epstein, W. L.; Shah, V. P., Riegelman, S. Griseofulvin levels in stratum corneum. Study after oral administration in man. Arch Dermatol. 1972; 106: 344–348.

58. Wallace, S. M., Shah, V. P., Epstein, W. L., Greenberg, J., Riegelman, S. Topically applied antifungal agents. Arch Dermatol 1977; 113:1539–1542.

59. Faergemann, J., Zehender, H., Denouel, J., Millerioux, L. Levels of terbinafine in plasma, stratum corneum, dermis-epidermis (without stratum corneum), sebum, hair and nails during and after 250 mg terbinafine orally once per day for four weeks. Acta Derm Venereol (Stockh). 1993; 73(4):305–309.

60. Faergemann, J., Zehender, H., Jones, T., Maibach, H. I. Terbinafine levels in serum, stratum corneum, dermis-epidermis (without stratum corneum), hair, sebum and eccrine sweat. Acta Derm Venereol (Stockh). 1991; 71(4):322–326.

61. Faergemann, J., Laufen, H. Levels of fluconazole in serum, stratum corneum, epidermis-dermis (without stratum corneum) and eccrine sweat. Clin Exp Dermatol. 1993; 18(2):102–106.

24

Interference with Stratum Corneum Lipid Biogenesis
An Approach to Enhance Transdermal Drug Delivery

Peter M. Elias
Veterans Affairs Medical Center and University of California School of Medicine, San Francisco, California

Vivien Mak and Carl Thornfeldt
Cellegy Pharmaceuticals Inc., San Carlos, California

Kenneth R. Feingold
Veterans Affairs Medical Center and University of California School of Medicine, San Francisco, California

I. INTRODUCTION

The traditional view of the stratum corneum (SC) regards this outer layer of the epidermis to be a relatively impermeable and highly resilient tissue. In fact, up to the mid-1970s, the SC was considered a metabolically inert, homogenous tissue, analogous to plastic wrap. According to this model, passive transdermal permeation is governed solely by the physical-chemical properties of this seemingly homogeneous tissue (Scheuplein and Blank, 1971). Based on this view, the permeability barrier can be reproduced in vitro with isolated stratum corneum sheets, regardless of whether cadaver-derived, frozen–thawed, or freshly obtained tissues are employed for permeability studies.

While well recognized for its species-related differences, within species the SC also displays site-related variations in the number of cell layers, the extent of hydration, and the overall quantities and distribution of its principal lipid species. Again, these differences can be integrated into the kinetics predicted by the plastic wrap model, as can the contribution of vehicle components in predicting transdermal delivery. Yet, while the plastic wrap model is able to explain the role of the SC as a competent physical-chemical barrier to the external environment, it incompletely explains the a) observed variable absorption of hydrophilic versus hydrophobic molecules, b) large variations in permeability of different topographic sites, such as palms/soles versus facial skin, and c) individual variations in the susceptibility to toxic insults or irritants.

II. TWO COMPARTMENT MODEL

The first development to cast doubt on this plastic wrap model, and its suppositions, was the discovery of the unique structural heterogeneity of the SC (reviewed in Elias and Menon, 1991). With tracer perfusion and freeze-fracture replication, the permeability barrier was shown to form coincident with the secretion of lamellar body (LB) contents at the stratum granulosum (SG)−SC interface. Most importantly, the delivery of LB contents to the SC interstices, an event observed concurrently with cornification, is associated with the formation of broad lamellar membranes, with the freeze-fracture characteristics of hydrophobic lipids (Williams, 1991). Under normal conditions, virtually all of the lipids in the SC are sequestered within the intercellular spaces, surrounding the corneocytes, an observation supported by histochemical cell fractionation, enzyme cytochemistry, and x-ray diffraction studies (reviewed in Elias and Menon, 1991). The cornified cells, with their surrounding lipid matrix, constitute the so-called "two-compartment" model of the SC (Elias, 1983).

Because of the segregation of lipids to the SC interstices, the two-compartment model also provides new insights into the pathogenesis of both disorders of cornification (DOC) and diseases associated with important barrier abnormalities. While not all DOC are linked to primary lipid abnormalities, selected abnormalities, such as recessive X-linked ichthyosis (RXLI), Sjögren−Larsson syndrome (SJS), Refsum's disease (RD), type II Gaucher's disease (GD), neutral lipid storage disease (NLSD), and certain forms of acquired ichthyosis, are associated with primary abnormalities in cholesterol sulfate (CS), fatty acid (FA), glucosylceramide (GlucCer), triacylglycerol (= triglyceride, TAG), and cholesterol (Chol) metabolism, respectively (reviewed in Williams, 1991). In most of these cases, the biochemical and/or structural abnormalities have been localized to the SC interstices. Furthermore, essential fatty acid deficiency (EFAD) is associated with certain lipid biochemical and structural abnormalities that result in a primary permeability barrier alteration, indicated by an increase in transepidermal water loss (TEWL). In addition, variations in intercellular lipid content and membrane structure also provide a structural basis for the wide variations in permeability of different human skin sites (e.g., palms/soles vs. leg vs. abdomen vs. face). Although there are some variations in SC lipid composition over different skin sites, regional differences in total lipid content alone can account for the differences in water-vapor permeability of these sites (Lampe et al., 1983). It follows that the lipid matrix in the SC interstices appears to comprise the permeability barrier for transdermal drug delivery, as well. Finally, in addition to the increase in the volume of the lipid fraction of the SC intercellular compartment, resulting from the delivery of LB contents to the SC interstices, this compartment can be expanded further via the incorporation of low-melting lipophilic agents following topical application to the skin surface. These results demonstrate that the SC interstices subserve both the putative intercellular transport route and the "reservoir function" of the SC (Nemanic and Elias, 1980).

III. EPIDERMAL LIPID BIOGENESIS

A. Biosynthetic Activities

Numerous studies have described the changes in lipid composition that accompany SC formation in murine, porcine, and human epidermis (reviewed in Yardley, 1969; Schurer and Elias, 1991). These studies demonstrate the loss of phospholipids and the emergence of neutral lipids, ceramides, and cholesterol in the SC during barrier formation. Moreover, a further gradient in lipid transformation occurs during the final stages of terminal differentiation and barrier formation; for example, small amounts of glucosylceramides and phospholipids persist in the lower SC (i.e., stratum compactum), but disappear from the outer SC (i.e., stratum disjunctum) (Elias and Menon, 1991). In addition, cholesterol sulfate content is maximal in the SG and stratum compactum, decreasing in the stratum disjunctum (Elias, 1984; Ramasinghe et al., 1986). Eventually, the three major classes of lipids in the SC are cholesterol, free fatty acids, and ceramides.

Studies have shown that epidermal lipid synthesis is both highly active and largely autonomous from systemic influences (reviewed in Feingold, 1991). The concept that lipid synthesis in the epidermis is relatively autonomous from circulating influences is supported by two further observations (reviewed in Feingold, 1991): a) epidermal lipid synthesis is not altered in animals with either very high or very low serum lipids, and b) only very small amounts of systemically administered lipids are incorporated into the epidermis, even when the barrier is disrupted. The synthesis of specific epidermal lipids, particular of the three key SC lipids—ceramides, free fatty acids, and cholesterol—is discussed next.

1. Cholesterol

In the early 1980s the laboratories of John Dietschy and Kenneth Feingold showed that mammalian skin is a highly active site of cholesterol synthesis. Subsequently, Feingold et al. (1983) demonstrated that abundant cholesterol is synthesized in the epidermis (accounting for about 30% of total cutaneous sterologenesis). Although most of this synthetic activity is localized to the basal layer, all of the nucleated epidermal cell layers retain the capacity to synthesize abundant cholesterol (reviewed in Feingold et al., 1983).

As noted earlier, epidermal cholesterol synthesis is relatively autonomous from changes in circulating lipids, reflecting its paucity of LDL receptors (Ponec et al., 1983), an important evolutionary adaptation to survival in a terrestrial environment. Although autonomous from the influence of circulating lipids, epidermal cholesterol synthesis nevertheless is regulatable by external influences, that is, changes in permeability barrier status (reviewed in Feingold, 1991). Acute barrier disruption provokes an increase in cholesterol synthesis, attributable both to a prior increase in mRNA for HMGCoA reductase and other key enzymes of cholesterol synthesis, and to changes in the phosphorylation (activation) state of HMGCoA reductase. The changes in mRNA for the key enzymes are, in turn, regulated by increased expression of sterol regulatory element binding protein-2 (SREBP-2), a transcription factor for both cholesterol and fatty acid synthesis (see also later discussion) (Harris, et al., 1998).

2. Free Fatty Acids (FFA)

The FFA in SC comprise both essential and nonessential species, with the latter comprising the bulk of the FFA in the intercellular lamellae (Lampe et al., 1983; Schurer and Elias, 1991). The essential FA, linoleic acid, comprises only a small proportion of the FA in the SC. Nonessential free fatty acids are synthesized in all epidermal cell layers and, as with cholesterol and ceramide synthesis, FFA synthesis is regulated by barrier requirements (reviewed in Feingold, 1991). Both of the key enzymes of FFA synthesis, FA synthase and acetyl CoA carboxylase (ACC), are upregulated. Moreover, as are the enzymes of cholesterol biosynthesis, their message expression appears to be transcriptionally regulated by SREBP-2 (Harris et al., 1998).

Keratinocytes also possess special mechanisms to take up and bind the essential FA linoleic acid and the prostaglandin precursor arachidonic acid from the blood (reviewed in Schurer and Elias, 1991). Whether one or more of these transport mechanism(s), in turn, is regulated by barrier requirements remains to be determined.

3. Ceramides

The ceramides in mammalian SC comprise a surprisingly well-conserved family of seven different species, derived from glucosylceramides, which are secreted from lamellar bodies into the SC interstices (reviewed in Elias and Menon, 1991). These molecules demonstrate minor species-to-species variations in sphingoid base structure, N-acyl chain length, and α/ω hydroxylation (reviewed in Schurer and Elias, 1991; Downing, 1992). Moreover, the ω-hydroxylated species contain an ω-esterified linoleate moiety (acylceramide) (Wertz, et al., 1984). Acylceramides have two putative functions: a) to form "molecular rivets," proposed to link adjacent, intercellular membrane structures in the SC (Downing, 1992), and b) to generate the deesterified ω-hydroxy ceramides, which are proposed to bind covalently to amino acids in the outer portion of the cornified envelope (Swartzendruber et al., 1987). The other major ceramide species comprise a mixture of ceramides with or without α-hydroxylated N-acyl groups, which interact with cholesterol and free fatty acids to form the bulk of the intercellular lamellae.

Like sterologenesis, ceramide synthesis also proceeds in all epidermal cell layers, peaking in the basal layer (Feingold, 1991). In contrast, glucosylceramide synthesis, catalyzed by the enzyme glucosylceramide (GC) synthase, peaks in the SG, coincident with the formation of lamellar bodies. As with cholesterol, ceramide synthesis is regulated by barrier requirements. Barrier-induced increases in ceramide synthesis are explicable by antecedent changes in both the activity and mRNA levels for serine palmitoyl transferase (SPT), the rate-limiting enzyme for ceramide synthesis (Harris et al., 1998). In contrast, GC synthase activity is not regulated by alterations in barrier function (Chujor et al., 1998). How SPT activity and ceramide synthesis are regulated remains unknown. Although SREBP-2 does not regulate SPT, it could regulate the availability of bulk FFA for sphingoid base and N-acyl group production.

B. Lamellar Body Secretion

As noted earlier, the unique two-compartment organization of the SC is attributable to the secretion of LB-derived lipids and colocalized hydrolases at the SG–SC interface (Elias and Menon, 1991). Under basal conditions, the rate of LB secretion appears to be slow, yet sufficient to provide for barrier integrity in the absence of stress. LB secretion requires active metabolism, since neither new organelle formation nor secretion occurs at 4°C. Moreover, calcium is an important regulator of LB secretion, with the epidermal calcium gradient restricting LB secretion to low, maintenance levels under basal conditions (Lee et al., 1992). Following acute barrier disruption, much of the preformed pool of LB in the outermost SG cell is quickly secreted (Menon et al., 1992a). LB contents are enriched in compressed, pleated sheets, which unfurl immediately after secretion (Elias and Menon, 1991; Menon et al., 1992a). Since occlusion blocks each of these steps, the lamellar body secretory response also is linked to barrier homeostasis. Pertinent to this review, the secretion inhibitors, monensin and brefeldin A block both LB formation and secretion after barrier disruption, leading to inhibition of barrier recovery (Man et al., 1995).

C. Extracellular Processing (ECP)

During the final stages of epidermal differentiation, a sequence of membrane transitions occurs within the SC extracellular domains. Extrusion of LB contents at the SG/SC interface is followed sequentially by unfurling, elongation, and processing into mature lamellar membrane unit structures (reviewed in Elias and Menon, 1991). As noted earlier, marked alterations in lipid composition occur, including depletion of glucosylceramides and phospholipids, with accumulation of ceramides and FFA in the SC. However, LB deliver not only glucosylceramides, phospholipids, and cholesterol, but also a family of hydrolytic enzymes to the SC interstices (Elias and Menon, 1991). Strong circumstantial evidence links these LB-derived hydrolases to extracellular processing (ECP) steps that are critical for barrier homeostasis. For example, the hydrolases that have been studied (glycosidase, phospholipase, steroid sulfatase, and neutral/acidic lipases) are localized specifically within SC membrane domains (reviewed in Elias and Menon, 1991).

1. GlucCer-to-Cer Metabolism by Beta-Glucocerebrosidase

The most compelling direct evidence for the central role of ECP in barrier formation comes from studies on glucosylceramide-to-ceramide processing. As with other lipid hydrolases, β-glucocerebrosidase (β-GlcCer'ase) is concentrated in the outer epidermis with the highest levels in SC (Holleran et al., 1992). In contrast, endogenous β-glycosidase activity is low in the SC of mucosal epithelia (Chang et al., 1991), which is about an order of magnitude more permeable. Moreover, glycosylceramides (types unspecified) predominate over ceramides; untransformed LB contents persist into the outer SC; and mature lamellar membrane structures do not form in mucosal epithelia. Recently, these correlative observations have been supported by direct evidence for the role of β-GlcCer'ase in the ECP of glucosylceramides-to-ceramides: a) Applications of specific conduritol-type inhibitors of β-GlcCer'ase both delay barrier recovery after acute perturbations and produce a progressive abnormality in barrier function when applied to intact skin (Holleran et al., 1993). b) In a transgenic murine model of Gaucher disease (GD), produced by targeted disruption of the β-GlcCer'ase gene, homozygous animals are born with an ichthyosiform dermatosis and a severe barrier abnormality (Holleran et al., 1994). c) In the severe type 2 neuron-opathic form of GD, infants present with a similar clinical phenotype, in-cluding an ichthyosiform erythroderma (Sidransky et al., 1996). In all three situations (inhibitor, transgenic murine, type 2 GD), the functional barrier deficit is accompanied by an accumulation of glucosylceramides and per-sistence of immature LB-derived membrane structures within the SC inter-stices. Together, these studies point to a critical role for glucosylceramide-to-ceramide processing in barrier homeostasis.

2. PL-to-FFA Catabolism by Phospholipases

Phospholipids, which are integral components of cell membranes, harbor within their structures bioactive moieties, such as arachidonic acid at the sn-2 position, whose generation is rate-limiting for the production of eicos-anoids and other lipid mediators. The fatty acids and lysophospholipids re-sulting from phospholipase A_2 (PLA_2) action can themselves act as second messengers or in membrane remodeling and phospholipid degradation. While the types of PLA_2 present in mammalian epidermis have not been extensively studied, recent studies have shown that both the $cPLA_2$ (85 kD cytosolic) and type 2 $sPLA_2$ (14 kD secretory) are constitutively expressed in epidermis under basal conditions (see references in Mao-Qiang et al., 1995, 1996). Since we have shown that bromphenacyl bromide (BPB) and MJ-33 (type 1 $sPLA_2$ inhibitors), but no MJ-45 (a type 2 $sPLA_2$ inhibitor), modulate barrier function, this suggests a role for the type 1 rather than the type 2 isoenzyme in barrier function (Mao-Qiang et al., 1995, 1996). That epidermal barrier function appears to require an $sPLA_2$ with characteristics of the type 1 form fits well with its putative role in barrier homeostasis.

3. Cholesterol Sulfate-to-Cholesterol Catabolism by Steroid Sulfatase

Just as with glucosylceramides, cholesterol sulfate content increases with epidermal differentiation, and then decreases quantitatively during passage from the inner to the outer SC (Elias et al., 1984). Both cholesterol sulfate and steroid sulfatase are concentrated in SC membrane domains, but not in LB (Elias et al., 1984). That cholesterol sulfate plays a critical role in des-quamation is demonstrated definitively by the accumulation of cholesterol sulfate in SC membranes in recessive X-linked ichthyosis (RXLI), and by the ability of topical cholesterol sulfate to induce hyperkeratosis (reviewed in Williams, 1991). Moreover, recent studies have demonstrated a barrier abnormality in RXLI (Zettersten et al., 1998).

Activity of steroid sulfatase, a close relative of arylsulfatase C, is ab-sent in RXLI (reviewed in Williams, 1991). This microsomal enzyme, which is expressed in high levels in the outer epidermis of mammals (Elias et al., 1984), hydrolyzes sulfate from the 3-β-hydroxy group on a variety of sterols and steroids. Whether cholesterol sulfate, in addition to its apparent roles in desquamation and signaling, also is an important participant in barrier homeostasis is less clear. An abundant and possibly sufficient pool of cholesterol is available from LB secretion directly without further modifications. However, in RXLI, cholesterol sulfate content increases to 10–12%, while cholesterol content decreases by 50% (Elias et al., 1984), leaving open the possibility that cholesterol sulfate could provide an important precursor pool for cholesterol in the SC. Recent studies have shown, however, that it is accumulation of cholesterol sulfate, rather than a paucity of cholesterol, that leads to the barrier abnormality in RXLI (Zettersten et al., 1998). Finally, cholesterol sulfate also could function in the barrier by additional or unrelated mechanisms, for example, in its capacity as a protease inhibitor (e.g., cholesterol sulfate regulates the acrosome reaction, a proteolytically triggered process, in spermatozoa).

4. pH and Ions

That the SC displays an acidic external pH ("acid mantle") is well documented, and such acid conditions are considered crucial for the resistance to microbial invasion. However, the origin of the acid mantle is not known— passive mechanisms, including net catabolic processes, sebaceous gland-derived FA, microbial metabolism, and apical proton flux due to the high electrical resistance of the SC have been proposed. Alternatively, protons could also be generated actively, such as by ion pumps inserted into the plasma membrane. If the limiting membrane of the LB contained such pumps, as suggested by recent inhibitor studies (Chapman and Walsh, 1989), then acidification of the extracellular space (ECS) could begin with insertion of such pumps coincident with LB secretion. Ongoing proton secretion at the SG–SC interface, perhaps coupled with one or more of the passive mechanisms described earlier, could explain the pH gradient across the SC interstices. The concept that acidification is required for sequential ECP is supported by the observation that barrier recovery is delayed when acutely perturbed skin is immersed in neutral pH buffers (Mauro et al., 1998). Thus, in addition to its potential antimicrobial function, the acid mantle may be important for permeability barrier homeostasis.

By a variety of independent methods, we and others have demonstrated a Ca^{2+} gradient in the epidermis, with the highest levels in the intercellular spaces and cytosol of the SG, followed by a rapid decline in Ca^{2+} from the SC (Menon et al., 1985). Although the epidermis displays a similar K^+ gradient, different gradients exist for other ions, including Na^+ and Cl^- (Mao-Qiang et al., 1997). Both acute and chronic barrier disruption dramatically alter the Ca^{2+} gradient, displacing Ca^{2+} outward through the SC interstices (Mao-Qiang et al., 1997; Menon et al., 1994). Finally, and pertinent to this chapter, exposure to high Ca^{2+} (and K^+) delays barrier recovery following acute perturbations, in a manner that is reversible by L-type Ca^{2+} channel inhibitor (Lee et al., 1992).

IV. BASIS FOR INTERFERING WITH EPIDERMAL LIPID BIOGENESIS TO ENHANCE TRANSDERMAL DRUG DELIVERY

The concept of a biochemical approach for enhancing cutaneous permeability derives from pharmacologic studies aimed at inhibiting the key enzymes of epidermal lipid synthesis, LB secretion, and ECP, thereby altering the critical molar ratio of the three key SC lipids. (See Table 1.) The first pharmacological study supporting this particular concept came from experiments in adult hairless mice where topical HMGCoA reductase inhibitors, such as lovastatin and fluvastatin, caused both a delay in barrier recovery (Feingold et al., 1990) and a barrier defect following repeated applications to intact skin (Feingold et al., 1991; Menon et al., 1992b). Since the ability of the inhibitors to alter barrier homeostasis could be reversed by coapplications of either mevalonate (the immediate product of HMGCoA reductase) or cholesterol (a distal product), the inhibitor effect could not be ascribed to nonspecific toxicity. Likewise, application of specific pharmacologic inhibitors of ACC (Mao-Qiang et al., 1993), SPT (Holleran et al., 1991), and GC synthase (Chujor et al., 1998), key enzymes of FA and ceramide synthesis, also provoked a delay in barrier recovery (Mao-Qiang et al., 1993b), as measured by TEWL. Yet, while the HMGCoA reductase and ACC inhibitors are additive in their capacity to alter barrier recovery, coapplications of HMGCoA reductase and SPT inhibitors are not additive, but instead paradoxically normalize the kinetics of barrier recovery (results of these studies are summarized in Table 2) (Mao-Qiang et al., 1993a). Although these studies were designed to determine which of the three key lipids is (are) required for permeability barrier homeostasis, these inhibitors also cause the "window to remain open longer," and/or they "open the window de novo." All of the pharmacological "knockout" studies support the concept that interference with the biosynthesis of each of the key SC lipids leads to a temporary increase in TEWL, which also could impact on the transdermal delivery of drug molecules.

The effects of these inhibitors on skin barrier function are explained in part by ultrastructural studies, which demonstrate that each of the single-enzyme, pharmacologic knockouts produces abnormalities of both LB contents and SC extracellular lamellae. In addition to abnormal extracellular

Table 1 Mechanistic Classification of Various
Biochemical Enhancers

Lipid synthesis inhibitors
   HMGCoA reductase
   Acetyl CoA carboxylase
   Serine palmitoyl transferase
   Glucosylceramide synthase
Lipid secretion inhibitors
   Organellogenesis
   Golgi processing
Extracellular acidification inhibitors
   Protonophores
   Proton pumps
Extracellular processing
   β-Glucocerebrosidase
   Secretory phospholipase A_2
   Acid sphingomyelinase
Others
   Lipids analogues[a]
   Other lipid hydrolases, e.g., acid lipase
   Complex lipids[b]
   Ion Pump inhibitiors, e.g., bafilamycin

---

[a]Nonphysiological, e.g., epicholesterol, *trans*-vaccenic acid.
[b]Some (e.g., cholesterol esters) but not all are effective.

Table 2 Targets and Examples of Lipid Synthesis and Processing Inhibitors That Modulate Barrier Homeostasis

Required Lipid	Enzyme Targets	Example
Cholesterol	HMGCoA reductase	Lovastatin, fluvastatin
Free fatty acids	Acetyl CoA carboxylase	TOFA
	Secretory phospholipase A_2	MJ33, BPB
Ceramides	Serine palmitoyl transferase	β-Chloralanine
		Morpholino agents (e.g., P4)
	Glucosylceramide synthase	Conduritols
	β-Glucocerebrosidase	

lamellae, these agents often produce the formation of separate lamellar and nonlamellar domains within the SC interstices. The basis for such domain separation presumably relates to changes in the critical mole ratio; that is, with deletion or excess of any one of the three key lipids, a portion of the excess species no longer can remain in a well-organized lamellar phase. For example, a 50% reduction in cholesterol would result in an excess of both ceramides and free fatty acids, with a portion of this excess forming a nonlamellar phase. The result of phase separation is SC interstices that are more permeable, due not only to deletion of a key hydrophobic lipid, but also to the creation of additional penetration pathways, distinct from the primary lamellar route.

Although vehicles, hydration, and conventional chemical enhancers also can produce domain separation, they do not significantly modify the structure or spacing of the extracellular lamellae themselves. Instead, the primary effect of standard chemical enhancers, vehicles, and physical methods appears to be an expansion of the fluid–lipid domains in the SC interstices (reviewed in Menon and Elias, 1997). In contrast, strategies that interfere with the synthesis, delivery, activation, or assembly/disassembly of the extracellular lamellar membrane interfere with permeability barrier homeostasis.

These biochemical approaches also can be viewed vectorially, that is, as operative within different layers of the epidermis. For example, most lipid synthesis occurs within the basal layer, while lamellar body formation, acidification and secretion occur in suprabasal, nucleated layers. Finally, ECP and membrane assembly occur within the SC interstices. Therefore, in theory, strategies could be put in place to take advantage of the localization and relative importance of the steps leading to the ultimate generation of SC extracellular lamellae in order to potentially enhance transdermal drug delivery.

V. ASSESSMENT OF EFFICACY OF BIOCHEMICAL APPROACH

To date the effectiveness of these biochemical approaches for transdermal drug delivery has been assessed mainly in adult hairless mouse epidermis. In our initial studies, caffeine and lidocaine were used as model permeants to assess whether their penetration characteristics parallel changes in TEWL measurements. The biochemical approaches here consisted of the topical application of either drug plus a cholesterol and/or fatty acid lipid synthesis inhibitor in two different, conventional enhancer/vehicle systems, dimethyl sulfoxide or propylene glycol:ethanol (7:3 vol), followed by assessment of both TEWL and drug delivery. Results from these studies showed that these biochemical enhancers enhanced lidocaine and caffeine delivery across intact skin beyond that found with the two conventional enhancer/vehicle systems alone (Tsai et al., 1996); that is, the net delivery of both drugs increased severalfold over either drug in the conventional enhancer system. Moreover, changes in TEWL correlated linearly with transdermal delivery of both drugs. This study is the first to show that biochemical enhancers can increase transdermal drug delivery in a widely used animal model. Additional work is needed to explore whether TEWL serves as an universal, accurate, and reproducible predictor for transdermal delivery of drugs, with a broad range of physical–chemical properties. Nevertheless, these preliminary studies suggest that great potential exists for the biochemical enhancer approach to increase transdermal drug delivery in humans.

REFERENCES

Chang, F., Wertz, P. W., and Squier, S. M. (1991). Comparison of glycosidase activities in epidermis, palatal epithelium, and buccal epithelium. *Comp. Biochem. Biophys. B* 150:137–139.

Chapman, S. J., and Walsh, A. (1989). Membrane-coating granules are acidic organelles which possess proton pumps. *J. Invest. Dermatol.* 93:466–470.

Chujor, C. S. N., Feingold, K. R., Elias, P. M., and Holleran, W. M. (1998). Glucosylceramide synthase activity in murine epidermis: Quantitation, localization, regulation, and requirement for barrier homeostasis. *J. Lipid Res.* 39: 277–285.

Downing, D. T. (1992). Lipid and protein structures in the permeability barrier of mammalian epidermis. *J. Lipid Res.* 33:301–313.

Elias, P. M. (1983). Epidermal lipids, barrier function, and desquamation. *J. Invest. Dermatol.* 80:44s–49s.

Elias, P. M., and Menon, G. K. (1991). Structural and lipid biochemical correlates of the epidermal permeability barrier. *Adv. Lipid Res.* 24:1–26.

Elias, P. M., Williams, M. L., Maloney, M. E., Bonifas, J. A., Brown, B. E., Grayson, S., and Epstein, E. H., Jr. (1984). Stratum corneum lipids in disorders of cornification: Steroid sulfatase and cholesterol sulfate in normal desquamation and the pathogenesis of recessive X-linked ichthyosis. *J. Clin. Invest.* 74: 1414–1421.

Feingold, K. R. (1991). The regulation and role of epidermal lipid synthesis. *Adv. Lipid Res.* 24:57–82.

Feingold, K. R., Brown, B. E., Lear, S. R., Moser, A. H., and Elias, P. M. (1983). Localization of *de novo* sterologenesis in mammalian skin. *J. Invest. Dermatol.* 81:365–369.

Feingold, K. R., Mao-Qiang, M., Menon, G. K., Cho, S. S., Brown, B. E., and Elias, P. M. (1990). Cholesterol synthesis is required for cutaneous barrier function in mice. *J. Clin. Invest.* 86:1738–1745.

Feingold, K. R., Mao-Qiang, M., Proksch, E., Menon, G. K., Brown, B., and Elias, P. M. (1991). The lovastatin-treated rodent: A new model of barrier disruption and epidermal hyperplasia. *J. Invest. Dermatol.* 96:201–209.

Harris, I. R., Farrell, A. M., Holleran, W. M., Jackson, S., Grunfeld, C., Elias, P. M., and Feingold, K. R. (1998). Parallel regulation of sterol regulatory element binding protein-2 and the enzymes of cholesterol and fatty acid synthesis but not ceramide synthesis in cultured human keratinocytes and murine epidermis. *J. Lipid Res.* 39:412–422.

Holleran, W. M., Mao-Qiang, M., Gao, W. N., Menon, G. K., Elias, P. M., and Feingold, K. R. (1991). Sphingolipids are required for mammalian barrier function: Inhibition of sphingolipid synthesis delays barrier recovery after acute perturbation. *J. Clin. Invest.* 88:1338–1345.

Holleran, W. M., Takagi, Y., Imokawa, G., Jackson, S., Lee, J. M., and Elias, P. M. (1992). β-Glucocerebrosidase activity in murine epidermis: Characterization and localization in relationship to differentiation. *J. Lipid. Res.* 33:1201–1209.

Holleran, W. M., Takagi, Y., Feingold, K. R., Menon, G. K., Legler, G., and Elias, P. M. (1993). Processing of epidermal glucosylceramides is required for optimal mammalian permeability barrier function. *J. Clin. Invest.* 91:1656–1664.

Holleran, W. M., Sidransky, E., Menon, G. K., Fartasch, M., Grundmann, J.-U., Ginns, E. I., and Elias, P. M. (1994). Consequences of β-glucocerebrosidase deficiency in epidermis: Ultrastructure and permeability barrier alterations in Gaucher disease. *J. Clin. Invest.* 93:1756–1764.

Lampe, M. A., Burlingame, A. L., Whitney, J., Williams, M. L., Brown, B. E., Roitman, E., and Elias, P. M. (1983). Human stratum corneum lipids: Characterization and regional variations. *J. Lipid Res.* 24:120–130.

Lee, S. H., Elias, P. M., Proksch, E., Menon, G. K., Mao-Qiang, M., and Feingold, K. R. (1992). Calcium and potassium are important regulators of barrier homeostasis in murine epidermis. *J. Clin. Invest.* 89:530–538.

Man, M.-Q., Brown, B. E., Wu-Pong, S., Feingold, K. R., and Elias, P. M. (1995). Exogenous nonphysiologic vs. physiologic lipids: Divergent mechanisms for correction of permeability barrier dysfunction. *Arch. Dermatol.* 131:809–816.

Mao-Qiang, M., Feingold, K. R., and Elias, P. M. (1993a). Inhibition of cholesterol and sphingolipid synthesis causes paradoxical effects on permeability barrier homeostasis. *J. Invest. Dermatol.* 101:185–190.

Mao-Qiang, M., Elias, P. M., and Feingold, K. R. (1993b). Fatty acids are required for epidermal permeability barrier function. *J. Clin. Invest.* 92:791–798.

Mao-Qiang, M., Feingold, K. R., Jain, M., and Elias, P. M. (1995). Extracellular processing of phospholipids is required for permeability barrier homeostasis. *J. Lipid Res.* 36:1925–1935.

Mao-Qiang, M., Jain, M., Feingold, K. R., and Elias, P. M. (1996). Secretory phospholipase A_2 activity is required for permeability barrier homeostasis. *J. Invest. Dermatol.* 106:57–63.

Mao-Qiang, M., Mauro, T., Bench, G., Warren, R., Elias, P. M., and Feingold, K. R. (1997). Calcium and potassium inhibit barrier recovery after disruption, independent of the type of insult in hairless mice. *Exp. Dermatol.* 6:36–40.

Mauro, T., Grayson, S., Gao, W. N., Mao-Qiang, M., Kriehuber, E., Behne, M., Feingold, K. R., and Elias, P. M. (1998). Barrier recovery is impeded at neutral pH independent of ionic effects: Implications for extracellular lipid processing. *Arch. Derm. Res.* 290:215–222.

Menon, G. K., and Elias, P. M. (1997). Morphologic basis for a pore-pathway in mammalian stratum corneum. *Skin Pharmacol.* 10:235–246.

Menon, G. K., Grayson, S., and Elias, P. M. (1985). Ionic calcium reservoirs in mammalian epidermis: Ultrastructural, localization by ion-capture cytochemistry. *J. Invest. Dermatol.* 84:508–512.

Menon, G. K., Feingold, K. R., and Elias, P. M. (1992a). The lamellar body secretory response to barrier disruption. *J. Invest. Dermatol.* 98:279–289.

Menon, G. K., Feingold, K. R., Man, M.-Q., Schaude, M., and Elias, P. M. (1992b). Structural basis for the barrier abnormality following inhibition of HMG CoA reductase in murine epidermis. *J. Invest. Dermatol.* 98:209–219.

Menon, G. K., Elias, P. M., and Feingold, K. R. (1994). Integrity of the permeability barrier is crucial for maintenance of the epidermal calcium gradient. *Br. J. Dermatol.* 130:139–147.

Nemanic, M. K., and Elias, P. M. (1980). In situ precipitation: A novel cytochemical technique for visualization of permeability pathways in mammalian stratum corneum. *J. Histochem. Cytochem.* 28:573–578.

Ponec, M., Hawkes, L., Kampanaar, J., and Vermeer, B. J. (1983). Cultured human skin fibroblasts and keratinocytes: Differences in the regulation of cholesterol synthesis. *J. Invest. Dermatol.* 81:125–130.

Ramasinghe, A. W., Wertz, P. W., Downing, D. T., and MacKenzie, I. C. (1986). Lipid composition of cohesive and desquamated corneocytes from mouse ears. *J. Invest. Dermatol.* 86:187–190.

Scheuplein, R. J., and Blank, I. H. (1971). Permeability of skin. *Physiol. Rev.* 51:702–747.

Schurer, N. Y., and Elias, P. M. (1991). The biochemistry and function of stratum corneum lipids. *Adv. Lipid Res.* 24:27–56.

Sidransky, E., Fartasch, M., Lee, R. E., Metlay, L. A., Abella, S., Zimran, A., Gao, W., Elias, P. M., Ginns, E. I., and Holleran, W. M. (1996). Epidermal abnormalities may distinguish Type 2 from Type 1 and Type 3 of Gaucher disease. *Pediatr. Res.* 39:134–141.

Swartzendruber, D. C., Wertz, P. W., Kitko, D. J., Madison, K. C., and Downing, D. T. (1987). Evidence that the corneocyte has a chemically bound lipid envelope. *J. Invest. Dermatol.* 88:709–713.

Tsai, J.-C., Guy, R. H., Thornfeldt, C. R., Feingold, K. R., and Elias, P. M. (1996). Metabolic approaches to enhance transdermal drug delivery. I. Effect of lipid synthesis inhibitors. *J. Pharm. Sci.* 85:643–648.

Wertz, P. W., Downing, D. T., Feinkel, R. K., and Traczyk, T. N. (1984). Sphingolipids of the stratum corneum and lamellar granules of fetal rat epidermis. *J. Invest. Dermatol.* 83:193–195.

Williams, M. L. (1991). Lipids in normal and pathological desquamation. *Adv. Lipid Res.* 24:211–262.

Yardley, H. J., and Sumerly, R. (1981). Lipid composition in normal and diseased epidermis. *Pharmacol. Ther.* 13:357–383.

Zettersten, E., Mao-Qiang, M., Sato, J., Farrell, A., Ghadially, R., Williams, M. L., Feingold, K. R., and Elias, P. M. (1998). Recessive X-linked ichthyosis: Role of cholesterol-sulfate accumulation in the barrier abnormality. *J. Invest. Dermatol.*, in press.

25
Percutaneous Drug Delivery to the Hair Follicle

Andrea C. Lauer
Boehringer Ingelheim–Roxane Laboratories, Columbus, Ohio

I. INTRODUCTION

Unique biochemical and immunological events dictate the complex cyclic growth and differentiation patterns of hair follicles and their associated sebaceous glands (1), structures that are collectively referred to as pilosebaceous units. Once regarded as mere evolutionary remnants, hair follicles and sebaceous glands have been recognized increasingly as significant pathways for percutaneous transport (2). Percutaneous transport routes via the lipoidal domains of the stratum corneum have been well established (3,4), whereas comparatively less is known about the specific roles of hair follicles and sebaceous glands. Determination of the roles of these structures is complicated by the lack of adequate animal models and methodologies that can distinctly distinguish follicular and stratum corneum pathways. Moreover, it may be possible that a topically applied compound traverses more than one pathway simultaneously.

The stratum corneum is acknowledged not only as the main barrier to skin penetration, but also as the major permeation pathway. The tightly packed, semicrystalline intercellular lipid domains and the extremely compact corneocytes of the stratum corneum create a barrier highly resistant to percutaneous transport (3,4). Modulation of stratum corneum lipid fluidity by topical agents has been thoroughly studied and is generally acknowledged as the major mechanism of percutaneous delivery (5). Passive percutaneous transport depends on several factors, including penetrant lipophilicity, charge, and molecular size (6). The upper limit of molecular size for permeation through the stratum corneum is still unknown.

Early suggestions of a follicular pathway were based on the hypothesis that hair follicles act as shunts, resulting in the rapid transport of ions and large polar molecules. Scheuplein (7,8) first described transient follicular delivery for small polar molecules and large polar steroids that ordinarily would not be expected to traverse the skin rapidly due to their charge or restrictive molecular size. Feldmann and Maibach (9,10) observed increases in percutaneous transport through skin areas with greatest follicular densities in both animals and humans, which also hinted at the possibility of follicular delivery. Later studies incorporated fluorescence microscopy, autoradiography, radiolabel deposition, confocal laser scanning microscopy, and other techniques, yielding data that have further supported this route for a wide range of drug molecules and vehicles (2). Specific particulate systems including liposomes (11–19) and synthetic microspheres (20,21) have been found to localize in follicular and sebaceous areas that act as drug reservoirs for compounds with varying physicochemical characteristics.

In several studies, hair follicles have been shown to act as channels, though not necessarily depots, for the iontophoretic flux of several molecules. Iontophoresis may be particularly useful in systemic delivery of ionic, polar compounds and high-molecular-weight peptides, which are slowly transported passively (22,23). Although the importance of follicular routes for iontophoretic flux should not be disregarded, a full discussion of iontophoresis is beyond the scope of this chapter, which is confined to passive follicular delivery.

II. CONSIDERATIONS FOR EXPERIMENTAL DESIGN

Careful consideration of several factors is necessary in designing an experimental approach to assess follicular permeation:

> Hair follicle and sebaceous gland anatomy (i.e., drug targets).
> Drug delivery goal: localized or systemic?
> Animal models.
> Quantitative and qualitative analysis methods.
> Physicochemical properties of permeants and vehicles.

III. HAIR FOLLICLE AND SEBACEOUS GLAND ANATOMY

The structure of the stratum corneum and its function in percutaneous transport are closely intertwined (3–5). Analogously, an understanding of the anatomy and physiology of the hair follicles and sebaceous glands is equally important, especially with regard to targeting specific therapeutic sites within the hair follicle. A simplified diagram of a vertically cross-sectioned mammalian pilosebaceous unit is illustrated in Fig. 1. The hair follicle is an invagination of the dermis and is continuous with the epidermis, which offers less resistance to transport than the stratum corneum. Dermal matrix cells overlying the follicular papilla give rise to the hair shaft and the inner root sheath, whereas the outer root sheath, which is continuous with the epidermis, is likely of epithelial origin (24,25). It is not unlikely that the tightly compacted cells of the inner root sheath beneath the hair follicle infundibulum (opening) may physically restrict transport deep within the hair follicle.

The pilosebaceous unit is a dynamic structure that undergoes cyclic growth phases: anagen (active), catagen (resorption), and telogen (resting) (24). Cotsarelis and coworkers (26) have described a mid-follicle "bulge area" that exhibits one of the fastest rates of cell division in mammals. It is thought that diffusable growth factors from the follicular papilla cause proliferation of the normally slow-cycling bulge cells in early anagen. However, during telogen, the bulge area is most responsive to topically applied carcinogens (26), suggesting that follicular delivery may vary depending on hair cycle stage. Hair cycle activity in humans occurs in a mosaic pattern, with a random distribution of neighboring hair follicles in varying growth

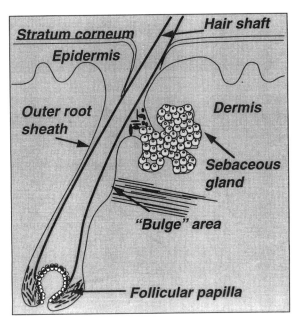

Figure 1 Simplified diagram of a mammalian hair follicle and sebaceous gland.

stages, whereas hair cycle activity in most animals is in a wave pattern of synchronized adjacent hair follicles.

Although located deep in the dermis, sebaceous glands are appendages of the epidermis that secrete sebum into the follicular canal, leading to the skin surface (24). Sebum flows into the follicular canal as a result of lysed sebocytes within sebaceous glands, creating a lipophilic environment that may favor transport for some molecules, but may also act as a chemical and physical barrier for others. Analysis of skin surface lipids of several species has shown tremendous variation in sebum composition among species. Human sebum is rich in squalene, wax esters, triglycerides, cholesterol, cholesterol esters, and free fatty acids (27). Androgenic hormones influence sebaceous gland size by stimulating rate of cell division and lipid accumulation (24,27). Targeting the sebaceous glands with agents to decrease androgen activity could be useful in the treatment of dermatologic disorders such as acne.

Several epithelial cell types, specialized structures, receptors, and immunocompetent cells reside within the pilosebaceous units. Hormones, growth factors, ultraviolet radiation, and drugs interact at various levels of the hair follicle (24). Greater understanding of the molecular signals that control the onset and duration of hair follicle growth and development, which still are not fully understood, may enable rational design of targeted follicular delivery systems. Obvious therapeutic targets exist within the pilosebaceous unit, which may offer promising treatments for acne, male-pattern baldness, alopecia areata, and some skin cancers (24). Besides localized delivery, systemic delivery via the hair follicle may also be achieved due to the large capillary networks associated with the pilosebaceous unit (28).

IV. ANIMAL MODELS

A. Hairless Animals

As stated previously, the choice of animal models for assessing follicular delivery is currently limited. A hairless rodent model may seem an obvious, convenient choice to represent the nonfollicular pathway; however, these animals only appear macroscopically hairless and do indeed possess abnormal hair follicles (19,29). In Fig. 2, the gross histological differences between CRL CD hairless rat skin and Sprague-Dawley hairy rat skin are shown by light-microscopy viewing of hematoxylin- and eosin-stained 5-μm vertical cross-sections of untreated dorsal skin (19). The stratum corneum of hairless rodent skin is typically hyperkeratinized, the hair follicles and sebaceous glands are enlarged and cysts frequently are present in the epidermis, dermis, and hair follicles (19,29).

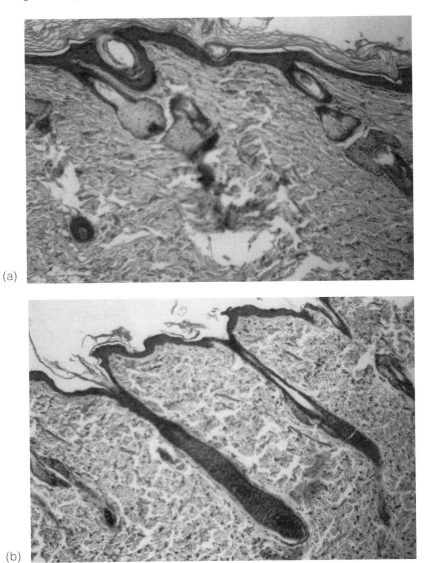

Figure 2 Light microscopy photographs (70×) showing hematoxylin- and eosin-stained 5μm vertical skin sections of (a) untreated hairless rat skin and (b) untreated hairy rat skin.

Despite the known histological differences, hairless rodents continue to be widely used as models for percutaneous penetration of a wide variety of molecules. In some studies, the hairless rat has been more accurately depicted as a follicular model, and has been compared to a follicle-free model developed by scarring hairless rat skin (20,21,30–34). Illel and Schaefer (30) induced follicle-free skin in hairless rats by immersing anesthetized rats into 60°C water for 1 min, followed by removal of epidermis and healing for 3 months. At this point, transepidermal water loss evaluations indicated normal barrier function, and histological analysis showed a complete absence of hair follicles and sebaceous glands. Preliminary in vitro studies by Illel and Schaefer (30) found that the steady-state flux and total diffusion of [^{3}H]hydrocortisone were 50-fold greater for hairless rat skin compared to follicle-free hairless rat skin after 24 h. Several additional studies followed comparing the percutaneous transport profiles of intact hairless rat skin and the scarred, follicle-free hairless rat skin (20,21,31–34). In vitro studies showed that permeation of a wide range of compounds—tritiated caffeine, niflumic acid and p-aminobenzoic acid—was approximately three times greater through hairless rat skin compared to the follicle-free model (31). Hueber and coworkers (32) confirmed these findings with in vivo studies using radiolabled hydrocortisone, progesterone, estradiol, and progesterone. The overall conclusion of these studies using intact hairless rat skin and scarred hairless rat skin was that follicular transport, especially via sebaceous glands, was significant for a wide range of topically applied compounds. Although this induced follicle-free model has offered some insight into follicular delivery, it is somewhat imperfect due to an incomplete assessment of the overall effects of heat-induced scarring on the hairless rat stratum corneum.

In addition, it has been shown that hairless rat skin does not exhibit typically normal absorption patterns, for both polar and nonpolar compounds, when compared to animals with fully developed hair follicles. In vivo permeability profiles of hairless rat (CRL CD) and hairy rat (Sprague-Dawley) skin have been investigated for up to 12 h using small polar and lipophilic molecules in various vehicles (19). Hairy rat skin was closely clipped with an electric clipper at least 12 h prior to the study in order to allow close skin contact with applied formulation. Aqueous [^{14}C]mannitol solution was applied for specified time periods to the dorsal surface of a 4-cm^2 skin area, followed by euthanization and preparation for radiolabel assay. Excised skin was tape-stripped to remove stratum corneum completely, followed by assay of tape strips and remaining epidermis/dermis (viable skin) for radiolabel by liquid scintillation counting. As shown in Fig. 3, in vivo topical application of aqueous [^{14}C]mannitol solution to hairless rat skin resulted in significant systemic accumulation in the urine, whereas systemic

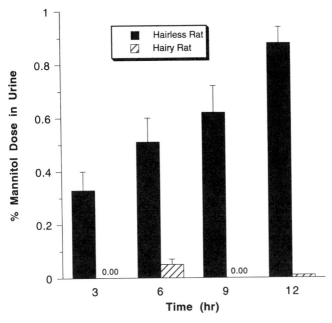

Figure 3 Urinary deposition of mannitol. In vivo deposition of ^{14}C-mannitol after topical application of aqueous solution to hairless and hairy rat skin ($n = 3-6$, mean $\pm$ SEM).

levels were negligible for hairy rat ($p < .05$). Mannitol deposition into both hairless and hairy rat viable skin (Fig. 4), which is defined as the epidermal and dermal layers remaining following complete tape stripping of the stratum corneum, was <0.2% of the applied dose at all time points. The results suggest that hairless rat skin acted as a sieve to aqueous mannitol solution, whereas hairy rat skin provided a sufficient barrier to this extremely polar molecule that is not expected to permeate the stratum corneum easily. In a similar study, a hydroalcoholic solution containing [^{3}H]progesterone was topically applied to hairless and hairy rat skin in vivo. Figure 5 shows that after 12 h, progesterone deposition into hairless rat skin was approximately 6-fold greater than that in hairy rat skin. Systemic levels of progesterone in the hairless rat were approximately twice the levels found in hairy rat urine. The data suggest that lipophilic compounds may be retained in the extensive depots of hairless rat skin provided by the numerous large sebaceous glands present in this species.

Hisoire and Bucks observed similar permeability differences for hairless and hairy guinea pig skin following in vitro topical application of

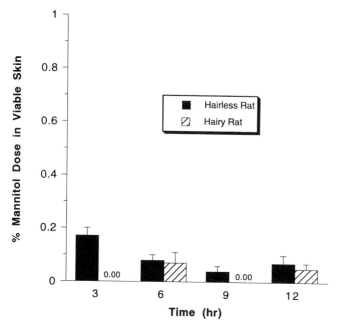

Figure 4 Viable skin deposition of mannitol. In vivo deposition of [^{14}C]mannitol after topical application of aqueous solution to hairless and hairy rat skin ($n = 3-6$, mean $\pm$ SEM).

0.025% retinoic acid in an ethanolic gel formulation (35). Histological analysis also indicated ultrastructural abnormalities in hairless guinea pig skin, including cysts and follicular aberrations. Although the authors concluded that the results were a consequence of stratum corneum differences between the two models, the possible influence of abnormal hair follicles and sebaceous glands cannot be excluded. Collectively, the results of these studies indicate that hairless animals are probably inappropriate models for topical absorption due to the presence of abnormal hair follicles, cysts, and other aberrations that result in a leaky barrier for polar compounds while acting as a lipophilic depot for nonpolar compounds (19,35).

The newborn rat also represents a follicle-free skin model until the age of 3–4 days, which is when hair follicles begin to develop. Illel and co-workers (31) found a 5-fold greater flux of [^{3}H]hydrocortisone through newborn rat skin 5 days after birth compared to 24 h after birth. Although the newborn rat lacks hair follicles, the possibility of barrier impairment due to underdeveloped stratum corneum cannot be eliminated since this skin has yet to be fully characterized.

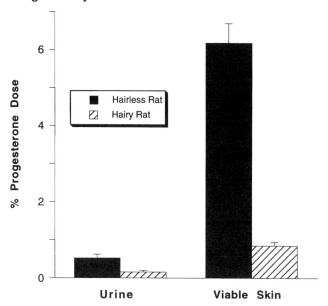

Figure 5 In vivo deposition of [³H]progesterone from hydroalcoholic solution (60% v/v) into hairless and hairy rat skin after 12 h of topical application (*n* = 3, mean ± SEM).

B. Follicle-Free Area of Guinea Pig Skin

The skin behind the ears of guinea pigs is completely devoid of hair follicles and sebaceous glands, which may provide a potential model that excludes follicular routes. However, investigations are limited by the extremely small surface area at this site and the possibility of site-specific lipid composition varying from other sites. Wahlberg (36) used surface radiation disappearance measurements to quantify in vitro and in vivo percutaneous absorption using aqueous and organic solutions of $HgCl_2$ and NaCl applied to hairy and hair follicle-free guinea pig skin. Although no clear differences were found in this study, it should be noted that systemic absorption was not assessed.

C. Syrian Hamster Ear

The ventral side of the Syrian hamster ear, which is rich in large sebaceous glands, was developed by Plewig and Luderschmidt (37) as a model for sebaceous gland deposition based on its structural similarity to human sebaceous glands. Matias (38) developed a method in which formulation is applied to the ventral surface of the ear and after a specified time point, the animal is euthanized and the ear is stratified into anatomical layers to allow for scraping of sebaceous contents. Sebaceous contents can then be quantitatively analyzed by several methods, depending on the applied permeant. A wide range of molecules in various vehicles, many of which were liposomal preparations, have been shown to be absorbed in appreciable amounts into hamster ear sebaceous glands (11,14,17).

D. Macaque Monkey

The perfect animal model for clearly defining contributions of the stratum corneum and follicular routes remains elusive, but at present, the study of an animal model possessing true, fully developed hair follicles may be the best option. The macaque monkey may be one of the most physiologically correct models, especially for studying follicular targeting to prevent androgenic baldness. This animal exhibits a species-specific frontal scalp baldness that coincides with puberty, and may be a particularly relevant model for the study of human androgenic alopecia since the hormonal and genetic factors that induce baldness closely parallel those causing baldness in humans. Uno and Kurata (39) found that topical application of hypertrichotic drugs, minoxidil and diazoxide, resulted in significant follicular enlargement and hair regrowth in bald macaques. In another study, topical application of an antiandrogen that inhibits 5α-reductase resulted in prevention of baldness in preadolescent macaques (40).

E. In Vitro Histocultures with Hair Follicles

Histocultured skin containing fully developed hair follicles may be a promising alternative to in vivo animal models. Li and coworkers (41) developed an in vitro three-dimensional histoculture using a collagen-sponge-gel support system to maintain tissue viability of mouse skin. Early anagen hair follicles in this model developed into full-grown anagen follicles and produced pigmented hair shafts in vitro. Similar mouse and human histocultures with well-developed hair follicles were used in vitro to assess follicular delivery of liposomal formulations containing calcein (12), melanin (13), or high-molecular-weight DNA (18). In limited cases, human skin can be assessed in vivo if follicular biopsy methods are used (42). Although in vitro methods have been improved, the potential effects of excessive hydration on excised skin and the lack of functioning circulatory networks surrounding the pilosebaceous units should still be considered.

V. METHODOLOGIES TO ASSESS FOLLICULAR PERMEATION

In the absence of a completely appropriate animal model, recently improved methodologies for quantifying and visualizing topical permeation routes may offer the best hope in more clearly defining these pathways. Following top-ical application of an agent for a specified time, the skin must be carefully excised and processed in at least one of several ways in order to assess drug deposition. If a thin cross-sectioning technique for microscopic visualization is chosen, care should be taken to avoid harsh fixatives and cross-contam-ination during sectioning.

A. Radiolabel Deposition

A frequently used method to assess topical drug deposition quantitatively, especially if a radiolabeled compound is available, is the tape-stripping method (43). At the end of the application time, excised skin is cellophane tape-stripped until stratum corneum removal is complete, as indicated by the glistening appearance of the residual viable skin (remaining epidermis and dermis) and lack of visible cellular components on tape strips. Tape strips and the residual viable skin can be analyzed for radiolabel by liquid scin-tillation counting after soaking in scintillation cocktail for several hours. Major organs should also be analyzed for radiolabel content to determine systemic absorption. The tape-stripping method is somewhat subjective, however, resulting in artificially high marker levels in the residual skin if stratum corneum removal is incomplete. Conversely, tape stripping may not be ideal for assessing follicular delivery since follicular contents may be stripped away, thereby underestimating follicular deposition. This method may be most useful for assessing follicular deposition when combined with other techniques such as visualization by microscopic methods. Deposition of topically applied radiolabeled agents can be assessed in hamster ear se-baceous glands by using the scraping technique, as described earlier in the animal model discussion, in combination with tape stripping.

B. Autoradiography

Autoradiography has been used previously to show localized drug deposition into the hair follicles (44,45). Nicolau and coworkers (44) combined micro-scopic autoradiography and liquid scintillation counting to assess in vivo follicular delivery of viprostol, a synthetic prostaglandin E_2 analog, in a petrolatum base. In mouse skin, distribution of radioactivity was evident in the stratum corneum and hair follicles 30 min after topical application. After 12 h, viprostol was localized in the stratum corneum, epidermis and lower hair follicle of monkey and mouse skin. Drug retention occurred only in follicular structures after 72 h, suggesting a depot function for this structure. Bidmon (45) found dose and vehicle dependencies using an improved dry-mount autoradiography technique to detect [^{3}H]estradiol-17β in hairy rat skin. Drug localization was visualized in the epidermis, sebaceous glands, and follicular papilla following a 2 h topical application of the drug in dimethyl sulfoxide (DMSO), ethylene glycol, or sesame oil formulations. After 24 h, autoradiograms indicated that hair follicles and sebaceous glands continued to act as drug depots.

C. Fluorescence Microscopy

Several investigators have visualized follicular deposition using simple fluorescence microscopy, and have combined these data with quantitative data for a follicular pathway (11,21,46). Topically applied fluorescent molecules can be visualized in skin structures after carefully cryosectioning skin into thin slices or in follicular biopsies. Fluorescence-labeled polymeric microspheres and microspheres loaded with fluorescent molecules have been visualized in follicular structures by simple epifluorescence microscopy and by scanning electron microscopy (20,21). The use of epifluorescence microscopy is limited by the requirement of extremely thin skin slices and the potential uncertain interpretation of data due to skin and hair follicle autofluorescence.

D. Confocal Laser Scanning Microscopy

Confocal laser scanning microscopy (CLSM) allows direct visualization of permeating fluorescent markers within hair follicles while minimizing detection skin and hair follicle autofluorescence. Importantly, CLSM does not require harsh fixation procedures, and relatively thick specimens can be viewed by collecting a series of images at regular focus intervals. CLSM minimizes the detection of autofluorescence by directly a single beam of laser light to the sample, resulting in a clearly defined fluorescent area that is much brighter than autofluorescent areas (47). With CLSM, a stack of optical sections from varying depths can be combined to reveal three-dimensional fluorescent marker distributions within the skin. Besides offering excellent qualitative visualization, the data can also be processed by computer interface to show quantitative site-dependent fluorescent intensities. CLSM has been used previously to visualize localization of topically applied fluorescent markers within the hair follicles and sebaceous glands (20,48).

E. Immunohistochemistry

Immunostaining and immunofluorescence techniques may be used to visualize topically applied agents that localize in follicular structures. Balsari and coworkers (15) used an immunohistochemistry technique to evaluate the in vivo percutaneous penetration of liposome-encapsulated monoclonal antibody to doxorubicin (MAD11) in 10-day-old rats. Following a 30-min application of formulation, the rats were euthanized and tissue samples were removed for cryosectioning. The immunoperoxidase technique was performed with an avidin–biotin–peroxidase complex (ABC) kit, followed by staining with diaminobenzidine and counterstaining with hematoxylin. Staining observed by light microscopy indicated localization of MAD11 in the stratum corneum, epidermis, and deep in the dermis surrounding follicular structures. The qualitative data obtained in this study via immunostaining were further supported with radiolabel penetration studies.

Yarosh and coworkers (16) quantitatively assessed the localization of topically applied DNA repair enzymes within mouse skin via immunocytology using antibodies against the DNA repair enzymes. After termination of the in vivo study, skin was sectioned and stained with antibodies against T4 endonuclease V and FITC linked secondary antibodies. After 1 h topical application period, immunostaining indicated that enzymes were found in the epidermis in the cells surrounding the follicular root sheath and the sebaceous glands.

F. Follicular Biopsies

Follicular casting methods have been used to remove follicular contents by applying a quick-setting cyanoacrylate adhesive on a glass slide to the skin surface (20,42,49,50). Drug deposition in the follicular casts pulled from the skin can be analyzed by microscopy or by various analytical techniques following extraction in solvents. Bojar and coworkers (50) investigated follicular deposition of a topically applied azelaic acid cream, 20% w/w, in human back and forehead skin. After 5 h, surface drug was removed followed by follicular biopsy and detection of azelaic acid in follicular cast supernatant by high-performance liquid chromatography (HPLC) analysis. The investigators explained the findings of lower drug concentrations in forehead follicular casts as a consequence of rapid removal from the richly vascularized sebaceous glands at that site. Follicular casting has previously been combined with microscopic analysis by fluorescence and scanning electron microscopy to assess deposition of drug-loaded microspheres into follicular structures (20).

G. Pharmacological Effect

The ability of a topically applied agent to elicit a pharmacological response that is specific to the pilosebaceous unit may be useful in supporting a follicular delivery hypothesis. Although this is not an absolute indicator of a follicular route, combined with additional data it can support the hypothesis. Balsari and coworkers (15) observed that a topically applied liposomal anti-doxorubicin monoclonal antibody, MAD11, was able to prevent alopecia in 31 of 45 young rats treated intraperitoneally with doxorubicin. While these data alone suggest a follicular route, the investigators also used radiolabel deposition and immunohistochemistry visualization techniques.

Lieb and coworkers (14) used a quantitative bioassay to assess in vivo follicular delivery of topically applied [^{3}H]cimetidine to hamster ear sebaceous glands. Various vehicles containing 3% cimetidine were applied to the ventral side of hamster ears twice daily for 4 weeks. Since cimetidine is known to have antiandrogenic activity, the basis of the assay was to measure changes in sebaceous gland sized as observed by phase-contrast microscopy. Drug deposition into sebaceous glands, as apparent by sebaceous gland size reduction, was greatest with ethanol and nonionic lipid-based formulations at pH values when cimetidine was mostly unionized.

Several pharmacological responses may be measured that suggest activity at the follicular level via transfollicular delivery, but systemic follicular delivery following transdermal delivery via the stratum corneum must be ruled out as an explanation. Hormonal effects on hair growth and acne are also examples of the therapeutic responses that occur at the follicular level.

H. Laser Doppler Flowmetry

Regional variations in cutaneous blood flow following topical application of agents can be measured by laser Doppler flowmetry (LDF). Tur and coworkers (51) measured skin blood flow by LDF following topical application of methyl nicotinate, a vasodilator, to human skin on the forearm, forehead and palm. The measured LDF responses indicated the greatest response at the forehead, an intermediate response at the forearm, and the least response through the palm. The investigators concluded that a preferential permeation route exists in areas of high follicular density, but conceded that other factors may have been influential such as potential skin lipid differences.

I. Gene Expression

Gene delivery to target sites within the hair follicles and sebaceous glands is currently an exciting development that may yield interesting new therapies. Investigators in recent studies have assessed follicular delivery of topically applied viral and nonviral vectors, as measured by gene expression at the follicular level. Li and Hoffman (52) demonstrated that a topically applied phospholipid (egg phosphatidylcholine) liposome preparation containing the lacZ reporter gene resulted in selective expression of the gene in hair bulb matrix cells and bulge area cells in mouse skin. Lu and coworkers (53) have shown that topically applied viral vectors can be used to transfer genes into various strata of the skin. Following a 2-day topical treatment of mouse skin in vivo with an adenovirus vector in phosphate-buffered saline, the skin was excised and stained with X-gal. The resulting blue staining of the sectioned skin indicated gene localization in the epidermis and some hair follicles. It may be important to note that ethanol pretreatment and tape stripping of the skin three to four times may have compromised the barrier function of the stratum corneum, thereby enhancing gene delivery from aqueous solution into the deeper skin strata.

VI. FORMULATION EFFECTS ON FOLLICULAR DEPOSITION

The presence of sebum in the follicular canal may restrict or enhance the transport of molecules into the follicle, depending on the physicochemical characteristics of the drug and vehicle. Dissolving the permeant in lipoidal solvents such as ethanol or acetone (3–5), which may reorganize or delipidize sebum, may thereby open the passageway for drug deposition within the follicle. Wetting agents (e.g., sodium lauryl sulfate) may be useful in decreasing the interfacial tension between polar drugs and sebum, which may promote mixing in the form of an emulsion, providing a more favorable environment for drug partitioning and absorption.

Bamba and Wepierre (33) assessed formulation effects on follicular deposition of [^{14}C]pyridostigmine bromine in a 72-h in vitro study using hairless rat and scarred (follicle-free) hairless rat skin. Greatest drug deposition into hair follicles was achieved by an ethanol vehicle during the first 24 h, whereas Azone and Nerol vehicles delivered pyridostigmine preferentially via a transepidermal route. The authors explained these results as a consequence of ethanol miscibility with sebum, which allowed more favorable drug deposition into sebaceous glands. The use of two other vehicles, dimethyl sulfoxide and propylene glycol, also resulted in significant drug delivery through the follicular routes.

Particulate delivery systems, including liposomes and polymeric microspheres, have yielded especially interesting data supporting a follicular route. These systems have been compared to conventional formulations and simple drug solutions, with particular emphasis on liposomal lipid composition and microbead particle size (11–21).

A. Liposomes

Liposomal systems, upon dehydration following topical application, may yield a fluid liquid-crystalline state in which bilayers containing drug can partition and pack into the follicular openings. Follicular delivery of hydrophilic molecules may be facilitated via association with polar head groups of liposomal bilayers. Results from recent studies have suggested that this mechanism of action is not dependent on the extent of hydrophilic drug entrapment (11), in that empty liposomes along with "free" aqueous solution exhibit similar absorption profiles. Following dehydration, extent of entrapment may be inconsequential due to the establishment of a new equilibrium between the drug and bilayers. Several studies have also shown facilitated follicular delivery of lipophilic molecules carried by liposomal formulations (11–19).

Lieb and coworkers (11) used an in vitro hamster ear model to quantitatively assess follicular deposition of a low-molecular-weight hydrophilic dye, carboxyfluorescein. Ears cut at the base were mounted ventral side up on Franz diffusion cells containing HEPES isotonic buffer, pH 7.4. Various formulations containing carboxyfluorescein, 100 µg/ml, were studied: a) isotonic HEPES buffer, b) 5% propylene glycol, c) 10% ethanol in HEPES, d) 0.05% sodium lauryl sulfate (SLS) in HEPES, e) multilamellar (MLV) egg phosphatidylcholine (PC):cholesterol (CH):phosphatidylserine (PS) liposomes, and f) a nonbilayer suspension of the same lipid mixture (MIX). After a 24-h in vitro study, quantitative fluorescence microscopy and fluorescence spectroscopy were used to visualize and quantitate carboxyfluorescein deposition into sebaceous glands. The dermal side of the ear was examined by epifluorescence microscopy, and images of several sebaceous glands were digitally processed and quantitated for fluorescence, as measured in radians. Phospholipid liposomes were superior to all formulations, as shown in Fig. 6. Data obtained from the scraping technique (not shown) correlated strongly with the quantitative fluorescence microscopy data. The data suggest that liposomes can enhance follicular delivery of small polar molecules, which are not ordinarily expected to penetrate the stratum corneum or sebaceous glands.

In a study of specific liposomal formulation effects, Niemiec and coworkers (17) compared peptide deposition from phospholipid (PC/CH/PS) liposomes and nonionic liposomes into hamster ear sebaceous glands. Nonionic liposomes consisted of glyceryl dilaurate, cholesterol, and polyoxyethylene 10-stearyl ether at a weight percent of 57:15:28, respectively. Topical application of $[^{125}I]$-α-interferon or $[^{3}H]$cyclosporine in various formulations was followed by the previously described scraping technique. Nonionic liposomes provided the greatest follicular deposition for both peptides, with deposition approximately fourfold greater than that achieved by phospholipid liposomes. Relatively low amounts of either drug were found in the dermis, suggesting that nonionic liposomes are capable of delivering peptides with varying physicochemical characteristics into sebaceous glands.

Li and coworkers (12) used confocal scanning laser microscopy to visualize the follicular localization of a liposome-entrapped fluorescent hydrophilic dye, calcein, in mouse skin histocultures. After a 20-min incubation

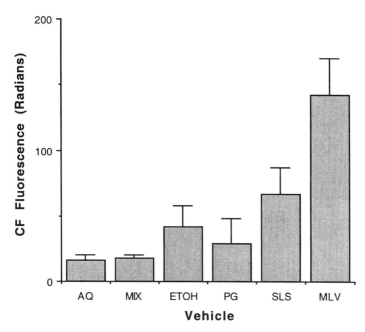

Figure 6 Formulation effects on deposition of carboxyfluorescein (CF) in hamster ear sebaceous glands, as determined by quantitative fluorescence microscopy after 24 h of in vitro topical application ($n = 4$, mean $\pm$ SEM). AQ = aqueous; MIX = lipid mixture; ETOH = ethanol; PG = propylene glycol; SLS = sodium lauryl sulfate; MLV = multilamellar liposomes. From Lieb (1992).

period, liposomes made with egg PC delivered calcein deeper into the hair follicle than those made with dipalmitoyl PC or aqueous calcein. The results suggested that formulation factors may have determined follicular deposition, based on differences in phase transition temperatures for the liposomal lipids. Egg PC liposomes were in a liquid-crystalline state at 37°C, whereas dipalmitoyl PC liposomes were in a gel phase. Another study by Li and coworkers (13), using liposomally entrapped melanin in a 12-h mouse skin histoculture, showed similar results. Follicular delivery of melanin from egg PC liposomes was superior to that from aqueous melanin solution, as indicated by fluorescence microscopy visualization of thin skin slices.

Enhancement of percutaneous transport of [^{14}C]mannitol via liposomes into follicular structures of the Sprague-Dawley hairy rat has been shown in vivo (19). As shown earlier in Fig. 2, hairless rat skin is a leaky barrier to aqueous mannitol solution, whereas hairy rat skin is relatively impenetrable to aqueous mannitol solution. However, as shown in Fig. 7, appreciable

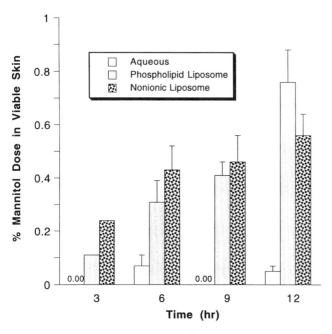

Figure 7 In vivo deposition of [^{14}C]mannitol into viable skin or hairy rats after topical application of phospholipid-based liposomes and nonionic lipid-based liposomes ($n = 3-6$, mean $\pm$ SEM).

amounts of liposomal mannitol are deposited into hairy skin following topical application for up to 12 h, with negligible systemic accumulation in the urine or feces (data not shown) (19). Conversely, hairless rat skin is also a leaky barrier to liposomal mannitol, with negligible amounts remaining in the viable skin, but relatively high urinary levels comparable to those attained following topical application of aqueous mannitol (data not shown). In a similar study, confocal laser scanning microscopy was used to visualize the deposition of a topically applied small polar fluorescent compound, carboxyfluorescein, at 100 μg/ml in aqueous or liposomal systems. With this technique, a 20-μm-thick vertical cryosection of treated hairy rat skin was viewed, showing upper follicular deposition of liposomal carboxyfluorescein up to 72 h (48), whereas aqueous solution resulted in fluorescence appearing only on the stratum corneum. Fluorescence was visualized only in the stratum corneum of hairless rat skin, regardless of formulation, at this time point. The results collectively suggest that fully developed follicular structures may act as depots for molecules in an appropriate carrier system, such as liposomes.

Most remarkably, liposomes have been shown to target macromolecules into the hair follicle. Li and coworkers (18) targeted liposomally entrapped high-molecular-weight DNA into hair follicles of histocultured mouse skin. Egg PC liposomes were loaded with [^{35}S]-dATP-labeled DNA that was isolated from a mouse genomic DNA library. Mouse skin histocultures were incubated with liposomal DNA or free DNA solution for 44 h. With liposomal DNA, autoradiography indicated radiolabel presence in the cell membranes and cytoplasm of hair follicle cells, whereas no radiolabel was found after incubation with free DNA solution. In another study discussed previously, Li and coworkers (52) also demonstrated follicular deposition of other macromolecules (genes) using liposomal formulations. Follicular delivery of such macromolecules seems unlikely via lipoidal pathways of the stratum corneum due to probable size restrictions.

As discussed previously, Balsari and coworkers (15) topically applied liposomal monoclonal antibodies to doxorubicin (MAD11), resulting in the prevention of systemic doxorubicin-induced alopecia in young rats. Phospholipid-based liposome formulations were compared with aqueous MAD11 solution and empty liposomes, using radiolabel deposition and immunochemistry histological studies. High dermal levels of radiolabeled antibody combined with immunostaining in follicular structures suggest that this macromolecule was able to be delivered via liposomes into the hair follicles, which exerted a protective effect against alopecia. Again, considering the large molecular size of MAD11 (>150 kD), a transport route through the stratum corneum lipids into the hair follicle seems unlikely.

B. Polymeric Microspheres

Results from a series of studies using topically applied polymeric microspheres have indicated that particle size and vehicle composition may be important factors determining the extent of follicular deposition. Schaefer and coworkers (20) investigated follicular deposition of fluorescent polystyrene microbeads, with diameters ranging from approximately 1 to 47 μm, suspended in aqueous or lipophilic vehicles. Microbeads were topically applied to hairless rat skin (in vivo) or human skin (in vitro) for 15 min, including a 5-min massage time period, after which follicular biopsies were performed and viewed with fluorescence microscopy. The 7-μm-diameter microbeads were found to be optimally sized, as indicated by fluorescence deep within hair follicles of both species. Suspension of these beads in a lipophilic vehicle, Miglyol, yielded best results for follicular deposition into hairless rat skin. Microbead localization deep within the hair follicle was also shown by scanning electron microscopy. Similar particle size and vehicle effects were also observed using poly-β-alanine microbeads labeled with dansyl chloride in an aqueous gel, hydroalcoholic gel, and a silicone oil. Microbeads with 5 μm diameter preferentially targeted the hair follicle, which was best achieved in the most lipophilic vehicle, silicone oil. Rolland and coworkers (21) used similar studies with polymeric microspheres loaded with adapalene, a fluorescent antiacne drug, and suspended in aqueous gel. Adapalene was visualized deep within hairless rat (in vivo) and human (in vitro) hair follicles by fluorescence microscopy following up to 5 h of topical application of the microsphere formulation. The microspheres were subsequently visualized by scanning electron microscopy deep within the hair follicle, with optimum delivery occurring with 5-μm microspheres. The results of these studies collectively suggest that particulate systems can be designed to specifically target follicular structures based on particle diameter and vehicle composition.

VII. SUMMARY

Passive follicular delivery may be a larger component of topical delivery than was previously assumed. As more knowledge is gained about the structure and function of hair follicles and sebaceous glands, the relevance of therapeutic targeting to specific areas may be better defined. The development of improved animal models and methodologies to assess follicular delivery may enable a better understanding of the kinetics of follicular transport. Follicular delivery is likely dependent on the physicochemical characteristics of both the drug and vehicle. Results from several studies have shown that particulate systems, including liposomes and polymeric microsphere systems, preferentially localize in pilosebaceous units. Manipulation of particle size and other formulation factors, such as lipid ratios and composition, may be useful in achieving a greater extent of delivery to the follicles and sebaceous glands. The potential of liposomes to deliver localized pharmacological agents and macromolecules such as genes into hair follicles and sebaceous glands may have a great impact on future treatments of follicular disorders.

REFERENCES

1. Hardy MH. The secret life of the hair follicle. Trends Genet 1992; 8:55–61.
2. Lauer AC, Lieb LL, Ramachandran C, Flynn GL, Weiner ND. Transfollicular drug delivery. Pharm Res. 1995; 12:179–186.
3. Elias PM. Epidermal lipids, barrier function and desquamation. J Invest Dermatol 1983; 80:44–49.
4. Downing DT. Lipid and protein structures in the permeability barrier of mammalian epidermis. J Lipid Res 1992; 33:301–313.

5. Williams AC, Barry BW. Skin absorption enhancers. Crit Rev Ther Drug Carrier Sys 1992; 9:305–353.

6. Kasting GB, Smith RL, Cooper ER. Effect of lipid solubility and molecular size on percutaneous absorption. In: Shroot B, Schaefer H, eds. Skin Pharmacokinetics. Basel: Karger, 1987:138–153.

7. Scheuplein RJ. Mechanism of percutaneous absorption. 1. Routes of penetration and the influence of solubility. J. Invest Dermatol 1965:334–346.

8. Scheuplein RJ, Blank IH, Brauner GJ, MacFarlane DJ. Percutaneous absorption of steroids. J Invest Dermatol 1969; 52:63–70.

9. Feldmann JR, Maibach HI. Regional variation in percutaneous penetration of ^{14}C-cortisol in man. J Invest Dermatol 1967; 48:181–183.

10. Maibach HI, Feldmann RJ, Hilby TH, Serat WF. Regional variation in percutaneous penetration in man. Arch Environ Health 1971; 23:208–211.

11. Lieb LM, Ramachandran C, Egbaria K, Weiner ND. Topical delivery enhancement with multilamellar liposomes via the pilosebaceous route: I. In vitro evaluation using fluorescent techniques with the hamster ear model. J Invest Dermatol 1992; 99:108–113.

12. Li L, Margolis LB, Lishko VK, Hoffman RM. Product-delivering liposomes specifically target hair follicles in histocultured intact skin. In Vitro Cell Dev Biol 1992; 28A:679–681.

13. Li L, Lishko VK, Hoffman RM. Liposomes can specifically target entrapped melanin to hair follicles in histocultured skin. In Vitro Cell Dev Biol 1993; 29A:192–194.

14. Lieb L, Ramachandran C, Flynn G, Weiner N. Follicular (pilosebaceous unit) deposition and pharmacological behavior of cimetidine as a function of formulation. Pharm Res 1994; 11:1419–1423.

15. Balsari AL, Morelli D, Ménard S, Veronesi U, Colnaghi MI. Protection against doxorubicin-induced alopecia in rats by liposome-entrapped monoclonal antibodies. FASEB J 1994; 8:226–230.

16. Yarosh D, Bucana B, Cox P. Localization of liposomes containing a DNA repair enzyme in murine skin. J Invest Dermatol 1994; 103:461–468.

17. Niemiec S, Ramachandran C, Weiner N. Influence of nonionic liposomal composition on topical delivery of peptide drugs into pilosebaceous units: An in vivo study using the hamster ear model. Pharm Res 1995; 12:1184–1188.

18. Li L, Lishko VK, Hoffman RM. Liposome targeting of high molecular weight DNA to the hair follicles of histocultured skin: A model for gene therapy of the hair growth processes. In Vitro Cell Dev Biol 1993; 29A:258–260.

19. Lauer AC, Elder JT, Weiner ND. Evaluation of the hairless rat as a model for in vivo percutaneous absorption. J Pharm Sci 1997; 86:13–18.

20. Schaefer H, Watts F, Brod J, Illel B. Follicular penetration. In: Scott RC, Guy RH, Hadgraft J, eds. Prediction of Percutaneous Penetration, Methods, Measurements, Modelling, London: IBC Technical Services, 1990:163–173.

21. Rolland A, Wagner N, Chatelus A, Shroot B, Schaefer H. Site-specific drug delivery to pilosebaceous structures using polymeric microspheres. Pharm Res 1993; 10:1738–1744.

22. Green PG, Hinz RS, Kim A, Szoka FC, Guy RH. Iontophoretic delivery of a series of tripeptides across the skin in vitro. Pharm Res 1991; 8:1121–1127.

23. Cranne-van Hinsberg WH, Verhoef JC, Bax LJ, Junginger HE, Boddé HE. Role of appendages in skin resistance and iontophoretic peptide flux: Human versus snake skin. Pharm Res 1995; 12:1506–1512.

24. Bertolino AP, Klein LM, Freedberg IM. Biology of hair follicles. In: Fitzpatrick TB, Eisen AZ, Wolff K, Freedberg IM, Austen KF, eds. Dermatology in General Medicine, 4th edition, Volume I. New York: McGraw-Hill, 1993:289–293.

25. Philpott MP, Sanders DA, Kealey T. Whole hair follicle culture. Dermatol Clin 1996; 14:595–607.

26. Cotsarelis G, Sun T, Lavker RM. Label-retaining cells reside in the bulge area of pilosebaceous unit: Implications for follicular stem cells, hair cycle and skin carcinogenesis. Cell 1990; 61:1329–1337.

27. Steward ME, Downing DT. Chemistry and function of mammalian sebaceous lipids. Adv Lipid Res 1991; 24:263–301.

28. Ryan TJ. Cutaneous circulation. In: Goldsmith LA, ed. Physiology, Biochemistry and Molecular Biology of the Skin. Oxford: Oxford Press, 1991:1019–1084.

29. Hanada K, Chiyoya S, Suzuki K, Hashimoto I, Hatayama I. Study of the skin of a new hairless rat mutant. J Dermatol 1988; 15:257–262.

30. Illel B, Schaefer H. Transfollicular percutaneous absorption: Skin model for quantitative studies. Acta Derm Vernerol 1988; 68:427–430.

31. Illel B, Schaefer H, Wepierre J, Doucet O. Follicles play an important role in percutaneous absorption. J Pharm Sci 1991; 80:424–427.

32. Hueber F, Wepierre J, Schaefer H. Role of transepidermal and transfollicular routes in percutaneous absorption of hydrocortisone and testosterone: In vivo study on hairless rat. Skin Pharmacol 1992; 5:99–107.

33. Bamba FL, Wepierre J. Role of the appendageal pathway in the percutaneous absorption of pyridostigmine bromine in various vehicles. Eur J Drug Metab Pharmacokinet 1993; 18:339–348.

34. Hueber F, Besnard M, Schaefer H, Wepierre J. Percutaneous absorption of estradiol and progesterone in normal and appendage-free skin of the hairless rat: Lack of importance of nutritional bloodflow. Skin Pharmacol 1994; 7:245–256.

35. Hisoire G, Bucks D. An unexpected finding in percutaneous absorption observed between haired and hairless guinea pig skin. J Pharm Sci 1997; 86:398–400.

36. Wahlberg JE. Transepidermal or transfollicular absorption? Acta Derm Venereol 1968; 48:336–344.

37. Plewig G, Lunderschmidt C. Hamster ear model for sebaceous glands. J Invest Dermatol 1977; 68:171–176.

38. Matias JR, Orentreich N. The hamster ear sebaceous glands. I. Examination of the regional variation by stripped skin planimetry. J Invest Dermatol 1983; 81:43–46.

39. Uno H, Kurata S. Chemical agents and peptides affect hair growth. J Invest Dermatol 1993; 101:143S–147S.

40. Uno H. Quantitative models for the study of hair growth in vivo. Ann NY Acad Sci 1991; 642:107–124.
41. Li L, Paus R, Slominski A, Hoffman RM. Skin histoculture assay for studying the hair cycle. In Vitro Cell Dev Biol 1992; 28A:695–698.
42. Mills OH, Kligman AM. The follicular biopsy. Dermatologica 1983; 167:57–63.
43. Rougier A, Lotte C. Predictive approaches. I. The stripping technique. In: Shaw VP, Maibach HI, eds. Topical Drug Bioavailability, Bioequivalence and Penetration. New York: Plenum Press, 1993:163–182.
44. Nicolau G, Baughman RA, Tonelli A, McWilliams W, Schiltz J, Yacobi A. Deposition of viprostol (a synthetic PGE_2 vasodilator) in the skin following topical administration to laboratory animals. Xenobiotica 1987; 17:1113–1120.
45. Bidmon HJ, Pitts JD, Solomon HF, Bondi JV, Stumpf WE. Estradiol distribution and penetration in rat skin after topical application, studied by high resolution autoradiography. Histochemistry 1990; 95:43–54.
46. Kao J, Hall J, Hellman G. In vitro percutaneous absorption in mouse skin: influence of skin appendages. Toxicol Appl Pharmacol 1988; 94:93–103.
47. Matsumoto B, Kramer T. Theory and applications of confocal microscopy. Cell Vision 1994; 3:190–198.
48. Lauer A. In Vivo Deposition of Polar Molecules from Various Topical Formulations into Hairless and Hairy Rat Skin. PhD dissertation, University of Michigan, Ann Arbor, 1996.
49. Mills OH, Kligman AM. A human model for assaying comedelytic substances. Br J Dermatol 1982; 107:543–548.
50. Bojar RA, Cutcliffe AG, Graupe K, Cunliffe WJ, Holland KT. Follicular concentrations of azelaic acid after a single topical application. Br J Dermatol 1993; 129:399–402.
51. Tur E, Maibach HI, Guy RH. Percutaneous penetration of methyl nicotinate at three anatomic sites: Evidence for an appendageal contribution to transport? Skin Pharmacol 1991; 4:230–234.
52. Li L, Hoffman RM. The feasibility of targeted selective gene therapy of the hair follicle. Nature Med 1995; 1:705–706.
53. Lu B, Federoff HJ, Wang Y, Goldsmith LA, Scott G. Topical application of viral vectors for epidermal gene transfer. J Invest Dermatol 1997; 108:803–808.

26
Relative Contributions of Human Skin Layers to Partitioning of Chemicals with Varying Lipophilicity

Tatiana E. Gogoleva, John I. Ademola, Ronald C. Wester, Philip S. Magee, and Howard I. Maibach
University of California School of Medicine, San Francisco, California

I. INTRODUCTION

The skin is an organized, heterogeneous, multilayered organ. The stratum corneum (SC), epidermis (SCE), and dermis (D), together with the appendages and the vasculatures, constitute the living protective system around the body. Percutaneous absorption of surface applied agents is the sum of the penetration and permeation of a chemical into and through the stratum corneum, epidermis, and some part of the dermis (1–5). In this investigation, the contribution of three skin layers to stratum corneum percutaneous absorption was evaluated by quantifying partitioning of model compounds between skin layers and vehicles (water and isopropyl myristate, IPM). The influences of drug concentrations, equilibration time, vehicle hydrophilicity, lipophilicity, and pH of the vehicle on partitioning behavior of model compounds were examined.

II. MATERIALS AND METHODS

A. Radioisotopes and Chemicals

The following chemicals were obtained from Radiochemical Center, Amersham, IL: [³H]propranolol hydrochloride (specific activity 17.1 Ci/mmol), 3-

[5(n)-³H]indolacetic acid (specific activity 29.6 Ci/mmol), and [³H]-8-meth-oxypsoralen (specific activity 85 Ci/mmol). Atrazine[ring-¹⁴C] (specific activity 7.8 mCi/mmol) and 4-acetamidophenol[ring-UL-¹⁴C] (specific activity 7.3 mCi/mmol) were purchased from Sigma Chemical Company (St. Louis, MO). [¹⁴C]Salicylic acid (specific activity 56.6 mCi/mmol) was received from NEN Research Products (Wilmington, DE) and theophylline [8-¹⁴C] from American Radiolabeled Chemicals, Inc. (St. Louis, MO). Propranolol base was prepared from [³H]propranolol HCl (specific activity 17.1 Ci/mmol) as follows: 0.1 ml of propranolol hydrochloride was evaporated, and water and sodium hydroxide were added (pH 8). Propranolol base was extracted three times by ethyl acetate. The radiochemical purity determined by TLC was >98% for all compounds. Unless otherwise specified the experiments were performed with the addition of cold compounds. 4-Acetamidophenol, indolacetic acid, 8-methoxypsoralen, propranolol hydrochloride, salicylic acid, and theophylline were purchased from Sigma Chemical Company (St. Louis, MO). Atrazine was received from Monsanto Agricultural Company (St. Louis, MO).

B. Skin Source

Skin samples were taken from the upper thigh of human cadavers (School of Medicine, University of California, San Francisco). The skin was der-matomed to targeted 0.5 mm thickness (Dermatome model B; Padgett, Kansas City, MO) and refrigerated at 4°C in phosphate-buffered saline prior to separation of skin layers. The skin was used within 2 days, postmortem.

C. Separation of Skin Layers

To separate dermis from epidermis, the tissue was submerged in phosphate buffered saline for 30 s at 60°C. The epidermis was peeled from the dermis with dissection forceps. Thin sheets of epidermis were placed dermal side down on a filter paper saturated with 0.0001% trypsin solution (pH 7.6, Sigma Chemical company, St. Louis, MO). After digestion of the viable epidermis, the sheets of stratum corneum were rinsed thoroughly with deionized water. The skin samples were dried at 37°C for 24 h in the incubator and stored in the desiccator (6,7).

Model compounds were placed in borosilicate glass tubes, secured with Teflon, and equilibrated for 3 h at room temperature. Aliquots taken from each vial were analyzed by liquid scintillation counting to obtain the initial concentration of compounds in each solution. Accurately weighed individual skin layer samples were then placed in each vial and the mixture equilibrated with occasional gentle agitation for 0.25–24 h at 20°C. The hydrated skin samples were then removed from the vials, blotted on filter paper, weighed, and dissolved in Soluene 350 (Packard Instrument Company, Downers Grove, IL). The weights of stratum corneum, epidermis, and dermis were 2.1 ± 0.3, 3.7 ± 0.7, and 12.2 ± 6.4 mg, respectively. The surface area of the skin samples was approximately 1 cm^2. The weight of the skin layers (mg) and the volume of the vehicle were chosen so that concentrations of the compounds in the vehicle would not exceed their solubility.

The amount of radiolabeled compound in the vehicle and dissolved skin sample was determined by liquid scintillation counting (TriCarb, model 1500; Packard Instrument Company, Downers Grove, IL). All experiments were performed in triplicate.

D. Skin Membrane/Water Partition Coefficient Determination

The partition coefficient (PC) of compounds between skin layers and vehicle was defined as the ratio of concentrations of radiolabeled chemicals revealed in the skin layers and in the vehicle (6,7). The skin membrane/vehicle partitioning was defined as

$$PC = \frac{\text{mg solute/1000 mg skin}}{\text{mg solute/1000 mg vehicle}}$$

E. Data Relations

The correlation matrix and linear relationships were derived using the MINITAB statistical package, version 6.1, from MINITAB, Inc. (State College, PA).

III. RESULTS

A. Partitioning Studies

The partitioning of model compounds in human stratum corneum, epidermis, and dermis as a function of equilibration time is shown in Figs. 1–3. In stratum corneum and epidermis the lipophilic chemicals—propranolol, atrazine, and salicylic acid—reached equilibrium in 3–6 h. The most hydrophilic compound, theophylline, required 12 h to achieve equilibrium. However, in the dermis/water system, equilibrium was reached in 3 h, which can be explained by the ease of dermal hydration. The partition coefficient time profile of MOP is different than for the other compounds. The skin layers were not saturated up to 24 h after the drug exposure, although the partition coefficient continued to increase with contact time.

The values of partition coefficient for lipophilic compounds are different in the three skin layers. The PCs of more lipophilic compounds such

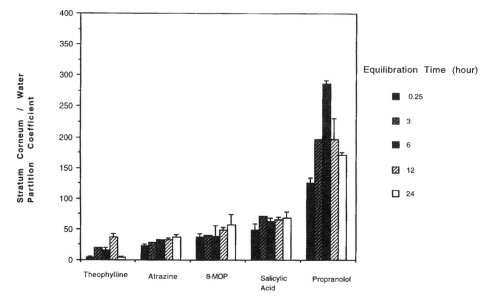

Figure 1   Stratum corneum/water partition coefficient (mean $\pm$ SD, $n = 3$) of model compounds as a function of equilibrium time. In the stratum corneum the lipophilic chemicals propranolol, atrazine, and salicylic acid reached equilibrium in 3–6 h. The hydrophylic compound theophylline required 12 h to achieve equilibrium. Solutions of chemicals used were at saturation solubility concentrations.

as propranolol, salicylic acid, and atrazine are higher in stratum corneum than epidermis and dermis. However the PC values of hydrophilic theophylline remained relatively constant in the three skin layers.

Table 1 shows the influence of drug concentration on PC of the model compounds. In all experiments the PC values decreased as drug concentration increased. This may be due to the saturation of available "binding sites" in the skin layers. As the number of moles available for binding and partitioning increases, the binding sites become saturated, while the concentration of unbound compound increases. Hence this process will continuously decrease the values of partition coefficient.

The PC values of these model compounds are comparable with the octanol/water PCs from the literature, summarized in Table 2. IPM is often used in PC studies because of its presumed similarities to the properties of skin lipids (8). In the skin layers the PCs of the model compounds in water and IPM are inversely related, while partition coefficients between octanol/water, skin layers/water, and IPM/water are related directly. Note that in the stratum corneum (SC)/IPM system the PC values for lipophilic compounds

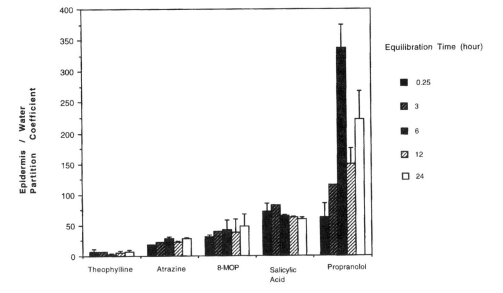

Figure 2 Epidermis/water partition coefficient (mean $\pm$ SD, $n = 3$) of model compounds as a function of equilibration time. Partitioning profiles of the tested chemicals follow the trend, observed in the stratum corneum. For the lipophilic compounds propranolol, atrazine, and salicylic acid it required 3–6 h to achieve equilibrium. Solutions of chemicals used were at saturation solubility concentrations.

(propranolol and atrazine) resemble the PC values of the more hydrophilic compounds (acetaminophen and theophylline). This is not surprising, given the complexity of the stratum corneum composition. It is likely that in SC in the presence of a lipophilic vehicle (IPM), the hydrophilic model compounds more readily partitioned into the protein domain of the stratum corneum.

The pH of the vehicle determines the degree of partitioning of salicylic acid. The data (Table 2) demonstrate that at pH 2.4 the PC values are higher than at pH 5.0 in all skin layers. The pH of the solvent determines the degree of ionization according to the Henderson–Hasselbach equation (9). The pK_a of salicylic acid is 3.0 (10). At pH 2.4 80% of molecules of salicylic acid exist in protonated or unionized form. Their permeation across the lipid membrane occurs more readily. At pH 5.0, 99% of the drug is unprotonated or in the ionized form, which is poorly soluble in lipids. The PC values in the skin decrease. The partitioning behavior of unionized and ionized salicylic acid between octanol and water is similar to partitioning of model compounds in the skin layers studied.

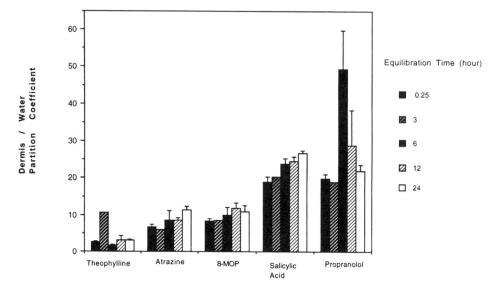

Figure 3   Dermis/water partition coefficient (mean ± SD, $n = 3$) of model compounds as a function of equilibration time. In the dermis/water system, equilibrium for hydrophilic theophylline was reached in 3 h. Solutions of chemicals used were at saturation solubility concentrations.

Table 1   Effect of Aqueous Phase Model Compounds Concentrations (C, $\mu g/cm^3$) on Skin Layers/Water Partition Coefficient (PC) (Mean ± SD; $n = 3$)

	Propranolol				Atrazine		
	PC					PC	
C	SC	SCE	D	C	SC	SCE	D
0.1	804 ± 86	691 ± 76	224 ± 60	1.5	110 ± 39	74 ± 16	45 ± 17
0.3	707 ± 88	433 ± 11	116 ± 59	2.9	111 ± 16	57 ± 17	12 ± 1.1
0.5	462 ± 103	321 ± 24	65 ± 11	5.8	70 ± 11	32 ± 1.8	16 ± 8.1
0.9	336 ± 21	235 ± 55	43 ± 6.3	12	59 ± 13	35 ± 1.8	10 ± 4.2
1.8[a]	271 ± 50	150 ± 17	22 ± 1.8	23[a]	37 ± 4	29 ± 1.1	11 ± 1.1

Note. The data demonstrate the influence of drug concentration on PC of the model compounds. As the concentration of the chemical increases, the PC values inversely decrease. Equilibration time was 24 h.
[a]Saturation solubility concentrations ($\mu g/cm^3$) are as follows: propranolol (1.8), atrazine (23), 8-MOP (17), salicylic acid (2.4×10^3), indolacetic acid (50), acetaminophen (1.7×10^3), theophylline (8.8×10^3).

Table 1 Continued

	8-MOP			$C \times$	Salicylic acid		
	PC				PC		
C	SC	SCE	D	10^2	SC	SCE	D
1.1	69 ± 12	60 ± 0.2	15 ± 1.5	1.3	150 ± 62	16 ± 4.3	6.6 ± 0.5
2.2	57 ± 14	56 ± 9.1	13 ± 3.8	2.6	194 ± 41	101 ± 32	7.0 ± 0.6
4.4	52 ± 11	40 ± 11	11 ± 1.4	5.8	182 ± 38	108 ± 31	9.8 ± 1.0
8.8	56 ± 3.2	46 ± 15	10 ± 1.7	12	91 ± 21	85 ± 41	16 ± 1.3
17[a]	56 ± 18	48 ± 19	11 ± 1.6	24[a]	68 ± 10	60 ± 3.2	27 ± 0.8

B. Data Relations

The correlation matrix was used to select the highest correlation for each of the experimental log P values. As experimental error is substantial and the data set is small ($n = 8$), no attempt was made to achieve higher correlations by multiple regression on two or more descriptors.

The simple linear relations are justified by the T value of the coefficient, the explained variance (r^2), and the Fisher distribution (F). The standard deviation around the equation line (point scatter) is represented by s. The correlation coefficient (r) is shown in the following equations in parentheses next to the explained variance, $r^2(r)$. The weakest expressions are significant at 96.3% confidence level (CL). The others range from 98.4 to 99.9% CL. The equations are arranged from highest to lowest significance, with w indicating water and o indicating oil.

$$\log P(SC/w) = 1.01 \log P(SCE/w) + 0.125$$
$$T = 8.05$$

$$n = 8 \quad s = 0.184 \quad r^2 = .915(.957) \quad F = 64.84$$

$$\log P(SC/IPM) = 1.00 \log P(SCE/IPM) + 0.164$$
$$T = 3.92$$

$$n = 7 \quad s = 0.505 \quad r^2 = .755(.869) \quad F = 15.40$$

$$\log P(SC/w) = 0.358 \log P(o/w) + 0.823$$
$$T = 3.32$$

$$n = 8 \quad s = 0.376 \quad r^2 = .648(.805) \quad F = 11.05$$

The equations show equal patterns of behavior of model compounds

Table 1 Continued

C	Indolacetic acid			Acetaminophen				Theophylline			
	SC	PC		$C \times 10^2$	SC	PC		$C \times 10^3$	SC	PC	
		SCE	D			SCE	D			SCE	D
3.2	31 ± 0.7	14 ± 1.4	11 ± 1.5	1.1	27 ± 2.2	13 ± 1.5	13 ± 0.53	0.6	12 ± 0.9	11 ± 0.9	4.4 ± 0.7
6.3	16 ± 3.1	9.7 ± 0.4	5.8 ± 0.8	2.2	15 ± 0.5	8 ± 0.9	7.8 ± 3.1	1.1	11 ± 0.3	9.4 ± 1.6	4.2 ± 0.6
13	11 ± 1.0	6.0 ± 1.4	5.6 ± 1.1	4.4	22 ± 8.6	8 ± 1.2	5.1 ± 0.7	2.2	8.3 ± 0.3	7.1 ± 2.9	5.1 ± 0.6
25	8.2 ± 0.8	4.2 ± 0.4	3.2 ± 0.8	8.8	14 ± 2.8	4.7 ± 0.8	5.8 ± 1.9	4.4	6.7 ± 1.3	8.1 ± 1.7	3.3 ± 0.5
50[a]	6.8 ± 0.3	4.4 ± 0.3	3.8 ± 0.6	17[a]	17 ± 2.6	6.3 ± 0.9	6.4 ± 0.7	8.8[a]	4.6 ± 0.5	7.1 ± 2.4	3.0 ± 0.4

Table 2 Partition Coefficient (log P) of Model Compounds

Model compound	log P octanol/water	log P IPM/water	Stratum corneum		Stratum corneum/ epidermis		Dermis	
			log P SC/water	log P SC/IPM	log P SCE/water	log P SCE/IPM	log P D/water	log P D/IPM
Propranolol	3.09	0.18	2.43	1.74	2.17	1.68	1.34	1.55
Atrazine	2.75	0.55	1.57	2.87	1.46	2.67	1.04	2.66
8-MOP	2.53	0.89	1.75	0.67	1.68	0.59	1.03	0.28
Salicylic acid, pH 2.4	2.26	0.68	1.83	0.14	1.77	0.16	1.43	0.72
Salicylic acid, pH 5	1.74	0.37	1.05	NA	0.91	NA	0.85	NA
Indolacetic acid	1.41	−0.68	0.83	1.05	0.64	1.29	0.58	1.44
Acetaminophen	0.32	−0.81	1.23	1.42	0.01	1.51	0.81	0.11
Theophylline	−0.78	−0.65	0.66	2.23	0.85	1.07	0.89	1.95

Note. The measured skin layers/water, skin layers/IPM and IPM/water partition coefficients are compared with the octanol/water PC values from the literature. Equilibration time was 24 h. Solutions of the chemicals used were at saturation solubility concentrations. NA, not available.

in isolated stratum corneum and epidermis in aqueous and organic vehicles. Apparently, there is no effect from water soluble epidermal tissue. Log $P(SC/w)$ versus log $P(o/w)$ has a lot of scatter; nevertheless, this relationship allows the approximation of log $P(SC/w)$ for any compound for which log $P(o/w)$ is known.

IV. CONCLUSION

The data presented here are compatible with physicochemical properties of the chemicals used in the study and previous literature observations. The calculated partition coefficients of model compounds as a function of equilibration time, chemical concentration, and nature of the solvent may be valuable in prediction of in vivo and in vitro transport of drugs and environmental agents through human skin.

Raykar et al. (11) suggested that two domains exist in the stratum corneum: the uptake of highly lipophilic compounds (log P values near 3.0) may be governed by the lipid domain of stratum corneum, while partitioning of the more hydrophilic solutes occurs in the protein domain. Theophylline partitions relatively equally into the stratum corneum, epidermis, and dermis; therefore, the retention of theophylline in those skin layers could be due to the binding to the protein domain of the skin. The data presented here are comparable with earlier observations and probably correspond to the 6–24 h period of hydration of stratum corneum (12–14).

REFERENCES

1. Schalla, W., Jamoulle, J., and Schaefer, H. 1989. Localization of compounds in different skin layers and its use as an indicator of percutaneous absorption. In *Percutaneous Absorption*, eds. R. L. Bronaugh and H. I. Maibach. New York: Marcel Dekker, 283–312.
2. Elias, P. M., Cooper, E. R., Korc, A., and Brown, B. E. 1981. Percutaneous transport in relation to stratum corneum structure and lipid composition. *J. Invest. Dermatol.* 76:297.
3. Noonan, P. K., and Wester, R. C. 1989. Cutaneous metabolism of xenobiotics. In *Percutaneous Absorption*, eds. R. L. Bronaugh and H. I. Maibach. New York: Marcel Dekker, 53–75.
4. Hansch, C., and Leo, A. 1979. *Substituent Constants for Correlation Analysis in Chemistry and Biology*, Chapter 4. New York: John Wiley and Sons.
5. Hansch C., and Dunn, W. J. III. 1972. Linear relationships between lipophilic character and biological activity of drugs. *J. Pharm. Sci.* 61:1.
6. Surber, C., Wilhelm, K. P., Hori, M., Maibach, H. I., and Guy, R. H. 1990. Optimisation of topical therapy: Partitioning of drugs into stratum corneum. *Pharm. Res.* 7:1320–1324.

7. Surber, C., Wilhem, K. P., Maibach, H. I., Hall, L. L., and Guy, R. H. 1990. Partitioning of chemicals into human stratum corneum: Implications for risk assessment following dermal exposure. *Fundam. Appl. Toxicol.* 15:99–107.

8. Hadgraft, J., and Ridout, G. 1987. Development of model membranes for percutaneous absorption measurements. I. Isopropyl myristate. *Int. J. Pharm.* 39: 149–156.

9. Lee, G., Swarbrick, J., Kiyohara, G., and Payling, D. W. 1985. Drug permeation through human skin. III. Effect of pH on the partitioning behavior of chromone-2 carboxylic acid. *Int. J. Pharm.* 23:43–54.

10. Katzung, B. G. 1982. *Basic and Clinical Pharmacology*, 4th ed., pp. 2–3. (city): Appleton and Lange.

11. Raykar, P. V., Fung, M., and Anderson, B. D. 1988. The role of protein and lipid domains in the uptake of solutes by human stratum corneum. *Pharm. Res.* 5:140–149.

12. Idson, B. 1983. Vehicle effects in percutaneous absorption. *Drug Metab. Rev.* 14:2, 207–222.

13. Bronaugh, R. L., and Congdon, E. R. 1984. Percutaneous absorption of hair dyes: Correlation with partition coefficients. *J. Invest. Dermatol.*, 83:124–127.

14. Chadrasekaran, S. K., Campbell, P. S., and Watanabe, T. 1980. 1. Application of the "dual sorption" model to drug transport through skin. *Polym. Eng. Sci.* 20:36–39.

27
Effect of Single Versus Multiple Dosing in Percutaneous Absorption

Ronald C. Wester and Howard I. Maibach
University of California School of Medicine,
San Francisco, California

I. INTRODUCTION

Standard pharmacokinetic practice is to first do a single dose application to determine bioavailability. This standard application is no different for percutaneous absorption, and most absorption values are for single doses. But think of topical application (or any drug dosing) in which the procedure is repeated, whether once per day for several days (or longer) or multiple times during the day, which also can go on for several days (or longer). There are few one-dose magic bullets in pharmaceutics. Therefore, it becomes important to view multiple topical dosing, especially if that is the standard procedure with which a topical drug is used, or if such exposure occurs for a hazardous environmental chemical.

II. SINGLE DAILY DOSE APPLICATION OVER MANY DAYS: HUMAN

Figure 1 illustrates the method used in this type of study (Wester et al., 1983). [^{14}C]Malathion was applied to the skin of human volunteers on day 1. For days 2–7, nonradioactive malathion was applied once per day to the same skin site. The radioactivity excretion curve for days 1–7 represents the single first daily dose. On day 8, the [^{14}C]malathion was applied again (note that malathion had been applied the previous 7 days). The radioactivity excretion curve for days 8–14 represents the multiple daily dose. Figure 1

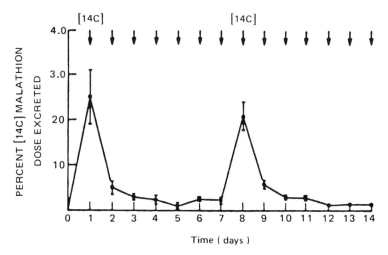

Figure 1 Percutaneous absorption of [^{14}C]malathion after single and repeated top-ical application (5 mg/cm^2) to the ventral forearm of man. Arrow represents appli-cation of malathion, and ^{14}C represents when [^{14}C]malathion was applied.

shows no difference in the percutaneous absorption of malathion in humans for single daily dose (exposure) over several days. This same method was used by Bucks et al. (1989) to study several steroids in humans. Table 1 shows no difference in the absorption of a single daily topical steroid dose over several days. The results are exactly like those with malathion (Fig. 2).

There is an exception to these clear results. Azone (1-dodecylazacy-cloheptan-2-one) is an agent that has been shown to enhance the percuta-neous absorption of drugs. Azone is believed to act on the stratum corneum (SC) by increasing fluidity of the lipid bilayers. Because Azone is nonpolar, it is thought to act by partitioning into the lipid bilayers, thereby disrupting the structure, and potentially allowing drug penetration to increase. Previous clinical studies with single-dose administration show neat Azone percuta-neous absorption to be <1%. A short-term, 4-day dosing sequence gave absorption of 3.5 ± 0.3%. However, the effect of long-term multiple dosing of Azone on the percutaneous absorption of Azone has never been assessed. A study such as this is important because the mechanism of Azone, disrup-tion of the lipid bilayer structure, suggests a potential for enhanced percu-taneous absorption with chronic administration.

Excretion from days 1–7 topical application gave a single-dose per-cutaneous absorption of 1.84 ± 1.56% dose. Percutaneous absorption from days 8–15 skin application was 2.76 ± 1.91%, and the absorption from days 15–21 skin application was 2.72 ± 1.21%. Statistical analysis showed

Table 1 Percutaneous Absorption of Steroids in Humans

	Mean % applied dose absorbed ($\pm$SD)	
	Nonprotected	Occlusion
Hydrocortisone		
Single application	2 ± 2	4 ± 2
Multiple application		
1st dose	3 ± 1	4 ± 1
8th dose	3 ± 1	3 ± 1
Estradiol		
Single application	11 ± 5	27 ± 6
Multiple application		
1st dose	10 ± 2	38 ± 8
8th dose	11 ± 5	22 ± 7
Testosterone		
Single application	13 ± 3	46 ± 15
Multiple application		
1st dose	21 ± 6	51 ± 10
8th dose	20 ± 7	50 ± 9

a significant difference for day 1 dosing versus day 8 dosing ($p < .001$) and for day 1 dosing versus day 15 dosing ($p < .008$). No difference was observed in percutaneous absorption for day 8 versus day 15 dosing (Fig. 3). The daily excretion patterns show that peak excretion occurred at 24 or 48 h following topical application. The results show that an increase occurs in the absorption of Azone with long-term multiple application, but that this enhanced self-absorption occurs early in use, and a steady-state absorption amount is established after the initial enhancement (Wester et al., 1994).

III. TRIPLE DAILY DOSE APPLICATION: HYDROCORTISONE

A. Study Design

The study was specifically designed to compare a single low dose (13.33 $\mu g/cm^2$) to a single larger dose (40.0 $\mu g/cm^2$; three times the amount) and to three multiple-application therapy (13.33 $\mu g/cm^2 \times 3 = 40.0$ $\mu g/cm^2$) treatments. Student's two-tailed, paired t-tests were employed to compare the percentage of the applied dose absorbed and observed mass absorbed per square centimeter between each of the treatments.

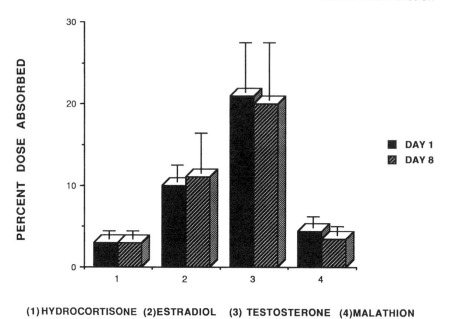

Figure 2 Skin absorption single (day 1) and multiple (day 8) dose in human.

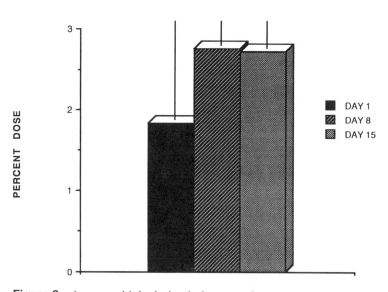

Figure 3 Azone multiple dosing in human volunteers.

Treatment 1: One bolus application of 1.0 μCi/13.33 μg/cm^2 on the right arm, 3 in from the antecubital fossa. The dose was exposed to the skin for 24 h, followed with removal by washing.

Treatment 2: One bolus application of 1.0 μCi/40.0 μg/cm^2 on the left arm, 3 in from the antecubital fossa. The dose was exposed to the skin for 24 h, followed with removal by washing.

Treatment 3: Three repeat applications of 1.0 μCi/13.33 μg/cm^2 on the left ventral forearm, 1 in from the antecubital fossa. One dose was applied, followed by identical doses 5 and 12 h after the initial dose. The site was washed 24 h after the initial dose was applied.

The dosing sequence is shown in Table 2.

Figures 4 and 5 show the predicted and observed hydrocortisone in vivo percutaneous absorption in acetone or cream vehicles dosed at 13.3 μg/cm^2 × 1 (single low dose), 40.0 μg/cm^2 × 1 (single high dose, and an amount three times that of the low dose), and 13.3 μg/cm^2 × 3 (multiple dose, which is three times the single low dose and equal in total amount to the 40 μg/cm^2 in the single high dose). The predicted amounts are multiples (three times) of that of the observed single dose value.

With acetone vehicle, 0.056 ± 0.073 μg/cm^2 hydrocortisone was absorbed for the low dose. The single-high-dose absorption was 0.140 ± 0.136 μg/cm^2, a value near its predicted linear amount of 0.168. The multiple-dose absorption should have been the same predicted 0.168; however, the absorption was 0.372 ± 0.034 μg/cm^2, a value statistically ($p < .05$) greater than that of the single high dose (Fig. 4).

With the cream vehicle, the same pattern emerged. The single high dose absorbed (0.91 ± 1.66 μg/cm^2) was three times that of the low dose absorbed (0.31 ± 0.43 μg/cm^2). The multiple dose absorbed (1.74 ± 0.93

Table 2 Hydrocortisone Dosing Sequence

Treatment	Dose per application (μg/cm^2)	Cumulative dose (μg/cm^2)	Total vehicle volume (μl)	
			Acetone	Cream
1[a]	13.33	13.33	20	100
2[b]	40.00	40.00	20	100
3[c]	13.33	40.00	60	100

[a]Single dose of 13.33 μg/cm^2, administered in 20 μl of vehicle.
[b]Single dose of 40.00 μg/cm^2, administered in 20 μl of vehicle.
[c]Three serial 13.33 μg/cm^2 doses, each administered in 20 μl of vehicle (total 60 μl).

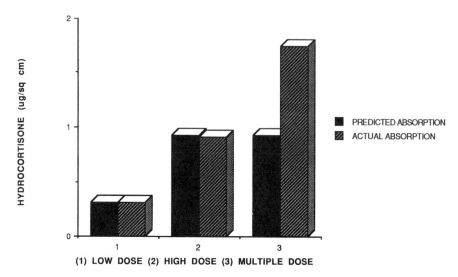

Figure 4   Hydrocortisone in vivo human skin absorption in cream vehicle.

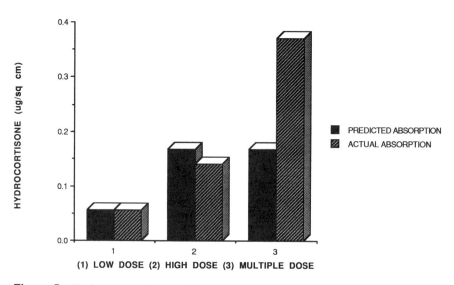

Figure 5   Hydrocortisone in vivo human skin absorption in acetone vehicle.

Table 3   Predicted and Observed Hydrocortisone Absorption In Vitro

Vehicle	Dosing sequence	Hydrocortisone (μg/cm^2)			
		Receptor fluid		Skin	
		Predict	Observe	Predict	Observe
Acetone[a]	13.3 μg/cm^2 $\times$ 1	—	0.13 $\pm$ 0.05	—	0.87 $\pm$ 0.23
	40.0 μg/cm^2 $\times$ 1	0.39[b]	0.35 $\pm$ 0.22	2.61[b]	2.21 $\pm$ 2.05
	13.3 μg/cm^2 $\times$ 3	0.39	0.55 $\pm$ 0.75	2.61	2.84 $\pm$ 2.05
Cream[a]	13.3 μg/cm^2 $\times$ 1	—	0.053 $\pm$ 0.023	—	0.30 $\pm$ 0.24
	40.0 μg/cm^2 $\times$ 1	0.16[b]	0.23 $\pm$ 0.03	0.90[b]	0.86 $\pm$ 0.53
	13.3 μg/cm^2 $\times$ 3	0.16	0.27 $\pm$ 0.21	0.90	1.19 $\pm$ 0.43

[a]$n = 3$; mean $\pm$ SD.
[b]0.39 μg/cm^2 is 3$\times$ the measured value of 0.13 μg/cm^2; 2.61 μg/cm^2 is 3$\times$ the measured value of 0.87 μg/cm^2; 0.16 μg/cm^2 is 3$\times$ the measured value of 0.053 μg/cm^2; 0.90 μg/cm^2 is 3$\times$ the measured value of 0.30 μg/cm^2.

μg/cm^2) exceeded the predicted amount and was statistically ($p < .006$) greater than that of the single high dose.

Table 3 gives the predicted and observed in vitro hydrocortisone percutaneous absorption. The receptor fluid accumulations (absorbed amounts) show the same trend as that seen in vivo. In vitro studies also allowed the human skin to be assayed for hydrocortisone content following the 24-h dosing interval. The skin content values markedly reflect those seen with the receptor fluid values. Only three observations were made per dosing sequence, so statistically no differences exist. The same human skin sources were used for both acetone and cream vehicles, so these absorption amounts can be compared. Hydrocortisone absorption is greater with the acetone vehicle (Melendres et al., 1992; Wester et al., 1995).

Little information is available on the most effective topical corticosteroid or other topical formulation dosing regimen regarding the number of skin applications in 1 day. Multiple applications for an ambulatory patient with a readily accessible skin site are common practice. However, for hospitalized patients, or patients where multiple dosing would be difficult, a single effective dose of hydrocortisone will eventually do the task.

IV. TRIPLE DAILY DOSE APPLICATION: DICLOFENAC

Diclofenac, a nonsteroidal anti-inflammatory drug, has been widely used in the treatment of rheumatoid arthritis and osteoarthritis. However, oral delivery of this drug poses certain disadvantages, such as fast first-pass metabolism and adverse side effects (including gastrointestinal reactions and idiosyncratic drug reactions). Therefore, alternative routes of administration have been sought. The skin has become increasingly important to this effect, and many drugs have been formulated in transdermal delivery systems, including diclofenac itself. However, diclofenac sodium is not easily absorbed through the skin due to its hydrophilic nature. Much work has concentrated on using percutaneous absorption enhancers or cosolvents to increase penetration. A new diclofenac sodium lotion named Pennsaid has been developed for topical application. Pennsaid includes the absorption enhancer dimethyl sulfoxide (DMSO). It is expected that the addition of DMSO may increase the in vivo permeation rate of diclofenac through the skin into the deeper target tissues beneath the skin.

Tables 4 and 5 show that multiple doses of Pennsaid lotion (2 μl/cm^2 and 5 μl/cm^2 $3\times$/day) delivered a total of 40.1 ± 17.6 μg and 85.6 ± 41.4 μg diclofenac, respectively at 48 h, compared to only 9.4 ± 2.9 μg and 35.7 ± 19.0 μg absorbed after topical application of diclofenac as an aqueous solution ($p < .05$). A single-dose study showed no statistical difference between diclofenac delivered in Pennsaid lotion or an aqueous solution. Over 48 h the total absorption for Pennsaid lotion was 10.2 ± 6.7 μg and 26.2 ± 17.6 μg (2 μl/cm^2 and 5 μl/cm^2, respectively), compared to 8.3 ± 1.5 μg and 12.5 ± 5.7 μg from an aqueous solution. Both single doses of Pennsaid lotion and aqueous diclofenac showed decreased diclofenac absorption into the receptor fluid between 12 and 24 h. However, when applied multiple times, absorption from Pennsaid lotion was continually increasing up to 48 h (Hewitt et al., 1998). Clinically, Pennsaid lotion has been shown to be effective in a multidose regimen.

Table 4 In Vitro Percutaneous Absorption of Diclofenac from Pennsaid Lotion and Aqueous Solution in Viable Human Skin

	Dosing regimen	Diclofenac absorbed (μg/cm^2)[a]
A.	30 μg/cm^2 single dose Pennsaid	10.2 ± 6.7
B.	75 μg/cm^2 single dose Pennsaid	26.2 ± 17.6
C.	30 μg/cm^2 single dose aqueous	8.3 ± 1.5
D.	75 μg/cm^2 single dose aqueous	12.3 ± 5.6
E.	240 μg/cm^2 multidose Pennsaid	40.1 ± 17.6
F.	600 μg/cm^2 multidose Pennsaid	85.6 ± 41.4
G.	240 μg/cm^2 multidose aqueous	9.4 ± 2.9
H.	600 μg/cm^2 multidose aqueous	35.7 ± 19.0

[a]Mean $\pm$ SD ($n = 4$ or 5). Absorbed = cumulative in receptor fluid plus skin.

Table 5 Statistical Summary: Diclofenac Absorption in
Viable Human Skin

Treatment	Statistic[a](P)
Pennsaid lotion	
A vs. B—single dose	0.09[b]
E vs. F—multiple doses	0.05[c]
A vs. E—single vs. multiple	0.007[c]
B vs. F—single vs. multiple	0.02[c]
Aqueous solution	
C vs. D—single dose	0.16[b]
G vs. H—multiple doses	0.02[c]
C vs. G—single vs. multiple	0.42[b]
D vs. H—single vs. multiple	0.03[c]
Pennsaid lotion vs. aqueous solution	
A vs. C—single dose	0.57[b]
B vs. D—single dose	0.13[b]
E vs. G—multiple doses	0.005[c]
F vs. H—multiple doses	0.04[c]

[a]Student's t-test.
[b]Not significant.
[c]Statistically significant.

These studies with hydrocortisone and diclofenac show enhanced human skin absorption from a multidose (defined as 3×/day application) regimen. The key for diclofenac is the Pennsaid formulation and, probably, the inclusion of the penetration enhancer DMSO. For hydrocortisone it may simply be the continuing application of vehicle that "washes" the drug through the skin.

V. ANIMAL MODELS

There are several multidose studies in the literature using animals (Roberts and Horlock, 1978; Wester et al, 1977, 1980a, 1980b; Courthheoux et al., 1986) that give mixed results when compared to subsequent human studies. A key animal study is that by Bucks et al. (1985), where percutaneous absorption of malathion multidose in the guinea pig is similar to multidose in the human except where skin washing is introduced into the study design. Skin washing enhances multidose malathion percutaneous absorption. In human studies skin washing is not a consideration because bathing is an everyday event. In animal experiments treatment of skin may compromise skin integrity (application of keratolytic agents, soap-and-water washing), and this can result in altered percutaneous absorption.

VI. DISCUSSION

Percutaneous absorption studies with single-dose regimens are generally the only types of studies done to determine typical bioavailability. Tradition and regulatory requirements are probably the driving forces. In actual clinical use, the multidose regimen is probably the more widely practiced treatment. Two types of experimental designs have been tested. The one of a single daily dose over the course of a few weeks probably does not affect percutaneous absorption unless the skin is compromised by disease (Wester and Maibach, 1992) or if some key ingredient in the formulation is an absorption enhancer, such as Azone. Azone seems to have only influenced the early part of the dosing period. The more intriguing multidose regimen is the multiple dose application during a single day. Both formulation (as seen with hydrocortisone) and formulation ingredients (DMSO) in Pennsaid lotion) can enhance skin absorption. The combined (and relevant) study of multiple doses within each day over several days dosing has not been done. Neither have longer term studies using multidose daily regimens been done to assess percutaneous absorption and diseased skin and the potential compromise of keratolytic agents such as salicyclic acid and hydrocortisone.

REFERENCES

Bucks D.A.W., Marty J.P.L., and Maibach H.I. (1985). Percutaneous absorption of malathion in the guinea pig: Effect of repeated skin application. *Food Chem. Toxicol.* 23:919–922.

Bucks D.A.W., Maibach H.I., and Guy R.H. (1989). *In vivo* percutaneous absorption: Effect of repeated application versus single dose. In *Percutaneous Absorption*, 2 ed. R. Bronaugh and H. Maibach, eds. Marcel Dekker, New York, pp. 633–651.

Courtheoux S., Pechnenot D., Bucks D.A., Marty J.P.L., Maibach H., and Wepierre J. (1986). Effect of repeated skin administration on in vivo percutaneous absorption of drugs. Br. J. Dermatol. 115:49–52.

Hewitt P.G., Poblete N., Wester R.C., Maibach H.I., and Shainhouse J.Z. (1998). *In vitro* cutaneous disposition of a topical diclofenac lotion in human skin: Effect of a multi-dose regimen. *Pharm. Res.*, 15:988–992.

Melendres J., Bucks D.A.W., Carnel E., Wester R.C., and Maibach H.I. (1992). *In vivo* percutaneous absorption of hydrocortisone: Multiple-application dosing in man. *Pharm. Res.* 9:1164–1167.

Roberts M.S., and Horlock E. (1978). Effect of repeated skin application on percutaneous absorption of salicylic acid. *J. Pharm. Sci.* 67:1685–1687.

Wester R.C., and Maibach H.I. (1992). Percutaneous absorption in diseased skin. In *Topical Corticosteroids*, H. Maibach and C. Surber, eds. Karger, Basel, pp. 128–141.

Wester R.C., Noonan P.K., and Maibach H.I. (1977). Frequency of application on percutaneous absorption of hydrocortisone. *Arch. Dermatol. 113*:620–622.

Wester R.C., Noonan P.K., and Maibach H.I. (1980a). Variations in percutaneous absorption of testosterone in the rhesus monkey due to anatomic site of application and frequency of application. *Arch Dermatol Res. 267*:229–235.

Wester R.C., Noonan P.K., and Maibach H.I. (1980b). Variations in percutaneous absorption of hydrocortisone increases with long-term administration: *In vivo* studies in the rhesus monkey. *Arch. Dermatol Res. 116*:186–188.

Wester R.C., Maibach H.I., Bucks D.A.W., and Guy R.H. (1983). Malathion percutaneous absorption after repeated administration to man. *J. Pharm. Sci. 68*: 116–119.

Wester R.C., Melendres J., Sedik L., and Maibach H.I. (1994). Percutaneous absorption of Azone following single and multiple doses to human volunteers. *J. Pharm. Sci. 83*:124–125.

Wester R.C., Melendres J., Logan F., and Maibach H.I. (1995). Triple therapy: Multiple dosing enhances hydrocortisone percutaneous absorption *in vivo* in humans. In *Percutaneous Penetration Enhancers*, E. Smith and H. Maibach, eds. CRC Press, Boca Raton, FL, pp. 343–349.

28
Measurement of Short-Term Dermal Uptake In Vitro Using Accelerator Mass Spectrometry

Garrett A. Keating, Kenneth T. Bogen, and John S. Vogel
Lawrence Livermore National Laboratory, University of California, Livermore, California

I. INTRODUCTION

Dermal absorption is commonly assessed in vitro with the two-chamber flow-through diffusion cell in which the chemical's diffusion through a skin membrane into unrecirculated receptor fluid is assessed by measuring chemical accumulation in collected receptor fluid. Typically, receptor fluid is collected for several hours or more until a sufficient amount of the chemical has collected in the receptor compartment for detection, for example, by scintillation counting of the radiolabeled chemical. Uptake into the receptor compartment is further monitored during the period in which chemical flux across the membrane remains constant, that is, during which total chemical mass collected in the receptor increases linearly over time. This experimental approach is designed to measure chemical flux between the compartments of the diffusion cell after the chemical concentration within the skin itself remains at an approximately constant "steady-state" equilibrium value. Under these conditions, in vitro data obtained using a diffusion cell model the integrated in vivo processes of chemical diffusion through skin and subsequent systemic absorption, but not the process of mass-loading within the skin itself. The rate constant for uptake into the receptor compartment during the "steady-state" or linear phase is then calculated using Fick's first Law of Diffusion (Scheuplein and Blank, 1971):

$$K_p = \frac{J_s}{A \, \Delta C} \qquad (1)$$

in which K_p (cm h^{-1}) is the permeability rate constant, J$_s$ (mg cm^{-2} h^{-1}) is the flux of solute passively diffusing across a given area, A is the area of exposed skin, and ΔC (mg cm^{-3} mg^{-1} cm^{-2}) is the concentration difference across that area. For drugs and cosmetics that come in contact with skin, exposure is expected to last several hours or longer. The linear phase of absorption is thus of greatest relevance for estimating chemical flux for this class of chemicals. However, human exposures to organic chemical contaminants in water typically last for <1 h (e.g., in bathing or showering), during which chemical flux through skin is not at steady state, but rather is dominated by partition-limited mass loading of the exposed skin itself. For exposures of this nature, an "effective K_p" or "time-weighted-average K_p" measure (K_p^{eff}) is defined as the quantity K_p in Eq. (1) *conditional* on the area A and a specified exposure duration t (McKone and Howd, 1992). K_p^{eff} thus characterizes an integration of time-dependent, non-steady-state processes that precede any potential steady-state permeability in skin (McKone, 1993).

K_p^{eff} values for short-term exposures are expected to be greater than K_p measures obtained under steady-state conditions, although just how much greater is unknown for most organic chemicals (Cleek and Bunge, 1993). Evaluation of this hypothesis currently is hampered by a lack of direct measures of short-term, non-steady-state uptake of chemicals into human skin. Such measurements require measurement of chemical concentration in skin rather than in receptor fluid. Scintillation-based measures of total dermal uptake of an organic chemical are complicated by the fact that recovery efficiencies need to be determined accurately, which is particularly difficult for volatile lipophilic chemicals dissolved in skin. Tape stripping is a well-established method for measuring chemical in the stratum corneum (Rougier et al., 1983; Pirot et al., 1997), but this method provides no information on chemical uptake into underlying skin layers. Autoradiography has been used for direct dermal-uptake measures using tritium-labeled organic chemicals at very low concentrations (Weber et al., 1991), but is labor- and time-intensive. Physiologically-based modeling studies using live animals (Jepson and McDougal, 1997) provide dynamic characterizations of dermal absorption, but this approach requires extensive acquisition of chemical-specific physiological values (partition coefficients and metabolism rates) in addition to model validation. We have developed a method for rapid, direct measure of short-term dermal uptake of organic compounds from water. The method involves highly sensitive accelerator mass spectrometry (AMS) analysis of human tissue exposed in vitro. We describe here how AMS was applied to study dermal uptake of the common water contaminant trichloroethylene during exposures up to 1 h.

II. MATERIALS AND METHODS

A. Test Chemical and Exposure Concentrations

1,2-^{14}C-Trichloroethylene (^{14}C-TCE, 5.4 Ci/mol, ≥98% purity, Sigma Chemical Co., St. Louis, MO), obtained in a sealed ampoule, was transferred to a 5-ml vial of methanol to yield a stock solution that was continuously refrigerated at 4°C. Aqueous ^{14}C-TCE exposure solution was prepared by pipetting 2–4 μl of stock solution into 250 ml distilled water prior to use. ^{14}C-TCE concentrations were measured prior to exposure or retention experiments by transferring four 500-μl samples of exposure solution with an automatic pipette to liquid scintillation vials containing 15 ml Sigma-Fluor universal liquid scintillation cocktail (Sigma Chemical Co., St. Louis, MO) for ^{14}C quantification in a Tri-Carb 4530 scintillation counter (Packard Instrument Co., Downers Grove, IL). The mean (±1 coefficient of variation, CV) of measured starting concentrations for 4 sets of dermal-exposure experiments was 4.9 μg/L (±34%, $n = 12$), and that for 4 sets of control-retention experiments was 4.7 μg/L (±5.2%, $n = 12$); CV = standard deviation (SD) divided by the mean.

B. Control-Retention Experiments

All experiments were conducted in a low-volume flow-through glass diffusion cell (LVDC) (RCR, Inc., Novato, CA; see Fig. 1). Some chemical loss during transfer of exposure solution to the LVDC was expected. Loss over a 1 hr period from a sealed LVDC containing dermal tissue was also possible. LVDC retention experiments, with and without exposed skin, were conducted to estimate the fraction, f_r, of aqueous chemical retained after transfer into the LVDC. These experiments were conducted for 1, 15, 30, or 60 min with liquid scintillation counting of solution recovered from the LVDC as described earlier.

C. Tissue

Experiments were conducted with full-thickness and dermatomed human skin. Human breast skin, obtained from three consenting Caucasian and Afro-American women (age 27–38 years) undergoing mammoplasty at the University of California San Francisco Medical Center, in accordance with an approved human-subjects protocol (LLNL IRB number 94-118), was used as the full-thickness skin. All skin was transported on ice, stored at 4°C, and

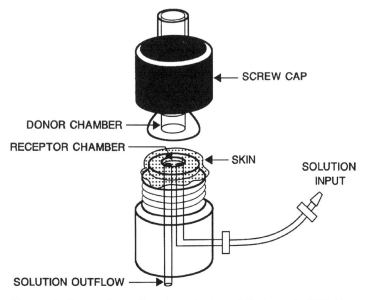

Figure 1 Low-volume flow-through glass diffusion cell (LVDC) used for short-term dermal uptake experiments (exposed-tissue diameter = 1.0 cm). The LVDC was modified by threading the glass at the top of the receptor compartment to accept a screw cap. The LVDC was not operated under flow-through conditions; skin was placed atop an aluminum-lined pad with stratum corneum facing the donor compartment.

used within 48 h of surgery. Skin samples were determined to be free of visually apparent abrasions and perforations, rinsed in distilled water, and separated from subcutaneous fat prior to use (having a mean thickness ±1 CV of 1.5 mm ± 10%). The dermatomed human tissue used was human leg skin (thigh or calf), obtained at an autopsy from a 43-year-old Caucasian female by the Northern California Transplant Bank with a dermatome set for a depth of 350 μm and cryopreserved for 2.5 years. The dermatomed skin sample was transported in liquid nitrogen and rapidly thawed in a 37°C water bath prior to use.

D. Dermal-Exposure Experiments

Tissue was exposed to solution at room temperature (20 ± 3.0°C) in the LVDC, which was not operated under flow-through conditions (i.e., the receptor compartment was empty, and access to it was blocked by an aluminum barrier as described later). For each experiment, a circular tissue piece was cut with a 18-mm-diameter aluminum cork bore and scissors, and placed on the metallic face of a 0.5-mm-thick, 18-mm-diameter cotton support pad lined with aluminum tape (0.10-mm A-25 tape, Lamart Corp., Clifton, NJ). The tissue/backing was inserted between the two LVDC compartments, with the donor compartment facing the outer (SC) surface of the skin area (0.785 cm^2) to be exposed. The two LVDC compartments were then tightly secured, a magnetic stir bar and a volume (3.7 ml) of exposure solution were placed in the donor compartment, and this compartment was then sealed with a 13-mm Mininert septum cap (VWR, San Francisco, CA) lined with aluminum tape (described earlier). In each experiment, the LVDC was placed on a magnetic stir plate together with an exterior magnet so as to elevate the interior stir bar above the exposed skin. The stir plate was then turned on, causing the stir bar to toggle slowly and thus circulate the exposure solution. Tissue samples were each exposed for 1, 5, 15, 30, or 60 min, except the 5-min experiment with the full-thickness skin, which was replicated three times for two of these samples. Control experiments measuring ^{14}C in unexposed samples were done using tissue in an LVDC that contained only distilled water and a magnet.

E. Tissue Sampling

Tissue was removed from the LVDC and blotted dry. Three to five core plugs (each with a surface area of 0.0483 cm^2) were taken by punch held perpendicular to the exposed tissue surface, dislodged from the punch, and placed into separate, precooled 6 $\times$ 50-mm glass sample tubes containing reactants used for AMS analysis (described later).

F. AMS Analysis

Tissue samples were analyzed for ^{14}C content with accelerator mass spectrometry (AMS), an isotope-ratio mass spectrometric method of quantifying radioisotopes independent of their decay times, as described more fully elsewhere (Vogel and Turteltaub, 1991). AMS uses mass selection and energy gain to separate the isotope of carbon (and other elements) so that ions of the rare isotope (in our case, ^{14}C) can be counted. Sensitivity of AMS for ^{14}C extends to 1 amol/mg carbon (i.e., 10^{-15}). The AMS unit comprises several parts (Fig. 2). At the ion source, the sample is bombarded by cesium ions, causing negatively charged ions to be ejected from the sample. These ions then pass through a low-energy mass spectrometer that selects for the desired atomic mass. As the ions pass through the accelerator they lose electrons and gain momentum. The second high-energy mass spectrometer removes the abundant isotope (^{13}C) from the ion stream. The remaining ^{14}C ions slow to a stop in the gas ionization detector with a characteristic pattern and are counted. AMS requires the samples be in solid form for placement

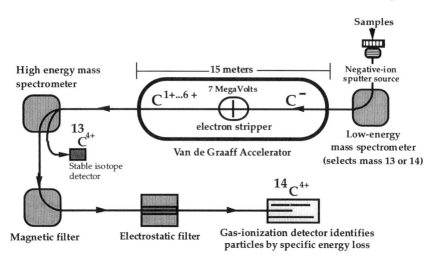

Figure 2 Schematic diagram of the accelerator mass spectrometer. Graphite samples are inserted in the ion source in the upper right of the diagram.

in the ion source, so samples were combusted and converted to graphite coated on a metallic powder (Vogel, 1992). To do this, the sample tubes were prebaked to oxidize and remove any carbon from the surfaces, and a few grams of CuO (Fisher Chemical Company, Springfield, NJ) and a small amount of pure Ag foil or powder (Alfa Products, Ward Hill, MA) were added to each tube prior to use. After the transfer of a tissue core to the sample tube, the tubes were flame-sealed within evacuated quartz tubes (9 × 150 mm) and an array of tubes was heated to 650°C for 3 h to oxidize the tissue to CO_2. Under vacuum, the CO_2 was transferred to a quartz tube (9 × 150-mm) containing reactants for graphitization. The tube was heat-sealed and heated to 650°C for 3 h, during which the CO_2 was catalytically reduced by Zn and by H_2 from TiH_3 (Alfa Products, Ward Hill, MA) to 1–2 mg of graphite (containing the ^{14}C from the tissue core) on 5–10 mg Co powder (Alfa Products, Ward Hill, MA), as described by Vogel (1992). The powder was removed and pressed into holes of 1.1 mm diameter and 1.5 mm deep in Al sample holders, which were inserted from a sample wheel in a computer controlled sequence into the Cs sputter source of the accelerator providing 35-keV negative ions from the graphite. Standards containing known amounts of ^{14}C and control samples to test for contamination were prepared and measured in exactly the same manner for comparison with the unknown samples.

Because AMS detects ^{14}C as a $^{14}C/^{13}C$ isotope ratio, the total carbon inventory of the combusted sample must be known in order to quantify ^{14}C

mass ($M^{14}C$, fmol) detected in each sampled plug. Therefore, three to six preweighed plugs from nonexposed skin samples from each subject were analyzed (Carlo Erba NA-1500 CNHS Analyzer, Fisons Instruments, Milan, Italy) for carbon content (M_C, μg). Measured concentrations of ^{14}C in exposed tissue samples were each expressed, after correction by standards and subtraction of concentrations of ^{14}C in control tissues, as a multiple (f_M) of "Modern" carbon, where "1 Modern" is defined as 5.9×10^{10} ^{14}C atoms per gram carbon (i.e., 98 fmol/g, 6.11 pCi/g, 73.3 pCi/mol). AMS measurements were converted from f_M to M_{14_C} using the relation: $M_{14_C} = f_M \times (1$ Modern/N) $\times$ (10^{-6} g/μg) $\times$ M_C, where N is Avogadro's number. AMS results obtained using a given starting exposure-solution concentration of W_o (μg/L) were normalized to a concentration of $W_0^* = 5$ ng/ml with the equation:

$$
M_{14_C} = \frac{\left(\dfrac{W_0^*}{W_0 f_r} \right)}{W_{14_C}}
\tag{2}
$$

where

$$
W_{14_C} = \frac{C_0}{MW_{TCE}} \times 10^6 \frac{\text{fmol}}{\text{nmol}} \times \frac{2 \text{ mol}^{14}\text{C}}{\text{mol TCE}} \times \frac{5.4 \text{ Ci/mol}}{124.8 \text{ Ci/mol}} = 3293 \frac{\text{fmol}}{\text{ml}}
$$

and where W_{14_C} is the aqueous ^{14}C concentration corresponding to W_0^*, 124 Ci/mol is the activity of pure ^{14}C-TCE, and f_r is the fraction of chemical retained (see Section II.B).

G. Data Analysis

Tissue ^{14}C contents were compared by Kruskal–Wallis one-way analyses of variance, where indicated using Hommel's Bonferroni-type adjusted p values (denoted p^*) to account for multiple tests (Wright, 1992). Means of the pooled data were plotted ± 1 standard deviation (SD). LVDC-retention data were normalized to 100% of the activity of exposure solution placed at $t = 0$ into the LVDC, corresponding linear regressions were analyzed and compared using appropriate F tests, and (based on the latter results) f_r in Eq. (2) was estimated from pooled data for $t > 0$.

H. Determination of Permeability Coefficient

Normalized ^{14}C uptake in full-thickness skin for the 5-min experiment were used with Eq. (1) to calculate corresponding K_p^{eff} values as follows. ^{14}C uptake per square centimeter at 5 min was calculated by dividing the tissue plug amount by the area of the tissue plug (0.0483 cm^2). The rate of ^{14}C

clearance from the exposure solution (cm/h) implied by this tissue amount was obtained using the relation: (fmol ^{14}C cm^{-2}/3293 fmol ^{14}C cm^{-3}) $\times$ 60 min/h.

All mathematical and statistical calculations were performed using Mathematica 3.0 (Wolfram, 1996).

III. RESULTS

The carbon content of tissue plugs sampled from the full-thickness skin and dermatomed skin had a mean (± 1 SD) of 31.3% ($\pm 8.5\%$, $n = 15$) and 28.1% ($\pm 12.5\%$, $n = 3$), respectively. Carbon contents differed significantly among subjects ($p = .0031$), which justified using individual-specific estimates of the ^{14}C content of tissue plugs analyzed by AMS (see Section II).

The LVDC retention data indicate that retained radiolabel was significantly negatively correlated with time after introduction of exposure solution into the LVDC, in experiments both with ($r = -.626$, $p = .00035$) and without ($r = -.627$, $p = .00037$) the presence of exposed skin. However, after excluding data collected at $t = 0$, linear models were found to be consistent with both the skin and no-skin data (F tests for linearity, $p > .11$). For these $t > 0$ data, the skin versus no-skin slopes do not differ significantly from zero ($p > .23$), and neither these slopes nor corresponding ordinate-axis intercepts differ significantly from each other (analysis of covariance, $p > .79$). For the pooled (skin + no-skin) data pertaining to $t > 0$, the estimated intercept (± 1 CV) is 92.3% ($\pm 1.2\%$), and the corresponding slope does not differ significantly from zero ($p = .089$). The LVDC retention data thus indicate that transfer of exposure solution into the LVDC resulted in a TCE loss of ~8% without discernible subsequent loss, regardless of tissue presence. An LVDC retention fraction of $f_r = .923$ was therefore assumed for AMS data analysis.

Comparison of ^{14}C detected by AMS revealed no significant ^{14}C difference between control tissue plugs that were exposed in the LVDC to distilled water only prior to experiments involving ^{14}C-TCE exposures, and control plugs similarly exposed after all ^{14}C-TCE exposures were completed ($p = .10$). Therefore, there was no evidence of sequential radiolabel contamination over the course of our experiments (which can be a problem using extremely sensitive AMS technology), and all the control data were pooled.

The ^{14}C measures obtained for tissue plugs obtained from the full-thickness skin at five exposure periods revealed significant ($p^* \leq .01$) heterogeneity among 5-min exposure data pertaining to three different individuals, which had corresponding mean uptake values equal to 98%, 60%, and 143% of the 5-min grand mean (3.2 fmol ^{14}C/plug $\pm 47\%$), and corresponding CVs all <30%. Interindividual differences at all other time points were nonsignificant ($p^* > .05$), so the data from all experiments with full-thickness skin were pooled for comparison with that from dermatomed skin. Coincidentally, 5-min exposures were also chosen for a detailed analysis of interexperiment variability among four replicate experiments (each yielding three to four AMS measurements), each done using tissue from each of two subjects. Significant interexperiment nonhomogeneity was found among measures involving only one of the subjects ($p^* = .017$). In the combined 5-min replicate data ($n = 39$), observed interindividual variability was significantly greater than observed interexperiment variability (F-ratio test, $F = 20.4$ with 1 and 3 df, $p = .020$). Figure 3 compares the increase in ^{14}C uptake of full-thickness with that of dermatomed skin. The only significant difference in ^{14}C uptake between full-thickness and dermatomed skin occurred at the 60 min exposure period ($p = .024$). ^{14}C uptakes at all time points in full-thickness skin were significantly greater from that of the pre-

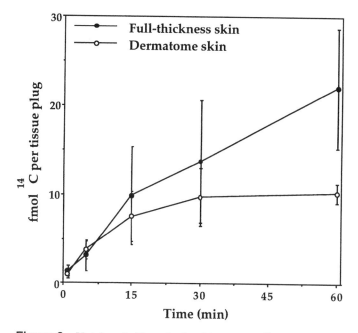

Figure 3 Uptake of dilute (~5 ppb) aqueous ^{14}C-radiolabeled trichloroethylene into full-thickness ($n = 3$, closed symbols) and dermatomed ($n = 1$, open symbols) human skin, plotted as fmol ^{14}C per tissue plug. Data points shown are means ± 1 SD of n_j measures made at each exposure duration. The number of measures obtained at 1, 5, 15, 30, and 60 min for full-thickness and dermatomed skin were 14, 44, 10, 9 and 9, and 5, 5, 4, 3, and 3, respectively.

ceding point, whereas in dermatomed skin, uptakes at 15, 30, and 60 min were not significantly different ($p \geq .18$).

The mean ($\pm$ SD) of the normalized ^{14}C uptakes in the 5-min experiments using full-thickness skin yielded a K_p^{eff} estimate of 0.24 ± 0.11 cm/h.

IV. DISCUSSION

We have applied highly sensitive AMS technology to a detailed investigation of short-term, non-steady-state, percutaneous uptake for full-thickness human skin in vitro for ≤ 1 h with dilute aqueous TCE. Replicate experiments using tissue from different individuals and exposure durations indicate that experimental error in the present study was due primarily to random measurement variability, but that such error may also reflect significant interindividual variability evident in data obtained for 5-min exposures. These data are unlikely to have been biased due to any lack of skin integrity, which is an important consideration in standard flow-through dermal-uptake experiments. Our diffusion system was operated in a static (non-flow-through) mode, using relatively fresh (48-h-old), apparently normal human skin tissue (Methods). Any unseen skin perforations could have enhanced measured uptake, but at most only proportional to their (undetected) surface area within plugs actually sampled from each exposed piece of tissue. In our experiments, we sampled randomly from the exposed skin, and no extraordinarily high measures were obtained in any of the experiments.

To obtain information on the depth of penetration of TCE within the skin, uptake in full-thickness skin was compared to that of skin dermatomed to a depth of 350 μm. Insofar as these two tissues were from different body regions and stored under different conditions, this comparison is limited. However, short-term absorption is expected to be controlled by the stratum corneum, which, as a nonviable skin layer, should not differ substantially between the two tissues. Similar uptakes in the two tissues for exposure periods up to 30 min suggest that TCE penetration was not affected by the thickness difference of the two tissues. At 60 min, membrane thickness may be the cause of the different uptakes in the two tissues. The full-thickness skin was expected to take considerably longer to reach equilibrium than the dermatomed skin. That uptake in dermatomed skin did not differ significantly after 15 min suggests that TCE partitioning was relatively complete in this tissue. Using a one-compartment pharmacokinetic model of skin to describe data obtained in these experiments (Bogen et al., 1998), the depth of penetration of TCE into the full-thickness skin was estimated to be in the range of 40 to 100 μm in these experiments, which indicates a diffusion rate for TCE in skin that is much faster than previously expected (U.S. EPA, 1992).

The K_p^{eff} estimate we obtained for TCE at 5 min is slightly greater than that of 0.20 cm h^{-1} ($\pm$ 11%) obtained in AMS experiments conducted using similarly exposed human cadaver skin (Bogen et al., 1996). Few other in vitro data are available with which to estimate an K_p^{eff} for TCE. A value of 0.11 cm h^{-1} is obtained for TCE using a physicochemical model for estimating dermal absorption from short-term exposures (U.S. EPA, 1992). In vivo data from other studies support the K_p^{eff} as well. An K_p^{eff} estimate of 0.23 ($\pm$ 17%) cm h^{-1} was obtained in a study in which immersed, sedated hairless guinea pigs were exposed to dilute aqueous ^{14}C-TCE at 32°C for 70 min (Bogen et al., 1992). Finally, the results of the present in vitro study of TCE uptake are consistent with in vivo data on TCE that were considered (along with in vivo data on eight other volatile organic compounds) by Bogen (1994) to indicate rapid dermal absorption for this compound. Using the K_p^{eff} value obtained in this study and assuming an exposed surface area of 18,000 cm^2, showering for 5 min will add 36% to the dose of TCE received through ingestion (0.24 cm h^{-1} $\times$ 18,000 cm^2 $\times$ 0.083 h).

This study illustrates the capability of AMS to measure total uptake of an organic chemical into human skin at extremely low concentrations and after relatively brief exposure durations, relevant to exposure scenarios of regulatory concern. AMS has distinct advantages over other methods for measuring chemical penetration into the skin. AMS sensitivity (1000-fold greater than scintillation counting) allows direct measures of uptake using test chemicals at ultra-low concentrations in exposure media, reducing potential vehicle effects on chemical partition into skin. AMS sensitivity can therefore provide uptake measures for a wide range of exposure durations (e.g., seconds to hours), facilitating studies of penetration kinetics. AMS can also be used with other methods to quantify chemical absorption into stratum corneum tape strips obtained after in vivo human dermal exposures (Pirot et al., 1997). Therefore, the capability of AMS to analyze small whole-tissue samples without extraction should facilitate short-term dermal-uptake studies, as well as biodistribution research in general.

ACKNOWLEDGMENTS

We are grateful to Kurt Haack (LLNL Center for AMS) for AMS sample preparation, Dr. William Reifenrath (RCR, Inc., Novato, CA) for LVDC assistance, Dr. Richard Guy (Centre Interuniversitaire de Recherche et d'Enseignement, Archamps, France) for supplying human surgical skin, and Dr. James Forsell (Northern California Transplant Bank) for supplying the dermatomed skin. This work was performed under the auspices of the U.S. Department of Energy at Lawrence Livermore National Laboratory under contract W-7405-ENG-48 and the U.S. Environmental Protection Agency, Interagency Agreement DW89937390-01-0.

REFERENCES

Bogen, K. T. Models based on steady-state in vitro permeability data underestimate short-term dermal exposures in vivo. J. Expos. Anal. Environ. Epidemiol. 1994; 4:457–476.

Bogen, K. T., Colston, B. W., and Machicao, L. K. Dermal absorption of dilute aqueous chloroform, trichloroethylene, and tetrachloroethylene in hairless guinea pigs. Fundam. Appl. Toxicol. 1992; 18:30–39.

Bogen, K. T., Keating, G. A., Meissner, S., and Vogel, J. S. Initial uptake kinetics in human skin exposed to dilute aqueous trichloroethylene in vitro. J. Expos. Anal. Environ. Epidemiol. 1998; 8:253–271.

Cleek, R. L., and Bunge, A. L. A new method for estimating dermal absorption from chemical exposure. 1. General approach. Pharm. Res. 1993; 10:497–506.

Jepson, G. W., and McDougal, J. N. Physiologically-based modeling of nonsteady state dermal absorption of halogenated methanes from aqueous solution. Toxicol. Appl. Pharmacol. 1997; 144:315–324.

McKone, T. E. Linking a PBPK model for chloroform with measured breath concentrations in showers: Implications for dermal exposure models. J. Expos. Anal. Environ. Epidemiol. 1993; 3:339–365.

McKone, T. E., and Howd, R. A. Estimating dermal uptake of nonionic organic chemicals from water and soil. Part 1. Unified fugacity-based models for risk assessments. Risk Anal. 1992; 12:543–557.

Pirot, F., Kalia, Y. N., Stinchcomb, A. L., Keating, G., Bunge, A., and Guy, R. H. Characterization of the permeability barrier of human skin in vivo. Proc. Natl. Acad. Sci. USA 1997; 94:1562–1567.

Rougier, A., Dupius, D., Lotte, C., Roguet, R., and Schafer, H. In vivo correlation between stratum corneum resevoir function and percutaneous absorption. J. Invest. Dermatol. 1983; 81:275–278.

Scheuplein, R. J., and Blank, I. H. Permeability of the skin. Physiol. Rev. 1971; 51:702–747.

U.S. Environmental Protection Agency. Dermal Exposure Assessment: Principles and Applications. EPA/600/8-91/011B. U.S. EPA Office of Environmental Assessment, Washington, DC, 1992.

Vogel, J. S. Rapid production of graphite without contamination for biomedical AMS. Radiocarbon 1992; 34:344–350.

Vogel, J. S., and Turteltaub, K. W. Biomolecular tracing through accelerator mass spectrometry. TRAC—Trends Anal. Chem. 1991; 11:142–149.

Weber, L. W., Zesch, A., and Rozman, K. Penetration, distribution and kinetics of 2,3,7,8-tetrachlorodibenzo-p-dioxin in human skin in vitro. Arch. Toxicol. 1991; 65:421–428.

Wolfram, S. The Mathematica Book, 3rd ed. Cambridge, UK: Cambridge University Press, 1996.

Wright S. P. Adjusted p-values for simultaneous inference. Biometrics 1992; 48:1005–1013.

29
New Developments in the Methodology Available for the Assessment of Topical Corticosteroid-Induced Skin Blanching

John M. Haigh
Rhodes University, Grahamstown, South Africa

Eric W. Smith
Ohio Northern University, Ada, Ohio

Howard I. Maibach
University of California School of Medicine, San Francisco, California

Since the publication of the previous edition of this book (1) there have been considerable developments and controversy in the field of topical corticosteroid bioequivalence assessment. There has been considerable discussion in the literature concerning the use of the Minolta chromameter for the measurement of corticosteroid-induced skin blanching (2), as it is believed this instrument would produce more objective results than the visual grading procedure. These efforts culminated in the release of a guidance document (3) from the Food and Drug Administration (FDA) detailing the procedures to be followed for the determination of topical corticosteroid bioequivalence using the chromameter. Since the promulgation of this document there have been challenges on the validity and scientific merit of the documented procedures (4), and recently the FDA itself conceded that it may be necessary to redefine some of the protocol evaluations (5). This chapter attempts to define the current standing of the two methods of response assessment.

I. INTRODUCTION

The human skin blanching assay (1) is unique in the field of bioequivalence testing of topical products in that the assessment methodology relies upon the production of a side effect of localized vasoconstriction and the consequent blanching (whitening) of the healthy skin following the topical application of corticosteroids. Since it has been shown (6) that the intensity of the induced blanching is directly proportional to the clinical efficacy of the corticosteroid, this procedure is particularly convenient as the induced response may be used as an indicator of the potency of a new corticosteroid drug moiety or the success of the topical vehicle delivery system in bioequivalence evaluations. In the past 15 years it has become patently obvious that formulations that are pharmaceutical equivalents (the same drug in the same type of formulation at the same concentration) may have markedly different clinical efficacy simply because of the differing potential of the compounded vehicle to release the drug to the stratum corneum. Considerable effort has, therefore, been applied to the research of delivery vehicle optimization and maximization of the thermodynamic leaving potential of drugs in topical formulations (7). Allied to this research effort has been the need to develop analytical systems capable of discriminating between the subtle drug delivery potentials of very similar formulations—especially for topical bioequivalence testing. Fortunately for corticosteroid products, this has been relatively facile because of the blanching phenomenon.

II. VISUAL ASSESSMENT

The human skin blanching assay is routinely practiced in laboratories throughout the world, as it is a valuable tool for the assessment of the topical availability of corticosteroid formulations (8). Since the degree of blanching is directly proportional to the clinical efficacy, it follows that if two formulations containing the same corticosteroid in the same concentration are being compared, the formulation that produces the greatest degree of blanching will be clinically most efficacious (9). If different corticosteroids are being compared, the one that produces the highest degree of blanching will be the more potent, and this has lead to the production of the corticosteroid formulation potency ranking tables, all of which are based on results obtained from the visually applied blanching assay.

This blanching effect has been successfully estimated over the last 35 years (10) by subjective visual assessments using an arbitrary grading scale.

This system has substantial merit as the human eye is an excellent discriminator of very small differences between colors, and numerous publications have documented the successful practical application, sensitivity, and remarkable reproducibility of the visual blanching methodology (11,12). The ability to parallel this color discrimination is still to be refined in optical analytical instruments, especially as the eye–brain combination has the ability to accurately and simultaneously assess skin color differences between the application site and surrounding (unmedicated) skin. To date this has required the manual subtraction of one value from another using instrumental measurements. Furthermore, the global visual assessment subconsciously accounts for skin factors such as inherent skin pigmentation, hirsuitism, and mottling, parameters that cannot easily be accounted for with instrumental measurements.

The methodology of the visual assay protocol was documented at length in the previous edition of this book (1), and only subsequent information is discussed here. This assay procedure is subjective in nature since the assessments are performed by the human eye. We have reported at length on the accuracy and reproducibility of this procedure (11,12) and have shown that, provided the assay is performed by trained observers, very similar results are obtained on repeated evaluation, and intertrial reproducibility was shown to be excellent (13). We routinely use three trained observers to assess the degree of skin blanching, because of the subjective nature of the observations, and normally 12 volunteers with at least three (and often more) application sites on each forearm being dosed with the same formulation.

To assess the reproducibility of the assay, three identical trials at 8-week intervals were mounted (12) comprising 18 volunteers, three observers, and two commercially available formulations (formulations A and B) containing the same concentration of the same steroid and each occupying six sites per arm. After a 6-h application period the formulations were removed from the skin and the blanching was estimated at intervals over an extended period of time. Since each formulation was applied to 12 sites per volunteer, it follows that at each time interval the blanching produced by a particular formulation was estimated 648 times. It can be seen from the area under the curve (AUC) values in Table 1 that, for each trial, each observer placed the two formulations in the same rank order, thus verifying the reproducibility of this assay procedure.

A further refinement of the assay protocol has been designed in our laboratories. A retrospective reanalysis of the data obtained from various blanching trials (14) showed quite clearly that the degree of blanching produced by the same formulation varies depending upon the position on the forearm which it occupies. Maximum blanching occurs in the middle of the forearm, with a slight reduction closer to the elbow and a dramatic reduction

Table 1 Area Under the Blanching Curves for Two
Pharmaceutically Equivalent Formulations Determined
by Three Independent Observers in Each of Three
Replicate Trials

Observer	Trial	Formulation A	Formulation B
1	1	803	619
	2	781	674
	3	767	624
2	1	815	656
	2	888	784
	3	799	639
3	1	731	558
	2	686	616
	3	605	516

closer to the wrist. For this reason we have suggested that more than one
site on each forearm should be demarcated for each formulation (preferably
a minimum of three sites) and these should be spread over the whole length
of the forearm. This finding is especially troublesome when assessed in terms
of the FDA guidance protocol, which makes no stipulation of the number
or positioning of application sites. In the light of the preceding discussion,
this would clearly produce incorrect results.

III. CHROMAMETER ASSESSMENT

The use of the eye to estimate corticosteroid-induced skin blanching has
been criticized (15) due to the subjectivity of visual assessments, which does
not allow for interlaboratory comparison of results. It has been suggested
(3,15) that the chromameter should be used to make these measurements,
as it is an objective method that quantifies the reflectance of a xenon light
pulse in terms of three indices, the a-scale (red–green), the b-scale (yellow–
blue), and the L-scale (light–dark). These three values define a point in
three-dimensional space that characterizes the color of the measured surface.

The FDA guidance suggests the use of the chromameter to measure
skin blanching in a complex protocol of pilot and pivotal trials with multiple
correction (baseline and unmedicated-site values) of the a-scale values only.
The a-scale values were chosen as they are the only set of values that appear
to show appreciable changes over the period of time during which blanching
is measured. One of the many contentious issues in the guidance document
that have been reported (4,16) is the requirement that the data from the pilot

investigation be modeled by suitable software to produce an effective dose at 50% response and establish a 33–67% range of response. The latter is used in the assessment of the results from the pivotal trial to exclude all subjects whose response does not fall within this range. This in itself does not seem logical, as no screening procedure is applied to the pilot study volunteers in order to ensure that they are a typical subsection of the population of responders. The inclusion/exclusion criteria that may be produced from an inappropriately-selected group of volunteers for the pilot study may skew the "acceptable" results generated in the pivotal study.

There are, similarly, several theoretical problems with the modeling procedure for the topical drug delivery data, since the exact dose of drug generating a specific response cannot be determined. The amount of formulation applied to the skin is a constant, known mass (arbitrarily chosen by the investigator), but the mass of drug that penetrates the skin and reaches the site of action is variable depending on the biological characteristics of the skin of each volunteer. In addition, the skin contact time of the dose of formulation is also left to the discrimination of the investigator, further confounding the concept of dose in the dose-response relationship. The guidance document suggests varying the skin contact time of the applied dose of formulation to generate different doses for modeling purposes; the relationship between skin contact time and the mass of drug penetrating the skin (especially when assessed in terms of stratum corneum reservoir formation) has not been sufficiently evaluated. Recent developments in the field of skin stripping methods for the determination of mass of drug in the skin may be beneficial in this regard. Moreover, the results of data modeling exercises are only as good as the raw data generated in the experimentation. In light of this discussion, there has been some comment on the accuracy and precision of the data generated by the chromameter for modeling purposes (4).

Our experiences with the chromameter (both the Minolta 200 and 300 models) have demonstrated (4,16) that the results obtained are far from objective. For the measurement of homogeneous, planar surface colors, the results obtained with the chromameter are reproducible and accurate and, therefore, objective. However, when it is applied to the manual measurement of the color of human skin, several problems arise:

1. The pressure of the head of the chromameter applied by the investigator to the skin of the volunteer can change the color of the skin due to vasocompaction.
2. The presence of hairs, moles, and skin mottling on the forearm can give artifactual readings.
3. The angle at which the chromameter is presented to the skin by the investigator can cause different values to be recorded at the same site.

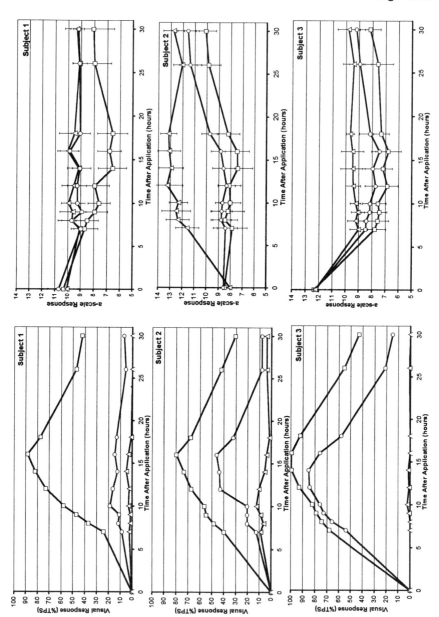

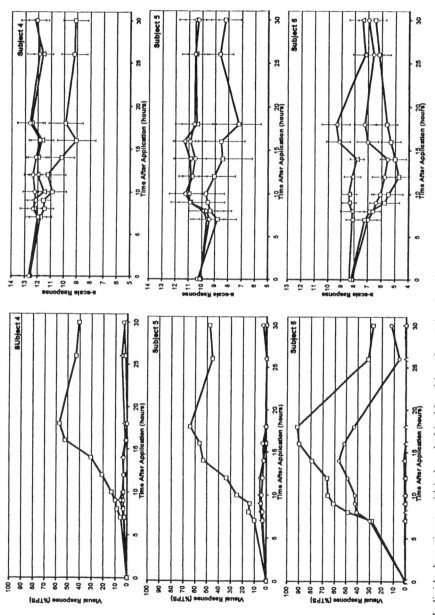

Figure 1 Individual corticosteroid-induced skin blanching data (visual and uncorrected a-scale chromameter) for six subjects using a typical study protocol (□ clobetasol propionate, ○ betamethasone 17-valerate, △ unmedicated control sites).

4. The chromameter is a hand-held instrument; thus the operator could become fatigued, as several hundred readings are taken per hour in a topical availability trial involving a full panel of subjects.

The inherent objectivity of the instrumental computations is reduced in overall value by the subjective manipulation of the chromameter head by the investigator; these are problems that would not arise if the instrument were being used in a nonbiological environment.

The results from six subjects in a typical blanching trial evaluation are depicted in Fig. 1. Here the blanching response profiles of two corticosteroid formulations from different potency groups (Dovate, 0.05% clobetasol propionate, Pharmacare-Lennon, South Africa; and Betnovate, 0.1% betamethasone 17-valerate, Glaxo-Wellcome, South Africa) were recorded by visual observation using four independent observers and by chromameter assessment. The visual profiles for each volunteer are typical, from our experience, and show a clear maximal response for the more potent formulation and, for four of the six subjects depicted, a lesser response for the lower potency formulation. There is negligible response recorded at the unmedicated (control) sites for all volunteers, and subjects four and five also show negligible response for the lower potency formulation. What is immediately apparent from the a-scale chromameter data is that there is less of a clear distinction between the control, lower, and higher potency formulation responses, although the rank order of response is appropriate. Especially troublesome is that the precision of the mean data values is such that there is no clear statistical difference between the responses of these two formulations from different potency groups. The period of peak blanching (12–14 h after initial application) is crucial in the determination of bioequivalence by visual observation. If we examine the profiles of the typical responders (subjects 1–3 and 6), there is no statistical difference between the data points of the two formulations at this region of peak response. This result is mirrored even for the "atypical" subjects four and five. In a blind assessment, one might conclude from these results that the two formulations are equivalent, whereas the visual methodology (and clinical usage) has proved these products to be of different potency. One may also debate the merits of attempting to model data with this degree of imprecision, especially when the maximal response of the most potent formulation produces a response profile that is only marginally different to that produced from untreated skin. These concerns reiterate those of a previous publication (4). We have observed that the differences between a-scale values for individual volunteers are so large that meaningful results are difficult, if not impossible, to obtain from commercially available modeling packages.

If one returns to the question of inclusion criteria for subjects, clearly only four of these six volunteers are typical responders and yet the guidance document does not make any provision for the exclusion of subjects from the pilot study pool. If this had been part of a pilot study pool, then all the results would have been used to determine the 50% effective dose and the 33–67% range. The mean data for all six subjects are shown in Fig. 2, which reinforces the discussion already presented: There is clear distinction between the products in the visual assessment, and in the chromameter assessment no statistical difference between the high-, moderate-potency, and unmedicated responses. Since the visual grading is based on an arbitrary

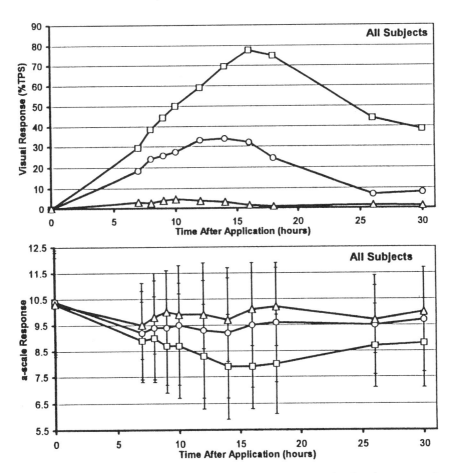

Figure 2 Mean corticosteroid-induced skin blanching data (visual and uncorrected a-scale chromameter) ± 1 standard deviation for six subjects using a typical study protocol (□ clobetasol propionate, ○ betamethasone 17-valerate, △ unmedicated control sites).

0–4 scale, standard deviation analysis of the visual means is inappropriate, but if this is conducted by assuming the scale is approximately linear, then very good precision is obtained about each point with clear statistical difference between the three profiles (16).

A further problem with the FDA guidance is the way in which it is suggested that the a-scale values should be corrected for both baseline and untreated site values. We have shown (16) that irrespective of the manner in which these values are corrected, the shapes of the blanching profiles remain essentially unchanged. More important from the view of determining bioequivalence, although some of the a-scale blanching profiles do show some similarity to the visually obtained curves, the standard deviations observed for the a-scale values are so large that there is overlap of the means of treated and untreated sites at all observation times.

The FDA guidance document suggests the use of a-scale values only. It seems to us that since the chromameter produces three values for the full definition of any color, it should be possible to utilize all three values for more accurate definition of skin color and therefore more appropriate conclusions to be drawn. We have measured the Euclidean distances in three-dimensional space (16) between untreated sites and treated sites, and plots of these values show greater similarity to the visually obtained blanching profiles than do the individual profiles of the a-, b-, and L-scale values. However, even the Euclidean distance plots of the reported initial assessment display unacceptably large standard deviations with overlapping of mean error bars of the treated and untreated sites at most observation points. Nevertheless, this mode of data analysis is worthy of further investigation.

IV. CONCLUSION

We concede that the visual assessment of corticosteroid-induced skin blanching is subjective. However, the proven reproducibility and accuracy of this assay procedure indicate that it still has a useful role to play in the assessment of topical corticosteroid availability and, especially, in providing the standard to which objective techniques must equate or surpass. The caveat that must be applied here is that the visual assay has to be performed correctly by experienced personnel. The most important factors in this regard are:

1. More than one observer should be used. In our laboratories we routinely use three or four observers.
2. The observers must be trained. We have found (17) that it takes inclusion in three full training trials before an observer can be

considered reliable and consistent for inclusion into a topical avail-ability trial.

3. More than one site per arm per formulation must be used, and these sites must be spread over the whole length of the forearm, avoiding wrist and elbow. We use three, four, or six sites per arm per formulation, depending on the trial structure.

4. Readings must be taken over a period of time after removal of the formulations to allow the construction of a blanching profile. We have often noted that a formulation that produces high blanching in the early part of a trial will produce lower AUC values than a formulation that produces lower blanching early in the trial.

We agree that it would be preferable to have an objective method for the measurement of corticosteroid-induced skin blanching, and we are work-ing toward the realization of this idea. We believe that the use of the chro-mameter is a step in the right direction, but we are not sure that this is the ideal instrument, used in the currently prescribed manner, for making these measurements. There is still much validation that has to be performed before this technique can achieve global acceptance.

REFERENCES

1. Smith EW, Meyer E, Haigh JM, Maibach HI. In: Bronaugh RL, Maibach HI, eds. Percutaneous Absorption. Vol. 2. New York: Marcel Dekker, 1989:443–460.

2. Pershing LK. Assessment of topical corticosteroid-induced skin blanching re-sponse using the visual McKenzie-Stoughton and colorimetric methods. Drug Infect J 1995; 29:923–934.

3. Guidance: Topical dermatologic corticosteroids: In vivo bioequivalence. Divi-sion of Bioequivalence, Office of Generic Drugs, Food and Drug Administra-tion, Rockville, MD. June 1995.

4. Demana PH, Smith EW, Walker RB, Haigh JM, Kanfer I. Evaluation of the proposed FDA pilot dose-response methodology for topical corticosteroid bio-equivalence testing. Pharm Res 1997; 14:303–307.

5. Smith EW, Walker RB, Haigh JM, Kanfer I. Is the topical corticosteroid bio-equivalence testing methodology accurate, precise and robust? Pharm Res 1998; 15:4–7.

6. Cornell RC, Stoughton RB. Correlation of the vasoconstrictor assay and clin-ical activity in psoriasis. Arch Dermatol 1985; 121:63–67.

7. Davis AF, Hadgraft J. Effect of supersaturation on membrane transport: 1. Hydrocortisone acetate. Int J Pharm 1991; 76:1–8.

8. Meyer E, Magnus AD, Haigh JM, Kanfer I. Comparison of the blanching activities of Dermovate, Betnovate and Eumovate creams and ointments. Int J Pharm 1988; 41:63–66.

9. Smith EW, Meyer E, Haigh JM. Blanching activities of betamethasone 17-valerate formulations: effect of the dosage form on topical drug availability. Arz Forsch 1990; 40:618–621.
10. McKenzie AW, Stoughton RB. Method for comparing percutaneous absorption of steroids. Arch Dermatol 1962; 86:608–610.
11. Smith EW, Meyer E, Haigh JM. Accuracy and reproducibility for the multiple reading skin blanching assay. In: Maibach HI, Surber C, eds. Topical Corticosteroids. Basel: Karger, 1992:65–73.
12. Haigh JM, Meyer E, Smith EW, Kanfer I. The human skin blanching assay for in vivo topical corticosteroid assessment. I. Reproducibility of the assay. Int J Pharm 1997; 152:179–183.
13. Haigh JM, Meyer E, Smith EW, Kanfer I. The human skin blanching assay for in vivo topical corticosteroid assessment. II. Subject- and observer-dependent variation in blanching responses. Int J Pharm 1997; 152:185–192.
14. Meyer E, Smith EW, Haigh JM. Sensitivity of different areas of the flexor aspect of the human forearm to corticosteroid-induced skin blanching. Br J Dermatol 1992; 127:379–381.
15. Shah VP, Peck CC, Skelly JP. Vasoconstriction-skin blanching assay—A critique. Arch Dermatol 1989; 125:1558–1561.
16. Smith EW, Haigh JM, Walker RB. Analysis of chromameter results obtained from corticosteroid-induced skin blanching. I. Manipulation of data. Pharm Res 1998; 15:280–285.
17. Smith EW, Meyer E, Haigh JM. The human skin blanching assay for topical corticosteroid bioavailability assessment. In: Shah VP, Maibach HI, eds. Cutaneous Bioavailability, Bioequivalence and Penetration. New York: Plenum Press, 1993:55–162.

30
Raman Spectroscopy

Adrian C. Williams and Brian W. Barry
University of Bradford, Bradford, West Yorkshire, England

I. INTRODUCTION

Numerous analytical techniques have been used to examine the nature of skin, and in particular the barrier structure of human stratum corneum, including thermal methods (1,2), electron spin resonance (3), and x-ray diffractometry (4,5). One approach has been to use vibrational spectroscopy to probe the molecular nature of the skin, with early work concentrating on the use of Fourier transform infrared (FTIR) spectroscopy (6). While the infrared technique has provided valuable insights into the concentration of water in the tissue (7) and has been used successfully to probe the tissue in vivo using attenuated total reflection methods (8,9), infrared is not ideally suited to examine the vibrational modes of biological samples that are naturally hydrated. This is because such materials show interference from water vibrational modes, while the water absorbs strongly the infrared radiation.

An alternative vibrational spectroscopic technique is Raman spectroscopy. Raman spectroscopy is both a qualitative and quantitative technique for characterizing materials at the molecular level in terms of molecular normal vibrational frequencies. If a sample is irradiated with an intense beam of monochromatic radiation (usually from a laser) operating at a wavenumber ν_0, most of the radiation is transmitted by the sample. However, a small portion of the exciting radiation (about 1 photon in 10^4) scatters elastically with a wavenumber equal to that of the incident radiation. An even smaller fraction (around 1 photon in 10^8) scatters inelastically with wavenumbers different to the incident radiation. The elastically scattered radiation is known as Rayleigh scattering, whereas the weaker inelastic scattered radiation is designated as Raman scattering. Raman scattering occurs when a

499

photon of the incident radiation at wavenumber ν_0 collides with a molecule causing one of the normal modes of vibration (ν_m) to increase its vibrational energy at the expense of the incident photon. The incident photon annihilates and a new photon scatters at a wavenumber of $\nu_0 \pm \nu_m$. The normal modes of vibration of a sample, ν_m, are characteristic vibrational wavenumbers of the molecular species undergoing excitation and thus give a measure of the molecular structure of the sample.

Thus, Raman scattering provides vibrational energy lower than the Rayleigh scattering (at $\nu_0 - \nu_m$) termed Stokes scattering and at higher wavenumbers ($\nu_0 + \nu_m$) termed anti-Stokes scattering, as Shown in Fig. 1. The proportion of Stokes or anti-Stokes scattering depends on whether the molecular mode occupied the ground state or first excited state energy level, respectively. This phenomenon can provide useful information as to the temperature of a sample; the relative intensities of the Stokes to anti-Stokes lines directly measure the ground or first excited energy levels and hence relate to temperature. Further theory and background can be found in the literature (10–12).

Raman and infrared spectroscopies are complementary techniques that differ in their interactions with the incident radiation. Briefly, for a vibrational mode to be infrared active, the dipole moment of a bond must change,

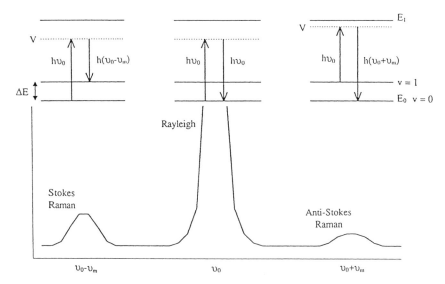

Figure 1   Transitions between energy levels with corresponding scatter bands (E_0, E_1; ground and excited electronic levels, respectively; ν, vibrational levels; V, virtual level, $\Delta E = h\nu_m$, vibrational energy; $E = h\nu_0$ excitation energy).

whereas for a mode to be Raman active the polarizability of a bond must alter. Thus, for bonds with an existing dipole, such as a C=O moiety, a change in the dipole moment can easily be induced and consequently the mode is strongly infrared active. However, such a dipole-distorted bond cannot easily have its polarizability altered and so only a very weak Raman effect arises. In contrast, homopolar bonds such as C=C moieties can easily have their electron cloud distorted and so are polarizable and thus Raman active, whereas they possess no dipole moments and hence are infrared inactive. Thus, the selection rules for activity in these two techniques differ, leading to complementarity of the two techniques. The practical consequences of this are that water is a very weak Raman scatterer but strongly absorbs infrared radiation, making Raman spectroscopy the usual vibrational spectroscopic technique of choice for examining the molecular structure of biological materials. Additionally, there is little sample preparation for Raman spectroscopy, samples can be analyzed in glass vessels, and the wavenumber range over which Raman spectroscopy provides molecular information is greater than that of infrared. However, the technique examines weak scattered radiation and hence is not as sensitive for low-concentration work as infrared. Some of the main practical differences between the two techniques are summarized in Table 1.

II. METHODS

The two main types of Raman spectrometers implement either dispersive or Fourier transform (FT) techniques, and both comprise four basic instrumental components: a radiation source (usually a laser), optics to illuminate the

Table 1 Comparison of Some of the Main Practical Differences Between FTIR and FT Raman Spectroscopies

Feature	IR	Raman
Polar bonds (C=O)	Good	Bad
Homopolar bonds (C=C)	Bad	Good
Water as a solvent	Bad	Good
Glass accessories	Bad	Good
Low concentration	Good	Bad
Spectral range	OK	Good
Fluorescence	Good	?
Linearity of signal with concentration	?	Good
Ease of sampling	OK	Good

sample and to collect and focus the scattered radiation, a monochromator (dispersive) or interferometer (FT), and a detector.

While (theoretically) Raman scattering can be excited anywhere in the electromagnetic spectrum, it is generally the ultraviolet–visible–infrared regions that are used. Selection of the radiation source, generally a laser, is influenced by sample color, stability, and molecular properties that may result in fluorescence, an emission of radiation while the sample is being irradiated. Fluorescence can often swamp the weaker Raman scattering. Generally, the fluorescence tends to weaken as the laser moves toward the infrared, and this has driven the current surge in manufacture of near-infrared Fourier transform Raman spectometers. The fluorescence effect is quite pronounced for biological materials, including skin.

Optics vary between FT and dispersive systems, notable with orientation of the sample. For human skin, a comparison of the two Raman techniques has been reported (13).

The use of a monochromator (dispersive) or interferometer (FT) is the main difference between the two Raman methods. In dispersive instruments, the scattered radiation focuses onto the narrow slits of a monochromator and then disperses into its constituent wavelengths via a diffraction grating. A camera mirror/lens arrangement collects the dispersed radiation and focuses it at the exit slit, where each spectral unit or spectral range presents to the detector sequentially. Recently, multiple or diode array detectors have been introduced that can detect a range of wavenumbers simultaneously, which greatly improves analysis speed. FT instruments simultaneously measure the signal of multiple data points by analysis of a single signal from the detector, providing a full scan (4000 to 50 cm^{-1}) from one scan using a Michelson interferometer. The absence of physical dispersion of radiation by diffraction gratings results in a more efficient use of the scattered radiation. Additionally, the data accumulation is superimposable, allowing accurate coaddition of scans.

Detector technology has improved rapidly over the last 10 years, which is another factor in the commercialization of FT systems. Choice of detector largely depends on the region of the electromagnetic spectrum being studied. Photographic plates, photomultipliers, and charge-coupled diode (CCD) array detectors are commonly used in the visible and ultraviolet regions. With infrared systems, mercury cadmium telluride (MCT) and liquid-nitrogen-cooled gallium arsenide (GaAs) detectors tend to be employed.

III. SPECTRA OF HUMAN STRATUM CORNEUM: INFRARED COMPARED TO RAMAN

Figure 2 gives a comparison of FTIR and FT Raman spectra of the same sample of human stratum corneum. This figure clearly illustrates the com-

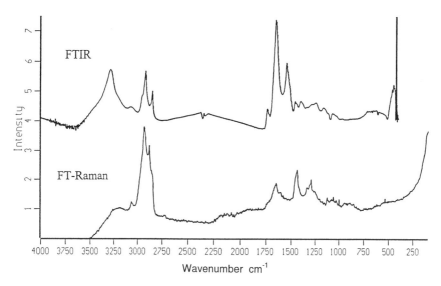

Figure 2 Comparison of FTIR and FT Raman spectra of human stratum corneum.

plementarity of the two techniques, although there are important differences. In the IR spectrum, there is a strong water feature around 3300 cm^{-1} that is largely absent in the Raman spectrum. Stratum corneum is naturally hydrated, and the presence of the strong water feature in the IR spectrum overlaps to some extent with the C-H stretching modes seen between 3100 and 2700 cm^{-1}. As described later, the position and relative intensities of these C-H stretching modes are important indicators of the degree of order/disorder in the stratum corneum lipids, so interference with these modes by water in the IR spectrum is problematic.

The fingerprint region of the spectra (1800 to 1000 cm^{-1}, where functional group modes are seen) also shows some important differences. While the Amide I band (largely C=O stretching) is strong in the infrared, it is a relatively weak feature in the Raman. Other modes, such as the CH$_2$ scissoring of lipid chains, are stronger in the Raman. However, most of the vibrational modes exhibit in both techniques.

As the spectra tend to lower wavenumbers (<1000 cm^{-1}) there is a marked difference in quality. Due to the strong absorbance of infrared radiation by water, the IR spectrum loses definition and peaks are difficult to observe. In the Raman spectrum, a series of weak but important bands show around 1200 to 1000 cm^{-1}. These are due to C-C stretching modes of the lipid chains, the position and intensity of which indicate the conformation of the C-C bonds. Further spectral features are discernible in the Raman spectrum down to around 400 cm^{-1}, including S-S stretching modes from proteins.

IV. ASSIGNMENT OF SPECTRAL FEATURES

Band assignments, consistent with the molecular modes, have been made for the Raman spectrum of human stratum corneum (14), and are compared with those from the infrared in Table 2. The origin of the bands has been determined by exhaustive lipid extraction of samples of human stratum corneum prior to analysis. As can be seen from the table, the Raman spectrum provides nearly twice the number of modes as the IR spectrum, in addition to some of the improvements in spectral quality described earlier.

There are numerous techniques available for fitting bands to the spectral profiles, although most of these tend to be iterative, with the operator entering the number and shape of the peaks in the spectral range to be fitted. Using a Fourier self-deconvolution technique, the main assignments and peak positions in Table 2 were confirmed (15).

V. RAMAN SPECTRA OF HUMAN AND ANIMAL SKINS

A comparison of Raman spectra from human and animal tissues has been made (16). Snake skin, a proposed model for human tissue (17), clearly possesses a molecular structure markedly different from that of human stratum corneum, notable due to the presence of the outer β-keratin sheet layer, whereas the spectrum for pig stratum corneum more closely resembles that of the human (Fig. 3). In snake skin, the lipids characteristic of human stratum corneum (ceramides) are replaced by phospholipids. The molecular differences have clear implications for spectroscopic or other analytical investigations of, for example, penetration enhancer mechanisms.

VI. IN VITRO/IN VIVO COMPARISONS

As with FTIR, spectra can be collected in vivo using a fiber-optic probe connected to the FT Raman spectrometer. Good quality data can be produced in this way, and a comparison of in vivo and in vitro data is given in Fig. 4. There are some minor differences in the spectra, notably the presence in the in vivo data of an additional feature around 3230 cm^{-1}. This is probably due to the presence of water—the probe tends to occlude the skin, and even though water is a weak Raman scatterer it is clearly seen. Probe technology is advancing, and it is now possible to sample the skin using noncontact probes, which minimize this effect. Other minor differences are seen in the relative intensities of spectral bands—for example, the C-C stretching

Table 2 FTIR and FT Raman Spectral Assignments Consistent with the Vibrational Modes from Human Stratum Corneum

Frequency (cm^{-1})		
Raman	Infrared	Assignment consistent with vibrational mode
424	—	δ(CCC) skeletal backbone
526		ν(SS)
600		ρ(CH) wagging
623		ν(CS)
644		ν(CS); amide IV
746		ρ(CH$_2$) in-phase
827		δ(CCH) aliphatic
850		δ(CCH) aromatic
883		ρ(CH$_2$)
931		ρ(CH$_3$) terminal; ν(CC) α-helix
956		ρ(CH$_3$); δ(CCH) olefinic
1002		ν(CC) aromatic ring
1031		ν(CC) skeletal cis conformation
1062	1076	ν(CC) skeletal trans conformation
1082		ν(CC) skeletal random conformation
1126		ν(CC) skeletal trans conformation
1155		ν(CC); δ(COH)
1172		ν(CC)
1207		
1244	1247	δ(CH$_2$) wagging; ν(CN) Amide III disordered
1274		ν(CN) and δ(NH) Amide III α-helix
1296	1298	δ(CH$_2$)
1336		
	1366	δ[C(CH$_3$)$_2$] symmetric
1385	1389	δ(CH$_3$) symmetric
	1401	δ[C(CH$_3$)$_2$] asymmetric
1421		δ(CH$_3$)
1438	1440	δ(CH$_2$) scissoring
	1451	δ(CH$_3$) asymmetric
	1460	δ(CH$_2$)
	1515	
1552	1548	δ(NH) and ν(CN) amide II
1585		ν(C=C) olefinic
1602		
1652	1650	ν(C=O) amide I α-helix
	1656	ν(C=O) amide I disordered
1743	1743	ν(C=O) lipid
1768		ν(COO)
2723		v(C-CH$_3$)
2852	2851	ν(CH$_2$) symmetric
	2873	ν(CH$_3$) symmetric
2883		ν(CH$_3$) symmetric
	2919	ν(CH$_2$) asymmetric
2931		ν(CH$_2$) asymmetric
2958	2957	ν(CH$_3$) asymmetric
3000		
3060		ν(CH) olefinic
	3070	1st overtone, amide II at 1548 cm^{-1}
	3287	ν(OH) of H$_2$O

Note. δ = deformation; ν = stretch; ρ = rock; v = nonfundamental vibration.

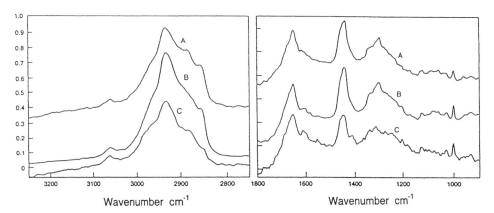

Figure 3 FT Raman spectra of (A) human stratum corneum, (B) pig stratum corneum, and (C) snake skin over the wavenumber ranges 3300 to 2700 and 1800 to 900 cm^{-1}.

modes tend to be more pronounced in vivo than in vitro. Portable Raman spectrometers are available and have been used in clinical situations for assisting in the complete removal of cancerous tissue, which has a spectral profile different from that of healthy skin (18).

VII. THERMAL PERTURBATION OF STRATUM CORNEUM LIPIDS

Using a novel environmental chamber (19), the thermal perturbation of human stratum corneum lipids has been researched (20). The positions of the CH stretching modes around 3100 to 2700 cm^{-1} indicate the conformation of the intercellular lipid chains and have been used to look at phase transitions in model lipid systems (21). In our work on heating stratum corneum it was found that a novel measure of separation between the CH$_2$ asymmetric and symmetric stretching modes was a valuable indicator of lipid order (20). Heating human stratum corneum from 25 to 100°C altered markedly the spectral profile (Fig. 5). From these data, a plot of the rate of change in band separation between the CH$_2$ asymmetric and symmetric stretching modes revealed three lipid transitions, which correlate well with transitions from thermal analysis (Fig. 6).

Other spectral features also alter on heating the tissue, and it can be seen from Fig. 5 that the profile of the CH$_2$ scissoring mode changes. This is a complex band with several overlapping vibrational modes. However,

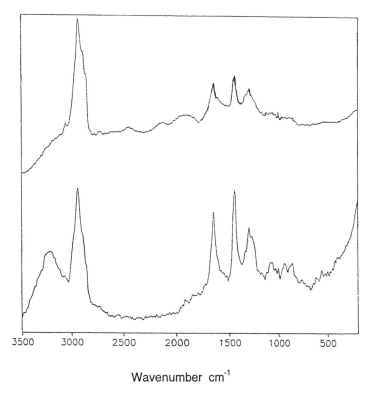

Figure 4 FT Raman spectra of human stratum corneum collected in vitro (top) and in vivo (bottom).

from curve fitting and analysis of the intensity (area) of the band centered around 1440 cm^{-1} relative to the total band profile area, it is again possible to discern transitions in the lipid domains. This analysis, illustrated in Fig. 7, shows the main thermal events as before, but with the addition of a further transition at around 55°C, which is occasionally seen in thermal studies. The spectroscopic evidence suggests that the 55°C event is a minor reorganization of the CH$_2$ moieties of the lipid alkyl chains.

Also of interest from Fig. 5 is the alteration of the C-C stretching modes on heating. Bands at 1126 and 1065 cm^{-1} arise from the C-C bonds in the trans conformation, and these are seen to decrease in intensity on heating. This indicates that free rotation about the C-C bond gradually develops within the lipid chains.

The thermal study was extended to examine more closely the events

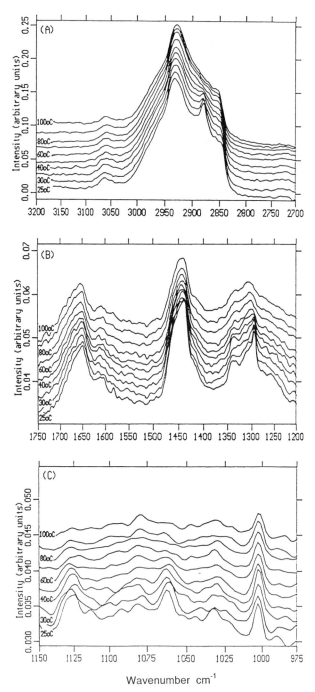

Figure 5 FT Raman spectra of human stratum corneum heated from 25 to 100°C: (A) 3200 to 2700 cm^{-1}, (B) 1750 to 1200 cm^{-1}, and (C) 1150 to 975 cm^{-1}.

508

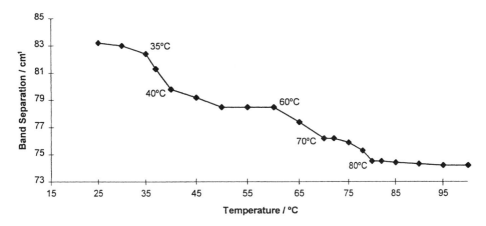

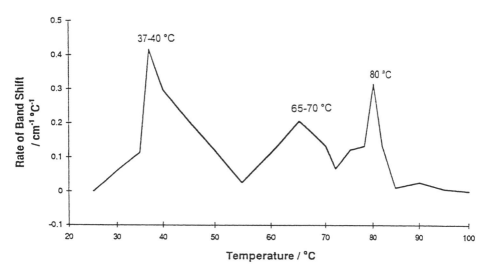

Figure 6 Plot of the band separation and the rate of change in band separation between the CH$_2$ asymmetric and symmetric stretching modes of human stratum corneum lipids with increasing temperature.

occurring during the main thermal transitions at ~70 and 80°C. Heating at 1°C intervals, it was seen that the lipid chains disordered, then reordered immediately after the melt, as molecules relaxed into a more stable packing arrangement before further heating again disordered the lipid chain organization (20).

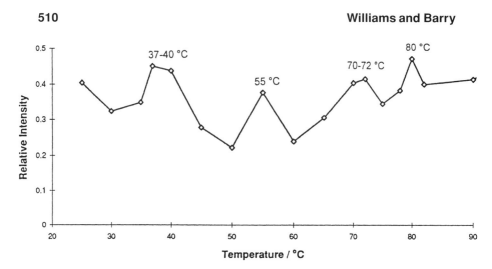

Figure 7   Plot of the variation in the CH_2 scissoring mode area of human stratum corneum lipids with increasing temperature.

VIII.   PENETRATION ENHANCER STUDIES USING FT RAMAN SPECTROSCOPY

Interference is a problem encountered using FTIR spectroscopy to examine the molecular interactions between penetration enhancers and stratum corneum constituents. Most enhancers possess vibrational modes that coincide with those of the tissue, notably the CH stretching modes discussed earlier. This has led to the use of deuterated enhancers, which shift the C-D modes to lower wavenumbers (22). Raman offers the advantage that linear subtractions can be made across the whole wavenumber range without the need for complex calibration curves. Thus, spectra can be generated with an enhancer present, and the signal for the enhancer can be removed by normalization on a unique band to leave just stratum corneum bands modified by the enhancer. This approach has been used for tissue treated with 1,8-cineole and with dimethyl sulfoxide (DMSO) (23,24). DMSO was shown to disturb the C-C lipid stretching modes, indicative of a trans gel to a trans–gauche liquid-crystalline phase. More clearly, the DMSO altered the α-keratin to β-sheets, an effect that could be semiquantified by curve fitting and intensity measurements. In contrast, the data from 1,8-cineole were more complex, with evidence that the enhancer existed in a discrete pool, in addition to direct modification of the lipid domain.

IX. FT RAMAN TO STUDY DISEASES TISSUE

FT Raman spectroscopy has been used to examine the molecular basis of many diseased states, notably cancerous tissue (18). It has also been employed to examine and compare the molecular nature of callus, psoriatic plaques, and verrucae (25–27). Clearly, in the callus and psoriatic samples, diminution of intercellular lipid was evident, and this may afford an opportunity to use the technique in clinical situations. In the study of verrucae, it was also possible to examine the interaction of salicyclic acid with the verrucae. The drug was shown to form monomers from the dimer when it interacted with the cells containing the human papillomavirus in the verruca. The use of FT Raman spectroscopy for investigation of skin diseases is still in its infancy, although its ease of sampling and noninvasive nature offer the potential for the technique to be enrolled as an aid to diagnosis and therapy.

X. OTHER USES OF FT RAMAN SPECTROSCOPY IN SKIN RESEARCH

FT Raman spectroscopy could be valuable in many other areas of skin research. In our laboratories, we have enlisted the technique to investigate the molecular events occurring during iontophoresis. Another area has been to use FT Raman spectroscopy to assess diffusion characteristics through human skin membranes for a mixture of permeants. As the permeants have different molecular structures, their Raman spectra also vary. Since the signal obtained is directly proportional to concentration, flux values can be obtained. Indeed, it has been possible to examine the concomitant flux of up to five permeants, allowing assessment of the flux of all components of a formulation in one experiment. The main limiting factor in this type of study is one of sensitivity.

A further use of the technique has been to examine archaeological samples. Skin from a 5200-year-old corpse from the Alps called the "Iceman" was examined, with Raman spectroscopy being the ideal tool due to its nondestructive aspect. The ancient stratum corneum was shown to be remarkably well preserved, with some expected degradation of olefinic bonds, but the stratum corneum structure had largely survived intact (15,28).

XI. CONCLUSIONS

FT Raman spectroscopy is a valuable tool in many areas of skin research. It offers clear advantages over infrared spectroscopy, but the two techniques should be viewed as complementary rather than competitive. For a full molecular picture of a sample, both techniques should be enlisted, as bands that are weak in the Raman (such as the Amide I mode) tend to be stronger in the infrared. A prime advantage of the Raman technique is the weakness of the water signal; indeed, Raman spectroscopy tends to be the vibrational spectroscopic technique of choice for most biological systems. Other advantages illustrated include nondestructive sampling, the ability to obtain in vivo data, and the ease of sample preparation. The main disadvantage of the procedure is the relative insensitivity compared to FTIR spectroscopy. However, advances in instrumentation methodology, and notably the improvements in detector design, mean that FT Raman in future may well supersede the use of FTIR in many areas of skin research.

REFERENCES

1. Van Duzee BF. Thermal analysis of human stratum corneum. J. Invest. Dermatol. 1975;65:404–408.
2. Goodman M, Barry BW. Differential scanning calorimetry of human stratum corneum: effects of penetration enhancers Azone and dimethyl sulphoxide. Anal. Proc. 1986;23:397–398.
3. Rehfeld SJ, Plachy WZ, Hou SYE, Elias PM. Localisation of lipid microdomains and thermal phenomena in murine SC and isolated membrane complexes: An electron spin resonance study. J. Invest. Dermatol. 1990;95:217–223.
4. Bouwstra JA, Gooris GS, van der Spek JA, Bras W. Structural investigations of human SC by small angle x-ray scattering. J. Invest. Dermatol. 1991;97:1005–1012.
5. Cornwell PA, Barry BW, Stoddart CP, Bouwstra JA. Wide angle x-ray diffraction of human stratum corneum: effects of hydration and terpene enhancer treatment. J. Pharm. Pharmacol. 1994;46:938–950.
6. Golden GM, Guzek DB, Harris RR, McKie JE, Potts, RO. Lipid thermotropic transitions in human SC. J. Invest. Dermatol. 1986;86:255–259.
7. Potts RO, Guzek DB, Harris RR, McKie JE. A non-invasive, in vivo technique to quantitatively measure water concentration of the stratum corneum using attenuated total-reflectance infrared spectroscopy. Arch. Dermatol. Res. 1985;277:489–495.
8. Mak VHW, Potts RO, Guy RH. Percutaneous penetration enhancement *in vivo* measured by attenuated total reflectance spectroscopy. Pharm. Res. 1990;7:835–841.
9. Bommannan D, Potts RO, and Guy RH. Examination of SC barrier function *in vivo* by infrared spectroscopy. J. Invest. Dermatol. 1990;95:403–408.
10. Long DA. Linear and non-linear Raman effects: the principles. Chem. Br. 1989;25:592–596.

11. Tu AT. Raman spectroscopy in biology: Principles and applications. New York: John Wiley & Sons, 1982.

12. Hendra P, Jones C, Warnes G. Fourier transform Raman spectroscopy: Instrumental and chemical applications. New York: Ellis Horwood, 1991.

13. Williams AC, Barry BW, Edwards HGM, Farwell DW. A critical comparison of some Raman spectroscopic techniques for studies of human stratum corneum. Pharm. Res. 1993;10:1642–1647.

14. Barry BW, Edwards HGM, Williams AC. Fourier transform Raman and infrared vibrational study of human skin: Assignment of spectral bands. J. Raman Spectrosc. 1992;23:641–645.

15. Edwards HGM, Farwell DW, Williams AC, Barry BW, Rull F. Novel spectroscopic deconvolution procedure for complex biological systems: Vibrational components in the FT-Raman spectra of Ice-man and contemporary skin. J. Chem. Soc. Faraday Trans. 1995;91:3883–3887.

16. Williams AC, Barry BW, Edwards HGM. Comparison of Fourier transform Raman Spectra of mammalian and reptilian skin. Analyst 1994;119:563–566.

17. Edwards HGM, Farwell DW, Williams AC, Barry BW. Raman spectroscopic studies of the skins of the Sahara sand viper, the carpet python and the American black rat snake. Spectrochim. Acta 1993;49A:913–919.

18. Gniadecka M, Wulf HC, Mortensen NN, Nielsen OF, Christensen DH. Diagnosis of basal cell carcinoma by Raman spectroscopy. J. Raman Spectrosc. 1997;28:125–129.

19. Edwards HGM, Farwell DW, Turner JMC, Williams AC. Novel environmental control chamber for FT-Raman spectroscopy: Study of *in situ* phase change of sulfur. Appl. Spectrosc. 1997;51:101–107.

20. Lawson EE, Anigbogu ANC, Williams AC, Barry BW, Edwards HGM. Thermally induced molecular disorder in human stratum corneum lipids compared with a model phospholipid system: FT-Raman spectroscopy. Spectrochim. Acta 1998;54A:543–558.

21. Gaber BP, Peticolas WL. On the quantitative interpretation of biomembrane structure by Raman spectroscopy. Biochim. Biophys. Acta 1977;465:260–274.

22. Ongpipattanakul B, Burnette RR, Potts RO, Francoeur ML. Evidence that oleic acid exists in a separate phase within stratum corneum lipids. Pharm. Res. 1991;8:350–354.

23. Anigbogu ANC, Williams AC, Barry BW, Edwards HGM. Fourier transform Raman spectroscopy in the study of interactions between terpene penetration enhancers and human skin. In: Brain KR, James V, Walters KA, eds. Prediction of Percutaneous Penetration; Methods, Measurements, Modelling. STS Publishing, Cardiff, 1993:27–36.

24. Anigbogu ANC, Williams AC, Barry BW, Edwards HGM. Fourier transform Raman spectroscopy of interactions between the penetration enhancer dimethyl sulfoxide and human stratum corneum. Int. J. Pharm. 1995;125:265–282.

25. Williams AC, Edwards HGM, Barry BW. Raman spectra of human keratotic biopolymers: skin, callus, hair and nail. J Raman Spectrosc. 1994;25:95–98.

26. Edwards HGM, Williams AC, Barry BW. Potential application of FT-Raman spectroscopy for dermatological diagnostics. J. Mol. Struct. 1995;347;379–388.

27. Lawson EE, Edwards HGM, Barry BW, Williams AC. Interaction of salicyclic acid with verrucae assessed by FT-Raman spectroscopy. J. Drug Target 1998; 5:343–351.

28. Williams AC, Edwards HGM, Barry BW. The "Iceman": Molecular structure of 5200-year-old skin characterised by Raman spectroscopy and electron microscopy. Biochim. Biophys. Acta 1995;1246:98–105.

31

Stratum Corneum and Barrier Performance
A Model Lamellar Structural Approach

Hamid R. Moghimi
School of Pharmacy, Shaheed Beheshti University of Medical Sciences, Tehran, Iran

Brian W. Barry and Adrian C. Williams
School of Pharmacy, University of Bradford, Bradford, West Yorkshire, England

I. INTRODUCTION

Transdermal drug delivery has been much researched over recent decades, yet there are still many problems to overcome. Percutaneous absorption is a complicated process, but a well-designed model system may allow us to isolate and examine individual processes and thus provide information about key barrier components. The intercellular lipid domain is the major pathway for permeation of most drugs through the stratum corneum (SC). Preparation of a model for SC intercellular lipids provides opportunities to probe the barrier nature of the horny layer and the drug and enhancer interactions with the SC intercellular lipids. As these lipids form lamellar structures, simple lamellar models may serve for this purpose and are examined here.

II. STRUCTURE AND FUNCTIONS OF SKIN

Human skin comprises a multilayer penetrated by hair shafts and gland ducts, consisting of the hypodermis (fatty subcutaneous layer), dermis, and epidermis. As the epidermis serves as the body's primary barrier against drug permeation, its structure is discussed here; further information is reviewed elsewhere (1–3).

The epidermis consists of five layers: stratum germinativum (basal layer), stratum spinosum (spinous layer), stratum granulosum (granular layer), stratum lucidum, and stratum corneum (horny layer). The epidermal cells (keratinocytes) rapidly divide, mature, and finally desquamate. Basal layer cells have typical intercellular organelles and are metabolically active. As they approach the SC they flatten, their metabolic activity decreases, and in the SC the cells are dead, anucleate, and metabolically inactive. Upward migration continues as cells at the surface desquamate.

A. Routes of Drug Permeation Across the Skin

Chemicals permeate the skin either across the intact epidermis (transepidermal pathway) or via the sweat glands and hair follicles (appendageal or shunt route). These routes provide barriers in parallel; therefore, the overall flux across the skin is the sum of the individual fluxes through each pathway, which itself depends on different physicochemical and geometrical properties (1).

1. The Appendageal Pathway

The available diffusional area of the shunt route is approximately 0.1% of the total skin area (4,5). Despite their small fractional area, the skin appendages may provide the main portal of entry into the subepidermal layers of the skin for ions (6,7) and large polar molecules (1,8,9).

The role of the appendageal pathway in the transport of low- to medium-molecular-weight nonelectrolytes is still not clear. It is argued that for small nonelectrolytes, the shunt route is dominant only in the transient (non-steady-state) phase of percutaneous absorption, but makes a negligible contribution to overall flux in the steady-state period (4,5). Thus, the contribution of shunt route in the steady-state transport of polar and non-polar steroids was less than 10% (10) and there was a poor correlation between appendageal density and percutaneous absorption when different skin regions were compared (11,12). However, recent studies have focused on the appendages as a potential pathway for steady-state permeation of a wide range of drugs (13,14). Such studies show that discussion of the issue is not yet closed and further investigations are required.

2. The Transepidermal Pathway

Transdermal permeation is a sequence of partitioning and diffusion in the SC, viable epidermis, and papillary layer of the dermis, with the microcirculation usually providing an infinite sink (1). A major problem in skin research has been the exact location of the barrier layer. Smith et al. (15) noted that dichloroethylsulfide is first rapidly taken up by some element on or adjacent to the skin surface, but cannot rapidly penetrate further. In 1951, the main barrier to permeation of water was found in the more external layers of the epidermis, particularly the SC (16). Subsequent experiments supported this finding, and it is now widely accepted that the SC is the rate-determining barrier to percutaneous absorption of most compounds in most conditions (5,17–21). However, for very lipophilic drugs, the rate-determining step may change from diffusion through, to clearance from, the SC (1).

B. Structure and Barrier Properties of the Stratum Corneum

The stratum corneum is a multilayered wall-like structure in which terminally differentiated keratin-rich epidermal cells (corneocytes) embed in an intercellular lipid-rich matrix. This two-compartment arrangement is usually simplified to a bricks (corneocytes) and mortar (intercellular domain) analogy (Fig. 1) (22,23). The SC comprises 15–20 layers of corneocytes and

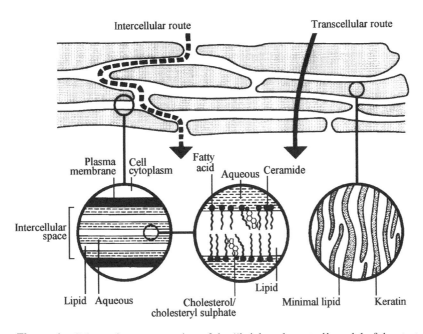

Figure 1 Schematic representation of the "brick and mortar" model of the stratum corneum and a very simplified lamellar organization of intercellular domain in which only major stratum corneum lipids are shown. Also illustrated are possible pathways of drug permeation through intact stratum corneum (62).

typically varies in thickness from 10 to 15 μm in its dry state (24–27). After hydration, the SC swells and its thickness may reach to around 40 μm, if well hydrated (4). Corns, calluses, soles, and palms are much thicker at hundreds of microns.

The anucleate corneocytes lack recognizable organelles and inclusions other than keratin filaments in an amorphous matrix. Corneocytes are polygonal, elongated, and flat, 0.2–1.5 μm thick with diameters varying from 34 to 46 μm (1,27–29).

SC lipids average 5–15% of the dry tissue weight, varying from 3 to 50% between individuals (30–32). They mainly comprise ceramides (Fig. 2), cholesterol, and fatty acids, together with smaller amounts of cholesteryl sulfate, sterol/wax esters, triglycerides, squalene, n-alkanes, and phospholipids (Table 1) (30,32–35). Lampe et al. (30,33) measured a high concentration of triglycerides (25%, w/w) in the SC, probably because of subcutaneous contamination.

Stratum corneum lipids localize mainly in the intercellular space with little in the corneocytes (23,36,37). Besides lipids, some proteins and enzymes inhabit the intercellular space (38,39).

1. Intercellular Lamellae

An organized bilayer structure for intercellular lipids was postulated by Michaels et al. (22). Freeze-fracture electron microscopy studies of the SC (40,41) confirmed that SC intercellular lipids arrange into bilayers (23). The lipids pack into lamellae, with the hydrocarbon chains mirroring each other and the polar groups dissolving in an aqueous layer (Fig. 1).

Ruthenium tetroxide staining electron microscopy provided information on the structure of intercellular bilayers and revealed that they arrange in pairs (42). The two bilayers that form each unit firmly bind together, possibly by the action of molecules such as acylceramides, which act as molecular rivets. Neighboring bilayer pairs may mutually contribute chains to an intermediate structure having the dimensions of half a bilayer, that is, a monolayer (42–44).

Small-angle x-ray studies (SAXS) of hydrated human SC (45,46) showed that SC lipids develop in two lamellar structures with repeat distances of 65 and 134 Å. SC intercellular lipids show a complex polymorphism. According to wide-angle x-ray diffraction studies (WAXS) of human SC, lipid alkyl chains arrange into orthorhombic perpendicular and (pseudo)hexagonal (gel-phase) packings, with some in the liquid state. Cholesterol is probably also present in the form of small crystals in human SC (47,48). That is, the SC intercellular domain comprises a mixture of crystalline, lamellar gel (L_β), and lamellar liquid crystalline (L_α) lipid phases.

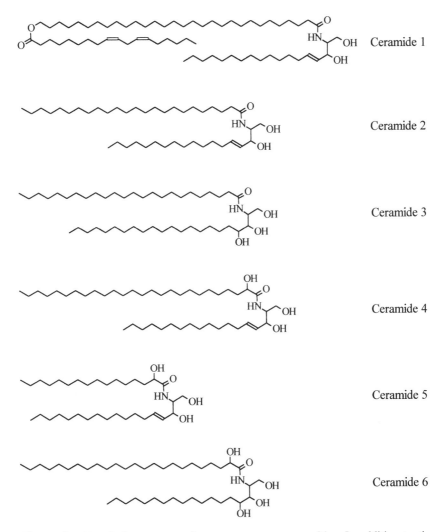

Figure 2 Chemical structures of stratum corneum ceramides. In addition to the structures shown, ceramide 6 also contains a minor component with an ester-linked hydroxyacid (35).

Intact intercellular lamellae are present at all levels in human SC and their appearance does not differ significantly between different anatomical locations (42,43). The number of intercellular lamellae between corneocytes varies from place to place within each SC sample and in some regions

Table 1 Composition of Human and Pig Stratum Corneum Lipids in Weight Percent

	Species (site)			
Constituent	Human (abdominal)[a]	Human (plantar)[b]	Human (?)[c]	Pig (?)[d]
Ceramides	18	35	41	39
Ceramide 1	14		3.2	6.0
Ceramide 2	4.3		8.9	13
Ceramide 3			4.9	5.2
Ceramide 4			6.1	4.1
Ceramide 5			5.7	3.5
Ceramide 6			12	7.2
Glucosylceramides	Trace		0.0	0.7
Fatty acids	19	19	9.1	11
Stearic acid	1.9			
Palmitic acid	7.0			
Myristic acid	0.7			
Oleic acid	6.3			
Linoleic acid	2.4			
Palmitoleic acid	0.7			
Others	<0.1			
Cholesterol	14	29	27	28
Cholesteryl sulphate	1.5	1.8	1.9	2.0
Sterol/wax esters	5.4	6.5	10	15
Di- and triglycerides	25	3.5	0.0	1.3
Squalene	4.8	0.2		
n-Alkanes	6.1	1.7		
Phospholipids	4.9	3.2		0.0

[a]Lampe et al. (30,33).
[b]Melnik et al. (34).
[c]Wertz and Downing (35).
[d]Wertz et al. (32).

lamellae are absent, allowing direct contact between the corneocyte lipid envelopes (43).

2. Origin of Stratum Corneum Intercellular Lamellae

SC lipids extrude into the intercellular space from secretory organelles—lamellar bodies or lamellar granules. Lamellar bodies first form in the upper stratum spinosum and lower stratum granulosum cells and contain lamellated material. When the keratinocytes reach the mid to upper part of the stratum granulosum, the lamellar granules aggregate in the peripheral cytoplasm; their membranes fuse with the plasma membrane and eventually discharge their lamellar contents into the intercellular space. During subsequent apical migration, at the granular–horny cell layer interface and between the first and second layer of the SC, the extruded material rearranges to form broad intercellular lamellae (40,41). Previously, it was thought that lamellar bodies contain flattened and stacked unilamellar vesicles that fuse after extrusion to form uninterrupted broad bilayers (49). However, Menon et al. (50) argued that the newly secreted lamellar body contents comprise pleated sheets (not discs) that unfold in the intercellular space to form continuous lipid bilayers.

3. Corneocyte Envelopes

An approximately 10-nm-thick protein envelope covers SC cells. Covalently bound to the outer surface of this envelope is an approximately 4-nm-thick lipid layer (the corneocyte lipid envelope). Thus, the intercellular lamellae do not contact the proteins of the corneocyte envelope, but a lipid layer (51,52). The lipids of the corneocyte envelope account for approximately 2% of SC dry weight and mainly comprise N-(ω-hydroxyacyl)sphingosines. It is postulated that the corneocyte lipid envelope resists the passage of water and other small polar molecules into and out of the corneocytes, plays an important role in natural SC cohesion, and acts as a scaffold for deposition of lamellar-body-derived, intercellular bilayers (44,51–53).

4. Permeation Pathways Across the Stratum Corneum

Figure 1 illustrates the intercellular and transcellular routes of drug permeation across the intact SC. Because the intercellular space of the SC was originally assumed to comprise a tiny portion of its overall volume (maximum 5% in dry SC and less than 5% in hydrated tissue), this space traditionally had been discounted as a possible pathway. It was assumed that water-soluble substances traverse the SC via diffusion within aqueous regions located near the outer surface of intracellular keratin filaments and lipid-soluble molecules through diffusion in the lipid matrix region between the filaments (5). Freeze-fracture studies showed, however, that the intercellular volume may be at least a factor of 3–7 times greater than was previously appreciated and accounts for 5–30% of the total tissue volume (40,54). Permeation studies of salicylic acid and water crossing the SC from different regions of human skin showed that while there is no clear correlation between either the penetration data and SC thickness or the number of cell layers, a distinct relationship emerges when these data for each site are compared to lipid composition (55). It is well known that lipid solubility is usually a prerequisite for good permeation through the horny layer. Histochemical studies revealed that SC lipids localize mainly in the intercellular space, and the cells are almost devoid of lipids (23,36). Tracer techniques elegantly demonstrated that this extracellular lipid material constitutes the morphological equivalent of the permeability barrier for most molecules (40,41). In the light of these considerations, Elias and coworkers (36,54) suggested that the intercellular (not the transcellular) is possibly the major pathway by which most molecules (both hydrophilic and hydrophobic) permeate the SC.

Further evidence supports an intercellular pathway. Albery and Hadgraft (56) showed that methyl nicotinate penetrates human skin through an intercellular pathway. Nemanic and Elias (57) perfused human and mouse SC with *n*-butanol followed by precipitation with osmium vapor and showed that although variable amounts precipitated within corneocytes, it was primarily found in intercellular domains. They concluded that the diffusant mainly permeates the SC through the intercellular pathway. However, caution is needed in such a conclusion as uptake and diffusion are independent phenomena and the domain with the higher uptake does not necessarily correlate with the rate-controlling route for permeation. The diffusion coefficient and diffusional pathlength of water penetrating porcine SC indicated that water permeates the SC through a tortuous pathway, that is, the intercellular lipids (58).

Boddé et al. (59) provided one of the most direct pieces of evidence in support of the intercellular route. Using a vapor fixation technique with electron microscopy, they visualized Hg^{2+} permeating down the SC through the intercellular pathway. However, with sufficient time, corneocytes also took up small amount of Hg^{2+}. Furthermore, they showed that during the washing process, most of the intercellular material went, while the small amount of Hg^{2+} present in the cells remained. This observation suggests that the intercellular lipid pathway is the main pathway for permeation of drugs on a usual time scale. Finally, the diffusion coefficients (D) of tiamenidine, 5-fluorouracil, and estradiol in lamellar lipid models are higher than the apparent SC values and are well correlated to the geometry (porosity and tortuosity) of SC intercellular channels, indicating that these drugs traverse the SC via intercellular lipids (60–62).

However, in spite of these considerations, the transcellular pathway should not always be dismissed as of no consequence. Contribution of intercellular or transcellular pathways in percutaneous absorption of drugs depends on the drugs' diffusivity in lipids and proteins, partitioning of drugs between these domains, and geometry of the SC (22,63,64). A micropore route may also be important for very polar compounds. However, at present, the majority of subject specialists consider the intercellular lamellar structure to be the major pathway for permeation of most drugs across the SC.

III. STRATUM CORNEUM INTERCELLULAR LIPID MODELS AND THEIR APPLICATIONS

A. Bilayer-Forming Ability of Stratum Corneum Intercellular Lipids: A Model Approach

The SC intercellular lipids that form lamellae are unusual, because, they are almost devoid of phospholipids (33). This is significant, as such molecules are generally the structural basis of biological bilayer membranes (65). The question is: How do SC lipids arrange themselves into bilayers in the absence of phospholipids?

Elias et al. (23) suggested that sphingolipids (glucosylceramides and ceramides) and cholesterol provide the amphiphilic properties required to form the membrane bilayers in the SC. Friberg and Osborne (66) provided evidence supporting an alternative hypothesis. They found that sphingolipids alone are unable to produce a lamellar structure with water, nor does a mixture of water and a lipid model containing phosphatidylethanolamine, ceramides, cholesterol, cholesteryl sulfate, fatty acids, oleic acid palmityl ester, squalene, triolein, and pristane (in almost the same proportions as exist in the SC). However, the condition changed when the pH of the skin (4.5–6.0), at which the fatty acids are partially neutralized, was considered. When about 40% of free fatty acids was neutralized, the system gave a lamellar mesomorphic structure with water. Combination of all the components with a proper acid/soap ratio produced a SAX pattern similar to that of the intact SC. These observations suggested that the fatty acids are key components in SC bilayer formation (66,67).

Cholesterol also seems to be important for structuring SC. Friberg et al. (68) prepared a lamellar mesomorphic phase from unsaturated fatty acids/soaps and water. Further addition of saturated fatty acids distorted the lamellar liquid crystal (microcrystals embedded into a liquid crystalline matrix), but adding cholesterol reformed the original liquid crystalline state. They concluded that disordering caused by cholesterol allowed that fraction of fatty acids residing between methylene group layers to penetrate the amphiphilic layer anchored at the water layers. Using a model lipid system and x-ray and differential scanning calorimetry techniques, Mizushima et al. (69) found that incorporation of cholesterol into anhydrous and hydrated equimolar mixtures of pseudoceramide and stearic acid enhances the mobility of hydrocarbon chain lipids, suggesting that cholesterol can regulate the mobility of hydrocarbon chains of the SC lipids. A mixture of cholesterol and ceramides forms a layered structure, indicating that such a combination is the basis of the double bilayer units of the SC intercellular domain (70). Further examples of lamellar systems prepared with ceramides and cholesterol (with or without other SC lipids) are provided later.

Another molecule apparently important in bilayer formation of SC lipids is cholesteryl sulfate. Wertz et al. (71) tried to prepare lamellar structures from SC containing lipids and found that a mixture of cholesterol and ceramides was unable to form liposomes in an aqueous medium of pH 7.5, but addition of one or both of palmitic acid and cholesteryl sulfate produced liposomes. Abraham and Downing (72) also prepared liposomes from ceramides, cholesterol, free fatty acids, and cholesteryl sulfate at pH 7.0. It was suggested that either free fatty acids or cholesteryl sulfate, ionized, is responsible for bilayer formation (71). Contradictory to these observations, a mixture of ceramides, cholesterol, and palmitic acid, with almost the same proportion and pH as employed previously (71), was unable to construct liposomes; however, addition of cholesteryl sulfate provided them (73).

These results indicate that no single component is responsible for the bilayer-forming ability of SC intercellular lipids; however, it seems that acids/soaps, cholesteryl sulfate, cholesterol, and ceramides provide the required amphiphilicity for such an arrangement.

B. Importance of the Lamellar Structure in the Barrier Performance of the Stratum Corneum Intercellular Lipids: A Model Approach

After the importance of the SC intercellular domain in the drug absorption was realized, the following question was raised: Which specific compound or structure is responsible for the barrier performance of SC intercellular lipids? To answer this query, Friberg et al. (74) studied the permeation of water through SC corneocytes reaggregated with model lipids. Their results revealed that membranes reaggregated with a model containing most of SC lipids show the same barrier performance toward water as those reaggregated by only a simple lamellar fatty acids/soaps system. The same group (75) then measured the permeation of water through reaggregated SC membranes using different aggregation elements of natural SC lipids, lecithin:water, sodium lauryl sulfate:decanol:water, Tween:water, and sodium lauryl sulfate: dioxyethylene dodecyl ether:water. These systems provided a lamellar (neat) mesomorphic structure except for Tween:water, which showed a hexagonal (middle) structure. Permeation experiments revealed that there was no drastic difference between the permeation of water through membranes using natural lipids and those reaggregated by other systems. They concluded that the presence of a mesomorphic structure is the primary factor in the barrier performance of SC intercellular lipids and that the differences in barrier property with specific lipid changes may be negligible (74,75). These experiments suggest that a hexagonal (middle) structure provides the same barrier properties as does a lamellar structure, which contradicts Lieckfeldt et al. (76). Using different mesomorphic systems containing fatty acids/soaps, cholesterol, and ceramides, they showed that as long as a lamellar gel structure existed, the diffusion coefficient of tiamenidine through different systems remained constant and did not alter with lipid composition, but when changes in the lipid mixture resulted in a hexagonal phase, the diffusional barrier decreased. Ward and du Reau (77) measured the permeation of water through reconstituted SC membranes using different lamellar structures of nonionic or anionic surfactants as aggregation lipids and found that the permeation of water is comparable in these systems.

The formation of crystalline instead of liquid crystalline phases in fatty acids:soaps:water systems reduces or entirely removes the water barrier (78). This mechanism was later related to the increased transepidermal water loss in essential fatty acid deficiency syndrome; essential-fatty-acid-deficient mice show increased transepidermal water loss (79). The oleic acid/sodium oleate system forms a crystalline structure when the water concentration reduces, but addition of linoleic acid/sodium linoleate prevents such crystallization, and the mixture of oleic acid/oleate and linoleic acid/linoleate forms a lamellar mesomorphic structure (80). The conclusion was that skin lipids should be mainly in a noncrystalline state to be an efficient barrier. Finally, as shown later, simple lamellar systems can model to some extent the barrier performance of SC toward permeation of drugs in presence or absence of enhancers.

In agreement with the preceeding conclusions, the fetal rat developed a competent barrier to water loss coincident with the organization of SC lipids into basic lamellar bilayer structure (81). Additionally, diminution in the SC intercellular lamellae plays an important role in barrier dysfunction after long-term use of topical corticosteroids (82).

In summary, no single component provides the barrier properties of the SC intercellular domain, which depend on the structure and physical state of intercellular lipids; it seems that a lamellar structure is essential.

C. Ability of Lamellar Systems to Model the Structure of Stratum Corneum Intercellular Lipids

Moghimi et al. (61) prepared a model matrix comprising 25% water, 20% cholesterol, and 55% fatty acids and their soaps in which fatty acids were in the same composition as they occur in abdominal SC (30). The matrix was studied by x-ray diffraction, polarized light hot-stage microscopy, and DSC, and results were compared with those of the SC intercellular structure.

The x-ray results of the model matrix are shown in Fig. 3. Analysis of SAXS ($2\theta < 10°$) determines the symmetry and dimension of the lattice. The sharp interference at $2\theta = 1.7°$ determines the dimension of the lattice

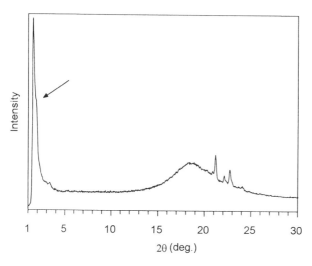

Figure 3 X-ray diffraction profile of the lamellar model matrix prepared by Moghimi et al. (61) at ambient temperature; θ is the scattering angle. The shoulder at $2\theta = 2.34°$ is illustrated by an arrow.

(repeat distance $d = 52.0$ Å), assuming that the main position of this peak is first order. The reflection with spacing of 52.0 Å and that at $2\theta = 3.38°$ ($d = 26.1$ Å) yield the ratio of 1:1/2 that is characteristic of lamellar periodicity. To confirm the x-ray results, the mesomorphic structure of the matrix was also assessed in cross-polarized light microscopy. The matrix showed an oily streaks texture in a planar matrix accompanied by a mosaic texture at 25 and 32°C (SC temperature), all representative of a lamellar structure. The shoulder on the primary reflection at $2\theta = 2.34°$ ($d = 37.8$ Å) in the x-ray diffraction profile is possibly a second-order diffraction peak of a unit cell with $d = 75.6$ Å. A lamellar model prepared from fatty acids/soaps, cholesterol, ceramides, and water (32%, w/w) produced a repeat distance of 65 Å at 21°C (83). This matrix also showed a shoulder on the primary reflection. Bouwstra et al. (84) prepared a matrix with ceramide I, ceramide II, cholesterol, palmitic acid, and water, and besides two lamellar phases with periodicities of around 120 and 55 Å, an additional phase with spacing of 37.7 Å, very similar to that of the shoulder on the primary reflection reported by Moghimi et al. (61), was also observed in their system.

A lamellar mesomorphic structure obtained from fatty acids/soaps:cholesterol:phosphatidylethanolamine and water was investigated by Friberg et al. (68). The interlayer spacing was studied as a function of water content (30–40%, w/w) by SAXS. Extrapolation of these to a water content of 25%

(that of the model matrix) gave an interlayer spacing of 53.6 Å (61), in reasonable correlation with their matrix results at small angle area ($d = 52.0$ Å). A lamellar matrix prepared from fatty acids/soaps and cholesterol (32:68 mol%) and water (30%, w/w) showed a periodicity of 56.7 Å at 25°C in further correlation (76). Also, a lamellar model prepared from fatty acids/soaps, cholesterol, ceramides (31:25:44%), and water (32%, w/w) showed a interlayer spacing of 65 Å at 21°C (83).

SAXS of hydrated human SC (45,46) provided a profile similar to that of the model matrix prepared by Moghimi et al. (61), showing that SC lipids are also arranged in two lamellar structures with repeat distances of 65 and 134 Å. Determinations on murine SC at 25°C also revealed a lamellar structure with a repeat period of 131 Å (85). Although the model lamellar systems described earlier show lamellar periodicities close to that of 65 Å reported for human SC, some of them do not produce the human and murine SC repeat distance of around 130 Å. This might be due to the absence of some SC lipids, including ceramides of very long chain length, in these model lipids. Parrott and Turner (70) prepared a lamellar structure from ceramide II with a repeat distance of 55 Å. On addition of cholesterol, a new reflection evolved at 104 Å. They suggested that the double bilayer formed by cholesterol and ceramides is the basis of the structure responsible for the 134-Å reflection seen in intact SC. They also proposed that proteins or polypeptides are not required to form the double bilayer. However, absence of the SC periodicity of about 130 Å in the model matrices is also reported for some ceramide- and cholesterol-containing systems. Other workers prepared model lamellar matrices comprising fatty acids/soaps, cholesterol, ceramides, and water and only observed the interlayer spacing of 65 (83) and 52.7 Å (76). This may be due to the ceramide type and/or cholesterol content of the model matrices. Bouwstra et al. (84) studied model systems comprising ceramide I, ceramide II, and cholesterol. At low cholesterol content (cholesterol/ceramide molar ratio of 0.4), the matrix showed one reflection with spacing of 67 Å. However, at an equimolar ratio two lamellar phases appeared with periodicities of 55 and 120 Å, in good agreement with human SC data. When they used ceramides isolated from pig SC (86), at a cholesterol molar ratio of 0.1, only one reflection was observed, while at 0.2, two reflections of 56 and 123 Å appeared. The lower cholesterol content needed for the pig lipid model than that for the ceramides I- and II-containing system to show the two bilayer periodicities indicates that at low cholesterol content, other ceramides of the SC (ceramides III–VI) play crucial roles in formation of the phases (86). McIntosh et al. (87) studied different model systems containing skin ceramides extracted from pig epidermis, cholesterol, and palmitic acid. Results showed that skin ceramide and cholesterol, in the presence or absence of palmitic acid, produce the 130-Å reflection, the intensity of which decreases after omission of ceramide I. The authors concluded that ceramide I is important in the formation of the 130-Å reflection.

The x-ray reflections or diffuse bands in the wide-angle region provide information concerning the conformation of hydrocarbon chains. The diffuse band at $\sim 2\theta = 19°$ (d of 4.6 Å, Fig. 3) shows that some of the hydrocarbon chains in the model matrix prepared by Moghimi and coworkers are highly disordered (liquid alkyl chains or type α). The same band (around 4.6 Å) is also reported for human SC and is attributed to both hydrocarbon chains in the liquid state (type α) and soft keratin located in the corneocytes (47). Note that there is no keratin in model systems. Sharp bands in the high-angle region indicate that some of the hydrocarbon chains of the model matrix are stiff and fully extended. Reflections of 4.20 Å ($2\theta = 21.1°$) and 4.11 Å ($2\theta = 21.6°$) may indicate the presence of hexagonal and/or pseudohexagonal nonspecific alkyl chain packings (gel phase or type β); the band at $2\theta = 22.7°$ ($d = 3.92$ Å) may arise from a specific tight alkyl chain packing (crystalline lipids), probably with triclinic or orthorhombic perpendicular arrangement (61). A model containing ceramide, cholesterol, and palmitic acid showed that ceramides can form complex crystalline phases, even in the presence of considerable amounts of cholesterol (88). Fourier transform infrared (FTIR) studies indicated that ceramides exist in conformationally ordered phases possibly with orthorhombic perpendicular subcells in such a model system (89). Also, Fenske et al. (93) studied a composition containing ceramide, cholesterol, and palmitic acid by deuterium nuclear magnetic resonance (NMR). The system was a mixture of solid-phase and lamellar liquid crystalline and gel phases at physiologic temperature.

These results correlate reasonably with those of SC. Based on WAXS of human SC, it has been suggested that alkyl chains of the SC lipids arrange in orthorhombic perpendicular and (pseudo)hexagonal (gel-phase) packings and that cholesterol is probably present in the form of small crystals (47,48). WAXS also revealed that crystalline and liquid alkyl chains coexist in the murine SC at 25°C (85). SAXS and WAXS in combination show that the SC intercellular domain exhibits a complex polymorphism and is a mixture of lamellar liquid crystalline (L_α), lamellar gel (L_β), and crystalline phases, a feature that can be modeled by simple lipid systems.

Moghimi et al. (61) studied the thermal behavior of their model matrix by DSC and hot-stage polarized light microscopy and compared the results with those of human SC. The matrix developed seven endothermic transitions (numbered T_1–T_7 for simplicity) from -30 to 120°C. The first transition (T_1, -11.5°C) may represent melting of some fatty acids chains (linoleic acid melts at -12°C) and correlates well with the lipid-based transition of human SC at -9°C (90).

Transition T_2 ($-1.7°C$) obviously arises from ice melting. This was confirmed by comparing the behavior of ice and that of the model matrix by hot-stage polarized-light microscopy. By measuring the enthalpy of fusion of ice and that derived from T_2 and also assuming that the free water in the matrix behaves like pure water, it was calculated that around 20% of the total water in the matrix was bound (61). This implies that intercellular lipids can also participate in the bound water-holding capacity of the stratum corneum, a phenomenon that is also reported for human SC intercellular lipids (91).

The fourth endothermic thermal transition (T_4) developed between approximately 25 and 45°C with a midpoint (peak) of around 36°C, which correlates well with that reported for human SC (92). The transition T_4 is possibly due to one or more of lamellar gel (L_β) to lamellar liquid crystalline (L_α), solid-to-liquid and lamellar mesomorphic-to-isotropic phase transitions. Fenske et al. (93) also prepared a model system containing ceramide, cholesterol, and palmitic acid, and showed it to be a mixture of solid-phase and lamellar liquid crystalline and gel phases at physiologic temperature. During heating, as the temperature increased, the solid phase decreased and disappeared entirely above 50°C, nearly the end of T_4 reported by Moghimi and corworkers. Anhydrous cholesterol shows an endothermic transition around 35°C, which is a polymorphic crystalline transition (94). To estimate if the T_4 transition of the model matrix arises in part from a cholesterol transition, the thermal behavior of anhydrous cholesterol was also studied by DSC (61). Cholesterol showed an endothermic transition at 36°C. By comparing the enthalpy of this transition and that of T_4 and considering the cholesterol content of the model matrix, it was concluded that contribution of the polymorphic crystalline transition of cholesterol to the T_4 could not be more than 10%.

Based on x-ray diffraction studies of human SC, the transition temperature at around 35°C was attributed to a change in the alkyl chain packing of the lipids within the intercellular bilayers from an orthorhombic to a hexagonal arrangement (46,47). However, from infrared spectroscopy data, it was suggested that the 35°C transition of human SC is a solid-to-fluid phase change for a discrete subset of SC lipids and not a change in the lattice packing of the lipids (95). It has been proposed that the difference in these observations may arise because x-ray and infrared detect different subpopulation of lipids within the SC. As explained earlier, the transition of the model systems (around 35°C) is also due to more than one structural change, including the solid-to-fluid transition, in good agreement with the SC data.

As the SC dehydrates below 20% (w/w), the temperature of the 35°C transition progressively increases and reaches 43°C in the dry stratum corneum (95). Our DSC studies on the lipid mixture (before hydration) reveal a transition at about 41°C that is higher than the T_4 transition temperature of the lipid mixture after hydration (the model matrix) at about 35°C (61)—again in good agreement with the SC data.

A minor transition (T_3) was seen at 22°C in the model matrix, which is possibly either a structural transformation in the gel (L_β) phase or arises from a gel (L_β) to a liquid crystal (L_α) transition (61). The same transition is also reported for porcine SC as studied by FTIR (96).

T_5 extended from around 50 to 60°C and showed a transition temperature around 56°C. A similar endothermic transition develops for a lamellar matrix model comprising fatty acids/soaps, cholesterol, ceramides, and water (83). Mizushima et al. (69) studied a mixture of pseudoceramide, stearic acid, and cholesterol and found that depending on the cholesterol content of the system, a transition between 57 and 62°C arises. A model system prepared with ceramide III, cholesterol, and palmitic acid was studied by FTIR spectroscopy (89). Results showed that both ceramides and palmitic acid exist in separate conformationally ordered phases. During the heating, conformational disordering of palmitic acid commenced at 42°C with a transition midpoint of 50°C. These model data agree well with a lipid-based transition at around 50°C (97) and 55°C (95) reported for human SC.

The sixth transition in the model matrix (T_6) extended from about 60 to around 90°C with a transition temperature of around 80°C. It possessed a very low enthalpy and was absent in some samples. It has also been reported that anhydrous nonhydroxy fatty acid containing ceramides undergo a broad endothermic transition over the range 50–90°C (98). Large unilamellar vesicles prepared with a lipid composition similar to that of the SC also showed an endothermic transition between 60 and 65°C (99). These transitions are also in partial agreement with two lipid-based SC transitions of 72 and 85°C reported by Goodman and Barry (92) and 65 and 80°C reported by Gay et al. (95).

The thermal behavior of the lipid mixture (before hydration) was also studied (61). The system provided a broad endothermic transition between approximately 50 and 90°C with a midpoint of 72°C. This window covers the T_5 and T_6 transitions of the matrix. Polarized-light microscopy studies revealed that while the model matrix retains its lamellar (neat) structure during heating from ambient to higher temperatures, this lamellar phase disorders continuously from around 35 to around 95°C. T_5 and T_6 transitions of the model matrices, therefore, possibly arise from melting of alkyl chains and correlate with the disordering of the lamellar phase and probably represent gel-to-liquid crystalline transitions. The lipid-based thermal transition of human SC at 55°C possibly represents disruption and a loss of crystalline orthorhombic lattice structure (89,95). Based on x-ray diffraction and FTIR studies, the lipid-based transitions of the SC at around 70 and 80°C are attributed to lipid alkyl chain melting and therefore disordering of the lamellar structure and change from hexagonal packing to a liquid phase (47,96), that is, from gel (L_β) to liquid crystalline (L_α) phase, in reasonable correlation with the model results mentioned earlier.

The last endothermic transition of the model matrix (T_7) showed transition temperatures of around 105°C. This transition also provided a very low enthalpy and was absent in some samples. The matrix texture changed from mosaic (lamellar structure) to fanlike and angular textures, representative of a hexagonal (middle) mesomorphic structure, at around 105°C. This feature confirms that T_7 arises from a lamellar-to-hexagonal (possibly reverse hexagonal) liquid crystalline transition. This transition cannot be seen in the SC (even if it exists) because the SC has a transition at around 100°C, with a range of approximately 90 to 110°C, related to proteins (92), which would obscure any T_7 transition. Such a phase transition (L_α to H_{II}) is also reported for a lamellar structure prepared from ceramides, cholesterol, palmitic acid, and cholesteryl sulfate above 60°C (100) and for a lamellar model comprising fatty acids/soaps, cholesterol, ceramides, and water above 50°C (83).

These results indicate that the model matrices discussed are not homogeneous and consist of different ultrastructural domains, correlating with the heterogeneous structure of SC intercellular lipids. They also indicate that simple lamellar structures can model some aspects of the SC intercellular architecture.

D. Stratum Corneum Geometrical Barrier and Application of Lamellar Models to Show the Permeation Pathway of Drugs Across the Stratum Corneum

Besides the continuous lipid bilayers, which provide a substantial diffusional barrier within the intercellular channels, a geometrical barrier (SC unique morphology) also exists within the SC and is additionally responsible for its low permeability in comparison to most other lipid membranes (22,56,60,63,64,101,102). The relatively impermeable corneocytes form this geometric barrier (58). There are three relevant geometrical properties: a) the intercellular pathway is of greater length than the simple thickness of the SC; b) the effective diffusional area, the cross-sectional area of the intercellular matrix normal to the flux, is smaller than the total cross-sectional area of the SC (Fig. 1) (e.g., see 22,60). c) The volume fraction of intercellular lipids, which is less than 1 (31), is the third geometrical property that is of importance in the calculation of partition coefficients (62) but is usually neglected. The geometrical values of human SC are summarized in Table 2. The importance of the SC geometrical barrier in the absorption of

Table 2 Geometrical Dimensions of the Stratum Corneum Intercellular Lipid Domain

Thickness (h) (μm)	Tortuous pathlength (τh) (μm)	Tortuosity factor (τ)	τ²	Porosity factor (ε)	τ/ε	Volume fraction (fᵥ, τε)[e]
15[a]	340[a]	22.7	500	0.007[a]	3200	0.16
				0.016[b]	1400	0.36
	500[c]	33.3	1100	0.007[a]	4800	0.23
				0.016[b]	2100	0.53
					900[d]	

[a]From Albery and Hadgraft (56).
[b]From Michaels et al. (22).
[c]From Potts and Guy (101).
[d]From Lieckfeldt and Lee (102).
[e]See text for detailed explanation.

drugs is reviewed here, focusing on the determination of the permeation pathways of drugs across the SC.

Simple passive transport across a membrane has two characteristic steps: a) a lag phase and b) a steady-state region (see Reference 1 for mathematical expressions). In passive transport, Fick's law is usually obeyed well for permeation across the skin whether the penetrant is a gas, an ion, or a nonelectrolyte (5). For the determination of D values of drugs through the SC, this membrane is normally simplified as isotropic and Eqs. (1) and (2) may be used. Lag time (L) and flux (J) values are measured practically, the drug's stratum corneum/vehicle partition coefficient (K_{sc}) is typically determined using the whole volume of the SC for calculation of the concentration of drug in the SC (the SC considered homogeneous), h is set to be the SC thickness, and A is the exposed surface area of the SC. In such an absence of geometric considerations, the calculated D values from lag time, Eq. (1), or from flux, Eq. (2), are actually apparent diffusion coefficients (D_{app}), which we can specify as $D_{app(L)}$ and $D_{app(J)}$, respectively:

$$D_{app(L)} = h^2/(6L) \tag{1}$$

$$D_{app(J)} = Jh/(K_{sc}CA) \tag{2}$$

in which J is the flux (the mass of diffusant which permeates the whole membrane per unit time) and C is the dissolved concentration of the penetrant in the vehicle.

However, if we take into account the geometry of the SC and assume that the intercellular channels are uniform in cross section, but tortuous, for drugs that permeate the SC only through intercellular lipids, Eqs. (1) and (2) will change to Eqs. (3) and (4) for calculation of the D of drugs through stratum corneum intercellular lipids (D_{lip}, the true diffusion coefficient):

$$D_{lip} = (\tau h)^2/(6L) \tag{3}$$

$$D_{lip} = J\tau h/(K_{lip}CA\varepsilon) \tag{4}$$

where τ is the tortuosity factor and τh the effective path length for transport via tortuous lipid channels, ε is the porosity factor (the fraction of the total area normal to the flow direction that is lipid), and K_{lip} is the drug's intercellular lipid/vehicle partition coefficient.

As we can see from Eqs. (1) and (3) for the lag-time method, the only geometrical property that directly affects the D is the tortuosity factor, by increasing the diffusional pathlength. However, if we use Eq. (2) to calculate the apparent D from flux data, $D_{app(J)}$, the porosity and, as explained later, the lipid volume fraction (f_v, fraction of stratum corneum volume that is lipid) will also indirectly affect the calculated apparent D. This means that the value of the apparent diffusion coefficient calculated from lag time will be different from that calculated from flux data.

Here we define a diffusivity ratio (DR), the ratio of true to apparent D (D_{lip}/D_{app}), to represent the error in calculation of the true D of a penetrant in the intercellular lipids. If we use the lag time to calculate the apparent D, $D_{app(L)}$, the DR can be derived from Eqs. (1) and (3) as:

$$D_{lip}/D_{app} = D_{lip}/D_{app(L)} = \tau^2 \tag{5}$$

If we use the steady-state flux to calculate the apparent D, $D_{app(J)}$, Eqs. (2) and (4) give a DR of:

$$D_{lip}/D_{app} = D_{lip}/D_{app(J)} = \tau K_{sc}/(\varepsilon K_{lip}) \tag{6}$$

As mentioned earlier, K_{sc} is normally measured using whole SC. The protein domain in SC is much more polar than the lipid domain and, depending on the lipophilicity of the penetrant, its uptake and therefore its partition coefficient may be governed by either the protein or the lipid domain or both of them (31). Therefore there are three different situations: a) $K_{sc} > K_{lip}$, which happens when the drug partitions mainly into the corneocytes; b) $K_{sc} = K_{lip}$, which is the situation when the drug partitions equally into corneocytes and intercellular lipids. These two cases apply to hydrophilic drugs. Case c), $K_{sc} < K_{lip}$, is applicable to lipophilic drugs where uptake by the intercellular lipids is greater than by corneocytes. According to these situations, the D_{lip}/D_{app} ratios in Eq. (6) will be:

$$K_{sc} > K_{lip} \qquad D_{lip}/D_{app} = D_{lip}/D_{app(J)} > \tau/\varepsilon \tag{7}$$

$$K_{sc} = K_{lip} \qquad D_{lip}/D_{app} = D_{lip}/D_{app(J)} = \tau/\varepsilon \tag{8}$$

$$K_{sc} < K_{lip} \qquad D_{lip}/D_{app} = D_{lip}/D_{app(J)} < \tau/\varepsilon \tag{9}$$

Equation (8), which defines the situation where $K_{sc} = K_{lip}$, is used by Lange-Lieckfeldt and Lee (60) to demonstrate the dependence of stratum corneum barrier performance toward tiamenidine on its geometry using a lamellar model matrix.

In the special case where the drug is taken up only by the intercellular lipids and instead of intercellular lipids volume ($f_v V$) the whole SC volume (V) is used in the calculation of partition coefficient, K_{sc}/K_{lip} will be equal to f_v (the volume fraction of intercellular lipids in the SC). Alternatively, if we assume that the intercellular channels of the stratum corneum are uniform in cross section, we can simply consider the channels as narrow tubes with a total surface area of $A\varepsilon$ and length of τh. Thus the volume of intercellular lipids will approximately be equal to $A\varepsilon\tau h$. Since the total volume of the stratum corneum (V) is equal to Ah, the volume fraction of lipids in the stratum corneum (f_v) will be $A\varepsilon\tau h/Ah$ or $\tau\varepsilon$. If in Eq. (6) we use $\tau\varepsilon$ instead of K_{sc}/K_{lip}, DR then will be:

$$D_{lip}/D_{app} = D_{lip}D_{app(J)} = \tau^2 \tag{10}$$

The DR terms calculated from lag times [Eq. (5)] and special case calculation from flux [Eq. (10)] both equal τ^2. This shows that for very lipophilic drugs that partition essentially into the intercellular lipids, the lag-time and flux methods should give the same DR and thus apparent D values.

Another parameter of permeation of molecules across a membrane that is normally measured and reported is the permeability coefficient (K_p). The permeability coefficients of drugs in the SC ($K_{p(sc)}$) can be calculated using the intercellular lipid data and SC geometrical parameters using Eq. (11) (102). See Reference 62 for more details.

$$K_{p(sc)} = K_{lip}D_{lip}\varepsilon/(\tau h) \tag{11}$$

Now we turn to application of the preceding equations in determining the permeation pathway across the SC using lamellar models. Lange-Lieck-feldt and Lee (60) prepared a lamellar matrix structure from fatty acids/soaps, cholesterol, ceramides; and water. They then measured the D of tiamenidine through this model matrix (D_{lip}) and human stratum corneum (D_{app}) at 33°C. Their results showed that the D of tiamenidine in the model matrix is around 7000 times greater than that of the SC, in good correlation with Eq. (7) and their theoretical calculation of τ/ε (order of 1000). These results imply that tiamenidine permeates the SC through the intercellular lipids and that the barrier property of the SC arises to a substantial degree from its internal geometry.

Moghimi et al. (62) measured the release of a model hydrophilic drug, 5-fluorouracil (5FU), and a model lipophilic drug, estradiol (OE), from a lamellar mesomorphic model structure (designated matrix here) at 32°C (SC surface temperature) and compared the results with those from human SC. The D of 5FU through matrices (D_{lip}) was calculated to be 5.97×10^{-4} cm^2 h^{-1}. That through human epidermis as calculated from flux data on the assumption of an isotropic structure for the SC, $D_{app(J)}$, is reported to be 0.95×10^{-7} cm^2 h^{-1} (103) and 0.81×10^{-7} cm^2 h^{-1} (104). From these results, the DR ($D_{lip}/D_{app(J)}$) was calculated (62) to be approximately 6000–7000. The calculated DR ($D_{lip}/D_{app(J)}$) is greater than τ/ε (900–4800, Table 2), which correlates well with Eq. (7), which predicts a $D_{lip}/D_{app(J)}$ of greater than τ/ε. This is reasonable, because the assumption underlying Eq. (7) favors the partitioning of drugs into corneocytes. 5FU is a hydrophilic drug and should thus have more affinity for corneocytes than for intercellular lipids. Good correlation of the DR with the theoretical expectation provides evidence indicating that 5FU permeates the SC through the intercellular route. We should bear in mind that the domain with the higher drug uptake does not necessarily correlate with the rate-controlling route for permeation. Thus although corneocytes play an important role in the uptake of water by the SC, water permeates the SC through the intercellular channels. Although corneocytes act as a sink for water, this water may not be available for net transport; that is, the corneocytes may be water-impermeable on a relevant time scale (58).

The D value of OE through the model lamellar matrices (D_{lip}) was calculated to be 2.19×10^{-5} cm^2 h^{-1} (62). The apparent D of OE in human epidermis ($D_{app(J)}$) was calculated from published permeability and partitioning data to be 8.48–34.1×10^{-8} cm^2 h^{-1}. The DR ($D_{lip}/D_{app(J)}$) was then estimated to be approximately 60–260. These values are less than τ/ε (900–4800) and are close to the lower range of τ^2 (500) and correlate well with Eq. (9), which predicts a $D_{lip}/D_{app(J)}$ of less than τ/ε, and Eq. (10), which predicts a $D_{lip}/D_{app(J)}$ equal to τ^2. An assumption underlying the derivation of Eqs. (9) and (10) is the favorable partitioning of the penetrant into the intercellular lipids, which applies to lipophilic drugs. Estradiol is such a drug, and good agreement of the DR ($D_{lip}/D_{app(J)}$) with theoretical expectation provides evidence indicating that OE permeates the SC through the intercellular route.

Moghimi et al. (62) also measured the permeability coefficients of 5FU and OE from their saturated aqueous solutions through a 1.35-mm-thick membrane prepared from the model lamellar matrix. Then using the permeability coefficients cited earlier and D values of drugs in the model matrix, the partition coefficients of 5FU and OE between the model matrix and water were calculated to be 0.64 and 23, respectively. Using the matrix data for the SC intercellular lipids, different reported geometrical dimensions of the SC (Table 2), and Eq. (11), they predicted permeability coefficients of

5FU and OE through human SC at 32°C. The predicted permeability coefficient for 5FU was $5.5-18 \times 10^{-5}$ cm h^{-1} which correlates well with reported average human epidermis data of $3.06-12.0 \times 10^{-5}$ cm h^{-1} (103–109). For OE, the predicted permeability coefficient was $0.07-0.24 \times 10^{-3}$ cm h^{-1} (62), which is somewhat lower than the mean reported values of $1.30-5.23 \times 10^{-3}$ cm h^{-1} (21,108–110). However, permeability coefficients of as low as 0.1×10^{-3} cm h^{-1} are measured for permeation of OE across human epidermal membrane (HEM), which is well correlated with the model matrix data (109).

The studies presented in this section clearly show that the main factor in the barrier performance of the SC intercellular lipids is the presence of a lamellar structure and that simple lamellar models can sometimes be used for the prediction of the permeation pathways of drugs across the SC, applying simple mathematical equations. These data also illustrate the importance of SC geometry in its barrier performance and that this geometrical barrier increases the diffusional resistance of the SC some 1000-fold.

E. Ability of Lamellar Structures to Model the Effects of Enhancers on the Barrier Performance of the Stratum Corneum

Different methods may increase the passage of drugs through the skin, of which one is addition of chemicals (called enhancers or accelerants) to this membrane (107,111). The enhancers usually change the lipid or protein structures of the SC (effect on D) and/or improve drug partitioning into the tissue (20). The lamellar intercellular domain is the main route for permeation of most drugs through the SC. Model lamellar structures, therefore, may be used to probe the mechanism of action of enhancers and to assess if they affect the intercellular lipids or not.

Kim et al. (112) prepared multilamellar liposomes from ceramides, cholesterol, cholesteryl sulfate, and palmitic acid. They added pyrollidone derivatives (enhancers) to a glucose solution that contained liposomes and studied their osmotic behavior. The pyrollidones perturbed the barrier function of liposomes. Using SC lipid liposomes and fluorescence lifetime studies, Yoneto and colleagues (113) showed that alkyl pyrrolidones can fluidize lipid bilayers.

Miyajima et al. (114) fixed liposomes of SC lipids (ceramides, palmitic acid, cholesterol, and cholesterol sulfate) onto a supporting filter. They studied the effects of Azone, decylmethyl sulfoxide, oleic acid, lauric acid, and capric acid on the permeation of model drugs ibuprofen (IB, a relatively lipophilic drug) and cyclobarbital (CB, a relatively hydrophilic drug) through the artificial membrane and guinea pig skin. For both membranes, the enhancers increased CB penetration, but had little effect on the lipophilic IB. These results indicated that the enhancers mainly acted on the intercellular lipid lamellae of the SC. However, the effects of enhancers on the model membrane and skin were not quantitatively the same. These enhancers increased the permeation of CB almost two to four times across the lipid membrane. The effects of the same enhancers on the permeation of CB through the skin were two to seven times greater than that of the lipid membrane. For IB, although the enhancer effects on the skin were greater than for the model membrane, the relative difference never exceeded two. This quantitative disagreement between model and skin was related to the different amount of enhancers in the model membrane and skin and also to possible effects of the enhancers on the other components of skin rather than the simple lipids used in the model. Other reasons for this quantitative disagreement are provided in the following sections.

Terpenes can be potent enhancers. Moghimi et al. (115,116) studied the effects of different concentrations (5–25%, w/w) of 1,8-cineole and (+)-limonene (monoterpenes) on the barrier performance of a lamellar model matrix toward permeation of OE and 5FU and compared the results with those of human SC.

Figure 4 summarizes the effects of cineole on the release of OE from the model matrix. The D of OE in the matrix containing 5% cineole was

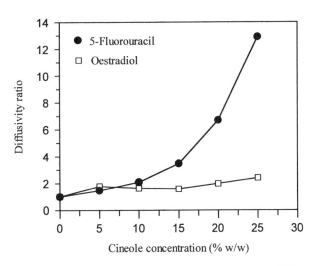

Figure 4 Effect of cineole concentration on the diffusivity of 5-fluorouracil and estradiol in a lamellar matrix model for stratum corneum intercellular lipids. Diffusivity ratio is the ratio of cineole-treated to untreated diffusion coefficients (115).

1.8-fold greater than its control value and further increased on increasing the cineole concentration to 25%, where the DR (diffusion coefficient after terpene treatment/diffusion coefficient before terpene treatment) reached 2.4 (115). Cineole increased the D of OE through human epidermis at 2.1-fold after 12 h of enhancer treatment (21). To compare the model data with those of human SC, the amount of enhancer in both systems should ideally be the same. It was calculated that the amount of enhancer in 12-h SC cineole-treated samples was equivalent to 40% cineole in the model matrix (115). However, due to matrix liquefaction, it was not practicable to perform release experiments using matrices containing 40% cineole. Extrapolation of data to a matrix cineole content of 40%, based on an assumed linear relationship between the D values of OE in the model matrix and cineole concentration in the range of 15–25% (Fig. 4) gave a DR of 3.6, which is in reasonable agreement with the HEM data.

Figure 4 also illustrates the effects of cineole on 5FU released from the matrix. Cineole increased the D of 5FU by about 1.5–13 times in matrices containing 5–25% enhancer relative to the untreated matrix (115). The DR of 5FU in 12-h cineole-treated human epidermal membrane was about 40 (104). Extrapolation of model data to a cineole concentration of 40% (equivalent to that of SC-treated samples) gave a DR of about 30–50 (115,117), in good agreement with HEM data.

Limonene increased the D of OE insignificantly (one to twofold) in matrices containing 5–20% terpene (Fig. 5). Increasing the limonene content to 25%, which liquefied the matrix, increased significantly DR to 4, in good agreement with the DR of 3.8 for HEM after 12 h of enhancer treatment (21). A matrix containing 25% limonene correlates with the amount of enhancer in 12-h SC terpene-treated samples (115). Good correlation of matrix results with those of HEM indicates that the matrix can model the effects of cineole and limonene on the D of OE and that of cineole on the D of 5FU through HEM and that these enhancers increase the D of OE and 5FU through the SC mainly via lipid interaction.

Limonene did not increase the matrix D of 5FU (Fig. 5). The differences between the D of 5FU in the untreated matrix and those of matrices containing 5 and 10% limonene (DR = 0.97 and 0.62, respectively) were not significant (p = .05). As the limonene concentration increased, the DR decreased significantly and reached values of 0.170, 0.015, and 0.066 in matrices containing 15, 20, and 25% limonene, respectively (115). Limonene is not a good penetration enhancer toward 5FU permeating through HEM. The DR of 5FU through HEM after 12 h of limonene treatment was only 1.3 (106) and 3.8 (104). Model data, however, show that at low concentrations limonene is not an enhancer toward 5FU releasing from the matrix and is a retardant at higher than 10% (the mechanism is discussed in Section

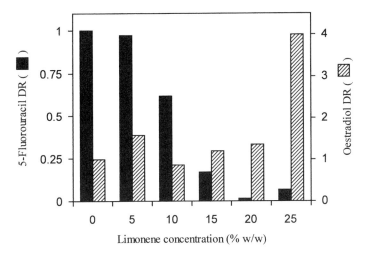

Figure 5 Effect of limonene concentration on the diffusivity of 5-fluorouracil and estradiol in a lamellar matrix model for stratum corneum intercellular lipids. Diffusivity ratio (DR) is the ratio of limonene-treated to untreated diffusion coefficients (115).

III.G), clearly indicating that the matrix does not model well the effect of limonene on the diffusion of 5-FU (a hydrophilic drug) through the SC. As shown earlier, there was no such problem for OE (a lipophilic drug) and the matrix modelled the effects of limonene on the D of OE through the SC. It was also shown that the efficiencies of different enhancers toward permeation of a hydrophilic drug (cyclobarbital) through a lamellar model matrix membrane were up to seven times less than for skin (114). However, the difference between the effects of the same enhancers toward permeation of a lipophilic drug (ibuprofen) in the model matrix and skin was only around twofold. A plausible reason for the underestimation of the effect of enhancers on the permeation of hydrophilic drugs by lamellar models might be the importance of the transcellular route for the permeation of these drugs through enhancer-treated SC, as discussed later.

The D in lipids and proteins and the partitioning of drugs between these domains may play important roles in permeation of drugs through the SC and the effects of enhancers on this process (20,22,56,63). For hydrophilic drugs, partitioning into the corneocytes should not be a rate-limiting step, and if such drugs do not permeate the SC through the transcellular route, the limiting step should be a diffusional barrier. It has been argued that the low permeability of corneocytes is possibly due to their durable lipid envelope (64). Suppose that an enhancer decreases the diffusivity barrier of the corneocytes and/or their envelopes. This reduction in the resistance of the transcellular pathway will improve the permeation of hydrophilic drugs but may not be useful for lipophilic drugs, as they find the intercellular domain favorable and do not, therefore, partition easily into the more hydrophilic corneocytes. Such a mechanism may explain why model lipid systems underestimate the effects of enhancers on the barrier performance of the SC toward hydrophilic (and not lipophilic) drugs. In support of this concept, it has been shown that limonene may interact with both the SC lipids and proteins (104,118).

Azone, 1-dodecanol, and dodecyl-2-(N,N-dimethylamino) propionate (DDAIP) affect the structural integrity of liposomes composed of SC lipids (119); the fluidizing effects of Azone and DDAIP were stronger than that of 1-dodecanol. Azone and DDAIP similarly affected the permeation of sotalol through human epidermis more than did 1-dodecanol.

Besides D, partition coefficient also plays an important role in the permeation of drugs and is a target for some penetration enhancers (20). Moghimi et al. (116) studied the effects of cineole and limonene on the partitioning of OE and 5FU from their aqueous solutions into a lamellar model membrane through release and permeation studies. Besides the increase in D, cineole also increased the partition coefficient of OE from water into the lipid model, in agreement with human SC data (21). Such data also revealed that limonene increased the permeation of OE through the model system partly by increasing the partition coefficient. For 5FU, however, cineole decreased the partition coefficient of this hydrophilic drug into the model matrix (116). In contrast, cineole increased the partitioning of 5FU from its aqueous solution into human SC sample (104). This difference relates to the increased equilibrium uptake of 5FU into the SC corneocytes (absent in the model system).

F. Ability of Lamellar Structures to Show the Mechanism of Action of Enhancers and to Model Their Effects on the Structure of the Stratum Corneum

The SC intercellular lipids models can be used to study the mechanism of action of enhancers. Moghimi et al. (120) monitored the effects of two terpenes, 5–40% cineole and 5–25% limonene, on the structure and thermal behavior of a lamellar model matrix by hot-stage polarized-light microscopy and DSC. An interesting change after cineole treatment was the replacement of most of the oily streaks and planar areas (representative of lamellar structure) of the untreated matrix by a network of positive and negative units (coarse mosaic texture, also representative of a lamellar structure) at 25 and 32°C, a transition induced by temperature in the untreated matrix. The mosaic texture represents the maximum possible disorder in the lamellar phase, the opposite extreme of a planar arrangement (121). This change indicates the disruption of lamellar structure by both enhancer and temperature, that is, their common mechanism of action. This may also suggest that enhancer efficacy around the transition temperatures (where the structure is already disordered by heat) is possibly less than that at lower temperatures. Thus the enhancement effect of cineole toward release of 5FU from the model matrix around the 35°C transition is less than that at lower temperatures, indicating that when the matrix is thermally disrupted, less structure remains for enhancer disruption (115). Similarly, preheating and sodium lauryl sulfate treatment of human SC produced the same effects (122), increasing the permeability of the SC to water and decreasing the intensity of x-ray diffraction peaks (i.e., disordering of the lamellar structure). A similar phenomenon was found when addition of Azone to a lecithin–water system induced a lamellar (L_α) to reversed hexagonal (middle) (H_{II}) phase transition at ambient temperature, a transition occurring above 200°C in an Azone-free system (123).

In model lamellar matrices containing 25% cineole, large lamellar liquid crystalline (L_α) particles (spherulites or vesicles) appeared, and at 40%, the matrix provided myelinic figures (120), representative of a lamellar liquid crystalline (L_α) structure (124). These textural changes, which indicate disordering of lamellar structure, might also be responsible for the enhancement action of cineole. Interestingly, electron microscopy indicated that water (a well-known penetration enhancer) creates vesicle-like structures in human SC intercellular lipids (125).

Besides mesomorphic structures, an isotropic liquid formed in cineole-treated matrices, with the proportion increasing with cineole concentration and becoming continuous in matrices containing 10% or more (120). The creation of such a continuous liquid pool, which has less resistance to diffusion than organized bilayers, is possibly one of the most important mechanisms for the enhancement effect of cineole. Based on x-ray diffraction studies, the same effect (pool formation) is also suggested for the effect of cineole on human SC (48,118). Such a phase separation developed when oleic acid was added to extracted human epidermal lipids (126). IR spectroscopy (in vivo in human) suggests that oleic acid disorders SC lipids and exists in a liquid phase at all levels of the SC, with phase separation predominant (127). These changes were held to be responsible for the enhancement effects of oleic acid. Interestingly, it was shown by freeze-fracture electron microscopy that water pools in human SC intercellular lipids (125). Data such as these show that simple lipid systems can model the effects of some enhancers on the SC intercellular lipid structure and are able to provide information about their mechanisms of action.

DSC studies were also performed on the 5-40% cineole-treated model matrices. The 35°C transition of the model matrix (T_4, see Section III.C) gradually increased from around 35°C in untreated samples to almost 39°C in matrices containing 25% cineole. This transition was absent in matrices containing 40% cineole (120). Because of the very low transition enthalpy at 35°C (equivalent to T_4) and, therefore, its absence in some SC samples, most investigations on effects of enhancers on the thermal behavior of the SC have involved higher transitions. One of the rare reports on the effects of enhancers on the 35°C transition is by Goodman and Barry (92), showing that N-methyl-2-pyrrolidone, propylene glycol, and dimethyl formamide increased the 35°C transition by about 1–5°C. The same enhancers decreased the temperature of SC transitions of around 70 and 80°C by about 5–13°C. Yamane et al. (128) studied the effects of cineole on SC transitions at 70 and 80°C and showed that cineole decreased these transitions by about 20 and 15°C, respectively. Considering that cineole, N-methyl-2-pyrrolidone, propylene glycol, and dimethyl formamide show the same effect on the higher transitions, their effects on the transition at 35°C are possibly similar. Cineole should, therefore, increase the transition temperature of 35°C in the SC, in good agreement with the model matrix results (120).

The enthalpy of T_4 (around 35°C) decreased significantly on addition of cineole to the model matrix, the mechanism of which was explained in detail (120). Based on DSC studies using human SC, Cornwell et al. (118) suggested that cineole is probably lipid disruptive around normal skin temperature (32°C), correlating with the decrease of the enthalpy of T_4 of model matrix and disruption of its lipids by cineole as observed by polarized-light microscopy (120).

Control samples (untreated model matrix) showed midpoint temperatures of around 79°C and enthalpies of around 1.0 J g^{-1} for matrix T_6 transition. Cineole decreased the transition of T_6 by about 8–19°C in matrices containing 5, 10, and 40% cineole and also decreased its enthalpies, providing enthalpy ratios (treated/untreated) of 0.66, 0.37, and 0.65, respectively (120). Using hydrated SC, 1–12 h of treatment with neat cineole decreased the temperatures of 70 and 80°C transitions (equivalent to model matrix T_6 transition) by about 15–20°C and also decreased the enthalpies of T_6 equivalent SC transitions, producing enthalpy ratios of 0.94–0.45 (128). Such results correlate well with model matrix data.

Limonene induced different textural changes in the matrix (120). At 25°C, with increasing limonene concentration, the streaks were gradually replaced by mosaic texture, which itself decreased with further terpene increase. The oily streaks were absent in matrices containing 15% or more limonene. Both oily streaks and mosaic textures indicate a lamellar structure (neat phase); however, mosaic texture represents the maximum degree of possible disorder (121). At 25% and room temperature, the matrix was a mixture of crystalline and mesomorphic phases [mixture of hexagonal (middle) and lamellar (neat) structures] dispersed in an isotropic liquid. SC takes up the equivalent of 25% limonene in the model matrix after 12 h of terpene treatment. X-ray diffraction studies showed that after 12 h of limonene treatment, lipid bilayers and areas of liquid enhancer coexist in the SC (48,97) in reasonable agreement with matrix data. Untreated lamellar systems also show a similar phase transition (L_α to H_{II}) at higher temperature: above 60°C (100) and about 105°C (61). This again shows that heating or enhancer treatment can induce the same alteration in lamellar structure.

A new apparently viscous isotropic (cubic) mesomorphic structure was also observed in lamellar matrices containing 15–25% limonene at 25°C (120). The same phenomenon arose when increasing the concentration of Azone in a monoolein–water system induced a lamellar-to-cubic and then a cubic-to-hexagonal (middle) phase transition (123). As discussed later, these phase transformations may be a main mechanism by which enhancers change the permeability of the SC.

Effects of limonene on the thermal behavior of the model matrix were also studied by DSC and results were correlated with the hot-stage polarized-light microscopy data (120). Results indicated that limonene interacts with the lamellar model at ambient and higher temperatures and affects all matrix transitions, including that around the normal skin temperature (T_4, around 35°C). To the best of our knowledge, there are no DSC data available on the effects of limonene on the thermal behavior of the SC lipids around the 35°C transition. However, from the effects of the limonene on the higher SC transitions (around 70 and 80°C) and classical thermodynamic analysis, it seems that limonene does not interact with the SC lipids to disorganize them around 35°C (118,129). This finding contradicts both the stratum corneum x-ray diffraction studies (48,118) and Moghimi and coworkers' model matrix results, which indicate that limonene interacts with SC lipids and its models; the conflict requires further investigation.

The preceding data show that simple lamellar structures can model some aspects of enhancers' effects on the structure of the SC, may be used to investigate their mechanism of action, and indicate that enhancers usually interact with the SC intercellular lipids to disorganize them.

G. Structure–Diffusivity Relationship

The effects of cineole and limonene on the barrier performance of a lamellar model matrix toward 5FU and OE at 32°C (SC temperature) and on the structure of the model matrix were investigated (115,120) and correlated here.

Addition of 5–25% cineole to the model matrix yielded DR values of 1.5–13 for 5FU and 1.6–2.4 for OE, revealing that as cineole concentration increases, the DR values increase. Polarized-light microscopy and DSC studies showed that cineole disrupted the lipid bilayers and created liquid pools, continuous in matrices containing 10% or more cineole. These observations suggest that 5FU and OE traverse the matrix through the pools and/or disturbed bilayers and/or interfacial regions, all of which have less resistance to diffusion than organised bilayers. From x-ray diffraction (48,118) and DSC (128) studies, the same mechanisms were suggested for the effects of cineole on the permeation of drugs through the SC. The progressive increase in D on increasing the cineole concentration should be due to the increase in the proportion of these more permeable structures.

Some enhancers may also increase the partitioning of drug into the intercellular lipids through solvent effects and complexation and thereby increase permeation of drugs. It was suggested that fatty acids and propranolol form complexes that readily partition into SC lipids and diffuse down to the viable epidermis, where the complex dissociates in the interfacial region and the freed propranolol partitions into water-rich viable tissue (130). Moghimi (117) measured the solubility of 5FU in isooctane (a simple model for the hydrocarbon interior of SC lamellar structure) and found that the effects of 26 different terpenes on 5FU solubility correlates well with the terpenes' enhancement activities toward permeation of 5FU through HEM. Cineole and 5FU also formed complexes.

Limonene decreased the D of 5FU successively in matrices containing 5–20% terpene (DR = 0.972–0.015), and then D slightly increased at 25% (DR = 0.066). The decrease was significant at 15–25% limonene. The effects of 5–20% limonene on D of OE (DR = 0.866–1.58) was not significant, but at 25%, the DR of OE increased significantly to 4.0 (115). With increasing limonene concentration, the consistency of the system also rose up to a concentration of 20% and then suddenly decreased in matrices containing 25% limonene where the matrix partially liquefied (120). The changes in consistency may suggest why D of 5FU in the matrix decreased up to 20% and then increased when the limonene concentration increased to 25%, but cannot explain why the diffusivity of 5FU in matrices containing 25% limonene, which is liquefied, is still around 15 times less than that of the ordered plain matrix. The diffusivity of OE in the model matrix did not fall as drastically as the limonene concentration increased. Therefore, another mechanism rather than consistency change could be involved.

The lamellar structure was replaced by an apparently viscous isotropic phase in limonene-treated matrices. At 25% limonene, the matrix also showed some crystalline lipids and hexagonal liquid crystalline (middle) phases dispersed in a liquid phase (120). If these mesomorphic structures are of reversed type (i.e., the matrices contain a continuous lipophilic phase), 5FU (a hydrophilic drug) would be expected to accumulate in the dispersed hydrophilic phase of the system and favorable partitioning toward this internal phase would render the drug somewhat unavailable to the continuous lipophilic phase. After liquefaction, D should increase. The DR of 5FU increased almost 4 times after liquefaction relative to matrices containing 20% limonene but still was 15 times less than that of untreated matrix (115). This shows that the rate-limiting step for the release of 5FU from the limonene-treated matrices (partitioning from an internal hydrophilic to an external lipophilic phase) cannot be compensated completely by liquefaction of the model matrix. The D of OE in the matrix is almost constant for matrices containing 5–20% limonene, which may arise from two opposing phenomena: a) the increase in the matrix consistency, which would decrease D, and b) replacement of lamellar with a reversed cubic phase, which creates a continuous lipophilic medium and consequently removes the OE partitioning step between hydrophilic and lipophilic layers of the lamellar structure. When the matrix liquefied at 25% limonene content, the DR of OE suddenly increased 3 times, presumably due to decreased consistency of the system.

IV. CONCLUSION

The stratum corneum (SC) is the main barrier to transdermal delivery of most drugs. The intercellular domain is considered as the major pathway for permeation of most drugs across the SC by the majority of the workers. The SC intercellular domain contains lipids with a lamellar structure (both in liquid crystalline and gel phases) at normal skin temperature. This domain also contains some crystalline material.

The main factor in the barrier performance of the SC intercellular lipids is the presence of the organized lamellar structure. Additionally, a substantial degree of SC barrier arises from its internal geometry (i.e., porosity and tortuosity), which increases the SC resistance 1000-fold.

Simple lamellar models can elucidate the barrier properties of SC, the effects of SC geometry, the permeation pathways of drugs, and the mechanism of action of penetration enhancers. These models can provide valuable information regarding the SC barrier–the type of information difficult to gain directly from the SC itself, which has a very complicated structure.

REFERENCES

1. Barry BW. Dermatological Formulations, Percutaneous Absorption. New York: Marcel Dekker, 1983.

2. Odland GF. Structure of the skin. In: Goldsmith LA, ed. Biochemistry and Physiology of the Skin. Vol. 1. New York: Oxford University Press, 1983: 3–63.

3. Bissett DL. Anatomy and biochemistry of skin. In: Kydonieus AF, Berner B, eds. Transdermal Delivery of Drugs, Vol. 1. Boca Raton: CRC Press, 1987: 29–42.

4. Scheuplein RJ. Mechanism of percutaneous absorption II. Transient diffusion and the relative importance of various routes of skin penetration. J Invest Dermatol 1967; 48:79–88.

5. Scheuplein RJ, Blank IH. Permeability of the skin. Physiol Rev 1971; 51: 702–747.

6. Grimnes S. Pathways of ionic flow through human skin in vivo. Acta Derm Venereol (Stockh) 1984; 64:93–98.

7. Cornwell PA, Barry BW. The routes of penetration of ions and 5-fluorouracil across human skin and the mechanisms of action of terpene skin penetration enhancers. Int J Pharm 1993; 94:189–194.

8. Tregear RT. The permeability of skin to albumin, dextrans and polyvinyl pyrolidone. J Invest Dermatol 1966; 46:24–27.

9. Scheuplein RJ, Blank IH, Brauner GJ, MacFarlane DJ. Percutaneous absorption of steroids. J Invest Dermatol 1969; 52:63–70.

10. Siddiqui O, Roberts MS, Polack AE. Percutaneous absorption of steroids: relative contribution of epidermal penetration and dermal clearance. J Pharmacokin Biopharm 1989; 17:405–424.

11. Rougier A, Dupuis D, Lotte C, Roguet R, Wester RC, Maibach HI. Regional variation in percutaneous absorption in man: Measurement by the stripping method. Arch Dermatol Res 1986; 278:465–469.

12. Rougier A, Lotte C, Maibach HI. In vivo percutaneous penetration of some organic compounds related to anatomic site in humans: Predictive assessment by stripping method. J Pharm Sci 1987; 76:451–454.

13. Illel B, Schaefer H, Wepierre J, Doucet O. Follicles play an important role in percutaneous absorption. J Pharm Sci 1991; 80:424–427.

14. Lauer AC, Lieb LM, Ramachandran C, Flynn GL, Weiner ND. Transfollicular drug delivery. Pharm Res 1995; 12:179–186.

15. Smith HW, Clowes GHA, Marshall EK. On dichloroethylsulfide (mustard gas) IV. The mechanism of absorption by the skin. J Pharmacol Exp Ther 1919; 13:1–30.

16. Berenson GS, Burch GE. Studies of diffusion of water through dead human skin: the effect of different environmental states and of chemical alterations of the epidermis. Am J Trop Med Hyg 1951; 31:842–853.

17. Blank IH. Further observations on factors which influence the water content of the stratum corneum. J Invest Dermatol 1953; 21:259–271.

18. Monash S, Blank H. Location and re-formation of the epithelial barrier to water vapor. Arch Dermatol 1958; 78:710–714.

19. Hadgraft J. Absorption of materials by or through the living skin. Int J Cosmet Sci 1985; 7:103–115.

20. Barry BW. Lipid-protein-partitioning theory of skin penetration enhancement. J Control Rel 1991; 15:237–248.

21. Williams AC, Barry BW. The enhancement index concept applied to terpene penetration enhancers for human skin and model lipophilic (oestradiol) and hydrophilic (5-fluorouracil) drugs. Int J Pharm 1991; 74:157–168.

22. Michaels AS, Chandrasekaran SK, Shaw JE. Drug permeation through human skin: Theory and in vitro experimental measurement. AIChE J 1975; 21:985–996.

23. Elias PM, Brown BE, Fritsch P, Goerke J, Gray GM, White RJ. Localization and composition of lipids in neonatal mouse stratum granulosum and stratum corneum. J Invest Dermatol 1979; 73:339–348.

24. Christophers E, Kligman A. Visualization of the cell layers of the stratum corneum. J Invest Dermatol 1964; 42:407–409.

25. Christophers E. Cellular architecture of the stratum corneum. J Invest Dermatol 1971; 56:165–169.

26. Anderson RL, Cassidy JM. Variations in physical dimensions and chemical composition of human stratum corneum. J Invest Dermatol 1973; 61:30–32.

27. Holbrook KA, Odland GF. Regional differences in the thickness (cell layers) of the human stratum corneum: An ultrastructural analysis. J Invest Dermatol 1974; 62:415–422.

28. Plewig G. Regional differences of cell sizes in the human stratum corneum. Part II. Effects of sex and age. J Invest Dermatol 1970; 54:19–23.

29. Plewig G, Marples RR. Regional differences of cell sizes in the human stratum corneum. Part I. J Invest Dermatol 1970; 54:13–18.

30. Lampe MA, Burlingame AL, Whitney J, Williams ML, Brown BE, Roitman E, Elias PM. Human stratum corneum lipids: Characterization and regional variations. J Lipid Res 1983; 24:120–130.

31. Raykar PV, Fung M-C, Anderson BD. The role of protein and lipid domains in the uptake of solutes by human stratum corneum. Pharm Res 1988; 5:140–150.

32. Wertz PW, Kremer M, Squier CA. Comparison of lipids from epidermal and palatal stratum corneum. J Invest Dermatol 1992; 98:375–378.

33. Lampe MA, Williams ML, Elias PM. Human epidermal lipids: Characterization and modulations during differentiation. J Lipid Res 1983; 24:131–140.

34. Melnik BC, Hollmann J, Erler E, Verhoeven B, Plewig G. Microanalytical screening of all major stratum corneum lipids by sequential high-performance thin-layer chromatography. J Invest Dermatol 1989; 92:231–234.

35. Wertz PW, Downing DT. Stratum corneum: biological and biochemical considerations. In: Hadgraft J, Guy RH, eds. Transdermal Drug Delivery: Developmental Issues and Research Initiatives. New York: Marcel Dekker, 1989: 1–22.

36. Elias PM, Goerke J, Friend DS. Mammalian epidermal barrier layer lipids: Composition and influence on structure. J Invest Dermatol 1977; 69:535–546.

37. Elias PM, Bonar L, Grayson S, Baden HP. X-ray diffraction analysis of stratum corneum membrane couplets. J Invest Dermatol 1983; 80:213–214.

38. Elias PM, Menon GK. Structural and lipid biochemical correlates the epidermal permeability barrier. Adv Lipid Res 1991; 24:1–26.

39. Sondell B, Thornell L-E, Egelrud T. Evidence that stratum corneum chymotryptic enzyme is transported to the stratum corneum extracellular space via lamellar bodies. J Invest Dermatol 1995; 104:819–823.

40. Elias PM, Friend DS. The permeability barrier in mammalian epidermis. J Cell Biol 1975; 65:180–191.

41. Elias PM, McNutt NS, Friend DS. Membrane alternations during cornification of mammalian squamous epithelia: A freeze-fracture, tracer, and thin-section study. Anat Rec 1977; 189:577–593.

42. Madison KC, Swartzendruber DC, Wertz PW, Downing DT. Presence of intact intercellular lipid lamellae in the upper layers of the stratum corneum. J Invest Dermatol 1987; 88:714–718.

43. Swartzendruber DC, Wertz PW, Kitko DJ, Madison KC, Downing DT. Molecular models of the intercellular lipid lamellae in mammalian stratum corneum. J Invest Dermatol 1989; 92:251–257.

44. Downing DT. Lipid and protein structures in the permeability barrier of mammalian epidermis. J Lipid Res 1992; 33:301–313.

45. Bouwstra JA, de Vries MA, Gooris GS, Bras W, Brussee J, Ponec M. Thermodynamic and structural aspects of the skin barrier. J Control Rel 1991; 15:209–220.

46. Bouwstra JA, Gooris GS, van der Spek JA, Bras W. Structural investigations of human stratum corneum by small-angle x-ray scattering. J Invest Dermatol 1991; 97:1005–1012.

47. Bouwstra JA, Gooris GS, de Vries MA, van der Spek JA, Bras W. Structure of human stratum corneum as a function of temperature and hydration: A wide-angle x-ray diffraction study. Int J Pharm 1992; 84:205–216.

48. Cornwell PA, Barry BW, Stoddart CP, Bouwstra JA. Wide-angle x-ray diffraction of human stratum corneum: effects of hydration and terpene enhancer treatment. J Pharm Pharmacol 1994; 46:938–950.

49. Landmann L. The epidermal permeability barrier. Comparison between in vivo and in vitro lipid structures. Eur J Cell Biol 1984; 33:258–264.

50. Menon GK, Feingold KR, Elias PM. Lamellar body secretory response to barrier disruption. J Invest Dermatol 1992; 98:279–289.

51. Swartzendruber DC, Wertz PW, Madison KC, Downing DT. Evidence that the corneocyte has a chemically bound lipid envelope. J Invest Dermatol 1987; 88:709–713.

52. Wertz PW, Downing DT. Covalently bound ω-hydroxyacylsphingosine in the stratum corneum. Biochim Biophys Acta 1987; 917:108–111.

53. Wertz PW, Swartzendruber DC, Kitko DJ, Madison KC, Downing DT. The role of the corneocyte lipid envelope in cohesion of the stratum corneum. J Invest Dermatol 1989; 93:169–172.

54. Elias PM. Epidermal lipids, membranes, and keratinization. Int J Dermatol 1981; 20:1–19.

55. Elias PM, Cooper ER, Korc A, Brown BE. Percutaneous transport in relation to stratum corneum structure and lipid composition. J Invest Dermatol 1981; 76:297–301.

56. Albery WJ, Hadgraft J. Percutaneous absorption: In vivo experiments. J Pharm Pharmacol 1979; 31:140–147.

57. Nemanic MK, Elias PM. In situ precipitation: a novel cytochemical technique for visualization of permeability pathways in mammalian stratum corneum. J Histochem Cytochem 1980; 28:573–578.

58. Potts RO, Francoeur ML. The influence of stratum corneum morphology on water permeability. J Invest Dermatol 1991; 96:495–499.

59. Boddé HE, van den Brink I, Koerten HK, de Haan FHN. Visualization of in vitro percutaneous penetration of mercuric chloride; Transport through intercellular space versus cellular uptake through desmosomes. J Control Rel 1991; 15:227–236.

60. Lange-Lieckfeldt R, Lee G. Use of a model lipid matrix to demonstrate the dependence of the stratum corneum's barrier properties on its internal geometry. J Control Rel 1992; 20:183–194.

61. Moghimi HR, Williams AC, Barry BW. A lamellar matrix model for stratum corneum intercellular lipids. I. Characterisation and comparison with stratum corneum intercellular structure. Int J Pharm 1996; 131:103–115.

62. Moghimi HR, Williams AC, Barry BW. A lamellar matrix model for stratum corneum intercellular lipids. II. Effect of geometry of the stratum corneum on permeation of model drugs 5-fluorouracil and oestradiol. Int J Pharm 1996; 131:117–129.

63. Tojo K. Random brick model for drug transport across stratum corneum. J Pharm Sci 1987; 76:889–891.

64. Heisig M, Lieckfeldt R, Wittum G, Mazurkevich G, Lee G. Non steady-state descriptions of drug permeation through stratum corneum. I. The biphasic brick-and-mortar model. Pharm Res 1996; 13:421–426.

65. Houslay MD, Stanley KK. Dynamics of Biological Membranes; Influence on Synthesis, Structure and Function. Chichester: John Wiley and Sons, 1982.

66. Friberg SE, Osborne DW. Small angle x-ray diffraction patterns of stratum corneum and a model structure for its lipids. J Dispersion Sci Technol 1985; 6:485–495.

67. Osborne DW, Friberg SE. Role of stratum corneum lipids as moisture retaining agent. J Dispersion Sci Technol 1987; 8:173–179.

68. Friberg SE, Suhaimi H, Goldsmith LB, Rhein LL. Stratum corneum lipids in a model structure. J Dispersion Sci Technol 1988; 9:371–389.

69. Mizushima H, Fukasawa JI, Suzuki T. Phase behaviour of artificial stratum corneum lipids containing synthetic pseudo-ceramide. A study of the function of cholesterol. J Lipid Res 1996; 37:361–367.

70. Parrott DT, Turner JE. Mesophase formation by ceramides and cholesterol: A model for stratum corneum lipid packing? Biochim Biophys Acta 1993; 1147: 273–276.

71. Wertz PW, Abraham W, Landmann L, Downing DT. Preparation of liposomes from stratum corneum lipids. J Invest Dermatol 1986; 87:582–584.

72. Abraham W, Downing DT. Preparation of model membranes for skin permeability studies using stratum corneum lipids. J Invest Dermatol 1989; 93: 809–813.

73. Kittayanond D, Ramachandran C, Weiner N. Development of a model of the lipid constituent phase of the stratum corneum: I. Preparation and characterization of "skin lipid" liposomes using synthetic lipids. J Soc Cosmet Chem 1992; 43:149–160.

74. Friberg SE, Kayali I, Beckerman W, Rhein LD, Simion A. Water permeation of reaggregated stratum corneum with model lipids. J Invest Dermatol 1990; 94:377–380.

75. Kayali I, Suhery T, Friberg SE, Simion FA, Rhein LD. Lyotropic liquid crystals and the structural lipids of the stratum corneum. J Pharm Sci 1991; 80: 428–431.

76. Lieckfeldt R, Villalain J, Gomez-Fernandez JC, Lee G. Diffusivity and structural polymorphism in some model stratum corneum lipid systems. Biochim Biophys Acta 1993; 1151:182–188.

77. Ward AJI, du Reau C. The essential role of lipid bilayers in the determination of stratum corneum permeability. Int J Pharm 1991; 74:137–146.

78. Friberg SE, Kayali I. Water evaporation rates from a model of stratum corneum lipids. J Pharm Sci 1989; 78:639–643.

79. Elias PM, Brown BE. The mammalian cutaneous permeability barrier; Defective barrier function in essential fatty acid deficiency correlates with abnormal intercellular lipid deposition. Lab Invest 1978; 39:574–583.

80. Friberg SE, Kayali I, Rhein L. Direct role of linoleic acid in barrier function: Effect of linoleic acid on the crystalline structure of oleic acid/oleate model stratum corneum lipid. J Dispersion Sci Technol 1990; 11:31–47.

81. Aszterbaum M, Menon GK, Feingold KR, Williams ML. Ontogony of the epidermal barrier to water loss in the rat: Correlation of function with stratum corneum structure and lipid content. Pediatr Res 1992; 31:308–317.

82. Sheu HM, Lee JYY, Chai CY, Kuo KW. Depletion of stratum corneum intercellular lipid lamellae and barrier function abnormalities after long-term topical corticosteroids. Br J Dermatol 1997; 136:884–890.

83. Schückler F, Bouwstra JA, Gooris GS, Lee G. An x-ray diffraction study of some model stratum corneum lipids containing Azone and dodecyl-L-pyro-glutamate. J Control Rel 1993; 23:27–36.

84. Bouwstra JA, Cheng K, Gooris GS, Weerheim A, Ponec M. The role of ceramide-1 and ceramide-2 in the stratum corneum lipid organisation. Biochim Biophys Acta Lipids Lipid Metab 1996; 130:177–186.

85. White SH, Mirejovsky D, King GI. Structure of lamellar lipid domains and corneocyte envelopes of murine stratum corneum. An x-ray diffraction study. Biochemistry 1988; 27:3725–3732.

86. Bouwstra JA, Gooris GS, Cheng K. Weerheim A, Bras W, Ponec M. Phase behavior of isolated skin lipids. J Lipid Res 1996; 37:999–1011.

87. McIntosh TJ, Stewart ME, Downing DT. X-ray diffraction analysis of isolated skin lipids: Reconstitution of intercellular lipid domains. Biochemistry 1996; 35:3649–3653.

88. Bouwstra JA, Thewalt J, Gooris GS, Kitson N. Model membrane approach to the epidermal permeability barrier: An x-ray diffraction study. Biochemistry 1997; 36:7717–7725.

89. Moore DJ, Rerek ME, Mendelsohn R. Lipid domains and orthorhombic phases in model stratum corneum: Evidence from Fourier transform infrared spectroscopy studies. Biochem Biophys Res Commun 1997; 231:797–801.

90. Tanojo H, Bouwstra JA, Junjinger HE, Boddé HE. Subzero thermal analysis of human stratum corneum. Pharm Res 1994; 11:1610–1616.

91. Imokawa G, Kuno H, Kawai M. Stratum corneum lipids serve as a bound-water modulator. J Invest Dermatol 1991; 96:845–851.

92. Goodman M, Barry BW. Action of penetration enhancers on human stratum corneum as assessed by differential scanning calorimetry. In: Bronaugh RL, Maibach HI. eds. Percutaneous Absorption, Mechanisms-Methodology-Drug Delivery. 2nd ed. New York: Marcel Dekker, 1989:567–593.

93. Fenske DB, Thewalt JL, Bloom M, Kitson L. Models of stratum corneum intercellular membranes. H-2 NMR of macroscopically oriented multilayers. Biophys J 1994; 67:1562–1573.

94. Loomis CR, Shipley GG, Small DM. The phase behavior of hydrated cholesterol. J Lipid Res 1979; 20:525–535.

95. Gay CL, Guy RH, Golden GM, Mak VHW, Francoeur ML. Characterization of low-temperature (i.e., <65°C) lipid transitions in human stratum corneum. J Invest Dermatol 1994; 103:233–239.

96. Ongpipattanakul B, Francoeur ML, Potts RO. Polymorphism in stratum corneum lipids. Biochim Biophys Acta 1994; 1190:115–122.

97. Cornwell PA. Mechanisms of Action of Terpene Penetration Enhancers in Human Skin. PhD thesis, University of Bradford, UK, 1993.

98. Hang CH, Sanftleben R, Wiedmann TS. Phase properties of mixtures of ceramides. Lipids 1995; 30:121–128.

99. Hatfield RM, Fung LWM. Molecular properties of a stratum corneum model lipid system, large unilamellar vesicles. Biophys J 1995; 68:196–207.

100. Abraham W, Downing DT. Lamellar structure formed by stratum corneum lipids in vitro: a deuterium nuclear magnetic resonance (NMR) study. Pharm Res 1992; 9:1415–1421.

101. Potts RO, Guy RH. The prediction of percutaneous penetration: a mechanistic model. In: Gurny R, Teubner A eds. Dermal and Transdermal Drug Delivery, New Insights and Perspectives. Stuttgart: Wissenschaftliche Verlagsgesellschaft mbH, 1993:153–160.

102. Lieckfeldt R, Lee G. Measuring the diffusional pathlength and area within membranes of excised human stratum corneum. J Pharm Pharmacol 1995; 47:26–29.

103. Cornwell PA, Barry BW. Sesquiterpene components of volatile oils as skin penetration enhancers for the hydrophilic permeant 5-fluorouracil. J Pharm Pharmacol 1994; 46:261–269.

104. Yamane MA. Terpene Penetration Enhancers in Human and Snake Skin; Permeation, Differential Scanning Calorimetry and Electrical Conductivity Studies. PhD thesis, University of Bradford, UK, 1994.

105. Kadir R, Barry BW. α-Bisabolol, a possible safe penetration enhancer for dermal and transdermal therapeutics. Int J Pharm 1991; 70:87–94.

106. Williams AC, Barry BW. Terpenes and the lipid-protein-partitioning theory of skin penetration enhancement. Pharm Res 1991; 8:17–24.

107. Williams AC, Barry BW. Skin absorption enhancers. Crit Rev Ther Drug Carrier Sys 1992; 9:305–353.

108. Goodman M, Barry BW. Action of penetration enhancers on human skin as assessed by the permeation of model drugs 5-fluorouracil and estradiol. I. Infinite dose technique. J Invest Dermatol 1988; 91:323–327.

109. Williams AC, Cornwell PA, Barry BW. On the non-Gaussian distribution of human skin permeabilities. Int J Pharm 1992; 86:69–77.

110. Megrab NA. Oestradiol Permeation Through Human Skin, Silastic and Snake Membranes; Effects of Supersaturation and Binary Co-solvent Systems. PhD thesis, University of Bradford, UK, 1994.

111. Walters KA, Hadgraft J. Pharmaceutical Skin Penetration Enhancement. New York: Marcel Dekker, 1993.

112. Kim C-K, Hong M-S, Kim Y-B, Han S-K. Effect of penetration enhancers (pyrrolidone derivatives) on multilamellar liposomes of stratum corneum lipids: A study by UV spectroscopy and differential scanning calorimetry. Int J Pharm 1993; 95:43–50.

113. Yoneto K, Li SK, Higuchi WI, Jiskoot W, Herron JN. Fluorescent-probe studies of the interactions of 1-alkyl-2-pyrrolidones with stratum corneum lipid liposomes. J Pharm Sci 1996; 85:511–517.

114. Miyajima K, Tanikawa S, Asano M, Matsuzaki K. Effects of absorption enhancers and lipid composition on drug permeability through the model membrane using stratum corneum lipids. Chem Pharm Bull 1994; 42:1345–1347.

115. Moghimi HR, Williams AC, Barry BW. A lamellar matrix model for stratum corneum intercellular lipids. III. Effects of terpene penetration enhancers on the release of 5-fluorouracil and oestradiol from the matrix. Int J Pharm 1996; 145:37–47.

116. Moghimi HR, Williams AC, Barry BW. A lamellar matrix model for stratum corneum intercellular lipids. IV. Effects of terpene penetration enhancers on the permeation of 5-fluorouracil and oestradiol through the matrix. Int J Pharm 1996; 145:49–59.

117. Moghimi HR. Modeling the Intercellular Lamellar Lipid Structure of Human Stratum Corneum for Drug Permeation Studies. PhD thesis, University of Bradford, UK, 1995.

118. Cornwell PA, Barry BW, Bouwstra JA, Gooris GS. Modes of action of terpene penetration enhancers in human skin; Differential scanning calorimetry, small-angle x-ray diffraction and enhancer uptake studies. Int J Pharm 1996; 127: 9–26.

119. Suhonen TM, Pirskanen L, Raisanen M, Kosonen K, Rytting JH, Paronen P, Urtti A. Transepidermal delivery of β-blocking agents; Evaluation of enhancer effects using stratum corneum lipid liposomes. J Control Rel 1997; 43: 251–259.

120. Moghimi HR, Williams AC, Barry BW. A lamellar matrix model for stratum corneum intercellular lipids. V. Effects of terpene penetration enhancers on the structure and thermal behaviour of the matrix. Int J Pharm 1997; 146: 41–54.

121. Rosevear FB. The microscopy of the liquid crystalline neat and middle phases of soaps and synthetic detergents. J Am Oil Chem Soc 1954; 31:628–639.

122. Ribaud C, Garson J-C, Doucet J, Lévéque J-L. Organisation of stratum corneum lipids in relation to permeability: influence of sodium lauryl sulfate and preheating. Pharm Res 1994; 11:1414–1418.

123. Engblom J, Engström S. Azone® and the formation of reversed mono- and bicontinuous lipid-water phases. Int J Pharm 1993; 98:173–179.

124. Benton WJ, Raney KH, Miller CA. Enhanced videomicroscopy of phase transitions and diffusional phenomena in oil-water-nonionic surfactant systems. J Colloid Interface Sci 1986; 110:363–388.

125. Vanhal DA, Jeremiasse E, Junginger HE, Spies F, Bouwstra JA. Structure of fully hydrated human stratum corneum. A freeze-fracture electron microscopy study. J Invest Dermatol 1996; 106:89–95.

126. Walker M, Hadgraft J. Oleic acid–A membrane "fluidiser" or fluid within the membrane? Int J Pharm 1991; 71:R1–R4.

127. Naik A, Pechtold LARM, Potts RO, Guy RH. Mechanism of oleic acid-induced skin penetration enhancement in vivo in humans. J Control Rel 1995; 37:299–306.

128. Yamane MA, Williams AC, Barry BW. Effects of terpenes and oleic acid as skin penetration enhancers towards 5-fluorouracil as assessed with time; Permeation, partitioning and differential scanning calorimetry. Int J Pharm 1995; 116:237–251.

129. Yamane MA, Williams AC, Barry BW. Terpene penetration enhancers in propylene glycol/water co-solvent systems; Effectiveness and mechanism of action. J Pharm Pharmacol 1995; 47:978–989.

130. Ogiso T, Shintani M. Mechanism for the enhancement effect of fatty acids on the percutaneous absorption of propranolol. J Pharm Sci 1990; 79:1065–1071.

32
Role of In Vitro Release Measurement in Semisolid Dosage Forms

Vinod P. Shah
Food and Drug Administration, Rockville, Maryland

Jerome S. Elkins
Food and Drug Administration, Dallas, Texas

Roger L. Williams
Food and Drug Administration, Rockville, Maryland

I. INTRODUCTION

A key aspect of any new drug product is its safety and efficacy as demonstrated in controlled clinical trials. The time and expense associated with such trials make them unsuitable as routine quality control methods to reestablish comparability in quality and performance following a change in formulation or method of manufacture. Therefore, in vitro and in vivo surrogate tests are often used to assure that product quality and performance are maintained over time in the presence of change. The focus of this chapter is the application of these surrogate approaches in the documentation of performance of semisolid dosage forms, specifically in the determination of bioavailability (BA) and bioequivalence (BE), defined as relative bioavailability.

For many dosage forms, in vivo approaches to document BA/BE rely on measurement of the active moiety/active ingredient in plasma or blood samples over time. This documentation is based on the use of pharmacokinetic metrics such as area under the curve (AUC) and C_{max}, which, with certain assumptions, are derived surrogates for safety and efficacy. In vitro approaches, such as in vitro dissolution, are also standard methods used to assess performance characteristics of a solid oral dosage formulation. In vitro dissolution may be used specifically to guide formulation development, monitor formulation quality from batch to batch, monitor control of the formulation manufacturing process, and possibly predict in vivo performance. When used as a quality control procedure, in vitro dissolution testing can signal an inadvertent change in drug and/or excipient characateristics or in the manufacturing process.

Extension of in vitro dissolution methodology to semisolid dosage forms (topical dermatological drug products such as creams, ointments, gels, and lotions) has been the subject of both substantial effort and debate. A simple, reliable, reproducible, relevant, and generally acceptable in vitro method to assess drug release from a semisolid dosage form would be highly valuable for the same reasons that such methodology has proved valuable in the development, manufacture, and batch-to-batch quality control of solid oral dosage forms.

Present quality control tests to assure the identity, strength, quality, purity, and potency for semisolid dosage forms include identification, assay, homogeneity, viscosity, specific gravity, particle size, microbial limits, and impurity profile. These tests may provide little or no information about drug release properties of the product, stability of the product, or effects of manufacturing and processing variables on the performance of the finished dosage form. A drug release test for topical products, analogous to a dissolution test for a solid oral dose form, is therefore of interest. Such a test should be able to detect formulation and manufacturing process changes that affect performance of the drug product in vivo. Ultimately, it would be desirable to demonstrate that in vitro performance of a topical formulation correlates in some way with in vivo performance of the formulation. Pending development of such data, a drug release test may be used to assess batch-to-batch drug release uniformity and quality of the product, just as in vitro dissolution is now used for solid oral dose forms.

II. IN VITRO RELEASE TESTING

In vitro release is one of several methods used to characterize performance characteristics of a finished topical dosage form. Important changes in the characteristics of a drug product or in the thermodynamic properties of the drug substance in the dosage form should be manifested as a difference in drug release. Drug release is theoretically proportional to the square root of time ($\sqrt{t}$) when the drug release from the formulation is rate limiting. A plot of the amount of drug released per unit area ($\mu g/cm^2$) against the square root of time yields a straight line, the slope of which represents the release rate. This release rate measure is formulation specific and can be used to monitor product quality. As summarized in an FDA guidance document (1), recommended methodology for in vitro release studies is as follows:

Diffusion cell system: a static diffusion cell system with a standard open cap ground glass surface with 15-mm-diameter orifice and total diameter of 25 mm (Fig. 1).

Synthetic membrane: appropriate inert, porous, and commercially available synthetic membranes such as polysulfone, cellulose acetate/nitrate mixed ester, or polytetrafluoroethylene 70 μm membrane

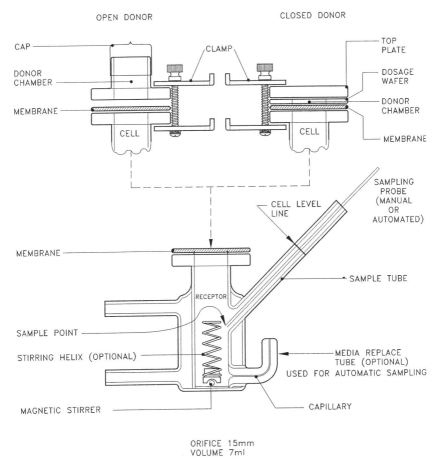

Figure 1 Schematic of diffusion cell assembly used for in vitro release measurement.

of appropriate size to fit the diffusion cell diameter (e.g., 25 mm in the preceding case).

Receptor medium: appropriate receptor medium such as aqueous buffer for water soluble drugs or a hydro-alcoholic medium for sparingly water soluble drugs or another medium with proper justification.

Number of samples: A minimum of six samples is recommended to determine the release rate (profile) of the topical dermatological product.

Sample applications: About 300 mg of the semisolid preparation is placed uniformly on the membrane and kept occluded to prevent solvent evaporation and compositional changes. This corresponds to an infinite dose condition.

Sampling time: Multiple sampling times (at least 5 times) over an appropriate time period to generate an adequate release profile and to determine the drug release rate (a 6-h study period with not less than five samples, i.e., at 30 min, 1, 2, 4, and 6 h) is suggested. The sampling times may have to be varied depending on the formulation. An aliquot of the receptor phase is removed at each sampling interval and replaced with fresh aliquot, so that the lower surface of the membrane remains in contact with the receptor phase over the experimental time period.

Sample analysis: Appropriate validated, specific, and sensitive analytical procedure, generally high-pressure liquid chromatography (HPLC), is used to analyze the samples and to determine the drug concentration and the amount of drug released.

In vitro release rate: A plot of the amount of drug released per unit membrane area ($\mu g/cm^2$) versus square root of time should yield a straight line. The slope of the line (regression) represents the release rate of the product. An X intercept typically corresponding to a small fraction of an hour is a normal characteristic of such plots.

Automation: The in vitro release test system can be completely automated (Fig. 2).

III. DISCUSSION

When drugs are applied topically, a pharmacologically active agent must be released from its carrier (vehicle) before it can contact the epidermal surface and be available for penetration in the stratum corneum and lower layers of the skin (2). A topical formulation thus is a complex drug delivery system, and the dynamics of drug release from a vehicle have been a subject of debate and investigation for many years (3–15). A simple and reproducible method, generally applicable to all topical dermatological dosage forms, has

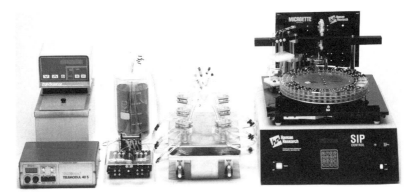

Figure 2 Automated setup for in vitro drug release experiment.

been developed to measure in vitro release of the drug from its vehicle using a diffusion cell and a synthetic membrane (3–5). This in vitro release test is gaining importance as a product performance and quality control test. Recent scientific workshops on scale-up of a semisolid disperse system (6) and on the value of in vitro drug release (7) have resulted in recommendations on the use of in vitro release tests as a measure of in-process control and also as a finished product specification for creams, ointments, and gels. In addition, the cited workshop report also recommends the use of in vitro release test for monitoring product reproducibility during component and compositional changes, manufacturing equipment and process changes, scale-up, and/or transfer to another manufacturing site (6).

A relatively simple methodology has been developed to assess drug release characteristics from topical dermatologic formulations. This methodology employs a static diffusion cell system, commonly referred to as the Franz cell, a commercially available polysulfone synthetic membrane, and an aqueous receptor phase (Fig. 1). To determine in vitro drug release, an infinite amount of drug is applied on the donor chamber. Aliquots of the receptor media are removed at 30-, 60-, 120-, 240-, and 360-min intervals, analyzed, and the cumulative amount of drug released, expressed in micrograms per square centimeter, is plotted against square root of time ($min^{0.5}$). The relationship between drug release and square root of time has been shown to be linear and valid for topical formulations as long as the percentage of drug release is less than 30% of the drug applied in the donor chamber (4,5). This relationship holds true for topical formulations with either fully dissolved or suspended drug. The release rate or the slope of the line is obtained by linear regression analysis and is considered to be the property of the formulation. The polysulfone synthetic membrane serves as a support membrane and should be chemically inert to the experimental formulations. It should not react with the drug or the receptor medium, should be permeable to the drug, and should not be rate limiting in the drug release process (14).

The methodology has been employed to measure in vitro drug release from marketed corticosteroid formulations (10). The release measurements from hydrocortisone cream with five different synthetic membranes were found to be nearly the same (3). Similarly, the release rates of other corticosteroid products using a cellulose acetate/nitrate mixed ester membrane and a polysulfone membrane were also found to be nearly the same (16). These findings confirm that the nature of the membrane does not contribute significantly to release rate of topical dosage forms, as long as sufficient porosity of the membrane is maintained. Using ointment corticosteroid formulations, the synthetic membrane has been shown not to contribute to drug release and is not a rate-limiting factor in in vitro release experiments (8). These findings confirm that the measured release rate is the property of the formulation under experimental conditions.

Selection of the receptor medium is an important and critical variable in an in vitro release test. The drug must have sufficient solubility in the receptor medium to maintain sink conditions. In addition, it must be determined that the receptor medium should not react with the membrane or alter the dosage form by back diffusion through the membrane. The medium is selected such that it is compatible with the analytical (HPLC) method, with the analysis carried out by direct injection of the receptor medium onto HPLC systems.

Selection of an appropriate receptor medium is important to maintain sink conditions during in vitro release studies for products containing water insoluble drugs. To promote drug release from topical preparations containing these drug substances, studies were carried out with receptor media containing surfactants and different organic/aqueous solvents. Use of surfactants resulted in foaming and formation of air bubbles during receptor mixing. The presence of air bubbles also interfered with required uniform contact between the receptor medium and the supporting membrane (9). Further studies with surfactants were discontinued for these reasons. Among the different water miscible organic solvents evaluated, a mixture consisting of ethanol and water resulted in an optimum receptor medium for the study of in vitro release of corticosteroid drug products (9).

The role of the receptor medium in in vitro release experiments is well illustrated using Diprolene (DI) and Diprosone (DO) cream products (Table 1). Although both of these products contain the same active drug, betamethasone dipropionate, in the same concentration, 0.05%, DI is assigned to a more potent category, Category 2, compared to DO, which is assigned

Table 1 Comparison of Diprolene and Diprosone Formulations

Ingredients[a]	Diprolene AF cream[b]	Diprosone cream[c]
Betamethasone dipropionate[d]	0.64 mg	0.64 mg
Chlorocresol	√	√
Propylene glycol	√	√
White petrolatum	√	√
White wax	√	
Cyclomethicone	√	
Sorbitol solution	√	
Glyceryl monooleate	√	
Ceteareth-30	√	√
Carbomer 940	√	
Sodium hydroxide	√	
Mineral oil		√
Ceteraryl alcohol		√
Phosphoric acid		√
Monobasic sodium phosphate		√
Water QS to 1 g	√	√

[a] As listed on the label/PDR.
[b] Emollient cream base.
[c] Hydrophilic emollient cream.
[d] Equivalent to 0.5 mg betamethasone.

to Category 3. The current compendial testing based on assay, content uniformity, and other tests cannot differentiate between DI and DO formulations. The drug release rates from DI and DO using different receptor media (60%, 40%, and 30% alcohol), are summarized in Fig. 3. Using 30% alcohol as a receptor medium, a "rank order" relation between the clinical potency and the release rate for two betamethasone dipropionate products has been reported (16). Based on these studies, 30% alcohol appears to be an appropriate receptor medium to use in assessing corticosteroid drug release in vitro.

IV. IN VITRO RELEASE—CORTICOSTEROIDS

In a series of studies, in vitro release of marketed corticosteroids was assessed using polysulfone membrane and a receptor medium consisting of 30% alcohol for creams and a receptor medium consisting of alcohol:isopropyl myristate:water in ratios of 85:10:5 (8–10,16). It was shown in these studies that this composition of receptor medium was needed to determine the release rate from corticosteroid ointments (8). All products re-

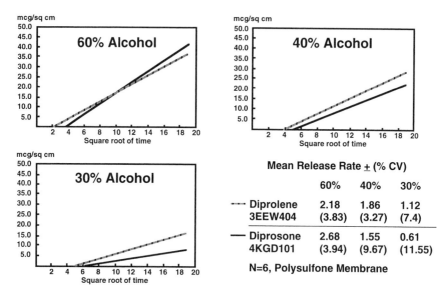

Figure 3 In vitro release: influence of receptor medium on release rate of diprolene and diprosone creams.

sulted in measurable, reproducible in vitro release rates. Comparisons of in vitro release rates of marketed drug products, both within a manufacturer and across manufacturers, provided useful information. In vitro release rates of triamcinolone acetonide products and the qualitative compositions of formulations from different manufacturers are summarized in Table 2 and Fig. 4. The release rates of different batches from a given manufacturer were found to be similar, indicating product release uniformity from batch to batch. In contrast, significantly different release rates, sometimes up to threefold and more ($4.2-11.6$ $\mu g/cm^2/min^{0.5}$, Table 3) were observed between products from different manufacturers. This observation is perhaps not unexpected because release rate is a property of the formulation and formulations are frequently significantly different between manufacturers. For betamethasone dipropionate, the release rates from different manufacturers were not as widespread ($0.51-1.29$ $\mu g/cm^2/min^{0.5}$) as that of triamcinolone acetonide (Fig. 5 and Table 4).

The improved performanace of betamethasone products compared to triamcinolone acetonide products may be explained by differences in regulatory requirements and recommendations. Prior to 1962, topical products containing triamcinolone acetonide (drug product approved before this date) and other agents had only to meet USP assay and content uniformity criteria.

Table 2 Comparison of Triamcinolone Acetonide Cream Formulations

Ingredients[a]	Manufacturer				
	A	B	C	D	E
Triamcinolone acetonide	5 mg	5 mg	5 mg	5 mg	5 mg
Benzyl alcohol		✓			
Ceteareph-20				✓	
Cetearyl alcohol				✓	
Cetyl alcohol	✓	✓	✓		✓
Cetyl esters			✓		
Emulsifying wax		✓			
Glycerine		✓	✓		
Glycerol monostearate			✓	✓	
Imidazolidinyl urea	✓				
Isopropyl palmitate	✓	✓	✓		✓
Lactic acid		✓			
Lanolin alcohol	✓				✓
Methyl and propyl paraben					✓
Mineral oil	✓				✓
Polyoxyl 40 stearate					✓
Polysorbate 40	✓				
Polysorbate 60			✓		✓
Propylene glycol	✓		✓	✓	✓
Propylene glycol stearate	✓			✓	✓
Simethicone				✓	
Sodium bisulfite	✓				
Sorbic acid	✓		✓	✓	✓
Sorbitan monopalmitate	✓				
Sorbitan monostearate			✓		✓
Sorbitol solution		✓		✓	
Stearyl alcohol	✓				
Water	✓	✓	✓	✓	✓
White petroleum				✓	

[a]Label declaration.

FDA did not recommend or require in vivo studies to document BA/BE for these products. Although all triamcinolone acetonide products studied will meet USP monograph criteria, the clinical relevance of the observed different release characteristics across products is not known. Approved after 1962, betamethasone dipropionate drug products from multiple manufacturers were required by FDA to show BE to the pioneer product via comparative clinical efficacy, pharmacokinetic, or pharmacodynamic studies. Because of this ad-

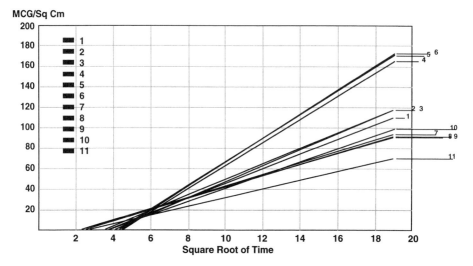

Figure 4 In vitro release of triamcinolone acetonide drug products. Line 1: manufacturer A, lot 1. Line 2: manufacturer A, lot 2. Line 3: manufacturer A, lot 3. Line 4: manufacturer B, lot 1. Line 5: manufacturer B, lot 2. Line 6: manufacturer B, lot 3. Line 7: manufacturer C, lot 1. Line 8: manufacturer C, lot 2. Line 9: manufacturer C, lot 3. Line 10: manufacturer D, lot 1. Line 11: manufacturer E, lot 1.

Table 3 Release Rates of Marketed Triamcinolone Acetonide Cream Drug Products

Manufacturer	Lot number	Release rate ($\mu g/cm^2/min^{0.5}$), mean[a] ± (%CV)
A	1	7.199 (15.7)
A	2	7.868 (5.06)
A	3	7.661 (6.28)
B	4	11.335 (3.37)
B	5	11.615 (7.25)
B	6	11.825 (5.98)
C	7	5.683 (6.06)
C	8	5.393 (4.83)
C	9	5.503 (3.37)
D	10	6.498 (6.51)
E	11	4.266 (6.20)

[a]Mean of six runs.

MCG/Sq Cm

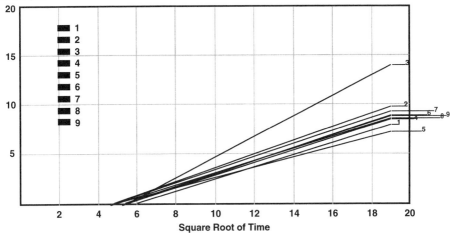

Figure 5 In vitro release of betamethasone dipropionate cream products. Line 1: manufacturer A, lot 1. Line 2: manufacturer B, lot 1. Line 3: manufacturer C, lot 1. Line 4: manufacturer D, lot 1. Line 5: manufacturer D, lot 2. Line 6: manufacturer D, lot 3. Line 7: manufacturer E, lot 1. Line 8: manufacturer E, lot 2. Line 9: manufacturer E, lot 3.

Table 4 In Vitro Release of Betamethasone Dipropionate Creams

Manufacturer	Lot number	Release rate ($\mu g/cm^2/min^{0.5}$), mean[a] $\pm$ (%CV)
A	1	0.611 (11.55)
B	2	0.687 (10.44)
C	3	1.036 (14.21)
D	4	0.619 (7.58)
D	5	0.515 (11.34)
D	6	0.659 (7.43)
E	7	0.649 (7.81)
E	8	0.611 (7.04)
E	9	0.626 (5.66)

[a]Mean of six runs.

ditional requirement, manufacturers may have performed better product development work, based on principles that have been termed reverse engineering, to assure that the components, composition, and in vitro performance of their products matched that of the pioneer product as closely as possible (see Table 5). Additional data also suggest the impact of method of manufacture on in vitro performance as well as on formulation, as exhibited by variability in in vitro release even when formulations are highly similar but manufacturing processes are different (17). This variability appears to be higher compared to the variability exhibited by identical formulations with similar or identical methods of manufacture.

In vitro release experiments using different agitation/mixing speeds and different lot numbers of synthetic membranes show no significant difference in release rate (16). In vitro release rate measurements performed over 2 years and from different batches of a given manufacturer also showed no important differences in results. These experiments establish the ruggedness and reproducibility of the in vitro release methodology and support its use in quality control/quality assurance. Unpublished work in our laboratory indicates that in vitro release methodology is also applicable to other groups

Table 5 Comparison of Betamethasone Dipropionate Cream Formulations

Ingredients[a]	Manufacturer				
	A	B	C	D	E
Betamethasone dipropionate	0.64 mg	0.64 mg	0.64 mg	0.64 mg	0.64 mg
Ceteareth-30	√				
Cetearyl alcohol	√				
Cetomacrogol 1000			√		
Cetostearyl alcohol		√	√	√	√
Chlorocresol	√	√	√	√	√
Mineral oil	√	√	√	√	√
Monobasic sodium phosphate	√	√	√	√	√
Phosphoric acid	√	√	√	√	√
Polyethylene glycol 1000 monocetyl ether		√		√	√
Propylene glycol	√		√	√	√
Sodium hydroxide		√	√	√	
Water	√	√	√	√	√
White petroleum	√	√	√	√	√

[a]Label declaration.

of dermatological dosage forms such as antifungal, antiviral, and antiacne drug products intended for topical administration. Overall, these results demonstrate the versatility and utility of the in vivo release test methodology.

V. APPLICATIONS

Application of in vitro release testing in drug development and its value in topical drug products quality assurance was discussed extensively in a recent scientific workshop entitled Assessment of Value and Applications of In Vitro Testing of Dermatological Drug Products (7). Consensus at this workshop was that in vitro release methodology is based on sound scientific principles and is of value in assessing product quality. It should not be used to compare fundamentally different topical formulations such as creams, ointments, and gels. Further, attendees agreed that release rate by itself should not be used as a measure of BA/BE. Nonetheless, in vitro release test can serve as a valuable tool for initial screening of experimental formulations in the product development area and can serve to signal possible bioinequivalence. In vitro release testing may find a future use as a quality control tool to assure batch-to-batch uniformity, just as the dissolution test is used to assure quality and performance of solid oral dosage forms.

A. SUPAC-SS

In May 1977, FDA released a guidance for industry entitled SUPAC-SS Nonsterile Semisolid Dosage Forms—Scale-Up and Postapproval Changes: Chemistry, Manufacturing, and Controls; In Vitro Release Testing and In Vivo Bioequivalence Documentation (1). A key public health objective of the SUPAC-SS document is the use of in vitro release testing to assure product sameness between prechange (approved, reference) product and postchange (test) product. The guidance document focuses on nonsterile creams, gels, lotions, and ointments intended for topical administration. The document describes recommendations for tests and filing recommendations for changes in the following four categories: a) components and composition; b) manufacturing equipment and process; c) scale (batch size); and d) site of manufacture. Changes are categorized as Level 1, Level 2, and Level 3, depending on the degree of change and the type of tests needed to document comparability in identity, strength, quality, purity, and potency of the drug product before and after the change. Level 1 changes are those that are unlikely to have any detectable impact on formulation quality and performance of the product. This degree of change does not require any additional testing for product approval beyond assurance that application and compendial specifications are met. Level 2 changes are those that could have a significant impact on formulation quality and performance of the product. For Level 2 changes, the guidance document recommends in vitro release testing in addition to assuring that application and compendial specifications are met. Level 3 changes are those that are likely to have a significant impact on formulation quality and performance of the product. This degree of change requires in vitro release testing for a site change or in vivo bioequivalence testing for changes in component and composition, in addition to assurance that application and compendial specifications are met. Release rates are considered similar when the ratio of the median release rate for the postchange (test) product over the median release rate for the prechange (reference) product is within the 90% confidence interval limits of 75% to 133.33%. The release rate is regarded as a "final quality control" test that can signal possible inequivalence in performance, thus comprising in the aggregate a number of physicochemical tests that might be performed individually.

B. Bioequivalence Waivers for Lower Strength

For solid oral dosage forms, biowaivers for generic products are generally granted in situations where the formulations of lower strength(s) product(s) are proportionately similar and the dissolution profile is also similar (21 CFR 320.22 (d) (2)). Using these same principles, bioequivalence waivers for lower strength(s) of topical dermatological drug products might also be granted based on in vitro release rate measurements. For a request of biowaiver for lower strength, the product must meet the following criteria:

> Formulations of the two strengths should differ only in the concentration of the active ingredient and equivalent amount of the diluent.
> No differences should exist in manufacturing process and equipment between the two strengths.
> For a generic application, that is, an abbreviated new drug application (ANDA), the reference listed drug (RLD) should be marketed at both higher and lower strengths.
> For an ANDA, the higher strength of the test product should be BE to the higher strength of RLD.

In vitro drug release rate studies should be measured under the same test conditions for all strengths of both the test and RLD products. The in vitro release rate should be compared between a) the RLD at both the higher (RHS) and lower strengths (RLS), and b) the test (generic) products at both higher (THS) and lower strengths (TLS). Using the in vitro release rate, the following ratios and comparisons should be made:

$$\frac{\text{Release rate of RHS}}{\text{Release rate of RLS}} \approx \frac{\text{Release rate of THS}}{\text{Release rate of TLS}}$$

The ratio of the release rates of the two strengths of the test products should be about the same as the ratio of the release rate of reference products.

VI. CONCLUSION

In vitro release rate methodology for topically applied locally acting drug products can reflect the combined effect of several physical and chemical parameters, including solubility and particle size of the active ingredient and rheological properties of the dosage form. In most cases, in vitro release rate is a useful test to assess product sameness between prechange and post-change semisolid products such as creams, gels, lotions, and ointments. The release test appears to be formulation specific and is a property of the formulation and its method of manufacture.

REFERENCES

1. Guidance for Industry: SUPAC-SS Nonsterile Semisolid Dosage Forms. Scale-Up and Postapproval Changes: Chemistry, Manufacturing, and Controls; In Vitro Release Testing and In Vivo Bioequivalence Documentation. US Department of Health and Human Services, Food and Drug Administration, Center for Drug Evaluation and Research, May 1997.
2. R.H. Guy, A.H. Guy, H.I. Maibach, and V.P. Shah. The bioavailability of dermatological and other topically administered drugs. Pharm. Res. 3:253–262, 1986.
3. V.P. Shah, J. Elkins, S.Y. Lam and J.P. Skelly. Determination of in vitro drug release from hydrocortisone creams. Int. J. Pharm. 53:53–59, 1989.
4. R. Guy and J. Hadgraft. On the determination of drug release rates from topical dosage forms. Int. J. Pharm. 60:R1–R3, 1990.
5. W.I. Higuchi. Analysis of data on the medicament release from ointments. J. Pharm. Sci. 51:802–804, 1962.
6. G.A. Van Buskirk, V.P. Shah, D. Adair, et. al. Workshop report: Scale-up of liquid and semisolid disperse systems. Pharm. Res. 11:1216–1220, 1994.
7. Workshop on Assessment of Value and Applications of In Vitro Testing of Topical Dermatological Drug Products. Sponsored by American Association of Pharmaceutical Scientists and FDA, September 1997.
8. V.P. Shah and J.S. Elkins. In vitro release from corticosteroid ointments, J. Pharm. Sci. 84:1139–1140, 1995.
9. V.P. Shah, J.S. Elkins, and J.P. Skelly. Relationship between in vivo skin blanching and in vitro release dose for betamethasone valerate creams. J. Pharm. Sci. 81:104–106, 1992.
10. V.P. Shah, J.S. Elkins, and R.L. Williams. In vitro drug release measurement of topical glucocorticoid creams. Pharm. Forum 19:5048–5059, 1993.

11. V.P. Shah, J. Elkins, J. Hanus, C. Noorizadeh, and J.P. Skelly. In vitro release of hydrocortisone from topical preparations and automated procedure. Pharm. Res. 8:55–59, 1991.

12. J.B. Li and P.C. Rahn. Automated dissolution testing of topical drug formulations using Franz cells and HPLC analysis. Pharm. Technol. 17(7):44–52, 1993.

13. M. Corbo, T.W. Schultz, G.K. Wong, and G.A. Van Buskirk. Development and validation of in vitro release testing methods for semisolid formulations. Pharm. Technol. 17(9):112–128, 1993.

14. J.L. Zatz. Drug release from semisolids: Effect of membrane permeability on sensitivity to product parameters. Pharm. Res. 2:787–789, 1995.

15. J.L. Zatz and J.D. Segers. Techniques for measuring in vitro release from semisolids. Dissolution Technol. 5(1):3–17, 1998.

16. V.P. Shah, J.S. Elkins, and R.L. Williams. Evaluation of the test system used for in vitro release of drugs for topical dermatological drug products. Pharm. Dev. Technol., in press.

17. FDA Research Contract 223-93-3016 with Professor Gordon L. Flynn, PhD, at University of Michigan, Ann Arbor, MI.

33
Percutaneous Absorption of Salicylic Acid, Theophylline, 2,4-Dimethylamine, Diethyl Hexyl Phthalic Acid, and *p*-Aminobenzoic Acid in the Isolated Perfused Porcine Skin Flap Compared to Humans In Vivo

Ronald C. Wester, Joseph L. Melendres, Lena Sedik, and Howard I. Maibach
University of California School of Medicine, San Francisco, California

Jim E. Riviere
North Carolina State University, Raleigh, North Carolina

I. INTRODUCTION

The ideal way to determine the percutaneous absorption potential of a compound in humans is to do the actual study in humans. Mechanisms and parameters of percutaneous absorption elucidated in vivo with human skin are most relevant to the clinical situation. Not all investigators will have access to human volunteers, and many compounds are potentially too toxic to test in vivo in humans, so their percutaneous absorption must be tested in animals. Mechanism studies and studies on factors affecting absorption must, therefore, be explored using animals and in vitro techniques.

One of the better partially validated animal models for percutaneous absorption in man is the miniature pig (Wester and Maibach, 1983). Bartek and La Budde (1975) and Bartek et al. (1972) showed for a variety of compounds that the percutaneous absorption in the pig ranged from 0 to 4 times that in humans in vivo. The pig has a potential disadvantage of a concentrated body fat. It is conceivable that a lipid-soluble compound upon in vivo absorption would concentrate in the fat rather than in the central compartment where biological fluids would be sampled (blood, urine). Additionally, it is possible that differences in patterns of systemic metabolism may also confound in vivo pig studies.

One way to take advantage of the permeability characteristics of pig skin and avoid the systemic/fat distribution problem is to use pig skin in an in vitro model. This led to the development of the isolated perfused porcine skin flap (IPPSF) (Riviere et al., 1986, 1995; Riviere and Monteiro-Riviere, 1991). The methodology is to surgically isolate a portion of skin (pig) with intact singular blood, which can be sampled for any chemical absorbed through skin. The skin flap can be used to study percutaneous absorption in vitro. The absorption of chemicals through skin and metabolism within the skin can be determined by assay of the perfusate.

This chapter compares percutaneous absorption prediction in the IPPSF system with that in humans in vivo. The study design utilized the same compounds with the same dose concentration and vehicle in both models. The compounds salicylic acid, theophylline, 2,4-dimethylamine, diethyl hexyl phthalic acid (DEHP) and p-aminobenzoic acid (PABA) were chosen because of scientific interest in the skin absorption of each compound, and because they represent a diverse selection of chemicals with different chemical and physical features.

II. MATERIALS AND METHODS

A. Chemicals

The following chemicals were used: [^{14}C]salicylic acid (0.46 mCi/mg, 99.9% pure; New England Nuclear, Boston) at 39.7 μg/cm^2; [^{14}C]theophylline (0.32 mCi/mg, 99.9% pure; Amersham, Arlington Heights, IL) at 3.2 μg/cm^2; [^{14}C]-p-aminobenzoic acid (0.05 mCi/mg, 99.9% pure; Sigma, St. Louis, MO) at 21.5 μg/cm^2; [^{14}C]-2,4-dichlorophenoxyacetic acid dimethylamine (2,4-dimethylamine) (0.022 mCi/mg, 99.9% pure; Sigma, St. Louis, MO) at 8.7 μg/cm^2; and [^{14}C]diethyl hexyl phthalic acid (9.7 mCi/mg, 98.0% or greater purity; Sigma, St. Louis, MO) at 18.5 μg/cm^2.

B. In Vivo Human Studies

The study included five or six normal volunteer outpatients per group (males, ages 18–85 years; and postmenopausal women, ages 50–65 years), from whom informed consent had been obtained. The subjects were topically dosed with ^{14}C-labeled chemical on the ventral forearm. Each ^{14}C-labeled chemical was solubilized in 50 μl ethanol and spread over 10 cm^2 skin surface area. The ethanol vehicle was allowed to air dry and the site was not occluded. Subjects were instructed to not touch or wash the study area for 24 h. They were allowed to wear clothing of their choice over the dosing area. The subjects were instructed to collect all urine in the containers provided for that day and the subsequent 6 days (7 days total urine collection period). After 24 h the dosing site was washed with 50% Liquid Ivory soap (1:1, v/v) and water. The washing sequence was soap (50%), water rinse, soap, followed by two water rinses using a cotton swab. At 7 days postapplication the skin dosing site was cellophane tape-stripped (Scotch Transparent; 3M Co., St. Paul, MN) 10 times for residual ^{14}C-labeled chemical. Percutaneous absorption was determined from the ^{14}C urinary excretion.

C. Isolated Perfused Porcine Skin Flap

Twenty-five IPPSFs (5/compound) were employed. The skin flaps were surgically prepared as previously described (Bowman et al., 1991), harvested and perfused in a nonrecirculating chamber (Riviere et al., 1986, 1995; Riviere and Monteiro-Riviere, 1991) and then dosed with radiolabeled compound 1 h after equilibration. A 10-cm^2 area on the IPPSF was dosed in a manner identical to the in vivo human studies. Skin flap viability was assessed by monitoring of glucose utilization (Riviere et al., 1986). Venous effluent was collected at 30-min increments for 8 h after which the perfusions were terminated. After termination, the flaps were weighed, and the dosing area swabbed two times with gauze dipped in a mild soap solution. The dosed skin was then stripped 12 times with cellophane tape. The treated area was then excised and removed from the subcutaneous tissue and weighed. The remainder of the skin flap was collected for mass balance purposes. All tissue samples were solubilized in BTS-450 (Beckman Instruments, Fullerton, CA) at 50°C overnight. Tissue digests, perfusate samples, and tape strips were then oxidized in an open-flame tissue oxidizer (Packard model 306, Packard Instruments Co., Downers Grove, IL) and trapped radiolabel was counted by scintillation spectroscopy (LKB-Wallac 122219 Rackbeta LSC, Wallac Co., Turke, Finland). Total penetration after 8 h was estimated as the radiolabel detected in perfusate plus treated skin samples.

D. Rhesus Monkey

The intravenous dose of 1.1 μCi ^{14}C]theophylline, ^{14}C]dimethyl hexyl phthalic acid, or ^{14}C]-*p*-aminobenzoic acid in 0.5 ml propylene glycol was administered into the saphenous vein of 4 female rhesus monkeys. Complete

24-h urine samples were collected for 7 consecutive days. A 5-ml aliquot of each urine sample was assayed for radioactivity to determine urinary dose disposition.

E. Sample Analysis (Human)

Urine samples were analyzed in duplicate for ^{14}C. A 5-ml aliquot of each urine sample was assayed in 10 ml scintillation cocktail (Universal ES, Costa Mesa, CA) with a Packard 4640 liquid scintillation spectrophotometer. The cotton swabs from the washing of the site of application were individually counted in 16 ml scintillation cocktail with the liquid scintillation spectrophotometer.

Background control samples and test samples were counted in a Packard model 4640 or model 15500 counter (Packard Instruments). Control and test sample counts were transferred to a computer program (Appleworks/ Apple IIE computer; Apple Computer Co., Mountain View, CA), which subtracted 1× background control samples and generated a spreadsheet, with the data reported under Results.

F. Percutaneous Absorption Determination

The in vivo percutaneous absorption in man was determined using this equation:

Percent dose absorbed

$$= \frac{\text{urinary } ^{14}\text{C excretion following topical}}{\text{urinary } ^{14}\text{C excretion following intravenous}} \times 100$$

The intravenous dosing is needed to account for that part of the dose that is excreted by other routes (feces) or retained in the body. The intravenous dose disposition for salicylic acid in humans was obtained from Feldmann and Maibach (1970) and that for 2,4-dimethylamine in the rhesus monkey was obtained from Moody et al. (1990). The urinary dose disposition for the other compounds were determined experimentally and are included in the Results.

The intravenous dose was given to rhesus monkeys instead of human volunteers to minimize human exposure. The need for relevant human in vivo exposure data and percutaneous absorption data is apparent, and human study approval committees are now recognizing this. Topical exposure is considered less risky than intravenous exposure, and thus approval for topical dosing can be obtained subject to certain safety guidelines. Intravenous dosing is considered more of a potential risk. Until this becomes feasible, the use of rhesus monkeys is a viable alternative (Wester et al., 1992a).

III. RESULTS

A. Salicylic Acid

Table 1 gives the in vivo data for salicylic acid in humans. A total of 5.8 ± 4.5% (mean ± SD) of applied dose was excreted over 7 days (Fig. 1). After 24 h some 53.4 ± 6.3% of dose could be recovered in the skin surface wash. The majority of the surface dose was recovered with the first soap application (Fig. 2). After 7 days, little (0.22 ± 0.25) residual material was recovered with the stratum corneum tape stripping.

Table 2 gives the percutaneous absorption of salicylic acid in the IPPSF. The perfusate gave 0.43 ± 0.44% of dose and the skin (after tape stripping) gave 7.1 ± 2.3% of dose. These combined give an experimental predicted absorption of 7.5 ± 2.6% of dose. Material balance recovered 48.2 ± 4.9% in the skin surface wash and 16.6 ± 5.3% in the skin tape strips, for an overall recovery of 75.4 ± 4.8% of applied dose.

B. Theophylline

Table 3 gives the in vivo data for theophylline in humans. A total of 13.0 ± 8.7% of applied dose was excreted in urine over 7 days (Fig. 1). Only 13.4 ± 10.9% of dose was recovered in the soap and water wash, and only 0.1 ± 0.08% of dose was recovered in the skin tape stripping.

In the IPPSF model, perfusate recovery was 0.6 ± 0.8% and skin recovery (after tape stripping) was 11.2 ± 3.6%, giving a predicted percu-

Table 1 In Vivo Urinary Excretion of Salicylic Acid in Humans

Subject	Percent dose		
	Urine excretion[a]	Skin surface wash[b]	Tape strip recovery[c]
1	8.9	49.3	0.04
2	3.9	56.7	0.09
3	13.6	49.3	0.02
4	2.6	46.1	0.35
5	3.7	63.2	0.67
6	2.3	56.0	0.17
Mean	5.8	53.4	0.22
SD	4.5	6.3	0.25

[a]For 2.4 μg/cm^2 ethanol vehicle, unoccluded ventral forearm.
[b]For 24-h postapplication soap-and-water wash.
[c]For 7 days postapplication/10 consecutive tape strips.

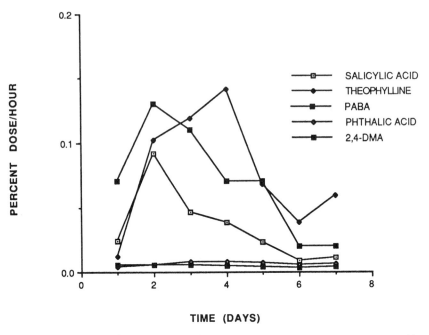

Figure 1   Urinary ^{14}C excretion in humans following topical application of ^{14}C-labeled chemical.

taneous absorption of 11.8 ± 3.8% (skin plus perfusate). The skin surface wash procedure recovered 42.4 ± 7.4% and tape stripping an additional 23.0 ± 7.0%, giving a material balance of 80.0 ± 4.1% (Table 4).

C. 2,4-Dimethylamine

Table 5 gives the in vivo data for 2,4-dimethylamine (2,4-DMA) in humans. Recovery of dose in urinary excretion was 0.77 ± 0.21% (Fig. 1). Only 2.0 ± 0.7% was recovered in the soap and water wash and only 0.06 ± 0.06% was recovered in the tape stripping.

Percutaneous absorption of 2,4-DMA in the IPPSF model is given in Table 6. The portion of dose in the perfusate was 1.5 ± 0.24% and skin content (after tape stripping) was 2.4 ± 0.5%, giving a predicted absorption of 2,4-DMA of 3.8 ± 0.6%. Only 21.4% of dose was recovered in the surface wash and 4.9 ± 2.6% in the tape strips, giving a material recovery of 31.2 ± 5.6%.

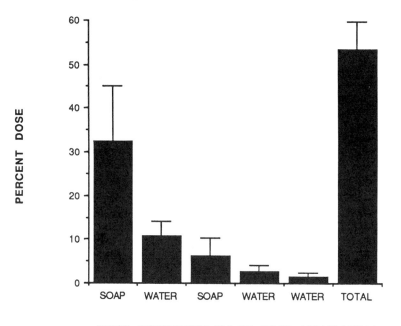

Figure 2 Removal of salicylic acid from human skin in vivo with soap-and-water wash 24 h after dosing.

Table 2 Percutaneous Absorption of Salicylic Acid in the Isolated Perfused Porcine Skin Flap

Subject	Tape strips	Skin	Perfusate	Skin surface wash	Experimental penetrated dose (skin + perfusate)	Recovery
1	15.3	8.8	0.79	51.4	9.6	80.0
2	16.5	4.8	0.11	46.5	4.9	69.0
3	24.9	9.4	1.00	40.6	10.4	79.9
4	10.2	7.8	0.06	48.9	7.9	72.4
5	16.2	4.5	0.18	53.2	4.6	75.8
Mean	16.6	7.1	0.43	48.2	7.5	75.4
(SD)	(5.3)	(2.3)	(0.44)	(4.9)	(2.6)	(4.8)

Table 3 In Vivo Urinary Excretion of Theophylline in Humans

		Percent dose	
Subject	Urine excretion[a]	Skin surface wash[b]	Tape strip recovery[c]
1	12.3	10.2	0.13
2	6.0	2.2	0.00
3	14.9	34.0	0.22
4	11.6	8.7	0.05
5	4.2	14.9	0.07
6	28.7	10.6	0.13
Mean	13.0	13.4	0.10
SD	8.7	10.9	0.08

[a]For 3.25 $\mu g/cm^2$ ethanol vehicle, unoccluded ventral forearm.
[b]For 24-h postapplication soap-and-water wash.
[c]For 7-days postapplication/10 consecutive tape strips.

D. Diethyl Hexyl Phthalic Acid (DEHP)

Table 7 gives the in vivo data for diethyl hexyl phthalic acid in humans. A total of $1.07 \pm 0.33\%$ of dose was recovered in the urine (Fig. 1). Skin surface wash recovered only $4.5 \pm 3.1\%$ and tape strip recovery was only $0.15 \pm 0.17\%$. With the IPPSF model (Table 8), some $0.14 \pm 0.1\%$ was recovered in the perfusate and an additional $3.8 \pm 0.1\%$ in the skin. Combined (skin plus perfusate), the predicted percutaneous absorption is $3.9 \pm 2.4\%$ of dose. Material balance was good, recovering $14.5 \pm 5.5\%$ in the

Table 4 Percutaneous Absorption of Theophylline in the Isolated Perfused Porcine Skin Flap

Subject	Tape strips	Skin	Perfusate	Skin surface wash	Experimental penetrated dose (skin + perfusate)	Recovery
1	24.9	10.2	0.4	41.7	10.6	81.0
2	19.4	7.9	0.3	42.6	8.2	73.0
3	33.2	12.4	1.9	33.5	14.4	83.5
4	22.8	16.8	0.0	40.2	16.9	82.1
5	14.4	8.3	0.2	53.9	8.6	80.3
Mean	23.0	11.2	0.6	42.4	11.8	80.0
(SD)	(7.0)	(3.6)	(0.8)	(7.4)	(3.8)	(4.1)

Table 5 In Vivo Urinary Excretion of 2,4-Dimethyl Amine in Humans

Subject	Percent dose		
	Urine excretion[a]	Skin surface wash[b]	Tape strip recovery[c]
1	0.18	2.8	0.13
2	0.92	2.1	0.03
3	0.88	2.7	0.02
4	0.61	1.3	0.15
5	0.45	1.9	0.01
6	0.99	1.0	0.02
Mean	0.77	2.0	0.06
SD	0.21	0.71	0.06

[a]For 39.7 $\mu g/cm^2$ ethanol vehicle, unoccluded ventral forearm.
[b]For 24-h postapplication soap-and-water wash.
[c]For 7 days postapplication/10 consecutive tape strips.

skin strips and 71.3 ± 10.3% in the skin surface wash, a total recovery of 94.1 ± 4.7%.

E. *p*-Aminobenzoic Acid

Table 9 gives the in vivo data for *p*-aminobenzoic acid (PABA) in humans. Urinary excretion was variable, ranging from 3.6 to 18.3% (mean of 11.5 ± 6.3%). Recovery in the surface wash was also variable, with a mean recovery of 24.8 ± 16.3%. Little residual PABA was recovered in the postapplication skin tape stripping (0.50 ± 0.47%). In the IPPSF model (Table

Table 6 Percutaneous Absorption of 2,4-Dimethyl Amine in the Isolated Perfused Porcine Skin Flap

Subject	Tape strips	Skin	Perfusate	Skin surface wash	Experimental penetrated dose (skin + perfusate)	Recovery
1	7.8	3.0	1.3	18.9	4.4	32.8
2	7.2	2.4	1.4	14.8	3.8	26.6
3	1.5	2.3	1.7	21.2	4.1	28.2
4	4.1	1.7	1.1	32.3	2.9	40.2
5	3.5	2.1	1.7	19.8	3.8	27.9
Mean	4.8	2.3	1.5	21.42	3.8	31.2
(SD)	(2.6)	(0.48)	(0.24)	(6.6)	(0.57)	(5.6)

Table 7 In Vivo Urinary Excretion of Diethyl Hexyl Phthalic Acid in Humans

| Subject | Percent dose | | |
	Urine excretion[a]	Skin surface wash[b]	Tape strip recovery[c]
1	1.1	2.5	0.03
2	1.2	0.9	0.01
3	0.8	6.0	0.04
4	0.6	6.9	0.38
5	1.6	8.5	0.36
6	1.2	2.0	0.10
Mean	1.1	4.5	0.15
SD	0.33	3.1	0.17

[a]For 18.5 $\mu g/cm^2$ ethanol vehicle, unoccluded ventral forearm.
[b]For 24-h postapplication soap-and-water wash.
[c]For 7 days postapplication/10 consecutive tape strips.

10) $0.73 \pm 0.53\%$ of dose was in the perfusate and $5.9 \pm 3.7\%$ was in the skin, giving a predicted penetrated dose of $5.91 \pm 3.7\%$. Skin surface wash recovered $56.2 \pm 10.1\%$ and tape stripping recovered $12.3 \pm 6.3\%$, giving a material balance of $76.9 \pm 5.2\%$.

F. Comparative Percutaneous Absorption

The data given in previous tables on in vivo human dosing show urinary excretion following topical application. Percutaneous absorption is determined from the ratio of urinary excretion following topical and intravenous

Table 8 Percutaneous Absorption of Diethyl Hexyl Phthalic Acid in the Isolated Perfused Porcine Skin Flap

Subject	Tape strips	Skin	Perfusate	Skin surface wash	Experimental penetrated dose (skin + perfusate)	Recovery
1	12.0	2.5	0.0	78.9	2.5	96.6
2	17.2	3.4	0.1	66.4	3.5	94.0
3	22.6	8.1	0.0	58.6	8.1	90.7
4	11.8	1.9	0.3	84.4	2.2	100.5
5	8.8	2.8	0.1	68.0	3.0	88.6
Mean	14.5	3.7	0.14	71.3	3.9	94.1
(SD)	(5.5)	(2.5)	(0.10)	(10.3)	(2.4)	(4.7)

Table 9 In Vivo Urinary Excretion of *p*-Aminobenzoic Acid in Humans

Subject	Percent dose		
	Urine excretion[a]	Skin surface wash[b]	Tape strip recovery[c]
1	7.1	36.2	1.2
2	11.8	25.8	0.23
3	3.6	37.6	0.19
4	18.3	8.6	0.10
5	16.9	39.3	0.76
Mean	11.5	29.5	0.56
SD	6.3	12.8	0.47

[a]For 21.5 μg/cm^2 ethanol vehicle, unoccluded ventral forearm.
[b]For 24-h postapplication soap-and-water wash.
[c]For 7 days postapplication/10 consecutive tape strips.

(where systemic availability is assumed 100% due to injection) dosing. Information on intravenous administration of salicylic acid in humans (89.8% excreted in urine) (Riviere et al., 1986) and of 2,4-dimethylamine in the rhesus monkey (72% excreted in urine) is available in the literature (Riviere and Monteiro-Riviere, 1991). Table 11 gives the intravenous urinary disposition of PABA, DHP and theophylline in the rhesus monkey. The majority of the dose for each compound was excreted in the urine, and excretion was essentially complete by day 7. Although the majority of the intravenous doses were excreted in the first day, only PABA showed this to be so with topical application (Fig. 1).

Table 10 Percutaneous Absorption of *p*-Aminobenzoic Acid in the Isolated Perfused Porcine Skin Flap

Subject	Tape strips	Skin	Perfusate	Skin surface wash	Experimental penetrated dose (skin + perfusate)	Recovery
1	15.4	5.8	0.07	58.12	5.9	83.7
2	6.4	1.7	0.42	67.4	2.1	77.2
3	17.3	5.2	0.89	54.0	6.0	79.7
4	4.6	2.9	0.77	62.0	3.6	72.7
5	17.6	10.4	1.48	39.6	11.9	71.0
Mean	12.3	5.2	0.73	56.2	5.9	76.9
(SD)	(6.3)	(3.4)	(0.53)	(10.1)	(3.7)	(5.2)

Table 11 Urinary Excretion of Theophylline, p-Aminobenzoic Acid, and Diethyl Hexyl Phthalic Acid Following Intravenous Administration to Rhesus Monkeys

Time (days)	Percent dose excreted[a]		
	Theophylline	PABA	DHP
1	59.4 ± 3.0	72.0 ± 17.7	52.6 ± 19.6
2	11.7 ± 1.6	1.3 ± 0.2	4.1 ± 1.0
3	3.4 ± 0.7	0.4 ± 0.3	1.3 ± 0.3
4	1.1 ± 0.1	0.4 ± 0.4	1.0 ± 0.1
5	0.6 ± 0.1	0.3 ± 0.1	0.7 ± 0.3
6	0.4 ± 0.1	0.3 ± 0.2	0.6 ± 0.2
7	0.4 ± 0.2	0.2 ± 0.1	0.4 ± 0.1
Total	77.1 ± 3.0	74.9 ± 18.2	60.8 ± 19.7

[a]Mean ± SD for four monkeys.

The in vivo percutaneous absorption of salicylic acid, theophylline, 2,4-DMA, DHP, and PABA is given in Table 12 and compared to the predicted penetrated dose from the IPPSF model. For a variety of chemical structures that exhibited a range of percutaneous absorption values, the IPPSF model was remarkably good in predicting what absorption in humans would be. The correlation coefficient equaled 0.78. The only difference in absorption between the two systems was with PABA, and with PABA the variability in the human absorption resulted in no statistical difference for absorption of PABA. Therefore, the IPPSF model appears a good predictor of percutaneous absorption in humans.

Table 12 In Vivo Percutaneous Absorption in Humans Compared to Penetrated Dose in the Isolated Perfused Porcine Skin Flap

Compound	Percent dose absorbed	
	Human	IPPSF
Salicylic acid	6.5 ± 5.0	7.5 ± 2.6
Theophylline	16.9 ± 11.3	11.8 ± 3.8
2,4-Dimethylamine	1.1 ± 0.3	3.8 ± 0.6
Diethyl hexyl phthalic acid	1.8 ± 0.5	3.9 ± 2.4
p-Aminobenzoic acid	15.3 ± 8.4	5.9 ± 3.7

IV. DISCUSSION

The isolated perfused porcine skin flap (IPPSF) was designed to take advantage of pig skin permeability characteristics, which resemble those of humans. For the variety of chemicals tested, the IPPSF system was fairly successful in predicting percutaneous absorption in humans.

To be successful, the IPPSF system required that prediction for humans be a summation of perfusate plus residual skin concentration after the dosing period. The most likely explanation relates to the time frame of the experiments. Since IPPSF studies were only conducted for 8 h, compounds that would have penetrated at later time points were still in the skin of the IPPSF. Thus, to precisely predict 7-day human absorption, both perfusate and residual skin concentrations are needed in the IPPSF.

The amount of material recovered in the tape strips was not included in the calculations for penetrated dose. The dosing areas on the human subjects were not covered to prevent loss of compound. The human subjects did wear clothing, and the data suggest that clothing and evaporation resulted in loss of compound. The amount of chemical in the IPPSF skin strips represents this loss of chemical. If an occluded/protected absorption prediction was desired, then these tape strippings should be included.

The skin surface wash recovery postapplication was similar for salicylic acid in humans (53.4 ± 6.3%) and the IPPSF system (48.2 ± 4.9%). For the other compounds, the similarity changes. 2,4-Dimethylamine is volatile, so its lower recovery in humans and the IPPSF system is understandable. For theophylline, DHP, and PABA, little chemical was recovered off human skin. Since the wash procedure was effective with pig skin, we can assume that these chemicals were lost to any clothing that the volunteers may have worn over the dosing site.

That the absorption in humans was not less than that in the IPPSF, and assuming the dose was lost in humans, it seems plausible that whatever compound was to penetrate human skin in solvent vehicle did so in a short period of time before the dosing was removed. Short-term skin exposure has been reported (Wester and Maibach, 1991). It is possible to use a ventilated plastic chamber in vivo to improve dose accountability (Bucks et al., 1991). This was tried with 2,4-dimethylamine (Moody et al., 1992). Accountability increased (73%); however, absorption also increased some 10-fold (12.5 ± 5.0%). Therefore, loss of compound from the skin through volatility and clothing/rubbing seems the natural process, except where the compound is encased in some transdermal unit. The use of nonocclusive covers to minimize skin surface loss is controversial. Human skin in the in vivo situation is not covered by protection, except for the natural wearing of clothing. Protective nonocclusive devices can remove compound from skin, thus affecting the applied dose (Wester et al., 1992b). With volatile compounds this can potentially create an artificial dosing situation. The need to see material balance needs to be strongly evaluated relative to any potential influence of the study results. Humans do not wear artificial protective devices, and a study intended to duplicate human exposure should use care in the study design.

The suggestion of the rapid establishment of a reservoir within some skin compartment is in accord with the observations of Feldmann and Maibach (1970) that it is difficult to wash several chemicals off skin after a waiting period of 15 min and with the development by Rougier et al. (1983; 1986) of a predictive model of penetration based on tape stripping of skin washed 30 min postdosing.

Of the five chemicals studied, the IPPSF was only significantly different with PABA. The large variability seen in the human data may explain this. Alternatively, since the pK_a of PABA is close to normal skin pH, difference in pH between human and pig skin may change the fraction of unchanged PABA available for penetration. This may also explain the variability seen between individuals.

Although these results and other studies (Riviere and Monteiro-Riviere, 1991; Riviere et al., 1992, 1995) are promising, the model compounds studied to date with the IPPSF do not allow us to overgeneralize until a broader group of diverse chemicals with varying physicochemical properties is examined.

ACKNOWLEDGMENT

We thank Canada Health and Welfare for their partial support. Technical assistance from Jim Brooks and Rick Rogers with the IPPSF system is appreciated.

REFERENCES

Bartek, M. J., La Budde, J. A., and Maibach, H. I. (1972). Skin permeability *in vivo*: Comparison in rat, rabbit, pig and man. J Invest Dermatol 58:114–123.

Bartek, M., and La Budde, J. A. (1975). Percutaneous absorption *in vitro*. In Animal Models in Dermatology (H. Maibach, Ed.), pp. 103–120. Churchill Livingstone, New York.

Bowman, K. F., Monteiro-Riviere, N. A., and Riviere, J. E. (1991). Development of surgical techniques for preparation of *in vitro* isolated perfused porcine skin flaps for percutaneous absorption studies. Am J Vet Res 52:75–82.

Bucks, D., Guy, R., and Maibach, H. (1991). Effects of occlusion. In In Vitro Percutaneous Absorption (R. Bronaugh and H. Maibach, Eds.), pp. 85–114. CRC Press, Boca Raton, FL.

Feldmann, R. J., and Maibach, H. I. (1970). Absorption of some organic compounds through the skin in man. J Invest Dermatol 54:399–404.

Moody, R. P., Franklin, C. A., Ritter, L., and Maibach, H. I. (1990). Dermal absorption of the phenoxy herbicides 2,4-D, 2,4-D amine, 2,4-D isooctyl, and 2,4,5-T in rabbits, rats, rhesus monkeys, and humans: A cross-species comparison. J Toxicol Environ Health 29:237–245.

Moody, R. P., Wester, R. C., Melendres, J. L., and Maibach, H. I. (1992). Dermal absorption of the phenoxy herbicide 2,4-D-dimethylamine in humans: Effect of DEET and anatomic site. J Toxicol Environ Health 36:241–250.

Riviere, J. E., and Monteiro-Riviere, N. A. (1991). The isolated perfused porcine skin flap as an *in vitro* model for percutaneous absorption and cutaneous toxicology. Crit Rev Toxicol 21:329–344.

Riviere, J. E., Bowman, K. F., Monteiro-Riviere, N. A., Carver, M. P., and Dix, L. P. (1986). The isolated perfused porcine skin flap (IPPSF). I. A novel *in vitro* model for percutaneous absorption and cutaneous toxicology studies. Fundam Am Appl Toxicol 7:444–453.

Riviere, J. E., Williams, P. L., Hillman, R., and Mishky, L. (1992). Quantitative prediction of transdermal iontophoretic delivery of arbutamine in humans using the *in vitro* isolated perfused porcine skin flaps (IPPSF). J Pharm Sci 81:504–507.

Riviere, J. E., Monteiro-Riviere, N. A., and Williams, P. L. (1995). The isolated perfused porcine skin flaps as an *in vitro* model for predicting transdermal pharmacokinetics. Eur J Pharm Biopharm 41:152–162.

Rougier, A., Dupuis, D., Lotte, C., Roguet, R., and Schaefer, H. (1983). Correlation between stratum corneum reservoir function and percutaneous absorption. J Invest Dermatol 81:275.

Rougier, A., Dupuis, D., Lotte, C., Roguet, R., Wester, R. C., and Maibach, H. I. (1986). Regional variation in percutaneous absorption in man: Measurement by the stripping method. Arch Dermatol Res 278:465.

Wester, R. C., and Maibach, H. I. (1983). Cutaneous pharmacokinetics: 10 Steps to percutaneous absorption. Drug Metab Rev 14:169–205.

Wester, R. C., and Maibach, H. I. (1991). Short-term *in vitro* skin exposure. In In Vitro Percutaneous Absorption (R. Bronaugh and H. Maibach, Eds.), pp. 231–236. CRC Press, Boca Raton, FL.

Wester, R. C., Maibach, H. I., Melendres, J., Sedik, L., Knaak, J., and Wang, R. (1992). *In vivo* and *in vitro* percutaneous absorption and skin evaporation of isofenphos in man. Fundam Appl Toxicol 19:521–526.

Wester, R. C., Melendres, J., and Maibach, H. I. (1992b). In vivo percutaneous absorption and skin decontamination of alachlor in rhesus monkey. J Toxicol Environ Health 36:1–12.

34
Transepidermal Water Loss Measurements for Assessing Skin Barrier Functions During In Vitro Percutaneous Absorption Studies

Avinash Nangia
ALZA Corporation, Palo Alto, California

Bret Berner
Cygnus, Redwood City, California

Howard I. Maibach
University of California School of Medicine, San Francisco, California

I. INTRODUCTION

In the course of performing in vitro percutaneous absorption studies, it is important to ensure the integrity of the stratum corneum. In vitro assessment of the permeability of chemicals is often preceded by studying the rate of penetration of tritiated water (Bronaugh et al., 1986; Scott et al., 1986). However, this technique has several limitations. For example, it results in hydrated stratum corneum, possibly affecting the penetration of the compound under investigation. This method cannot be used if the test compound is tritium labeled. In addition, it is time-consuming and requires a license and facilities to handle radioactive materials. In practice, the repetitive measurement of tritiated water in laboratories is accomplished by determining the fraction of the dose applied absorbed at a fixed time. Selection of that fixed time is difficult without a priori knowledge of the permeability, and improper selection can lead to an erroneous assessment of the barrier properties of skin. Recently an alternate technique, which measures the electrical resistance across the skin slices, has also been proposed for this purpose (Lawrence, 1997). Based on the initial results, this technique appears to have some advantages, but more studies are required to validate it.

Transepidermal water loss (TEWL) measurements with an evaporimeter, under controlled environmental conditions, are predictive of altered barrier functions (Pinnagoda et al., 1990). Measurement of TEWL with an evaporimeter has been used to evaluate the competency of the skin barrier in vivo (Leveque et al., 1979; Roskos and Guy, 1989). Investigators have also addressed the possibility of using this technique in vitro (Wilson et al., 1982; Moloney, 1988). In a study by Moloney, an evaporimeter was used to investigate the structural requirements of lipids, which are capable of altering the barrier functions and thus promoting skin permeability. Subtle changes in the skin barrier were observed with this device, suggesting TEWL usefulness in detecting damaged skin. TEWL is also a relevant and sensitive indicator in vivo and in vitro for prediction of percutaneous absorption of various chemicals (Murahata et al., 1986; Lotte et al., 1987).

Measurement of TEWL by an evaporimeter is a continuous measurement of water permeation through skin under gradient conditions. In contrast, tritiated water permeation is a cumulative measurement over time. With increasing time, finite-dose tritiated water permeation methods reflect the dose absorbed and not the permeability of the tissue. Assuming a negligible time lag for water permeation (10–15 min), from mass balance we find that

$$\frac{dC}{dt} = -APC/V$$

where C is the concentration of tritiated water in the donor solution, t is the time, A is the surface area, V is the volume of tritiated water applied and is assumed to be constant, and P is the permeability of skin. By integrating, the fraction absorbed may be obtained as

$$\text{Fraction absorbed} = 1 - \exp\left[-(A/V)\int_0^t dt\, P\right]$$

For constant P at short times ($APt/V \ll 1$),

$$\text{Fraction absorbed} = APt/V$$

and the tritiated water permeation reflects the permeability of skin. However, in general, even for constant P,

$$\text{Fraction absorbed} = 1 - \exp(-APt/V)$$

At long times, particularly, the fraction absorbed of tritiated water equals unity because it does not depend on the permeability. In contrast, TEWL is a measure of the skin permeability at all times.

II. METHODS

In our laboratory, we have examined the relationship between TEWL and skin barrier functions that have been damaged to varying extents by different techniques. After altering the skin barrier, tritiated water flux measurements were also obtained and correlated with the TEWL measurements. The TEWL results were then further analyzed to study the relationship between water loss and in vivo primary skin irritation caused by various chemicals.

In vitro skin permeation studies were performed using the flow-through diffusion cell through the heat-separated human epidermis. Using physiological saline (containing 0.01% w/v gentamicin sulfate) and a 1.0-cm^2 permeation area (maintained at 32°C), the donor side of the epidermis was subjected to various physical and chemical pretreatments to create various damaged skin models. In the first category, epidermis was delipidized with a chloroform:methanol (C:M) mixture (2:1% v/v) for different time intervals under occlusion. In the second category, different concentrations of sodium lauryl sulfate (SLS) were exposed for various time intervals. In the third category, five basic compounds (norephedrine, antipyrine, imipramine, naphazoline, and mecamylamine) of different basicity (pK_a) were applied for in different concentration for 24 h. In the last category, mechanical insults to the skin slices were induced by either abrading it with sandpaper or a sharp needle or by stripping with adhesive tape. The extent of barrier function perturbation was assessed by observing for changes in TEWL values from the baseline as well as by quantitating the tritiated water flux. TEWL measurements were made by placing a collard probe on top of the upper half of the permeation cell (Fig. 1) and leaving it there until a constant value was established. An unstirred column of 2.5 cm above the skin provided little resistance to water permeation. Only skin samples with a baseline TEWL reading between 1.0 and 6.0 g/m^2/h were used. Details of these procedures have been previously reported by Nangia et al. (1993, 1996).

An in vivo skin irritation study was also conducted to correlate with the in vitro data and to establish that the in vitro technique has clinical application. To 12 healthy female volunteers were applied plastic chambers containing aqueous or ethanolic solutions of 5 basic permeants in a random order, in longitudinal rows on both the left and right interscapular areas of the back. Chambers containing water and ethanol served as vehicle controls. After 24 h, the patches were removed and the test sites were marked. Thirty minutes later, the sites were graded for erythema and edema, using a 0–5

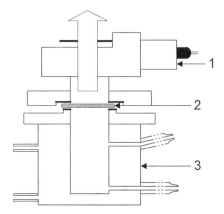

Figure 1   Schematic diagram of the diffusion cell coupled to an evaporimeter: (1) evaporimeter probe; (2) skin slice; (3) diffusion cell.

visual irritation scoring system (Nangia et al., 1993). TEWL values were measured using the evaporimeter.

III.   RESULTS AND DISCUSSION

Exposure to the C:M mixture resulted in a rapid and irreversible damage to the barrier properties. The extent of damage increased with the contact time (Fig. 2). An exposure of 3 min had a pronounced effect and resulted in a TEWL of 22.1 ± 8.0 g/m^2/h, which was approximately 5 times higher than

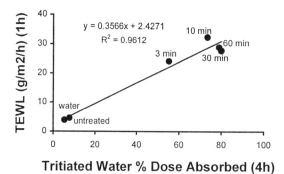

Figure 2   Linear correlation between TEWL (measured after 1 h) and the fraction of the dose of tritiated water absorbed (in 4 h) after chloroform–methanol treatment applied to excised human skin.

the baseline values. Damage to the water barrier of the excised skin was most effective when the skins were exposed for 10 min with a TEWL value of 29.9 ± 8.0 g/m²/h. This value was similar to that of an uncovered cell filled with water (27.5 ± 5.7 g/m²/h), suggesting that 10 min of exposure to C:M mixture is adequate to destroy completely the water barrier of the skin. Treatment for 60 min did not exhibit further damage. C:M treatment also resulted in an increased fraction of tritiated water absorbed in 4 h; these results were consistent with the trend for TEWL. The correlation between the effects of C:M treatment on TEWL and the fraction of tritiated water absorbed was found to be excellent ($r^2 = .96$). In a similar type of study in our laboratory, Abrams et al. (1993) evaluated the effect of topical exposure to various organic solvents for various time intervals on the barrier functions of excised human skin. After exposure to various solvents, the contents of the donor compartment of the cell were analyzed for various lipids. By measuring the TEWL with an evaporimeter, various solvents were ranked for their ability to alter skin barrier integrity. C:M exposure for 12 min was found to extract the maximum quantity of stratum corneum lipids and also induced greatest TEWL change, further confirming our results and demonstrated that evaporimeter is a useful tool to evaluate skin barrier functions.

The concentration-dependent increases in TEWL and tritiated water absorbed by SLS treatment are shown in Fig. 3. A good correlation between the two was observed ($r^2 = .98$). Treatment with SLS resulted in a significant increase in TEWL at all concentrations. A concentration below the critical micelle concentration (CMC) (i.e., 0.125% w/v) caused the least damage to the skin. However, the extent of damage observed with concentrations higher than the CMC was more severe. A linear dependence of TEWL on surfactant

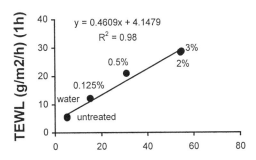

Tritiated Water % Dose Absorbed (4h)

Figure 3 Linear correlation between TEWL (measured after 1 h) and the fraction of the dose of tritiated water absorbed (in 4 h) at various concentrations of SLS (after 4 h) applied to excised human skin.

concentration between 0.125% and 2% w/v was observed. No further increase in TEWL was evident at 3% w/v, suggesting that 2% w/v SLS solution caused maximum possible damage to the skin.

At a fixed concentration (2% w/v) of SLS, the time dependence of the effect of SLS on the barrier function was characterized. The extent of damage, as reflected by either TEWL or the fraction of tritiated water absorbed, increased linearly with contact time (Fig. 4) ($r^2 = .82$ and $r^2 = .93$ for TEWL and tritiated water measurements, respectively).

Of the three acute physical injuries created, abrasion with sandpaper inflicted the greatest damage, while a needle had little effect on the increase in TEWL and tritiated water permeation (Fig. 5). For these acute injuries, TEWL correlated well with results for water permeation ($r^2 = .89$; Fig. 5).

Table 1 shows the TEWL values after exposure to five basic permeants. Application of the antipyrine solution ($pK_a = 1.4$, nonirritant control) for 24 h did not cause any damage to the skin, and the TEWL was similar to that for control water-treated skin. All the remaining four basic compounds, norephedrine ($pK_a = 9$), imipramine ($pK_a = 9.5$), naphazoline ($pK_a = 10.9$), and mecamylamine ($pK_a = 11.2$) resulted in an increase in TEWL and the fraction of tritiated water absorbed, with the maximum effect seen with mecamylamine, followed by naphazoline, imipramine, norephedrine, and an-

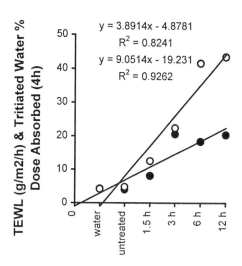

Time SLS Treatment (hours)

Figure 4   Linear correlation between TEWL (measured after 1 h) and the fraction of the dose of tritiated water absorbed (in 4 h) after different durations of SLS treatment applied to excised human skin. ●, TEWL; ○, tritiated water absorbed.

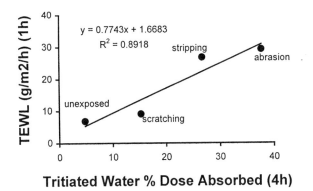

Tritiated Water % Dose Absorbed (4h)

Figure 5 Linear correlation between TEWL (measured after 1 h) and the fraction of the dose of tritiated water absorbed (in 4 h) after various physical injuries applied to excised human skin.

tipyrine. The rank for the bases was in accordance with their pK_a values; that is, TEWL increases with higher pK_a values. Mecamylamine, with the highest pK_a of 11.2, perturbed the skin barriers completely. The maximum percentage of tritiated water absorbed through cadaver skin also ranked the permeants in a manner similar to that of TEWL, and the correlation between TEWL and water permeation was found to be reasonably good ($r^2 = .71$).

Table 1 Irritation Induced by Basic Compounds with Increasing pK_a Values

Compound[a]	pK_a	TEWL values[b] ($g/m^2/h$)	Dose ^3H_2O absorbed (% absorbed in 20 h)	Erythema score[c]
Antipyrine	1.4	4.8 ± 0.8	30.0	0.00 ± 0.00
Norephedrine	9.0	5.2 ± 1.6	45.2	1.03 ± 0.43
Imipramine	9.5	9.8 ± 0.6	55.0	2.30 ± 1.70
Naphazoline	10.9	13.8 ± 4.1	90.5	2.50 ± 1.70
Mecamylamine	11.2	28.4 ± 6.2	88.7	2.83 ± 1.14
Water	—	4.6 ± 1.3	28.2	0.08 ± 0.19
Ethanol	—	7.7 ± 1.2	43.6	0.50 ± 0.61

[a] Antipyrine and norephedrine solutions were in distilled water, while the remaining bases were in ethanol.
[b] TEWL values were meaured in vitro after 1 h, using the evaporimeter.
[c] Scored using a 0–5 visual irritation scoring system: (0) no erythema; (1) very slight erythema—barely perceptible; (2) well-defined uniform erythema; (3) moderate to severe erythema; (4) severe erythema to slight eschar formation; (5) severe erythema with edema.

Deviations occurred for large fractions of water absorbed, because in that regime this measurement no longer reflects water permeation. Comparison of the results between in vitro TEWL values and the visual irritation score further validates this technique for in vitro toxicological studies.

From these studies, it is quite clear that TEWL measurement with an evaporimeter, under controlled environmental conditions, is a simple and rapid method of screening the integrity of the barrier functions of skin in vitro and can thus be routinely used as an alternative to tritiated water permeation.

REFERENCES

1. Abrams K, Harvell JD, Shriner D, Wertz P, Maibach H, Maibach HI, Rehfeld SJ. Effect of organic solvents on in vitro human skin water barrier function. J Invest Dermatol 1993; 101:609–613.
2. Bronaugh RL, Stewart RF. Methods for in vitro percutaneous absorption studies. VI. Use of excised human skin. J Pharm Sci 1986; 75:1094–1097.
3. Lawrence JN. Electric resistance and tritiated water permeability as indicators of barrier integrity of in vitro human skin. Toxicol In Vitro 1997; 11:241–249.
4. Leveque JL, Garson JC, de Rigal J. Transepidermal water loss from dry and normal skin. J Soc Cosmet Chem 1979; 30:333–343.
5. Lotte C, Rougier A, Wilson DR, Maibach HI. In vivo relationship between transepidermal water loss and percutaneous penetration of some organic compounds in man: Effect of anatomic site. Arch Dermatol Res 1987; 279:351–356.
6. Moloney SJ. Effect of exogenous lipids on in vitro transepidermal water loss and percutaneous absorption. Arch Dermatol Res 1988; 280:67–70.
7. Murahata RI, Crowe DM, Roheim JR. The use of transepidermal water loss to measure and predict the irritation response to surfactants. Int J Cosmet Sci 1986; 8:225–231.
8. Nangia A, Camel E, Berner B, Maibach HI. Influence of skin irritants on percutaneous absorption. Pharm Res 1993; 10:1756–1759.
9. Nangia A, Anderson PH, Berner B, Maibach HI. High dissociation constants (pK_a) of basic permeants are associated with in vivo skin irritation in man. Contact Dermatitis 1996; 34:237–242.
10. Pinnagoda J, Tupker RA, Agner T, Serup J. Guidelines for transepidermal water loss (TEWL) measurement. Contact Dermatitis 1990; 22:164–178.
11. Roskos K, Guy RH. Assessment of skin barrier functions using transepidermal water loss: Effect of age. Pharm Res 1989; 6:949–953.
12. Scott RC, Dugard PH, Doss AW. Permeability of abnormal rat skin. J Invest Dermatol 1986; 86:201–207.
13. Wilson DR, Maibach HI. A review of transepidermal water loss. In: Maibach HI, Boisits EK, eds. Neonatal Skin: Structure and Function. New York: Marcel Dekker, 1982:83–100.

III
DRUG AND COSMETIC ABSORPTION

35

Percutaneous Penetration as a Method of Delivery to Skin and Underlying Tissues

Parminder Singh
Novartis Pharmaceuticals Corporation, Suffern, New York

I. INTRODUCTION

The skin is the outermost organ of the body. It is a complex membrane comprising of three major layers—epidermis, dermis, and hypodermis. The epidermis is an avascular stratifying layer of epithelial cells that overlies the connective tissue layer, the dermis. The outermost layer of epidermis, the stratum corneum, is primarily responsible for the barrier properties of the skin. The underlying viable epidermis is different from the stratum corneum in being physiologically more akin to other living cellular tissues. The dermis comprises the largest fraction of the skin and is responsible for providing its structural strength. The dermis also provides an environment for nerve and vascular networks and appendages required to support the epidermis. The dermis is also a primary site in which cutaneous inflammation occurs in response to skin injury (1). The blood supply to the dermis is important in the systemic absorption of substances applied to the skin. The epidermis and dermis are supported by an internal layer of adipose tissue, the hypodermis. This layer provides a cushion between the external skin layers and internal structures such as bone and muscle. It also provides an energy source, allows for skin mobility, molds body contours, and insulates the body.

In recent years there has been a great deal of interest in the dermal and transdermal delivery of drugs (2). Most of the interest has focused on systemic (transdermal) blood levels of drugs with successful commercialization of transdermal patches, such as scopolamine, nitroglycerin, clonidine, estradiol, testosterone, and nicotine. Transdermal drug delivery offers advantages of avoiding local gastrointestinal irritation and the hepatic first-pass effect, providing controlled plasma levels of potent drugs and improved patient compliance (3). A second use of dermal application is to treat diseases of the upper skin layers, which is the field of external dermatotherapy. A third use of dermal application is to reach deeper underlying tissues with minimal systemic exposure.

Various conditions may occur at different sites within the skin and require therapeutic drug levels within the particular region(s). For instance, the use of sunscreens should limit the penetration of active ingredient only to the depth of stratum corneum, conditions like acne require drug targeting of pilosebaceous structures, in psoriasis therapeutic drug levels should be attained in deeper epidermal layers, certain local musculo-skeletal inflammations and osteoarthritis require drug levels as deep as underlying muscle, etc. Extensive literature is available regarding topical therapy of superficial skin structures. A literature has also now been established in the last two decades in the field of transdermal drug delivery. The area of deep tissue penetration after topical application has, however, received limited attention. It will be clinically useful if topical penetration can concentrate more active compound in local affected muscular tissue than can alternative administration routes. Better therapy may thus be possible with minimal systemic distribution of drug and associated side effects (4).

The concept of using the transdermal route to advantageously reach deeper tissue, such as the muscle below the applied site, is contrary to the conventional thinking. It is generally believed that cutaneous microcirculation acts as a sink, so that following its passage through stratum corneum, epidermis, and dermis, a penetrant will be effectively removed by the dermal blood supply. However, reports have appeared in literature showing evidence of direct penetration of solutes below topical site of application. Local subcutaneous drug delivery has been demonstrated for a diverse array of chemical substances such as dimethyl sulfoxide, steroids, and commonly used organophosphorous pesticides (4). Topical preparations of diclofenac, ibuprofen, piroxicam, indomethacin, and felbinac are commercially available for local treatment of rheumatism and pain.

The topical delivery of local anesthetics to deeper layers of skin will be most useful when patients are averse to the use of hypodermic needles, particularly in the pediatric patient population where even relatively minor procedures are venipuncture and insertion of catheters can be source of considerable difficulty (5). Percutaneous local anesthesia may also permit excision of abscesses and split-skin grafting without any need for general or infiltration anesthesia (6).

II. IN VIVO DERMAL PENETRATION STUDIES

We have recently studied the tissue penetration of a number of solutes below a topically applied site (7–13). Tissue concentrations at various target site(s) including skin, underlying subcutaneous tissue, fat, muscle, blood, and contralateral (similar tissues on an untreated site) tissues were quantified.

III. EXPERIMENTAL MODEL

In these studies, solutions of solutes were applied directly on the exposed rat dermis in vivo, in the absence of stratum corneum barrier. The epidermis was removed by electrodermatome and a glass cell was then adhered to the exposed dermis and warmed to 37°C by means of an external heating device. A solution of solute was introduced into the dermal glass cell and the solution was stirred by a glass stirrer driven by an external motor. Samples were removed from the dermal cell at various times and analyzed for solute concentration. The glass cell containing drug solution was removed from the rat dermis at predetermined times and a blood sample was taken from the tail vein. The animals were then sacrificed with an overdose of anesthetic ether and tissues below the treated site—that is, dermis, subcutaneous tissue, fascia, muscle lining or superficial muscle, muscle, fat pad, and deep muscle (Fig. 1)—were dissected. Similar tissues from the contralateral side were also removed. Tissues and plasma samples were analyzed for radiolabel (7–13).

The relative importance of the diffusion and perfusion processes in the dermis are most easily studied by placing the solute directly on the dermis in the absence of stratum corneum barrier (14,15). The application of the drug solution directly on the dermis mimics perfect delivery through epidermis and overcomes the inherent differences in the stratum corneum permeability to solutes related to vehicle–skin, vehicle–drug, and drug–skin interactions (14). In practice, compounds are applied on the intact stratum corneum for their local or systemic effects. Dermal absorption studies combined with penetration fluxes through isolated human epidermis have been used to examine the concentrations in dermis after topical application (14,16). This approach was evaluated for certain selected solutes whereby underlying tissue levels of drugs were determined after dermal application and then combined with isolated human epidermal studies (10).

IV. TISSUE PENETRATION OF SOLUTES AFTER DERMAL APPLICATION

Using the model just described, target tissue penetration of a variety of solutes was evaluated. The solutes studied were salicylate (9), nonsteroidal

DERMIS

SUBCUTANEOUS

FASCIA

SUPERFICIAL
MUSCLE

MUSCLE

FAT PAD

DEEP MUSCLE

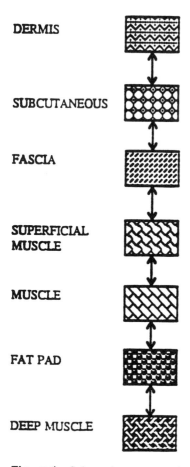

Figure 1   Schematic representation of dermis and underlying tissues in rat.

anti-inflammatory drugs (NSAIDs, including diclofenac, piroxicam, naproxen, and indomethacin) (10), bases (including lidocaine, antipyrine, iodoantipyrine, haloperidol, and diazepam) (11,12), steroids (including hydrocortisone, fluocinolone acetonide, progesterone, and testosterone) (12), and
the polar nonelectrolytes water and sucrose (8). The deep tissue penetration
potential of solutes was evaluated by comparing the tissue concentration of
solutes below the applied site with concentrations in similar tissues on the
contralateral side and plasma concentrations. Figure 2 shows the time course
of salicylic acid in target tissues and plasma after dermal application. The
concentrations of salicylic acid (Fig. 2) and other nonsteroidal anti-inflammatory agents (not shown) were higher in underlying dermis and subcuta-

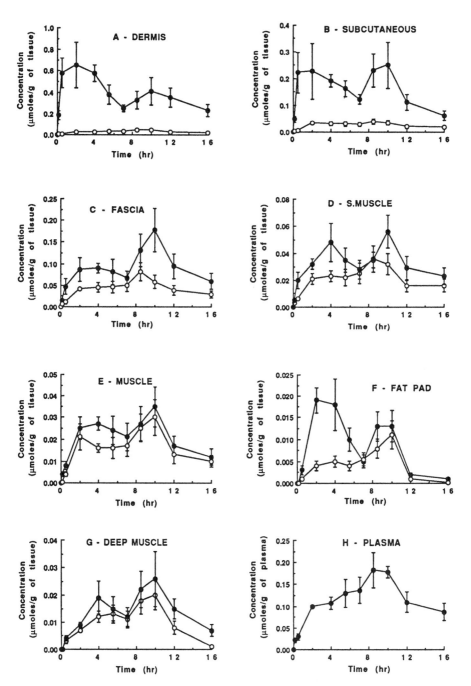

Figure 2 (A–G) Tissue concentrations versus time profiles of salicylic acid applied to anesthetized rat dermis: (●) underlying tissues, (○) similar tissues from a contralateral site. (H) Concentration–time profile in plasma. S. muscle, superficial muscle; D. muscle, deep muscle.

neous tissue as compared to plasma concentrations and concentrations in similar tissues on the contralateral side (9,10). At greater tissue depths (3–4 mm), the concentrations in underlying tissues were always less than the plasma concentrations and comparable to contralateral tissue concentrations (9,10). Figure 3 shows time course of lidocaine in tissues and plasma after dermal application. Lidocaine (Fig. 3), other bases, and steroidal compounds (not shown) achieved concentrations in underlying tissues that were higher than in contralateral tissues and plasma concentrations up to a depth of about 8–10 mm below the applied site (11,12).

Salicylic acid and other nonsteroidal anti-inflammatory compounds are predominantly ionized at the dermal pH of 7.4 and are therefore effectively removed by the dermal blood supply. Their depth of penetration below the applied site is limited (9,10). The bases and steroids used were predominantly unionized at the dermal pH. The more lipophilic solutes, such as lidocaine, haloperidol, progesterone, testosterone, and diazepam, penetrated to a greater depth as compared to the less lipophilic ones, antipyrine, iodoantipyrine, and hydrocortisone (11,12). Since deep tissue penetration of solutes was estimated in terms of differential concentrations in tissues below the applied and contralateral sites, the higher deep tissue potential of certain compounds may also be interpreted in terms of their higher total body clearance. Other lipophilic solutes such as thyroxine, triiodothyronine, and estradiol have been shown to penetrate directly into local subcutaneous structures after topical application, while dexamethasone (a polar steroid) and diisopropyl phosphate were effectively removed by the dermal blood supply (17,18). Hydrocortisone, a polar steroid, shows direct penetration to a lesser extent than the more lipophilic progesterone. Triethanolamine salicylate, a lipophilic salt of salicyclic acid, effectively penetrates underlying tissues after topical application (19,20) as compared to ionized salicylic acid. The subcutaneous tissue/muscle accumulation of other lipophilic solutes such as diisopropyl fluorophosphate, malathion, parathion, estradiol, and progesterone has also been shown in the mouse and rat (17,18). It has been suggested that the blood supply to the dermis is not capable of resorbing certain chemicals, and thus the substrate accumulates with time and is able to diffuse to deeper tissues (17).

Rabinowitz et al. have shown high levels of salicylate in local tissues after topical application of its triethanolamine salt to knees of dogs (19). Baldwin et al. applied trolamine salicylate to biceps femoris in pigs and observed levels in underlying muscle (~4 cm) that were manifold higher than those observed in contralateral muscle and in the blood (20). Riess et al. have reported significant concentrations of diclofenac in synovial fluid and tissue after topical administration on the hands of arthritic patients (21). In contrast, Radermacher et al. observed that distribution of topically applied

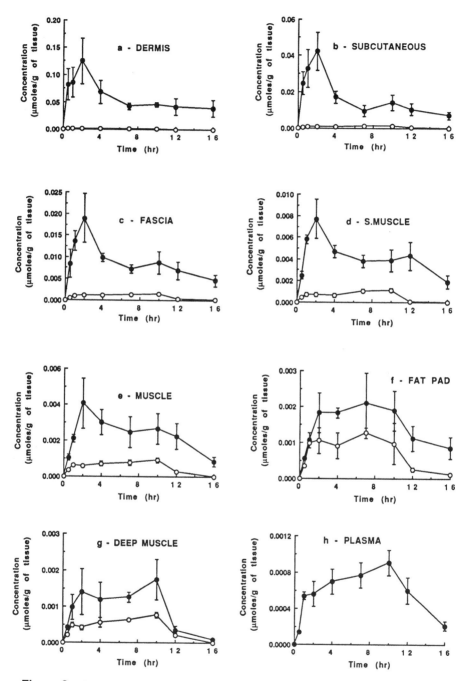

Figure 3 (a–g) Tissue concentrations versus time profiles of lidocaine applied to anesthetized rat dermis: (●) underlying tissues, (○) similar tissues from a contralateral site. (h) Concentration–time profile in plasma.

diclofenac to synovial fluid occurred mainly via systemic blood supply and direct penetration was minimal (22). A similar conclusion was drawn for biphenylacetic acid (23). It was suggested that relatively higher diclofenac concentrations in synovial fluid observed by Riess et al. (21) were measured in smaller joints such as finger and wrist joints, probably reflecting a shorter diffusion distance (22). The percutaneous application of the antifungal agent flutrinazole in mini-pigs resulted in the compound accumulation only in the superficial layers of skin (100–200 μm), with very low deep tissue penetration potential (24).These observations suggest that the depth of target tissue penetration is an important parameter and should be stated in local tissue penetration studies. The depth of penetration may also be altered by biological variables such as the presence or absence of adipose tissue or fat beneath the area of application (25). The observed differences in various studies may also be due to variables such as differences in formulation, method of application—solution, ointment, or cream, with or without rubbing—duration, application site, species studied, etc.

V. INTEGRATION OF EPIDERMAL PENETRATION AND DERMAL ABSORPTION STUDIES

In practice, all topically used NSAIDs are applied to intact skin with the stratum corneum barrier present. Consider a solute of concentration C_v applied to the epidermis and removal of that solute from the epidermis by a dermal clearance CL_D. The rate of change of concentration in epidermis dC_e/dt) can then be expressed as

$$V_e \frac{dC_e}{dt} = k_p A(C_v - C_{ss}) - CL_D C_{ss} \tag{1}$$

where V_e is the volume of epidermal compartment, k_p is the epidermal (stratum corneum) permeability coefficient, C_{ss} is the steady state concentration in epidermis, and A is the area of application. Equation (1) strictly applies only after the lag time for permeation from the stratum corneum. When the uptake process approaches a steady state, $dC_e/dt \to 0$ and Eq. (1) reduces to

$$\frac{C_{ss}}{C_v} = \frac{k_p A}{k_p A + CL_D} \tag{2}$$

The epidermal concentration fraction after topical application of various NSAIDs was estimated from Eq. (2) using permeability coefficients from excised skin flux measurements and an area of 2.54 cm^2. Expressions analogous to Eq. (2) can be written for other tissues (10). Hence the expression for the dermis following the application of an aqueous solution under pseudo-steady-state conditions is:

$$\frac{Cd_{ss}}{C_s} = \frac{k_a V_s}{k_a V_s + CL_{sc}} \tag{3}$$

where k_a is the disappearance rate constant from the dermis, C_s is the concentration of solution applied to dermis, Cd_{ss} is the concentration in the dermis at steady-state, V_s is the volume of solution applied to dermis, and CL_{sc} is the clearance into subcutaneous tissue (10,26).

Figure 4 shows estimated target tissue concentrations following epidermal delivery of the NSAIDs at their maximum fluxes from applied vehicle. Salicylic acid was found to give highest local tissue levels, followed by piroxicam, naproxen, indomethacin, and diclofenac. Maximal fluxes are based on the product of permeability constants of solutes from aqueous solutions and aqueous solubility (10,14). Such fluxes are, however, only approximate, due to possible changes in the activity coefficient within the vehicle or vehicle–skin interactions (10). The preceding results should be

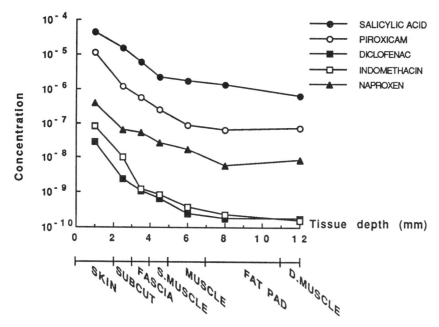

Figure 4 Concentration (fraction of initial donor concentration) of NSAIDs in underlying tissues estimated at maximum epidermal fluxes of NSAIDs across human epidermis from an aqueous vehicle.

placed in the context that aqueous solutions were used for estimating tissue concentrations, whereas in clinical practice NSAIDs are usually administered topically in more viscous and partially nonaqueous formulations such as creams, ointments, and jellies. Release characteristics will therefore vary depending on vehicle–skin, vehicle–drug, and drug–skin interactions (14). Thus the use of a particular formulation may favor the penetration of one particular compound relative to another across epidermis. Higher fluxes across epidermis can also be obtained by use of either physical (iontophoresis, phonophoresis, etc.) or chemical (penetration enchancers) means.

VI. FACILITATION OF DEEP TISSUE PENETRATION

A. Use of Vasoactive Chemicals

Epinephrine and phenylephrine are often used to decrease systemic absorption of local anesthetics from subcutaneous and other injection sites by decreasing local blood flow (27). The peak systemic concentrations of timolol after ocular administration in rabbits have also been shown to decrease significantly in the presence of phenylephrine and epinephrine (28,29). The effects of phenylephrine on underlying tissue (vs. systemic) distribution of salicylic acid and lidocaine after dermal application were recently investigated (13). The experimental model was as described earlier. Appropriate concentrations of phenylephrine were coapplied with the solutes and their effects on tissue distribution and plasma uptake evaluated as already described.

Figure 5 shows the fraction (of the initial concentration) of salicylic acid in underlying tissues, contralateral tissues, and plasma at various PE concentrations after dermal application. The concentrations of salicylic acid in underlying tissues (except for fat pad) increase with an increase in PE concentration up to 0.01%. At higher PE concentrations, the concentrations of salicylic acid approach a constant value. The plasma and contralateral tissue concentrations of salicylic acid decrease with increasing PE concentration, suggesting decreased systemic availability and distribution of salicylic acid (13). The concentrations of lidocaine in the tissues below the applied site also increased in the presence of phenylephrine, whereas those in plasma and contralateral tissues decreased (13). The addition of PE significantly increased the treated tissue:contralateral tissue and treated tissue:plasma ratio of salicylic acid and lidocainein in underlying dermis, subcutaneous tissue, fascia, and muscle (13). These results show that the addition of PE significantly increases the quantities of solutes delivered to local subcutaneous structures with reduced systemic uptake. Altering local tissue blood flow is likely to affect tissue concentration most when drug

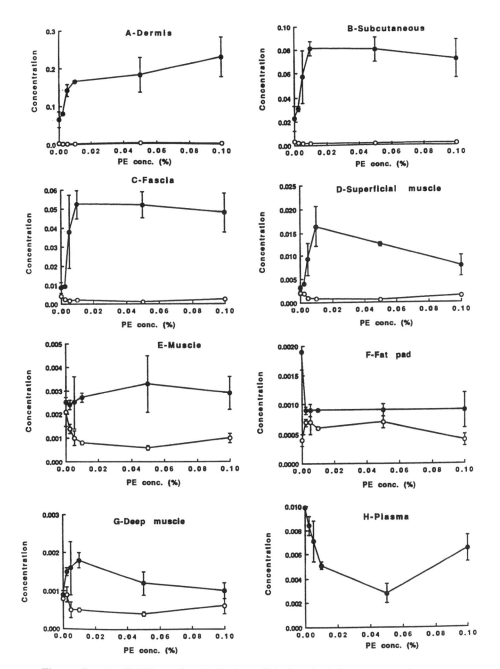

Figure 5 (A–G) Effect of applied (dermally) phenylephrine concentration on the observed concentrations of salicylic acid (expressed as fraction of initial donor concentration) in underlying tissues (●), contralateral tissues (○), and plasma (H) in rats.

absorption is blood-flow limited and solute penetrates the biological membranes relatively easily. The tissue concentrations of these and various other solutes in sacrificed animals have been observed to be higher as compared to anesthetized animals, showing the importance of viable blood supply (8–13,16,30).

B. Iontophoresis

Iontophoresis is a novel noninvasive drug delivery system used to facilitate transdermal delivery. It shares the advantages of passive transdermal delivery including bypass of first-pass effect and the variability associated with gastrointestinal absorption, provides controlled plasma levels of potent drugs with short biological half-lives, and increased patient compliance. Iontophoresis minimizes trauma, risk of infection, and damage to the wound and is an important alternative to parenteral therapy. Transdermal iontophoresis is useful for delivering drugs both to the skin for topical therapy and to the blood for systemic therapy (31). Passive transdermal delivery is usually limited to the delivery of small, nonpolar and lipophilic solutes. Iontophoresis facilitates the transport of charged molecules and has great potential in the delivery of peptide and protein drugs.

The rationale of using iontophoresis to deliver drugs to their local site of action is based on the expected achievement of high tissue concentrations of the solutes in target areas with minimal side effects (31). Compounds such as salicylate, phenylbutazone, ketoprofen, indomethacin, bufexamac, piprofen, and diclofenac have been claimed to be effective in various rheumatic conditions and soft-tissue rheumatisms such as epicondylitis, capsulitis, and tendonitis, after transdermal iontophoresis (31).

Using the model just described, epidermal iontophoresis yielded significant concentrations of salicylic acid in the underlying tissues, with levels comparable to those obtained when salicylic acid was applied to exposed rat dermis. The concentrations of salicylic acid in the skin and subcutaneous tissue below the treated site were higher than in plasma and contralateral tissues up to a depth of 3–4 mm (Fig. 6). Concentrations in deeper tissues were always lower than concentrations in plasma but higher than, although comparable to, concentrations in contralateral tissues, suggesting that the drug is first absorbed into the bloodstream and then distributed to deeper tissues (7). Consistent with dermal application studies, the direct penetration of salicylic acid is limited to a depth of 3–4 mm below the applied site.

Application of iontophoresis yielded high concentrations of lidocaine in all underlying tissues to an approximate depth of 1.2 cm (7). The tissue lidocaine levels after iontophoretic epidermal application were higher than the levels in similar tissues and plasma after passive dermal application, a

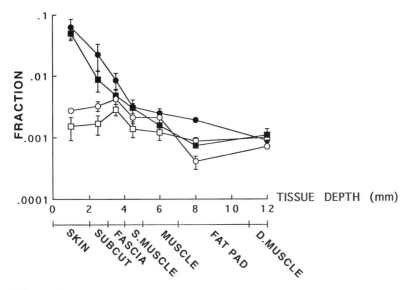

Figure 6a Tissue distribution of salicylic acid in rats after application to dermis [treated site (●) and contralateral site (○)] and iontophoresis through epidermis [treated site (■) and contralateral site (□)].

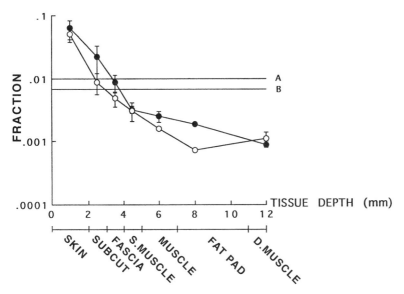

Figure 6b Tissue concentrations of salicylic acid in rats: (●) application to dermis; (○) iontophoresis through epidermis; (A) levels in plasma after dermal application; (B) levels in plasma after iontophoresis through epidermis.

result highlighting the efficacy of iontophoretic delivery (Fig. 7) (7). Russo et al. previously reported the penetration of iontophoretically applied lidocaine to the depth of subcutaneous tissue (observed by the placement of sutures) in humans (32). Murray et al. also demonstrated the safety of steroid iontophoresis in diseases of subcutaneous tissue at an estimated depth of 1.25 cm (33). Lidocaine could not be detected in contralateral tissues after iontophoresis, and very low concentrations were observed in the blood (7). Earlier studies by Glass et al. revealed that <0.4% dexamethasone was absorbed into the total blood volume after iontophoretic application in monkeys (34). Percutaneous iontophoresis of prednisolone has also been shown to yield levels in plasma that are about one-third those produced by oral administration (35). It is therefore apparent that transdermal iontophoresis is a useful technique for enhancing local delivery of drugs without exposing the rest of the body to unwanted side effects.

The combined use of iontophoresis and vasoactive chemicals was effectively used to selectively target local tissues or systemic circulation (36). Epinephrine was found to increase the local skin concentrations of lidocaine and elevate lidocaine mass-depth profiles while the vasodilator tolazoline decreased the cutaneous reservoir of lidocaine after transdermal iontophoresis (36).

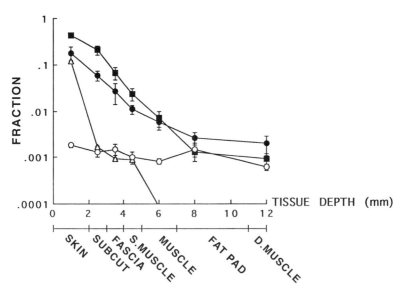

Figure 7a Tissue distribution of lidocaine in rats after application to dermis measured at treated site (●) and contralateral site (○), iontophoresis through epidermis measured at treated site (■), and application to epidermis measured at treated site (△).

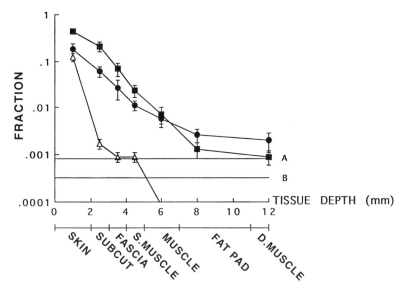

Figure 7b   Tissue concentrations of lidocaine in rats: (●) application to dermis; (■) iontophoresis through epidermis; (△) application to epidermis; (A) levels in plasma after dermal application; (B) levels in plasma after iontophoresis through epidermis.

The underlying skin and target tissue penetration of solutes after topical application involves a complex interplay of physiological variables such as tissue diffusion, tissue blood supply and its distribution, local/systemic metabolism, reversible/irreversible tissue binding, tissue/plasma partitioning, etc., and physicochemical variables such as lipophilicity, molecular size, steric factors, state of ionization, protein binding, etc. The work discussed in this chapter provides an insight in defining the target sites and ways of estimating target site concentrations within the skin and underlying tissues as well as systemic uptake of compounds after topical application. Strategies to enhance dermal distribution relative to systemic uptake have been outlined.

REFERENCES

1. Bucks DAW. Skin structure and metabolism: Relevance to the design of cutaneous therapeutics. Pharm Res 1984; 1:148–153.
2. Nowack H, Marin U, Reger R, Bohme H, Schriever K-H, Bocionek, Elbers R, Kampffmeyer H-G. Cutaneous absorption of indomethacin from two topical preparations in volunteers. Pharm Res 1985; 2:202–206.

3. Ranade VV. Drug delivery systems. 6. Transdermal drug delivery. J Clin Pharmacol 1991; 31:401–418.

4. Guy RH, Maibach HI. Drug delivery to local subcutaneous structures following topical administration. J Pharm Sci 1983; 72:1375–1380.

5. Hallen B, Uppfeldt A. Does Lidocaine-Prilocaine cream permit painfree insertion of IV catheters in children? Anesthesiology 1982; 57:340–342.

6. McCafferty DF, Woolfson AD, McClelland KH, Boston V. Comparative in vivo and in vitro assessment of the percutaneous absorption of local anesthetics. Br J Anaesth 1988; 60:64–69.

7. Singh P, Roberts MS. Iontophoretic transdermal delivery of salicylic acid and lidocaine to local subcutaneous structures. J Pharm Sci 1993; 82:127–131.

8. Singh P, Roberts MS. Blood flow measurements in skin and underlying tissues by microsphere method—Application to dermal pharmacokinetics of polar non-electrolytes. J Pharm Sci 1993; 82:873–879.

9. Singh P, Roberts MS. Dermal and underlying tissue pharmacokinetics of salicylic acid after topical application. J Pharmacokinet Biopharm 1993; 21:337–373.

10. Singh P, Roberts MS. Skin permeability and local tissue concentrations of non-steroidal anti-inflammatory drugs (NSAIDs) after topical application. J Pharmacol Exp Ther 1994; 268:141–151.

11. Singh P, Roberts MS. Dermal and underlying tissue pharmacokinetics of lidocaine after topical application. J Pharm Sci 1994; 83:773–782.

12. Singh P, Roberts MS. Deep tissue penetration of bases and steroids after dermal application in rat. J Pharm Pharmacol 1994; 46:956–964.

13. Singh P, Roberts MS. Effects of vasoconstriction on dermal pharmacokinetics and local tissue distribution of compounds. J Pharm Sci 1994; 83:783–791.

14. Roberts MS. Structure-permeability considerations in percutaneous absorption. In: Scott RC, Guy RH, Hadgraft J, Bodde H, eds. Prediction of Percutaneous Penetration. IBC Technical Services Ltd: London, 1991:210–228.

15. Roberts MS, Singh P, Yoshida NH. Targeted drug delivery after topical application. Proc Int Symp Cont Rel Bioact Mater 1991; 18:295–296.

16. Siddiqui O, Roberts MS, Polack AE. Topical absorption of methotrexate: Role of dermal transport. Int J Pharm 1985; 27:193–203.

17. Marty JP, Guy RH, Maibach HI. Percutaneous penetration as a method of delivery to muscle and other tissues. In: Bronaugh RL, Maibach HI, eds. Percutaneous Absorption. Marcel Dekker: New York, 1989:511–529.

18. Marty JP, James M, Hajo N, Wepierre J. Percutaneous absorption of oestradiol and progesterone: Pharmacokinetic studies. In: Jarvais-Mauvais P, ed. Percutaneous penetration of steroids. Academic Press: New York, 1980:205–218.

19. Rabinowitz JL, Feldman ES, Weinberger A, Schumacher HR. Comparative tissue absorption of oral ^{14}C-salicylate in humans and canine knee joints. J Clin Pharmacol 1982; 22:42–48.

20. Baldwin JR, Carrano RA, Imondi AR. Penetration of trolamine salicylate into the skeletal muscle of the pig. J Pharm Sci 1984; 73:1002–1004.

21. Riess VW, Schmid K, Botta L, Kobayashi K, Moppert J, Schneider W, Sioufi A, Strusberg A, Tomasi M. Die perkutane Resorption von Diclofenav. Arzneim-Forsch/Drug Res 1986; 36:1092–1096.

22. Radermacher J, Jentsch D, Scholl MA, Lustinetz T. Frolich JC. Diclofenac concentrations in synovial fluid and plasma after cutaneous application in inflammatory and degenerative joint disease. Br J Clin Pharmacol 1991; 31:527–541.

23. Dawson M, McGee CM, Vine JH, Watson TR, Brooks PM. The disposition of biphenylacetic acid following topical application. Eur J Clin Pharmacol 1988; 33:639–642.

24. Conte L, Ramis J, Mis R, Forn J, Vilaro S, Reina M, Vilageliu J, Basi N. Percutaneous absorption and skin distribution of [^{14}C] flutrimazole in minipigs. Arzneim-Forsch/Drug Res 1992; 42:847–853.

25. Rosenberg PH, Kytta J, Alila A. Absorption of bupivacaine, etidocaine, lignocaine and ropivacaine into n-hepatane, rat sciatic nerve, and human extradural and subcutaneous fat. Br J Anaesth 1986; 58:310–314.

26. Singh P, Roberts MS. Local deep tissue penetration of compounds after dermal application: Structure-tissue penetration relationships. J Pharmacol Exp Ther 1996; 279:908–917.

27. Tucker GT, Mather LE. In: Cousins MJ, Bridenbaugh PO, eds. Neural blockade in clinical anesthesia and management of pain. J. B. Lippincott: Philadelphia, 1988:47–110.

28. Kyyronen K, Urtti A. Effect of epinephrine pretreatment and solution pH on ocular and systemic absorption of ocularly applied timolol in rabbits. J Pharm Sci 1990; 79:688–691.

29. Urtti A, Kyyronen K. Ophthalmic epinephrine, phenylephrine, and pilocarpine affect the systemic absorption of ocularly applied timolo. J Ocul Pharmacol 1989; 5:127–132.

30. Cross SE, Roberts MS. Importance of dermal blood supply and epidermis on the transdermal iontophoretic delivery of monovalent cations. J Pharm Sci 1995; 84:584–592.

31. Singh P, Maibach HI. Iontophoresis in drug delivery: Basic principles and applications. Crit Rev Ther Drug Carr Sys 1994; 11:161–213.

32. Russo J, Lipman AG, Vomstock TJ, Page BC, Stephen RL. Lidocaine anesthesia: Comparison of iontophoresis, injection and swabbing. Am J Hosp Pharm 1980; 37:843–847.

33. Murray W, Lavine LS, Seifter E. The iontophoresis of C_{21} esterified glucocorticosteroids: Preliminary report. J Am Phys Ther Assoc 1983; 43:579–581.

34. Glass JM, Stephen RL, Jacobsen SC. The quantity and distribution of radiolabeled dexamethasone delivered to tissues by iontophoresis. Int J Dermatol 1980; 19:519–525.

35. James MP, Graham RM, English J. Percutaneous iontophoresis of prednisolone–A pharmacokinetic study. Clin Exp Dermatol 1986; 11:54–61.

36. Riviere JE, Riviere-Monteiro NA, Inman AO. Determination of lidocaine concentrations in skin after transdermal iontophoresis; Effects of vasoactive drugs. Pharm Res 1992; 9:211–214.

36
Phonophoresis

Joseph Kost
Ben-Gurion University, Beer-Sheva, Israel

Samir Mitragotri and Robert Langer
Massachusetts Institute of Technology, Cambridge, Massachusetts

I. INTRODUCTION

In spite of major research and development efforts in transdermal systems and the many advantages of the transdermal route, low permeability of the human skin is still a major problem that limits the usefulness of the transdermal approach. It is well accepted that the stratum corneum is the major rate-limiting barrier to molecular diffusion through the mammalian epidermis (1–3). Due to the fact that most drugs do not permeate the skin in therapeutic amounts, chemical and physical approaches have been examined to lower the stratum corneum barrier properties and enhance transdermal permeation.

Although chemical approaches using molecules such as dimethyl sulfoxide (DMSO), 1-dodecylazacycloheptan-2-one (Azone), surfactants, solvents, and binary polar and apolar systems have been shown to provide enhancement, many have not been widely accepted because of either suspected pharmacologic activity or unresolved questions about safety. In addition to the chemical approach, several physical approaches for skin penetration enhancement, such as stripping of the stratum corneum, thermal energy, iontophoresis, electroporation, and ultrasound, have also been evaluated.

Phonophoresis or sonophoresis is defined as the movement of drugs through intact skin into soft tissue under the influence of an ultrasonic perturbation (1,4–54). This chapter attempts to present phonophoresis, experimental variables and possible mechanisms of action, and the most representative experimental and clinical studies.

615

II. ELEMENTARY PHYSICS OF ULTRASONIC WAVES

Ultrasound is defined as sound having a frequency beyond 20 kHz. Most modern ultrasound equipment is based on the piezoelectric effect. When pressure is applied to crystals (quartz), and to some polycrystalline materials such as lead–zirconate–titanium (PZT) or barium titanate, electric charges develop on the outer surface of the material. Thus, rapidly alternating potential when applied across opposite faces of a piezoelectric crystal will induce corresponding alternating, dimensional changes and thereby convert electrical into vibrational (sound) energy.

There are three distinctly different sets of ultrasound conditions based on their frequency ranges and applications (55–58):

1. High-frequency or diagnostic ultrasound (2–10 MHz).
2. Medium-frequency or therapeutic ultrasound (0.7–3 MHz).
3. Low-frequency or power ultrasound (20–100 kHz).

The therapeutic ultrasound most frequently applied for transdermal drug delivery enhancement consist of a high-frequency generator that is connected to a piezoelectric crystal (the treatment head). The resonant frequency of the crystal is partly determined by the thickness of the piezoelectric material, and consequently the frequency of the ultrasound is so determined as well.

A. Properties of the Ultrasound Beam

The ultrasound beam has two distinctive areas: the near field (Fresnel zone) and the distant field (Fraunholer zone). The near field is characterized by interference phenomena in the ultrasound beam, which may lead to marked variations in intensity, expressed as the beam nonuniformity ratio (BNR). The distant field is characterized by the near absence of interference phenomena, so that the sound beam is uniform and the intensity gradually decreases with increasing distance. The length of the near field depends on the diameter of the treatment head and the wavelength. With the common treatment heads for therapeutic applications, the near field is about 10 cm long for a 5-cm^2 head and 1 cm long for a 1-cm^2 head at 1 MHz. At 3 MHz the near field is 3 times as long. Because the depth effect of ultrasound is limited, the therapeutic effects occur mainly in the near field (55).

B. Nature of the Ultrasound Wave

The ultrasound wave is of a longitudinal nature; that is, the direction of propagation is the same as the direction of oscillation (59,60). The longitudinal sound waves cause compression and expansion of the medium at half a wavelength's distance, leading to pressure variations in the medium. The wavelength of ultrasound is expressed by the relationship

$$\lambda f = c$$

where λ is the wavelength, f the frequency, and c the speed of propagation.

In soft tissue and in water the wavelength at 1 MHz is approximately 1.5 mm and in bony tissue it is about 3 mm. The medium is compressed and expanded at the same frequency as that of the ultrasound. At 1 MHz the resultant pressure changes are fairly large. For instance, at an intensity of 1 W/cm² the pressure variation is about 1.7 bar. At a wavelength of 1.5 mm this implies a pressure gradient of 3.4 bar over a distance of 0.75 mm, as the high and low pressures are half a wavelength apart.

C. Mass Density and Acoustic Impedance

The mass density of a medium (ρ) and the specific acoustic impedance (Z) determine the resistance of the medium to sound waves. The mass density also partly determines the speed of propagation (c). The higher the mass density, the higher is the speed of propagation. The specific acoustic impedance, which is a material parameter, depends on the mass density and the speed of propagation:

$$Z = \rho c$$

The specific acoustic impedance for skin, bone, and air are 1.6×10^6, 6.3×10^6, and 400 kg/m² s, respectively.

D. Absorption and Penetration of Ultrasound

As ultrasound energy penetrates into the body tissues, biologic effects can be expected to occur if the energy is absorbed by the tissues. The absorption coefficient (a) is used as a measure of the absorption in various tissues. For ultrasound consisting of longitudinal waves with perpendicular incidence on homogeneous tissues, the following formula applies:

$$I(x) = I_0 \times e^{-ax}$$

where $I(x)$ is the intensity at depth x, I_0 the intensity at the surface, and a the absorption coefficient.

A different value relating to absorption is the half-value depth ($D_{1/2}$), defined as the distance in the direction of the sound beam at which the intensity in a certain medium decreases by half. For skin the $D_{1/2}$ is 11.1 mm at 1 MHz and 4 mm at 3 MHz (60). In air $D_{1/2}$ is 2.5 mm at 1 MHz and 0.8 mm at 3 MHz.

To transfer the ultrasound energy to the body it is necessary to use a contact medium, because of the complete reflection of the ultrasound by air. The many types of contact media currently available for ultrasound transmission can be broadly classified as oils, water–oil emulsions, aqueous gels, and ointments.

III. SELECTION OF ULTRASOUND PARAMETERS FOR SONOPHORESIS

Proper selection of ultrasound parameters is required to ensure safe and efficacious sonophoresis. Ultrasound parameters such as frequency, intensity, duty cycle, and distance of transducer from the skin influence the efficiency of sonophoresis. We here present a general discussion of the role played by various ultrasound parameters in sonophoresis. The objective of this discussion is not to point out the exact values of ultrasound parameters to be selected, but rather to present information regarding the dependence of sonophoretic enhancement on each parameter.

A. Ultrasound Frequency

Ultrasound at various frequencies in the range of 20 kHz to 16 MHz has been used for sonophoresis. These studies of sonophoresis can be classified into three categories based on the ultrasound frequency used: therapeutic, high-frequency, and low-frequency ultrasound.

Therapeutic frequency ultrasound (frequency between 1 and 3 MHz). This is the most commonly used ultrasound frequency range for sonophoresis. Mitragotri et al. (35) reported that the sonophoretic enhancement in the therapeutic frequency range varies inversely with ultrasound frequency. They found that while 1-MHz ultrasound enhances transdermal transport of estradiol across human cadaver skin in vitro by 13-fold, 3-MHz ultrasound at the same intensity induces an enhancement of only 1.5-fold. They further hypothesized that the observed inverse dependence of sonophoretic enhancement on ultrasound frequency occurs since cavitational effects, which are primarily responsible for sonophoresis, vary inversely with ultrasound frequency (61).

High-frequency ultrasound (frequency above 3 MHz). Bommanan et al. (7,8) performed sonophoresis of salicylic acid and lanthanum tracers across hairless rat skin in vivo using high-frequency ultrasound ($f = 2$ MHz, 10 MHz, and 16 MHz). They investigated the dependence of sonophoresis on ultrasound frequency in the high-frequency region and found that 10-MHz ultrasound is more effective in enhancing transdermal transport of salicylic acid than that at 16 MHz, which in turn is more effective than that at 2 MHz. They proposed that the sonophoretic enhancement in the high-frequency region should vary directly with ultrasound frequency, though the anomalously high efficiency of sonophoresis at 10 MHz was due to higher efficiency of the transducer operating at that frequency.

Low-frequency ultrasound (frequency below 1 MHz). Tachibana et al. (50–52) have reported use of low-frequency ultrasound (48 kHz) to enhance transdermal transport of lidocaine and insulin across hairless mice skin. Low-frequency ultrasound has also been used by Mitragotri et al. (36,37) to enhance transport of various low-molecular-weight drugs including salicylic acid and corticosterone as well as high-molecular-weight proteins including insulin, γ-interferon, and erythropoeitin across human cadaver skin in vitro.

B. Ultrasound Intensity

Various ultrasound intensities in the range of 0.1 to 3 W/cm^2 have been used for sonophoresis. In most cases, use of higher ultrasound intensities is limited by thermal effects. Several investigations have been performed to assess the dependence of sonophoretic enhancement on ultrasound intensity. Miyazaki et al. (40) found a relationship between the plasma concentrations of indomethacin transported across the hairless rat skin by sonophoresis (therapeutic conditions) and the ultrasound intensity used for this purpose. Specifically, the plasma indomethacin concentration at the end of 3 h after sonophoresis (0.25 W/cm^2) was about threefold higher than controls at the same time. However, increasing intensity by threefold (to 0.75 W/cm^2) further increased sonophoretic enhancement only by 33%. Mortimer et al. (43) found that application of ultrasound at 1 W/cm^2 increased transdermal oxygen transport by 40% while that at 1.5 and 2 W/cm^2 induced an enhancement by 50% and 55%, respectively.

C. Pulse Length

Ultrasound can be applied either in a continuous or a pulsed mode. A pulsed mode of ultrasound application is used commonly because it reduces the severity of adverse side effects of ultrasound, such as thermal effects. However, pulsed application of ultrasound may have a significant effect on the efficacy of sonophoresis. As will be discussed later, cavitational effects, which play a crucial role in sonophoresis, vary significantly with the pulse length (62). Mitragotri et al. (35) reported that while a continuous application of therapeutic ultrasound (1 MHz, 2 W/cm^2) increased human skin permeability to estradiol by 13-fold, a pulsed application (2-ms pulses applied every 10 ms) did not significantly enhance transdermal estradiol flux. In very low-frequency ultrasound region, Kost et al. (29,63) reported that urea permeability of cuprophane membranes increased from 6% to 56% as the ultrasound (20 kHz) pulse length increased from 100 ms to 400 ms (applied every second).

D. Distance of the Transducer From the Skin

The ultrasound pressure (or intensity) field around a transducer is quite complex. As described earlier, the intensity of the ultrasound passes through a series of maxima and minima in a region near the transducer and beyond a certain distance decreases monotonically with distance. The region in which the ultrasound intensity passes through the series of minima and maxima is referred to as the near field, and the region beyond the near field is referred to as the far field. Julian et al. (22) studied the effect of transducer distance on the sonophoretic permeability of benzoic acid through a polydimethyl-siloxane membrane under low-frequency conditions. They observed that the effect of 20-kHz ultrasound on the permeability of the membrane is insensitive to the distance of the transducer from the membrane. This probably occurs because of the successive reflections of ultrasound waves in the diffusion cell, which prevents any systematic pressure pattern from forming in the diffusion cell.

IV. CLINICAL STUDIES OF SONOPHORESIS

Over the past 40 years, numerous clinical reports have been published concerning phonophoresis. The technique involves placing the topical preparation on the skin over the area to be treated and massaging the area with an ultrasound probe. Some of the earliest studies done with phonophoresis involve hydrocortisone. Fellinger and Schmidt (13) reported successful treatment of polyarthritis of the hand's digital joints using hydrocortisone ointment with phonophoresis. Newman et al. (45) and Coodley (64) showed improved results of hydrocortisone injection combined with ultrasound "massage" compared to simple hydrocortisone injection for a bursitis treatment.

In addition to joint diseases and bursitis, phonophoresis has been tested for its ability to aid penetration in a variety of drug–ultrasound combinations, mainly for localized conditions. The major medications used include the anti-inflammatories such as cortisol, dexamethasone, and salicylates, and local anesthetics (65). Cameroy (9) reported success using carbocaine phonophoresis before closed reduction of Colle's fractures. Griffin et al. (17) treated 102 patients with diagnoses of elbow epicondylitis, bicipital tendonitis, shoulder osteoarthritis, shoulder bursitis, and knee osteoarthritis with hydrocortisone and ultrasound: 68% of patients receiving drug in conjunction with ultrasound were rated as "improved," demonstrating a pain-free normal functional range of motion, while only 28% of patients receiving placebo with ultrasound were rated as "improved." Similar effects were presented by Moll (41), who published a double blind study with three groups of patients receiving either lidocaine/Decadron with ultrasound, a placebo with ultrasound, or a placebo with ultrasound at zero intensity. Percentage improvement for the three groups were 88.1, 56.0, and 23.1%, respectively.

McElnay et al. (31) evaluated the influence of ultrasound on the percutaneous absorption of lignocaine from a cream base. Mean data indicated that there was a slightly faster onset time for local anesthesia when ultrasound was administered when compared with control values (no ultrasound). However, the differences were not statistically significant.

Benson et al. (4) reported on the influence of ultrasound on the percutaneous absorption of lignocaine and prilocaine from Emla cream. The local anesthetic cream formulation Emla was chosen because it requires a relatively long contact time with the skin before the application site becomes anesthetized (60 min). The authors evaluated three frequencies (0.75, 1.5, and 3 MHz) at a continuous intensity of 1.5 W/cm^2 or frequencies of 1.5 or 3.0 MHz, 1:1 pulsed output, at an intensity of 1.0 W/cm^2. The 1.5-MHz (1:1 pulsed output) and 3.0-MHz (continuous output) ultrasound appeared to be the most effective frequencies in improving the rate of percutaneous absorption, while the 1.5-MHz and 3-MHz (1:1 pulsed output) ultrasound treatments were the most effective frequencies in improving the extent of drug absorption. Benson et al. (5,6) demonstrated that ultrasound is also capable of enhancing the percutaneous absorption of methyl and ethyl nicotinate. For the lipophilic hexyl nicotinates no effect of ultrasound on its percutaneous absorption could be detected. The pharmacodynamic parameter of vasodilation caused by nicotinates was used to monitor percutaneous absorption. McEnlay et al. (33) also evaluated the skin penetration enhancement effect of phonophoresis on methyl nicotinate in 10 healthy volunteers in a double-blind, placebo-controlled, crossover clinical trial. Each treatment consisted of the application of ultrasound massage (3.0 MHz, 1.0 W/cm^2 continuous output) or placebo (0 MHz) for 5 min to the forearm, followed by a standardized application of methyl nicotinate at intervals of 15 s, 1 min, and 2 min ultrasound massage. Ultrasound treatment applied prior to methyl nicotinate led to enhanced percutaneous absorption of the drug. The authors suggest that the ultrasound affects the skin structure by disordering the structured lipids in the stratum corneum.

Similar experiments were performed by Hofman and Moll (20), who studied the percutaneous absorption of benzyl nicotinate. For recording the reddening of the skin a reflection photometry was employed. Ultrasonic treatment of the skin at levels lower than 1 W/cm^2 reduced the lag time as a function of ultrasound power. Repetition of the ultrasonic treatment confirmed the reversibility of the changes in skin permeability.

Kleinkort and Wood (24) compared phonophoretic effects of a 1% cortisol mixture to that of a 10% mixture. Although an improvement of approximately 80% of the patients receiving 1% cortisol was demonstrated, the group treated with the 10% mixture showed improvement in 95.7% of the patients, while treatment of 16 patients with subdeltoid bursitis showed 100% improvement. In all groups, the 10% compound was more effective. The transmission characteristics of a number of topical proprietary preparations containing drugs suitable for use with ultrasound have been investigated by Benson et al. (4) Gel formulations were found to be the most suitable coupling agents.

Williams (53) developed an electrical sensory perception threshold technique for use with human volunteers in order to evaluate the effect of phonophoresis on three commonly available topical anesthetic preparations. Low intensities (0.25 W/cm^2) of 1.1 MHz ultrasound had no detectable effects upon the rate of penetration of either one of the three anesthetic preparations through human skin under conditions where temperature increases had been minimized.

V. SAFETY

Safety of sonophoresis involves two main issues: a) the reversibility of the skin barrier properties after turning ultrasound off, and b) the effect of ultrasound on the living parts of the skin and underlying tissues. Many reports exist in literature describing preliminary assessment sonophoresis with respect to these two issues. We next present a summary of these studies.

A. Recovery of the Skin Barrier Properties After Sonophoresis

Numerous reports exist to suggest that application of therapeutic ultrasound (1–3 MHz, 0–2 W/cm^2) does not induce any irreversible change in the skin permeability to drugs in vivo. Quantitative measurements of estradiol transport across human skin (in vitro) have also shown that application of therapeutic ultrasound (1 MHz, 2 W/cm^2) does not induce any statistically significant irreversible change in skin barrier properties (35). Similar studies have also been performed using very low-frequency ultrasound (20 kHz, 125 mW/cm^2, 100-ms pulses applied every second) to assess whether application of low-frequency ultrasound results in any permanent loss of the barrier properties of skin measured in terms of water permeability (37). It has been found that in the case of a 1-h-long ultrasound exposure, the skin permeability to water measured within 2 h postexposure was comparable to the passive skin permeability to water. In the case of a 5-h-long ultrasound exposure, the skin permeability 2 h postexposure was about 6 times higher than the passive permeability to water. However, this value continued to decrease, and was within a factor of 2 of the passive skin water permeability at 12 h postexposure. Studies have also been performed (7,8,34) to assess whether application of high-frequency ultrasound induces any irreversible damage to the barrier properties of the skin measured in terms of transepidermal water loss (TEWL) across hairless mice skin exposed to high-frequency ultrasound (16 MHz). No significant difference in TEWL values of the skin exposed to ultrasound and that not exposed to ultrasound was found (34).

B. Biological Effects of Ultrasound

Ultrasound over a wide frequency range has been used in medicine over last century. For example, therapeutic ultrasound (1–3 MHz) has been used for massage, low-frequency ultrasound has been used in dentistry (23–40 kHz) (57,58), and high-frequency ultrasound (3–10 MHz) has been used for diagnostic purposes (60). In view of this, significant attention has been dedicated to investigating the effects of ultrasound on biological tissues. However, no conclusions have been reached regarding the limiting ultrasound conditions required to ensure safe exposure.

Ultrasound affects biological tissues via three main effects: thermal effects, cavitational effects, and acoustic streaming. A summary of qualitative guidelines to determine under what conditions these effects become critical is presented (56):

Thermal effects may be important when:

1. The tissue has a high protein content.
2. A high-intensity continuous-wave ultrasound is used.
3. Bone is included in the heated volume.
4. Vascularization is poor.

Cavitation may be important when:

1. Low-frequency ultrasound is used.
2. Gassy fluids are exposed.
3. Small gas-filled spaces are exposed.
4. The tissue temperature is higher than normal.

Streaming may be important when:

1. The medium has an acoustic impedance different from its surroundings.

2.   The fluid in the biological medium is free to move.
3.   Continuous-wave application is used.

Various investigators have reported histological studies of animal skin exposed to ultrasound under various conditions in order to assess the effect of ultrasound on living skin cells. Levy et al. (29) performed histological studies of hairless rat skin exposed to therapeutic ultrasound using hematoxylin and eosin and reported that application of ultrasound (1 MHz, 2 W/cm^2) induced no damage to the skin. Tachibana et al. (51) performed similar studies of rabbit skin expose to low-frequency ultrasound (48 kHz) and reported no damage to the skin upon ultrasound application. Mitragotri et al. (36,37) also performed histological studies of hairless rat skin exposed to low-frequency ultrasound (20 kHz) and found no damage to the epidermis and underlying living tissues. Note that although these histological studies indicate no adverse effects of ultrasound, further research focusing on safety issues is required to evaluate limiting ultrasound parameters for safe exposure.

VI. MECHANISM

A possible mechanism of improved percutaneous absorption by ultrasound suggested by several groups (6–8,25,29,35,37) is that ultrasound may interact with the structured lipids located in the intercellular channels of the stratum corneum. This is similar to the postulated effects of some chemical transdermal enhancers that act by disordering lipids (66). The ultrasound energy may act to facilitate diffusion through lipid domains.

Alternatively, hydrophilic and lipophilic layers are suggested to constitute two major pathways for permeation across skin: one transcellular, the other via the tortuous but continuous path of intercellular lipids. Kost et al. (30) found that ultrasound enhanced the permeability through both the lipophilic and hydrophilic routes, suggesting that the effect of ultrasound on the lipoidal configuration is not the only contributor of ultrasound to the enhanced transdermal permeability. Differential scanning calorimetry (DSC) and Fourier transform infrared (FTIR) studies suggested that there were no irreversible morphological changes in the stratum corneum due to the ultrasound exposure.

Levy et al. (29) also performed in vitro studies to gain insight into the mechanism of ultrasonically enhanced transdermal delivery. The authors considered three factors that might contribute to the ultrasound enhanced permeability: mixing, cavitation, and temperature. To examine whether ultrasound might affect a boundary layer in the neighborhood of the skin and therefore cause higher permeabilities, experiments were performed in vitro under controlled mixing rates. The temperature of the skin exposed to ultrasound was monitored. Cavitation effects were evaluated through in vitro permeability experiments in degassed buffer, where cavitation was minimized. The authors concluded that the small increase in surface skin temperature observed after ultrasound application $(1-2°C)$ is not likely to cause dramatic changes in skin permeability. The ultrasound enhancing phenomenon was mainly attributed to mixing and cavitation effects.

Mortimer et al. (43) showed that ultrasound exposure led to an increase in the rate of oxygen diffusion through frog skin. The authors found that the oxygen permeability increase is dependent upon the ultrasound average intensity but does not depend on the peak acoustic pressure. Based on this finding they concluded that it is not likely that cavitation is the dominant mode of action, since cavitation mechanisms are a function of acoustic pressure rather than average intensity. In addition, transient cavitation was not observed through the measurement of OH radicals. Since diffusion increased with increasing average intensity, the most likely mechanism proposed by the authors is acoustic streaming (quartz wind), leading to stirring action in the vicinity of the membrane, affecting the boundary layer (reducing the concentration gradient in the immediate neighborhood of the membrane (42).

Tachibana and Tachibana (50,51) postulated that the energy of ultrasonic vibration enhanced transdermal permeability through the transfollicular and transepidermal routes. The microscopic bubbles (cavitation) produced at the surface of the skin by ultrasonic vibration might generate a rapid liquid flow when they implode, thereby increasing skin permeability.

Simonin suggested the sweat ducts as the main hydrophilic molecules pathways affected by sonophoresis. The author concluded that there is not enough space to allow for transient cavitation in the inter- or intracellular routes, so cavitation is unlikely to disrupt the orderly structure of the stratum corneum. In contrast, bubbles generated by ultrasound cavitation in the sweat duct lumen having a nominal diameter of about 5 μm can be formed and grow, possibly attaining their resonant size. Thereby a vigorous mixing may be formed by microstreaming and bubble collapse, resulting in enhanced transport (67).

Bommannan et al. (7) examined the effects of ultrasound on the transdermal permeation of the electron-dense tracer lanthanum nitrate. The results demonstrate that exposure of the skin to ultrasound can induce considerable and rapid tracer transport through an intercellular route. Prolonged exposure of the skin to high-frequency ultrasound (20 min, 16 MHz), however, resulted in structural alterations of epidermal morphology.

Mitragotri et al. (35) evaluated the role played by various ultrasound-related phenomena, including cavitation, thermal effects, generation of convective velocities, and mechanical effects. The authors' experimental findings suggest that among all the ultrasound-related phenomena evaluated, cavitation plays the dominant role in sonophoresis using therapeutic ultrasound (frequency $1-3$ MHz, intensity $0-2$ W/cm^2). Confocal microscopy results indicate that cavitation occurs in the keratinocytes of the stratum corneum upon ultrasound exposure. The authors hypothesized that oscillations of the cavitation bubbles induce disorder in the stratum corneum lipid bilayers, thereby enhancing transdermal transport. The theoretical model developed to describe the effect of ultrasound on transdermal transport predicts that the sonophoretic enhancement depends most directly on the passive permeant diffusion coefficient and not on the permeant diffusion coefficient through the skin.

Significant attention has also been given to understanding the mechanisms of ultrasonic transport enhancement across polymeric systems. Although it is difficult to be certain whether the data obtained in the in vitro experiments with synthetic polymeric membranes are extendible to the in vitro situation with skin, it is likely that since both involve diffusion through membranes, those factors that ultrasound affects in vitro with synthetic polymeric membranes also play a role in vivo.

Three possible mechanisms were proposed by Howkins (21) by which ultrasound could influence the rate of permeability through a polymeric membrane: a) the direct heating of the membrane; b) the sinusoidal pressure variations across the membrane producing some rectification of flow and thus small net dissolved permeant; and c) the reduction of the effect of weak forces between the membranes and the molecules diffusing through it. However, the results indicated that none of the suggested mechanisms is feasible, and the authors concluded that the major effect was due to stirring of fluid layers next to the membranes.

Fogler and Lund (14) proposed that the enhancement of mass transport by ultrasound was due to ultrasonically induced convective transport created by acoustic streaming in addition to diffusional transport. Acoustic streaming is a secondary flow that produces time-independent vortices when an acoustic wave is passed through the medium (60). The formation of these vortices or cells inside ducts, tubes, and pores can increase the rate of mass transfer through these enclosures. Between adjacent cells, molecular diffusion is the only means of mass transport; however, within each cell, transport is primarily by convection. A differential mass transport equation was coupled with the second-order time-independent streaming equation in a rectangular membrane duct and was solved by finite-difference techniques. The acoustic streaming strongly modifies the concentration field, which would be present when only diffusional mass transfer takes place. An analytical solution of the proposed model showed that with the application of ultrasound an increase of up to 150% above the normal diffusive transport could be obtained.

Kost et al. (68) showed that application of ultrasound enhances protein blotting. Three minutes of ultrasound exposure (1 MHz, 2.5 W/cm^2) was sufficient for a very clear transfer of proteins from a polyacrylamide gel to nitrocellulose or nylon 66 membrane. The authors proposed the enhancement of mass transport by ultrasound in the polymeric synthetic membranes is due to ultrasonically induced convective transport created by acoustic streaming.

Lenart and Auslander (28) found that ultrasound enhances the diffusion of electrolytes through cellophane membranes. They proposed the mechanism to be diminution of the hydration sphere surrounding the electrolytes, thus increasing the electrolyte mobility and diffusion coefficient. The authors also proposed a local temperature effect due to the implosion of cavitation bubbles to be a possible mechanism.

Julian and Zentner (22) systematically investigated the effect of ultrasound on solute permeability through polymer films. In these studies the known parameters of permeation were controlled. Diffusivity of benzoic acid in polydimethylsiloxane films and hydrocortisone in cellulose films was increased 14% and 23%, respectively, with 23 W ultrasound. The increase in permeability was unique to the ultrasonic perturbation and was not attributed to disruption of stagnant aqueous diffusion layers, increased membrane/solution temperature, or irreversible changes in membrane integrity. The authors suggested that ultrasonically enhanced diffusion is a result of a decrease of the activation energy necessary to overcome the potential energy barriers within the solution-membrane interfaces.

Kost et al. (68,69) suggested the feasibility of ultrasonic controlled implantable polymeric delivery systems in which the release rates of substances can be repeatedly modulated at will from a position external to the delivery system. Both bioerodible and nonerodible polymers were found to be responsive to the ultrasound; enhanced polymer erosion and drug release were noted when the delivery systems were exposed to ultrasound. It was proposed (70,71) that cavitation and acoustic streaming are responsible for the enhanced polymer degradation and drug release.

In summary, cavitation has been shown to play a major role in ultrasonic enhancement of drug diffusion across the skin as well as polymeric membranes. Further studies investigating the nature of cavitation and its effects on the skin should be performed to advance our knowledge of the mechanisms of sonophoresis.

VII. CONCLUSIONS

The possibility of using ultrasound to mediate and enhance transdermal delivery of diverse substances of wide-ranging molecular size and chemical composition has appealing therapeutic and commercial aspects. As most of the reported results applied ready-made ultrasonic units that were not designed for this specific application, there is no doubt that specifically designed units will enable higher transdermal permeability mediation, which will lead to the preparation of transdermal delivery patches linked to miniature power sources that can be externally adjusted for a wide range of clinical applications. The recent results on the enhancing effect of low-frequency ultrasound on protein permeability suggest the possible future application of this approach for needle-free injections of proteins such as insulin, which is of very high interest. Such efforts at developing miniature and relatively inexpensive power sources will also be important for patient use and convenience.

REFERENCES

1. Kost, J., Levy, D., Langer, R., Ultrasound effect on transdermal drug delivery. Proc. Int. Control. Rel. Bioact. Mater. 13, Control. Rel. Soc. 1986, 177–178.
2. Elias, J. J., Microscopic structure of the epidermis and its derivative. In Percutaneous Absorption: Mechanisms-Methodology-Drug Delivery, Bronaugh, R. L., Maibach, H. I., Eds. 1989, Marcel Dekker, New York, pp. 3–12.
3. Bronaugh, R. L., Determination of percutaneous absorption by in vitro techniques. In Percutaneous Absorption: Mechanisms-Methodology-Drug Delivery, Bronaugh, R. L., Maibach, H. I., Eds. 1989, Marcel Dekker, New York.
4. Benson, H. A. E., McElnay, J. C., Harland, R. Phonophoresis of lingocaine and prilocaine from Emla cream. Int. J. Pharm. 1988, 44:65–69.
5. Benson, H. A. E., McElnay, J. C., Harland, R., Use of ultrasound to enhance percutaneous absorption of benzydamine. Phys. Ther. 1989, 69(2):113–118.
6. Benson, H. A. E., McElnay, J. C., Hadgraft, J., Influence of ultrasound on the percutaneous absorption of nicotinate esters. Pharm. Res. 1991, 9:1279–1283.
7. Bommannan, D., Menon, G. K., Okuyama, H., Elias, P. M., Guy, R. H., Sonophoresis. II. Examination of the mechanism(s) of ultrasound-enhanced transdermal drug delivery. Pharm. Res. 1992, 9(8):1043–1047.
8. Bommannan, D., Okuyama, H., Stauffer, P., Guy, R. H., Sonophoresis. I. The use of high-frequency ultrasound to enhance transdermal drug delivery. Pharm. Res. 1992, 9(4):559–564.
9. Cameroy, B. M. Ultrasound enhanced local anesthesia. Am J. Orthoped. 1966, 8:47.
10. Byl, N. N., McKenzie, A., Halliday, B., Wong, T., O'Connell, J., The Effects of phonophoresis with corticosteroids: A controlled pilot study. J. Orthoped. Sports Phys. Ther. 1993, 18(5):590–600.
11. Ciccone, C. D., Leggin, B. Q., Callamaro, J. J., Effects of ultrasound and trolamine salicylate phonophoresis on delayed-onset muscle soreness. Phys. Ther. 1991, 71(9):666–678.

12. Davick, J. P., Martin, R. K., Albright, J. P., Distribution and deposition of tritiated cortisol using phonophoresis. Phys. Ther. 1988, 68(11):1672–1675.

13. Fellinger, K., Schmidt, J., Klinik and Therapies des Chromischen Gelenkreumatismus. Maudrich Vienna, Austria, 1954, pp. 549–552.

14. Fogler, S., Lund, K., Acoustically augmented diffusional transport. J. Acoust. Soc. Am. 1973, 53(1):59–64.

15. Griffin, J. E., Touchstone, J., Ultrasonic movement of cortisol into pig tissue. Am. J. Phys. Med. 44(1):20–25.

16. Griffin, J. E., Physiological effects of ultrasonic energy as it is used clinically. J. Am. Phys. Ther. Assoc. 1966, 46:18–26.

17. Griffin, J. E., Echternach, J. L., Proce, R. E., Touchstone, J. C., Patients treated with ultrasonic driven hydrocortisone and with ultrasound alone. Phys. Ther. 1967, 47(7):600–601.

18. Griffin, J. E., Touchstone, J. C., Low-intensity phonophoresis of cortisol in swine. Phys. Ther. 1968, 48(12):1136–1344.

19. Griffin, J. E., Touchstone, J. C., Effects of ultrasonic frequency on phonophoresis of cortisol into swine tissues. Am. J. Phys. Med. 1972, 51(2):62–78.

20. Hofman, D., Moll, F., The effect of ultrasound on in vitro liberation and in vivo penetration of benzyl nicotinate. J. Control. Rel. 1993, 27:187–192.

21. Howkins, S. S., Diffusion rates and the effect of ultrasound. Ultrasonics 1969, 8:129–130.

22. Julian, T. N., Zentner, G., Ultrasonically mediated solute permeation through polymer barriers. J. Pharm. Pharmacol. 1986, 38:871–877.

23. Julian, T. N., Zentner, G. M., Mechanism for ultrasonically enhanced transmembrane solute permeation. J. Control. Rel. 1990, 12:77–85.

24. Kleinkort, J. A., Wood, F., Phonophoresis with 1 percent versus 10 percent hydrocortisone. Phys. Ther. 1975, 55(12):1320–1324.

25. Kost, J., Levy, D., Langer, R., Ultrasound as a transdermal enhancer. In Percutaneous Absorption: Mechanisms-Methodology-Drug Delivery, Bronaugh, R. L., Maibach, H. I., Eds. 1989, Marcel Dekker, New York, p. 595–601.

26. Kost, J., Langer, R., Ultrasound-mediated transdermal drug delivery. In Topical Drug Bioavailability, Bioequivalence, and Penetration, Maibach, H. I. Shah, V. P., Eds. 1993, Plenum, New York, pp. 91–103.

27. Kost, J., Pliqueet, U., Mitragotri, S., Yamamoto, A., Weaver, J., Langer, R., Enhanced transdermal delivery: Synergistic effect of ultrasound and electroporation. Pharm. Res. 1996, 13:4,633–638.

28. Lenart, I., Auslander, D. The effects of ultrasound on diffusion through membranes. Ultrasonics 1980, September: 216–217.

29. Levy, D., Kost, J., Meshulam, Y., Langer, R., Effect of ultrasound on transdermal drug delivery to rats and guinea pigs. J. Clin. Invest. 1989, 83:2074–2078.

30. Machluf, M., Kost, J., Ultrasonically enhanced transdermal drug delivery. Experimental approaches to elucidate the mechanism. J. Biomater. Sci. 1993, 5: 146–156.

31. McElnay, J. C., Matthews, M. P., Harland, R., McCafferty, D. F., The effect of ultrasound on the percutaneous absorption of lingocaine. Br. J. Clin. Pharmacol. 1985, 20:421–424.

32. McElany, J. C., Kennedy, T. A., Harland, R., The influence of ultrasound on the percutaneous absorption of fluocinolone acetonide. Int. J. Pharm. 1987, 40: 105–110.

33. McEnlay, J. C., Benson, H. A. E., Harland, R., Hadgraft, J., Phonophoresis of methyl nicotinate: A preliminary study to elucidate the mechanism. Pharm. Res. 1993, 4:1726–1731.

34. Menon, G., Bommanon, D., Elias, P., High-frequency sonophoresis: Permeation pathways and structural basis for enhanced permeability. Skin Pharmacol. 1994, 7(3):130–139.

35. Mitragotri, S., Edwards, D., Blankschtein, D., Langer, R., A mechanistic study of ultrasonically enhanced transdermal drug delivery. J. Pharm. Sci. 1995, 84(6):697–706.

36. Mitragotri, S., Blankschtein, D., Langer, R., Ultrasound-mediated transdermal protein delivery. Science 1995, 269:850–853.

37. Mitragotri, S., Blankschtein, D., Langer, R., Transdermal drug delivery using low-frequency sonophoresis. Pharm. Res. 1996, 13:3,411–420.

38. Mitragotri, S., Blankschtein, D., Langer, R., An explanation for the variation of the sonophoretic transdermal transport enhancement from drug to drug. J. Pharm. Sci. 1997, 86:1190–1191.

39. Mitragotri, S., Blankschtein, D., Langer, R., Sonophoresis: Enhanced transdermal drug delivery by application of ultrasound. In Encyclopedia of Pharmaceutical Technique, Swarbrick, J., Boylan, J., Eds., 1996, Marcel Dekker, New York, pp. 103–122.

40. Miyazaki, S., Mizuoka, O., Takada, M., External control of drug release and penetration: Enhancement of the transdermal absorption of indomethacin by ultrasound irradiation. J. Pharm. Pharmacol., 1990, 43:115–116.

41. Moll, M. A., New approaches to pain. US Armed Forces Med. Serv. Dig., 1979, 30:8–11.

42. Mortimer, A. J., Maclean, J. A., A dosimeter for ultrasonic cavitation. J. Ultrasound Med. 1986, 5(suppl.):137.

43. Mortimer, A. J., Trollope, B. J., Roy, O. Z., Ultrasound-enhanced diffusion through isolated frog skin. Ultrasonics 1988, 26:348–351.

44. Novak, E. J., Experimental transmission of lidocaine through intact skin by ultrasound. Arch. Phys. Med. Rehab. 1964, May:231–232.

45. Newman, J. T., Nellermo, M. D., Crnett, J. L., "Hydrocortisone phonophoresis: A literature review. J. Am. Podiatr. Med. Assoc. 1992, 82(8):432–435.

46. Oziomek, R. S., Perrin, D. H., Herold, D. A., Denegar, C. R., Effect of phonophoresis on serum salicylate levels. Med. Sci. Sports Exercise. 1990, 23(4): 397–401.

47. Pottenger, J. F., Karalfa, L. B., Utilization of hydrocortisone phonophoresis in United States Army physical therapy clinics. Milit. Med. 1989, 154(7):355–358.

48. Quillen, W. S., Phonophoresis: A review of the literaure and technique. Athelet. Train., 1980, 15:109–110.

49. Skauen, D. M., Zentner, G. M., Phonophoresis. Int. J. Pharm. 1984, 20:235–245.

50. Tachibana, K., Tachibana, S., Transdermal delivery of insulin by ultrasonic vibration. J. Pharm. Pharmacol. 1991, 43:270–271.

51. Tachibana, K., Transdermal delivery of insulin to alloxan-diabetic rabbits by ultrasound exposure. Pharm. Res. 1992, 9(7):952–954.

52. Tachibana, K., Tachibana, S., Use of ultrasound to enhance the local anesthetic effect of topically applied aqueous lidocaine. Anesthesiology 1993, 78(6): 1091–1096.

53. Williams, A. R., Phonophoresis: An in vivo evaluation using three topical anaesthetic preparations. Ultrasonics 1990, 28:137–141.

54. Wing, M., Phonophoresis with hydrocortisone in the treatment of temporomandibular joint dysfunction. Phys. Ther. 1981, 62(1):32–33.

55. Hoogland, R., Ultrasound Therapy, 1986, Ernaf Nonius, Delft.

56. Suslick, K. S., Ultrasound: Its Chemical, Physical and Biological Effects, 1989, VCH, New York.

57. Walmsley, A. D., Applications of ultrasound in dentistry. Ultrasound Med. Biol. 1988, 14(1):7–14.

58. Walmsley, A. D., Potential hazards of the dental ultrasonic descaler. Ultrasound Med. Biol. 1988, 14(1):15–20.

59. Hueter, T. F., Bolt, R. H., Sonics: Techniques for the Use of Sound and Ultrasound in Engineering and Science. 1962, John Wiley & Sons, New York.

60. Wells, P. N. T., Biomedical Applications of Ultrasound, 1977, Plenum Press, New York.

61. Gaertner, W., Frequency dependence of acoustic cavitation. J. Acoust. Soc. Am. 1954, 26:977–980.

62. Crum, L. A., Folwlkes, J. B., Acoustic cavitation generated by microsecond pulses of ultrasound. Nature 1986, 52:319.

63. Kost, J., Ultrasound for controlled delivery of therapeutics. Clin. Mater. 1993, 13:155–161.

64. Coodley, G. L., Bursitis and post-traumatic lesions. Am. Pract. 1960, 11:181–187.

65. Antich, T. J., Phonophoresis: The principles of the ultrasonic driving force and efficacy in treatment of common orthopedic diagnoses. J. Orthoped. Sports Phys. Ther. 1982, 4(2):99–102.

66. Johnson, M. E., Mitragotri, S., Patel, A., Blankschtein, D., and Langer, R., Synergistic effect of ultrasound and chemical enhancer on transdermal drug delivery. J. Pharm. Sci. 1996, 85(7):670–679.

67. Simonin, J.-P., On the mechanisms of in vitro and in vivo phonophoresis. J. Control. Rel. 1995, 33:125–141.

68. Kost, J., Liu, L. S., Ferreira, J., Langer, R., Enhanced protein blotting from PhastGel media to membranes by irradiation of low-intensity ultrasound. Anal. Biochem. 1994, 216:27–32.

69. Kost, J., Liu, L. S., Gabelnick, H., Langer, R., Ultrasound as a potential trigger to terminate the activity of contraceptive delivery implants. J. Control. Rel. 1994, 30:77–81.

70. D'Emanuele, A., Kost, J., Hill, J., Langer, R., An investigation of the effect of ultrasound on degradable polyanhydride matrices. Macromolecules 1992, 25: 511–515.

71. Liu, L. S., Kost, J., D'Emanuele, A., Langer, R., Experimental approach to elucidate the mechanism of ultrasound-enhanced polymer erosion and release of incorporated substances. Macromolecules 1992, 25:123–128.

37
Facilitated Transdermal Delivery by Iontophoresis

Parminder Singh and Puchun Liu
Novartis Pharmaceuticals Corporation, Suffern, New York

Steven M. Dinh
Lavipharm Inc., Hightstown, New Jersey

I. INTRODUCTION

Transdermal iontophoresis may be defined as the facilitated movement of ions of soluble salts across the skin under an externally applied potential difference. It was first introduced as early as 1740 to treat arthritis, and general systemic effects of this technique were demonstrated in 1879 (1). The use of iontophoresis briefly gained momentum in the 1930s when iontophoresis was employed for the treatment of hyperhidrosis, a condition characterized by excessive sweating, and for the transport of compounds into the skin (2). In 1959, Gibson and Cooke used iontophoretic application of pilocarpine to induce sweating, and the technique has since been used for the diagnosis of cystic fibrosis (3). The facilitation of ion transport by the process of electrophoresis was also demonstrated by Strohl et al. (4). The use of iontophoresis in the field of dentistry has been extensively evaluated by Gangarosa and coworkers (5). It has also been much studied in the areas of dermatology (6) and has recently been extended to ophthalmology (7) and otolaryngological indications. In addition to local indications, the present approach of iontophoresis research and development for the first time is more focused toward exploiting this technique for systemic delivery of drugs, particularly the effective and controlled systemic delivery of proteins and peptide drugs (8).

II. BASIS FOR IONTOPHORESIS

Surface tissues, including, skin, oral mucosa, tympanic membrane, ocular epithelium, etc., consist of membrane barriers composed predominantly of lipids and proteins. Following passive application, the unionized form of any compound is better absorbed than its ionized counterpart (9). However, the penetration rates of drug molecules across skin or other epithelial surfaces are usually low due to their excellent barrier properties. In addition, many drug candidates exist in their ionized form, thus rendering them virtually ineffective for membrane permeation. The rate of membrane penetration of drugs may be increased by means of an external energy source. The technique of iontophoresis provides an external energy source to drug ions in their passage across the membrane. Thus in order to deliver a negatively charged drug across any epithelial surface, it is placed under the negative electrode, where it is repelled, and is attracted toward a positive electrode placed elsewhere on the body, with the opposite being true for positively charged ions. The transport of neutral or uncharged molecules is also facilitated by the process of electro-osmosis (10–13).

III. BENEFITS OF IONTOPHORESIS

Iontophoresis shares the advantages of transdermal drug delivery, which include bypass of hepatic first-pass effect and gastrointestinal vagaries, controlled plasma levels of potent drugs with short biological half-lives, increased patient compliance, and ease of terminating drug delivery. It also minimizes trauma, risk of infection, and damage to the wound and is an important alternative to the needle. Passive transdermal drug delivery is usually limited to delivery of small, nonpolar, lipophilic solutes. Iontophoresis facilitates the transport of charged and high-molecular-weight compounds, which cannot be normally delivered by passive delivery. Iontophoresis provides quick onset of action, since the lag time is of the order of minutes, as compared to hours in passive diffusion. The inter- and intrasubject variability is considerably reduced for relatively small compounds in iontophoresis, since the rate of drug delivery is proportional to the applied current (14–17). The dosage regimen can be tailored on an individual basis to deliver drug at preprogrammed rates (18). In addition, the proper selection of formulation/electric variables allows plasma profiles to be tailored analogous to bolus injection/infusion therapy but without using needle.

IV. IN VITRO METHODOLOGY

Both the horizontal side-by-side (14,19) and vertical glass diffusion cells (17,20) have been employed in studying in vitro iontophoretic drug transport. In most cases, a system of two electrodes is employed, one in each compartment, and voltage or current measurement is made between the two electrode locations across the skin (21). Some investigators have also used four electrodes in one system (22,23), which decouples the reference electrodes from the working electrodes to provide a more accurate measurement of the voltage drop across the skin surface. The mechanisms and penetration enhancements for the two systems should, however, be similar (21). The skin from different species is sandwiched between the donor and receptor compartments and the diffusion assembly is maintained at 32°C by immersing in water bath. The two compartments (only the receiver in vertical glass cells) are magnetically stirred and their volumes may vary depending on experimental design. Platinum wires or silver/silver chloride electrodes are generally employed. The anode is placed in the donor compartment and the cathode in the receptor compartment during anodal iontophoresis (i.e., delivery of positively charged drug ions), while the electrode orientation is reversed in cathodal iontophoresis. A constant or pulsed current is passed between the two electrodes for a given period of time.

The flux of an analogue of growth hormone releasing factor across hairless guinea pig skin has been observed to be independent of cell design (vertical vs. horizontal) and electrode design (platinum wire was placed in receiver side and served as cathode in both horizontal and vertical assemblies; a platinum wire also served as anode in horizontal cells, whereas in the vertical cell, a platinum plate electrode embedded in Teflon was used as anode) (24). Alterations in the diffusion cell configurations have also been shown to have no significant impact on the transport of benzoate ions across hairless mouse skin (14). The iontophoresis fluxes of lithium and pyridostigmine through human, pig, and rabbit skin have been found to be comparable (17). Burnette and Ongpipattanakul observed comparable iontophoretic fluxes of sodium, chloride, and mannitol across human thigh skin obtained from male and female cadavers and also from areas with high and low density of hair follicles (16). In another study, the results of five successive iontophoretic treatments (each treatment 10 mA, 20 min) of potassium iodide to the same knee of a human volunteer indicated constant uptake of iodine for each application, and only a modest interindividual effect was observed (25). Triplicate treatments using pulsed current (0.3 mA, 40 min each) were found to give comparable skin permeation rates of vasopressin across excised rat skin (26). Insulin delivery through hairless rat skin was comparable to fuzzy rat (27), and only a moderate enhancement of alkanoic acid permeation was observed in nude rat as compared to furry rat following transdermal iontophoresis (19). These studies, though limited, led to the suggestion that iontophoretic delivery may be independent of the type of skin studied, since drug delivery depends on applied current. However, factors such as skin metabolism, nature of the skin (normal, diseased), and dermal clearance can affect drug transport in vivo. The influence of dermal blood flow on transdermal iontophoresis has been evaluated by monitoring lidocaine flux in pigs in the presence of vasoactive chemicals. The vasoconstrictor epinephrine decreased the lidocaine flux due to a vasoconstricting effect, while the vasodilator tolazoline increased the iontophoretic lidocaine flux due to a vasodilatory effect on the dermal vasculature (28,29).

The isolated perfused porcine skin flap (IPPSF) has been proposed as an alternate in vitro animal model for cutaneous absorption and transdermal drug delivery studies (30) and was shown to predict human absorption data for a variety of compounds applied passively to skin (31). More recently, the IPPSF has been used to predict human plasma concentration–time profiles for arbutamine administered by transdermal iontophoresis (32).

V. IN VIVO STUDIES

Blood level–time profiles following iontophoresis have been generated in some cases (27,31,33–36) and compared with other administration routes. The response to iontophoretic administration of insulin has been determined by estimating blood glucose levels in diabetic animal models (26,37). Local tissue concentrations after transdermal iontophoresis of lidocaine (29,38), salicylic acid (38), etidronate (34), and corticosteroids (39) have also been determined. The in vivo human studies have been predominantly of a clinical nature, involving both controlled and uncontrolled trials. The clinical or subjective responses to various iontophoretic treatments have included pain-prick sensation for testing anesthesia, and improved mobility of joints or reduction in pain (as determined by pain score) after iontophoresis of anti-inflammatories.

VI. ELECTRODE CHOICE IN IONTOPHORESIS

The desirable properties of electrodes used in transdermal iontophoresis include good conductive material, shape and form easily adaptable to the contour of skin surface, produce minimal changes in pH, no gaseous by-products, and produce no skin burns. Silver/silver chloride electrodes are most commonly used because they resist the changes in pH that often accompany the use of inert electrodes such as platinum. Silver/silver chloride or so-called reversible electrodes are made from a metal in contact with a solution of its own ions (40). Silver at the anode oxidizes under the applied potential and reacts with chloride to form insoluble silver chloride. At the cathode, the reduction of the silver chloride to silver metal liberates chloride ion. Thus electrochemical reactions do not involve electrolysis of water, and pH changes are eliminated. Such electrodes display the smallest bias potentials and are practically unpolarizable. In order to prevent damage on the Ag/AgCl layer, the electrodes should be cleaned carefully. The electrodes should be washed under flowing water immediately after use and rinsed additionally with distilled water to avoid any calcium deposits from tap water. Dry conditions are recommended for prolonged storage without usage. Continuous use of Ag/AgCl electrodes may result in a black deposit of silver chloride on top of the grey Ag/AgCl layer as a result of chlorination. The electrodes will continue to function normally as long as bias potentials and polarization tendencies do not increase too much. Mechanical removal of deposits is generally not recommended. A 5:1 dilution of dilute ammonium hydroxide with distilled water has been used with success (40). Given the significance of electrodes in iontophoresis with regard to drug delivery and system design, more work needs to be done to modify existing or explore newer forms of electrodes.

The importance of electrode material selection in optimizing drug delivery was demonstrated by Phipps et al. by studying the transport of lithium across polyvinyl alcohol (PVA) hydrogel membrane (17). The use of a platinum anode in the donor compartment caused a pH decrease from 5.9 to 2.6 after 6 h at 1 mA, which was attributed to the oxidation of water to form oxygen gas and hydronium ion. The increased presence of hydronium ions coupled with their faster mobility as compared to lithium ion resulted in a lithium delivery efficiency of $\sim20\%$. The use of silver anode instead of platinum electrode caused virtually no change in pH since the oxidation reaction in this case ($Ag^+ + Cl^- = AgCl + e^-$) produced insoluble silver chloride, which precipitated on the anode surface, and therefore there was no net migration of silver ions into the donor compartment. The resulting lithium delivery efficiency increased significantly to $\sim37\%$ (17). Proper choice of the electrode system is also important in order to avoid false negative results. Dexamethasone sodium phosphate is most effectively delivered from a negative electrode, although it is possible to deliver it by positive electrode by the process of electro-osmosis (41). However, more dexamethasone is delivered per milliampere-minute by iontophoresis than by electro-osmosis, implying that the same amount of the drug can be administered in a shorter time from the negative electrode. It may also be possible to deliver compounds of opposite polarity from the same mixture by reversing the polarity of the electrode (41), although the total time of iontophoresis treatment may increase in the latter case.

VII. IONTOPHORESIS UNITS/DEVICES

Most conventional systems for iontophoresis use a steady direct current in which the active (or reservoir) electrode is often a moist pad or a gauze in direct contact with skin on which a metal plate somewhat smaller in surface area is placed (42). The dispersive (or indifferent) electrode is placed on a convenient body region some distance from the active electrode (43). The position of the dispersive electrode does not strongly affect the efficacy of iontophoresis since the electrophoretic diffusion is mainly determined by the current density in the region of application of the substance (43). Iontophoresis devices are generally designed to deliver small amounts of therapeutically active materials at a constant current for a given period of time. The salient considerations for an iontophoretic device include safety, convenience, reliability, cost, and portability. Three types of iontophoresis units are commercially available: line-operated units, simple battery-operated units, and rechargeable power sources (44). Line-operated devices are used for the iontophoresis of pilocarpine to diagnose cystic fibrosis. Phoresor is a multifunctional battery-operated device (manufactured by Iomed Corp., Salt Lake City, UT that generates a precise dose of direct current irrespective of the changes in skin resistance and has an automatic control for shutoff if the skin/electrode resistance exceeds preset limits (45). Drionic, another battery-operated device (General Medical Company, Los Angeles, CA), has recently been approved by the Food and Drug Administration and is commercially available for the treatment of hyperhidrosis. It is designed for home self-use and has been shown to be clinically safe and effective (44,46). Hidrex (Gesellschaft für Mediani and Technic, Wuppertal, FRG) is a device used for treating palmoplantar hyperhidrosis and can be operated by a rechargeable energy source or by batteries (47). It is also designed for home use and is characterized by safety equipment, automatic timing, and remote control for amperage adjustment.

Two pulsed DC generating iontophoretic delivery systems—the Advance Depolarizing Pulse Iontophoresis System (ADIS-4030) and Transdermal Periodic Iontotherapeutic System (TPIS)—have recently been described. The TIPS is capable of delivering pulsed DC with various combinations of waveform, frequency, on/off ratio, and current density for a specific duration of treatment (26). The ADIS is designed to give continuous drug delivery under constant application of pulse current (48).

VIII. THEORY

By definition, iontophoresis is the increased movement of ions under the applied electric current. The introduction of electric current can also interfere with the morphology of skin, leading to changes in skin permeability and in a drug's physical, chemical, and biological properties. The skin is believed to have an isoelectric point around pH 3–4, so that it carries a net positive charge in contact with a solution of pH lower than 3 and a net negative charge at pH higher than 4 (49,50). The presence of a negative charge on the skin during iontophoresis will affect the electro-osmotic movement of water into the body from anode toward the outer surface of the skin at cathode. This water movement during iontophoresis results in the pore shrinkage at the anode and pore swelling at the cathode (49). Cation transfer may be facilitated during anodal iontophoresis since it occurs in the same direction as water flow.

Due to the complex behavior of an iontophoretic setup, a precise relationship defining the observed rate of iontophoretic delivery has not yet been established. Abramson and Gori derived an equation to correlate the iontophoretic flux to electric mobility, electroosmosis, and simple diffusion (51). According to the Nernst–Planck flux equation, the flux of an ion across a membrane under the influence of an electric field is due to three components: a diffusive component, an iontophoretic component, and an electro-osmotic component.

$$J_{isp} = J_p + J_e + J_c \tag{1}$$

where J_p is the flux due to passive delivery and is given by

$$J_p = K_s D_s \frac{dC}{h_s} \tag{2}$$

J_e is the flux due to electric current facilitation and is given by

$$J_e = \frac{Z_i D_i F}{RT} C_i \frac{dE}{h_s} \tag{3}$$

and J_c is the flux due to convective transport and is given by

$$J_c = k C_s I_d \tag{4}$$

where K_s is the partition coefficient between donor solution and stratum corneum, D_i is the donor concentration of the ionic species i, D_s is the concentration in the skin tissue, dC/h_s is the concentration gradient across the skin, dE/h_s is the electric potential gradient across the skin, D_s is the diffusivity across the skin, D_i is the diffusivity of ionic species i in the skin, I_d is the current density, Z_i is the electric valence of ionic species i, k is the proportionality constant, F is the Faraday constant, T is the absolute temperature, and R is the gas constant (26).

For relatively small ionized solutes, the contribution of the electro-osmotic component to overall iontophoresis-facilitated transport is small and may be neglected (52). In order to calculate the flux of a given ion, Eq. (1) should be written for every ionic species in the system and solved with appropriate boundary conditions. Often the Goldmann constant field approximation is used to facilitate the integration of Eq. (1). With this assumption, Eq. (1) can be integrated to give an enhancement factor E (relative to passive flux), which is given by (52):

$$E = (P_{\text{ionto}}/P_{\text{pass}}) = \frac{-K}{1 - \exp(K)} \qquad (5)$$

where

$$K = \frac{Z_i F \, \Delta E}{RT} \qquad (6)$$

However, deviations from predictions based on Eq. (5) have been noted, particularly at higher voltages (52,53).

While the Nernst–Planck flux equation describes drug flux through a membrane using an applied potential, Faraday's law describes flux in terms of electric current flowing in the circuit. According to Faraday's law, the mass of substance transported in an aqueous solution is proportional to the charge (product of current and time) applied. The migration flux of a compound (J_{migr}) across the skin is given by:

$$J_{\text{migr}} = \frac{t}{zF} I \qquad (7)$$

where z is the charge on the compound, F is Faraday's constant and is equal to 96,500 C/equiv, I is the current density (A/cm^2), and t is the transport number of the compound. In iontophoresis, every ion in the formulation and endogenous ions in the skin will carry a fraction of applied current. The number of interest is the transport number of the active ingredient (t), which may be defined as the fraction of total current carried by the drug and is given by:

$$t = \frac{z^2 uc}{\sum_i z_i^2 u_i c_i} \qquad (8)$$

where u is the ionic mobility, c is the concentration of the ion, and i is all ions in the system. Thus iontophoretic flux of a compound may be predicted if the skin mobility of the drug and other competitive ions is known. It is difficult to estimate the skin mobility of a compound. As a first approximation, free solution mobility of a compound may be used to get a ballpark number of iontophoretic flux of a compound across the skin (54).

IX. PATHWAYS OF ION TRANSPORT

The prominent role of appendageal pores in iontophoretic transport including the sweat ducts and hair follicles has been suggested (55,56). The use of pilocarpine in the diagnosis of cystic fibrosis is in itself a suggestion toward electric current traveling down the sweat ducts. A dot-like pattern over the sweat gland openings following iontophoresis of charged dyes in human skin in vivo has been observed (51). Evidence suggesting that sweat ducts are an important pathway in iontophoretic transport also comes from the work of Papa and Kligman, who observed a direct correlation between methylene blue staining of the skin and the location of sweat ducts (57). The reduction in blood glucose levels following iontophoretic delivery of insulin in hairless and regular rats was found to be not significantly different, suggesting that the drug was being transported mainly through sweat ducts and apocrine and sweat glands (27). Ion transport through sweat duct units using electrodes was shown by Grimnes (55).

Cullander and Guy used a vibrating-probe electrode to identify the largest currents in the vicinity of residual hairs (58). The appendageal pathways for iontophoretic transport have also been shown for mercuric chloride and ferric and ferrous ions by scanning electron microscope (60), and by the observed localization of fluorescein dye at the pores following cathodal iontophoretic transport and by obtaining maximal responses over the pores using microelectrodes (56). A nonappendageal pore pathway has also been suggested (58,60) and attributed to "artificial shunts" as a result of temporary disruption of the organized structure of stratum corneum. A potential-dependent pore formation in the stratum corneum has also been reported and attributed to the flip-flop movements in polypeptide helices in the stratum corneum. The repulsion between neighboring dipoles forms pore openings through which water and charged ions flow and neutralize the dipole moments (61).

The occurrence of paracellular transport for charged and polar solutes has also been demonstrated by confocal microscopy of iontophoretically driven fluorescent ions in the skin (60). Iontophoresis has been shown to significantly increase the penetration of water and glutamic acid across snake skin, which is devoid of shunt routes, thus implicating other pathways for iontophoresis facilitated transport (62). An intercellular pathway has also been suggested by Burnette for ions such as Hg^{2+}, Ni^{2+}, and Na^+ that find their way between keratinized cells (63).

Evidence of the "aqueous" nature of iontophoretic pathways is indicated by the apparent similarities between the absorption and transport kinetics of charged compounds following either transdermal iontophoresis or passive delivery through dermis both in vitro and in vivo, suggesting a similar rate-controlling mechanism (38,64). These "aqueous" pathways may include the skin appendages (hair follicles, sweat glands) and/or transient pores that open and close momentarily and/or hydrated polar regions of the bipolar lipid laminates that exist between the corneocytes (64).

Using alkali metals, it has been demonstrated that the skin is a cation-selective membrane at a physiological pH of 7.4, facilitating the transport of positively charged ions in comparison to negatively charged ions (15,16,65). This negative charge is a result of greater number of protein amino acid residues carrying negative charges (e.g., carboxylic groups) as opposed to positive charges (e.g., amine moieties) (66). The permselectivity of the skin induces a net volume flow during iontophoresis, and it has been demonstrated by Pikal and Shah that this induced volume flow during iontophoresis is in the direction of positive ion transport, supporting the belief in cation selectivity of skin (11–13).

X. FACTORS AFFECTING IONTOPHORETIC TRANSPORT

A. pH

By definition, iontophoresis is the movement of ions under an applied electric field. Thus the optimum pH for iontophoretic delivery is where the compound exists predominantly in ionized form. The effect of pH of the aqueous vehicle on the rate and extent of lidocaine iontophoresis through human stratum corneum was investigated by Siddiqui et al. (67). The rate of penetration was greatest at the pH where lidocaine exists mainly in the ionized form. The importance of pH in the enhancement of solute transport by iontophoresis has also been shown for other solutes (68). The pH changes become particularly significant for protein and peptide drugs, since the pH of the solution not only determines the charge on these molecules but can also affect the chemical stability of these molecules. For example, insulin has been shown to have greater skin permeability at a pH below its isoelectric pH than at a pH at or above its isoelectric point (27). Membrane group ionization induced by pH changes has been implicated to explain anomalous behavior of increased verapamil iontophoretic flux when pH is increased from 4 to 6, although the degree of ionization is fairly constant over pH range 4–6.

B. Current Strength

A linear relationship has been observed between the apparent flux of a number of compounds and the applied current. Figure 1 shows a linear relationship between applied current density and methylphenidate flux across split-thickness human skin (54.) The slope (also termed iontophoretic efficiency) was found to be 700 $\mu g/cm^2$-h. The permeability coefficient of butyrate was also shown to increase linearly when the current density was increased from 0.08 to 1.6 mA/cm^2 (14). A linear increase in steady-state flux with current has also been reported for sodium, chloride, and mannitol (16), lithium (17),

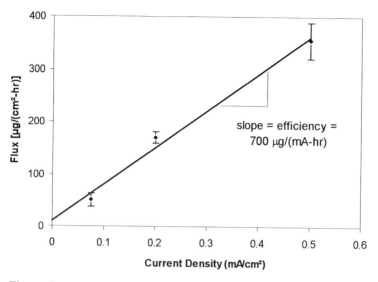

Figure 1 Relationship between methylphenidate skin flux and applied current density at a methylphenidate concentration of 0.5 M ($n = 4$).

tryptophanamide HCl (17), thyrotropin-releasing hormone (15), amphotericin (69), gonadotropin-releasing hormone (70), certain alkanols and alkanoic acids (19), verapamil (23), and diclofenac sodium (71). The pyridostigmine plasma concentration after its iontophoretic administration in pigs in vivo was also shown to be dependent on the magnitude of current (17). The tissue distribution of radioactive phosphorus after iontophoretic application was also proportional to the applied current density (72).

However, the maximum current that can be used is limited by patient safety considerations and the upper limiting value of current has been suggested to be 0.5 mA/cm^2 (73). The maximum tolerable current increases with electrode area, which may help in achieving therapeutic levels of certain drugs. This relationship is, however, curvilinear and may not apply to all solutes (17,74).

C. Ion Competition

The electroneutrality condition requires that there be an equal quantity of negative and positive charge within a given volume. Therefore, migration of a particular ion requires that an ion of the opposite charge be in close proximity. This ion of opposite charge is generally referred to as the *counterion*. An ion of like charge but of a different type is termed a *co-ion*. The pH of the donor medium is often controlled by the addition of buffering agents. The use of buffers adds to the buffer medium co-ions that are usually more mobile than the drug itself. This results in reduction of the fraction of current carried by the drug ion and consequent reduction in transdermal flux of the compound. The drug transport number of a drug in the skin will always be less than unity due to the presence of relatively small and more mobile endogenous ions (e.g., sodium and chloride). The effect of competitive ions on solute transport has been demonstrated by Bellantone et al., who observed reduction in benzoate flux by more than half when an approximately equimolar amount of sodium chloride was added to the donor compartment (14). A detailed influence of changing the ionic strength of the donor solution in iontophoresis has been discussed by Lelawongs et al. (75). Arginine-vasopressin (AVP) has an isoelectric point of 10.9, so that >99% of the drug is protonated in a buffer solution of pH lower than 9. A greater flux of AVP across hairless mouse skin was observed at pH 5.0 as compared to pH 7.4 when ionic strength was a variable, and comparable flux of AVP was observed at both pH 5.0 and 7.4 at a fixed ionic strength. A likely explanation for this phenomenon is the competition between the drug ions and buffer ions for the current applied. Since most of the current will be carried by buffer ions with relatively high mobilities, the actual fraction of the applied current carried by AVP ions would be proportionally reduced as the concentration of buffering agent in the donor solution increases, resulting in lower skin penetration rate of AVP (75). In addition to ionic competition effects, the activity coefficient of the drug can also change with the variation of ionic strength in the donor solution. Within a range of ionic strengths, a linear decline in skin permeation rate of AVP was observed with increase in ionic strength (expressed in terms of its activity coefficient) (75). It is, therefore, imperative to state the experimental conditions in order to make a meaningful comparison of the fluxes or permeability coefficients from different studies, considering a diverse array of buffer systems and varying ionic strengths being employed in different studies. Ideally, the use of buffer system should be avoided in iontophoresis, or alternate buffers consisting of ions with relatively low mobility or conductivity should be preferred.

D. Concentration

The apparent steady-state flux of a number of solutes has been shown to increase with increase in applied concentration. An increase in concentration of butyrate (19), diclofenac (71), and AVP (75) in the donor compartment was found to produce a proportional increase in their fluxes across the skin. An apparent, though modest, linear increase in benzoate and gonadotropin-releasing hormone (leutinizing hormone releasing hormone, LHRH) flux with increasing concentrations of sodium benzoate and LHRH in the donor

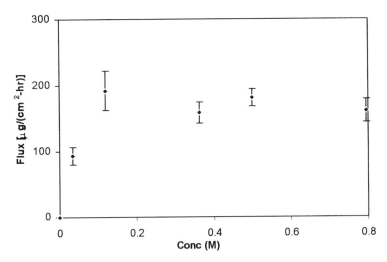

Figure 2 Relationship between applied concentration and iontophoretic skin flux of methylphenidate at an applied current density of 0.2 mA/cm^2 ($n = 4$).

compartment has also been reported (14,70). An increase in tissue levels of phosphorus after iontophoresis was also observed with an increase in phosphorous concentration (72). The steady-state flux of methylphenidate was found to increase with concentration up to 0.1 M. At higher drug concentrations, there was little change in drug flux with increasing drug concentration (Fig. 2) (54), probably due to charge saturation of the aqueous conducting pathways of the skin. Similar behavior has also been shown for lidocaine (76) and sodium (77).

E. Molecular Size

A plot of the iontophoretic fluxes of monovalent anions and cations through excised human skin against molecular weight (M) yielded the following relationships (78,79).

Anions: log iontophoretic flux = log $0.73 - 0.0032M$

Cations: log iontophoretic flux = log $0.87 - 0.0018M$

The difference found in the intercepts and slopes between anions and cations was attributed to the negative fixed charge on the stratum corneum, which facilitates the transport of positively charged cations relative to negatively charged anions. The cathodal iontophoretic data of Green et al. (80,81) yielded a negative slope of 0.0017, while a similar regression on Del Terzo's (alkanoic acids) and Phipps's (for cations) data yielded negative slopes of 0.0062 and 0.0032, respectively (17,19).

A mean slope of 0.0032 was interpreted in terms of free volume model to give an average free volume equivalent to the molecular volume of an ionized solute with a molecular weight of 135 or an equivalent radius of about 0.3 nm (78,79). However, solutes of higher molecular weight, such as insulin, growth hormone, etc., have been shown to penetrate across the skin, suggesting different mechanisms/pathways of transport.

F. Convective or Electro-Osmotic Transport

An electrically driven flow of ions across a membrane having a net charge can induce a coupled flow of solvent, called electro-osmosis (11–13,82). For relatively small compounds, the electro-osmotic contributions to the overall ion transport have been observed to be small compared to the direct electrical field effects (52,68,69,83,84). The penetration of uncharged substances like thymidine (10), and 9β-arabinofuranosyladenine (10), mannitol (16,64,85–87), neutral thyrotropin-releasing hormone (15), and glycine, glucose, tyrosine, carboxyinulin and bovine serum albumin (11,12), and inulin (64) has been shown to be facilitated by the volume flow effect induced by an applied potential difference across the skin. Pikal and Shah measured the current-induced volume flow across hairless mouse skin and observed that it is related to the charge flow through the skin (11–13). The concept of volume flow is supported by the experimental evidence presented by Burnette and Ongpipattanakul. They measured anodic iontophoretic flux of mannitol in the presence and absence of Ca^{2+}. The addition of Ca^{2+} in place of Na^+ in the donor compartment markedly decreased the mannitol flux across excised human skin (16), probably due to binding of Ca^{2+} with tissue's fixed negative binding sites. This electrostatic binding results in a decrease in monovalent anion/monovalent cation mobility ratio (88) and the accompanying volume flow.

G. Continuous Versus Pulsed Current

Use of continuous DC current may result in skin polarization, which can reduce the efficiency of iontophoretic delivery proportional to the length of DC application (89). The build up of this polarizable current can be overcome by using pulsed DC, a direct current that is delivered periodically (18). During the "off time" the skin gets depolarized and returns to near its initial electric condition. The optimum selection of frequency can eliminate any residual polarization from the previous cycle. Enhanced skin depolarization can, however, decrease the efficiency of pulsed transport if the frequency is high (90).

Enhanced iontophoretic transport of peptide and protein drugs using pulsed DC as compared to conventional DC has been reported (26). Two pulsed DC generating iontophoretic delivery systems—the Advance Depolarizing Pulse Iontophoresis System (ADIS-4030) and the Transdermal Periodic Iontotherapeutic System (TPIS)—have recently been described. The TIPS is capable of delivering pulsed DC with various combinations of waveform, frequency, on/off ratio, and current density for a specific duration of treatment. The in vivo transdermal iontophoresis of vasopressin in rabbits was facilitated twofold using pulse current from the TPIS as compared to simple DC generated from the Phoresor system using the same current density (0.22 mA/cm^2) for the same duration of application (40 min). The efficiency of pulsed DC was also demonstrated in diabetic rabbits by the attainment of early and higher plasma levels of insulin and greater reduction in blood glucose levels as compared to simple DC for the same current density and using the same device. Pulse current (1 mA, 40 min) generated by TPIS was also superior to DC generated by the Phoresor system (4 mA, 80 min) in obtaining peak plasma levels of insulin in 30 min, as compared to 1–2 h for the Phoresor system (26). However, blood glucose reduction observed with pulsed-mode iontophoresis (using constant voltage of 1.5 V at 1000 and 2000 kHz) was no better than simple DC-mode iontophoresis (constant voltage of 1 V) in diabetic rats (91). While Liu et al. observed better blood glucose reduction at 2 kHz frequency as compared to 1 kHz frequency (92), no significant difference in the decrease in blood glucose levels was observed in the studies of Haga et al. when the frequency was changed from 1 to 2 kHz (91). Using a pulse current with 20% duty cycle (4 μs), followed by 80% depolarizing period (16 μs) generated by the device ADIS-4030, metoprolol was effectively delivered in 5 human subjects with therapeutically relevant concentrations without any polarization-induced skin damage (48). However, comparable fluxes of Na$^+$ across excised nude mouse skin have been obtained using either pulsed constant current or an equivalent continuous current (90). More systematic studies comparing direct and pulsed current or voltage mode are needed to explain the conflicting results from different studies.

XI. PROGRAMMABLE DRUG DELIVERY

Figure 3 shows the flux of methylphenidate as a function of time when current is increased in a stepwise manner. The "on" portion signifies when the current was turned on, and "off" signifies when the current was switched off. As can be seen, iontophoresis significantly facilitates the transport of protonated methylphenidate relative to passive transport, and the delivery is proportional to applied current. These results suggest that iontophoretic de-

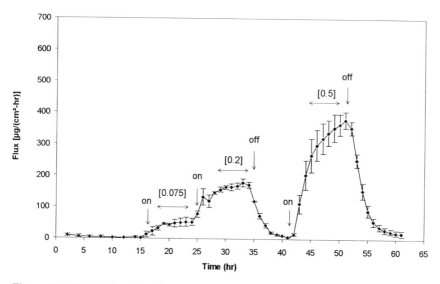

Figure 3 Methylphenidate flux as a function of time at different current densities ($n = 4$). Square brackets indicate current density (mA/cm^2).

livery can be preprogrammed by switching the current on or off and higher doses may be delivered by increasing the current density. This regimen may be particularly useful for compounds such as methylphenidate that require dose adjustments on an individual basis (54).

XII. IN VITRO/IN VIVO CORRELATION

We recently studied systemic disposition of diclofenac sodium in rabbits after transdermal iontophoresis and iv infusion (unpublished results). Figure 4 shows plasma concentrations of diclofenac sodium during 6 h of iv infusion of 1.25 mg diclofenac sodium (0.2 mg/h). An average clearance of 0.90 ± 0.22 L/h ($n = 4$) was calculated from the plasma level data. Figure 5 shows plasma concentrations of diclofenac sodium during 6 h of iontophoresis (first hour was passive) at 0.5 and 0.2 mA/cm^2. The plasma levels of diclofenac increased with applied current density, reached a maximum at 1–2 h, and declined thereafter (probably due to competing chloride ions liberated at the cathode). This decline was less pronounced at lower current densities. Using C_{max} values from the iontophoresis data and clearance values from iv infusion studies, iontophoretic delivery rates were estimated to be 0.074 ± 0.028 mg/(cm^2-h) and 0.027 ± 0.012 mg/(cm^2-h) at 0.5 and 0.2 mA/cm^2, respectively. The efficiency of diclofenac transport in vivo in rab-

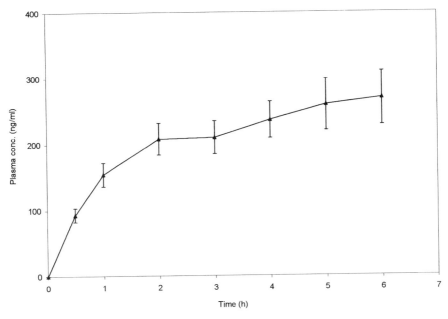

Figure 4 Plasma concentrations of diclofenac during 6-h intravenous infusion of diclofenac sodium at the rate of 0.2 mg/h ($n = 4$).

bits of 0.149 mg/(h-mA) was about twofold higher than estimated from in vitro experiments using hairless mouse skin (unpublished results). Given the limited nature of in vivo study and associated variability and the different skin species used in vitro and in vivo work, the agreement between in vitro and in vivo studies appears reasonable and shows that a reasonable estimate of in vivo iontophoretic delivery rates may be obtained from in vitro experiments. No significant skin irritation in rabbits was observed during iontophoretic treatment.

XIII. DISADVANTAGES OF IONTOPHORESIS

A slight feeling of tingling or itch generally accompanies all iontophoresis procedures (93–97). Transient erythema and local vasodilatation are the other two common side effects associated with the use of iontophoresis (93,96,97). Minor skin burns have also been reported (96). The actual current density across the pores may be higher than the current per unit area applied, depending on the density of pores in a given region. These high current densities increase the possibility of current-induced skin damage (56). Isolated cases of contact sensitization to the drug ketoprofen (98) and to com-

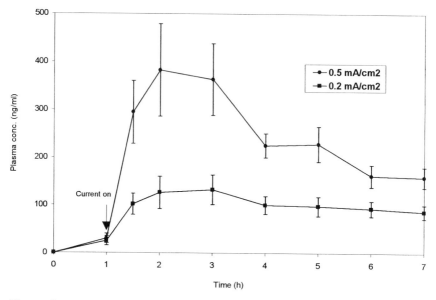

Figure 5 Plasma concentrations of diclofenac during 6-h cathodal iontophoresis
($n = 4$). Donor concentration = 7 mg/ml; pH 7.4.

ponents of electrodes and electrode gels—propylene glycol (99), nickel
(100)—have been reported. Lidocaine iontophoresis in pigs was shown to
induce a reversible dose-dependent non-immune-mediated epidermal alteration of medium duration (59).

Contraindications for iontophoresis include patients with higher susceptibility to applied currents, patients carrying electrically sensitive implanted devices such as cardiac pacemakers, and hypersensitivity to the applied drug. Iontophoresis is not recommended for use over broken or
damaged skin (101). Iontophoretic drug delivery can be costly because of
the electrodes and the number of batteries required to sustain a therapeutic
delivery rate (102).

XIV. FUTURE OF IONTOPHORESIS

With rapid and significant advances in the area of biotechnology resulting
in increased number of peptide and protein pharmaceuticals, iontophoresis
provides a unique opportunity for noninvasive, safe, convenient, effective,
and patient-controlled delivery of such drugs. Given the limitations of the
maximum current that can be tolerated and the diffusional area considerations, the judicious use of iontophoresis is determined by a compound's

physicochemical properties (e.g., size) and its dosage, dosage schedule, and PK/PD profile. The combined use of iontophoresis and penetration enhancers may further facilitate delivery of certain compounds. The challenges in system development lie in designing an easy to use and relatively inexpensive system while maintaining the compound's physical and chemical stability during storage. Disposable or rechargeable batteries could be used based on number of treatments, cost, and environmental considerations. In terms of pharmacoeconomics, a relatively high initial investment in developing such systems will eventually be offset by patient acceptance of the device and applicability to a large number of other compounds for various indications. Since the patient can self-administer the drug and eliminate regular visits to hospitals, physicians, and other health-care outlets, iontophoresis therapy in the long term has the potential to be highly cost-effective—particularly for treatment of chronic ailments such as diabetes, osteoporosis, and cancer, and for pain control and administration of biotechnology-derived drugs. Successful commercialization of the first available system should generate significant interest among the health-care providers and the health-care recipients.

REFERENCES

1. Licht S. History of electrotherapy. In: Therapeutic Electricity and Ultraviolet Radiation. Stillwell, GK, ed. Baltimore: Williams & Wilkins, 1983.
2. Schaefer H, Zzest A, Stuttgen G. Skin Permeability. Berlin: Springer-Verlag, 1982.
3. Gibson LE, Cooke RE. A test for the concentration of electrolytes in sweat in cystic fibrosis of the pancreas utilizing pilocarpine by iontophoresis. Pediatrics 1959; 223:545–549.
4. Strohl A, Verne J, Roucayrol JC, Ceccaldi PF. CR Soc Biol 1950; 144: 819–824.
5. Gangarosa LP. Iontophoresis in Dental Practice. Chicago: Quintessence, 1983.
6. Sloan JB, Soltani K. Iontophoresis in dermatology. J Am Acad Dermatol 1986; 15:671–684.
7. Friedberg ML, Pleyer U, Mondino BJ. Device drug delivery to the eye: Collagen shields, iontophoresis and pumps. Ophthalmology 1991; 98:725–732.
8. Singh P, Maibach HI. Iontophoresis in drug delivery. Basic principles and applications. Crit Rev Ther Drug Carr Sys 1994; 11:161–213.
9. Swarbrick J, Lee G, Brom J, Gensmantel P. Drug permeation through skin. II. Permeability of ionizable compounds. J Pharm Sci 1984; 73:1352–1354.
10. Gangarosa LP, Park NH, Wiggins CA, Hill JM. Increased penetration of nonelectrolytes into hairless mouse skin during iontophoretic water transport (iontohydrokinesis). J Pharmacol Exp Ther 1980; 212:377–381.
11. Pikal MJ, Shah S. Transport mechanisms in iontophoresis. I. A theoretical model for the effect of electroosmotic flow on flux enhancement in transdermal iontophoresis. Pharm Res 1990; 7:118–126.

12. Pikal MJ, Shah S. Transport mechanisms in iontophoresis. II. Electroosmotic flow and transference number measurements for hairless mouse skin. Pharm Res 1990a; 7:213–221.

13. Pikal MJ, Shah S. Transport mechanisms in iontophoresis. III. An experimental study of the contributions of electroosmotic flow and permeability change in transport of low and high molecular weight solutes. Pharm Res 1990b; 7:222–229.

14. Bellantone NH, Rim S, Francoeur ML, Rasadi B. Enhanced percutaneous absorption via iontophoresis I. Evaluation of an in vitro system and transport of model compounds. Int J Pharm 1986; 30:63–72.

15. Burnette RR, Marreo D. Comparison between the iontophoretic and passive transport of thyrotopin releasing hormone across excised nude mouse skin. J Pharm Sci 1986; 75:738–743.

16. Burnette RR, Ongpipattanakul B. Characterization of the permselective properties of excised human skin during iontophoresis. J Pharm Sci 1987; 76: 765–773.

17. Phipps JB, Padmanabhan RV, Lattin GA. Iontophoretic delivery of model inorganic and drug ions. J Pharm Sci 1989; 78:365–369.

18. Banga A, and Chein YW. Iontophoretic delivery of drugs: Fundamentals, developments and biomedical applications. J Control Rel 1988; 7:1–24.

19. DelTerzo S, Behl CR, Nash RA. Iontophoretic transport of a homologous series of ionized and nonionized model compounds: Influence of hydrophobicity and mechanistic interpretation. Pharm Res 1989; 6:89–90.

20. Glikfeld P, Cullander C, Hinz RS, Guy RH. A new system for in vitro studies of iontophoresis. Pharm Res 1988; 5:443–446.

21. Behl CR, Kumar S, Malick W, DelTerzo S, Higuchi WI, Nash RA. Iontophoretic drug delivery: Effects of physicochemical factors on the skin uptake of nonpeptide drugs. J Pharm Sci 1989; 78:355–359.

22. Masada T, Higuchi WI, Srinivasan V, Rohr U, Fox J, Behl C, Pons S. Examination of iontophoretic transport of ionic drugs across skin: Baseline studies with the four-electrode system. Int J Pharm. 1989; 49:57–62.

23. Wearly LL, Liu JC, Chein YW. Iontophoresis facilitated transdermal delivery of verapamil. I. In vitro evaluation and mechanistic studies. J Control Rel 1989; 8:237–250.

24. Kumar S, Char H, Patel S, Piemontese D, Iqbal K, Malick AW, Neugroschel E, and Behl CR. Effect of iontophoresis on in vitro skin permeation of an analogue of growth hormone releasing factor in the hairless guinea pig model. J Pharm Sci 1992; 8:635–639.

25. Puttemans FJM, Massart DL, Gilles F, Lievens PC, Jonckeer MH. Iontophoresis: Mechanism of action studied by potentiometry and x-ray fluorescence. Arch Phys Med Rehabil 1982; 63:176–180.

26. Chien YW, Lelawongs P, Siddiqui O, Sun Y, Shi WM. Facilitated transdermal delivery of therapeutic peptides and proteins by iontophoretic delivery devices. J Control Rel 1990; 13:263—278.

27. Siddiqui O, Sun Y, Liu JC, Chien YW. Facilitated transport of insulin. J Pharm Sci 1987; 76:341–345.

28. Riviere JE, Sage B, Williams PL. Effects of vasoactive drugs on transdermal lidocaine iontophoresis. J Pharm Sci 1991; 80:615–620.
29. Riviere JE, Riviere NA, Inman AO. Determinantion of lidocaine concentrations in skin after transdermal iontophoresis: Effects of vasoactive drugs. Pharm Res 1992; 9:211–214.
30. Riviere JE, Bowman KF, Monteiro-Riviere NA, Dix LP, Carver MP. The isolated perfused porcine skin flap (IPPSF) I. A novel in vitro model for percutaneous absorption and cutaneous toxicology studies. Fundam Appl Toxicol 1986; 7:444–453.
31. Riviere JE, Monteiro-Riviere NA. The isolated perfused porcine skin flap as an in vitro model for percutaneous absorption and cutaneous toxicology. Crit Rev Toxicol 1991; 21:329–344.
32. Riviere JE, Williams PL, Hillman RS, Mishky LM. Quantitative prediction of transdermal iontophoretic delivery of arbutamine in humans with the in vitro isolated perfused porcine skin flap. J Pharm Sci 1992; 81:504–507.
33. Heit MC, Williams PL, Friederike LJ, Chang SK, Riviere JE. Transdermal iontophoretic peptide delivery: In vitro and in vivo studies with luteinizing hormone releasing hormone. Pharm Res 1993; 82:240–243.
34. Slough CL, Spinelli MJ, Kasting GB. Transdermal delivery of etidronate (EHDP) in the pig via iontophoresis. J Membr Sci 1988; 35:161–165.
35. Kumar S, Char H, Patel S, Piemontese D, Waseem W, Iqbal K, Neugroschel E, Behl CR. In vivo transdermal iontophoretic delivery of growth hormone releasing factor GRF (1–44) in hairless guinea pig. J Cont Rel 1992; 18:213–220.
36. Meyer BR, Kreis W, Eschbach J, O'Mara V, Rosen S, Sibalis D. Transdermal versus subcutaneous leuprolide: A comparison of acute pharmacodynamic effect. Clin Pharmacol Ther 1990; 48:340–345.
37. Kari B. Control of blood glucose levels in alloxan-diabetic rabbits by iontophoresis of insulin. Diabetes 1986; 35:217–221.
38. Singh P, Roberts MS. Iontophoretic transdermal delivery of salicylic acid and lidocaine to local subcutaneous structures. J Pharm Sci 1993; 82:127–131.
39. Glass JM, Stephen RL, and Jacobsen SC. The quantity and distribution of radiolabeled dexamethasone delivered to tissue by iontophoresis. Int J Dermatol 1980; 19:519–525.
40. Boucsein W. Electrochemical Activity. New York: Plenum Press, 1992.
41. Petelenz TJ, Buttke JA, Bonds C, Lloyd LB, Beck JE, Stephen RL, Jacobsen SC, Rodriguez P. Iontophoresis of dexamethasone: laboratory studies. J Control Rel 1992, 20:55–66.
42. Stillwell GK. Electrical stimulation and iontophoresis. In: Krussen FH, ed. Handbook of Physical Medicine and Rehabilitation. 2nd ed. St. Louis: W. B. Saunders, 1971.
43. Lekas MD. Iontophoresis treatment. Otolaryngol Head Neck Surg 1979; 87:292–298.
44. Tyle P. Iontophoretic devices for drug delivery. Pharm Res 1986; 3:318–326.
45. Phoresor Iontophoretic Drug Delivery System. Package insert, Iomed, Inc., Salt Lake City, UT.

46. Holzle E. Ruzicka T. Treatment of hyperhidrosis by a battery-operated iontophoretic device. Dermatologica 1986; 172:41–47.

47. Holzle E, Alberti N. Long-term efficacy and side effects of tap water iontophoresis of palmoplantar hyperhidrosis-the usefulness of home therapy. Dermatologica 1987; 175:126–135.

48. Okabe K, Yamaguchi H, Kawai Y. New iontophoretic transdermal administration of the betablocker metoprolol. J Control Rel 1986; 4:79–85.

49. Harris R. Iontophoresis. In: Stillwell GK, ed. Therapeutic Electricity and Ultraviolet Radiation. 3rd ed. Baltimore: Williams & Wilkins, 1967:156–178.

50. Rosendal T. Studies on the conducting properties of the human skin to direct current. Acta Physiol Scand 1943; 5:130–151.

51. Abramson HA, Gorin MH. Skin reactions. IX. The electrophoretic demonstration of the patent pores of the living human skin; Its relation to the charge of the skin. J Phys Chem 1940; 44:1094–1102.

52. Srinvasan V, Higuchi WI, Sims SM, Ghanem AH, Behl CR. Transdermal iontophoretic drug delivery: Mechanistic analysis and application to polypeptide delivery. J Pharm Sci 1989; 78:370–375.

53. Kasting GB, Keister JC. Application of electrodiffusion theory for a homogeneous membrane to iontophoretic transport through skin. J Control Rel 1989; 8:195–210.

54. Singh P, Boniello S, Liu P, Dinh S. Iontophoretic transdermal delivery of methylphenidate hydrochloride. Pharm Res 1997; 14(suppl):309–310.

55. Grimnes S. 1984. Pathways of ionic flow through human skin in vivo. Acta Dermato-Venereol (Stockh) 198; 64:93–98.

56. Burnette RR, Ongpipattanakul B. Characterization of the pore transport properties and tissue alteration of excised human skin during iontophoresis. J Pharm Sci 1988; 77:132–137.

57. Papa CM, Kligman AM. Mechanism of eccrine anhidrosis. J Invest Dermatol 1966; 47:1–9.

58. Cullander C, Guy RH. Sites of iontophoretic current flow into the skin: Identification and characterization with the vibrating probe electrode. J Invest Dermatol 1991; 97:55–64.

59. Monteiro-Riviere NA. Altered epidermal morphology secondary to lidocaine iontophoresis: In vivo and in vitro studies in porcine skin. Fundam Appl Toxicol 1990; 15:174–185.

60. Cullander C. What are the pathways of iontophoretic current flow through mammalian skin? Adv Drug Rel Rev 1992; 9:119–135.

61. Jung G, Katz H, Schmitt H, Voges KP, Menestrina G, Boheim G. Conformational requirements for the potential dependent pore formation of the peptide antibiotics alamethicin, suzukacillin and trichotoxin. In: Spach G, ed. Physical Chemistry of Transmembrane Ion Motion. New York: Elsevier, 1983.

62. Millard J, Barry BW. The iontophoresis of water and glutamic acid across full thickness human skin and shed snake skin. J Pharm Pharmacol 1988; 40(suppl):41P.

63. Burnette RR. Iontophoresis. In: Hadgraft J, Guy RH, eds. Transdermal Drug Delivery. New York: Marcel Dekker, 1988, Ch. 11.

64. Singh P, Anliker M, Smith G, Zavortink D, Maibach H. Transdermal ionto-phoresis and solute penetration across excised human skin. J Pharm Sci 1995; 84:1342–1346.

65. Lakshminarayanaiah N. Transport Phenomenon in Membranes. New York: Academic Press, 1969.

66. Stutten G, Spier HW, Schwarz G. Handbuch Der Haut-und Geschlecht-skrankheiten. Berlin: Springer-Verlag, 1981.

67. Siddiqui O, Roberts MS, Polack AE. The effect of iontophoresis and vehicle pH on the in vitro permeation of lignocaine through human stratum corneum. J Pharm Pharmacol 1985; 37:732–735.

68. Siddiqui O, Roberts MS, Polack, AE. Iontophoretic transport of weak elec-trolytes through the excised human stratum corneum. J Pharm Pharmacol 1989; 41:430–432.

69. Roberts MS, Singh J, Yoshida NH, Currie KI. Iontophoretic transport of selected solutes through human epidermis. In: Scott RC, Hadgraft J, Guy RH, eds. Prediction of Percutaneous Absorption. Oxford: IBS, 1990:231–241.

70. Miller LL, Kolaskie CJ, Smith GA, Riviere J. Transdermal iontophoresis of Gonadotropin releasing hormone (LHRH) and two analogues. J Pharm Sci 1990; 79:490–493.

71. Koizumi T, Kakemi M, Katayama K, Inada H, Sudeji K, Kawasaki M. Trans-fer of diclofenac sodium across excised guinea pig skin on high-frequency pulse iontophoresis II. Factors affecting steady-state transport rate. Chem Pharm Bull 1990; 38:1022–1023.

72. O'Malley EP, Oester YT. Influence of some physical chemical factors on iontophoresis using radio isotopes. Arch Phys Med Rehabil 1955; 36:310–316.

73. Abramson HA, Gorin MH. Skin reactions. X Preseasonal treatment of hay fever by electrophoresis of ragweed pollen extracts into the skin: preliminary report. J Allergy 1941; 12:169–175.

74. Sanderson JE, Riel DS, Dixon R. Iontophoretic delivery of nonpeptide drugs: Formulation optimization for maximum skin permeability. J Pharm Sci 1989; 78:361–364.

75. Lelawongs P, Liu JC, Siddiqui O, Chien YW. Transdermal iontophoretic de-livery of arginine-vasopressin (I): Physicochemical considerations. Int J Pharm 1989; 56:13–22.

76. Sage B. Iontophoresis. In: Smith EW, Maibach HI, eds. Percutaneous Pene-tration Enhancers. Boca Raton, FL: CRC Press, 1995:351–368.

77. Yoshida NH, Roberts MS. Role of conductivity in iontophoresis. 2. Anodal iontophoretic transport of phenylethylamine and sodium across excised human skin. J Pharm Sci 1994; 83:344–350.

78. Yoshida NH, Roberts MS. Structure-transport relationships in transdermal ion-tophoresis. Adv Drug Del Rev 1992; 9:239–264.

79. Yoshida NH, Roberts MS. Solute molecular size and transdermal iontopho-resis across excised human skin. J Control Ref 1993; 25:177–195.

80. Green PG, Hinz RS, Cullander C, Yamane G, Guy RH. Iontophoretic delivery of amino acids and amino acid derivatives across the skin in vitro. Pharm Res 1991a; 8:1113–1119.
81. Green PG, Hinz RS, Kim A, Szoka FC, Guy RH. Iontophoretic delivery of a series of tripeptides across the skin in vitro. Pharm Res 1991b; 8: 1121–1127.
82. Aveyard R, Haydon DA. An Introduction to the Principles of Surface Chemistry. New York: Cambridge University Press, 1973.
83. Sims MS, Higuchi WI, Srinivasan V. Skin alteration and convective solvent flow effects during iontophoresis II. Monovalent anion and cation transport across human skin. Pharm Res 1992; 9:1402–1409.
84. Inada H, Endoh M, Katayama K, Kakemi M, Koizumi T. Factors affecting sulfisoxazole transport through excised rat skin during iontophoresis. Chem Pharm Bull 1989; 37:1870–1873.
85. Sims SM, Higuchi WI, Srinivasan V. Skin alterations and convective solvent flow effects during iontophoresis I. Neutral solute transport across human skin. Int J Pharm 1991; 69:109–121.
86. Kim A, Green PG, Rao G, Guy RH. Convective solvent flow across the skin during iontophoresis. Pharm Res 1993; 10:1315–1320.
87. Delgado-Charro MB, Guy RH. Characterization of convective solvent flow during iontophoresis. Pharm Res 1994; 11:929–935.
88. Barry PH, Diamomd JM, Wright EMJ. The mechanism of cation permeation in rabbit gall bladder: Dilution potentials and biionic potentials. J Membr Biol 1971; 4:358–394.
89. Lawler JC, Davis MJ, Griffith E. Electrical characteristics of the skin: The impedance of the surface sheath and deep tissues. J Invest Dermatol 1960; 34:301–308.
90. Bagniefski T, Burnette RR. A comparison of pulsed and continuous current iontophoresis. J Control Rel 1990; 11:113–122.
91. Haga M, Akatani M, Kikuchi J, Ueno Y, Hayashi M. Transdermal iontophoretic delivery of insulin using a photoetched microdevice. J Control Rel 1997; 43:139–149.
92. Liu JC, Sun O, Siddiqui O, Chien YW, Shi WM, Li J. Blood glucose control in diabetic rats by transdermal iontophoretic delivery of insulin. Int J Pharm 1988; 44:197–204.
93. Ledger PW. Skin biological issues in electrically enhanced transdermal delivery. Adv Drug Del Rev 1992; 9:289–307.
94. Kellog DL, Johnson JM, Kosiba, WA. Selective abolition of adrenergic vasoconstrictor responses in skin by local iontophoresis of bretylium. Am J Physiol 1989; 257:H1599.
95. Jarvis CW, Voita DA. Low voltage skin burns. Pediatrics 1971; 48:831–832.
96. Maloney JM. Local anesthesia obtained via iontophoresis as an aid to shave biopsy. Arch Dermatol 1992; 128:331.
97. Branda RM, Singh P, Aspe-Carranza E, Maibach HI, Guy RH. Acute effects of iontophoresis on human skin in vivo; Cutaneous blood flow and trans-epidermal water loss measurements. Eur J Phar Biopharm 1997; 43:133–138.

98. Teyssandier MJ, Briffod P, Ziegler G. Interêt de la dielectrolyse de ketoprofene en eheumatologie et en petite traumatologie. Sci Med 1977; 8:157–162.

99. Zugerman C. Dermatitis from transcutaneous electrical nerve stimulation. J Am Acad Dermatol 1982; 6:936.

100. Fisher AA. Dermatitis associated with transcutaneous electrical nerve stimulation. Cutis 1978; 21:24.

101. Lark MR, Gangarosa LP. Iontophoresis: An effective modality for the treatment of inflammatory disorders of the temporomandibular joint and myofascial pain. J Craniomandibular Pract 1990; 8:108–119.

102. Parasrampuria D, Parasrampuria J. Percutaneous delivery of protein and peptides using iontophoretic techniques. J Clin Pharm Ther 1991; 16:7–17.

38

Percutaneous Penetration as It Relates to the Safety Evaluation of Cosmetic Ingredients

Jeffrey J. Yourick and Robert L. Bronaugh
Food and Drug Administration, Laurel, Maryland

I. INTRODUCTION

Exposure of consumers to cosmetic products mainly occurs via the dermal route. Once a chemical contacts skin, absorption begins. Diffusion into and through the stratum corneum typically is the rate-limiting step in percutaneous absorption. However, the rate-limiting barrier to absorption is dependent upon the specific chemical. For a cosmetic chemical that is applied to skin, the accuracy of the risk assessment can be improved by basing the potential systemic exposure on an estimate of the dermal exposure that has been corrected with skin absorption data (1). Dermal exposure to a cosmetic ingredient is a function of the concentration of chemical contacting skin and the duration of skin contact. Leave-on cosmetic products represent the category of products resulting in large dermal exposures, while rinse-off product use results in dermal exposures that are brief and discontinuous.

Percutaneous absorption of cosmetic ingredients, including fragrances, can represent a major route of ingredient uptake and subsequent systemic exposure. There are many factors that alter the extent of percutaneous absorption of cosmetic ingredients, such as physicochemical properties of the ingredient, hydration state of skin, duration of product contact, vehicle/formulation effects, and area of application, to name a few. To generate data that will be useful in the risk assessment, the absorption study design should attempt to incorporate testing conditions that approximate consumer use conditions.

In this chapter, we discuss how percutaneous absorption data can be used to refine the risk assessment process, especially the exposure estimate, for cosmetic ingredients. We discuss various issues regarding the safety/risk assessment for both noncarcinogenic and carcinogenic cosmetic ingredients.

II. HAZARD IDENTIFICATION

The evaluation of cosmetic safety begins with the identification of a hazard. Hazard identification may be defined as a determination of whether exposure to a cosmetic ingredient or impurity in the ingredient can lead to an increased incidence of an adverse health effect and the relative strength of the evidence for biologic causation. Hazard identification can arise from many different sources. Consumer use of products can result in adverse health effects that are reported to physicians, the Food and Drug Administration (FDA), or the company that produced the product. Animal toxicity testing of chemicals can also raise concerns about the safety of chemicals. Toxicology (short-term and subchronic) and carcinogenesis testing, such as that performed at the National Toxicology Program, is a means to identify potential hazardous chemicals. Reports published in the open literature are another source to identify hazardous chemicals. These types of reports may range from human patch testing of cosmetic chemicals for sensitization or irritation to animal toxicity testing.

During the process of hazard identification it is important to identify toxicity studies that define doses of the chemicals that are toxic and ideally doses at which no toxicity was found. The dosage level at which no adverse effects were observed is referred to as the no-observed-adverse-effects level (NOAEL). It is not usually the case that a human NOAEL for a cosmetic finished product or cosmetic raw ingredient will be available. For most toxicity testing, a NOAEL is derived from an animal study that administered the chemical by one of several possible routes, such as feed, gavage, drinking water, or skin painting. Therefore, to evaluate human safety from animal data requires some method of extrapolation. For cosmetics the extrapolation from animal data to human safety involves many different factors, including the species of animals, body weights, body surface area, quality and length of animal test, and sensitive human populations, just to name a few (2).

III. EXPOSURE ESTIMATE

Once a potential hazard has been identified, the next step in the process of safety evaluation is to estimate human exposure. Generally, exposure to a cosmetic ingredient is via the dermal route. The dermal exposure to a cosmetic ingredient is the amount of that ingredient that is applied to the skin.

However, the systemic exposure to that cosmetic ingredient may be much lower, given the barrier properties of skin. Furthermore, systemic exposure will also be dependent upon the duration of skin contact. For an accurate estimate of systemic exposure for a specific dermally applied chemical it is important also to know the extent of skin absorption. Once the percent of applied dermal dose absorbed is determined experimentally, refinement of the systemic exposure estimate can be made from the applied dermal dose. In the absence of any skin absorption data, it should be assumed that all (i.e., 100%) of the chemical that is applied to skin is absorbed (conservative approach).

A. Dermal Exposure

To estimate dermal exposure to a cosmetic ingredient, it is important to know the use conditions for the specific product(s) containing the ingredient of interest. From the product use instructions, it is possible to determine many factors pertinent to the exposure calculation, such as the approximate frequency of product use, volume/weight of product used, duration of exposure, and other conditions of exposure (e.g., apply heat or cover application area with plastic). An ingredient could be contained in one product intended for leave-on usage (e.g., moisturizing lotion) or in a different product intended as a rinse-off formulation (e.g., hair dye product or shampoo). All these use conditions must be defined and utilized in the exposure estimate.

An important piece of information necessary for the dermal exposure estimate is the concentration of the ingredient in the finished product. However, this product information is proprietary and difficult to obtain. FDA's Cosmetic Technology Branch occasionally conducts surveys that directly measure ingredient concentrations in finished cosmetic products. These data are used in the exposure estimate.

The initial dermal exposure estimate is calculated from as many of the already mentioned exposure conditions as are deemed pertinent to include. When actual data are unavailable for a specific parameter, an estimate of the parameter must be used. However, any estimates used in the calculation should be clearly noted in the exposure estimate summary. This is important since uncertainties in the exposure estimate will affect the final safety assessment. It is desirable to calculate an exposure in units of ingredient weight per kilogram body weight per day. This facilitates comparison of dermal exposure to a NOAEL obtained from a dietary or parental administration toxicology study.

B. Percutaneous Absorption

Percutaneous absorption is measured in fresh, viable human and/or animal skin using in vitro flow-through diffusion cell methodology. These techniques are described in detail in Chapter 12 of this book and by Bronaugh and Collier (3). A variety of receptor fluids may be used, the composition of which will depend upon the specific chemical being tested. The application of chemicals should be made in a manner that approximates consumer use conditions as closely as possible to generate data providing realistic exposure estimates.

Skin absorption is dependent on many factors, including lipid solubility of the chemical, duration of skin contact, location of skin contact, vehicle for the chemical, environmental conditions, occlusion of the dosing area by clothing, surface area of skin application, and the age of the individual. These factors must be considered when attempting to accurately define the exposure estimate.

Skin absorption studies should be conducted under conditions that approximate specific product use/abuse conditions. A well-designed absorption study should take into account the relevant experimental conditions necessary to replicate consumer use conditions. A typical experimental design should consider the dosing vehicle, the dosing concentration, and the duration of exposure, to simulate use conditions. To simulate a leave-on product the dosing solution may be left on the skin for 24 h, whereas for a rinse-off product the dosing solution may be removed after $1-2$ min. The percent of applied dose absorbed is determined experimentally and is used in estimating systemic exposure to the chemical.

IV. SAFETY ASSESSMENTS

A. Noncarcinogenic Cosmetic Ingredient Safety Evaluation: Assume a Threshold for Toxicity

One approach for extrapolating data from animal studies to human hazard/safety is the safety factor approach. The safety factor approach implies that there is a threshold dose for a toxic effect. If the NOAEL is considered as the threshold dose, then the NOAEL is divided by a safety factor (usually 100; 10 for interspecies and 10 for intraspecies variability) to determine a safe human dose or an acceptable daily intake (ADI) (4).

1. Exposure Estimate: Acute and Chronic Exposure

The systemic exposure estimate for a single exposure to a cosmetic ingredient is a function of the amount applied to skin, the concentration of the ingredient in the product, the duration of skin contact, and the extent of percutaneous absorption. If no data are available for these specific exposure conditions, then these parameters must be estimated. The volume of the product used is the typical amount of product applied to the skin. If the volume of the product applied is not available, then it is possible to use a typical application rate, such as milligrams product applied to skin per square centimeter, for specific cosmetic product categories (e.g., a lotion is typically applied at 2 mg/cm^2). The amount of a cosmetic ingredient systemically absorbed can be estimated as follows:

Amount absorbed/kg body weight = (application rate)

$\times$ (concentration of ingredient in product)

$\times$ (duration of exposure) $\times$ (surface area exposed)

$\times$ (% skin absorption)/(weight of individual)

The estimated daily dose is an exposure estimate based on chronic usage of a specific product. The amount absorbed per kilogram body weight is multiplied by the estimated frequency of use throughout a lifetime, divided by the total number of days represented by a lifetime of use:

Estimated daily dose = (amount absorbed/kg body weight)

$\times$ (estimated frequency of use over a lifetime)/(total number of

days represented by a lifetime of use)

2. Safety Assessment:

a. If a NOAEL for the Ingredient Is Available Applicable equations are:

Acceptable daily intake (ADI) = NOAEL/Safety Factor

Margin of safety = NOAEL/Estimated Daily Dose

b. If a NOAEL for the Ingredient Is Not Available It is sometimes the case when a toxicology study is completed that all experimental doses caused an adverse effect such that a NOAEL is not determined, but the lowest experimental dose was identified as the lowest observed adverse effect level (LOAEL). It is possible to divide the LOAEL by an additional safety factor to estimate the NOAEL. This estimated NOAEL can then be used in the calculation of acceptable daily intake and margin of safety as presented earlier.

c. Assessment The safety assessment is reviewed and after a qualitative evaluation of the uncertainties inherent in the exposure estimate, a decision is made by the risk managers as to whether there is a potential safety problem with exposure to a specific cosmetic ingredient or finished product. Furthermore, risk managers will ultimately be responsible for making decisions on any corrective actions that might be necessary.

B. Carcinogenic Cosmetic Ingredient Safety Evaluation

The safety assessment of a cosmetic ingredient suspected to be carcinogenic (determined, e.g., from open literature publications, National Toxicology Program reports, IARC reports, unpublished studies, etc.) in either animals or humans is currently conducted using a different approach. An initial exposure estimate is completed as outlined in the preceding section to derive an estimated daily dose. Since it may not be appropriate to consider a threshold for carcinogenic potential, quantitative risk assessment calculations are used to determine the relative lifetime cancer risk resulting from the potential exposure. A mathematical model is used to perform the linear regression for the low-dose estimate of cancer risk. The procedure for the cancer risk estimation has been outlined (4). Briefly, a direct method (linear-at-low dose approach) for low-dose cancer risk estimation is used. A straight line is drawn from a point on the dose-response curve to the point of origin. The slope of the straight line is used as an index of carcinogenic potency. The upper limit estimate of relative cancer risk (4) is then:

$$\text{Risk} \leq \text{slope} \times \text{dose}$$

C. Risk Management

Risk management deals with identifying and considering the range of regulatory options and then making a decision about which approach to use. Risk managers review the risk assessment and consider all other legal, economic, social, ethical, and political issues that may arise from a risk management decision (5). A set of decision options is formulated to address the specific risk assessment findings. The risk managers must then decide on the nature of any corrective actions required to protect the public health.

V. CASE STUDY—EXPOSURE ESTIMATE FOR THE DERMALLY APPLIED FRAGRANCE MUSK XYLOL

Musk xylol is a fragrance ingredient used in a variety of cosmetic and household products. It accumulates in the environment, and levels have been measured in worldwide bodies of water (6). The fragrance was found to be carcinogenic in a rodent bioassay (7). Mutagenicity testing, both mammalian and bacterial, has indicated that musk xylol is not genotoxic (8). We have reported that 22% of musk xylol was absorbed through human skin after topical application in a cosmetic emulsion vehicle (9). In addition to absorption data, a determination of the daily amount applied to skin is needed to estimate systemic exposure. The amount of musk xylol applied to skin is difficult to determine because the fragrance is found in a number of product categories that might be used simultaneously by consumers.

Musk xylol, as a component of the perfume, is contained in many different types of cosmetic products. Therefore for any determination of the total amount of musk xylol applied to skin, it is necessary to estimate the amount of musk xylol on skin from use of each product type. There are many exposure parameters that need to be defined and estimated prior to calculating musk xylol on skin for each product type. First, it is necessary to estimate the total amount of each product type that will be used daily by consumers. This is represented as "product usage" and is estimated in grams of product applied per day (Table 1). However, only a small percentage of the actual finished cosmetic product is perfume. Furthermore, the perfume component of most finished cosmetic products is composed of many different fragrances. Therefore, the perfume concentration (Perfume Conc. (%), see Table 3) (RIFM survey, unpublished, 1993) and the percentage of the fragrance that is musk xylol (Table 2) can be estimated. The "retention factor" (Table 3) (RIFM survey, unpublished, 1993) adjusts the exposure estimate to musk xylol based on either the volatility of the applied product or whether the product is a wash-off or leave-on type product. The amount of perfume retained (mg/day) (Table 3) is estimated from the perfume concentration, retention factor, and product usage. The musk xylol on skin (μg/

Table 1 Product Usage (g/day) from Cosmetic Toiletry and Fragrance Association Survey[a]

Product type	50th Percentile	90th Percentile
Bath preparations	1.81	9.50
Colognes	0.55	0.93
Perfumes	0.10	0.40
Shampoos, rinses	12.02	29.40
Hair sprays	0.93	2.13
Other hair preparations	5.09	13.04
Soaps	2.47	3.72
Deodorants	0.42	0.67
Cleansing creams	1.04	2.60
Face, body preparations	3.08	5.99
Moisturizers	0.45	0.87
Other skin preparations	2.57	6.13
Suntan preparations	2.00	4.00
Air fresheners	2.50	5.74
Household detergents	61.12	140.31

[a]Survey by the Cosmetic, Toiletry and Fragrance Association (unpublished), 1983.

Table 2 Musk Xylol in Perfumes (%) from Research Institute for
Fragrance Materials Survey[a]

Product Type	50th Percentile	90th Percentile
Bath preparations	0.3	2.5
Colognes	0.1	1.5
Perfumes	0.2	0.5
Shampoos, rinses	1.5	2.0
Hair sprays	1.5	1.7
Other hair preparations	1.5	2.3
Soaps	1.5	3.8
Deodorants	0	1.5
Cleansing creams	0.4	1.5
Face, body preparations	0.4	1.5
Moisturizers	0.4	1.4
Other skin preparations	0.4	1.3
Suntan preparations	0.4	2.5
Air fresheners	0	1.1
Household detergents	0.8	4.1

[a]RIFM (unpublished), 1993.

day) for each product type (Table 3) is then calculated from the perfume
retained and the percentage of perfume attributed to the fragrance, musk
xylol. The musk xylol in perfume (%) (Table 3) was estimated by Monte
Carlo simulation (see following discussion) using the data from Table 2. The
musk xylol on skin (μg/day) (Table 3) is estimated from perfume retained
and musk xylol in perfume. The total amount of musk xylol applied to skin
(mg/day) (Table 4) was estimated by 2000 iterations of the Monte Carlo
simulation (see following discussion). Finally, the amount of musk xylol
available for systemic absorption is estimated from musk xylol on skin (Table 4, 90th percentile) and the percent of musk xylol absorbed through skin
(22%).

A. RIFM Dermal Exposure Estimate

An exposure estimate for musk xylol was prepared by the Research Institute
for Fragrance Materials (RIFM) in 1993 (RIFM, unpublished data, 1993)
using the point estimate approach. Estimate of total daily musk xylol applied
to skin was based on product usage (Table 1) and percent musk xylol (Table
2) in 15 different categories of commercial products (13 cosmetic and 2
household). For the point exposure estimate, RIFM summed the exposures
to all categories of products and established four exposure levels based on

Table 3 Musk Xylol Retained on Skin

Product type	Product usage (g/day)[a]	Perfume conc. (%)[b]	Retention factor[b]	Perfume retained (mg/day)	Musk xylol in perfume (%)[a]	Musk xylol on skin (μg/day)
Bath preparations	1.78	2	0.01	0.36	0.57	2.04
Colognes	0.67	5	0.9	30.1	0.31	92.0
Perfumes	0.088	18	0.9	14.2	0.28	39.9
Shampoos	6.47	0.5	0.2	6.47	0.44	28.3
Hair sprays	0.88	0.15	0.3	0.40	1.22	4.85
Other hair preparations	10.9	0.5	0.2	10.9	0.93	101.2
Soaps	1.81	1.2	0.2	4.33	2.85	123.7
Deodorants	0.47	0.75	1	3.51	0.27	9.44
Cleansing creams	0.87	0.5	1	4.37	0.79	34.4
Face/body preparations	1.90	0.5	1	9.52	0.47	45.1
Moisturizers	0.18	0.5	1	0.91	0.58	5.27
Other skin preparations	4.28	0.5	1	21.4	0.58	123.5
Suntan preparations	1.43	0.4	1	5.72	1.07	61.2
Air fresheners	1.35	1.75	0.01	0.24	0.097	0.23
Household detergents	48.7	0.25	0.001	0.12	2.26	2.75

[a]Monte Carlo simulation estimates from one iteration of the modeling.
[b]Data from RIFM (unpublished), 1993.

Table 4 Total Amount of Musk
Xylol Applied to Skin (mg/day)[a]

Minimum	0.34
Mean	0.84
Maximum	1.5
50th Percentile	0.82
90th Percentile	1.1

[a]Final results from the Monte Carlo simulation (2000 iterations) of musk xylol dermal exposure (compilation of results like those found in Table 3).

the amount of product usage and the musk xylol concentration in the product. RIFM selected, as a more realistic exposure estimate, the 50th percentile level, which was 0.5 mg/day, not the 90th percentile value of 3.9 mg/day. To provide the exposure estimate a margin of safety, RIFM doubled the 0.5-mg/day exposure estimate and used 1.0 mg/day as the estimated daily exposure for its safety assessment.

B. Dermal Exposure Estimates Using Monte Carlo Simulation

A different approach for estimating certain exposure parameters is to estimate product usage and percent musk xylol in perfume using survey data and Monte Carlo simulation techniques. The Chemistry Review Branch of the Office of Premarket Approval at FDA has adapted Monte Carlo simulation techniques to estimate exposures to food additives contained in multiple food sources. We have applied this simulation approach to estimate the total amount of musk xylol applied to skin, since musk xylol is contained in multiple cosmetic product categories and these products might be used simultaneously by consumers. The Monte Carlo simulation approach is a quantitative method that is able to summarize an exposure condition (i.e., product usage or percent musk xylol in perfume for our example) as a probability distribution. This simulation method can generate a distribution of possible outcomes by a number of model iterations, and for each iteration uses a different randomly selected set of input values (e.g., the distribution input values for "Bath Preparations" from the CTFA Survey of Product Usage of Table 1 were 0 at the zero percentile, 1.81 at the 50th percentile, and 9.5 at the 90th percentile). The Monte Carlo simulation results will give the range of probabilities of possible outcomes for the exposure parameter. It is then possible to use the simulation results to examine estimates of probable exposure parameters at any desired exposure level (e.g., 90th percentile estimate of total amount of musk xylol applied to skin) from a number of exposure pathways, that is, cosmetic product categories.

Monte Carlo simulation of musk xylol exposure parameters were conducted using the PC-based software program @RISK (Palisade Corporation, Newfield, NY) within the Excel (Microsoft Corporation, Redmond, WA) spreadsheet program. Monte Carlo sampling techniques with 2000 iterations were used to generate probability distributions of product usage and percent musk xylol in perfume for each cosmetic or household product. Each distribution (TRIANG [min, most likely, max]) was created by input of the minimum (0), most likely (50th percentile), and maximum values (90th percentile) for a) product usage (Table 1) and b) percent musk xylol in perfume (Table 2). The Monte Carlo derived estimates of product usage (column 1, Table 3) and percent musk xylol in perfume (column 5, Table 3) were multiplied by RIFM estimates of perfume concentration and retention of topically applied product on skin to estimate musk xylol on skin (column 6, Table 3). The 90th percentile estimate of total amount of musk xylol applied to skin (1.1 mg/day) (Table 4) was selected as the dermal exposure for use in the systemic exposure estimate.

C. Potential Systemic Exposure Estimate

Potential systemic exposure is the total amount of musk xylol applied to skin multiplied by the fraction of musk xylol absorbed through the skin (from percutaneous absorption study, 0.22) (9). The potential systemic exposure to musk xylol based on the Monte Carlo simulation estimate follows:

$$1.1 \text{ mg/day} \times 0.22 = 0.24 \text{ mg musk xylol/day}$$

D. Conclusion

Determination of ingredient usage in products can be difficult when the ingredient is contained in many different commercial product categories at various concentrations within each product category. A consumer is likely to be exposed to a cosmetic ingredient by using products in several different categories. An individual might use a product from each musk xylol-containing product category over a short time period, but it is probably not realistic to assume an individual would use all categories of products over a lifetime. An overestimate of exposure will occur by assuming that the consumer is simultaneously using products from all product categories. This was the case with the exposure estimate for musk xylol done by RIFM. They assumed that a consumer would use a product daily from each product category that contains musk xylol. RIFM calculated both the 50th and 90th percentile levels of exposure. RIFM chose an intermediate exposure level of 1.0 mg/day (50th percentile value of 0.5 mg/kg/day $\times$ 2) for use in its safety assessment.

A second approach for estimating musk xylol exposure was done by using Monte Carlo simulation techniques to estimate certain exposure parameters. The Monte Carlo simulation method is useful in simulating exposure from a number of pathway sources. This simulation approach can evaluate models where at least one input value (e.g., product usage) can be defined by a distribution of values. Each iteration of the simulation uses a set of input values sampled from the distribution of possible input values to calculate the outcome of the model as a single result. @RISK generates an output probability distribution by consolidating single results from each individual iteration. The advantage of this approach is that an exposure at any desired percentile can be derived from the outcome probability function. It was found by using the Monte Carlo simulation approach that the 90th percentile exposure estimate for musk xylol was 1.1 mg/day.

A comparison of the RIFM and Monte Carlo simulation approaches found that the estimates of daily dermal exposure to musk xylol with the two approaches were quite close (1.0 mg/day for the RIFM estimate vs. 1.1 mg/day for the estimate using Monte Carlo simulation). However, to obtain its estimate, RIFM chose to use the 50th percentile exposure value (0.5 mg/day) multiplied by 2, without providing any theoretical basis for selecting these parameters, rather than using the 90th percentile exposure. The RIFM approach (using a summation of all product categories) estimated a 90th percentile exposure to musk xylol of 3.9 mg/day, while the approach using Monte Carlo simulation estimated a 90th percentile exposure of 1.1 mg/day. This represents a greater than threefold difference in the 90th percentile exposure estimates to musk xylol using the two different approaches. RIFM's assertion that use of the 50th percentile exposure value and a safety factor of 2 provided a more realistic exposure estimate is confirmed by the exposure estimate obtained using Monte Carlo simulation. However, without a theoretical basis to justify its selection of parameters (50th percentile exposure and a safety factor of 2), the extrapolation of the RIFM approach to other exposure estimates is questionable. Monte Carlo simulation provides modeling of the available exposure data for product usage and ingredient concentration, rather than simply summing the exposure to the ingredient across all product categories, which probably leads to an overestimate of exposure. As shown by the musk xylol exposure estimate, use of this technique provides a more realistic estimate of exposure parameters, especially for the higher percentile exposures, than using a single point estimate approach. Therefore, the use of Monte Carlo simulation to estimate certain exposure parameters represents an improvement in our ability to provide realistic estimations of consumer exposure to cosmetic ingredients, especially for ingredients that are contained in multiple product categories.

REFERENCES

1. ECETOC. Percutaneous Absorption. Monograph No. 20. Brussels. 1993.
2. Faustman EM, Omenn GS. Risk assessment. In: Klaassen CD, ed. Casarett & Doull's Toxicology: The Basic Science of Poisons, 5th ed. New York: McGraw-Hill, 1996:75–88.
3. Bronaugh RL, Collier SW. Protocol for *in vitro* percutaneous absorption studies. In: Bronaugh RL, Maibach HI, eds. *In Vitro* Percutaneous Absorption: Principles, Fundamentals, and Applications. Boston: CRC Press, 1991:237–241.
4. Kokoski CJ, Henry SH, Lin CS, Ekelman KB. Methods used in safety evaluation. In: Branen AL, Davidson PM, Salminen S, eds. Food Additives. New York: Marcel Dekker, Inc., 1990:579–615.
5. National Research Council. Understanding Risk: Informing Decisions in a Democratic Society. Stern PC, Fineburg HV, eds. Washington, DC: National Academy Press. 1996:33–34.
6. Rimkus GG, Wolf M. Nitro musk fragrances in biota from freshwater and marine environment. Chemosphere 1995; 30:641–651.
7. Maekawa A, Matsushima Y, Onodera H, Shibutani M, Ogasawara H, Kodama Y, Kurokaea Y, Hayashi Y. Long-term toxicity/carcinogenicity of musk xylol in B6C3F$_1$ mice. Food Chem Toxicol 1990; 8:581–586.
8. Nair J, Ohshima H, Malaaveille C, Friesen M, O'Neill IK, Hautefeuille A, Bartsch H. Identification, occurance and mutagenicity in *Salmonella typhimurium* of two synthetic nitromusks, musk ambrette and musk xylene, in Indian chewing tobacco and betal quid. Chem Toxicol 1986; 24:27–31.
9. Hood HL, Wickett RR, Bronaugh RL. *In vitro* percutaneous absorption of the fragrance ingredient musk xylol. Food Chem Toxicol 1996; 34:483–488.

39
Percutaneous Absorption
of Fragrances

Jeffrey J. Yourick, Harolyn L. Hood,* and Robert L. Bronaugh
Food and Drug Administration, Laurel, Maryland

I. INTRODUCTION

A fragrance is defined as a sweet or delicate odor of flowers or other growing things. A fragrance can also be derived from scents given off by animals, as would be the case in the family of musk-related chemicals. Fragrance chemicals are often extracted directly from plant or animal material, but as specific fragrance chemical structures are identified, fragrance manufacturers are producing synthetic fragrance chemicals. There is an advantage to using synthetic fragrance chemicals because of limitations on the supply of some natural (plant- and animal-derived) fragrances and the lot-to-lot variability in their composition. New synthetic fragrances can be developed by simple changes to the parent molecule. This provides the fragrance chemist with an almost unending supply of novel fragrance ingredients.

Fragrance chemicals are the second largest group of skin allergens and the most frequent cause of cosmetic allergies (1,2). Existing and newly synthesized fragrance chemicals should undergo proper safety testing because of these concerns. Fragrance chemicals commonly found to induce allergic reactions are eugenol, isoeugenol, oak moss, geraniol, cinnamic aldehyde, amyl cinnamaldehyde, hydroxycitronellal, and cinnamic alcohol (3). An extensive study examining the reaction rates of patch-tested patients to 48 fragrances found mixed results, with the top 25 most commonly used fragrances causing few allergic reactions (4). In a fragrance-induced dermal allergic reaction, a fragrance must be absorbed through the skin before it can serve as a hapten that initiates an allergic response.

---

*Present affiliation: Bristol-Myers Squibb Worldwide Beauty Care, Stamford, Connecticut.

673

Children are sensitive to fragrances and can develop allergic reactions easily. The causative chemicals for inducing contact dermatitis in 125 children under the age of 12 years have been investigated (5). These children were patch tested over a period of 7 years. The most frequently identified allergens were metals, fragrances, and rubber chemicals. Forty children were also tested for contact urticaria reactions and 20 children exhibited positive reactions (benzoic acid, 14; cinnamaldehyde, 12). The data suggest that exposure of children to fragrance chemicals poses a real potential for allergic reactions. Hence, percutaneous absorption of fragrances is also a factor to consider in the development of allergic reactions in children.

The safety of fragrance ingredients poses other concerns. For instance, musk ambrette has been found to cause neurotoxic effects and testicular atrophy after dermal application to rats (6,7), and it can also induce photocontact sensitization. The International Fragrance Association has recommended that it not be used in products applied to the skin. The ability of fragrances to induce carcinogenesis in animals is another concern; one specific example is coumarin. In a National Toxicology Program (NTP) carcinogenicity bioassay, coumarin, given by gavage, was found to cause an increase in the incidence of renal tubule adenomas in male rats, an increase in the incidence of alveolar and bronchiolar adenomas in male mice, and an increase in the incidence of alveolar and bronchiolar adenomas and carcinomas and hepatocellular adenomas in female mice (8).

The annual use of fragrance chemicals in the United States is enormous. Estimates on the use of benzyl acetate alone total nearly 1 million kilograms per year (9). Fragrances are used in a variety of consumer products such as cosmetics, perfumes, household soaps and detergents, air fresheners, and as flavorings in foods. Human exposure to these fragrances can be assumed to be mainly via two routes, either dermal or inhalation. This chapter summarizes some of the existing literature regarding the percutaneous absorption of fragrance chemicals and discusses absorption of fragrances from a variety of experimental systems, animal species, and both in vivo and in vitro studies.

II. EXPOSURE CONDITIONS FOR FRAGRANCES

A. Skin

The main route of human exposure to fragrances is via the dermal route for cosmetics and perfumes. Many cosmetic products containing fragrances are applied directly to either skin or hair. Upon application of fragrance-containing products to skin, the fragrances, aided by the higher body temperature, volatilize. Once the fragrance is detected by smell, then we can assume that exposure is occurring via both the dermal and inhalation routes. For the purposes of this discussion we focus on factors that would affect the extent of fragrance absorption.

Dermal exposure to fragrances involves both nonoccluded and occluded conditions. Fragrance-containing products applied directly to skin without the cover of clothing could be considered as nonoccluded applications. A covering of clothing over the application site could represent an absorption scenario under occluded conditions. Occlusion of the application site typically results in a higher extent of absorption when compared with nonoccluded dosing applications (10). Occlusion is thought to enhance skin absorption by either increasing the hydration of the stratum corneum, increasing the surface skin temperature, or simply reducing surface evaporation of the test chemical (see Chapter 4).

The integrity of skin determines the extent of fragrance absorption. If fragrances are applied to damaged skin, it is likely that more absorption will occur. Bronaugh and Stewart (11) observed an increased absorption of chemicals in damaged skin. In Sweden, the most common contact allergen in men is perfume (12). Edman (12) showed that shaving with a razor blade may damage skin, thereby resulting in enhanced fragrance absorption, leading to an increase in the risk of developing a perfume allergy.

B.   Categories of Dermally Applied Fragrance Containing Products

Fragrances are constituents of many different types of cosmetic and perfume products. Fragrances can be found in leave-on products such as lotions, creams, perfumes, colognes, leave-in conditioners, and aftershaves. The duration of skin contact with this type of product maximizes the exposure to fragrance ingredients. Fragrances are also found in wash-off products that include soaps, detergents, shampoos, and hair conditioners, to name a few. Human exposure is minimal to fragrance ingredients in wash-off products.

A typical daily human exposure to fragrance ingredients likely includes exposure to multiple fragrances contained in many different product categories. An example of characterizing an exposure to musk xylol in multiple products is described in Chapter 38. Musk xylol was found to be in 15 different product categories. The estimate of musk xylol exposure is a function of the daily use of each of these 15 products and the concentration of the musk xylol in each product.

C.   Effect of Fragrance Formulation on Absorption

The absorption of a fragrance is dependent upon the vehicle or formulation in which it is contained. The majority of perfumes, colognes, and aftershaves are based on formulations containing a high percentage of ethanol. Ethanol and acetone applied to ex vivo pig skin flaps evaporated rapidly (1–2 min), with more acetone absorbed than ethanol (13). The evaporation of ethanol increases the relative concentration of the fragrance on the skin surface, although the fragrance is also evaporating at the same time. For example, a study of the volatility of safrole found that in 1 min approximately 93% of the safrole evaporated from a waxed paper surface (10).

Many fragrance-containing formulations are emulsions. Although the majority of fragrance ingredients are lipophilic, fragrances tend to dissolve in the lipid phase of emulsions. This may minimize evaporative loss of the fragrance and increase skin absorption. This was found to be the case for the dermal absorption of coumarin. Coumarin absorption was greater from an oil-in-water emulsion when compared to an ethanol vehicle in both human and rat skin (14). Hence, the formulation will have significant effects on the extent of fragrance absorption.

D. Children's Exposure to Fragrance-Containing Products

It is common practice to use enticing fragrances to attract children to certain products. Fragrances are recognized as inducers of allergic reactions in children (5). Any consideration of risk should include exposure to fragrances in this potentially sensitive population. A child's exposure to a product is greater than an adult's, just given the relative relationship between body weight and body surface area. Even the typical wash-off products such as soaps and shampoos may be more of an exposure problem in children, given the higher frequency and longer duration of baths.

E. Use of Skin Absorption Data to Refine Exposure Assessments

The determination of skin absorption for a specific fragrance chemical can influence the subsequent estimation of exposure. When planning the experimental design for an absorption study, the study should be designed with experimental conditions that closely approximate or simulate actual consumer use conditions. This will generate relevant skin absorption data that can be used to estimate realistic exposures to fragrances. This subject is discussed in more detail in Chapter 38. The other issue to consider is the type of skin to be used for the absorption study (see Chapters 11 and 12). The desired choice is to use fresh, viable human skin for the testing.

III. SPECIFIC EXAMPLES OF FRAGRANCE PERCUTANEOUS ABSORPTION

A. Coumarin

Coumarin is widely used as a fragrance in cosmetics, perfumes, and soaps. Coumarin is a natural plant product contained in at least 60 plants. It is frequently associated with the fragrance of sweet clover and new-mown hay (15). An estimated annual usage of coumarin in fragrances in the United States is approximately 250,000 lb (9). An estimated daily human coumarin exposure of approximately 1.2 mg was calculated based solely on annual production figures (14).

Coumarin is rapidly absorbed through both viable split-thickness human and rat skin (14). Absorption of [^{14}C]coumarin from both ethanol and an oil-in-water emulsion was followed over 24 h using flow-through diffusion cells and a receptor fluid of HEPES-buffered Hanks balanced salt solution (HHBSS). The total applied dose absorbed (skin + receptor fluid) of coumarin in rat skin over a 24-h period was 55 and 87% for the ethanol and emulsion vehicles, respectively. The total applied dose absorbed (skin + receptor fluid) of coumarin in human skin over a 24-h period was 64 and 98% for the ethanol and emulsion vehicles, respectively. Coumarin rapidly penetrated both rat and human skin, with the majority of the absorbed dose found in the receptor fluid within the first 6 h. Percutaneous absorption of coumarin was greater from the emulsion vehicle than from the ethanol vehicle. Coumarin metabolism in the skin was also investigated in this study, although no evidence of metabolism was found. Absorption of coumarin in skin was concluded to be significant. Therefore, dermally applied coumarin would be expected to result in a systemic exposure.

Coumarin absorption was also investigated in viable full-thickness human, rat, and mouse skin (16). This study measured [^{14}C]coumarin absorption over a 72-h period from an ethanol vehicle using flow-through diffusion cells with HHBSS as a receptor fluid. Coumarin absorption was rapid; maximum absorption occurred at 4 h in mouse skin, 4–6 h in rat skin, and 8 h in human skin. The extent of coumarin absorption through human, rat, and mouse skin into the receptor fluid was enhanced by occluding the diffusion cells. The amount of coumarin remaining in the skin after 72 h ranged from 8.5 to 13% in unoccluded and occluded rat and human skin. In mouse, there was approximately 10 and 32% remaining in the skin with occlusion and no occlusion, respectively. A linear relationship between the dose of coumarin applied to rat skin and the amount absorbed into the receptor fluid was observed. The metabolism of coumarin upon absorption was also investigated in this study. The lack of coumarin metabolism in skin in this study agrees with the findings of Yourick and Bronaugh (14). Because there is significant absorption of coumarin and no metabolism of coumarin upon absorption, a high systemic coumarin exposure is likely in humans after consumer products containing coumarin are applied to skin. Once coumarin is systemically absorbed, it is postulated that it will circulate to the liver, where extensive biotransformation will occur (15).

B. Musk Xylol

Musk xylol is a nitro musk that is used as a fixative for fragrance ingredients in cosmetic products such as soaps, perfumes, and lotions. Musk xylol is typically used in cosmetic formulations at concentrations of 0.01 to 0.5% (9). The annual use of musk xylol as a fragrance in the United States has been estimated at 150,000 lb (9).

Absorption of [^{14}C]musk xylol in human and hairless guinea pig skin was determined from both a methanol and oil-in-water emulsion vehicle (17). Absorption was measured for 24 h or 7 days using an in vitro flow-through diffusion cell system with a receptor fluid of HHBSS + 4% bovine serum albumin (BSA). In hairless guinea pig skin, total absorption (skin + receptor fluid) of musk xylol was 55% from the emulsion vehicle and 45% from the methanol vehicle. Approximately 20% of the absorbed material was found to remain in the skin after 24 h. In human skin, the total absorption was approximately 22% for both vehicles, with the majority (18 to 21%) of the material found in skin. No metabolism of musk xylol was noted through the course of the 24-h study. In separate experiments a permeability constant, K_p (steady-state conditions), for musk xylol was determined to be 6.86×10^{-5} cm/h in hairless guinea pig skin.

The vast majority of absorbed musk xylol was retained in skin at the end of 24 h. Whether the musk xylol remained in skin or if it diffused slowly out into the receptor fluid over time was unclear; thus human skin studies were extended for 6 additional days (i.e., 7 days total). After 7 days only 6% of the applied dose remained in the skin. This suggests that the musk xylol content of skin is absorbed material that will eventually be systemically absorbed. It was concluded that cosmetic products, such as lotions, that are applied to a large body surface area and remain on the skin for a long duration will result in the greatest absorption of musk xylol.

C. Safrole

Safrole is the principal component of sassafras, and it is a constituent of more than 70 other essential oils. Safrole is used in fragrances at an annual rate of about 50,000 lb (9).

The percutaneous absorption of safrole was measured in vivo in the rhesus monkey using [^{14}C]safrole (10). Absorption of safrole through human skin in vitro was determined by using both static diffusion cells and flow-through diffusion cells with normal saline supplemented with 6% oleth 20 (used to facilitate partitioning of lipophilic chemicals from the skin into the receptor fluid) as the receptor fluid (10). Safrole absorption through monkey skin in vivo over 5 days was about 4% of the applied dose in unoccluded skin and about 13% in occluded skin. Approximately 15 and 38% of the safrole applied dose was absorbed in unoccluded and occluded cells, respectively, in in vitro studies with human skin. Occlusion of the application site resulted in a greater amount of safrole absorbed in both the in vitro and in vivo studies.

There was less absorption of safrole by monkey skin in vivo as compared to its absorption by human skin in vitro. Safrole was found to be relatively volatile, with 93% of it evaporating within 1 min from a waxed paper surface, and it is possible that the high walls of the diffusion cell used in the in vitro human skin study prevented the evaporation of safrole from the human skin. Other possible explanations of this difference in skin absorption might also include differences in barrier properties of human and monkey skin, biological variability of the skin, or the accuracy of measurements made in the in vivo and in vitro studies.

D. Cinnamic Acid

Cinnamic acid is a constituent of basil and Chinese cinnamon essential oils. Its estimated annual usage as a fragrance in the United States is less than 1000 lb (9). The percutaneous absorption of cinnamic acid was measured in vivo in the rhesus monkey using [^{14}C]cinnamic acid (10). Absorption of cinnamic acid through human skin in vitro was determined by using both static diffusion cells and flow-through diffusion cells, with normal saline supplemented with 6% oleth 20 as the receptor fluid (10). Cinnamic acid absorption through monkey skin in vivo over 5 days was about 39% of the applied dose in unoccluded skin and about 84% in occluded skin. Approximately 18 and 61% of the safrole applied dose was absorbed in unoccluded and occluded cells, respectively, in in vitro studies with human skin. Permeation of cinnamic acid occurred rapidly within the first 8 h. Occlusion of the application site resulted in a significantly greater amount of cinnamic acid absorbed in both the in vitro and in vivo studies. Cinnamic acid was found to be relatively volatile, with 40% evaporating within 1 min from a waxed paper surface.

E. Benzyl Acetate

Benzyl acetate is widely used as a fragrance and flavoring agent, with an estimated annual usage in the United States of about 1 million kg (9). Benzyl acetate is a major constituent of jasmine, hyacinth, and gardenia essential oils. The percutaneous absorption of benzyl acetate and five other benzyl derivatives (benzophenone, benzoin, benzamide, benzyl benzoate, and benzyl alcohol) was determined in monkeys in vivo (18). When the application site on the monkey skin was occluded, approximately 70% of the applied dose was absorbed in 24 h for all the chemicals tested. When absorption studies were performed with the application sites unoccluded, the absorption of the fragrances was reduced and there were differences between the relative amounts of chemicals absorbed. It was concluded that the observed differences in absorption for the chemicals could be explained by the rates of evaporation from the application site. These results indicate that a systemic exposure to this family of fragrance chemicals is likely after dermal application.

Skin absorption of benzyl acetate was also investigated in other studies. The percutaneous absorption of neat benzyl acetate was determined in vitro in flow-through diffusion cells by using full-thickness rat skin (19). It was found that under occluded conditions, absorption of benzyl acetate was rapid (1 h), and approximately 50% of the applied dose was absorbed into the receptor fluid after 48 h. The content of benzyl acetate in skin after the 48-h period amounted to nearly 19% of the applied dose. It was also shown that there was a linear relationship between the amount of benzyl acetate applied to skin and the amount found in the receptor fluid at 24 h. A good correlation ($r = 0.993$) was found in rat between benzyl acetate absorption in vivo and in vitro. Various dosing vehicles were tested for their effect on benzyl acetate absorption (under occlusion) in vitro using rat skin. In these studies it was found that, when compared to neat benzyl acetate, there was no increase in absorption from an ethanol vehicle, but there were modest increases in absorption from a phenylethanol or a dimethyl sulfoxide vehicle (20).

A comparison of benzyl acetate absorption in human and rat skin was investigated. Human and rat skin in vitro absorption of neat benzyl acetate were compared by using full-thickness skin mounted in flow-through diffusion cells (21). Absorption of benzyl acetate was rapid in rat skin, with approximately 56% of the applied dose absorbed at 72 h. Absorption of benzyl acetate in human skin was not as rapid, and only 18% of the applied dose was absorbed at 72 h. Absorption of benzyl acetate was greater in rat skin at all time points measured when compared with human skin. These data suggest that absorption is likely to lead to human exposure following dermal application of benzyl acetate-containing products. If only data from rat skin studies were used to estimate human exposure, an overestimate of human exposure would be likely. The absorption data from rat skin could be used as a conservative estimate of human exposure.

F. Cinnamaldehyde and Cinnamyl Alcohol

Cinnamaldehyde is a natural constituent of essential oils of cinnamon leaves, cinnamon bark, myrrh, and hyacinth. Cinnamaldehyde is used in fragrances in the United States at an annual rate of about 100,000 lb (9). Cinnamyl alcohol is a natural component of cinnamon leaves and hyacinth. Its annual usage as a fragrance in the United States is estimated to be 150,000 lb (9). Cinnamaldehyde and cinnamyl alcohol have been reported to cause sensitization reactions in humans (4).

In vitro absorption of cinnamaldehyde and cinnamyl alcohol in human skin was measured in static diffusion cells (22). The K_p for cinnamaldehyde and cinnamic acid was calculated to be 3.8×10^{-5} and 6.9×10^{-5} cm/h, respectively. When cinnamaldehyde was applied to skin, little parent cinnamaldehyde was found in the receptor fluid. It was found that cinnamaldehyde was metabolized in skin to cinnamyl alcohol and cinnamic acid, with their subsequent diffusion into the receptor fluid. Further investigation indicated that metabolism of cinnamaldehyde is facilitated by binding to protein in the skin.

G. Diethyl Maleate

Diethyl maleate is a manufactured fragrance chemical. Its use in the United States has been estimated to be approximately 1000 lb per year (9). Skin absorption of diethyl maleate was determined in vivo in human and monkey skin (18). These studies found that application of diethyl maleate under occlusion resulted in 54% of the applied dose absorbed in human skin in 24 h compared with 69% absorption in monkey skin. As with other previously discussed fragrance chemicals, occlusion of the application site significantly increased the amount of absorbed diethyl maleate in monkey skin. It was concluded that for diethyl maleate, monkey skin serves as a good model for human skin.

IV. CONCLUSIONS

We have presented skin absorption results for various fragrance ingredients. These studies included many different types of skin, thicknesses of skin, experimental approaches (in vivo and in vitro), vehicles, and dosing conditions. The majority of information reviewed suggests that fragrance chemicals are well absorbed. This is caused, in part, by the lipophilic character of most fragrance chemicals. The information also suggests that occlusion of the dosing site enhanced the extent of absorption. This is not surprising, given the relative volatility of fragrance ingredients. Evaporation of the fragrance from the surface of the skin has the effect of decreasing exposure to the fragrance, which in turn decreases its skin absorption. For most cosmetic uses involving fragrance-containing products, the most realistic use situation would be represented by unoccluded site of application conditions. Hence, the net amount of fragrance absorbed will be dependent upon the rate of evaporative loss and the rate of penetration into skin.

Fragrance ingredients are commonly found in a multitude of cosmetics, perfumes, and household products. Many consumers have experienced allergic reactions to certain products, and these reactions have been due in part to the fragrances. In the cosmetics industry certain product lines are formulated and marketed as "fragrance-free," implying to the consumer that these products are safer. Other safety concerns ranging from neurotoxicity to carcinogenicity have been raised with specific fragrance ingredients. When a question of safety arises, it is necessary to know the extent of percutaneous absorption. Determination of skin absorption using experimental conditions that approximate consumer use conditions as closely as possible will provide realistic exposure estimates.

REFERENCES

1. Adams RM, Maibach HI. A 5-year study of cosmetic reactions. J Am Acad Dermatol. 1985; 13:1062–1069.
2. DeGroot AC, Bruynzeel DP, Weyland JW, et al. The allergens in cosmetics. Arch Dermatol. 1988; 124:1525–1529.
3. Johansen JD, Menné T. The fragrance mix and its constituents: a 14-year material. Contact Dermatitis. 1995; 32:1823.
4. Frosch PJ, Pilz B, Andersen KE, Burrows D, Camarasa JG, Dooms-Goossens A, et al. Patch testing with fragrances: Results of a multicenter study of the European environmental and contact dermatitis research group with 48 frequently used constituents of perfumes. Contact Dermatitis. 1995; 33:333–342.
5. Rademaker M, Forsyth A. Contact dermatitis in children. Contact Dermatitis. 1989; 20:104–107.
6. Spencer PS, Bischoff-Fenton MC, Moreno OM, Opdyke DL, Ford RA. Neurotoxic properties of musk ambrette. Tox Appl Pharmacol. 1984; 75:571–575.
7. Ford RA, Api AM, Newberne PM. 90-Day dermal toxicity study and neurotoxicity evaluation of nitromusks in the albino rat. Food Chem Toxicol. 1990; 28:55–61.
8. National Toxicology Program (NTP). Toxicology and Carcinogenicity Studies of Coumarin (CAS No. 91-64-5) in F344/N Rats and B6C3F$_1$ Mice, Technical Report Series No. 422, NIH Publication No. 93-3153. US Department of Health and Human Services, Public Health Service, National Institutes of Health, Research Triangle Park, NC, 1993.
9. Opdyke DLJ. Monographs on Fragrance Raw Materials. Oxford: Pergamon Press, 1979.
10. Bronaugh RL, Stewart RF, Wester RC, Bucks D, Maibach HI. Comparison of percutaneous absorption of fragrances by humans and monkeys. Food Chem Toxicol. 1985; 23:111–114.
11. Bronaugh RL, Stewart RF. Methods for in vitro percutaneous absorption studies. V: Permeation through damaged skin. J Pharm Sci. 1985; 74:1062–1066.

12. Edman B. The influence of shaving method on perfume allergy. Contact Dermatitis. 1994; 31:291–292.

13. Williams PL, Brooks JD, Inman AO, Monteiro-Riviere NA, Riviere JE. Determination of physicochemical properties of phenol, p-nitrophenol, acetone and ethanol relevant to quantitating their percutaneous absorption in porcine skin. Res Commun Chem Pathol Pharmacol. 1994; 83:61–75.

14. Yourick JJ, Bronaugh RL. Percutaneous absorption and metabolism of coumarin in human and rat skin. J Appl Toxicol. 1997; 17:153–158.

15. Cohen AJ. Critical review of the toxicology of coumarin with special reference to interspecies differences in metabolism and hepatotoxic response and their significance to man. Food Cosmet Toxicol. 1979; 17:277–289.

16. Beckley-Kartey SAJ, Hotchkiss SAM, Capel M. Comparative in vitro skin absorption and metabolism of coumarin (1,2-benzopyrone) in human, rat, and mouse. Toxicol Appl Pharmacol. 1997; 145:34–42.

17. Hood HL, Wickett RR, Bronaugh RL. In vitro percutaneous absorption of the fragrance ingredient musk xylol. Food Chem Toxicol. 1996; 34:483–488.

18. Bronaugh RL, Wester RC, Bucks D, Maibach HI, Sarason R. In vivo percutaneous absorption of fragrance ingredients in rhesus monkeys and humans. Food Chem Toxicol. 1990; 28:369–373.

19. Hotchkiss SAM, Chidgey MAJ, Rose S, Caldwell J. Percutaneous absorption of benzyl acetate through rat skin in vitro. 1. Validation of an in vitro model against in vivo data. Food Chem Toxicol. 1990; 28:443–447.

20. Hotchkiss SAM, Miller JM, Caldwell J. Percutaneous absorption of benzyl acetate through rat skin in vitro. 2. Effect of vehicle and occlusion. Food Chem Toxicol. 1992; 30:145–153.

21. Garnett A, Hotchkiss SAM, Caldwell J. Percutaneous absorption of benzyl acetate through rat skin in vitro. 3. A comparison with human skin. Food Chem Toxicol. 1994; 32:1061–1065.

22. Weibel H, Hansen J. Penetration of the fragrance compounds, cinnamaldehyde and cinnamyl alcohol, through human skin in vitro. Contact Dermatitis. 1989; 20:167–172.

40
Hair Dye Absorption

William E. Dressler
Bristol-Myers Squibb Worldwide Beauty Care,
Stamford, Connecticut

I. INTRODUCTION

Because of their chemical characteristics, methods of formulation, application, and usage patterns, hair dyes represent a relatively unique class of cosmetic ingredients with respect to their percutaneous penetration potential. It is therefore important to consider these factors carefully in the design and interpretation of experiments intended to provide reliable data for the purpose of risk assessment. This chapter provides an overview of the chemistry relevant for major classes of hair dyes and discusses available historical data on hair dye absorption measured in human volunteers under actual dyeing conditions. It also summarizes more recent studies using in vitro techniques and discusses their relationship to the prior in vivo data as well as their utility for assessing the relative percutaneous penetration potential of various hair dyes. It also considers the versatility of in vitro studies in isolating and controlling key experimental variables that may be systematically evaluated for their potential to influence hair dye penetration.

II. UNIQUE CHARACTERISTICS OF HAIR DYES AND HAIR DYE EXPOSURE

Some considerations for hair dye percutaneous absorption that, taken together, set them apart from other skin permeants including many other cosmetic materials, are listed in Table 1.

First, oxidative hair coloring, the class most widely used by consumers both at home and in salons, involves chemical complexation of primary dye

Table 1   Hair Dye Absorption—Special Considerations

Permeants have intended affinity for (hair) keratin.
Presence (and abundance) of hair allows for excessive "loading" doses of hair
  dye formulation on scalp.
Hair provides a large surface area and competes with scalp for dyestuff
  absorption.
Substantial differences in follicular density between skin (scalp) and nonhairy
  skin.
Certain permeants react chemically and are consumed and altered during the
  dyeing process.
Alkaline vehicle pH (9–10) intended to soften hair keratin; formulation
  components intended to retard uptake and promote even color distribution.
Substantial shade-related variations in applied doses.
Products are rinsed off after a brief exposure and are used on a discontinuous
  basis.

intermediates with dye couplers, themselves uncolored, to form colored polymeric materials, which subsequently become entrapped within the hair-shaft. Thus, the concentration of potential skin permeant is rapidly and continually reduced during the exposure period. Further, for oxidative hair colors, the initial dye concentration in the product is reduced, usually by one-half, by mixing with hydrogen peroxide (developer) immediately prior to dye application.

The usefulness of hair dyes, either oxidative or direct preformed colors, directly relates to their ability to bind to hair keratin. The amino acid content of hair (and nails) is similar to that of skin, with the exception of the former showing a higher cysteine content at the expense of glycine (1). Additionally, nondestructive Raman spectroscopy has demonstrated a higher stability of the disulfide bond and a higher folding of proteins in hair (and nail) as compared to skin (2). Intended targeting of hair dyes to keratin may serve to limit the mobility of the dyestuff to the uppermost layers of the stratum corneum, thereby diminishing the potential for systemic absorption. Bronaugh and Congdon (3) have noted that prediction of percutaneous absorption may be confounded for compounds such as hair dyes, which are capable of binding to skin.

Skin staining provides obvious evidence of such an effect, and the removal of such stains by moderate scrubbing illustrates the potential for loss of dye by desquamation of the outer layer of the stratum corneum. Conversely, and less obvious, there is the potential for certain dye materials to be leached from the hair over time following repeated washing, resulting in potential skin reexposure during showering or bathing, albeit at very low concentrations.

Dose, an important consideration for any potential skin permeant, also merits special discussion for this class of cosmetic materials. Except in certain instances where dye is selectively applied to the hairshaft as in highlighting procedures, scalp exposure usually is an unavoidable consequence of hair dye use. However, the bulk of the material is used to coat the hair, and only a portion of the applied amount is in actual scalp contact and, hence, available for percutaneous absorption. Typical product use levels of 60 ml (direct dyes) to 100 ml (oxidative dyes after mixing) result in apparent "loading doses" of 100 to 165 mg/cm^2, based on an approximate scalp surface area of 600 cm^2. This is clearly greater than the amount of material that could be realistically applied to nonhairy skin. In practice, hair coloring often involves application of about one-third of the product to the root (and scalp) area, which is left on for 20 to 30 min, followed by application of the balance of the material to the shaft and ends of the hair for the final 5–10 min. Based on surface area considerations using a figure of about 50,000 cm^2 for a medium (4-in) hair length (4), this would result in a scalp exposure fraction of only about 1.4% if the applied material were to distribute evenly on the basis of surface area. These considerations are important, particularly when percentage absorption values are cited based on different in vivo and in vitro exposure scenarios.

In addition to providing a competitive surface area, the presence of hair may modify the reactive chemistry of oxidative hair coloring, changing the species and persistence of potential permeants (5). This may be due to the hair protein or melanin, or to metal ions, either free or bound to protein.

Another important factor for products such as hair color applied to the scalp is the potential for exposure by the transfollicular route. Scheuplein (6) postulated that "shunt" diffusion through follicles and ducts was dominant in the initial transient diffusion stage, particularly for poorly absorbed materials. Currently, some controversy exists regarding the importance of the relatively large surface areas of deeply invaginated follicles to the surface area of the stratum corneum, particularly if the route of exposure still requires entry via the limited surface area provided by the follicular orifice itself (7,8). Further, demonstration of skin permeants within the follicle may reflect only a depot of material that, in fact, has not been absorbed. For certain materials, lipophilicity and vehicle have been shown to influence the relative transappendageal/transepithelial penetration, based on studies conducted using follicle-free scar skin (9). While demonstration of the potential for hair dyes to penetrate via a transfollicular route awaits detailed experiments using appropriate methodology, comparison of in vitro data obtained using nonhairy skin surfaces with actual in vivo scalp use data can at least provide inferential information for dye materials evaluated under the same exposure conditions. In this regard, Meidan et al. (10) have recently proposed the use of low-intensity ultrasound, which discharges sebum from the sebaceous glands and blocks the shunt pathway for hydrophilic molecules, as a probe to elucidate the relative follicular contribution to total transdermal absorption.

An understanding of hair dye formulation characteristics is also important. Unlike therapeutic materials applied to the skin in relatively narrow concentration ranges, there are substantial shade-related variations in dye exposure concentrations. For certain hair dyes, particularly oxidative colors, this may be as large as four orders of magnitude (e.g., 0.003 to 3.0%). Additionally, some dyes may be specific to a particular shade or narrow range of shades, and individual products may be formulated using 2–10 dyes selected from a larger palette of 10–12 materials used to provide the full shade range. Further, hair color products are typically formulated in alkaline vehicles (pH 9–10), which are intended to soften hair keratin to facilitate dye uptake. In some circumstances, it is beneficial to incorporate formulation ingredients that retard the rate of dye uptake, in order to promote even color distribution in between the worn ends of the hair cuticle and newer cuticle near the hair root.

III. HAIR DYE PRODUCT TYPES AND CLASSIFICATION

While hair color product types may be described to the consumer in terms of performance characteristics such as permanent, demipermanent (or long-lasting semipermanent), semipermanent, and temporary, the chemical distinction of oxidative versus direct dyes is more relevant with respect to percutaneous penetration potential.

Oxidative hair dyes undergo chemical reactions within the cortex of the hairshaft and impart a permanent change to the color of the hair, which does not wash out. While the color may fade somewhat over time, regrowth of new hair is required to remove the dye completely. Typically, such products may be applied every 6–8 weeks to recolor the root area and refresh the shade. The use of alkaline peroxide, which bleaches the natural melanin in the cortex of the hair, affords the ability to produce shades lighter than the original color. This is often referred to as "lift." Oxidative hair color technology is also used in demipermanent or "long-lasting" semipermanent products, which generally maintain color through up to 24 shampoos.

In contrast, direct dyes are preformed colors that diffuse into the outer hair cuticle and, depending upon the size of the molecule, may persist through several shampoos (semipermanent) or rinse completely after a single washing (temporary).

Oxidative and direct semipermanent hair colors account, respectively, for about 88–90% and 10–12% of the U.S. market. Temporary hair dyes or gradual colors utilizing lead acetate, which cause progressive darkening of the hair with successive uses, together with natural (henna) colors, account for the small remaining fraction of hair dye use. Only the major classifications of oxidative and direct hair dyes for which some generalizations with respect to percutaneous penetration potential can be made are discussed further in this chapter.

IV. HAIR DYE CHEMISTRY

A. Oxidative Hair Dyes

This hair coloring process involves coupling of amino- and hydroxyaromatic primary intermediates, typically with meta-aminophenols or meta-diaminobenzene couplers, themselves uncolored, to form colored reaction products. This chemical process, which occurs under alkaline (pH 9.5 to 10.5) peroxide conditions, is illustrated in Figs. 1 and 2 for the common primary intermediate p-phenylenediamine (PPD). A reactive imine, which forms initially, combines with various couplers to produce a leuco dye, which undergoes further oxidation to the basic indo dye chromophore. Further coupling to form trinuclear or higher polynuclear dyes can occur if the initial indo dye still contains a reactive site. In the absence of alternate couplers, p-phenylenediamine can self-couple to form a trimer known as Bandrowski's base. Thus, the potential for skin absorption of oxidative hair dyes is influenced not only by the nature and initial concentrations of the primary intermediates and couplers, but also by the rate of formation and nature of the higher molecular weight species formed. The array of colors that may be formed by interactions between various oxidative primary intermediates and

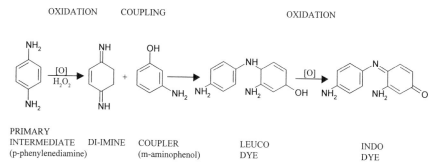

Figure 1   Representative chemical reactions involved with oxidative hair dyes.

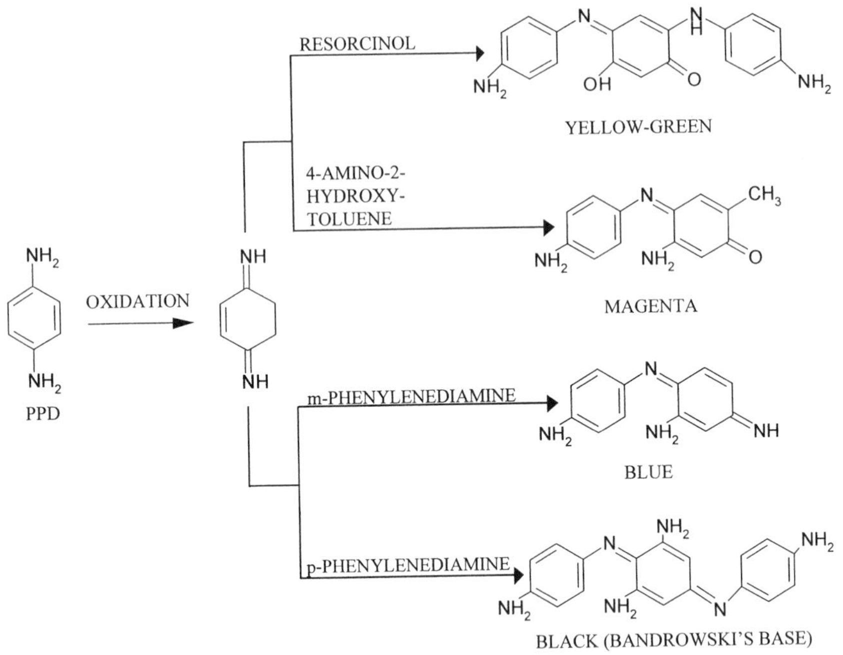

Figure 2   Reaction of *p*-phenylenediamine (PPD) with various couplers to produce colored indo dyes.

couplers is shown in Table 2. Commonly used oxidative dyes are listed in Table 3.

B. Direct Hair Dyes

These preformed colors are often derivatives of nitrobenzene, di- or tri-substituted with amino, hydroxy, and alkoxy groups. Higher molecular weight compounds including nitrodiphenylamines, amino- or hydroxysub-stituted anthraquinones, and, occasionally, azo dyes may be used. Representative chemical structures are shown in Fig. 3, and commonly used direct dyes are listed in Table 3. In order to promote an even color distribution between the worn ends of the hair cuticle and the new cuticle near the hair root, products may be formulated with pairs of different molecular weight dyes of the same basic color (e.g., HC Red No. 1 and HC Red No. 3). Direct hair dyes are slowly removed from the hair cortex by successive shampooing and usually require reapplication after 6–12 shampoos.

Table 2 Colors Formed by Oxidative Dye Couples

Coupler	Primary intermediate		
	PPD	PAP	PTD
m-Phenylenediamine	Blue violet	Violet red	Blue violet
1-Naphthol	Warm violet	Red-orange blonde	Light purple
4-Amino-2-hydroxytoluene	Violet	Red	Warm violet
Resorcinol	Medium yellow-green to brown	Gray yellow blonde	Very light brown
2-Methylresorcinol	Medium warm brown	Light warm blonde	Orange golden brown
m-Aminophenol	Reddish brown	Yellow orange	Light magenta

Source: From Brown and Pohl (27).

In some circumstances, direct hair dyes may be incorporated into oxidative hair color formulations as toners, but they do not undergo the same reactive color processes as do the oxidative dye materials themselves.

V. HISTORICAL IN VIVO DATA

A. Actual or Simulated Dyeing Conditions in Humans

There are only a few studies evaluating the systemic absorption of hair dyes in humans under actual conditions of use. Maibach and Wolfram (11) and Wolfram and Maibach (12) conducted detailed investigations in volunteers with a number of ^{14}C-ring-labeled oxidative and direct dyes that can, however, serve as useful benchmarks. The oxidative formulations were mixed with 6% aqueous hydrogen peroxide, and the mixture (approximately 110 g) was worked into dry hair for over a period of 5 to 8 min and left on the hair for an additional 20 min. Direct dyes (approximately 88 g) were applied in a similar manner, after which a plastic turban was wrapped around the hair for the additional 30-min exposure period. Following rinsing, towel blotting, and drying, the hair was either clipped off with an electric clipper (application-only studies) or left in place (application and wear studies).

^{14}C levels were determined in rinse, in digested and decolorized hair, and in each subject's urine via liquid scintillation counting. Urinary values were corrected for incomplete excretion based on parenteral or oral data. The data from the application-only studies, expressed as cumulative 4-day absorption in $\mu g/cm^2$ of scalp area (assuming 700 cm^2 of scalp) as calculated

Table 3 Hair Dyes Used Widely in the United States and Europe

Oxidative dyes	Direct dyes
Primary intermediates	2-Amino-3-nitrophenol
2-Aminophenol (*o*-aminophenol)	4-Amino-3-nitrophenol
4-Aminophenol (*p*-aminophenol)	Cl Disperse Black 9
N,*N*-Bis(2-hydroxyethyl)-*p*-	Cl Disperse Blue 3
phenylenediamine	
2-Chloro-*p*-phenylenediamine	Cl Disperse Violet 1
1,4-Diaminobenzene (*p*-phenylenediamine)	HC Blue No. 2
2,5-Diaminotoluene	HC Orange No. 1
N-Methyl-4-aminophenol	HC Red No. 1
N-Phenyl-*p*-phenylenediamine	HC Red No. 3
2-(2-Hydroxyethyl)-*p*-phenylenediamine	HC Red No. 7
N,*N*[1]-bis(2-hydroxyethyl)*N*,*N*[1]-bis (4'-	HC Violet No. 2
aminophenyl)-1,3-diamino-2-	HC Yellow No. 2
hydroxypropane	
3-Methyl-4-aminophenol	HC Yellow No. 4
	HC Yellow No. 6
Couplers	HC Yellow No. 7
5-Amino-2-methylphenol	HC Yellow No. 10
3-Aminophenol (*m*-aminophenol)	HC Yellow No. 11
4-Chlororesorcinol	2-(*N*-Hydroxyethylamino)-5-
	nitroanisole
1,3-Diaminobenzene (*m*-phenylenediamine)	4-β-Hydroxyethylamino-3-nitrophenol
2,4-Diaminophenol	3-Methylamino-4-nitrophenoxyethanol
2,4-Diaminophenoxyethanol	3-Methylamino-4-
	nitrophenoxypropan-1,2-diol
3-*N*-Ethylamino-4-methylphenol	2-Nitro-*p*-phenylenediamine[a]
7-Hydroxybenzomorpholine	4-Nitro-*o*-phenylenediamine[a]
5-*N*-(2-Hydroxyethyl)-amino-2-	Sodium picramate
methylphenol	
2-Methylresorcinol	
1-Phenyl-3-methylpyrazolone-5	
Resorcinol	

[a]May be used as toners in oxidative hair dye formulations.
Source: Adapted from IARC (28).

by Kalopissis (4), as well as data from the application and wear studies, are shown in Table 4.

In the application-only studies, the oxidative primary intermediate *p*-phenylenediamine (PPD) showed a cumulative absorption of 4.47 μg/cm^2, while the values for the two couplers (resorcinol and 2,4-diaminoanisole)

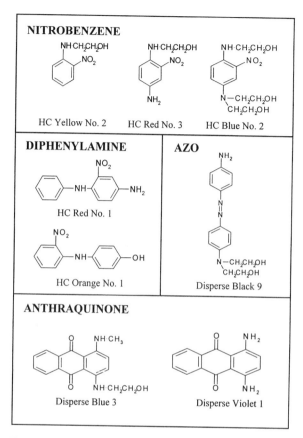

Figure 3 Representative chemical structures of some direct hair dyes.

were 0.21 and 0.46 µg/cm^2. Cumulative absorption for the three direct dyes (4-amino-2-nitrophenol, 2-nitro-*p*-phenylenediamine and HC Blue No. 1) evaluated in the application-only studies ranged between 0.67 and 1.55 µg/cm^2. In all cases these amounts represented no more than 0.2% of the dye amounts applied when 126 to 157 mg/cm^2 (i.e., 88 to 110 g to 700 cm^2 of scalp) of formulation was used under these conditions.

Data for PPD and three direct dyes studied under application and wear conditions show distinctly different patterns of cumulative absorption over time. For PPD, cumulative absorption increased between days 1 and 10, with most of the change occurring within 2 days of application. In contrast, two of the direct dyes (2-nitro-*p*-phenylenediamine and HC Blue No. 1) showed incremental absorption over the entire 30-day period, while the third

Table 4 Systemic Absorption of Hair Dyes from Commercial Formulations in Human Volunteers Under Actual Conditions of Use

Dye	Classification[b]	On-head[c] (%)	Application[d] only	Absorbed, $\mu g/cm^{2a}$ Application and wear			
				Day 1	Day 10	Day 20	Day 30
p-Phenylenediamine	O	1.35	4.47	4.35	7.12	7.81	7.81
Resorcinol	O	0.61	0.46	—	—	—	—
2,4-Diaminoanisole	O	0.87	0.21	—	—	—	—
4-Amino-2-nitrophenol	(D)	0.21	0.67	—	—	—	—
2-Nitro-p-phenylenediamine	(D)	0.68	1.33	2.20	4.86	7.18	8.68
HC Blue No. 1	D	1.48	1.55	2.81	5.26	5.63	9.38
HC Blue No. 2	D	1.77	—	0.23	1.57	2.09	2.00

[a]Calculated from percentage absorption data assuming a 700-cm^2 (human) or 32-cm^2 (rhesus) scalp surface area.
[b]O—oxidative; D—direct; (D)—direct dyes used as toners in oxidative hair color formulations.
[c]After 1:1 dilution with 6% hydrogen peroxide.
[d]Expressed as cumulative 4-day absorption data.

(HC Blue No. 2) showed a relatively large increase by day 10 and little further change after day 20. Such results are entirely consistent with the chemistry involved. The formation of larger polymeric dye molecules in the hair shaft subsequent to the uptake and chemical reaction of the monomeric PPD with couplers would serve to entrap dye, thereby limiting subsequent bioavailability. In contrast, deposition of direct dye materials into the outer cortex of the hair allows for the possibility of subsequent leaching and reexposure of the scalp and body to the incremental dye amounts during wear, hair shampooing, and bathing. Observations that over the 30-day wear period, only about 10% of the original dye content was lost from hair with the oxidative materials, while about 60% was lost from hair colored with direct dyes, are also consistent with the evidence for incremental reexposure from the hair reservoir.

In addition to these radiolabel studies, there appear to be only two other published evaluations conducted in humans under actual use conditions. Goetz et al. (13) used a gas chromatography/mass spectrometry (GC-MS) method to monitor the excretion of PPD metabolites in urine following application of commercial oxidative hair dyeing products containing 0.54% to 2.52% PPD. In 19 individual experiments, five different subjects were each evaluated on one to six occasions. Aliquots of 24- to 48-h urinary samples were flash-hydrolyzed in the injection port of a GC, resulting in cleavage of PPD from its acetylated and conjugated metabolites. N^1,N^4-Di-acetyl-PPD was also measured independently using high-performance liquid chromatography (HPLC).

In one subject using a product containing 0.6% PPD, 43 μg of PPD was detected within the first 30-min urine sample, while approximately 500 μg was determined after 36 h. The amounts of PPD recovered in urine ranged from 340 to 698 μg in 24 h (15 experiments in five subjects) and from 463 to 671 μg in 48 h (six experiments in four subjects), although these data showed no obvious trends indicating that the extent of absorption was related to the dye concentrations applied. The 24–48 h data corresponds to a cumulative absorption of about 1 μg/cm^2 of scalp area, somewhat less than the value of about 4.5 μg/cm^2 calculated from the Maibach and Wolfram (11) data cited earlier. However, after normalizing for differences in applied dosages and expressing data on a percentage basis, values of 0.04 to 0.25% for Goetz (13) are in close accord with values of 0.07 to 0.21% for Maibach and Wolfram (11).

Kiese and Rauscher (14) evaluated the percutaneous absorption of p-toluenediamine (2.5%), another common oxidative hair color primary intermediate, from a hair color base containing a single coupler, resorcinol (2.5%), following mixing with 6% hydrogen peroxide and application to the hair of four female and one male volunteers for 40 min. The appearance of the N^1,N^4-diacetyl p-toluenediamine derivative in the urine was calculated from absorption curves of the derivative, isolated via thin-layer chromatography, and corrected for incomplete excretion using subcutaneous injection recovery data. Analysis showed the highest excretion within 5 to 8 h after dye application, with an average total recovery equivalent to 4.6 mg p-toluenediamine. This would correspond to about 6.6 μg/cm^2, based on a 700-cm^2 human scalp area. While this is clearly in the range of values reported for PPD, the cold analytical methods employed for this study have been subsequently critiqued and may overestimate exposure (15).

B. Other Data

The percutaneous absorption of ^{14}C-radiolabeled hair dyes has also been evaluated in vivo under simulated use conditions in other species, notably rhesus monkeys and rats.

Maibach and Wolfram (11) and Wolfram and Maibach (12) found reasonable accord with human data in rhesus monkey experiments in which application amounts were normalized for the relative differences in scalp surface area. Dye formulations containing PPD, 2,4-DAA, HC Blue No. 1, resorcinol, and 2-nitro-p-phenylenediamine, identical to those used for the human studies, were evaluated.

In in vivo studies using rats, Beck et al (16) found 72-h cumulative absorption values ranging from 0.05 to 2.23 μg/cm^2 for 10 direct dyes, and values ranging between 0.87 to 4.82 μg/cm^2 for eight oxidative dyes mixed with hydrogen peroxide.

Also in rats, Howes and Black (17) evaluated 2-nitro-p-phenylene-diamine under a variety of exposure conditions including dye base and ethanol vehicles, varied exposure times (including 48 h under occlusion) and concentrations. With 5-min exposures at concentrations ranging between 0.06 to 0.48%, they found cumulative absorption values ranging between 0.1 to 3.2 μg/cm^2, and found a value of 6.1 μg/cm^2 for a 30-min exposure, most reflective of a use situation.

VI. IN VITRO METHODS

Methodological limitations associated with the complexity of cold analytical techniques as well as reticence about the routine use of radiolabel studies for nontherapeutic materials in human volunteers provide impetus to utilize alternative approaches to obtain skin absorption data appropriate for risk assessment purposes. This is especially true for novel synthetic cosmetic ingredients, where it may be advantageous to estimate the potential for systemic exposure at an early point in their toxicological evaluation. In vitro methods, particularly those conducted using human skin (or pig skin as a preferred alternative), offer an excellent opportunity to evaluate the percutaneous absorption potential, particularly if results can be benchmarked against existing in vivo data. Such in vitro studies may also be helpful in evaluating the relative permeability of individual hair dyes and may allow precise control of a number of key experimental variables (e.g., dye concentration, presence of other dye materials, formulation constituents) in order to evaluate their potential influence on the rate and/or extent of percutaneous penetration.

A. In Vitro Human Skin Protocols

Typical human in vitro experiments that have been conducted in our own and in collaborators' laboratories involve the use of female abdominal and/ or breast skin obtained at autopsy or from cosmetic surgery. Full-thickness skin is heat-separated in deionized water ($60 \pm 1°C$ for 45 s) and the epidermal membranes are placed between two halves of Franz-type static diffusion cells. The cells are placed in a water bath set at $37 \pm 1°C$, which maintains a skin surface temperature of approximately $32°C$. Receptor chambers are stirred continuously with a magnetic bar. Prior to application of hair dye formulations, the integrity of the epidermal membranes is routinely evaluated using ^3H_2O. Only those membranes with calculated permeation rates of less than 1.5 $mg/cm^2/h$ are selected, based on analysis of ^3H in receptor fluid (phosphate-buffered saline, PBS) after a 1-h exposure. Prior to dye exposure, the receptor fluid is replaced with fresh PBS containing 6% Volpo20.

^{14}C-Ring-labeled dye materials together with cold dye are incorporated stepwise into the hair dye formulation process, rather than spiking finished formulations. Dyes are studied singly and in combinations representative of commercial formulations, depending on the purpose of the experiment. Oxidative formulations are usually mixed with an equal volume of 6% hydrogen peroxide, while direct dye formulations are applied neat. Products are applied using a positive displacement micropipette at target doses ranging from about 7.5 mg/cm^2 to 100 mg/cm^2, with the higher dose most closely emulating the scalp use situation. Formulations are spread evenly over the exposed skin surface (approximately $1.0-1.5$ cm^2) using a tared glass rod, and actual applied doses are calculated by subtracting any material removed during spreading as determined gravametrically. Following a 30-min exposure period, residual material is removed from the skin surface by rinsing with distilled water (8×500 μl) and the presence of radiolabel in receptor fluid is monitored by liquid scintillation counting, typically up to 48 h postexposure. Mass balance is determined as the sum or counts from 30-

min rinsates; solubilized skin samples (Soluene), terminal 48-h skin rinse (2×500 µl), and blots made using small strips of filter paper; and receptor fluid.

Some experiments were designed to study the absolute dye permeation rates for comparison with historical in vivo data when such data were available for specific compounds, including a few that are no longer in commercial use. Such studies utilized identical formulations and exposure conditions as the original human volunteer studies. Other studies were designed to evaluate the relative permeation of a series of dyes when applied at the same concentration (typically 1%) and in identical hair dye vehicles. In some instances, studies were designed to evaluate the influence of various factors such as dye concentration, applied product dose, vehicle, the presence of other dyes or developer, and influence of hair on the permeation process. The objectives and results of these experiments are described here, grouped according to the major dye type in Sections VII and VIII. Preliminary reports of some of these studies have been published previously (18–21).

B. In Vitro Studies Using Pig or Rat Skin

A number of investigators have used pig skin to study hair dye absorption in vitro (16,22,23). Skin was obtained from 3- to 5-month-old male castrated pigs and, dermatomed to a thickness of 1 mm, either used directly or held at $-20°C$ for up to 5 weeks. Skin was mounted in flow-through chambers with an exposed surface area of 27.3 cm^2 to which hair dye formulations with or without hydrogen peroxide were applied to a defined 10-cm^2 area at a level of 100 mg/cm^2. Dye was rinsed with a warm shampoo solution after 30 min, and receptor fluid was collected in 8- or 10-h intervals for a total of 72 h at a flow rate of 10 ml/h.

Beck et al. (16) also employed the clipped abdominal skin obtained from Wistar rats within 3 h of sacrifice after careful removal of subcutaneous fat. Skin was mounted on glass permeation chambers, and formulations were applied at a level of 20 mg/cm^2 and rinsed after 30 min with water and shampoo, or else left in contact for the 20-h measurement period.

VII. HUMAN SKIN, IN VITRO: OXIDATIVE HAIR DYES

A. Primary Intermediates

1. PPD

This key primary intermediate was evaluated under a variety of exposure conditions in order to assess the influence of dosage volume, the presence or absence of other dyes and developer, and the presence or absence of hair on the extent of percutaneous absorption.

a. Dosage Volume Application of a hair dye base containing 2.7% PPD alone at target levels of 100 mg/cm^2 (reflective of use conditions) and 20 mg/cm^2 (reflective of typical in vitro studies) for 30 min resulted in a similar deposition on or in skin after 48 h, as shown in Fig. 4, A and B. Absorption reached a plateau within 4 h, after which there was an apparent slight decrease in receptor fluid radiolabel over the remaining 48 h (see Fig. 6A for time course). While the reasons for this apparent decrease are unclear, it is possible that PPD or its polymerization products, which have already permeated into the receptor chamber, bind either to the inner surface of the epidermal membrane or to the receptor chamber glass. Binding to glass was noted with the coupler 2,4-diaminophenol, as described later in this section.

Despite a fivefold difference in applied dosage volume, the amount of dye present in the receptor fluid at 4 h increased only twofold from 2.98 to 6.00 µg/cm^2. This indicated that under these exposure conditions an infinite dose was being approached at the lower applied dosage level.

b. Presence of Other Dyes and Developer The formulation containing 2.7% PPD alone was diluted 1:1 with distilled water, and another formulation containing PPD in the presence of other dyes (total dye load of 9.2%), representative of a commercial black shade, was mixed 1:1 with 6% hydrogen peroxide developer. Mixed formulations were applied at a target level of 100 mg/cm^2 and rinsed after a 30-min exposure. Penetration was monitored up to 48 h. As shown in Fig. 4, C and D, there was about a fourfold (0.69 vs. 2.83 µg/cm^2) increase in radiolabel remaining on or in the skin after 48 h when the dye was applied as the complete shade mixed with peroxide. Despite this variation in amounts on or in the skin through 48 h, which probably reflected differences in both rates and types of color reactions involved as described earlier in this chapter, there were only slightly greater amounts (1.96 to 2.36 µg/cm^2) of dye penetrated into receptor fluid at the 4-h plateau with the complete shade, as compared to PPD alone.

c. Presence of Hair In another series of experiments 5 mg/cm^2 of finely chopped hair was added to the skin prior to application of a 1.35% (after mixing) PPD-containing complete hair dye formulation with hydrogen peroxide developer. While the hair could certainly compete with the skin for absorptive surface, the excess of hair color formulation applied (100 mg/cm^2) was considered adequate to compensate for this variable. The purpose of this specific comparison was, however, to evaluate whether faster and more complex color reactions than would be expected to occur in simple aqueous solution, would influence the extent of penetration. As shown in Fig. 4; E and F, in the presence of hair, there was almost twice the amount

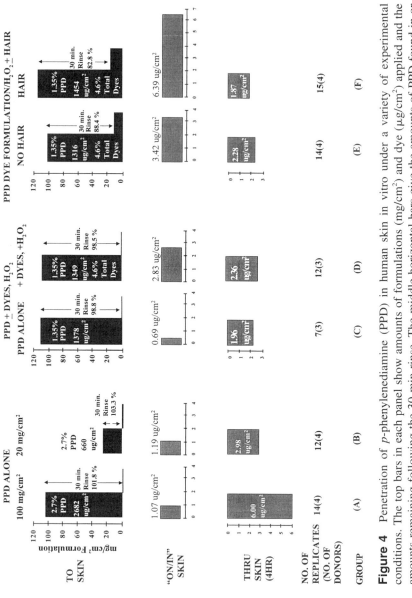

Figure 4 Penetration of *p*-phenylenediamine (PPD) in human skin in vitro under a variety of experimental conditions. The top bars in each panel show amounts of formulations (mg/cm²) and dye (μg/cm²) applied and the amounts remaining following the 30-min rinse. The middle horizontal bars give the amounts of PPD found in or on skin after solubilization at 48 hr. The bottom bars indicate the cumulative amounts of PPD in receptor fluid at 4 hr (maximum).

of radiolabel left in or on the skin surface after 48 h (6.39 vs. 3.42 $\mu g/cm^2$), while the amount penetrated into receptor fluid at the 4-h plateau was, in fact, slightly less (1.87 vs. 2.28 $\mu g/cm^2$) than that seen without hair.

Further comparison of the influence of mimicking more closely the in-use situation of other dyes, developer, and hair can be made by contrasting the data in Fig. 4C with that in Fig. 4F. Despite an approximately 10-fold difference in the amount of dye (0.69 vs. 6.39 $\mu g/cm^2$) persisting in or on the skin through 48 h after the 30-min rinse, the maximal amount of dye penetrated under these different exposure scenarios was remarkably similar (1.96 vs. 2.28 $\mu g/cm^2$). Thus, the presence of hair, other dyes, and developer impacted color deposition only on the skin surface and did not remarkably influence the amount penetrated. In the course of these studies, differences in the amount of color formation on the skin surface under the simulated dyeing conditions were apparent from the relative amount of skin staining present on the epidermal membranes at the conclusion of the experiment. It was also noteworthy that maximal penetration was noted at 4 h in these experiments, suggesting that there was little potential for the remaining colored polymeric material to continue to penetrate, despite the apparent persistence of residual material on the skin surface until the termination of experiments at 48 h.

2. *p*-Aminophenol (PAP)

PAP, another primary intermediate, was evaluated at a concentration of 0.42% (after mixing) in the presence and absence of other dyes and developer in another series of experiments analogous to those conducted with PPD. The formulation containing only PAP was diluted with distilled water prior to application. As shown in Fig. 5, with PAP alone there were higher amounts of radiolabel on or in skin after 48 h (2.21 vs. 1.40 $\mu g/cm^2$) as well as greater penetration (0.46 ± 0.14 vs. 0.13 ± 0.03 $\mu g/cm^2$) into receptor fluid than with PAP combined with other dyes and developer. Comparison of the receptor fluid values between 24 and 48 h indicated that, in addition to the reduction in extent of absorption in the presence of other dyes and couplers, the formation of colored reaction products apparently reduced the rate of absorption over the second 24-h interval (see also Fig. 6A).

B. Oxidative Dye Couplers

1. 2,4-Diaminophenol (DAP)

This material was studied as a coupler in complete dye formulation at a level of 0.27% (after mixing), although DAP is sometimes used alone in auto-oxidative hair color formulations intended to produce a gradual dark-

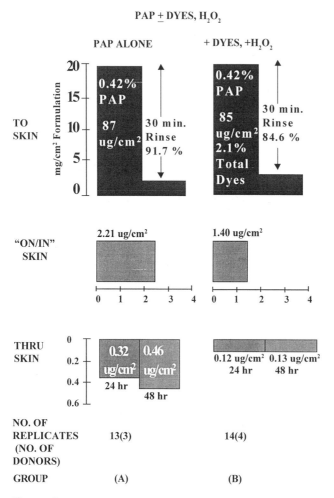

Figure 5 Penetration of *p*-aminophenol (PAP) in human skin in vitro as a single dye and as a component of a representative hair dye shade. The top bars in each panel show amounts of formulations (mg/cm²) and dye (μg/cm²) applied and the amounts remaining following the 30-min rinse. The middle horizontal bars give the amounts of PAP found in or on skin after solubilization at 48 hr. The bottom bars indicate the cumulative amounts of PAP in receptor fluid at 24 and 48 hr.

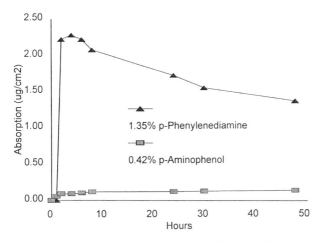

(A) PRIMARY INTERMEDIATES

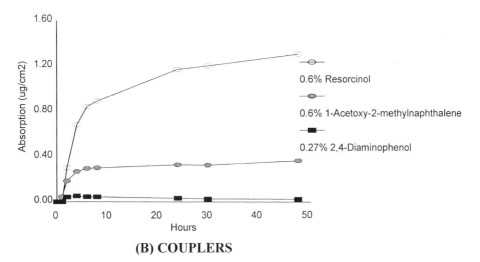

(B) COUPLERS

Figure 6 Cumulative penetration of several hair dye primary intermediates (A) and couplers (B) from oxidative hair dye formulations in human skin, in vitro. Dyes were evaluated as components of complete hair dye shades at the stated concentrations (after mixing with hydrogen peroxide).

ening of the hair with successive uses. As illustrated in Fig. 6B, this material showed a very low overall penetration of 0.052 ± 0.01 $\mu g/cm^2$ (mean of 4 donors and 15 replicates), which was maximal at 4 h postexposure despite the persistence of material in or on the skin at a level equivalent to 0.85 $\mu g/cm^2$ through 48 h. An apparent decrease in receptor fluid concentration of radiolabel from 4 through 48 h may have been due to absorption of the dye onto the surface of the glass receptor chamber, as determined by separate experiments in which skin was untreated but the receptor fluid was spiked with known amounts of ^{14}C-2,4-DAP. Receptor fluid as well as glass scintillation vials spiked with radiolabel showed 21% and 33% losses, respectively, after 28 h.

2. 2-Methyl-1-naphthol (2-MN)

This material is formed in hair color formulations in situ from a stable precursor, 1-acetoxy-2-methylnaphthalene, within 5–15 days of formulation (24). It then serves as a coupler that may react with primary intermediates such as PPD or PAP, in which case it would form, respectively, violet or orange-red indo dyes. The percutaneous penetration potential of 2-MN was evaluated using the ^{14}C-radiolabeled precursor in a prototype oxidative hair dye formulation at a level of 0.6% (total dyes 1.6%) after dilution with 6% hydrogen peroxide. As shown in Fig. 6B, cumulative absorption values remained relatively flat after 8 h, as reflected by values of 0.30 ± 0.04, 0.33 ± 0.04, and 0.36 ± 0.05 $\mu g/cm^2$ after 8, 24, and 48 h, respectively. At 48 h, 0.15 ± 0.001 $\mu g/cm^2$ of material was present on or in the skin.

3. Resorcinol

This coupler was evaluated from a representative hair dye formulation that contained 0.61% resorcinol (total dyes 2.7%) after dilution with developer. Mean data for 3 donors and 16 replicates shown in Fig. 6B indicate a plateau in receptor fluid concentration between 24 and 48 h, as reflected by cumulative absorption values of 1.17 ± 0.24 and 1.30 ± 0.36 $\mu g/cm^2$. At 48 h, 2.66 ± 0.001 $\mu g/cm^2$ of material was present on or in the skin.

VIII. HUMAN SKIN, IN VITRO: DIRECT DYES

A. HC Yellow No. 4

A series of experiments analogous to those conducted with PPD were performed using the direct dye HC Yellow No. 4 to evaluate the influence of dosage volume and the presence or absence of other dyes on the extent of absorption. Experiments using hair were not conducted with direct dyes since there was no reactive color chemistry involved and the competing surface area offered by the hair could readily be overcome by adding excess formulation. Additionally, the influence of hair dye vehicle was also evaluated for this dye.

1. Dosage Volume

A formulation containing 1% HC Yellow No. 4 alone was applied at actual levels of 7.6 ± 0.5, 14.6 ± 0.7, and 22.2 ± 2.1 mg/cm^2 (target doses of 10, 20, and 30 mg/cm^2). Skin from three different donors was used with three to six replicates per group. The resulting permeation over 48 h shown in Fig. 7A indicated absorption at the mid-dose level was approximately twice that of the low dose and equivalent to that of the high dose, indicating that an infinite dose had been achieved at an applied dose of approximately 15 mg/cm^2 of formulation.

In these experiments the ratio of wash volume to formulation was kept constant. Thus, 4 × 500 μl, 8 × 500 μl, and 16 × 500 μl were used for the low, mid, and high doses, respectively. Inspection of the mass balance data shown in Table 5 indicates that at the higher dosages, proportionately less material was found in the terminal (48 h) rinse and blot and in or on the skin at the termination of the experiment. This may have been a result of greater serial dilution of residual material on the skin surface because of the greater number of 500-μl rinsing aliquots used at the higher dosages.

2. Presence of Other Dyes

Although no chemical color reactions occur with direct dyes, it was of interest to compare the percutaneous absorption of HC Yellow No. 4 from a formulation containing 0.28% dye alone with that of absorption from a representative dark shade containing the same concentration of HC Yellow No. 4 in combination with seven other direct dyes at a total dye load of 1.0%. Seven different donors were used with a total of 17–18 replicates per group. The resulting data shown in Fig. 7B indicate that the percutaneous absorption of HC Yellow No. 4, when applied in a representative shade, was only about 37% of that when applied alone as the sole dye constituent in the formulation. The basis for the observed difference may relate to the formation of aggregates through hydrogen bonding, which would influence the molecular volume and partition coefficients of the individual dyes and, hence, their penetration. That dyes in combination are known to affect the solubilities of individual components supports the potential for such binding to occur. While it is clear, therefore, that studies involving a single dye material may conservatively overestimate the exposures to be anticipated from complete formulations, it would be difficult to generalize on the magnitude of such effects, which may depend on the relative permeabilities of

A. FORMULATION AMOUNT: HC YELLOW NO. 4 (1%)

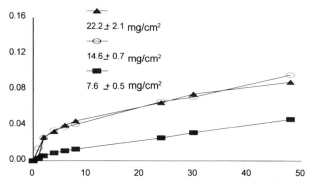

B. DYE ALONE (0.28%) vs. COMPLETE SHADE (1% TOTAL DYES)

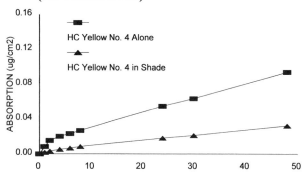

C. HC YELLOW NO. 4 (1%) IN DIFFERENT VEHICLES

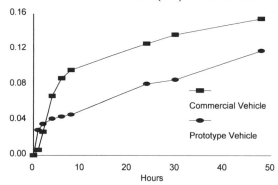

Figure 7 Cumulative penetration of HC Yellow No. 4 in human skin, in vitro. (A) Dosage volume effects. (B) Dye alone vs. complete shade. (C) Vehicle (formulation) effects. For (A) the amounts of dye applied, corresponding to the stated formulation amounts, were (mean $\pm$ SD) 76 $\pm$ 5, 146 $\pm$ 7, and 222 $\pm$ 21 $\mu g/cm^2$, respectively.

Table 5 Mass Balance Data, HC Yellow No. 4

Percent applied dose	Applied dose of dye, μg/cm^2		
	76 ± 5	146 ± 7	222 ± 2
30-min Rinse	82.75 ± 2.08	82.56 ± 3.75	79.34 ± 4.18
48-hr Rinse & blot	1.92 ± 0.79	0.79 ± 0.47	0.05 ± 0.03
Skin	1.50 ± 0.87	1.03 ± 0.87	0.18 ± 0.09
Receptor fluid	0.07 ± 0.03	0.06 ± 0.02	0.05 ± 0.02
Total	86.23 ± 2.11	84.45 ± 2.89	79.62 ± 4.12

the individual dyes and their presence and concentration in individual shades of hair dye formulations.

3. Vehicle Effects

Variations in the rate and/or extent of skin permeation are often anticipated among formulations of varying composition. To determine such influences for hair dye formulations, HC Yellow No. 4 was evaluated at a level of 1% as the sole dye constituent of a commercial and prototype formulation, which utilized different thickeners and other excipients. Four donors with 10–12 replicates per group were used. The data shown in Fig. 7C indicate that while the extent of absorption after 48 h was similar, there were apparent differences in the rates of absorption particularly at the earlier time points (2, 4, and 8 h). At these intervals, there was less absorption from the prototype base, with differences achieving a marginal level of statistical significance ($p < .1$).

B. Other Direct Dyes

In addition to HC Yellow No. 4, a number of other direct dyes were evaluated under similar protocols, all but one at a concentration of 1.0% as a single component in an identical or very similar dye base. These dyes included HC Yellow No. 2 and HC Yellow No. 15; HC Orange No. 1; HC Red No. 1, HC Red No. 3 (1.5%), and HC Red No. 14; and Disperse Black 9. Design features and selected data from these experiments are compiled in Table 6, and the cumulative absorption time courses are illustrated in Fig. 8.

The percutaneous absorption data for these dyes tested under very similar exposure conditions appear to fall into three groupings. Disperse Black 9 and HC Yellow No. 15 demonstrated the lowest absorption, with similar values of 0.035 ± 0.012 and 0.039 ± 0.016 μg/cm^2, respectively.

Table 6 Design Features and Results of In Vitro Percutaneous Absorption Studies in Human Skin with Various Individual Direct Dyes

Dye	Chemical class	MW (g/mol)	No. donors (no. replicates)	^3H_2O perm. (mg/cm²/hr)	Formulation dose (mg/cm²)	% Recovers		Absorbed (ug/cm²), 48 hr
						30-min Rinse	Skin	
Disperse Black 9	Azo	300	4 (13)	0.81 ± 0.39	9.3 ± 1.5	100.2 ± 2.0	0.42 ± 0.08	0.035 ± 0.01
HC Yellow No. 15	Nitrobenzene	227	3 (11)	0.67 ± 0.17	9.4 ± 2.2	91.1 ± 1.6	5.70 ± 0.82	0.039 ± 0.02
HC Yellow No. 4	Nitrobenzene	242	4 (10)	0.52 ± 0.14	12.4 ± 0.9	91.0 ± 2.2	2.23 ± 0.83	0.12 ± 0.04
HC Red No. 3[a]	Nitrobenzene	197	3 (19)	0.52 ± 0.05	12.2 ± 0.6	99.7 ± 1.1	0.81 ± 0.20	0.24 ± 0.04
HC Yellow No. 2	Nitrobenzene	182	4 (19)	0.30 ± 0.03	15.3 ± 0.9	89.5 ± 1.8	0.27 ± 0.05	1.36 ± 0.15
HC Red No. 14	Nitrobenzene	195	3 (13)	0.52 ± 0.09	18.7 ± 0.9	88.4 ± 1.7	0.71 ± 0.09	1.47 ± 0.19
HC Orange No. 1	Diphenylamine	230	3 (11)	0.75 ± 0.16	11.2 ± 0.6	96.4 ± 1.3	0.44 ± 0.14	1.49 ± 0.30
HC Red No. 1	Diphenylamine	229	3 (10)	0.75 ± 0.16	9.9 ± 0.7	98.5 ± 5.1	0.37 ± 0.08	1.57 ± 0.28

[a]1.5%; all others were 1%.

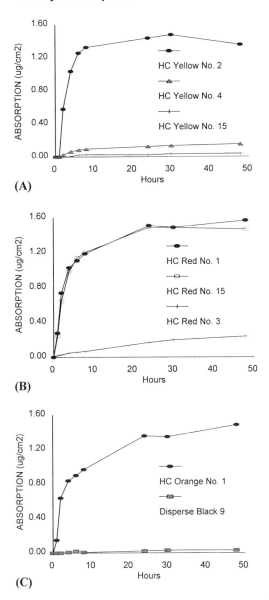

Figure 8 Cumulative penetration of several individual direct hair dyes in human skin in vitro. (A) yellow dyes, (B) red dyes, and (C) other dyes. All dye concentrations were 1% except for HC Red No. 3, which was evaluated at 1.5%. The numbers of donors (numbers of replicates) and additional experimental details are given in Table 6.

HC Yellow No. 4 and HC Red No. 3 (evaluated at 1.5%) were next, with values of 0.12 ± 0.04 and 0.24 ± 0.04 μg/cm^2, respectively. HC Yellow No. 2, HC Red No. 14, HC Orange No. 1, and HC Red No. 1 all showed very similar absorption values in the narrow range of 1.36 ± 0.15 to 1.57 ± 0.28 μg/cm^2. Notably, HC Yellow No. 4 and particularly HC Yellow No. 15 showed low cumulative absorption values relative to the other dyes, despite the fact that more substantial levels of dye were found on or in the skin after 48 h. The time-course data for these materials generally showed a leveling off of penetration within 8–20 h with some of the more poorly absorbed dyes showing small but continual incremental absorption through 48 h. Some of this increase could reflect changes in barrier function of the epidermal membrane over time.

IX. IN VIVO/IN VITRO CORRELATIONS

For comparison purposes, in vivo and in vitro data from hair dye studies involving actual or simulated use conditions conducted in our own and in collaborators' laboratories as well as cited literature are compiled in Table 7A (oxidative dyes) and Table 7B (direct dyes). The human volunteer data are limited by the few available observations (three direct and three oxidative dyes), but the ranges for cumulative absorption values in the various systems show relatively good agreement. In particular, comparisons between pig skin in vitro and in vivo data involving identical dye formulations evaluated under similar exposure conditions showed excellent statistical correlation as reflected by R values of .92 for 8 oxidative dyes (including one studied in a water vehicle which gave values stated in the footnote to Table 7A) and .95 for 10 direct dyes.

Several of the identical dye formulations evaluated in human volunteers have also been evaluated using rhesus monkeys and are also currently under investigation in our own and others' (16) laboratories using in vitro systems with human and pig skin. Available data are summarized in Table 8. Where possible, the values that reflect cumulative 4-day absorption from the application-only studies by Wolfram and Maibach (12) are cited, particularly since the in vitro systems would not be expected to reflect the incremental absorption for the direct dyes, attributed to leaching from the hair reservoir. For HC Blue No. 2, which was studied only under application and wear conditions, the range for the 1–10 day cumulative absorption figures is cited. There was a substantial increase in absorption over this time (0.23 to 1.57 μg/cm^2) for this material.

Although the sample sizes are relatively small (generally two to five subjects per group), there appears to be reasonable accord for the in vivo human and in vivo rhesus monkey data, particularly for the relative rank

Table 7 Surveys of Hair Dye Absorption In Vivo and In Vitro

System	Exposure and sampling	No. of dyes	Range of cumulative absorption values ($\mu g/cm^2$)	Reference
A. Oxidative dyes				
Human, in vivo	30-min Application in commercial formulation ($+H_2O_2$; 100 mg/cm^2; 144 hr sampling	3	0.21–4.47	Wolfram and Maibach (12); Kalopissis (4)
Human, in vitro	30-min Application in commercial formulation ($+H_2O_2$; 100 mg/cm^2; 48 hr sampling	5	0.02–2.1	Current data
Pig, in vitro	30-min Application in formulation ($+H_2O_2$); 72 hr sampling	7	0.14–2.46 $\uparrow$ $R = .92^a$ $\downarrow$	Beck (16)
Rat, in vivo	30-min Application in formulation ($+H_2O_2$); 72 hr sampling	7	0.87–4.82	
B. Direct dyes				
Human, in vivo	30-min Application in commercial formulation (100 mg/cm^2); 96 hr sampling	3	0.67–1.55	Wolfram and Maibach (12); Kalopissis (4)
Human, in vitro	30-min Application as single dye in base (10–15 mg/cm^2); 48 hr sampling	8	0.03–1.56	Current data
Pig, in vitro	30-min Application in formulation (100 mg/cm^2); 72 hr sampling	10	0.04–1.14 $\uparrow$ $R = .95$ $\downarrow$	Beck et al. (16)
Rat, in vivo	30-min Application in formulation (20–100 mg/cm^2); 72 hr sampling	10	0.05–2.23	

[a]This correlation value includes one dye studied in an aqueous vehicle that gave cumulative absorption values of 4.48 and 12.54 $\mu g/cm^2$ for pig skin in vitro and rat skin in vivo, respectively.

Table 8 Comparison of In Vivo and In Vitro Hair Dye Absorption Data Obtained Under Actual or Simulated Use Conditions

Dye	Classification[c]	On-head[b] %	Absorbed, $\mu g/cm^{2a}$ In vivo Human[d]	In vivo Rhesus[d]	In vitro Human
p-Phenylenediamine	O	1.35	4.47	4.05	2.10
Resorcinol	O	0.61	0.46	1.68	1.30
2,4-Diaminoanisole	O	0.87	0.21	0.47	—
4-Amino-2-nitrophenol	(D)	0.21	0.67	—	—
2-Nitro-p-phenylenediamine	(D)	0.68	1.33	5.12	1.55
HC Blue No. 1	D	1.48	1.55	2.79	0.42
HC Blue No. 2	D	1.77	$(0.23-1.57)^e$	—	0.20

[a]Calculated from percentage absorption data assuming a 700-cm^2 (human) or 32-cm^2 (rhesus) scalp surface area.
[b]After 1:1 dilution with 6% hydrogen peroxide.
[c]O—oxidative; D—direct; (D)—direct dye used as toner in oxidative hair dye formulation.
[d]Expressed as cumulative 4-day absorption data.
[e]Range for Day 1 to Day 6.

order of absorption for the three oxidative dyes, which was PPD > resorcinol > 2,4-diaminoanisole. A similar comparison for the in vitro human data set awaits completion of ongoing work with 2,4-diaminoanisole. For the direct dyes, the in vivo and in vitro human data for 2-nitro-p-phenylenediamine show excellent accord (1.33 vs. 1.55 $\mu g/cm^2$), while the absolute values for the direct blue dyes appear lower than the human data.

Completion of additional experiments in human skin and of ongoing in vitro work in pig skin with these materials may serve to further clarify the ability of the in vitro models to mirror the in vivo results. In the interim, the in vitro models would at least appear useful to evaluate the relative percutaneous potential of a series of related materials evaluated under similar exposure conditions, such as the work cited here for eight direct dyes that showed approximately a 50-fold variation between the highest and lowest value (0.03–1.56 $\mu g/cm^2$) with essentially the same maximum as the available human data. Such information provides confidence in the utility of in vitro data for risk assessment purposes. It is also important here to acknowledge the caution expressed by Bronaugh (25) and others that a lack of agreement between in vivo and in vitro data may not necessarily imply that the in vitro data are in error.

X. RISK ASSESSMENT

Using the ranges of cumulative percutaneous absorption data determined in vivo and in vitro cited in the preceding section, it is feasible to estimate the potential for systemic exposure with representative direct and oxidative dyes. For example, at the higher end of the observed ranges, cumulative absorption values of 5–10 $\mu g/cm^2$ would yield a single application exposure of 3000–6000 μg based on a scalp surface area of 600 cm^2. This level of exposure would correspond to 50–100 $\mu g/kg$ using a 60-kg body weight. Considering the discontinuous use of such products, typically once every 4 weeks (28 days), this would reflect average daily exposures in the range of only about 2–4 $\mu g/kg/day$. This would be for the more highly absorbed materials and for darker dye shades containing higher dye loads. Actual typical exposures to individual dyes contained in lighter shades may, in fact, be substantially lower than such worst-case analyses might suggest.

Corresponding safety factors, derived as the ratio of the systemic exposure dose to the no-observed-adverse-effect level (NOAEL) determined in animal experiments, would, in general, greatly exceed a 100-fold margin nominally required in such circumstances. For typical NOAELs in the range of 10–100 mg/kg/day observed with most hair dye materials in 90-day subchronic toxicity or teratology studies, such safety factors would be in the range of 2500–25,000 (i.e., 10,000–100,000 $\mu g/kg/day$ NOAEL divided by 4$\mu g/kg/day$ exposure). This is clearly adequate for toxic effects associated with thresholds below which no adverse effects are anticipated. In certain circumstances, the systemic exposure may be sufficiently low to provide de minimus risks when extrapolations are based on assuming absence of a threshold where any dose is associated with a small but finite risk. Details of such quantitative risk assessments have been published for the oxidative coupler 2,4-diaminoanisole (4) and for the direct dye Disperse Blue No. 1 (26).

XI. SUMMARY AND CONCLUSIONS

Hair dyes represent a relatively unique class of skin permeants for which the most meaningful data is obtained in studies that reflect their actual use conditions. Overall, the extent of absorption for hair dyes appears to be relatively low, representing up to only a few tenths of a percent of the amounts applied to the hair and scalp. Available in vitro data suggest absorption values may vary by about 50-fold when individual direct dyes are evaluated at similar concentrations and by about 100-fold when oxidative dyes present at their typical use percentages are also considered. In vivo human data exist for a limited number of materials representative of the chemical types of oxidative and direct dyes in current use. These data serve as useful benchmarks for safety assessment purposes as well as for the validation of in vitro alternative methods that hold promise as useful tools either for evaluating novel synthetic compounds or for isolating the experimental variables that may influence absorption. Maximum observed cumulative hair dye absorption for materials evaluated in vivo and in vitro under use conditions in a variety of test systems appears to be in the range of no more than $5-10$ $\mu g/cm^2$ even when incremental exposure during wear is considered. Limitation of percutaneous absorption within this range would generally be expected to offer a very wide margin of safety for most hair dye materials, particularly with respect to those toxicity endpoints considered to be associated with thresholds.

ACKNOWLEDGMENTS

The valuable contribution of Richard Grabarz in conducting the in vitro studies in our laboratories and of the An-eX group (Drs. Kenneth A. Walters, Adam C. Watkinson, Valerie J. James, and Darren M. Green) in Cardiff, Wales, in performing collaborative studies with several of the dyes is gratefully acknowledged. The skillful assistance of Elizabeth Kantor in the preparation of the manuscript is also appreciated.

REFERENCES

1. Baden HP, Goldsmith LA, Fleming B. A comparative study of the physico-chemical properties of human keratinized tissues. Biochim Biophys Acta. 1973; 322:169–278.

2. Gniadecka M, Nielsen OF, Christensen DH, Wulf CH. Structure of water, proteins, and lipids in intact human skin, hair and nail. J Invest Dermatol. 1998; 110:393–398.

3. Bronaugh RL, Congdon ER. Percutaneous absorption of hair dyes: Correlation with partition coefficients. J Invest Dermatol. 1984; 83:124–127.

4. Kalopissis G. Toxicology and hair dyes. In: Zivak C, ed. The Science of Hair Care. New York: Marcel Dekker, 1986:287–308.

5. Brown KC. Hair coloring. In: Johnson D, ed. Hair and Hair Care. New York: Marcel Dekker, 1997:191–215.

6. Scheuplein RJ. Mechanism of percutaneous absorption. II. Transient diffusion and the relative importance of various routes of skin penetration. J Invest Dermatol. 1967; 48(1):79–88.

7. Schaefer H, Watts F, Brod J, Illel B. Follicular penetration. In: Scott RC, Guy RH, Hadgraft J, eds. Prediction of Percutaneous Absorption: Methods, Measurements, Modelling. London: IBC Technical Services, 1990:163–173.

8. Lauer AC, Lieb LM, Ramachandran C, Flynn GL, Weiner ND. Review: Transfollicular drug delivery. Pharm Res. 1995; 12(2):179–185.

9. Wepierre J, Doucet O, Marty JP. Percutaneous absorption of drugs in vitro: Role of transepidermal and transfollicular routes. In: Scott RC, Guy RH, Hadgraft J, eds. Prediction of Percutaneous Absorption: Methods, Measurements, Modelling. London: IBC Technical Services, 1990:129–134.

10. Meidan V, Docker M, Walmsley A, Irwin W. Low intensity ultrasound as a probe to elucidate the relative follicular contribution to total transdermal absorption. Pharm Res. 1998; 15(1):85–92.

11. Maibach HI, Wolfram LJ. Percutaneous penetration of hair dyes. Cosmet J Soc Chem. 1981; 32:223–229.

12. Wolfram LJ, Maibach HI. Percutaneous penetration of hair dyes. Arch Dermatol Res. 1985; 277:235–241.

13. Goetz N, Lasserre P, Bore P, Kalopissis P. Percutaneous absorption of p-phenylene diamine during actual hair dyeing procedure. Int J Cosmet Sci. 1988; 10:63–73.

14. Kiese M, Rauscher E. The absorption of p-toluenediamine through human skin in hair dyeing. Toxicol Appl Pharmacol. 1968; 13:325–331.

15. Corbett JF. Application of oxidative coupling reactions to the assay of p-phenylenediamines and phenols. Anal Chem. 1975; 47(2):308–313.

16. Beck H, Bracher M, Faller C, Hofer H. Comparison of in vitro and in vivo skin permeation of hair dyes. Cosmet Toilet. 1993; 108:76–83.

17. Howes D, Black JG. Percutaneous absorption of 2-nitro-p-phenylenediamine. Int J Cosmet Sci. 1983(5):215–226.

18. Dressler WE, Azri-Meehan S, Grabarz R. Utilization of *in vitro* percutaneous penetration data in the safety assessment of hair dyes. In Lisanski SG, Macmillan R, Dupuis J, eds. Proceedings of an International Scientific Conference Organised by the European Cosmetic Industry (COLIPA), Brussels, Belgium, 1995: Alternatives to Animal Testing. Newbury UK: CPL Press, 1996, 269.

19. Dressler WE, Grabarz R, Sharma R, Brain KR, Green DM, James VJ, Walters KA. Percutaneous penetration of oxidative hair dyes from human cadaver skin, in vitro. In Brain KR, James VJ, Walters KA, eds. Perspectives in Percutaneous Penetration. Cardiff: STS Publishing, 1997; 5A:21.

20. Azri-Meehan S, Dressler WE, Grabarz R. Percutaneous penetration of semi-permanent (direct) hair dyes determined in vitro using human cadaver skin. Toxicologist 1996; 30:125.

21. Sharma RK, Dressler WE, Grabarz R, Watkinson AC, Green DM, Walters KA. In vitro skin permeation of para-phenylenediamine (PPD). Toxicologist 1997; 36:324.

22. Bracher M, Faller C, Grötsch W, Marshall R, Spengler J. Studies on the potential mutagenicity of p-phenylenediamine in oxidative hair dye mixtures. Mutat Res. 1990; 241:313–323.

23. Noser F, Bracher M, Faller C. In vitro permeation with pig skin: instrumentation and comparison of flow-through versus static-diffusion protocol. J Appl Cosmetol. 1988; 6:111–122.

24. Lim MI, Pan YG, Stasaitis LR, O'Donoghue JD. Storage stable 2-Methyl-1-naphthol couplers. US Patent 5,529,583, June 25, 1996.

25. Bronaugh RL. Methods for *in vitro* percutaneous absorption. Toxicol Methods. 1995; 5(4):265–273.

26. Haws LC, Jackson BA, Turnbull D, Dressler WE. Two approaches for assessing human safety of Disperse Blue 1. Regul Toxicol Pharmacol. 1994; 19: 80–96.

27. Brown KC, Pohl S. 1996. Permanent Hair Dyes: Society of Cosmetic Chemists Monograph, 1–41.

28. IARC. Occupational exposures of hairdressers and barbers and personal use of hair colourants; some hair dyes, cosmetic colourants, industrial dyestuffs and aromatic amines. IARC Monogr Eval Carcinogen Risks Hum. 1993; 57: 43–118.

41
Percutaneous Absorption of Alpha-Hydroxy Acids in Human Skin

Margaret E. K. Kraeling and Robert L. Bronaugh
Food and Drug Administration, Laurel, Maryland

I. INTRODUCTION

Alpha-hydroxy acids (AHAs) are hydroxycarboxylic acids that are widely used in many cosmetic products as exfoliants, moisturizers, and emollients. Dermatologists have used AHAs to treat various skin disorders, including ichthyosis (1), acne, photoaging (2,3), and other hyperkeratotic conditions (4,5). Because of their reported effects in the skin, such as increased cell turnover in the stratum corneum (6,7), AHAs have become increasingly popular for cosmetic use, and are marketed for their antiaging effects (8,9). The ingredients, concentration, and pH of AHAs vary in cosmetic products. The small, short-chain compounds glycolic acid and lactic acid are most commonly used in cosmetic products; however, longer chain AHAs have been found to increase stratum corneum extensibility and flexibility (10) and have been used in some products. AHAs have been shown to decrease the thickness of the stratum corneum (11), and increase the thickness of the viable epidermis (3,12). AHAs have also been shown to affect the deeper dermal layer of skin with an increase in thickness, acid mucopolysaccharides (3), glycosaminoglycan formation (13), collagen deposition (3,13), and other cellular changes (14).

The mechanism for AHA effects on skin is still unknown. Van Scott et al. have suggested that AHAs reduce stratum corneum corneocyte cohesion by interference with ionic bonding (4,5). Structural changes in the epidermis and dermis after dermal exposure to AHAs suggest, however, that effects on the stratum corneum could originate from AHA effects in these deeper layers. Therefore, understanding the extent of absorption of AHAs is important, particularly with regard to localization in the various layers of skin. The effects of AHAs on skin are likely influenced by their ability to be absorbed into the different layers of skin. AHAs are weak acids; thus, their absorption should also be sensitive to pH changes in formulations due to ionization. Cosmetic product pH could be an important factor affecting the skin absorption of AHAs; therefore, we investigated AHA skin absorption using an oil-in-water emulsion vehicle, adjusted to a low pH (3.0) and a neutral pH (7.0). Product formulation changes, such as the use of a stronger surfactant, on the skin absorption of glycolic acid and on permeability changes in the skin barrier were also studied. Other factors that could influence AHA skin absorption include chemical structure and lipophilicity of the AHA compound. Therefore, we evaluated the effect of AHA chemical structure on absorption by studying the absorption of a homologous series of AHA compounds.

The percutaneous absorption of five AHAs was measured through viable excised human skin in diffusion cells: glycolic acid (GA), lactic acid (LA), 2-hydroxyhexanoic acid (2-hydroxycaproic acid), 2-hydroxyoctanoic acid (2-hydroxycaprylic acid), and 2-hydroxydecanoic acid (2-hydroxycapric acid). AHA absorption was assessed by determining levels of absorbed material in skin layers and in the receptor fluid beneath the skin.

II. MATERIALS AND METHODS

A. Materials

$[1\text{-}^{14}C]$Glycolic acid (specific activity, 55 mCi/mmol; 99% purity) was obtained from American Radiolabeled Chemicals, Inc. (St. Louis, MO) and $[1\text{-}^{14}C]$-DL-lactic acid (specific activity, 50 mCi/mmol; 98% purity) was obtained from Sigma Chemical Co. (St. Louis, MO). $[1\text{-}^{14}C]$-2-Hydroxyhexanoic acid (specific activity, 17.6 mCi/mmol; 96% purity), $[1\text{-}^{14}C]$-2-hydroxyoctanoic acid (specific activity, 19.2 mCi/mmol; 97% purity), and $[1\text{-}^{14}C]$-2-hydroxydecanoic acid (specific activity, 16.4 mCi/mmol; 92% purity) were synthesized by Research Triangle Institute (Research Triangle Park, NC). $[^{3}H]$Water (specific activity, 55.5 mCi/mmol; 97% purity) was purchased from New England Nuclear Corp. (Boston). Nonlabeled glycolic acid (98% purity), 2-hydroxyoctanoic acid (<98% purity), and 2-hydroxydecanoic acid (98% purity) were obtained from Sigma Chemical Co. Nonlabeled 2-hydroxyhexanoic acid (99% purity) was obtained from Aldrich Chemical Co. (Milwaukee, WI). Commercial product 1 (5% GA, pH 2.5) and commercial product 2 (10% GA, pH 3.5) were obtained from a local cosmetic supplier.

B.   Oil-in-Water Emulsion Formulations

Percutaneous absorption of glycolic acid was studied by using two oil-in-water emulsion formulations (Formulations A and B). The composition of Formulation A is given in Table 1 (15). It contained two nonionic emulsifying agents: polyethylene glycol (PEG) 100 stearate (2%) and PEG-4 lauryl ether (Laureth-4) (1%). Formulation B had the same composition as Formulation A, except that 1% ammonium laureth sulfate (ALS), an ionic surfactant, was used in place of the Laureth-4. Formulation A was the vehicle utilized for studying the absorption of all AHAs (glycolic, lactic, 2-hydroxyhexanoic, 2-hydroxyoctanoic, and 2-hydroxydecanoic acids). Formulation B was used for studying the effect of surfactant changes on the absorption of glycolic acid.

Table 1   Composition of the Oil-in-Water Emulsions for 5% Alpha-Hydroxy Acids

Formulation A	g/100 g Emulsion
Phase A	
Polyoxyethylene (100) glycerol stearate	2.0
(ICI Surfactants, Wilmington, DE)	
Mineral oil (light)	10.0
(Penreco, Karns City, PA)	
Cetearyl alcohol	3.0
(Henkel Corp., Hoboken, NJ)	
Phase B	
Laureth-4	1.0
(Lipo Chemicals, Paterson, NJ)	
Propylene glycol	5.0
(Aldrich Chemical Co., Milwaukee, WI)	
Alpha-hydroxy acid	5.0
Buffer	73.0
(phthalate-HCl[a] or potassium phosphate-NaOH[b])	
Phase C—Preservative	
Methyl-p-hydroxybenzoate	0.5
(Pfaltz & Bauer, Inc., Stamford, Ct)	
Propyl-p-hydroxybenzoate	0.5
(Pfaltz & Bauer, Inc., Stamford, Ct)	

[a]Phthalate-HCl buffer (pH = 3): 50 ml of 0.1 M potassium biphthalate + 22.3 ml of 0.1 M HCl diluted with water to 100 ml.
[b]Potassium phosphate–NaOH buffer (pH = 7): 50 ml of 0.1 M potassium phosphate + 29.1 ml 0.1 M NaOH diluted with water to 100 ml.
Source: From Kraeling and Bronaugh (15).

Emulsions containing 5% AHAs were prepared by dissolving the acid in either the pH 3 or pH 7 buffer, readjusting the buffer to the proper pH, and then mixing with the other ingredients in phase B. Phases A and B were heated separately to 75–80°C; then phase B was added to phase A and mixed at high shear in an Omni-Mixer homogenizer (Omni International, Warrenton, VA) for 1 h. Mixing was continued at a lower shear until the temperature of the emulsion reached room temperature. Phase C, the preservative, was then added and the emulsion stirred for an additional 30 min. For the 0.5% emulsions, a stock emulsion (containing no AHA) was prepared and then appropriate amounts of AHA were added to aliquots of the stock emulsion to give the desired concentration of AHA.

C. Percutaneous Absorption Experiments

Skin absorption studies were conducted by using human skin freshly obtained from abdominoplasty procedures. The skin was placed in a saline solution or in HEPES-buffered Hanks balanced salt solution (HHBSS) at the clinic and kept in cool packing as it was transported to the laboratory. Tissue transported in saline was immediately transferred to HHBSS at the laboratory. Subcutaneous fat was removed from the skin, and the surface was gently cleaned with a 10% soap solution and rinsed with distilled water. The skin was mounted on a Styrofoam block and cut with a Padgett dermatome (Padgett Instruments, Dermatome Division, Kansas City, MO) to a thickness of 200–340 μm. Skin discs were prepared with a punch and placed epidermal side up in Teflon flow-through diffusion cells (16). Prior to assembly, the flow-through diffusion cell system was disinfected with 70% ethanol and rinsed with receptor fluid. The diffusion cells were maintained at 35°C in an aluminum holding block heated by a circulating water bath; this maintained the surface temperature of the stratum corneum at 32°C. The skin was perfused with HHBSS, pH 7.4, receptor fluid at a flow rate of 1.5 ml/h to maintain the viability of the skin in the diffusion cells for the duration of the 24-h study (17).

A 20-min skin barrier integrity check, using [^{3}H]water, was conducted prior to the application of the AHA test formulations to ensure that the permeability of the human skin was in the normal range and that the skin was not damaged (18). Cells in which the percent of the applied dose of [^{3}H]water absorbed through the skin was greater than the historical limit of 0.35% were discarded.

The AHA test formulations were prepared to give an average dose of 0.55 μCi of ^{14}C-radiolabeled AHA per cell. The emulsion was applied to the skin at 3 mg/cm^2 of exposed skin in the diffusion cells (exposed skin = 0.64 cm^2). At the end of each experiment the skin surface was washed three times with 0.3 ml of a 10% soap solution and rinsed three times with 0.3 ml of distilled water to remove unabsorbed material remaining on the surface of the skin. The skin was removed from the diffusion cell and tape-stripped with Scotch Magic cellophane tape (3M Commercial Office Supply Division, St. Paul, MN) 10 times to remove the stratum corneum.

The remaining epidermis was separated from the dermis by wrapping the skin in Saran Wrap plastic wrap (DowBrands L.P., Indianapolis, IN) and submerging it in a 60°C water bath for 40 s. The skin was unwrapped and the epidermis was then slowly peeled from the dermis. The epidermis and dermis were cut into thin strips with a razor and digested with tissue solubilizer.

III. ALPHA-HYDROXY ACID ANALYSIS

The absorbed radiolabeled material in the 6-h receptor fluid fractions and the skin layers was measured by liquid scintillation counting (Minaxiβ Tri-Carb 4000 series liquid scintillation counter, Packard Instrument Co., Downers Grove, IL) using Ultima Gold (Packard Instrument Co., Meriden, CT) liquid scintillation cocktail.

IV. BARRIER INTEGRITY DETERMINATIONS

Barrier integrity of hairless guinea pig skin after 24-h exposure to glycolic acid formulations was assessed by measuring the steady-state rate of penetration of [^{3}H]water and then calculating a permeability constant (K_p). Skin from 4- to 6-month-old male hairless guinea pigs [strain Crl:AF/HA (hr/hr)Br](Charles River Laboratories, Wilmington, MA) was dermatomed to a thickness of 200–300 μm and assembled into flow-through diffusion cells. Glycolic acid formulations were applied to the surface of the skin (3 mg/cm^2), while some diffusion cells containing skin were left untreated (control skin). After 24 h, the surface of the skin (including untreated control skin) was washed three times with 0.3 ml of a 10% soap solution, rinsed three times with distilled water, and blotted dry with a cotton-tipped applicator. [^{3}H]Water (2.34 to 2.88 μCi) was applied in excess (800 μl) to the surface of the skin, the diffusion cell was covered, and effluent from the flow cell was collected every half hour until a steady-state rate of permeation was established (about 4 to 4.5 h). Permeability constants were calculated by dividing the rate by the initial concentration of [^{3}H]water.

V. SKIN SURFACE pH MEASUREMENTS

The pH profile of human skin in flow-through diffusion cells was determined 24 h after application of an oil-in-water (O/W) emulsion (Formulation A, without AHA) at pH 3.0. The O/W emulsion was applied to the surface of the skin (3 mg/cm^2) and after 24 h the skin was washed, rinsed, and dried in a manner described previously. The skin was removed from the diffusion cell, and the pH of the skin surface was measured on a Corning pH meter model 320 (Corning Inc., Science Products Division, Corning, NY) using an MI-404 flat membrane pH electrode with an MI-402 micro-reference electrode (Microelectrodes, Inc., Bedford, NH). The layers of the stratum corneum were removed by stripping 15 times with cellophane tape. After each tape strip, the pH of the skin surface was measured.

VI. STATISTICAL ANALYSIS

Total absorption values of AHA, representing the combined absorption values for receptor fluid and skin (stratum corneum, viable epidermis, and papillary dermis), were compared by Student's *t*-test or a one-way analysis of variance (ANOVA, SigmaStat Statistical Software, Jandel Scientific Software, San Rafael, CA). The permeability constant (K_p) determinations for the barrier properties of skin were compared statistically by performing Student's *t*-test and ANOVA. The Student–Newman–Keuls test was used as the method for multiple pairwise comparisons at a significance level of $p <$ .05 (SigmaStat Statistical Software, Jandel Scientific Software, San Rafael, CA).

VII. RESULTS

The in vitro percutaneous absorption of glycolic acid was measured from an oil-in-water (O/W) emulsion (Formulation A) at a concentration of 5% at pH 3.0 and 7.0. Greater glycolic acid absorption was observed in all locations with the emulsion adjusted to pH 3.0. Total absorption of glycolic acid in 24 h decreased from 27.2% at pH 3.0 to 3.5% with the pH 7.0 emulsion (Table 2) (15). Significant amounts of glycolic acid were found in the receptor fluid at pH 3.0 (2.6%), but larger amounts were found in the skin layers (24.6%). Glycolic acid was found in the surface layer (the stratum corneum) (5.8%); however, greater amounts (18.8%) were found in the deeper skin layers (the viable epidermis and dermis).

In order to study the effects of surfactants on the percutaneous absorption of glycolic acid, a second emulsion (Formulation B) was prepared with 1% ammonium laureth sulfate (ALS). ALS is an ionic surfactant contained in some AHA rinse-off formulations. Use of Formulation B did not change the total absorption of glycolic acid at either pH (Tables 2 and 3) (15). However, the absorbed glycolic acid was distributed differently in the skin and receptor fluid at the end of the 24-h studies. A greater amount of

Table 2 Percent Applied Dose Absorbed of 5% AHA in Formulation A

Location	5% Glycolic acid		5% Lactic acid		5% 2-OH-hexanoic acid	
	pH 3	pH 7	pH 3	pH 7	pH 3	pH 7
Receptor fluid	2.6 ± 0.7^a	0.8 ± 0.3	3.6 ± 1.2^b	0.4 ± 0.1	$32.9 \pm 2.6^{a,b}$	1.0 ± 0.2
Stratum corneum	5.8 ± 2.8	1.2 ± 0.4	6.3 ± 1.4	3.2 ± 0.8	3.4 ± 0.4	2.8 ± 0.3
Viable epidermis	6.6 ± 2.5	0.8 ± 0.3	6.6 ± 0.9	3.2 ± 0.8	2.8 ± 1.4	3.7 ± 1.3
Dermis	12.2 ± 1.4^a	0.6 ± 0.2	13.9 ± 2.3^b	2.9 ± 1.3	$4.0 \pm 1.8^{a,b}$	2.0 ± 0.3
Total in skin	24.6 ± 4.0^a	2.6 ± 0.6	26.8 ± 4.5	9.4 ± 2.1	10.2 ± 3.3^a	8.4 ± 1.0
Total absorption	27.2 ± 3.3	3.5 ± 0.9	30.4 ± 3.3	9.7 ± 2.0	43.1 ± 5.9	9.4 ± 1.1

Values are the mean $\pm$ SEM of two to five determinations in each of three subjects. [a,b]Values obtained at pH 3.0 in each location with similar superscripts are significantly different from each other (ANOVA, $p < .05$).
Source: From Kraeling and Bronaugh (15).

Table 3 Percent Applied Dose Absorbed of 5% Glycolic Acid in Formulation B[a]

| | Percent applied dose absorbed | |
Location	pH 3.0	pH 7.0
Receptor fluid	12.2 ± 5.3	1.4 ± 0.7
Stratum corneum	2.4 ± 1.3	0.1 ± 0.0
Viable epidermis	11.6 ± 2.5	0.4 ± 0.2
Dermis	8.6 ± 2.0	0.4 ± 0.1
Total in skin	22.6 ± 3.2	0.9 ± 0.0
Total absorption	34.8 ± 3.9	2.3 ± 0.8

[a]Values are the mean ± SEM of two to six determinations from five donors (pH 3.0) and three determinations from three donors (pH 7.0).
Source: From Kraeling and Bronaugh (15).

the absorbed material was found in the receptor fluid with the use of Formulation B (Table 3), with approximately 12% of the applied dose completely penetrating skin at pH 3.0, as compared to 2.6% from the use of Formulation A, pH 3.0 (Table 2).

The effect of product formulation on the barrier properties of hairless guinea pig skin was studied in Formulations A and B (5% GA, pH 3.0), as well as two commercial AHA products (5% GA, pH 2.5, and 10% GA, pH 3.5). A [³H]water permeability constant was determined after 24-h exposure to each of the formulations (Fig. 1) (15). The average of K_p values for all formulations was higher than the untreated control value. However, one-way analysis of variance showed that none of the formulations were significantly different from each other ($p < .05$).

The correlation between GA absorption (Formulation B) through skin and the barrier properties of skin was determined in skin from the five human donors (Fig. 2) (15). The skin from all donors was within our historical normal limits of [³H]water absorption (≤0.35% of the applied dose) as assessed by the 20-min test prior to application of the GA formulations (18). Glycolic acid absorption through donor skin varied from 24 to 44% of the applied dose. A high correlation was observed between the water and glycolic acid absorption values ($r^2 = .92$) from each donor, indicating that the variability in glycolic acid absorption was associated with the normal variability in the barrier properties of human skin.

Formulation A was used to study the effects of chemical structure on the percutaneous absorption of a homologous series of AHA ingredients. The absorption of lactic acid (LA) and 2-hydroxyhexanoic acid was determined from the O/W emulsion at an AHA concentration of 5%, at pH 3.0

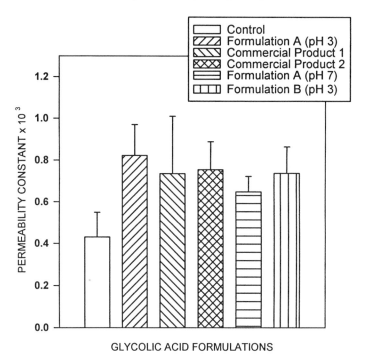

Figure 1 Effect of various glycolic acid formulations on the barrier properties of hairless guinea pig skin. The values are the mean ± SEM of three to four determinations in each of two to five animals. A one-way ANOVA indicated that none of the formulations were significantly different from each other ($p < .05$). From Kraeling and Bronaugh (15).

and 7.0 (Table 2). The pH-related differences in total absorption observed for these compounds were substantial but were less than the difference seen for GA. Total AHA absorption at pH 3.0 did not differ significantly between the 3 AHAs (ANOVA, $p = .087$). However, at pH 3.0, receptor fluid levels of 2-hydroxyhexanoic acid (32.9%) were significantly higher than for GA and LA (2.6% and 3.6%, respectively), and skin levels of 2-hydroxyhexanoic acid (10.2%) were significantly lower than the values for GA and LA (24.6% and 26.8%, respectively).

The longer chain AHAs were tested at 0.5% concentration because of solubility limitations and to simulate product usage (Table 4) (15). 2-Hydroxyhexanoic acid absorption studies were repeated at this lower concentration for comparison of absorption with chemical dose. The percentage of the applied dose absorbed into the receptor fluid is significantly lower at the

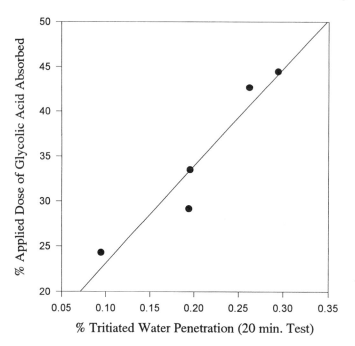

Figure 2   Human skin variability: correlation between barrier integrity and glycolic acid absorption. The values are the mean of two to six determinations in each of five subjects. From Kraeling and Bronaugh (15).

Table 4   Percent Applied Dose Absorbed of 0.5% AHA in Formulation A

Location	0.5% 2-OH-hexanoic acid, pH3	0.5% 2-OH-octanoic acid, pH3	0.5% 2-OH-decanoic acid, pH 3
Receptor fluid	10.1 ± 2.7	15.4 ± 3.1	8.8 ± 2.5
Stratum corneum	3.2 ± 0.9	1.4 ± 0.3	2.6 ± 0.6
Epidermis	8.4 ± 1.1[a]	2.8 ± 0.4[a,b]	5.8 ± 0.9[b]
Dermis	6.7 ± 0.7[a,b]	1.4 ± 0.2[a]	2.1 ± 0.3[b]
Total in skin	18.3 ± 2.6[a,b]	5.5 ± 0.9[a]	10.5 ± 1.0[b]
Total absorption	28.4 ± 3.9	21.0 ± 2.5	19.3 ± 3.1

Values are the mean ± SEM of two to four determinations in each of three subjects. [a,b]Values in each location with similar superscripts are significantly different from each other (ANOVA, $p < .05$).
Source: From Kraeling and Bronaugh (15).

0.5% dose level, but the percentage absorbed into skin and the total percentage absorbed are not statistically different (*t*-test, $p < .05$). There was no significant difference between 0.5% 2-hydroxyhexanoic, 2-hydroxyoctanoic, and 2-hydroxydecanoic acids with regard to total absorption (ANOVA, $p = .19$) or receptor fluid levels (ANOVA, $p = .28$). However, total in-skin values for 2-hydroxyhexanoic acid were significantly higher than corresponding values for the other AHAs.

The pH profile of human skin assembled in diffusion cells was measured (Fig. 3) (15). Skin from two donors was used. Initial skin surface pH values were approximately 5.3 for both donors. However, the pH of the stratum corneum increased gradually to 6.5 and 7.3 for the two donors as the stratum corneum layers were completely removed.

VIII. DISCUSSION

The percutaneous absorption of GA (Tables 2 and 3) was found to be dependent on the pH of the formulation. This finding was expected because the ionized molecule is more polar and therefore would be less readily absorbed. The effect of pH on the ionization of GA ($pK_a = 3.8$) can be calculated from the Henderson–Hasselbach equation (Fig. 4) (15). At pH 3.0, the GA remains mostly unionized (87%), and even at pH 3.8, 50% of the compound is in the unionized form. We evaluated the absorption of GA and other AHAs at pH 3.0 to simulate the acidic pH of some commercial cosmetic products containing these ingredients. The effect of product pH on the levels of AHA absorbed into various locations (receptor fluid and skin layers) is shown in Tables 2–4. The magnitude of reduction in absorption from a pH 7.0 product differed among the AHAs in these locations. Formulation pH made less of a difference in the skin levels obtained with LA and 2-hydroxyhexanoic acid than with GA. Even at pH 7.0, between 9 and 10% of the applied lactic and 2-hydroxyhexanoic acid was absorbed.

Compared with the two commercial AHA products, our formulations (Formulation A, pH 3 and pH 7, and Formulation B, pH 3) did not differ significantly in their effects on hairless guinea pig barrier integrity (Fig. 1). However, the differing GA absorption profiles with Formulations A and B (Tables 2 and 3) illustrate the potential effects of cosmetic vehicles on AHA absorption. Differences in skin response to AHA products with the same pH and AHA concentration may result from the effects of the vehicle on absorption.

The variability in absorption properties of normal human skin is illustrated in Fig. 2. Although all human skin samples used in these studies were within the historical limit for the normal range of human skin permeability, there was still a measurable amount of variability in GA skin absorption.

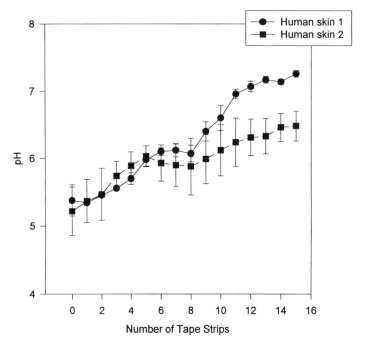

Figure 3 The pH profile of human skin in flow-through diffusion cells 24 h after application of an O/W emulsion pH 3.0. The values are the mean ± SEM of three determinations in each of two subjects. From Kraeling and Bronaugh (15).

Absorption values obtained from the skin of only one or two donors can be misleading, not only because of human skin barrier variability, but also if skin is damaged in harvesting or is stored before use. The use of a standard compound such as tritiated water to check barrier integrity of the skin aids in the assessment of the accuracy of absorption.

At pH 3.0, there were no significant differences in the total absorption of the AHA ingredients studied (Tables 2 and 4). However, there were some differences in levels found in the receptor fluid or skin locations. Levels of 2-hydroxyhexanoic acid are much higher in the receptor fluid and lower in skin than LA and GA (Table 2). The skin levels of 2-hydroxyoctanoic acid are lower than the other AHAs (Table 4). Comparison of the percent of the absorbed dose remaining in the skin (Fig. 5) (15) shows a tendency toward a decrease in the percent of the absorbed AHA remaining in skin and stratum corneum (data not shown) with the longer chain compounds. This is opposite to what might be expected for the more lipophilic compounds, such as 2-hydroxydecanoic acid, which could be retained in the skin. The octanol/

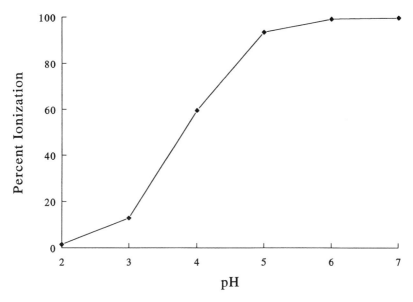

Figure 4 Effect of pH on glycolic acid ionization as determined by the Henderson–Hasselbach equation. From Kraeling and Bronaugh (15).

water (pH 3.0) partition coefficients for 2-hydroxyhexanoic, 2-hydroxyoctanoic, and 2-hydroxydecanoic acids were determined to be 3.7, 30.6, and 71.1, respectively.

However, these acids ionize to polar compounds at physiological pH as they enter and are absorbed through the stratum corneum. The pH of human stratum corneum in the diffusion cells was determined to range from initial surface values of approximately 5.3 up to values ranging from 6.5 to 7.3 at the stratum corneum–viable epidermal interface (Fig. 3). These values are in agreement with in vivo stratum corneum stripping studies that found a pH gradient in human stratum corneum ranging from pH 4.5 to 5.3 on the skin surface to a pH of about 7 at the viable epidermal layer (19). Therefore, these longer chain AHAs are not expected to form a reservoir in skin based on their lipid solubility properties.

The percutaneous absorption of GA has previously been reported from an aqueous solution through animal skin (20). Absorption values of 0.7 and 0.9% were reported in 8 h through minipig and hairless mouse skin, respectively, from a pH 3.8 aqueous solution. However, an infinite dose applied to skin of over 100 μl/cm^2 makes the values not relevant to consumer use conditions.

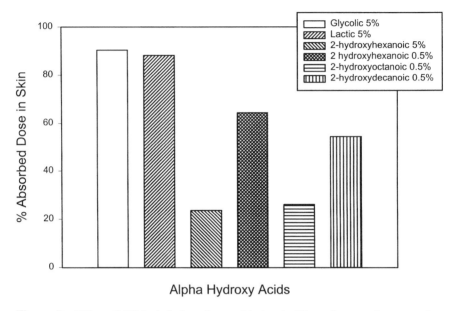

Figure 5 Effect of AHA chain lengths on skin levels. The values are the mean ± SEM of two to five determinations in each of three subjects. From Kraeling and Bronaugh (15).

We have found that AHAs are extensively absorbed into and through human skin from a relevant dose of an O/W emulsion adjusted to pH 3.0. Approximately 27% of the applied dose of glycolic acid was absorbed in 24 h, and there was no significant difference in total absorption when compared with values obtained from longer chain AHAs.

REFERENCES

1. Van Scott EJ, Yu RJ. Control of keratinization with alpha hydroxy acids and related compounds: I. Topical treatment of ichthyotic disorders. Arch Dermatol. 1974; 110:586–590.
2. Ridge JM, Siegle RJ, Zuckerman J. Use of alpha hydroxy acids in the therapy for "photoaged" skin. J Am Acad Dermatol. 1990; 23:932.
3. Ditre CM, Griffin TD, Murphy GF, Sueki H, Telegan B, Johnson WC, Yu RJ, Van Scott EJ. Effects of alpha-hydroxy acids on photoaged skin: A pilot clinical, histologic, and ultrastructural study. J Am Acad Dermatol. 1996; 34: 187–195.
4. Van Scott EJ, Yu RJ. Hyperkeratinization, corneocyte cohesion and alpha hydroxy acids. J Am Acad Dermatol. 1984; 11:867–879.

5. Van Scott EJ, Yu RJ. Alpha hydroxy acids: Procedures for use in clinical practice. Cutis 1989; 43:222–228.

6. Smith WP. Hydroxy acids and skin aging. Soap Cosmet Chem. 1993; 69:54–76.

7. Smith WP. Hydroxy acids and skin aging. Cosmet Toilet. 1994; 109:41–48.

8. Sargisson S. The AHA phenomenon continues. Drug Cosmet Ind. 1995; 156:34–36.

9. Kurtzweil P. Alpha hydroxy acids for skin care: smooth sailing or rough seas? FDA Consumer. 1998; 32:30–35.

10. Hagan DB, Parrott DT, Taylor AP. A study of the structure-activity relationships present in skin active agents. Int J Cosmet Sci. 1993; 15:163–173.

11. Leyden JJ, Lavker RM, Grove G, Kaidbey K. Alpha hydroxy acids are more than moisturizers. J Geriatr Dermatol. 1995; 3 (Suppl A):33A–37A.

12. Lavker RM, Kaidbey K, Leyden JJ. Effects of topical ammonium lactate on cutaneous atrophy from a potent topical corticosteroid. J Am Acad Dermatol. 1992; 26:535–544.

13. Bernstein EF, Uitto J. Connective tissue alteration in photoaged skin and the effects of alpha hydroxy acids. J Geriatr Dermatol. 1995; 3:7A–18A.

14. Griffin TD, Murphy GF, Sueki H, Telegan B, Johnson WC, Ditre CM, Yu RJ, Van Scott EJ. Increased factor XIIIa transglutaminase expression in dermal dendrocytes after treatment with alpha-hydroxy acids: Potential physiologic significance. J Am Acad Dermatol. 1996; 34:196–203.

15. Kraeling MEK, Bronaugh RL. In vitro percutaneous absorption of alpha hydroxy acids in human skin. J Soc Cosmet Chem. 1997; 48:187–197.

16. Bronaugh RL, Stewart RF. Methods for *in vitro* percutaneous absorption studies IV: The flow-through diffusion cell. J Pharm Sci. 1985; 74:64–67.

17. Collier SW, Sheikh NM, Sakr A, Lichtin JL, Stewart RF, Bronaugh RL. Maintenance of skin viability during in vitro percutaneous absorption/metabolism studies. Toxicol Appl Pharmacol. 1989; 99:522–533.

18. Bronaugh RL, Stewart RF, Simon M. Methods for in vitro percutaneous absorption studies VII: Use of excised human skin. J Pharm Sci. 1986; 75:1094–1097.

19. Öhman H, Vahlquist A. In vivo studies concerning a pH gradient in human stratum corneum and upper epidermis. Acta Derm Venerol (Stockh). 1994; 74:375–379.

20. Goldstein M, Brucks R. Evaluation of glycolic acid permeation through skin. Pharm Res. 1994; 11:S–180.

42
Optimizing Patch Test Delivery
A Model

Ronald C. Wester and Howard I. Maibach
University of California School of Medicine,
San Francisco, California

I. INTRODUCTION

A positive allergic or irritant response from a chemical applied to skin during patch testing requires that the chemical inherently be able to produce the toxicological response, and that the chemical be absorbed into the skin at a concentration sufficient to produce this response. Various test systems are available with which to apply a chemical to skin. If, for some reason, a test system decreased the absorption of chemical into the skin, then the changes for the chemical to produce a toxicological response are also reduced. This may result in a false-negative response (Wester and Maibach, 1983). The following is a brief explanation of a patch test, followed by the effect that a patch test system might have on the percutaneous absorption of the test chemical. It is paramount that the test chemical produce the results, not the test system.

II. METHODOLOGY

A. Conventional Patch Test

For routine use, a 4-h exposure for predictive testing for irritation was suggested by the National Academy of Sciences (NAS) panel. Although materials may be tested as described for the Draize-type test, it may be desirable to test dilutions of new materials and unknowns. If the test is intended to supplement Draize-type animal studies, 1-in gauze squares can be used with the same amount of material applied as in the animal study. Commercial patches, chambers, gauze squares, or cotton bandage material such as Webril (Curity) may be used. If commercial patches/devices are used, the volume of test material may need to be adjusted. Pads of these devices should be saturated with liquid, and/or a sufficient amount of solid/slurry should be used to cover the surface of the pad. Because of variations in patch size, it is helpful to express the dose in the form of mg/cm² and to report both the area of exposure and amount of material applied. Patches are secured in place with hypoirritating surgical tape without wrapping the trunk or arm. The degree of occlusion of the patch/device and of the surgical tape will affect the degree of irritation that develops. Increasing the degree of occlusion by use of Duhring or Hilltop chambers or occlusive tapes such as Blenderm will generally increase the severity of the response. When volatiles are tested under patch conditions (after screening in an open application test), a relatively nonocclusive tape should be used to minimize extreme irritation.

After the desired period of exposure, test patches/devices are removed and the area is rinsed with water to remove any residue. Because some tapes may produce skin damage that could be falsely attributed to the test material, test sites are marked to facilitate locating treatment sites for evaluation. Evaluation of the response is usually deferred for 30 min to 1 h after patch removal to allow hydration and pressure effect of the patch to subside. Test sites are reevaluated 24 h after patch removal. A scale is used for grading skin responses (Patil et al., 1996). Table 1 outlines the patch test system (Amin et al., 1996), Table 2 gives the patch test grading scale (Patil et al., 1996) and Table 3 gives variables influencing patch test sensitivity.

B. Percutaneous Absorption and Patch Testing

Table 3 is adapted from Patil et al. (1996) in that percutaneous absorption is added to both intrinsic and extrinsic listing. This is because the test chemical must get from the test system into the skin to create a biological response. This is illustrated with the chemical [^{14}C]paraphenylenediamine in a bioavailability study on patch test systems (Kum et al., 1987). [^{14}C]Paraphenylenediamine (PPDA; 6.25 µCi/mg) was the test chemical, formulated as 1% petrolatum (USP) and applied to each test system at a concentration of 2 mg/mm². Thus, the amount of PPDA was normalized to the surface area of each test system. The [^{14}C]PPDA was dissolved in propylene glycol for intraperitoneal injection. Percutaneous absorption was determined by urinary excretion of ^{14}C after intraperitoneal and topical administration to guinea pigs. Two days before chemical application, guinea pigs (Hartley strain, 300–500 g) were clipped and depilated (NEET; Wahlberg,

Table 1 Patch Test for Allergic Contact Dermatitis

Materials	(1) Allergens in vehicle (e.g., petrolatum, ethanol, water).
	(2) Vehicles.
	(3) Aluminum chambers (Finn chamber), Scanpor tape, and filter papers (for solutions).
	OR:
	(1) Ready-made patch test series (TRUE test).
Method	Patches on tape or ready-made patch test series are applied on intact skin of the back. Filter papers are used for solutions; 17 μl of allergen in vehicle is used for each patch. Ready-made patch series is applied as on similar skin sites. The patches are removed after 48 h.
Reading time	48 h and 96 h
Interpretation	Erythema and edema or more is positive. Distinguishing between allergic and irritant responses is critical: If the reaction spreads across the boundaries of the patch site, the reaction is more likely to be allergic. If the reaction peaks at 48 h and starts to fade rapidly after that, it may be irritant.
Precautions	Intense skin reactions possible; these can be treated with topical glucocorticosteroids. Active sensitization possible.
Controls	Required.

1972) on the dorsal midlumbar region. The PPDA in the test systems was placed on the skin and wrapped with gauze for 48 h. After 48 h, the chemical and test system were removed, the skin washed and soap and water, and the animal returned to the metabolic cage. Guinea pigs were housed in individual metabolic cages. Urine was collected at 24-h intervals for 5 days. Each urine sample was assayed for radioactivity content by liquid scintillation counting.

The chemical, concentration, and amount per area of skin were the same for each patch test system. Because the systems used were of different surface area, the dose applied was normalized at 2 mg PPDA/mm^2. The PPDA was applied to each test system, put on skin, and then held in place with gauze wrap. The chemical in the patch test system could then either diffuse into the skin or stay in the system (not move or diffuse into patch

Table 2 Human Patch Test Grading Scales

Number	Indication
	Detailed
0	No apparent cutaneous involvement
½	Faint, barely perceptible erythema or slight dryness (glazed appearance)
1	Faint but definite erythema, no eruptions or broken skin, *or* no erythema but definite dryness; may have epidermal fissuring
1½	Well-defined erythema or faint erythema with definite dryness; may have epidermal fissuring
2	Moderate erythema, may have a few papules or deep fissures, moderate to severe erythema in the cracks
2½	Moderate erythema with barely perceptible edema *or* severe erythema not involving a significant portion of the patch (halo effect around the edges), may have a few papules, *or* moderate to severe erythema
3	Severe erythema (beet redness), may have generalized papules *or* moderate to severe eythema with slight edema (edges well defined by raising)
3½	Moderate to severe erythema with moderate edema (confined to patch area) *or* moderate to severe erythema with isolated eschar formations or vesicles
4	Generalized vesicles or eschar formations *or* moderate to severe erythema and/or edema extending beyond the area of the patch
	Simple
0	Negative, normal skin
±	Questionable erythema not covering entire area
1	Definite erythema
2	Erythema and induration
3	Vesiculation
4	Bullous reaction

test structure). Factors such as conformity to skin surface and degree of occlusion could also influence the absorption from the system into skin.

Table 4 gives the percutaneous absorption of PPDA from various patch test systems. Analysis of variance was statistically significant for differences in absorption resulting from the test patch systems ($F = 4.83$; $p < .02$). The Newman–Keuls multiple-range test gave specific statistical differences ($p < .05$) between the Hill Top chamber and AL test, and between the Teflon

Table 3 Variables Influencing Test Sensitivity

Extrinsic	Intrinsic
Degree of occlusion	Anatomical site
Choice of vehicle	
Frequency of dosing	Concomitant disease
Duration of exposure	Species differences
Dose (concentration)	Age
Temperature	
Environmental conditions	Gender (effect disputed)
Altered barrier function	Race
Abrasion	Percutaneous absorption
Chemical damage	
Tape stripping	
Test system chemical delivery	

patch and AL test. Figure 1 shows PPDA absorption for the patch test systems and verifies the complete excretion for each. In decreasing order, the percentage absorption from the various systems were Hill Top chamber (53.4 ± 20.6) > Teflon patch (48.6 ± 9.3) > small Finn chamber with paper disk insert (34.1 ± 19.8), small Finn chamber (29.8 ± 9.0) > large Finn chamber > (23.1 ± 7.3) > AL test (8.0 ± 0.8). Note that there is a sixfold difference in the range of absorption. The highest efficiency of skin absorption was with the Hill Top chamber. Polikandritou and Conine (1985) performed comparative studies using the Hill Top chamber system and Webril patch system to compare delayed contact hypersensitivity. Reactions were induced at sig-

Table 4 Percutaneous Absorption of PPDA from Patch Test Systems

Patch test system	Total load in chamber (mg)	Concentration in chamber (mg/mm^2)	Absorption %	Total (mg)
Hill Top chamber	40	2	53.4 ± 20.6	21.4
Teflon (control)	16	2	48.6 ± 9.3	7.8
Small Finn chamber	16	2	29.8 ± 9.0	4.8
Large Finn chamber	24	2	23.1 ± 7.3	5.5
AL test chamber	20	2	8.0 ± 0.8	1.6
Small Finn chamber (with paper disk insert)	16	2	34.1 ± 19.8	5.5

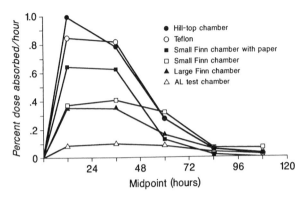

Figure 1   Rate of ${}^{14}C$ excretion after skin application of $[{}^{14}C]$-PPDA in various patch test systems to guinea pig.

nificantly lower concentrations for samples tested with the Hill Top chamber. The reason for this may have been higher skin absorption, as suggested by the results in this study. Further note that the two chambers that are identical except for size (Finn chamber: small and large) had similar delivery—quite different from what might be expected on the basis of the description.

An aluminum eye patch nonocclusive system was compared to a polypropylene nonocclusive patch system (Hill Top Co., Cincinnati, OH) for dose recovery and possible dose interference (Wester and Maibach, 1993). Table 5 shows that there were no differences in ${}^{14}C$ urinary excretion or ${}^{14}C$ total recovery ($p > .05$), but that a difference existed in ${}^{14}C$ recovery between

Table 5   Comparison of Topical Nonocclusive Systems for Alachlor Balance Study in Rhesus Monkey

	Percent applied dose	
System	Patch I	Patch II
Urinary excretion (7 days)	9.1 ± 6.3	10.1 ± 7.2
Nonocclusive cover	31.3 ± 9.0	4.3 ± 1.6[a]
Soap-and-water wash (24 h)	29.3 ± 8.3	62.3 ± 1.4[a]
Tape strip (day 7)	0.7 ± 0.1	2.1 ± 0.6
Total recovery	70.4 ± 5.8	78.9 ± 8.2

Note. Patch I, polypropylene patch with Goretex; Patch II, aluminum eye patch with Goretex; $n = 3$ animals per group.
[a]Significant difference between Patch I and Patch II ($p < .05$).

Table 6 Relationship of Percutaneous Absorption and Erythema for Several Oils Used in Cosmetics

Absorbability (greatest to least)	Erythema
Isopropyl myristate	++
Glycerol trioleate	−
n-Octadecane	±
Decanoxydecane	+
2-Hexyldecanoxyoctane	−

nonocclusive covers and soap-and-water skin surface wash ($p < .05$). The plastic system touched the topically applied dose and removed a large portion of that dose upon removal. However, this surface contact did not influence urinary ^{14}C excretion (skin absorption) or total recovery (Wester et al., 1992).

Table 6 gives the relationship of percutaneous absorption and erythema for several oils used in cosmetics (Wester and Maibach, 1994). There seems to be an agreement between absorbability and erythema, suggesting that skin absorption will play a part in patch test response. Note that these oils also can serve as vehicles and their biological response can be part of the toxicological response.

III. DISCUSSION

Of the two types of errors that can be produced in a study, the false-negative error is more serious than the false-positive, because it leads to the assumption that a chemical is not reactive when, in fact, it may be. Because a chemical must penetrate the skin at some base-level concentration to elicit a response, the skin absorption must be considered important. We therefore suggest that the bioavailability (rate and extent of absorption) be considered as important as the intrinsic activity of the chemical when a patch test (or any type of biological/toxicological response) is performed.

The differences noted here presumably add to the many other variables affecting patch test reactivity. This variable is readily measured and, if performed for some of the common antigens in the routine allergy series, might clarify how important a factor this represents. At the moment, we can state that it is a potentially large and clinically relevant variable that requires addressing, so that appropriate standards can be promulgated. In the interim, each patch test must assume that the negative and positive threshold levels must be made available for a given diagnostic patch unit.

The principles observed with these patch test units must also be relevant to the various types of transdermal delivery systems. The technology utilized here (^{14}C excretion in the urine) may be a more quantitative and clinically relevant method of assessing such patches than some of the currently employed methods (plasma area under the curve). This dermatopharmacokinetic approach is facile and robust; studies of this type should accelerate refinement of biologically relevant patch test systems.

REFERENCES

Amin, S. K., Lauerma, A. I., and Maibach, H. I. (1996). Diagnostic tests in dermatology: Patch and photopatch testing and contact urticaria. In Dermatotoxicology, 5th ed., eds. F. Marzulli and H. Maibach, pp. 505–513. Washington, DC: Taylor & Francis.

Kim, H. O., Wester, R. C., McMaster, J., Bucks, D. A. W., and Maibach, H. I. (1987). Variable skin absorption from toxicological patch test systems. Contact Dermatitis 17:178–180.

Patil, S. M., Patrick, E., and Maibach, H. I. (1996). Animal, human, and *in vitro* test methods for predicting skin irritation. In Dermatotoxicology, 5th ed., eds. F. Marzulli and Maibach, pp. 411–436. Washington, DC: Taylor & Francis.

Polikandritou, M., and Conine, D. L. (1985). Enhancement of the sensitivity of the Buehler method by use of the Hill Top chamber. J. Soc. Cosmet. Chem. 36: 159–168.

Wahlberg, J. E. (1972). Impairment of skin barrier function by depilatories. J. Invest. Dermatol. 59:160–162.

Wester, R. C., and Maibach, H. I. (1983). Cutaneous pharmacokinetics: 10 Steps to percutaneous absorption. Drug Metab. Rev. 14:169–205.

Wester, R. C., and Maibach, H. I. (1993). In vivo topical bioavailability and skin decontamination: example alachlor. In: Topical Drug Bioavailability, Bioequivalence, and Penetration, eds. V. Shah and H. Maibach, pp. 197–207. New York: Plenum Press.

Wester, R. C., and Maibach, H. I. (1994). Percutaneous absorption and cosmetics. In Cosmetic Dermatology, eds. R. Baran and H. Maibach, pp. 55–60. London: Martin Duntz.

Wester, R. C., Melendres, J., and Maibach, H. I. (1992). In vivo percutaneous absorption and skin decontamination of alachlor in rhesus monkey. J. Toxicol. Environ. Health 36:1–12.

43
Permeation of *N*-Nitrosamines Through Human Skin

Keith R. Brain
An-eX Analytical Services Ltd., Cardiff, Wales

Stephen D. Gettings
Avon Products, Inc., Suffern, New York

Kenneth A. Walters
An-eX Analytical Services Ltd., Cardiff, Wales

I. INTRODUCTION

Although there has been no clear demonstration of the contribution of ni-trosamines to the occurrence of cancer in humans, awareness of the carci-nogenic potential of *N*-nitrosamines in animals has existed for over 30 years and therefore the importance of minimizing human exposure to these compounds is well recognized (1). A number of *N*-nitroso compounds (nitrosamines) have been detected as low-level (parts per billion) contami-nants in a wide range of matrices, including industrial, pharmaceutical, food, and cosmetic materials (2). Cosmetics containing nitrosamines may be con-sidered adulterated and therefore subject to regulatory action by the U.S. Food and Drug Administration (FDA) (3) and to the enforcement pro-visions of California Proposition 65 (4).

Increasing sophistication in analytical technology has resulted in the ability to detect very low levels of analytes and has allowed the determi-nation of trace contaminants in cosmetic preparations. In most cases the presence of extremely low levels of contaminants in such products is of minimal relevance to product safety, but it is important to consider the im-plications of such contamination in the context of human exposure to a product during normal everyday use by the consumer.

Table 1   ECETOC Classification of Nitrosamines (5)

Group I	Steam-volatile, short-chain (e.g., N-dimethylnitrosamine [DMN]) or heterocyclic (e.g., N-nitrosopyrrolidone) compounds
Group II	Other volatile, long-chain compounds (e.g., N-nitroso-N-methyldodecylamine [NDOMA])
Group III	Nonvolatile, high-polarity compounds (e.g., N-nitrosodiethanolamine [NDELA])

Nitrosamines display a wide range of physical characteristics and have been categorized by ECETOC into three broad classifications (Table 1) (5). Physicochemical characteristics (6) of representative compounds from each category are given in Table 2, and corresponding structures are shown in Fig. 1. From a consideration of the wide variation in physicochemical properties of these nitrosamines, large differences in percutaneous penetration can be expected between individual compounds.

The contamination of personal care products with nitrosamines has been reviewed and methods of detection and analysis described elsewhere (5,6,7–13). N-Nitrosamines are formed by the interaction of a nitrogen-containing organic compound, such as an amine or amide, and a nitrosating agent, such as nitrogen oxide. Although all types of amines are nitrosable (14), the rate of nitrosation of tertiary amines is reported to be 10,000 times slower than that of related secondary amines (15). Products containing amines or amino derivatives may form nitrosamines if they also contain an ingredient that can act as a nitrosating agent [e.g., 2-bromo-2-nitropropane-1,3-diol, 5-bromo-5-nitro-1,3-dioxane, or tris(hydroxy-methyl)nitromethane], or if they are contaminated with a nitrosating agent (e.g., sodium nitrite). Other possible precursors in cosmetic products are quaternary amines, amine oxides, and secondary and tertiary amides (14). Amines and their derivatives

Table 2   Physicochemical Characteristics of Representative Nitrosamines (ECETOC Groups I, II, and III) (2)

	Boiling point (°C)	cLogP[a]
DMN	148	-0.64 ± 0.23
NDOMA	290	5.21 ± 0.23
NDELA	125	-1.89 ± 0.36

[a]Calculated using ACD/LogP (ACD, Inc., Toronto, Canada).

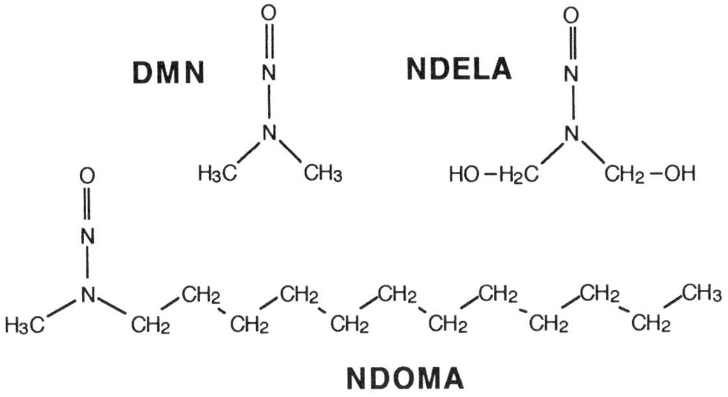

Figure 1 Structures of representative nitrosamines (ECETOC groups I, II, and III). DMN, *N*-dimethylnitrosamine; NDELA, *N*-nitrosodiethanolamine; NDOMA, *N*-nitroso-*N*-methyldodecylamine.

are commonly present in creams, cream lotions, hair shampoos, and cream hair conditioners. Nitrosation may occur during manufacture or during product storage. NDELA (*N*-nitrosodiethanolamine) has been detected in cosmetic formulations at levels ranging from the limit of detection to 47 mg/kg (1), while measureable levels of DMN (*N*-dimethylnitrosamine) are much lower (9) possibly due to the high volatility of the latter). Formation of DMN may result from the nitrosation of elimination products during manufacture of formulations containing, for example, quaternary ammonium compounds, betaines, and amine oxides (10,11). Long-chain nitrosamines, such as NDOMA (*N*-nitroso-*N*-methyldodecylamine), may occur as contaminants in raw materials used in cosmetic and personal care products or be formed by nitrosation of amine oxides or contaminating amines during formulation and storage (10,11). In addition to cosmetic products, NDOMA has also been detected in household products (16).

II. ASSESSMENT OF CONSUMER EXPOSURE TO *N*-NITROSAMINES

Under appropriate conditions, manufacturers of personal care and related products must perform risk assessments in support of product use, and anticipated misuse, to ensure consumer safety and to comply with government regulations. Risk assessment comprises hazard identification, determination of dose-response, exposure assessment, and risk characterization. In order to be meaningful, any risk assessment must be based upon determination of exposure that truly reflects actual use conditions. In practice, the types of exposure encountered during the normal use of topically applied products by consumers involve application of small amounts of formulation (which may possibly contain very low levels of contaminants) to the skin. In addition, many products are of the "rinse-off" type, where actual exposure is markedly reduced as a consequence of the fact that the bulk of the applied material is washed off after a short time period.

Until recently, very little information on the penetration of nitrosamines through human skin (information essential for the estimation of exposure following topical application of cosmetic products) was available. Edwards et al. (17) detected the excretion of ~2% of the dose of NDELA in the urine of a volunteer after application of 12.7 g of a cosmetic containing 77 ppm NDELA to a 2090-cm^2 area for 7.75 h followed by washing. Marzulli et al. (18) examined in vivo permeation of NDELA from a cosmetic vehicle in pigs and monkeys and reported absorption of ~4–23% over 24 h. Bronaugh et al. (19) demonstrated that cosmetic vehicles had significant effects on the penetration of NDELA through excised human skin. However, more recent studies conducted by ourselves and others (20–22) have now considerably extended our knowledge base in this area.

III. DETERMINATION OF THE PENETRATION OF NITROSAMINES THROUGH HUMAN SKIN IN VITRO

Franz et al. (20) compared the permeation of NDELA through excised human skin in vitro under both finite and infinite dose conditions from both isopropyl myristate (IPM) and generic representative personal care formulations 10% shampoo dilution and sunscreen; Table 3). While absorption from any particular formulation will be ultimately dependent upon precise composition, the "generic" cosmetic formulations used in this study were generally representative of products of this type.

When an infinite dose of NDELA (500 $\mu g/cm^2$ at 1 $\mu g/\mu l$) was applied, the total percent absorption was greatest from the shampoo formulation (~35%) and very considerably higher from IPM (~25%) than from the sunscreen formulation (~4%) (Table 4). In the case of the shampoo formulation, there was evidence that the rate of absorption increased with time, indicative that this formulation (under these exposure conditions) caused skin barrier damage. This was not an unexpected consequence of unreasonably long-term exposure to such a shampoo product.

For the three finite doses of NDELA evaluated (0.6, 2.0, and 6.0 $\mu g/cm^2$ as 10 μl of 0.06, 0.2, and 0.6 $\mu g/\mu l$, respectively), the percent absorption was found to be largely independent of the concentration, that is, the total dose absorbed ($\mu g/cm^2$) was directly proportional to the concentration

Table 3 Generic Representative Personal Care Formulations Used in Infinite vs. Finite Study

Sunscreen (anionic oil/water emulsion)	
Water	57.90
Carbomer 940, 2% dispersion	10.00
Methylparaben	0.30
Glycerin	3.00
Stearic acid	5.00
Glyceryl stearate SE	2.00
Polysorbate 20	2.00
Cetyl alcohol	0.50
Stearyl alcohol	0.50
Dimethicone 200 cts	0.50
Octyl dimethyl PABA	8.00
Benzophenone-3	3.00
Octyl methoxycinnamate	5.00
Triethanolamine, 99%	2.00
DMDM hydantoin	0.30
Total	100.00
Shampoo	
Water	81.55
Sodium lauryl sulfate	15.00
Sodium chloride	3.00
Perfume	0.25
Preservative	0.20
Dyes	Trace
Total	100.00

Source: Reference 20.

Table 4 Percent Total Absorption from Finite and Infinite Doses of NDELA

	Infinite dose 500 μl/cm^2	Finite dose		
		10 μl/cm^2	10 μl/cm^2	10 μl/cm^2
Total dose (μg/cm^2)	500	0.6	2.0	6.0
Conc. (μg/μl)	1.0	0.06	0.20	0.6
IPM	24.9 ± 5.2	41.8 ± 10.9	65.2 ± 9.5	61.9 ± 13.9
Shampoo	35.4 ± 7.6	42.7 ± 15.1	42.9 ± 13.7	46.4 ± 12.4
Sunscreen	3.9 ± 1.6	37.3 ± 12.3	35.3 ± 8.3	41.4 ± 14.5

Source: Reference 20.

of NDELA in the vehicle) (Fig. 2). The total absorption of NDELA at 48 h was from 35 to 65% of the applied dose and was dependent upon the type of formulation. Absorption was greatest from IPM ($\sim$56%), and higher from the shampoo ($\sim$44%) than the sunscreen ($\sim$38%). This equated to a total absorption of $\sim$4–6 μg/cm^2 over 48 h from a NDELA dose of 10 μg/cm^2.

With infinite doses, total absorption from the sunscreen formulation was much less than from the IPM and shampoo vehicles, whereas at low finite doses there was little difference between total absorption from the formulations.

Recovery of NDELA in the various compartments at the end of these experiments (mass balance) is shown in Figs. 3 and 4. Total recovery was 81–87% for the infinite and 91–101% for the finite dose studies. In both cases a considerable quantity of NDELA was recovered by surface washing at the end of the experiment, although this amount appeared to be somewhat dependent upon the nature of the formulation. In the infinite dose study (Fig. 3) the recovery of NDELA from the skin fraction was much higher from the IPM vehicle ($\sim$20%) than from the generic cosmetic formulations (<2%).

In the finite dose studies (Fig. 4), epidermal and dermal compartments

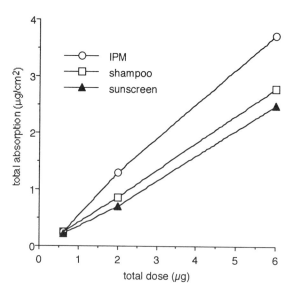

Figure 2 Linear relationship between total dose (μg) and total absorption (μg/cm^2) for three low finite doses of NDELA. From Reference 20.

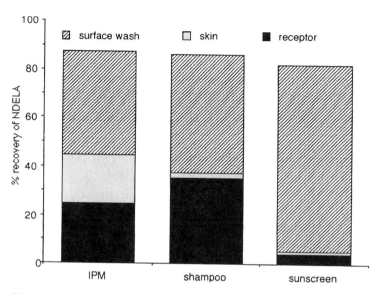

Figure 3 Mass balance of NDELA from infinite doses in three vehicles. From Reference 20.

were assessed separately. In all cases recovery from the epidermis was greater than from the dermis, although, under these finite dose conditions, there was little difference between the total recovery from the skin after treatment with the three different vehicles.

Infinite dose studies use conditions where the concentrations of neither the compound of interest, nor any of the vehicle components, are appreciably reduced during the experimental time frame. The advantage of this approach is that steady-state conditions are normally reached, which simplifies analysis of the data, although a major shortcoming is that the results cannot be directly extrapolated to "in-use" conditions. There were clear differences between the results obtained by Franz et al. (20) using infinite and finite dose conditions. It is the results from the finite dose experiments that are relevant for risk assessment in humans exposed to NDELA in cosmetics where it only occurs as a trace contaminant. Under "in-use" conditions the dose is clearly finite and dose depletion will commence as soon as percutaneous absorption occurs. The finite dose exposure values reported (20) were conservative estimates as they did not take into account practical factors, such as rinsing off shampoo shortly after use and loss of sunscreen by mechanical abrasion, swimming, etc.

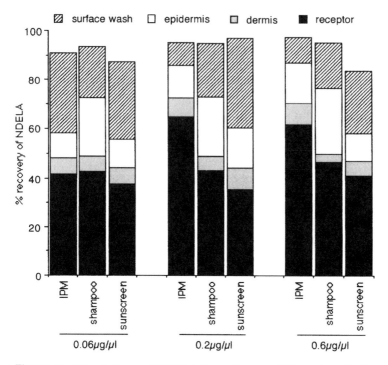

Figure 4   Mass balance of NDELA from three small finite doses in three vehicles. From Reference 20.

IV.   STRUCTURE–ACTIVITY RELATIONSHIPS

The percutaneous absorption of a compound is well known to be related to structural features. A comparison between the penetration of three possible contaminant nitrosamines (DMN, NDELA, and NDOMA), representing each of the ECETOC categories, is therefore of interest (20–22). Investigation of the effect of structure on fundamental permeation parameters requires the use of infinite doses, while assessment of the effect of structure on potential consumer exposure requires the use of finite doses under "in-use" conditions.

A.   Permeation from Infinite Doses of DMN, NDELA, and NDOMA in IPM

The data obtained from infinite doses (1 μg/μl) in IPM are shown in Table 5. The mean total absorption of NDELA over 48 h was 23.6 ± 6.4% (103.9

Table 5 Absorption and Recovery of DMN, NDELA, and NDOMA After Application as Infinite Doses (1 μg/μl) in IPM

Parameter	DMN	NDELA	NDOMA
Total absorption at 48 h (μg/cm^2)	11.3 $\pm$ 5.52	103.9 $\pm$ 28.4	0.445 $\pm$ 0.036
Total absorption at 48 h (%)	2.57 $\pm$ 1.22	23.6 $\pm$ 6.40	0.098 $\pm$ 0.008
Recovery in wash (%)	5.04 $\pm$ 4.52	40.5 $\pm$ 8.69	90.1 $\pm$ 1.05
Recovery in skin (%)	0.78 $\pm$ 0.48	45.9 $\pm$ 5.33	0.79 $\pm$ 0.21
Total recovery (%)	8.39 $\pm$ 6.04	109.9 $\pm$ 5.04	90.9 $\pm$ 1.18
Permeability coefficient (cm/h)	nd	4.1×10^{-3}	9.0×10^{-6}
Lag time (h)	nd	5.4	4.5

Note. All data = mean $\pm$ standard error; nd = not determinable.
Source: References 20 and 21.

$\pm$ 28.4 μg/cm^2), which correlated well with the value of 24.9 $\pm$ 5.2% reported by Franz et al. (20). However, mean total absorption from the shorter chain compound DMN was unexpectedly an order of magnitude lower at 2.57 $\pm$ 1.22% (11.3 $\pm$ 5.52 μg/cm^2), while that from the long-chain compound NDOMA was, as anticipated, almost three orders of magnitude lower at 0.098 $\pm$ 0.008% (0.445 $\pm$ 0.036 μg/cm^2). The majority of the dose of the long-chain NDOMA was recovered in the postexposure skin wash (90.1 $\pm$ 1.05%), whereas less than half (40.5 $\pm$ 8.69%) of the more polar NDELA was removed by washing, and only a small proportion (5.04 $\pm$ 4.52%) of the short-chain DMN was recovered in this fraction (Fig. 5).

B. Permeation of Finite Doses of DMN from Representative Formulations

Permeation of DMN was investigated from finite doses of vehicles typical of those which might be contaminated with this material (Table 6). The average total absorption of DMN from a finite dose in a representative o/w (oil-in-water) emulsion over 48 h, expressed as a percent of the applied dose, was 3.96 $\pm$ 0.25% (Table 7). While the flux rate over the first 3 h of exposure was rapid (Fig. 6), this rate declined thereafter to close to zero. Similar results were observed for absorption of DMN from a shampoo vehicle that was rinsed off 10 min after dosing. Much less DMN was recovered from the skin fraction after dosing with the shampoo ($\sim$0.08%) than after treatment with the o/w emulsion ($\sim$1.2%). Total recovery of DMN after dosing with sunscreen was very low ($\sim$7.3%), and even recovery from the

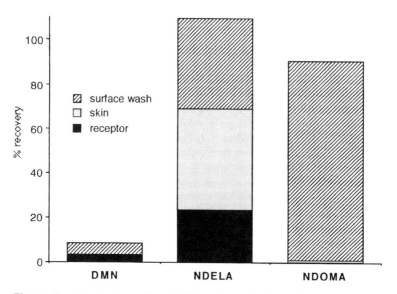

Figure 5 Mass balance from infinite doses of three nitrosamines in IPM. From References 20 and 21.

shampoo vehicle rinsed off after 10 min was relatively low (~73%). Consideration of the high volatility of DMN led to the conclusion that the rate of evaporative loss of DMN from the skin surface was far higher than the rate of absorption, resulting in an overall low uptake. Thus, although DMN has a very high potential for skin penetration, this is negated in practice by rapid evaporative loss.

C. Permeation of Finite Doses of NDOMA from Representative Formulations

Permeation of NDOMA was investigated from finite doses of vehicles typical of those which might be contaminated with this material (Table 8). The average total absorption of NDOMA after application as a finite dose in the o/w formulation over 48 h, expressed as a percent of the applied dose, was $4.66 \pm 0.76\%$ (Table 9 and Fig. 7), whereas total absorption from the shampoo which was rinsed off after 10 min was considerably lower ($0.75 \pm 0.17\%$). Recovery of NDOMA from the skin fractions was similar in this case (~5.4% from o/w emulsion and 6.7% from shampoo). The data were consistent with the hypothesis that NDOMA rapidly partitions into the skin, but only partitions out slowly into an aqueous environment.

Table 6 Composition of Vehicles Used in DMN Study

Vehicle	%
o/w Emulsion	
Amphisol	3.00
Stearic acid	4.00
Cetyl alcohol	2.00
Dimethicone 200	0.30
Cetiol LC	8.00
Water	75.40
Propylene glycol	5.00
Carbomer 940	0.15
Water + triethanolamine (99%)	1.15
Germaben IIE	1.00
Total	100.00
Shampoo	
Water	48.00
Sodium lauryl sulfate	50.00
Sodium chloride	1.50
Perfume	0.25
Preservative	0.20
Dyes	0.05
Total	100.00

Source: Reference 21.

Table 7 Absorption and Recovery of DMN After Application as Finite Doses in o/w Emulsion and Shampoo Formulations

Parameter	o/w Emulsion	Shampoo
Total absorption at 48 h ($\mu g/cm^2$)	0.0042 ± 0.0003	0.0003 ± 0.00002
Total absorption at 48 h (%)	3.98 ± 0.25	1.10 ± 0.08
Recovery in 10-min rinse	n/a	71.82 ± 2.53
Recovery in wash (%)	2.09 ± 0.04	0.04 ± 0.00
Recovery in skin (%)	1.20 ± 0.02	0.08 ± 0.00
Total recovery (%)	7.27 ± 0.85	73.04 ± 2.74

Note. All data = mean ± standard error.
Source: Reference 21.

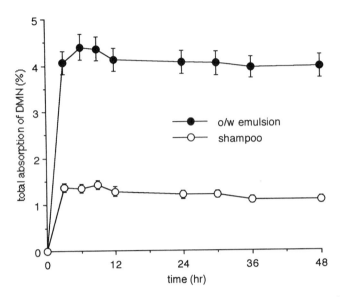

Figure 6 Cumulative absorption of finite doses of DMN from representative vehicles over 48 h. From Reference 21.

Table 8 Composition of Vehicles Used in NDOMA Study

Vehicle	%
o/w Emulsion	
Methyl glucose sesquistearate	0.50
PEG-20 methyl glucose sesquistearate	1.50
Mineral oil	6.00
C12/15 alkyl benzoate	1.00
Cetyl alcohol	0.80
Methyl paraben	0.15
Hydroxyethylcellulose	0.50
Distearyldimonium chloride	1.00
Quaternium-15	0.20
Water	88.35
Total	100.00
Shampoo: as for DMN study	

Source: Reference 22.

Table 9 Absorption and Recovery of NDOMA After Application as Finite Doses in o/w Emulsion and Shampoo Formulations

Parameter	o/w Emulsion	Shampoo
Total absorption at 48 h (μg/cm^2)	0.0363 $\pm$ 0.0069	0.0122 $\pm$ 0.0028
Total absorption at 48 h (%)	4.66 $\pm$ 0.76	0.75 $\pm$ 0.17
Recovery in 10-min rinse	n/a	71.72 $\pm$ 3.63
Recovery in wash (%)	80.44 $\pm$ 4.27	13.15 $\pm$ 1.53
Recovery in skin (%)	5.43 $\pm$ 0.68	6.67 $\pm$ 0.57
Total recovery (%)	90.53 $\pm$ 4.20	92.29 $\pm$ 2.29

Note. All data = mean $\pm$ standard error.
Source: Reference 22.

V. DISCUSSION

The data published to date demonstrate that while dermal absorption of *N*-nitrosamines is related to structure, the situation is more complex than might have been expected. Permeation from a compound with a long alkyl chain (NDOMA) was extremely low, but the short-chain DMN also showed unexpectedly low uptake due to rapid loss by evaporation. Clearly, devel-

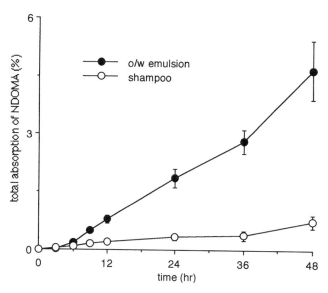

Figure 7 Cumulative absorption of NDOMA from representative vehicles over 48 h. From Reference 22.

opment of a meaningful generic predictive model for nitrosamine absorption requires the generation of further human skin permeation data on additional pertinent compounds.

It is clear from the studies discussed here that permeation of nitrosamines under "in-use" conditions was low. Under finite conditions the percentage absorption was independent of the dose. It follows that human systemic exposure to N-nitrosamines as low-level impurities in products will be dependent upon the level of contamination. Although short-chain compounds, such as DMN, have a relatively high intrinsic permeation rate, their high volatility results in such high evaporative loss that the actual skin absorption is very low. Long-chain compounds, such as NDOMA, partition into the skin relatively readily, but partition out into an aqueous environment only slowly, which significantly reduces their systemic absorption. The rate and extent of absorption were dependent on the application vehicle and time of exposure.

Appropriate use of in vitro percutaneous absorption data for risk assessment should consider a number of critical factors, including the frequency and extent of exposure (i.e., how often a product is used and the amount of body surface area exposed), the application site (e.g., head and neck vs. trunk), the conditions of use (e.g., neat or diluted, rinse-off or leave-on), the concentration of the permeant, and the nature of the vehicle (23). Products such as sunscreens are applied at levels of ~ 2 mg/cm^2, whereas preparations such as shampoos may have an application dose as high as 30 mg/cm^2 (15 g applied to ~ 500 cm^2 of scalp). However, in the latter case most of the applied material coats the hair and is unavailable for absorption.

For risk assessment purposes it may be assumed that, following finite-dose exposure to a leave-on product, ~ 4–5% of any DMN or NDOMA present may penetrate the skin (21,22). For a rinse-off product this is reduced to $\sim 1\%$. Absorption from a finite dose of NDELA from a leave-on product was an order of magnitude higher. Because the initial NDELA studies (20) described here were designed to estimate the total possible percutaneous absorption, actual exposures from cosmetic products are likely to be grossly overestimated by relying solely on the value of 40% of the applied dose absorbed reported. In contrast to the procedure in the other studies described (21,22), the NDELA shampoo formulation was not removed (and the skin was not rinsed) after 10 min of exposure. In addition, absorption kinetics indicated a likely perturbation of skin-barrier function (and hence increased absorption of penetrants) produced by the protracted (48-h) exposure to high levels of surfactants in the formulation (20). Although no data are available on absorption of NDELA after exposure to a rinse-off product under "in-use" conditions, it can reasonably be extrapolated that a similar reduction in absorption will occur.

In conclusion, it is important to remember that when data derived from in vitro percutaneous penetration studies are used for risk assessment purposes, other considerations must be incorporated into the final estimate of exposure. For example, removal and rinsing (as in normal product use) substantially reduces the amount of applied material subsequently available and hence absorption. If the applied dose of NDELA were 100 ng, the data reported by Franz et al. (20) suggest that very little (less than 1%) would be absorbed in the first 10 min. Based on data obtained for DMN and NDOMA, which suggest that at least 72% of the applied dose is readily removed by gently rinsing the skin after 10 min of exposure, it is reasonable to assume that the actual amount of NDELA available for absorption would be reduced to 30 ng. If 40% (12 ng) of this amount was subsequently absorbed, this represents only 12% of the total "applied dose." Under conditions of actual product use, briefer periods of initial exposure (less than 10 min) and more vigorous rinsing than is feasible for fragile in vitro test systems would further tend to minimize consumer exposure.

Similarly, formulations that interfere with evaporative loss could well yield different absorption profiles for DMN than those evidenced here. Moreover, exposure conditions resulting in partial or full occlusion could result in a greater degree of penetration of DMN than has been reported (21). On the other hand, loss of applied or deposited contaminants does not occur to any meaningful extent under the experimental conditions that have been used in vitro, whereas loss due to friction (e.g., from clothing) or bathing may be significant under conditions of actual product use. Each of these factors should be carefully considered when using data derived from human skin penetration studies in vitro for purposes of comprehensive risk assessment.

REFERENCES

1. ECETOC. (1990). Technical Report No. 41. Human exposure to N-nitrosamines, their effects, and a risk assessment for N-nitrosoethanolamine in personal care products. European Chemical Industry Ecology and Toxicology Centre, Brussels.
2. Klein, R. G. (1982). Calculations and measurements on the volatility of N-nitrosamines and their aqueous solutions. Toxicology, 23, 135–147.
3. 44 Fed. Reg. 21365 (April 10, 1979).
4. California Code of Regulations, Title 22, Section 12000.
5. ECETOC. (1991). Technical Report No. 42. Critical evaluation of methods for the determination of N-nitrosamines in personal care and household products. European Chemical Industry Ecology and Toxicology Centre, Brussels.
6. Havery, D. C., and Chou, H. J. (1994). N-Nitrosamines in cosmetic products. An overview. Cosmet. Toilet., 109, 53–62.

7. Fan, T. Y., Goff, U., Sing L., Fine, D. H., Arsenault, G. P., and Biemann, K. (1977). N-Nitrosodiethanolamine in cosmetics, lotions and shampoos. Food Cosmet. Toxicol., 16, 423–430.

8. Westin, J. B., Spiegelholder, B., Preussmann, R., and Shani, J. (1990). Assay of suntan lotions for the carcinogenic, non-volatile N-nitrosamine N-nitroso-diethanolamine. Cancer Lett., 50, 157–160.

9. Spiegelholder, B., and Preussmann, R. (1984). Contamination of toiletries and cosmetic products with volatile and non-volatile N-nitroso carcinogens. J. Cancer Res. Clin. Oncol., 108, 160–163.

10. Hecht, S. S., Morrison, J. B., and Wenninger, J. A. (1982). N-Nitroso-N-methyldodecylamine and N-nitroso-N-methyltetradecylamine in hair-care products. Food Chem. Toxicol., 20, 165–169.

11. Morrison, J. B., Hecht, S. S., and Wenninger, J. A. (1983). N-Nitroso-N-methyloctadecylamine in hair-care products. Food Chem. Toxicol., 21, 69–73.

12. Ikeda, K., and Migliorese, K. G. (1990). Analysis of nitrosamines in cosmetics. J. Soc. Cosmet. Chem., 41, 285–333.

13. Chou, H. J., and Yates, R. L. (1998). A rapid and selective method for determining potential nitrosating agents in cosmetic products by chemiluminescence detection of nitric oxide, J. AOAC Int., 81, 368–372.

14. Douglass, M. L., Kobacoff, B. L., Anderson, G. A., and Cheng, M. C. (1978). The chemistry of nitrosamine formation, inhibition and destruction. J. Soc. Cosmet. Chem., 29, 581–606.

15. Mirvish, S. S. (1975). Formation of N-nitroso compounds. Chemistry, kinetics and in vivo occurrence. Toxicol. Appl. Pharmacol., 31, 325–351.

16. Kemp, E., and Eisenbrand, G. (1991). Long-chain N-nitroso-N-methylalkyl-amines in commercial cosmetics, light-duty dishwashing liquids and household cleaning products. Food Chem. Toxicol., 29, 203–209.

17. Edwards, G. S., Peng, M., Fine, D. H., Spiegelholder, B., and Hann, J. (1979). Detection of N-nitrosodiethanolamine in human urine following application of a contaminated cosmetic. Toxicol. Lett., 4, 217–222.

18. Marzulli, F. N., Anjo, D. M., and Maibach, H. I. (1981). In vivo skin penetration studies of 2,4-toluenediamine, 2,4-diaminoanisole, 2-nitro-p-phenylene-diamine, p-dioxane and N-nitrosodiethanolamine in cosmetics. Food Cosmet. Toxicol., 19, 743–747.

19. Bronaugh, R. L., Congdon, E. R., and Scheuplein, R. J. (1981). The effect of cosmetic vehicles on the penetration of N-nitrosoethanolamine. J. Invest. Dermatol., 76, 94–96.

20. Franz, T. J., Lehman, P. A., Franz, S. F., Demetrulias, J. L., North-Root, H., Kelling, C. K., Moloney, S. J., and Gettings, S. D. (1993). Percutaneous penetration of N-nitrosoethanolamine through human skin (in vitro); Comparison of finite and infinite dose applications from cosmetic vehicles. Fundam. Appl. Toxicol., 21, 213–221.

21. Brain, K. R., Walters, K. A., James, V. J., Dressler, W. E., Howes, D., Kelling, C. K., Moloney, S. J., and Gettings, S. D. (1995). Percutaneous penetration of dimethylnitrosamine through human skin in vitro: Application from cosmetic vehicles. Food Chem. Toxicol., 33, 315–322.

22. Walters, K. A., Brain, K. R., Dressler, W. E., Green, D. M., Howes, D., James, V. J., Kelling, C. K., Watkinson, A. C., and Gettings, S. D. (1997). Percutaneous penetration of N-nitroso-N-methyldodecylamine through human skin in vitro: Application from cosmetic vehicles. Food Chem. Toxicol., 35, 705–712.
23. Gettings, S. D., Azri-Meehan, S., Demetrulias, J. L., Dressler, W. E., North-Root, H., Kelling, C. K., Moloney, S. J., and Gettings, S. D. (1993). The use of in vitro skin penetration data in the safety assessment of cosmetic formulations. In Prediction of Percutaneous Penetration, Vol. 3b, eds. K. R. Brain, V. J. James, and K. A. Walters, pp. 621–637. Cardiff: STS Publishing.

44
Nail Penetration
Focus on Topical Delivery of Antifungal Drugs for Onychomycosis Treatment

Ying Sun, Jue-Chen Liu, and Jonas C. T. Wang
Johnson & Johnson Consumer Products Worldwide, Skillman, New Jersey

Piet De Doncker
Janssen Research Foundation, Beerse, Belgium

I. PHYSICAL AND CHEMICAL PROPERTIES OF HUMAN NAIL PERTAINING TO TOPICAL DRUG DELIVERY

Nail properties pertaining to topical drug delivery are briefly discussed here. For more detailed information, several excellent reviews are available (1–3). The human nail plate is composed of approximately 25 layers of flattened, keratinized cells fused into a dense, hard, yet slightly elastic plate, with a typical thickness of 0.5–1.0 mm. These cells have their origin in the nail matrix, a living, highly proliferative epithelial tissue. The growth rate of normal fingernails varies from less than 1.8 mm to more than 4.5 mm/month (2). There are wide interpersonal differences in nail growth rate, although the difference is much less among members of the same family.

Nail growth is faster for the dominant hand, and faster for the male than the female (4). Nail growth also relates to age and environment; for example, nail growth slows down in older age, as well as in cold climates. Nails grow faster when regenerating after avulsion, and in persons with psoriasis and onycholysis. Accelerated nail growth has also been reported to be induced by treatment with drugs such as benoxaprofen, biotin, cystine, etretinate, methionine, levodopa, and itraconazole (5). The rate of nail growth is less than normal in persons with fever, decreased circulation, malnutrition, during lactation, with yellow nail syndrome, in acute infection, and during therapy with antimitotic drugs (2). The average rate of growth of 0.1 mm/day (3 mm/month) is generally used to predict nail replacement time. Toenails grow at one-half to one-third the rate of fingernails. A normal fingernail grows out completely in about 6 months, but a normal toenail takes 12 to 18 months (2). The nail plate overlays the nail bed, a soft and normally noncornified tissue. Although it appears visually uniform, a nail plate is made of at least two discernible macroscopic strata: the dorsal nail plate and intermediate nail plate. The dorsal nail plate is harder and thinner than the intermediate nail plate. A third layer is sometime observed, the ventral plate, which is very thin and contains only one or two layers of cells (1). A schematic diagram of a human nail with its surrounding tissues is shown in Fig. 1.

Although the nail plate is similar to stratum corneum of the skin in that it is also derived from epidermis, it is mainly composed of hard keratin with high content of disulfide linkages (cystine), and is approximately 100-fold thicker than stratum corneum. Table 1 summarizes some physical and chemical properties of nails pertaining to drug permeation. Although both are composed of keratinized dead cells, there are marked differences between nail plate and stratum corneum in their physical and chemical characteristics (Table 2). In comparison to the thin stratum corneum, the much thicker nail plate means a much longer diffusion pathway for drug delivery through the nail. In contrast with the elastic and pliable stratum corneum,

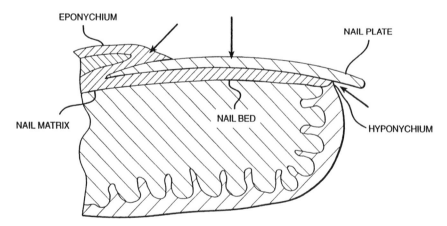

Figure 1 Schematic diagram of a human nail. The target sites of topical onychomycosis treatment are nail plate, nail bed, and nail matrix, where a topically administered antifungal drug should be able to reach with a concentration above its fungicidal level. Arrows indicate the potential points of entry for drug delivery.

Table 1 Physical Properties of Nails at Target Sites for Antifungal Drug Delivery

Target site	Characteristics
Nail plate	Thickness: approximately 0.5 mm for fingernails, up to 1.3 mm for great toenails
	Consists of three cornified keratin layers:
	Dorsal nail plate: dense and hard (approx. 200 μm thick)
	Intermediate nail plate: homogeneous, thick and less hard
	Ventral nail plate: very thin with only one or two layers of cells
Nail bed	Noncornified soft tissue underneath the nail plate
Nail matrix	A thick, highly proliferative epithelium tissue, the "root" of a nail plate.

the nail plate is dense and hard, especially the dorsal nail plate layer. Based on the biochemical composition of the nail, one would expect the permeation characteristics of the nail to be different from those of the stratum corneum. This turns out to be true from the evidence available to date. Some physical and chemical differences between a nail plate and stratum corneum are:

1. The total lipid content of the nail lies between 0.1 and 1%, as opposed to approximately 10% lipid content of stratum corneum (1), suggesting the role of the lipoidal pathway in the nail plate is probably much less important than that in the stratum corneum.
2. The nail has a high sulfur content (cystine) in its hard keratin domain, whereas the stratum corneum does not.
3. Under average conditions, the nail contains 7 to 12% water, in comparison to 25% water in the stratum corneum. At 100% relative humidity, the maximal water content in the nail is approximately 25%, in sharp contrast to the stratum corneum, which can take up water by 200–300% (6,7).

Table 2 Comparison of Nail Plate and Stratum Corneum

Composition	Nail	Stratum corneum
Thickness	500–1000 μm	10–40 μm
Disulfide linkage	10.6%	1.2%
Maximal swelling (in water)	25%	200–300%
Lipid domain	0.1–1%	10–20%

Water permeability through the nail plate was reported to be in the range of 1.2–5.3 mg/cm²/h, as determined from the nail plates with a dry dorsal side (6,8–10). When the thickness difference is taken into consideration, the water permeation rate of nail is approximately severalfold to 10 times higher than that of stratum corneum. Using the same evaporimeter on 21 healthy volunteers, Jemec et al. obtained the value of transonychial water loss to be 1.94 mg/cm²/h, with a value of transepidermal water loss of 0.56 mg/cm²/h. If the thickness of the nail plate is normalized, the intrinsic nail permeability for water is two to three orders higher than stratum corneum. The markedly different water permeability exhibited by the nail plate from the stratum corneum is probably due to the difference in the lipid content. It is probably also due to the different composition of amino acids in nail keratin. A complete amino acid composition of the nail is listed in Table 3

Table 3 Amino Acid Composition (per 100 Residues) of Human Nail Compared to Stratum Corneum, Hair, and Sheep Horn, Hoof, and Wool

Amino acid	Human[a]			Sheep[b]		
	Stratum corneum	Nail	Hair	Horn	Hoof	Wool
Lys	4.2	3.1	2.5	3.8	4.0	2.7
His	1.5	1.0	0.9	1.3	0.9	0.8
Arg	3.8	6.4	6.5	6.7	7.2	6.2
Asp	7.9	7.0	5.4	7.8	8.4	5.9
Thr	3.0	6.1	7.6	4.8	5.0	6.5
Ser	13.6	11.3	12.2	9.6	9.5	10.8
Glu	12.6	13.6	12.2	12.9	13.7	11.1
Pro	3.0	5.9	8.4	3.8	4.0	6.6
Gly	24.5	7.9	5.8	11.1	9.1	8.6
Ala	4.4	5.5	4.3	5.9	6.4	5.2
Val	3.0	4.2	5.5	5.2	5.7	5.7
Met	1.1	0.7	0.5	0.8	0.8	0.5
Ile	2.7	2.7	2.3	3.3	3.6	3.0
Leu	6.9	8.3	6.1	9.1	9.5	7.2
Tyr	3.4	3.2	2.2	5.0	4.0	3.8
Phe	3.2	2.5	1.7	2.6	2.7	2.5
$\frac{1}{2}$Cys	1.2	10.6	15.9	6.2	5.7	13.1
S	1.4%	3.2%	4.5%	—	—	—

[a]Baden et al. (6).
[b]Marshall and Gillespie (11).

in comparison to human hair and stratum corneum. The chemical compositions of sheep horn, hoof, and wool are also included for comparison (11). It can be seen that the amino acid composition of human nail is much closer to hair and wool than to the stratum corneum, because of the lower glycine and higher half-cystine content in the nail keratin.

The high disulfide linkages in nail keratin are directly responsible for the toughness of the nail. On the other hand, they also represent a problem when drug permeation through the nail is of concern. Porcine and bovine hooves have been used as a barrier membrane model in place of a human nail plate to study drug penetration (12–16). Because of the composition difference between a human nail plate and a hoof membrane, such as the half-cystine content, caution should be used in data interpretation and making predictions, especially when chemical penetration enhancement is involved. More discussion is devoted to this subject in a later section.

Walters et al. studied the nail permeation pattern of homologous alcohols and compared it with that of the stratum corneum (17). The *n*-alkanols tested for nail permeation included methanol, ethanol, propanol, butanol, pentanol, hexanol, heptanol, octanol, decanol, and dedecanol. The permeability of water in the nail was also included (designated as alkyl chain length equal to zero, i.e., $n = 0$). Their results suggested that, for the polar and semipolar alcohols, the hydrated human nail plate behaves somewhat like a hydrogel membrane of high ionic strength: The alkanol permeability in the nail decreased as the alkyl chain length increased from 1 to 8 with a gradual increase in its lipophilicity. When the alkyl chain length further increased to 10 and 12, however, the permeability increased, indicating the permeability had become sensitive to oil/water partitioning.

The most striking feature of the results is the total departure from the patterns seen with the same permeants in other body membranes such as human and animal skin, and vaginal and gastrointestinal mucoses. The permeability patterns indicate that those membranes act like lipid barriers, whereas the nail plate behaves exactly opposite (1). The observation that the nail permeability decreased from water to *n*-octonol indicates a lipid pathway was not involved for these permeants (17). In addition, unlike the stratum corneum, there is no significant difference in the nail permeability between the ionic and nonionic forms of a drug (18).

These differences between the nail and stratum corneum, both physical and chemical, are probably responsible for the lack of efficacy of topical nail antifungal formulations presently on the market (19), as well as for the ineffectiveness of some well-known skin penetration enhancers, such as dimethyl sulfoxide and homologous alcohols, on nail permeation enhancement (18).

II. CLINICAL EXPERIENCES IN TOPICAL ANTIFUNGAL TREATMENT OF ONYCHOMYCOSIS

Fungal infection of the nails, onychomycosis, makes up about 30% of fungus infections of the skin. Onychomycosis involves about 2–5% of general population, increasing to 8–10% in persons over 50 years of age. Of the 245 million people in the United States, between 4.9 and 12.3 million are infected with onychomycosis. Data from the National Diagnostic and Therapeutic Index estimates only 371,000 people were treated in 1993. Although there are many potent antifungal drugs currently available, nail fungal infections remain a disease very difficult to treat. In the United States, only oral administration of three antifungal drugs is approved by the U.S. Food and Drug Administration (FDA) as effective therapy for nail fungal infection: itraconazole (Sporanox), terbinafine (Lamisil), and an old drug, griseofulvin. Because of the relatively longer duration of oral treatment for onychomycosis (1 week pulse for 2–4 months or continuous therapy for 3–6 months), potential adverse effects, such as hepatotoxicity and drug interactions, are often of concern. To minimize the undesirable side effects associated with systemic treatment, it is highly desirable to have an efficacious topical product to treat onychomycosis. Since the newer antifungal agents have shown a better risk–benefit ratio, topical therapy may in combination with oral therapy help increase overall efficacy, reduce the duration of oral therapy, and decrease side effects associated with the oral agent. In addition, topical therapy has more and more been used to prevent recurrences.

Traditionally, topical treatments alone for onychomycosis are only able to inhibit the growth of fungal nail infections and to contain them, but not able to cure them (20,21). The lack of efficacy for topical treatment alone is probably due to the fact that the antifungal drugs cannot penetrate the nail plate to reach the infection sites: the nail plate, nail bed (underneath the nail plate), and nail matrix (nail "root") (Fig. 1).

The thick nail plate and its dense keratin nature make it a very difficult barrier for a topically applied drug to penetrate. It has been shown, however, that once the nail plate has been removed by either surgical or chemical nail avulsion, topical treatments of onychomycosis are effective in treating onychomycosis, using miconazole (22) or ketoconazole (23).

Keratolytic agents, such as urea and salicylic acid, are often used to soften nail plates for nail avulsion. Urea and a combination of urea and salicylic acid were reported to be used for nonsurgical avulsion of nail dystrophies in clinical studies prior to topical treatment of onychomycosis, with satisfactory results (24–26). The urea concentrations used by these investigators ranged from 20% to 40% in a hydrophobic base. In one study, the formulation also contained 10% salicylic acid (26). It took about 1 week under occlusion to accomplish the dystrophic nail avulsion.

In order to deliver a therapeutically sufficient quantity of an antifungal drug to the fungus-infected sites (i.e., nail plate, nail bed, and nail matrix), certain methodologies for penetration enhancement may need to be employed to overcome the nail barrier. Both physical and chemical methods have been used to achieve this goal. As mentioned earlier, physically removing the entire diseased nail plate would be a logical and effective way of eliminating the nail barrier. However, the requirement of health care professionals for the nail avulsion procedure would certainly limit its use and reduce patient compliance. Nevertheless, the partial removal of a diseased nail plate, or thinning of the diseased nail plate with a file-like tool by the patient or with the assistance of a family member, prior to topical drug application, would improve the efficacy of topical onychomycosis treatment.

Hardjoko et al. reported the results of a clinical trial for onychomycosis treatment, using a combination of 1% bifonazole and 40% urea ointment (27). The combination ointment was applied to the fungus-affected nail with an occlusive dressing every day for 2 weeks, followed by 1% bifoconazole cream topically for 1 month. Before changing the dressing, the affected nail was immersed in warm water for about 10 min to facilitate easy removal of the nail. Then the nail and nail bed were scraped with a plastic nail file. It was observed that thinning and total avulsion of the nails occurred in 97% of 30 patients with candidal onychomycosis. Regrowth of a normal-appearing nail was found in 92% of 26 patients. Failure rate was 4% and recurrence rate was 12% in 26 patients. Thinning and avulsion of the affected nail plates probably contributed a great deal to the treatment efficacy.

Pittrof et al. and Lauharanta reported the use of an amorolfine-containing nail lacquer (Loceryl nail lacquer) to treat onychomycosis in clinical trials (12,28). The patients were instructed to clean the surface of each affected nail with lacquer remover, and to file and cut away as much of the nail as possible before the application of the amorolfine nail lacquer. Clinical trials demonstrated that this procedure was essential. Again, the physical elimination of a part of the nail barrier certainly facilitated the drug penetration for a better efficacy.

Filing of the diseased nails has also been used for Ciclopirox formulated in a 8% lacquer base to improve penetration of the drug into the nail (29). Topical ciclopirox has been reported to be beneficial in treatment onychomycosis in acquired immunodeficiency syndrome patients (30).

III. RECENT PROGRESS ON THE RESEARCH OF TOPICAL DRUG DELIVERY TO THE NAIL

As mentioned in the previous section, the pioneer work on the research of topical drug delivery to the nail was made by Walters and his colleagues in the early and mid 1980s (1,17,18,31). They reported that the nail plate was a very stable barrier. For example, its permeability to methanol remained unchanged over 49 days of aqueous immersion. Some of the organic solvents (e.g., dimethyl sulfoxide [DMSO] and alcohols) that tend to promote diffusion through the skin horny layer did not show much effect on permeation enhancement through the nail.

The most significant finding of their research is probably that, as a permeation barrier, a human nail plate does not mimic the behavior of a lipophilic membrane, which has been the case for almost all the other body membranes, such as the skin and the vaginal and gastrointestinal mucosal membranes. Instead, a hydrated nail plate behaves more like a hydrogel membrane in its barrier properties (17). This conclusion was confirmed recently by Mertin and Lippold (13–15). The permeation behavior of homologous nicotinic acid esters was studied in vitro using human nail and a keratin membrane from bovine hooves with modified Franz diffusion cells. Despite significantly different lipophilicity among the penetration substances investigated (i.e., the partition coefficient octanol/water [$PC_{oct/w}$] ranging from 7 to 51,000], the permeability (P) of the compounds in the nail and hoof membrane did not increase with increasing $PC_{oct/w}$ or lipophilicity of these compounds. This result indicates that both barriers behaved like a hydrophilic gel membrane, rather than as a lipophilic membrane as in the case of the stratum corneum (13).

Mertin and Lippold compared the permeation results of chloramphenicol through the human nail and bovine hoof membrane from three vehicles saturated with the drug: phosphate buffer (pH 7.4), n-octanol, and medium-chain triglyceride (14). There was about a 10-fold difference in the chloramphenicol solubilities in the vehicles. However, the maximum fluxes from these vehicles through each barrier membrane were virtually identical, after data normalization to the same membrane thickness. It was concluded that when swelling of the nail plate or hoof membrane is not altered by the lipophilic vehicles, the maximum flux of the drug is independent of its solubility in the vehicle and is the same as that from the saturated aqueous solution.

Chloramphenicol penetration from nail lacquers of various concentrations through human nail and bovine hoof membrane was studied. The composition of the lacquers contained 0.5–20% chloramphenicol, which produced the drug concentrations of 2.2–47.6% in the dry lacquers. The lacquers were composed of quaternary poly(methyl methacrylates) (Eudragit RL) as film former, dibutyl sebacate as plasticizer, and methanol as solvent. The results in Fig. 2 and Table 4 show that there was a linear increase of maximum flux from the lacquers through the hoof membrane as the drug concentrations in dry lacquers increased up to 31%. This was attributed to

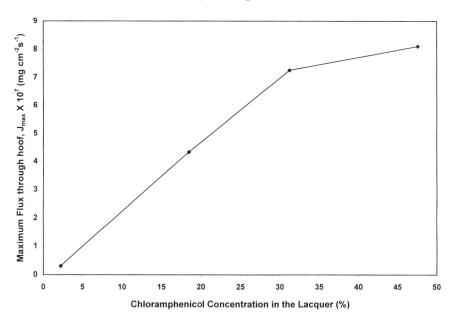

Figure 2 The data show the relationship between maximum flux of chloramphenicol through bovine hoof membrane and human nail plate and the drug concentrations in the dry lacquer. The maximum fluxes have been standardized to the same thickness of 1 mm. There was a linear increase of the flux from the lacquers up to 31% of the drug. Plotted from the data in Mertin and Lippold (14).

the formulation of supersaturated drug solution in the lacquers. At concentrations higher than 31%, the supersaturated solution might no longer be stable, thus leading to the plateau of the flux. Table 4 shows that the chloramphenicol flux through bovine hoof membrane was 54-fold higher than that through the human nail plate with a nail lacquer (43.4% drug), and 65-fold higher with the saturated aqueous chloramphenicol solution (phosphate buffer pH 7.4), when the membrane thickness was normalized to be the same. A similar phenomenon was also observed with the homologous nicotinic acid esters: The fluxes were about 40-fold higher in the bovine hoof membrane than the human nail (13). This observation was important since hoof membranes from animals are used sometimes as a model membrane to study drug penetration through human nail (12). The interpretation for the observed difference in the barrier property between human nail and bovine hoof may lie in the different biochemical composition of the two barriers (Table 3). It can be seen that sheep hoof has only about one half of the half-cystine content as compared to the human nail. The high sulfur-containing

Table 4　Chloramphenicol Penetration at 32°C from Nail Lacquers and from Saturated Drug Solution

Chloramphenicol concentration in the lacquer (%)	Maximum flux of chloramphenicol, $J_{max} \times 10^7$ (mg cm^{-2} s^{-1})		J_{max} (hoof)/J_{max} (nail)
	Bovine hoof membrane	Human nail plate	
2.2	3.1	—	—
18.5	43.4	0.80	54
31.3	72.6	—	—
47.6	81.1	—	—
Saturated drug solution in phosphate buffer, pH 7.4 (4.5 g L^{-1})	53.2	0.82	65

Note. Fluxes are standardized to 1 mm thickness of barriers.
Source: Modified from Mertin and Lippold (14).

keratin is responsible for the toughness of the nail, and presumably for the good barrier property as well.

　　The impact of molecular size of a penetrating substance through the human nail and bovine hoof membrane was also investigated by Mertin and Lippold (15). Figure 3 is a plot extracted from their experimental results (Table 5). As can be seen in Fig. 3 and the top portion of Table 5, the ratio of the drug permeability coefficient through bovine hoof to that through human nail is linearly proportional to the molecular weight (MW) of the penetrating substance investigated. In other words, the difference between the human nail and bovine hoof as a penetrating barrier increases as the molecular weight of penetrating drugs increase. This leads to a potential danger of overestimating the drug penetration into human nail using an animal hoof membrane as a model barrier.

　　In order to predict drug permeation in the human nail with hoof membrane data, a quantitative correlation between the human nail and the hoof membrane may need to be established first in a similar molecular weight range of the penetrating substances, and preferably, with the similar solvent compositions. It should be pointed out that the prediction of drug permeability through human nails based on hoof results is only possible when no significant changes occurred in the barrier membranes. If the use of certain chemical penetration enhancers results in such a change—for example, cleavage of disulfide bonds in the keratin membrane as discussed in a later section—a significant deviation may render the prediction nonmeaningful.

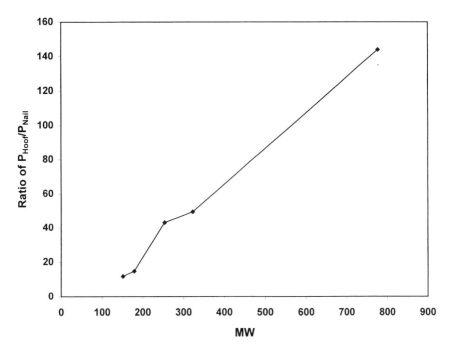

Figure 3 The ratio of the drug permeability coefficient through bovine hoof to that through human nail is linearly proportional to the molecular weight (MW) of the penetrating substance investigated. The difference between the human nail and bovine hoof as a penetrating barrier increases as the molecular weight of penetrating drugs increases. Extracted from the data in Mertin and Lippold (15).

The data in Fig. 3 and Table 5 also indicate that drug penetration through nail plate is quite sensitive to the molecular weight of the penetrating compound. Consequently, a drug with a lower molecular weight might be a better candidate than higher molecular weight for a topical product, assuming similar potency and solubility in the nail.

Mertin and Lippold rightly point out that for onychomycosis treatment, not only is the flux of an antimycotic drug through the nail plate of importance, but also the antifungal potency, which is commonly expressed as the minimum inhibitory concentration (MIC). An efficacy coefficient E is therefore introduced, which should be maximized for high therapeutic effectiveness (15):

$$E = \text{Flux/MIC}$$

This simple equation allows an estimation and comparison of the relative efficacy among various antifungal agents.

Table 5 Molecular Weights (MW) of Antimycotic and Other Drugs, with Their Permeability Coefficients

Drug	MW	$P_{nail} \times 10^8$ (cm^2 s^{-1})	$P_{Hoof} \times 10^8$ (cm^2 s^{-1})	P_{hoof}/P_{nail}
Medium: aqueous phosphate buffer, pH 7.4				
Paracetamol	151.2	1.78	20.97	11.8
Phenacetin	179.2	1.40	20.78	14.8
Diprophylline	254.3	0.142	6.14	43.2
Chloramphenicol	323.1	0.182	9.01	49.5
Iopamidol	777.1	0.010	1.44	144
Medium: ethanol-containing phosphate buffer, pH 8.1				
Ciclopirox	207.3	—	2.03	—
Naftifine	287.4	—	4.08	—
Tolnaftate	307.4	—	3.44	—
Bifonazole	310.4	—	3.05	—
Amorolfine	317.5	—	2.03	—
Clotrimazole	344.8	—	2.30	—
Griseofulvin	352.8	—	1.00	—
Econazole	381.7	—	3.37	—
Ketoconazole	531.4	—	0.84	—
Nystatin	926.1	—	0.19	—

Source: Modified from Mertin and Lippold (15).

It was shown by Franz that the use of different solvents had an influence on the drug uptake and penetration through human nails (16). The uptake of amorolfine into human nail was determined by soaking small pieces of nail plates for 48 h in a nail lacquer formulation containing 5% amorolfine with either ethanol or methylene chloride as solvent. The drug uptake was found to be greater from the methylene chloride lacquer than from the ethanol lacquer (2.9 vs. 1.2 μg drug/mg nail, respectively). The uptake result was consistent with the result of amorolfine penetration through the nail (16).

The lower portion of Table 5 shows a general trend of decreasing permeability coefficient with increasing molecular weight of antimycotic drugs investigated. Based on such a relationship, Mertin and Lippold made a prediction of the permeabilities of the antimycotic drugs through the human nail (15). Table 5 also shows that ciclopirox and econazole exhibit a noticeable deviation from the general trend. The deviation of ciclopirox was explained by the effect of the Donnan equilibrium (13,15). Because approximately 50% of ciclopirox dissociated into anions at pH 8.1 (the medium pH), its penetration into the negatively charged nail keratin membrane was inhibited. However, this explanation can not be used for the exceptionally high econaozole permeability. With a pK_a around 5.4 in the solvent composition, almost all the drug molecules are nonionized at the testing pH. Either an experimental error or an unknown interaction between the econazole and the keratin membrane might be the reason for the observed deviation.

The existing clinical evidence supports the notion that the key to successful treatment of onychomycosis by a topical antifungal product lies in effectively overcoming the nail barrier, and in successfully developing a novel drug delivery system to transport a therapeutically sufficient quantity of antifungal drugs to the target sites. Physical methods mentioned earlier have demonstrated this point to a certain extent. Alternatively, a chemical method, the use of nail penetration enhancers of chemical nature, may also achieve this goal. The advantages of chemical nail penetration enhancers include: a) improved drug delivery; b) elimination of the need for whole nail avulsion or partial nail removal; c) better patient compliance; and d) better control in clinical efficacy because of a more consistent mode of drug applications.

In sharp contrast to the abundant literature available on chemical skin penetration enhancers, information on chemical nail penetration is rather scarce. Due to the different nature of the barriers, as discussed in the previous section, the likelihood of success would not be very high if one simply used a proven-effective skin penetration enhancer for nail delivery. The working mechanisms for a vast majority of skin penetration enhancers involve the lipid domains or pathways in the stratum corneum, either by increasing the lipid fluidity or by increasing the drug partitioning into it. These skin penetration enhancers are unlikely to have the same penetration enhancement effect on the nail simply because the nail contains much less lipid, so probably has much less well-developed lipid pathways.

The chemical composition of nail, as well as some experimental evidence, indicates that the aqueous pathway plays a dominant role in drug penetration into nail (1,31). Water is the principle plasticizer for the nail (1). Once hydrated, a hard nail plate becomes more elastic, and hence probably more permeable to a topically applied drug. Nail hydration is influenced by many factors, such as solution pH, temperature, and the presence of certain chemicals.

Sun et al. reported the use of chemical enhancers that act directly on the nail keratin to increase the nail hydration and elasticity and consequently resulted in enhanced drug permeation through the nail (32). Certain sulfhydryl compounds, such as cysteine, a natural amino acid, and its derivative acetylcysteine, cleave the disulfide bonds of nail keratin. Urea was used to

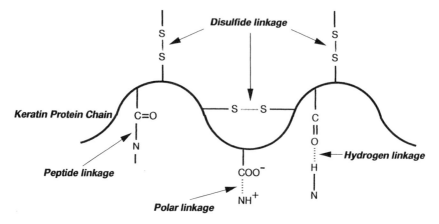

Figure 4 Simplistic presentation of chemical bonds involved in a nail keratin chain.

interact with hydrogen bonds of nail keratin to facilitate the disulfide bond breaking (Fig. 4). The chemical reaction that occurs between the cystine linkages in nail keratin and cysteine is shown in Scheme 1. Both cysteine and acetylcysteine have been used in Japan as hair-waving agents (33).

N-Acetyl-L-cysteine has long been used as mucolytic agent with an excellent safety record. It is therefore logical to incorporate nail penetration enhancers into topical antifungal formulations to enhance the drug delivery

$$
\begin{array}{ccc}
\}\text{NH} & \}\text{NH} & \\
| & | & \\
\text{CHCH}_2-\text{S}-\text{S}-\text{CH}_2\text{CH} & + & 2\ \text{HS}-\text{CH}_2-\text{CH}(\text{NH}_2)-\text{COOH} \\
| & | & \\
\text{CO} & \text{CO} & \\
\} & \} &
\end{array} \quad \rightleftharpoons
$$

$$
\begin{array}{cc}
\}\text{NH} & \}\text{NH} \\
| & | \\
\text{CHCH}_2-\text{SH} & + & \text{HS}-\text{CH}_2\text{CH} & + & \text{HOOC}-\text{CH}(\text{NH}_2)-\text{CH}_2-\text{S}-\text{S}-\text{CH}_2-\text{CH}(\text{NH}_2)-\text{COOH} \\
| & | \\
\text{CO} & \text{CO} \\
\} & \}
\end{array}
$$

Scheme 1 The chemical reaction that occurs between the disulfide bonds in nail keratin and a sulfhydryl compound (in this case, cysteine).

into and through the nail. The antifungal nail formulations containing itraconazole and miconazole nitrate were evaluated by in vitro methodologies such as a nail swelling test, drug partitioning test, and drug permeation through human nails. The ability of a drug in a formulation to penetrate the nail plate should be reflected by the rate and extent of the nail uptake of the formulation (i.e., nail swelling in the formulation), as well as by the amount of the drug migration into the nail (drug partitioning). Nail swelling and drug partitioning tests were conducted by immersing human nail clippings in a topical drug formulation at 32°C for 2 days. After washing off the surface-bond formulation, the nail clippings were first examined for weight gain, then digested for drug contest analysis with a high-pressure liquid chromatography method.

Figures 5 and 6 show the effect of N-acetyl-L-cysteine and urea as penetration enhancers on nail swelling and itraconazole partitioning. All four itraconazole formulations contained 1% itraconazole and different enhancer compositions: Formulation A was the control, containing neither N-acetyl-L-cysteine nor urea; Formulation B contained 10% urea but no N-acetyl-L-cysteine; Formulation C contained 5% N-acetyl-L-cysteine but no urea; and Formulation D contained both 5% N-acetyl-L-cysteine and 10% urea. The numbers over the bars are the enhancement factors, that is, the ratio of the

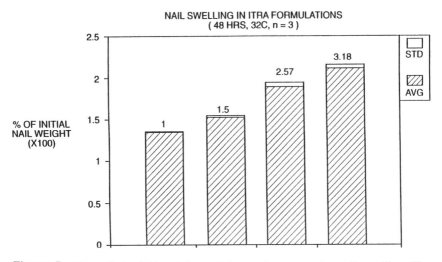

Figure 5 The effect of N-acetyl-L-cysteine and urea on the nail swelling. The numbers above the bars are the enhancement factor, that is, the ratio of percentage weight gain in the nail sample in the testing formulation to that in the control formulation without the enhancer and/or urea.

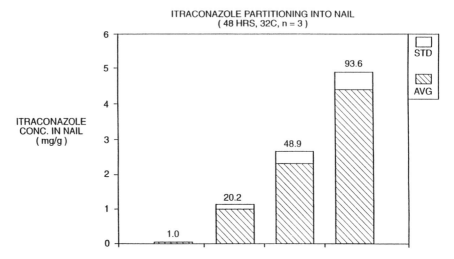

Figure 6 Effect of *N*-acetyl-L-cysteine and urea on itraconazole partitioning into nail. The numbers above the bars are the enhancement factor, that is, the ratio of the itraconazole concentration in the nail sample in the testing formulation to that in the control formulation without the enhancer and/or urea.

nail weight gain or nail itraconazole content from the test formulations (Formulations B–D) to those from the control (Formulation A). Figure 5 shows that urea alone resulted in 0.5-fold increase in nail swelling in Formulation B; *N*-acetyl-L-cysteine alone produced 1.6-fold (Formulation C); and both *N*-acetyl-L-cysteine and urea produced 2.2-fold (Formulation D). Higher formulation uptake into the nail means higher nail drug content. Indeed, Fig. 6 shows a 20.2-fold increase of itraconazole partitioning into nail from Formulation B over the control, 48.9-fold from Formulation C, and 93.6-fold from Formulation C. The benefit of incorporating nail penetration enhancers (*N*-acetyl-L-cysteine, urea, and salicylic acid) into the formulations is demonstrated by a clear trend of rapid increase of itraconazole partitioning into nail.

As described earlier, nail swelling and drug partitioning tests are good screening methods for more penetrating formulations. However, in vitro nail penetration studies are necessary to confirm that the formulation is really capable of delivering the drug through the nail plate. A nail penetration experiment was conducted with human cadaver nails using custom-designed nail permeation cells. The nail permeation cells resembled the Franz diffusion cell, but with a nail clamping mechanism to hold the nail in place. The

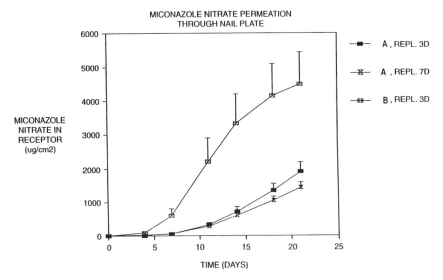

Figure 7 Permeation profile of miconazole nitrate through human nail plates ($n = 3$). Miconazole nitrate penetrated through the nail plates into the receptor fluid, and reached the detection levels within the first few days. The frequency of replenishing Formulation A (every 3 days vs. every 7 days) did not result in much difference in miconazole nitrate permeation through the nail.

nail permeation cells were maintained at 32°C during the 3-week experiment. Figure 7 shows the permeation profiles of miconazole nitrate from two formulations. The cumulative drug content in the nail and receptor was shown in Fig. 8. Both formulations contained 2% miconazole nitrate. Formulation A contained 5% N-acetyl-L-cysteine and 20% urea. Formulation B contained 10% N-acetyl-L-cysteine and 20% urea. As can be seen, miconazole nitrate penetrated through the nail plates into the receptor fluid, and reached the detectable range within the first few days. The frequency of replenishing Formulation A (every 3 days vs. every 7 days) did not result in much difference in miconazole nitrate permeation through the nail, nor in the amount of the drug retained in the nail (Fig. 8). On the other hand, when the concentration of N-acetyl-L-cysteine was doubled, from 5% in Formulation A to 10% in Formulation B, both the miconazole nitrate permeation rate and drug retention in the nail doubled (Figs. 7 and 8). The data indicate that the nail permeation of a drug formulation can be controlled by adjusting the penetration enhancer level. The substantial amount of the drug retained in the nail reflects the keratinophilic nature of miconazole.

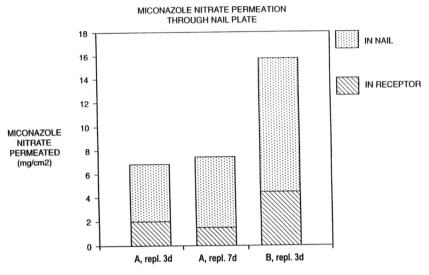

Figure 8 Amount of miconazole nitrate penetrated through the nail plate and amount retained in the nail plate. Substantial amount of the drug retained in the nail reflects the keratinophilic nature of iconazole.

IV. CONCLUSIONS

In summary, the chemical composition of human nail is significantly different from the other body membranes such as the skin and the vaginal and gastrointestinal membranes. The nail plate is composed of keratin molecules of high disulfide bond linkage and low lipid content. As a penetration barrier, the nail plate does not mimic the behavior of a lipophilic membrane, which has been the case for almost all the other body membranes. Instead, a nail plate behaves more like a hydrogel membrane in its barrier properties.

Due to the markedly different properties between the nail and skin, many skin absorption enhancers cannot be used effectively to promote drug delivery into the nail. Consequently, the mechanism of certain nail penetration enhancers may be quite different from those commonly employed in the transdermal and topical drug delivery, as illustrated by the example of disulfide-bond cleavage. Although we cannot neglect the issue of patient compliance with long-time topical application and the difficulty for the elderly to reach their extremities, topical treatment of onychomycosis remains primarily a drug delivery problem. Therefore the lack of effective topical products in the market represents a serious challenge to the scientists in this field. Recent research findings begin to lay a theoretical foundation for successful development of efficacious topical products to treat nail diseases.

REFERENCES

1. Walters KA, Flynn FL. Permeability characteristics of the human nail plate. Int J Cosmet Sci 1983; 5:231–246.
2. Fleckman P. Basic science of the nail unit. In: Scher RK, Daniel CR, eds. Nails: Therapy, Diagnosis, Surgery. Philadelphia: WB Saunders, 1990:36–51.
3. Gonzalez-Serva A. Structure and function. In: Scher RK, Daniel CR, eds. Nails: Therapy, Diagnosis, Surgery. Philadelphia: WB Saunders, 1990:11–30.
4. Hamilton JB, Terada H, Mestler GE. Studies of growth throughout the lifespan in Japanese: growth and size of nails and their relationship to age, sex, hereditary, and other factors. J Gerontol 1955; 10:401.
5. André J. Illustrated Nail Pathology. Diagnosis and Management. CD-ROM. Publ. Lasion Europe N.V., Belgium. 1995.
6. Baden HP, Goldsmith LA, Fleming B. A comparative study of the physicochemical properties of human keratinized tissues. Biochim Biophys Acta 1973; 322:269–278.
7. Scheuplein RJ, Morgan LJ. Bound water in keratin membranes measured by a microbalance technique. Nature 1967; 214:456.
8. Burch GE, Windsor T. Diffusion of water through dead plantar, palmer and torsal human skin and through toenails. Arch Dermatol 1946; Syph. 51:39–41.
9. Spruit D. Measurement of water vapor loss through human nail in vivo. J Invest Dermatol 1971; 56:359–361.
10. Jemec GBE, Agner T, Serup J. Transonychial water loss: Relation to sex, age and nail plate thickness. Br J Dermatol 1989; 121:443–446.
11. Marshall RC, Gillespie JM. The keratin proteins of wool, horn and hoof from sheep. Aust J Biol Sci 1977; 30:389–400.
12. Pittrof F, Gerhards J, Erni W, Klecak G. Loceryl nail lacquer—Realization of a new galenical approach to onychomycosis therapy. Clin Exp Dermatol 1992; 17(Suppl 1):26–28.
13. Mertin D, Lippold BC. In vitro permeability of the human nail and a keratin membrane from bovine hooves: Influence of the partition coefficient octanol/water and the water solubility of drugs on their permeability and maximum flux. J Pharm Pharamcol 1997; 49:30–34.
14. Mertin D, Lippold BC. In vitro permeability of the human nail and a keratin membrane from bovine hooves: Penetration of chloramphenicol from lipophilic vehicles and a nail lacquer. J Pharm Pharmacol 1997; 49:241–245.
15. Mertin D, Lippold BC. In vitro permeability of the human nail and a keratin membrane from bovine hooves: prediction of the penetration rate of antimycotics through the nail plate and their efficacy. J Pharm Pharmacol 1997; 49:866–872.
16. Franz TJ. Absorption of amorolfine through human nail. Dermatology 1992; 184(Suppl 1):18–20.
17. Walters KA, Flynn GL, Marvel JR. Physicochemical chemical characterization of the human nail: permeation pattern for water and the homologous alcohols

and difference with respect to the stratum corneum. J Pharm Pharmacol 1983; 35:28–33.

18. Walters KA, Flynn GL, Marvel JR. Physicochemical characterization of the human nail: solvent effects on the permeation of homologous alcohols. J Pharm Pharmacol 1985; 37:771–775.

19. Haneke E. Fungal infections of the nail. Semin Dermatol 1991; 10:41–53.

20. Cohen PR, Scher RK. Topical and surgical treatment of onychomycosis. J Am Acad Dermatol 1994; 31:S74–S77.

21. Korting HC, Schäfer-Korting M. Is tinea unguium still widely incurable? Arch Dermatol 1992; 128:243–248.

22. Rollman O. Treatment of onychomycosis by partial nail avulsion and topical miconazole. Dermatologica 1982; 165:54–61.

23. Hettinger DF, Valinsky MS. Treatment of onychomycosis with nail avulsion and topical ketoconazole. J Am Podiatr Med Assoc 1991; 81:28–32.

24. Farber EM, South DA. Urea ointment in the nonsurgical avulsion of nail dystrophies. Cutis 1978; 22:689–692.

25. South DA, Farber EM. Urea ointment in the nonsurgical avulsion of nail dystrophies—A reappraisal. Cutis 1980; 25:609–612.

26. Buselmeier TJ. Combination urea and salicylic acid ointment nail avulsion in nondystrophic nails: A follow-up observation. Cutis 1980; 25:397–405.

27. Hardjoko FS, Widyanto S, Singgih I, Susilo J. Treatment of onychomycosis with a bifonazole-urea combination. Mycoses 1990; 33(4):167–171.

28. Lauharanta J. Comparative efficacy and safety of amorolfine nail lacquer 2% and 5% once weekly. Clin Exp Dermatol 1992; 17(Suppl. 1):41–43.

29. Ceschin-Roques CG, Hanel H, Pruja-Bougaret SM, et al. Ciclopirox nail lacquer 8%: In vivo penetration into and through the nails and in vitro effect on pig skin. Skin Pharmacol 1991; 4:89–94.

30. Dompmartin D, Dompmartin A, Deluol AM, Coulaud JP. Onychomycosis and AIDS: Treatment with topical ciclopirox olamine. Int J Dermatol 1990; 29:3, 233.

31. Walters KA, Flynn FL, Marvel JR. Penetration of the human nail plate: The effects of vehicle pH on the permeation of miconazole. J Pharmacol 1985; 37:498–499.

32. Sun Y, Liu JC, Kimbleton E, Wang J. Antifungal treatment of nails. US patent No. 5,696,164, 1997.

33. Iwasaki A. Cysteine waving lotions—Past improvement and future prospects. Cosmet Toilet 1994; 109:69–78.

45

Topical Dermatological Vehicles
A Holistic Approach

Eric W. Smith
Ohio Northern University, Ada, Ohio

Christian Surber
University Hospital, Basel, Switzerland

Howard I. Maibach
University of California School of Medicine, San Francisco, California

I. INTRODUCTION

The broad spectrum of topical preparations in use today vary in their physicochemical nature from powders through semisolids to liquids. While early topical formulations were often crude mixtures of chemicals, the optimized drug delivery systems common today achieve a balance between the physicochemical requisites for stability of active and inactive constituents, preservation against microbial spoilage, and, most importantly, presentation of the drug to the skin in a system that will allow appropriate release of the active to the stratum corneum. This is in addition to ensuring the elegance and user acceptability of the formulation. It is clear, therefore, that topical vehicle formulation is not a facile process, and as many of the desirable properties of the formulation as possible must be fulfilled in this process. Obviously, the synthetic and semisynthetic formulation constituents that are available today make this task somewhat easier than the formulation techniques of yesterday that utilized predominantly natural products.

In dermatology, the drug is rarely applied to the skin in the form of a pure chemical but, instead, is normally incorporated into a suitable carrier system—the vehicle. The term *vehicle* in this context is relatively new and

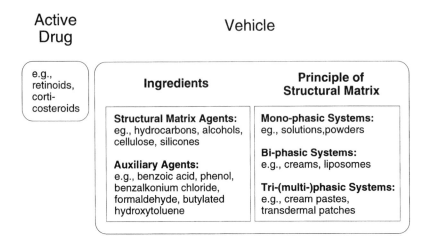

Figure 1 Principle of topical preparations.

was developed only when it became possible to designate a specific (ther-apeutic) effect to a chemical substance. Thereby it became possible to dis-tinguish between "active" and "inactive" ingredients in a formulation. In crude terms, the vehicle or base may be regarded as the sum of the ingre-dients in which the drug is presented to the skin. In practical terms, the vehicle is not only the sum of the formulation ingredients but also (in most cases) represents the existence of a physical, structured matrix comprised of the vehicle constituents. This structural matrix can be monophasic (e.g., a simple lipid), biphasic (e.g., a cream), or tri- or multiphasic systems (e.g., multiple emulsions, pastes, patch systems) into which a drug is incorporated (Fig. 1).

II. CLASSIFICATION OF TOPICAL VEHICLES

The simplest classification of topical vehicles consists of an initial division of the preparation into liquid, semisolid, or solid classes (Table 1). It is obvious that this is a simplification of the diversity of external formulations and does not account for many of the newer formulations (e.g., liposomes, microcapsules). Although a comprehensive classification has been proposed for isolated dermatological vehicles (1), clinical textbooks tend to combine the pharmaceutical nomenclature, the character of the structural matrix, and the formulation performance when attempting to classify the topical vehicles available for therapy.

Table 1 Simple Classification System for Topical Dermatological Vehicles

System	Monophasic	Diphasic	Multiphasic
Liquid	Nonpolar solution, often designated as oil	Emulsion (o/w, w/o), often designated as milk, lotion, shake, etc.	Emulsion (o/w/o, w/o/w), often designated as milk, lotion, shake, etc.
	Polar solution, often designated as paint, lotion, etc.	Suspension, often designated as paint, shake, etc.	Suspension, often designated as paint, shake, etc.
Semisolid	Anhydrous, nonpolar ointment, polar ointment	Emulsion (o/w, w/o), often designated as washable (o/w), nonwashable (w/o), or amphiphilic (o/w, w/o) creams	Emulsion (o/w, w/o), with powder, often designated as cream pastes
	Hydrous, nonpolar gel, polar gel	Suspension, often designated as paste	
Solid	Powder	Transdermal patch	Transdermal patch

Modern vehicles are frequently tailor-made and developed as carefully as the drug that they are intended to contain. Formulator research and development are extensive in terms of the stability, compatibility, and patient or consumer acceptability of the vehicle. More recently, it has become obvious that the type of vehicle or the nature of the excipients can markedly affect (or negate) the percutaneous absorption of a drug. This realization has added another, essential, dimension to the formulation development process. Several studies have demonstrated how markedly different drug delivery potentials are achieved by incorporating the same concentration of the same drug into topical vehicles of different physicochemical composition (2) or by incorporation into identical vehicles where the microstructure of the vehicle matrix is altered by micronization (3). Typically for most topically delivered drugs, the greatest delivery rates of the actives are achieved from alcoholic solutions, which have marked drug dissolution and skin barrier modification potential, and from lipid ointments, which tend to be occlusive in nature. Recent scientific research in this field has demonstrated that the delivery vehicle may have a marked effect on the barrier potential of the stratum corneum. Thus, modern pharmaceutical (and cosmetic) formulation development is based upon the stability and compatibility of excipients and active drug(s), cosmetic or aesthetic acceptability of the vehicle, and bioavailability of the drug(s).

III. THE IDEAL VEHICLE

The "ideal" vehicle should fulfill many different functions, all of which need to be addressed in the development process. It should be easy to apply and remove, and be nontoxic, nonirritant, nonallergenic, chemically stable, homogeneous, bacteriostatic, cosmetically acceptable, pharmacologically inert, and should readily release the drug to the stratum corneum (see Table 2). In topical dermatological treatment, formulators, producers, legislators, prescribers, and consumers perceive both the performance of the structural matrix and the performance of the ingredients of the topical preparations to be important. These requirements have led to extensive discussions on the "vehicle effect," where different interest groups have different conceptions of the delivery vehicle and different expectations of its role. The effect of the vehicle on the topical drug availability is probably much greater in topical drug delivery than in any other route of drug administration. It is common knowledge that potent molecules may be made clinically ineffective or that enhanced efficacy may be generated with weaker molecules, depending

Table 2 Selection Criteria for Topical Vehicles

Pharmaceutical, technological criteria for pharmaceutical semisolid
    Stability of active drugs
    Stability of ingredients
    Rheological properties—consistency, extrudability
    Loss of water and other volatile components
    Phase changes—homogeneity/phase separation, bleeding
    Particle size and particle size distribution of dispersed phase
    Apparent pH
    Microbial contamination/sterility (in unopened containers and under conditions
      of use)
    Enhanced or controlled drug release from the vehicle

Cosmetic and usage criteria for topical vehicles
    Visual appearance of product
    Odor, color
    Sampling and dispensing characteristics: ease of removal from container
    Application properties, texture (stiffness, grittiness, greasiness, adhesiveness)
    Residual impression after application, permanency on the skin

Biopharmaceutical criteria for topical vehicles
    Enhanced drug delivery and drug retention in the skin
    Controlled drug delivery and drug retention in the skin
    Targeted drug delivery and drug retention in the skin

upon the vehicle used (4,5). Despite the wishes of many formulators, there is no universal vehicle; each drug, at each concentration, requires a unique vehicle for optimized therapy.

In attempting to design a topical delivery vehicle, one should bear in mind that this is a situation of constant dynamic equilibrium; the constituents of the formulation will interact with one another and will interact with the skin once the product has been applied. Katz and Poulsen (6) have listed the interactions that may occur, between the vehicle, drug and skin as:

Vehicle–drug interactions.
Vehicle–skin interactions.
Drug–skin interactions.
Vehicle–drug–skin interactions.

Vehicle–drug interactions include all physical and chemical reactions that may take place between the drug molecules and the combined molecules of all the other formulation constituents. Solubility is of primary importance here, in that this parameter will define the physical state of the drug molecule in the vehicle (solution, suspension, ionized or nonionized) and will define the magnitude of the concentration gradient for passive diffusion.

Vehicle–skin interactions include the broad spectrum of physical and chemical events that may take place once the vehicle comes into contact with the stratum corneum. The formulation constituents all have potential to partition from the applied vehicle and enter the stratum corneum. If one considers the holistic, classical approach to dermatological therapy, then the partitioning of the ancillary, formulation substances into the skin have an important function in fulfilling the emollient, tactile, and rubifacient functions of the formulation. Modern topical delivery theory may suggest that the partitioning of the ancillary formulation chemicals is unimportant in the therapeutic goal unless such partitioning interferes with the delivery or diffusion of the active drugs. Recent technology has made use of this phenomenon in the form of penetration enhancer chemicals that are specifically designed to penetrate the stratum corneum from the applied vehicle and facilitate transport of the active drug principal through the barrier layer. Generally, penetration enhancers are vehicle components that interact with the stratum corneum to bring about changes in drug solubility or drug diffusion or both (7,8).

Drug–skin interactions include cutaneous metabolism and binding of the drug by the skin strata. Although it was classically believed that the stratum corneum was a "dead" layer of cells, we are now only beginning to understand the complex metabolic activity that is possible in this tissue and the potential for using this phenomenon in terms of prodrug diffusion through the skin. Similarly, it has been long established that a depot of topically delivered drug is rapidly established in the stratum corneum and that the active agent may be delivered from this binding site for days or weeks after initial application of the vehicle to the skin (9).

Formulations in which significant *vehicle–drug–skin interactions* occur are probably the most common. Many pharmaceutical solvents, propylene glycol for example, are known to have modest effects on reducing the skin barrier function (thereby altering the diffusivity of the drug through the skin) as well as influencing the solubility of the drug in the delivery vehicle (increasing the concentration gradient across the barrier layer) and changing the partitioning of the drug from the vehicle into the stratum corneum.

IV. THE CHOICE OF VEHICLE

Depending on environmental conditions, ethnic origin (10,11), gender, age, localization, and state of disease, different skin conditions are treated with respect to oiliness or humidity. The choice of a vehicle in a particular disease or in a specific patient often follows recommendations that are either based on a classification of topical preparations (12–15) or follow a few simple assessment parameters. It is a basic dermatological concept that the more acute the dermatosis, the blander the treatment should be. The principle of "wet-on-wet" and the use of occlusive ointments for dry or chronic dermatoses has become commonplace. As the condition improves, a "wet" dermatosis may subsequently be treated with either a drying paste or an oil-in-water cream, and a "dry" dermatosis may have a hydrous ointment, water-in-oil cream, or even an occlusive ointment applied (12,16). The capability of vehicles to alter the condition of the skin surface can be attributed to their influence on the lipid and water content of the skin or lipid composition (17–19). Vehicles with hydrophilic properties are suitable for oily and normal skin conditions, whereas vehicles with lipophilic properties are more suitable for dry skin conditions because of their emollient action.

There are several anatomical issues that need to be considered in the choice of an appropriate vehicle. Chemicals applied to different regions of the body permeate to varying extents. Most importantly from a clinical perspective, the reactivity of the same dermatosis to the same dose of formulation at different anatomical locations may vary markedly. Table 3 gives an overview of current prescribing practices. Similarly, topical preparations are tested in the pharmaceutical and cosmetic industries with respect to stability under various carefully defined climatic conditions (20). Practical assessments of the stability problems of vehicles or vehicle constituents encountered during use are rarely available. Preparations formulated for certain geographical climates are often inappropriate in more extreme climatic con-

Table 3 Localization, Skin Status, and Vehicle Used in Different Locations

Localization	Status	Dosage forms used
Hairy skin	Dry	Solutions, w/o cream
	Oily	Solution, gel
Face	Dry	w/o Cream
	Oily	Solution, o/w cream
Ear	Oily	o/w Cream
Body/extremities	Dry	Ointment, w/o cream
	Oily	o/w Cream
Intertrigenous area	Humid	Drying pastes, o/w cream
Hand/feet	Dry	During the day: o/w or w/o cream
	Oily	During the night: ointment, w/o cream
Nail	Dry	Solutions, lacquer

ditions. Hence, the final conditions of usage of the topical preparation should be kept in mind when selecting an appropriate vehicle.

Cosmetic or aesthetic criteria such as visual appearance, odor, application properties, or permanency of vehicle and drug on the skin are also important factors that influence patient acceptance of a topical preparation and compliance with a prescribed regimen of therapy. Commercially, these aspects are also important in maintaining customer loyalty. Ointments are often associated with adhesiveness by users and prescribers and are usually used for more severe diseases; creams or lotions are easier to apply and have a cooling effect and are usually preferred by patients.

V. CONCLUSION

As outlined here, many factors influence the development, the choice, the performance, and hence the clinical effectiveness of the vehicle. These factors are often viewed from different perspectives by patients, chemists, formulators, and clinicians. Due to this complexity, no uniform and comprehensive recommendations and guidelines are available for the development and the use of semisolid formulations, and it seems unlikely that this can be achieved in the future. Dermatological vehicle treatment may therefore remain an "art" gleaned by personal experience, rather than a defined, textbook science.

REFERENCES

1. Juch, R. D.; Rufli, T.; Surber, C. Pastes: What do they contain? How do they work? Dermatology. 1994; 189:373–377.
2. Wall, D. S.; Abel, S. R. Therapeutic-interchange algorithm for multiple drug classes. Am J Health-Syst Pharm. 1996; 53:1295–1296.
3. Wilhelmsen, L. Ethics of clinical trials—The use of placebo. Eur J Clin Pharmacol. 1979; 16:295–297.
4. Hadgraft, J. Formulation of anti-inflammatory agents. In: Hensby, C.; Lowe, N. J., editors. Nonsteroidal Anti-Inflammatory Drugs. New York: S. Karger; 1989:21–43.
5. Polano, M. K.; Bonsel, J.; Van der Meer, B. J. The relation between the effect of topical irritants and the ointment bases in which they are applied to the skin. Dermatologica. 1950; 101:69–80.
6. Katz, M.; Poulsen, B. J. Absorption of drugs through skin. In: Brodie, B. B.; Gillete, J. R., editors. Handbook of Experimental Pharmacology. New York: Springer; 1971; 28:103–174.
7. Barry, B. W. Mode of action of penetration enhancers in humans skin. J Control Release. 1987; 6:85–97.
8. Walters, K. A.; Hadgraft, J. Pharmaceutical Skin Penetration Enhancement. New York: Marcel Dekker; 1993.
9. Woodford, R.; Haigh, J. M.; Barry, B. W. Possible dosage regimens for topical steroids, assessed by vasoconstrictor assays using multiple applications. Dermatologica. 1983; 166:136–140.
10. Schlossmann, M. L. Formulating ethnic makeup products. J Cosmet Toilet. 1995; 110:59–63.
11. Sugino, K.; Imokawa, G.; Maibach, H. I. Ethic difference of stratum corneum lipid in relation to stratum corneum function. J Invest Dermatol. 1993; 100: 597–601.
12. Griffiths, W. A. D.; Wilkinson, J. D. Topical therapy. In: Rook, A.; Wilkinson, D. S.; Ebling, F. J. G. In: Champion, R. H.; Burton, J. L.; Ebling, F. J. G., editors. Textbook of Dermatology. London: Blackwell Scientific Publications; 1992:3037–3084.
13. Thoma K. Dermatika. München: Werbe- und Vertriebsgesellschaft Deutscher Apotheker; 1983.
14. Hundeiker, M. Grundlagen der Therapie mit äusserlichen Arzneimittelzubereitungen. Zentrlbl Hautkr. 1982; 148:683–697.
15. Katz, M. Design of topical drug products: Pharmaceutics. In: Ariëns, E. J., editor. Drug Design. New York: Academic Press; 1973:93–148.
16. Barry, B. W. Dermatologic Formulations. New York: Marcel Dekker; 1983.
17. Gabard, B. Testing the efficacy of moisturizers. In: Elsner, P.; Berardesca, E.; Maibach, H. I., editors. Bioengineering of the Skin: Water and Stratum Corneum. Boca Raton, FL: CRC Press; 1994:147–170.
18. Lodén, M. The increase in skin hydration after application of emolients with different amounts of lipids. Acta Derm Venereol (Stockh). 1992; 72:327–330.

19. Choudhury, T. H.; Marty, J. P.; Orecchini, A. M.; Seiller, M.; Wepierre, J. Factors in the occlusivity of aqueous emulsions. Influence of humectants. J Soc Cosmet Chem. 1985; 36:255–269.
20. Grimm, W. International harmonization of stability tests for pharmaceuticals. The ICH tripartite guideline for stability testing of new substances and products. Eur J Pharm Biopharm. 1995; 41(3):194–196.

46

Human In Vivo and In Vitro Hydroquinone Topical Bioavailability, Metabolism, and Disposition

Ronald C. Wester, Joseph L. Melendres, Xiaoying Hui, Rebecca Cox, Steffany Serranzana,* Hongbo Zhai, Danyi Quan,† and Howard I. Maibach
University of California School of Medicine, San Francisco, California

I. INTRODUCTION

Hydroquinone is a ubiquitous chemical, used in medicine and in industry. The efficacy of hydroquinone (1,4-dihydroxybenzene) as a skin lightening agent has been established in both human and animal studies. Clinically, hydroquinone is applied topically in the treatment of hypermelanosis (Bucks et al., 1988). Hydroquinone is available over the counter (OTC) at 2% concentration, in many OTC cosmetic products, and by prescription at higher concentrations. Hydroquinone is an important industrial chemical, used primarily as a developer in black-and-white photography, as an antioxidant in the rubber industry, and as a polymerization inhibitor. Thus, hydroquinone is readily applied to skin for medical and cosmetic reasons and through industrial use.

Hydroquinone's acute and chronic toxicity toward higher terrestrial organisms appears to be minimal in humans (Carlson and Brewer, 1953; Friedlander et al., 1982; Pifer et al., 1994; Anderson, 1974), in Sprague-Dawley and Carworth rats (Carlson and Brewer, 1953; Christensen et al.,

Present affiliations:
*Penederm Inc., Foster City, California.
†Theractech Corporation, Salt Lake City, Utah.

1976; Topping et al., 1988), and in dogs (Carlson and Brewer, 1953). The reported nephropathy and cell proliferation, as evidence of carcinogenicity, observed in Fischer 344/N rats (NTP, 1991) appears to be strain and sex specific (English et al., 1994a, 1994b). Hydroquinone was negative in the Ames/*Salmonella* and *Drosophila* genotoxicity assays (Goeke et al., 1981). Others suggest carcinogenic and teratogenic potentials have been at present inadequately studied (Diviliers et al., 1990), and that both hydroquinone and benzoquinone produce cytotoxic effects on human and mouse bone marrow cells (Colinas et al., 1994).

Thus, hydroquinone is an ever-present chemical with human exposure at or through the skin. Risk assessment thus requires a relevant assessment of hydroquinone bioavailability into and through the skin. Bucks et al. (1988) describe hydroquinone in vivo percutaneous absorption in humans to be high. Barber et al. (1995) sets a value for the absorption rate using human stratum corneum at about one sixth as reported by Bucks et al. (1988), and Barber et al. (1995) expound a slow permeability constant for hydroquinone. The study described in this chapter was designed to determine the in vivo percutaneous absorption of hydroquinone in humans and in vitro in human skin, using the same formulation in both conditions to assess the true topical bioavailability state of hydroquinone in humans.

II. MATERIALS AND METHODS

A. Test Article and Formulation

Hydroquinone, a white, crystalline solid, was received from Kiwi Brands, Inc. (Douglassville, PA). Eastman Chemical Company (Kingsport, TN) was the manufacturer. The compound specification and analysis were conducted by Van Waters and Rogers, Inc. (Kirkland, WA), and the purity by element analysis was 99.6%. The compound was stored in its original container, at room temperature and ambient humidity. [UL-^{14}C]-hydroquinone was obtained from Wizard Laboratories (Davis, CA). It was stored in its original container, at 0–5°C and ambient humidity, and protected from light. The specific activity was 20 mCi/mmol (MW 110), and radiochemical purity 98.8%.

The formulation was 2% hydroquinone cream containing 200 μCi radioactivity and 0.02 g hydroquinone/g cream. The topical formulation was stored at approximately 0–5°C and ambient humidity, and was protected from light. The stability of each formulation was verified immediately after preparation, once a week, 1 day before dosing, and 1 day after dosing, by a validated high-performance liquid chromatography (HPLC) method.

B. In Vivo Study Design

The study was divided into three groups. Group A was to determine the percutaneous absorption of hydroquinone after topical application of radiolabeled hydroquinone in human volunteers. The forehead was chosen as the dosing area because it is the skin site most clinically relevant to hydroquinone use, and it is one of the higher absorbing skin sites. A 2% [^{14}C]hydroquinone cream was applied on the forehead of six volunteers. The treated skin site was washed with soap and water after 24 h. Four-day urine samples were collected to determine hydroquinone percutaneous absorption.

Group B was to determine the concentration of hydroquinone and metabolites within the stratum corneum over time. Five sites on the ventral forearm, 1 cm^2 skin area per site, were treated with [^{14}C]hydroquinone cream. At set time intervals (0, 1, 3, 6, and 24 h), the skin surface was washed with soap and water followed by 10 cellophane tape strips. These samples were analyzed for radioactivity. The value of the skin washing technique is that all the radiolabeled hydroquinone not recovered from the treated skin site is considered absorbed. The passage of hydroquinone through the stratum corneum can be shown.

Group C used the ipsilateral and contralateral technique to capture hydroquinone and metabolites (formed within the skin) as they emerged into the blood stream. A 25-cm^2 skin area on the left ventral forearm was treated with 2% [^{14}C]hydroquinone cream. Blood was sampled from both right and left forearms at the same time. This was to determine if there were differences between the "emerging/ipsilateral" blood samples, which might give a picture of what is happening to hydroquinone in vivo at the skin:blood junction (Karim, 1987).

Human volunteers were recruited from the University of California, San Francisco, and surrounding San Francisco Bay Area community. Fourteen normal, healthy males between 18 and 80 years of age, without preexisting dermatologic disease, were selected for this study. Six volunteers were used for Group A, four for Group B, and four for Group C. The study was approved by UCSF Committee on Human Research, and each volunteer signed a consent form.

C. In Vivo Percutaneous Absorption

[^{14}C]Hydroquinone (25 μCi containing 2.5 mg hydroquinone in 0.125 g cream spread over 25 cm^2 skin surface area) was administered, using a 100-μl glass syringe, on the forehead of six volunteers. The delivered dose was quantitated by weight difference of the syringe before and after dosing. The volunteers were requested not to touch or wash the treated area for 24 h; however, no provisions were made to cover or protect the dosed site (as the product is actually used). It could possibly be rubbed off during sleep. The volunteers were requested to collect all of their urine in the provided containers.

The next day, the volunteers returned and the treated site was washed with cotton balls with liquid soap and water (50:50, v/v). The washing procedure was 50% Ivory soap solution, followed by a rinse with distilled deionized water. Soap was again applied, followed by two water rinses. These cotton balls were individually placed in a borosilicate glass vial with 10 ml distilled deionized water and then shaken for 24 h. Radioactivity of these samples was counted to quantify removable material.

Urine samples were collected every 24 h from time 0 until 96 h after dosing in the provided containers. Each urine vessel was pretreated with 0.5 g ascorbic acid, kept in a cold condition with blue ice packs during the collection, and changed every 24 h. The volume of each urine sample was measured and recorded. Aliquots of urine sample at each time point were frozen immediately at $-20°C$ until further analysis.

D. In Vivo Skin Tape Stripping

A 1×1 cm^2 area per skin site, with a total of 5 sites on the ventral forearm, was used for topical application. Each site was treated with [^{14}C]hydroquinone (1 μCi containing 0.1 mg hydroquinone in 0.005 g cream spread over 1 cm^2 skin surface area) as 2% cream. The delivered dose was quantitated by weight difference of the pipette before and after dosing. The volunteers were requested not to touch or wash the treated area for 24 h. At set time intervals (0, 1, 3, 6, and 24 h), the surface of treated skin was washed 5 times by cotton swab Q-Tips with the same procedure used for in vivo percutaneous absorption. The washed skin site then was tape stripped 10 times with cellophane tape (3M Commercial Office Supply Division, St. Paul, MN).

E. In Vivo Blood Concentration

Onto human volunteers was applied [^{14}C]hydroquinone (25 μCi containing 2.5 mg hydroquinone in 0.125 g cream spread over 25 cm^2 skin surface area) as 2% cream on the left forearm. The skin dosing site was not covered or protected, as the product is normally used. A catheter was placed in a vein draining the treated area (ipsilateral site) near the elbow flexure. A second catheter was placed in the same area on the right ventral forearm (contralateral site). Blood samples (10 ml) were taken from ipsilateral and contralateral sites at 0 h (before dosing) and at 0.5, 1, 4, and 8 h after dosing. Each blood sample was immediately placed in a sodium heparin Vacutainer (Becton Dickinson Vacutainer Systems, Rutherford, NJ) and then centrifuged by an Eppendorf microcentrifuge at 10,000 × g for 1 min to separate plasma and packed red blood cells.

The dosed skin site was washed at 8 h after dosing. The dosed site was washed by cotton balls with liquid soap and water (50:50, v/v). The washing procedure was as in the percutaneous absorption study.

Urine samples were collected in the provided containers at 24, 48, 72, and 96 h after dosing.

F. In Vitro Human Skin

Skin samples, obtained from UCSF tissue bank or hospitals, were from amputation, tummy tuck, abdominoplasty, or from cadaver. Six human skin sources were used. Skin samples were dermatomed to 500 μm thickness, stored in Eagle's minimum essential medium (MEM) at 5°C (refrigerator), and used within 48 h after removal from donor or death of donor. Obtaining human skin was approved by the UCSF review board. Informed consent was obtained by contributing hospitals.

Each skin sample was used for three treatments: a control of base cream without hydroquinone, skin dosed with 2% [^{14}C]hydroquinone cream, and skin pretreated with 2% sodium azide for 30 min and then dosed with 2% [^{14}C]hydroquinone cream. Sodium azide is a metabolic inhibitor that inhibits cytochrome oxidase (Casarett, 1975).

The in vitro diffusion assembly consisted of flow-through design glass penetration cells. Fresh human skin was dermatomed to a targeted thickness of 500 μm and mounted in penetration cells, with an available diffusion area of 1 cm^2. The receiver side consisted of 3.5 ml phosphate-buffered saline (PBS) containing 0.1% ascorbic acid solution (pH 7.4). The receiver cells were placed over magnetic stirrers, and all the permeation experiments were performed at 37°C using a recirculating constant-temperature water bath. Before pretreatment or dosing, the system was equilibrated for 30 min. For pretreatment with a metabolic inhibitor, 200 μl of 2% sodium azide solution was added to the donor side for 30 min; then all residual solution was removed with a water rinse. The donor side consisted of 10 μl of 2% hydroquinone cream containing [^{14}C]hydroquinone, or 10 μl of control cream. At 4, 8, 12, 16, 20, and 24 h after dosing, 5 ml of receiver solution was automatically collected into separate scintillation vials. The in vitro experiment was performed in a dark place to minimize potential oxidation of hydroquinone. After each 24-h permeation experiment, the residual receiver solution, skin surface washing solution, and the skin itself were collected for assay.

At the end of the experiment, the skin surface was washed three times with a 50% commercial detergent solution (Ivory Liquid, Procter & Gamble) and rinsed with distilled water to remove unabsorbed materials. All washing solutions were kept for scintillation counting. Skin from each diffusion cell was removed and homogenized by a Brinkman homogenizer (Littau, Switzerland) for 30 s (30 rpm) with 2 ml of PBS containing 0.1% ascorbic acid. The homogenate was centrifuged for 5 min, and the supernatant and solid precipitate separated. The solid precipitate was washed twice by using the same buffer solution.

The steady-state flux was calculated from the slope of plots of the amount of hydroquinone penetrated versus time:

$$\text{Flux} = (dc/dt)/A \qquad (\mu g/h\text{-}cm^2)$$

where dc/dt is the steady-state slope, which is usually obtained at $12-24$ h after the beginning of the permeation study, and A is the effective diffusion area of hydroquinone. The lag time (h) was obtained by extrapolating the linear portion to the time axis.

G. Extraction and Thin-Layer Chromatography

The thin-layer chromatography (TLC) system was designed to separate and identify hydroquinone and benzoquinone in the receptor fluid and skin extracts. Identification was by cochromatography with hydroquinone and benzoquinone standards. Benzoquinone was visualized by (NH_4OH) ammonia vapor and hydroquinone by ammoniacal ($AgNO_3$) silver nitrate spray (Gasparic and Churacek, 1978).

One milliliter of receiver fluid (receptor fluid or skin homogenate) was transferred into a covered conical glass tube. Two milliliters ethyl acetate was added to the conical tube and the tube was shaken three times, then kept for $10-15$ min to let the aqueous and ethyl acetate layers separate. The top ethyl acetate layer was transferred to another conical glass tube. Two more extractions were done and the pooled ethyl acetate extracts were evaporated under nitrogen. The dried extract was reconstituted in double-distilled water or methanol, and the vial purged with nitrogen and mixed for TLC. An aliquot of the residual receiver fluid was assayed for ^{14}C content, and this amount was expressed as nonextracted. The extraction procedure was done in a dark place, and tubes were covered with foil.

TLC plates (20×20 cm silica gel 60) were divided into lanes. Standards of benzoquinone and hydroquinone were spotted in separate lanes. The water-reconstituted extraction solution was spotted on the plate. The plates were developed in the solvent system chloroform:ethyl acetate:acetic acid (60:30:10). After developing, the plates were air dried to remove solvent. The solvent front was marked. Standard spots of hydroquinone and benzoquinone were visualized by ammonia and silver nitrate treatments.

Each plate lane was scraped separately in 1-cm units from 0.5 cm below the origin through the solvent front. Each 1-cm scraping was placed into a scintillation vial and 10 ml Universol scintillation fluid was added for ^{14}C determination. To test for stability, a sample of [^{14}C]hydroquinone was subjected to the extraction and TLC procedures. The TLC profile showed only one radioactive peak, which corresponded with the R_f of hydroquinone.

Background control samples and liquid test samples were combined with 10 ml liquid scintillation cocktail (Universal ES, Costa Mesa, CA) and counted in a Packard 1500 liquid scintillation counter (Packard Instruments). Before each measurement, the count was audited for accuracy using sealed samples of quenched and unquenched standards as detailed by the instrument manual.

Control and test sample counts were transferred by hand to a computerized spreadsheet (Microsoft Excel 4.0 for Macintosh) and the percentage of labeled material recovered from the applied dose was determined. From these values, the concentration of ^{14}C-labeled hydroquinone was calculated, based on the specific activity of unmetabolized material. The samples were extracted and chromatographed in duplicate.

H. Urine Sample Extraction/Enzyme Treatment

Urine samples were extracted without enzyme (glucuronidase, sulfatase) treatment and compared to duplicates samples with enzymatic treatment. One milliliter of urine was transferred into a covered conical glass tube that had 1 ml buffer solution (KH_2PO_4, pH 6.9, for glucuronidase; Tris-HCl, pH 7.1, for sulfatase) and then enzyme was added (final amount of β-glucuronidase and sulfatase for 96 u/ml and 6 u/ml, respectively). Control negative and positive of enzyme reactions were done with substrate to test enzyme activity. Sample was incubated for 2 h at 37°C in a shaker bath, then purged with nitrogen and covered with foil. The same extraction and TLC procedures were used as with the in vitro samples noted earlier.

I. Radioactivity Analyses

The scintillation cocktail used for radioactivity determination was Universal-ES (ICN Biomedicals, Costa Mesa, CA). Background control samples and test samples were counted in a Packard model 1500 counter (Packard Instruments). The counter was audited for accuracy using sealed samples of quenched and unquenched standards as detailed by the instrument manual. Control and test sample counts were transferred by hand to a computerized spreadsheet (MS Excel 5.0 for Windows) and the percentage of labeled material recovered from the applied dose determined. From these values, the concentration of ^{14}C-labeled hydroquinone was calculated, based on the specific activity of unmetabolized material. Statistical analyses (Student's *t*-test where $p < .05$ is significant) were done using Primer (McGraw-Hill, 1988).

III. RESULTS

A. In Vivo Human

Table 1 presents the radioactivity recovery following 2% [^{14}C]hydroquinone cream application to the forehead. Dose recovered in urine was 45.3 ± 11.2%, of which the majority was excreted in the first 24 h. Recovery from the unprotected treated skin site at 24 h was 5.2 ± 3.2%. It is assumed that the remainder of the unprotected topical dose was lost, probably by rubbing contact during sleep. Figure 1 shows the decrease in topical hydroquinone from skin and subsequent rise in hydroquinone levels in the stratum corneum (tape stripping). Skin surface hydroquinone quickly and constantly transfers into the stratum corneum. Table 2 shows that hydroquinone on skin can be decontaminated early after exposure, although the total dose cannot be removed with soap-and-water wash. Figure 2 shows ipsilateral (blood draw site near dosed area on one forearm) and contralateral (blood draw from opposite, nondosed forearm) plasma levels. The ipsilateral concentration should be initially higher than the contralateral because of body dilution of drug in getting to the opposite arm (Karim, 1987). This is initially the case with hydroquinone (0.5 h), but the two sites are subsequently nearly the same. Peak plasma concentration was at 4 h, at 0.04 μg-Eq/ml for the dosed hydroquinone (2500 μg/25 cm^2). Urinary excretion for the 8-h period was

Table 1 Radioactivity Recovery Following [^{14}C]Hydroquinone Topical Application In Vivo in Humans: Forehead Site

Time (h)	Percent administered dose	
	Urine	Skin surface wash
24	37.0 ± 9.8	5.2 ± 3.2[a]
48	7.1 ± 2.1	
72	0.9 ± 0.6	
96	0.4 ± 0.2	
Total	45.3 ± 11.2	5.2 ± 3.2

Note. Values are mean ± SD ($n = 6$).
[a]Dosed skin site washed with soap and water after 24-h dosing period.

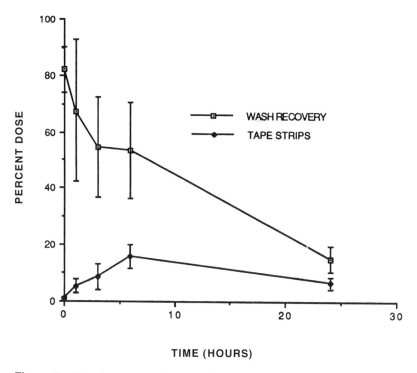

Figure 1   Timed sequence of soap-and-water wash recovery and stratum corneum tape stripping from human volunteers dosed with 2% hydroquinone cream on the ventral forearm.

Table 2   Radioactivity Recovery Following [^{14}C]Hydroquinone Topical Application In Vivo in Humans: Forearm Site

Time (h)	Percent administered dose	
	Skin wash recovery	Skin tape strips
0	82.3 ± 8.1	1.2 ± 0.4
1	67.5 ± 25.3	5.4 ± 2.5
3	54.8 ± 18.0	8.6 ± 4.5
6	53.6 ± 17.3	15.8 ± 4.2
24	15.0 ± 4.5	6.6 ± 2.1

Note. Skin absorption is occurring, and that this contributes to decreased skin wash recovery, as well as that which is lost to the environment (e.g., rubbing during sleep). Values are mean ± SD ($n = 4$).

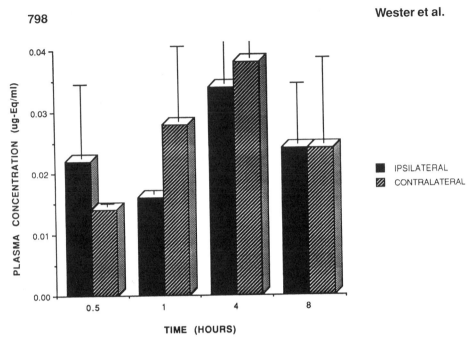

Figure 2 Plasma radioactivity in blood samples taken ipsilateral (near dosing area on forearm) and contralateral (opposite forearm) from human volunteers dosed with 2% hydroquinone cream on the ventral forearm.

8.0 ± 4.1% dose. Thus, there is a picture of hydroquinone quickly disappearing from the dosed surface of skin, filling and passing through the stratum corneum, into the body circulation, and being excreted in the urine. The amount absorbed is significant (45.3%) and the absorption process is continuous.

B. In Vitro Human Skin Absorption

Table 3 shows the permeation profiles of hydroquinone through six sources of human skin without or pretreatment with sodium azide. This in vitro study shows that after 24 h application, 34.0 ± 4.9% of hydroquinone permeated through the skin, and 9.3 ± 4.4% of hydroquinone was in the skin. The total permeation of hydroquinone without treatment was 43.3%. When the test skins were pretreated with 2% sodium azide for 30 min, there was no significant change in the permeation profiles compared with no pretreatment. The test showed that 29.8 ± 10.0% of hydroquinone penetrated through the skin, and 13.0 ± 7.8% of hydroquinone was in the skin. The total permeation of hydroquinone with a pretreatment was 42.7%. Dose accountability was 100%.

Table 3   In Vitro Percutaneous Absorption of Hydroquinone in Human Skin

Treatment	Percent dose			
	Receptor fluid	Skin	Surface wash	Total
None	34.0 ± 4.9	9.3 ± 4.4	55.8 ± 6.2	99.2 ± 2.6
Azide[a]	29.8 ± 10.0	13.0 ± 7.8	58.0 ± 7.6	100.8 ± 9.2

Note. Values are mean ± SD ($n = 6$).
[a]Sodium azide, a metabolic inhibitor, was dosed on skin for 30 min and then the skin surface
  was rinsed with buffered saline, and hydroquinone dosed on the skin.

The permeation parameters calculated from 8 to 20 h of penetration after dosing show that no treatment and inhibitor treatment did not show significant difference in their flux (2.94 and 2.93 μg/h-cm^2) and lag time (7.99 and 8.03 h). The lag time indicates that it takes 8 h for hydroquinone to penetrate through human skin into the receptor fluid in this in vitro study.

C.   Human Skin In Vitro and In Vivo Metabolism

Figure 3 shows the result of a control extraction and TLC of hydroquinone. No peaks other than hydroquinone were seen. Figure 4 is a TLC metabolic profile for extraction and TLC of skin following hydroquinone topical application. The second peak had the same R_f and was tentatively identified as benzoquinone. The profile was converted to percent dose by zone, and subjected to statistical analyses. The hydroquinone (HQ) and benzoquinone (BQ) zones were identified by cochromatography with authentic standards. The other zones are identified as origin (O), zone 1 (Z1) between origin and HQ, zone 2 (Z2) between hydroquinone and benzoquinone, and zone 3 (Z3) between benzoquinone and (including) solvent front.

Table 4 indicates the percent dose radioactivity for not extracted and for each TLC zone of the extracted skin homogenate for no treatment and azide pretreatment. The percent doses not extracted were 14.3 ± 10.2 for no treatment and 19.8 ± 12.4 for azide treatment. Peaks of radioactivity were isolated for hydroquinone with no treatment (28.5 ± 12.9%) and with azide treatment (28.1 ± 18.4%), and for benzoquinone with no treatment (10.8 ± 5.7%) and for azide treatment (4.1 ± 2.0%). Statistically there was no difference in hydroquinone with and without azide treatment. However, the percent radioactivity for benzoquinone was significantly decreased with azide treatment. There were no outstanding radioactivity peaks or differences with and without azide treatment for the other radioactivity zones. The major peak of radioactivity over all time periods was hydroquinone; the other TLC

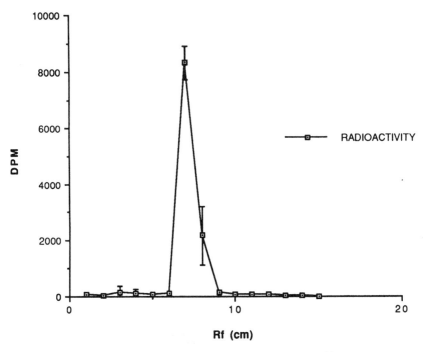

Figure 3   Control extract and thin-layer chromatography of [^{14}C]hydroquinone.

zones showed less percent radioactivity. Statistically, there was no effect of azide on percent radioactivity associated with hydroquinone and benzoquinone. The percent radioactivity not extracted ranged from 6.4 ± 5.6% to 20.2 ± 17.2% for no treatment and a similar 8.0 ± 5.6% to 24.8 ± 22.7% for azide treatment. Both showed a decrease in percent radioactivity at 24 h. The levels of not extracted radioactivity from homogenized skin (Table 5) fell within the range of receptor fluid not extracted (Fig. 5).

Table 5 presents the percent dose radioactivity for extraction from urine samples (0–24 and 24–48 h, respectively) with and without enzyme treatment. For 0–24 h samples the percent doses extracted were 3.5 ± 0.8 for no treatment; 30.7 ± 5.5 for glucuronidase treatment, and 2.7 ± 1.2 for sulfatase treatment. For 24–48 h samples the percent doses extracted were 5.6 ± 3.4 for no treatment, 28.5 ± 6.6 for glucuronidase treatment, and 2.3 ± 1.3 for sulfatase treatment. Statistically there was no difference in the percentage dose radioactivity for extraction with and without sulfatase treatment at both 0–24 and 24–48 h; however, the percentage dose radioactivity for extraction with and without glucuronidase treatment was statistically different at both 0–24 and 24–48 h.

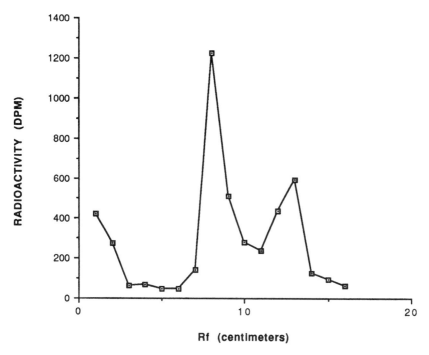

Figure 4 Extract and thin-layer chromatography of homogenized human skin in vitro percutaneous absorption with [^{14}C]hydroquinone.

Table 4 Thin-Layer Chromatography After Hydroquinone Topical Application to Human Skin In Vitro: Skin Homogenate

TLC zone	Percent dose[a]	
	No treatment	Azide treatment
Not extracted	14.3 ± 10.2	19.8 ± 12.4
Origin	3.8 ± 2.5	3.6 ± 2.5
Zone 1	1.6 ± 0.8	1.7 ± 1.0
Hydroquinone	28.5 ± 12.9	28.1 ± 18.4
Zone 2	2.7 ± 1.3	1.6 ± 0.8
Benzoquinone	10.8 ± 5.7	4.1 ± 2.0*
Zone 3	0.3 ± 0.1	0.1 ± 0.1

[a]Mean ± SD ($n = 6$).
*Statistically different ($p < .02$; paired t-test).

Table 5 Extraction After Hydroquinone Topical Application to Human Skin In Vivo: Urine Samples With and Without Enzyme Treatment

	Percent dose[a]		
Extraction	No treatment	Glucuronidase treatment	Sulfatase treatment
0–24 h	3.5 ± 0.8	30.7 ± 5.5*	2.7 ± 1.2
24–48 h	5.6 ± 3.4	28.5 ± 6.6*	2.3 ± 1.3

[a]Mean ± SD (n = 6).
*Statistically significant (p < .001; paired t-test between glucuronidase treatment and no treatment).

Table 6 contains the percent doses of radioactivity for urine samples at 0–24 and 24–48 h for not extracted amounts and for the various zones following glucuronidase and sulfatase treatment, extraction, and TLC. No unidentified radioactive peaks were seen, other than hydroquinone and a small amount of benzoquinone.

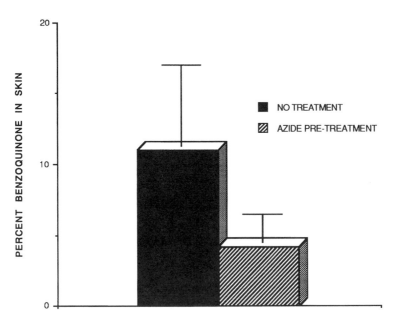

Figure 5 Metabolic inhibition by azide on skin metabolism of hydroquinone to benzoquinone.

Table 6 Thin-Layer Chromatography After Hydroquinone Topical Application to Human Skin In Vivo: Urine Samples with Glucuronidase Treatment

	Percent dose[a]	
TLC zone	0–24 h	24–48 h
(Not extracted)	(56.1 ± 8.3)	(58.5 ± 2.8)
Origin	1.9 ± 0.9	2.6 ± 1.5
Zone 1	1.6 ± 0.4	1.7 ± 1.3
Hydroquinone	61.2 ± 4.4	58.3 ± 13.2
Zone 2	0.0 ± 0.0	0.0 ± 0.0
Benzoquinone	1.3 ± 0.8	1.4 ± 0.7
Zone 3	0.1 ± 0.2	0.3 ± 0.6

[a]Mean ± SD ($n = 6$).

D. Flux

Table 7 gives the flux rates for humans in vivo when hydroquinone is applied to the forehead (1.9 μg/cm^2/h) and to the forearm (1.0 μg/cm^2/h) compared to human skin in vitro flux rate of 2.8 and 2.9 μg/cm^2/h. Note that the dose in vitro per square centimeter was double that of the in vivo applications, and this is reflected in the larger flux (approximately doubled).

IV. DISCUSSION

For a topical hydroquinone concentration in cream formulation of 100 μg/cm^2 to the forehead of humans, urinary excretion was 45.3% dose for a 24-h duration. Bucks et al. (1988) applied hydroquinone in a variety of alcohol vehicles at 125 μg/cm^2 to the forehead of humans for 24 h and reported a range of 35–66% urinary excretion, depending upon vehicle. Thus, the Bucks et al. (1988) study and this current study agree that hydroquinone readily penetrates human forehead skin in vivo. It should be noted that in the Bucks et al. (1988) study and this current study, complete urinary ^{14}C excretion is assumed. This is the case for the rat, but not for the rhesus monkey (Hui et al., personal communication). Only an intravenous administration in humans would determine this for sure. If excretion in humans was not complete, then a correction factor for incomplete recovery (feces, body retention) would need to be applied. This would increase the estimated absorption for humans.

In the present study, urinary excretion for the 8-h forearm dosing was 8% of dose. Assuming linear absorption, this would project to 24% of dose

Table 7 Flux Rate for Topical Hydroquinone (2% Cream)

Species	Dose HQ/surface area	Calculation
Human in vivo (forehead)	2500 μg HQ/25 cm^2	= 100 μg/cm^2 × 0.4530 bioavailability[a]
		= 45.3 μg/cm^2/24 h
		= 1.9 μg/cm^2/h
Human in vivo (forearm)	2500 μg HQ/25 cm^2	= 100 μg/cm^2 × 0.08 bioavailability[a]
		= 8.0 μg/cm^2/8 h
		= 1.0 μg/cm^2/h
Human skin (in vitro)	200 μg HQ/1 cm^2	= 200 μg/cm^2 × 0.34 bioavailability[a]
		= 68 μg/cm^2/24 h
		= 2.8 μg/cm^2/h
	2% Cream graphic method	2.9 μg/cm^2/h[b]

[a]Bioavailability calculated from total urinary ^{14}C excretion or receptor fluid accumulation.
[b]The steady-state flux is calculated from the slope of plots of the amount of hydroquinone penetrated vs. time:

$$\text{Flux} = (dc/dt)A \qquad (\mu g/h\text{-}cm^2)$$

where dc/dt is the steady-state slope, which is usually obtained at 12–24 h after the beginning of the permeation study, and A is the effective diffusion area of hydroquinone.

for 24 h. The forehead at 45% dose absorption would thus be a higher absorbing skin site than the forearm, and this is in agreement with the literature (Wester and Maibach, 1989).

The recovered dose in vivo from the forehead after the 24-h treatment period was only 5.2% of dose. The soap-and-water washing procedure is sufficient to remove most of the applied dose, as evidenced by the 82.3% control recovery (0 h) in Table 2. Thus, it can be assumed that the remainder of the dose (remembering that 45.3% of dose is absorbed) was probably lost by rubbing during sleep. In animal studies with the same hydroquinone formulation, but where the animal's movement is restricted during the 24-h dosing period, dose accountability for topical application was 100.1 ± 6.1% in the rhesus monkey. Thus, in regular human use, some of the topically applied or environmentally derived hydroquinone should be expected to be lost from the surface of the skin.

For this 2% hydroquinone cream dose, the calculated flux was 1.9 μg/cm^2/h for the in vivo forehead, 1.0 μg/cm^2/h for the in vivo forearm, and

2.8 or 2.9 μg/cm^2/h for human skin in vitro, depending on the method of calculation. Barber et al. (1995), using a higher 5% aqueous dose, reported lower flux values of 0.52 μg/cm^2/h for isolated human stratum corneum and 1.1 μg/cm^2/h for rat skin. Isolated human stratum corneum is considered a low-confidence method and thus probably not a valid model compared to fresh skin for estimating hydroquinone percutaneous absorption (Howes et al., 1996).

There is another cautionary note to be made regarding in vitro absorption studies, and that is the excess time needed for chemical to diffuse through the artificial structure of skin in a diffusion cell. In the current study, a lag time of 8 h was seen before sufficient steady-state absorption was achieved. It is obvious from the in vivo plasma data that hydroquinone and/or metabolites are readily systemically available within the first 0.5 h of dosing. Thus, any concept that hydroquinone absorption is slow (Barber et al., 1995) probably is due to the in vitro system. In vivo absorption of hydroquinone is relatively rapid, appearing in the first 0.5 h in vivo plasma samples. Any "lag time" or "slow" concept from in vitro studies should be considered as an artifact of the system (Wester and Maibach, 1996).

Hydroquinone was shown to have some metabolism to a second peak, which corresponded to benzoquinone in this TLC system. This was in isolated human skin; however, extraction and subsequent TLC showed systemic/urinary metabolite(s) to be mostly conjugated, mainly the glucuronide of hydroquinone. Skin does have the capacity to do both phase I and II reactions (Noonan and Wester, 1983), but no conjugation was seen with the in vitro skin, the main metabolism sequence being conjugation of hydroquinone systemically. Monteiro-Riviere (1990) suggests that 70–80% of viable skin energy metabolism is anaerobic. Perhaps, although viable skin has many metabolism potentials (Noonan and Wester, 1989), passage through skin may not involve all metabolism possibilities. Skin absorption can occur through a variety of routes, such as diffusion through cells, diffusion around cells, and diffusion by other pathways such as hair shaft. Metabolizing enzymes may be present, but the absorbing chemical may not encounter the enzyme in passage through the skin. It should also be noted that Barber et al. (1995) did not detect benzoquinone in their studies. This may be related to the condition of the skin used, or it may be related to a difference in assay procedure. Wester et al. (1998) have shown that dermatomed human skin remains viable during the storage/use regimen that was used in this study.

Thus, with hydroquinone, most of these studies show a chemical that continuously partitions from the skin surface into and through the stratum corneum, viable epidermis, and into the systemic circulation. Systemic blood levels are measurable within the first half hour, and continue as long as hydroquinone is in contact with the skin. Barber et al. (1995), using their estimate of hydroquinone flux, suggested exposure to be approximately four orders of magnitude lower than the dose of hydroquinone required to produce nephrotoxicity in the rat. However the Barber et al. (1995) flux rate is at least 4-fold less than that reported by the Bucks et al. (1988) study and this current study. The major metabolism of hydroquinone is conjugation for systemic removal. A TLC peak with the R_f of benzoquinone was found, but this identification is tentative until further chemical characterization is done.

REFERENCES

Anderson, B. (1974). Corneal and conjunctival pigmentation among workers engaged in manufacture of hydroquinone. Arch. Ophthalmal. 38:812–826.

Barber, E. D., Hill, T., and Schum, D. B. (1995). The percutaneous absorption of hydroquinone (HQ) through rat and human skin *in vitro*. Toxicol. Lett. 80: 167–172.

Bucks, D. A. W., McMaster, J. R., Guy, R. H., and Maibach, H. I. (1988). Percutaneous absorption of hydroquinone in humans: effect of 1-dodecylazacyloheptan-2-one (Azone) and the 2-ethylhexyl ester of 4-(dimethylamino)benzoic acid (Escalol 507). J. Toxicol. Environ. Health 24:279–289.

Carlson, A. J., and Brewer, N. R. (1953). Toxicity studies on hydroquinone. Proc. Soc. Exp. Biol. 84:684–688.

Casarett, L. J. (1975). Toxicology: The Basic Science of Poisons. Macmillan, New York.

Christensen, R. T., Clark, C. S., Cody, T. E., Witherup, S., Gartside, P. S., Elia, V. J., Eller, P. M., Lingg, R., and Cooper, G. P. (1976). The development of a test for the potability of water heated by a direct reuse system. University of Cincinnati.

Colinas, R. J., Burkart, P. T., and Lawrence, D. A. (1994). *In vitro* effects of hydroquinone, benzoquinone, and doxarubicin on mouse and human bone marrow cells at physiological oxygen partial pressure. Toxicol. Appl. Pharmacol. 129: 95–102.

Diviliers, J., Boule, P., Vassfur, P., Prevot, P., Steman, R., Sfigue-Murandi, F., Benoit-Guyod, J. L., Nendza, M., Grioni, C., Dive, D., and Chambon, P. (1990). Environmental and health risk of hydroquinone. Ecotoxicol. and Environ. Safety 19:327–354.

English, J. C., Hill, T., O'Donogue, J. L., and Reddy, M. V. (1994a). Measurement of nuclear DNA modification by ^{32}P-postlabeling in the kidneys of male and female Fischer 344 rats after multiple gavage doses of hydroquinone. Fundam. Appl. Toxicol. 23:391–396.

English, J. C., Perry, L. G., Vlaovic, M., Moyer, C., and O'Donogue, J. L. (1994b). Measurement of cell proliferation in the kidneys of Fischer 344 and Sprague-Dawley rats after gavage administration of hydroquinone. Fundam. Appl. Toxicol. 23:397–406.

Friedlander, B. R., Hearne, T., and Newmann, B. J. (1982). Mortality, cancer incidence, and sickness in photographic processors: An epidemiologic study. J. Occup. Med. 24:605–613.

Gasparic, J., and Churacek, J. (1978). Laboratory Handbook of Paper and Thin-layer Chromatography. John Wiley and Sons, New York.

Goeke, E., King, M. T., Eckardt, K., and Wild, D. (1981). Mutagenicity of cosmetics ingredients licensed by the European community. Mutat. Res. 90:91–109.

Howes, D., Guy, R., Hadgraft, J., Heylings, J., Hoeck, U., Kemper, F., Maibach, H., Marty, J-P., Merk, H., Parra, J., Rekkas, D., Rondelli, I., Schaefer, H., Tauber, U., and Verbiese, N. (1996). Methods for assessing percutaneous absorption. ECVAM Workshop Report 13. ATLA 24:81–106.

Karim, A. (1987). Transdermal absorption of nitroglycerin via a microsealed drug delivery system. In Transdermal Delivery of Drugs, Vol 1, A. Kydonieus and B. Berner, eds. CRC Press, Boca Raton, FL, pp. 131–143.

Monteiro-Riviere, N. A. (1990). Specialized technique: The isolated perfused porcine skin flap (IPPSF). In Methods for Skin Absorption, B. Kemppainen and W. Reifenrath, eds. CRC Press, Boca Raton, FL, pp. 175–189.

National Toxicology Program. (1991). Toxicology and carcinogenesis studies of hydroquinone in F-344 rats and B6C3F mice. NIH Publication No. 90-2821, Research Triangle Park, NC.

Noonan, P. K., and Wester, R. C. (1989). Cutaneous metabolism of xenobiotics. In Percutaneous Absorption, 2nd edition, R. Bronaugh and H. Maibach, eds. Marcel Dekker, New York, pp. 53–75.

Pifer, J. W., Hearne, F. T., Swanson, F. A., and O'Donoghue, J. L. (1994). Mortality study of employees engaged in the manufacture and use of hydroquinone. Int. Arch. Occup. Environ. Health 67:267–280.

Topping, D., Bernard, L., Blacker, A., Murphy, S., and O'Donogue, J. (1988). Subchronic functional observational battery (FOB) and neuropathology studies on hydroquinone. Chemical Manufacturers Association, Washington, DC.

Wester, R. C., and Maibach, H. I. (1989). Regional variation in percutaneous absorption. In Percutaneous Absorption, 2nd edition, R. Bronaugh and H. Maibach, eds. Marcel Dekker, New York, pp. 111–119.

Wester, R. C., and Maibach, H. I. (1996). Percutaneous absorption: Short-term exposure, lag time, multiple exposures, model variations, and absorption from clothing. In Dermatotoxicology, 5th edition, F. Marzulli and H. Maibach, eds. Taylor & Francis, Washington, DC, pp. 35–48.

Wester, R. C., Christoffel, J., Hartway, T., Problete, N., Maibach, H., and Forsell, J. (1998). Human cadaver skin viability for in vitro percutaneous absorption: Storage and detrimental effects of heat-separation and freezing. Pharm. Res. 15:82–84.

47

Human Skin In Vitro Percutaneous Absorption of Gaseous Ethylene Oxide from Fabric

Ronald C. Wester, Tracy Hartway, Steffany Serranzana,*
and Howard I. Maibach
University of California School of Medicine, San Francisco, California

I. INTRODUCTION

Ethylene oxide, a colorless gas at ordinary room temperature and pressure, is used as a fumigant for textiles and foodstuffs, and to sterilize surgical instruments and hospital gowns. It is a highly reactive alkylating agent that can react directly with cellular macromolecules, including DNA, RNA, and protein, without prior metabolic activation (Brown et al., 1996). Data have been collected on the genotoxicity of ethylene oxide in somatic and germ cells (Preston et al., 1995). Major et al. (1996) have reported genotoxicological changes in nurses occupationally exposed to low dose ethylene oxide. Studies done to assess ethylene oxide risk assessment use inhalation and intravenous routes of administration (Ehrenberg and Törnquist, 1995). Since ethylene oxide exposure to humans can occur through exposure to fabric/skin in the workplace (Wester et al., 1996), a study was done to determine the percutaneous absorption of [^{14}C]ethylene oxide from fabric into and through human skin. This study also serves as a model for exposure of any potential hazardous gas to fabric/skin, be it war (Wester et al., 1996), the workplace, or some terrorist act as recently occurred in Japan.

II. MATERIALS AND METHODS

The test chemical was [1,2-^{14}C]ethylene oxide with a specific activity of 1.0 mCi/mmol and 1800 ppm (Sigma Chemical Co., St. Louis, MO). The fabric

---

Present affiliation: Penederm Inc., Foster City, California.

was 100% polyester cuff material from a surgical gown, the fabric cut into 1-cm^2 round disks. The fabric disks were placed into a break seal glass container, [^{14}C]ethylene oxide was added into the container, and the container was sealed.

Two separate human skin sources were used. Cadaver skin was dermatomed to a targeted 500 μm and stored refrigerated at 4°C in Eagle's minimal essential medium to preserve skin viability. The skin was used within 5 days of donor death. This preservation/use regimen follows that used by the human skin transplant bank (Hurst et al., 1984) and the work of Bronaugh et al. (1989).

Glass diffusion cells with 1 cm^2 surface area were filled with phosphate-buffered saline. Each cell reservoir was sealed to the outside to contain potential absorbed chemical, and each contained a stirring bar to ensure continual contact between receptor fluid and the under dermis side of skin.

Human skin was mounted on the diffusion cells. The sealed containers containing fabric/[^{14}C]ethylene oxide was opened and the fabric disks were removed. Four disks were placed directly into scintillation vials containing scintillation fluid to determine how much [^{14}C]ethylene oxide had penetrated the fabric. The remaining four fabric disks were placed on the surface of the mounted human skin. Two fabric disks were covered immediately with double-layer latex glove material to avoid loss of gaseous [^{14}C]ethylene oxide, and two fabric disks were left open to the air. The diffusion system was mounted in a hood and the procedure was done in the hood to protect technicians from exposure.

Every 4 h the receptor fluid in the diffusion cell was removed for assay and replaced with fresh receptor fluid. At the end of 24 h the skin surface was washed with liquid soap (50% Ivory Liquid, Procter & Gamble, Cincinnati, OH), and rinsed with water. The skin itself was completely solubilized in Soluene 350 (Packard Instruments, Downers Grove, IL). The receptor fluid, solubilized skin, latex cover material, residual fabric, and skin washings were assayed for ^{14}C content by liquid scintillation counting. The counting process and computer program have been verified to be accurate by a quality assurance officer.

III. RESULTS

Table 1 gives the in vitro percutaneous absorption of [^{14}C]ethylene oxide from fabric through human skin. In cells where the fabric/skin surface area was open to the air, 0.72 and 0.85% of dose was accumulated in receptor fluid and 0.62 and 0.32% of dose was in skin. This gives percutaneous absorption values of 1.34 (receptor fluid accumulation plus skin content) and

Table 1 In Vitro Percutaneous Absorption of [^{14}]Ethylene Oxide from Fabric Through Human Skin

	Percent dose	
Item assayed	Skin source 1	Skin source 2
Surface open to air		
Receptor fluid (RF)	0.72	0.85
Fabric	7.65	5.65
Skin wash	0.03	0.03
Skin rinse	0.001	0.001
Skin content	0.62	0.32
Average absorbed (RF + skin)	1.3	
Surface Occluded		
Receptor fluid (RF)	46.16	39.54
Fabric	3.62	3.62
Latex cover	0.23	0.21
Skin wash	0.07	0.05
Skin rinse	0.01	0.02
Skin content	3.25	3.12
Average absorbed (RF + skin)	46.0	

1.17% of dose (average 1.3% of dose). Less than 10% residual radioactivity was recovered in residual fabric and washings.

In cells where the fabric/skin was occluded with double latex material, 46.16 and 39.54% of dose was recovered in receptor fluid and 3.25 and 3.12% of dose was in skin. This gives percutaneous absorption values of 49.41 and 42.66% of dose (average 46.0%). Less than 5% residual radioactivity was recovered in residual fabric and washings. It is assumed that the remainder of the [^{14}C]ethylene oxide was lost to the surrounding air.

Figure 1 shows the accumulation of radioactivity in receptor fluid. Almost all of the skin-absorbed radioactivity was in the first 0–4 h interval.

Some receptor fluid accumulations from a nonradioactive study were quickly frozen and shipped to NAmSA laboratories (Irvine, CA) for assay. At a dose of 600 ppm, no ethylene oxide, ethylene chlorhydrin, or ethylene glycol was detected. At 3892 ppm, ethylene oxide represented 4.5–8.3% of dose in receptor fluid, and no ethylene chlorhydrin or ethylene glycol was detected. Potential loss of ethylene oxide during analysis is not known; however, ethylene oxide is able to be absorbed intact across human skin.

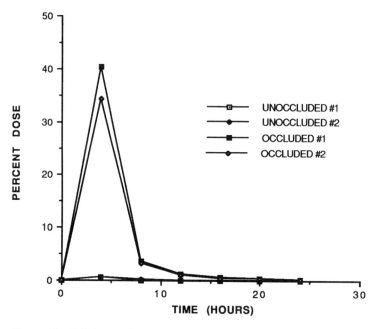

Figure 1 Ethylene oxide can be absorbed through human skin rapidly in the first
0–4 h sampling period.

IV. DISCUSSION

Risk assessment studies of ethylene oxide have been done using inhalation
and intravenous administration (Ehrenberg and Törnquist, 1995). This study
shows that ethylene oxide in contact with fabric/skin will absorb into and
through human skin. A value of 1.26% of dose absorbed was obtained when
the fabric/skin was open to the surrounding air, where it is assumed most
of the ethylene oxide dissipated. However, when the fabric/skin is occluded
to limiting ethylene oxide air dissipation, a large 46.0% of dose was absorbed through the human skin. This absorption was rapid, occurring in the
first sampling period (1–4 h). The cost and nature of the study limited the
number of replicates. The specific activity of the [14C]ethylene oxide and
the size of the break seal glass container limited the number of fabric disks
that could be exposed.

   Thus, ethylene oxide percutaneous absorption through human skin can
be rapid, and ethylene oxide can be absorbed across human skin as intact
chemical. This also serves as a model for potential hazardous gas absorption
into and through fabric and human skin, be it war (Wester et al., 1996) or
the workplace.

REFERENCES

Bronaugh, R. L., Stewart, R. F., and Storm, J. E. (1989). Extent of cutaneous metabolism during percutaneous absorption of xenobiotics. Toxicol. Appl. Pharmacol. 99:534–543.

Brown, C. D., Wong, B. A., and Fennell, T. R. (1996). *In vivo* and *in vitro* kinetics of ethylene oxide metabolism in rats and men. Toxicol. Appl. Pharmacol. 136: 8–19.

Ehrenberg, L., and Törnquist, M. (1995). The research background for risk assessment of ethylene oxide: aspects of dose. Mutat. Res. 330:41–54.

Hurst, L. N., Brown, D. H., and Murray, K. A. (1984). Prolonged life and improved quality for stored skin grafts. Plast. Reconstr. Surg. 73:105–109.

Major, J., Jakob, M. G., and Tompa, A. (1996). Genotoxicological investigation of hospital nurses occupationally exposed to ethylene oxide: 1. Chromosome aberrations, sister-chromated exchanges, cell cycle kinetics, and UV-induced DNA synthesis in peripheral blood lymphocytes. Environ. Mol. Mutagen. 27: 84–92.

Preston, R. J., Fennell, T. R., Leber, A. P., Sielken, R. L., Jr., and Swenberg, R. L. (1995). Reconsideration of the genetic risk assessment for ethylene oxide exposures. Environ. Mol. Mutagen. 26:189–202.

Wester, R. C., Quan, D., and Maibach, H. I. (1996). *In vitro* percutaneous absorption of model compounds glyphosate and malathion from cotton fabric into and through human skin. Food Chem. Toxicol. 34:731–735.

48

In Vivo Percutaneous Absorption of Acetochlor in Rhesus Monkey
Dose Response and Exposure Risk Assessment

Ronald C. Wester, Joseph L. Melendres, and Howard I. Maibach
University of California School of Medicine, San Francisco, California

I. INTRODUCTION

Acetochlor, recently approved by the U.S. Environmental Protection Agency (EPA) as an herbicide for corn crops, has a median lethal dose of 1063 mg/kg in male rats and 768 mg/kg in female rats (Antov et al., 1991; Khalkova et al., 1990). It exhibits toxicological effects at 50–60 mg/kg subchronic dosing in dog and rat. Dermal dosing at 100 mg/kg for 3 weeks did not produce systemic toxicity in the rat, and acetochlor has been labeled as carcinogenic.

Acetochlor was found to be clastogenic in the in vitro (human lymphocyte) cytogenetic assay. In the Ames test, a weak response was observed in *Salmonella typhimurium* TA1538 strain only with S9. When tested in vivo, it was negative in the mouse micronucleus assay, but produced a positive, albeit weak, response in an in vivo unscheduled DNA synthesis assay. Acetochlor has therefore shown genotoxicity activity in vitro but only weak activity in vivo (Zeneca, personal communication, 1992).

Support for acetochlor approval was because farmers only need to use half as much acetochlor per acre as other herbicides. As risk assessment endpoint is with the human user and not acreage, the bioavailability (rate and kinetic extent of absorption) is critical in determining how much acetochlor will be absorbed into the body. This bioavailability value relative to the herbicides it is to replace should be a pivotal point in determining relative safety. Thus the objective of this study was to determine the in vivo percutaneous absorption of acetochlor at relevant user dose levels in the rhesus monkey, an animal model relevant to humans for percutaneous absorption (Wester and Maibach, 1989a).

II. MATERIALS AND METHODS

A. Formulation

Four dose preparations containing [^{14}C]acetochlor were made: Three of these doses represented preparations equivalent to a commercial formulation concentrate and nominal 1/70 and a 1/1000 aqueous dilution thereof and were used for topical administration. Application of 40 μl of these three preparations represented nominal doses of 30.7 (3070 μg/cm^2), 0.43 (43 μg/cm^2), and 0.03 (3 μg/cm^2) mg acetochlor, respectively. The doses were applied over 10 cm^2 skin surface area. The fourth dose, equivalent to the preceding intermediate dose, was prepared in propylene glycol for intravenous administration. The analyzed specific activities of [^{14}C]acetochlor in the four dose preparations were 0.26 μCi/mg for the formulation concentrate, 18.03 μCi/mg for the 1/70, and 250.7 μCi/mg for the 1/1000 aqueous dilutions, respectively, and 18.3 μCi/mg for the intravenous dose solution. The [^{14}C]acetochlor in each dose preparation was shown to be stable for longer than the interval between preparation and dosing.

B. Test Animals

The in vivo percutaneous absorption of acetochlor was determined in four female rhesus monkeys with body weight range of 7.4 to 9.5 kg. These monkeys have permanent identification numbers (tattoo). The same animals were used for each dose application. The animals are under care of veterinary staff at UCSF, and undergo routine physical examinations and other health assessments and care. Samples of urine were collected prior to dosing to establish the background radioactivity level for these samples. All animals were housed in an environmentally controlled room with a temperature of 23 $\pm$ 6°C and a humidity of 50 $\pm$ 20%. Fluorescent lighting provided illumination 12 h/day. During the study, monkeys were housed individually in metabolic cages. The animals were acclimated (as needed) in the metabolic cages. Food (Purina monkey chow, Ralston Purina Company) and tap water were provided daily. Fruit and fruit juice were provided.

C. Study Design

The study design was as follows:

Number of animals*			
Topical nominal dose level (mg/10 cm^2)			Parenteral nominal dose levels (mg),
0.03	0.43	30.7	0.43
4	4	4	4

*The same group of animals was used for each dosing phase.

Percutaneous absorption and applied dose accountability were determined by assaying the following samples: a) predose urine excretion; b) blood sample collection; c) daily urine and fecal excretion for 7 days; and d) skin surface washings (50% soap and water) at the 24-h interval following topical dosing.

D. Procedure

Rhesus monkeys were placed in specially designed metabolic chairs for the designated time period. A dose of 40 μl acetochlor was administered to 10 cm^2 of monkey's abdominal skin. The dosing site was not covered. The site of application was washed with 50% soap (Ivory Liquid) and water after the 24-h skin application time. The monkeys were transferred to metabolic cages for continued blood, urine, and feces collection. Each study was concluded after 7 days, at which time the animals were placed in an appropriate room to allow for the depuration of any remaining radioactivity for a period of 3 weeks (4 weeks between skin application periods). The animals were observed for mortality, morbidity, presence of emesis and/or diarrhea, and skin irritation. None of these events occurred.

E. Sample Analysis

Blood samples were separated into plasma and red blood cells by centrifugation. Plasma radioactivity was measured by direct liquid scintillation counting.

Total urine excretion was collected by metabolic chairs and cages at room temperature throughout the study. Total urine volume was recorded and a 15-ml aliquot saved for analysis. All urine samples were well mixed and aliquoted in duplicate for direct counting in scintillation cocktail. Five-milliliter aliquots of urine were counted.

Total fecal excretion was determined. Fecal samples were homogenized, then an aliquot combusted for ^{14}C content.

The scintillation cocktail was Universal$_{ES}$ (ICN Biomedicals, Costa Mesa, CA). Background control and test samples were counted in a Packard model 4640 or model 1500 counter (Packard Instruments). Control and test sample counts were transferred to a computer program, which subtracted 1 × background control samples and generated a spreadsheet, with the data reported in the next section.

III. RESULTS

Table 1 gives radioactivity excretion following intravenous and topical acetochlor administration. After intravenous administration, 69.1 ± 8.3% dose was excreted in urine and 12.5 ± 1.1% was excreted in feces over the 7-day collection period. Total dose accountability was 81.6 ± 9.3%. Figure 1 shows that most of the administered dose was excreted in the first few days.

After topical administration of the low dose (0.03 mg), 16.6 ± 6.7% was excreted in urine and 2.1 ± 0.9% excreted in feces. Skin surface wash 24 h after dosing removed 42.4 ± 11.0% dose, giving a dose accountability of 61.1 ± 7.0%. The mid dose (0.43 mg) was 10.0 ± 1.5% excreted in urine, 3.8 ± 4.6% excreted in feces, and 55.7 ± 2.9% recovered in skin wash after administration (69.5 ± 3.4% dose accountability). The high dose (30.7 mg) was 2.6 ± 1.3% excreted in urine, 1.4 ± 0.7% excreted in feces, and 86.0 ± 5.8% recovered in the skin wash (90.0 ± 5.2% dose accountability). As the topical dose mass increased, the efficiency of absorption (percent) decreased. In addition, more dose was recovered on the skin surface, which improved dose accountability.

Table 1 Radioactivity Excretion Following Intravenous and Topical [^{14}C]Acetochlor Administration to Rhesus Monkeys

Route	Dose (mg/cm^2)	Percent of dose[a]			
		Urine	Feces	Surface wash	Total
Intravenous	0.43	69.1 ± 8.3	12.5 ± 1.1	—	81.6 ± 9.3
Topical	0.03	16.6 ± 6.7	2.1 ± 0.9	42.4 ± 11.0	61.1 ± 7.0
Topical	0.43	10.0 ± 1.5	3.8 ± 4.6	55.7 ± 2.9	69.5 ± 3.4
Topical	30.7	2.6 ± 1.3	1.4 ± 0.7	86.0 ± 5.8	90.0 ± 5.2

[a]Mean ± SD (n = 4).

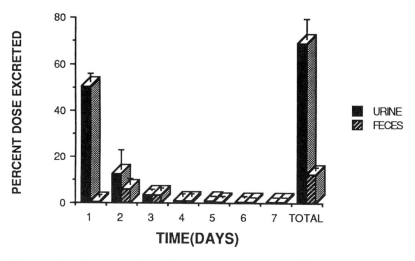

Figure 1 Urinary and fecal ^{14}C excretion following [^{14}C]acetochlor intravenous administration to rhesus monkeys. Note that administered dose is completely excreted by day 7.

Table 2 gives the individual percent acetochlor dose excreted in urine and feces following intravenous and topical administration. From the equation given in calculated percent, doses absorbed were 23.1 ± 8.7% for 0.03 mg dose, 17.3 ± 5.9% for 0.43 mg dose, and 4.9 ± 1.4% for 30.7 mg dose. In Table 3, these percent doses absorbed are converted to daily and hourly flux rates.

Thus, as the topical dose was increased a 1000-fold (0.03 mg, 0.4 mg, 30.7 mg), the efficiency/percent dose absorption decreased (23.1%, 17.3%, 4.9%); however, the flux continued to increase (0.03, 0.3, 6.3 μg/cm^2/h) with increasing dose.

Table 4 gives the plasma radioactivity as μg-Eq/ml following intravenous and topical acetochlor administration. After intravenous administration, plasma concentrations rapidly decreased during the first day and were at background level by day 7. Plasma levels for topical 0.03 mg and 0.43 mg were extremely low. With the high 30.7-mg dose, plasma levels were detectable in 1 h and continued at a somewhat steady level through the 24-h dosing period. After the skin surface wash (24 h), plasma levels decreased but were still detectable at the 168-h sampling period (Fig. 2). Radioactivity from acetochlor administration penetrated red blood cells. Figure 3 shows the concentrations and ratios of radioactivity in plasma and red blood cells following intravenous administration.

Table 2 Calculation of Percent Dose [^{14}C]Acetochlor Absorbed Following Topical Application to Rhesus Monkeys

Monkey number	Percent of dose excreted, urine and feces				Percent of dose absorbed (calculated)[a]		
	0.03 mg	0.43 mg	iv (0.43 mg)	30.7 mg	0.03 mg	0.43 mg	30.7 mg
I	18.1	10.24	85.60	3.46	21.1	12.0	4.0
II	22.7	12.63	92.74	3.52	24.5	13.6	3.8
III	24.45	13.25	72.14	4.91	33.9	18.4	6.8
IV	9.77	19.08	75.99	3.78	12.9	25.1	5.0
Mean	18.76	13.80	81.62	3.92	23.1	17.3	4.9
Standard deviation	6.56	3.75	9.33	0.68	8.7	5.9	1.4

Note. The iv dose is assumed 100% bioavailable (all injected). Other route (topical) of administration is then determined relative to the iv dose.
[a]Percent dose absorbed = (percent dose excreted [urine and feces] following topical application)/(percent dose excreted [urine and feces] following intravenous administration) × 100.

Table 3 Acetochlor Flux Rates

Dose ($\mu g/cm^2$)	Percent of dose absorbed	Daily flux ($\mu g/cm^2/24$ h)	Hourly flux ($\mu g/cm^2/h$)
3	23.1 ± 8.7	0.70 ± 0.27	0.03 ± 0.01
43	17.3 ± 5.9	7.4 ± 2.7	0.3 ± 0.1
3070	4.9 ± 1.4	150.4 ± 42.9	6.3 ± 1.8

Note. Daily and hourly flux rates can be used to determine potential skin absorption when specified exposure time is known. Values are mean ± SD.

IV. DISCUSSION

The in vivo percutaneous absorption of acetochlor in rhesus monkey varied with topical dose. As the dose was increased, the efficiency of absorption (percent) decreased; however, the mass of chemical (flux) absorbed always increased with increased topical dose. This percutaneous absorption relative to concentration was first reported by Wester and Maibach (1976), and the relationship of dose and absorption remains part of the dose response for percutaneous absorption (Wester and Maibach, 1989b). Absorption results for varying concentrations give relevant data for a chemical where handlers are exposed to different concentrations. Pesticides are supplied at commercial concentrations and diluted for use.

The approval for acetochlor was in part based on the premise that only half as much acetochlor was needed per acre as other herbicides. However, risk assessment is for exposed people, and in vivo percutaneous absorption determines the in vivo bioavailability of exposure. Alachlor is one of the herbicides that acetochlor could replace. The in vivo percutaneous absorption of alachlor in the rhesus monkey is 15.3 ± 3.9% for a topical dose of 46 $\mu g/cm^2$ (Wester et al., 1992a). This present study shows that the corresponding acetochlor absorption is 17.3 ± 5.9% for a topical dose of 43 $\mu g/cm^2$. The two herbicides are bioequivalent and, therefore, use of acetochlor in place of alachlor at one-half the amount per acre should, if all else were equal, decrease risk assessment proportionally.

Other human exposure ensues from residual acetochlor in environmental water and soil. Chemicals are readily absorbed from water and soil by children and adults through play, recreation, and work (Wester et al., 1990, 1992b, 1992c, 1993a, 1993b, 1993c). It is appropriate that comparative in vivo skin bioavailability from soil be done for acetochlor and the chemicals that acetochlor is intended to replace.

The rhesus monkey is an animal model predictive of humans for percutaneous absorption (Wester and Maibach, 1976, 1989a). The standard tox-

Table 4   Plasma Radioactivity Following Intravenous and Topical [^{14}C]Acetochlor Administration to Rhesus Monkeys

Time (h)	Plasma concentration (μg-Eq/ml)[a]			
	Intravenous (0.43 mg)	Topical (0.03 mg)	Topical (0.43 mg)	Topical (30.7 mg)
0	0.00 ± 0.00	0.00 ± 0.00	0.00 ± 0.00	0.00 ± 0.00
0.25	0.0469 ± 0.0066	—	—	—
1	0.0282 ± 0.0065	0.0000 ± 0.0000	0.0011 ± 0.0020	0.0254 ± 0.0508
2	0.0179 ± 0.0045	0.0001 ± 0.0002	0.0000 ± 0.0000	0.0940 ± 0.1869
4	0.0101 ± 0.0041	0.0001 ± 0.0002	0.0005 ± 0.0010	0.0670 ± 0.0921
6	0.0073 ± 0.0024	0.0001 ± 0.0003	0.0000 ± 0.0000	0.0257 ± 0.0486
10	0.0081 ± 0.0004	0.0001 ± 0.0002	0.0000 ± 0.0000	0.0811 ± 0.1504
24	0.0032 ± 0.0021	0.0002 ± 0.0004	0.0002 ± 0.0004	0.0550 ± 0.0659
48	0.0024 ± 0.0022	0.0002 ± 0.0003	0.0001 ± 0.0002	0.0123 ± 0.0156
96	0.0030 ± 0.0044	0.0002 ± 0.0004	0.0001 ± 0.0002	0.0146 ± 0.0194
168	0.0001 ± 0.0002	0.0001 ± 0.0003	0.0001 ± 0.0001	0.0142 ± 0.0166

[a]Mean ± SD ($n = 4$).

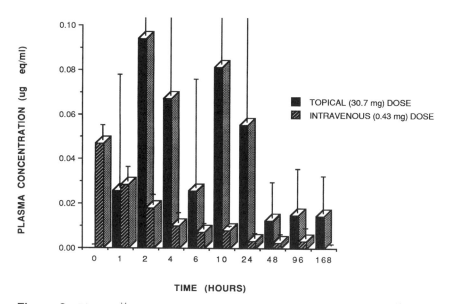

Figure 2 Plasma ^{14}C radioactivity following topical and intravenous [^{14}C]acetochlor administration. Note that plasma ^{14}C is detectable 1 h after topical administration, and that plasma ^{14}C decreases after 24 h (when topical dose site was washed with soap and water).

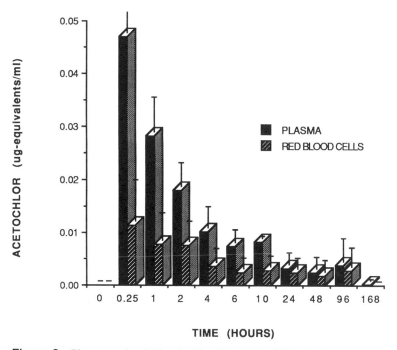

Figure 3 Plasma and red blood cell radioactivity following intravenous administration.

icological test animal, the rat, varies greatly in percutaneous absorption relative to humans (Wester and Maibach, 1993). Thus, for risk assessment where skin bioavailability is factor, it is appropriate to have a relative animal model such as the rhesus monkey.

REFERENCES

Antov G, Zaikov C, Bouzidi A, Mitova S, Michaelova A, Halkova J, Choumkov N (1991). Biochemical and histological changes after acute oral poisoning with the acetanilide herbicide acetochlor. J. Toxicol. Clin. Exp. 11:349–356.

Khalkova Z, Zaikov K, Bouzidi A, Antov G (1990). Biochemical changes in the brain tissue of white rats exposed to the herbicide acetochlor under acute and subacute oral experimental conditions. Probl. Khigien. 15:96–102.

Wester R, Maibach H (1976). Relationship of topical dose and percutaneous absorption in rhesus monkey and man. J. Invest. Dermatol. 67:518–520.

Wester R, Maibach H (1989a). In vivo animal models for percutaneous absorption. In Percutaneous Absorption, R. Bronaugh and H. Maibach, eds. Marcel Dekker, Basel, pp. 221–238.

Wester R, Maibach H (1989b). Interrelationship in the dose-response of percutaneous absorption. In Percutaneous Absorption, R. Bronaugh and H. Maibach, eds. Marcel Dekker, New York, pp. 343–355.

Wester R, Maibach H (1993). Animal models for percutaneous absorption. In Topical Drug Bioavailability, Bioequivalence, and Penetration. V. Shah and H. Maibach, eds. Plenum Press, New York, pp. 333–349.

Wester R, Maibach H, Bucks D, Sedik L, Melendres J, Liao C, DiZio S (1990). Percutaneous absorption of [^{14}C]-DDT and [^{14}C]-benzo(a)pyrene from soil. Fundam. Appl. Toxicol. 15:510–516.

Wester R, Melendres J, Maibach H (1992a). In vivo percutaneous absorption and skin decontamination of alachlor in rhesus monkey. J. Toxicol. Environ. Health 36:1–13.

Wester R, Maibach H, Sedik L, Melendres J, DiZio S, Wade M (1992b). In vitro percutaneous absorption of cadmium from water and soil into human skin. Fundam. Appl. Toxicol. 19:1–5.

Wester R, Maibach H, Sedik L, Melendres J, Liao C, DiZio S (1992c). Percutaneous absorption of [^{14}C]chlordane from soil. J. Toxicol. Environ. Health 35:269–277.

Wester R, Maibach H, Sedik L, Melendres J, Wade M (1993a). In vivo and in vitro percutaneous absorption and skin decontamination of arsenic from water and soil. Fundam. Appl. Toxicol. 20:336–340.

Wester R, Maibach H, Sedik L, Melendres J, Wade M, DiZio S (1993b). Percutaneous absorption of pentachlorophenol from soil. Fundam. Appl. Toxicol. 20: 68–71.

Wester R, Maibach H, Sedik L, Melendres J, Wade M (1993c). Percutaneous absorption of PCBs from soil: In vivo rhesus monkey, in vitro human skin, and binding to powdered human stratum corneum. J. Toxicol. Environ. Health 39: 375–382.

49

The Development of Skin Barrier Function in the Neonate

Lourdes B. Nonato
University of California, San Francisco, California

Yogeshvar N. Kalia and Aarti Naik
*Centre Interuniversitaire de Recherche et d'Enseignement,
Archamps, France, and Laboratoire de Pharmacie Galénique,
University of Geneva, Geneva, Switzerland*

Carolyn H. Lund
Children's Hospital, Oakland, California

Richard H. Guy
*Centre Interuniversitaire de Recherche et d'Enseignement,
Archamps, France, and Laboratoire de Pharmacie Galénique,
University of Geneva, Geneva, Switzerland*

I. INTRODUCTION

The skin, one of the largest organs of the body, functions as a protective and regulatory barrier between the body and the external environment. At birth, infants must make the transition from a fluid intrauterine environment to dry, extrauterine life. Full-term infants, born at 40 weeks of gestation, make the transition with a competently formed barrier. However, the skin of the premature neonate comprises as much as 13% of the body weight, compared to only 3% of the body weight of an adult (1). In addition, the surface area to body weight ratio of the neonate is four times that of adults (2). Thus, the immaturity of the skin has numerous ramifications for the neonate, including ineffective thermoregulation (3), fluid imbalance (4), percutaneous absorption of toxins (2), tissue injury (5), infection (6), and delayed healing (7).

Within the last decade, new technologies have changed the caretaking practices of the premature neonate; advances in respirator design, monitor technology, reduction of blood volume required for specialized tests, and the sophistication of new diagnostic techniques have all contributed to increasing the chances of survival of the premature neonate (8). Nevertheless, much remains to be learned about the complex extrauterine skin development of these neonates.

On the other hand, the increased permeability of premature neonatal skin may be advantageous for drug delivery and monitoring via the skin. About 99% of neonates weighing less than 1500 g are routinely treated with multiple drug therapy in the intensive care nursery (9). The most commonly used drugs include antibiotics, nutritional agents, analgesics, and cardiovascular-respiratory agents (10), and require drug dosing that is precise and safe. The predominant use of intravenous drug administration suffers from several disadvantages. The introduction of an intravenous cannula is painful and stressful and may cause severe bruising at the site of insertion. Furthermore, an inappropriate infusion rate can result in under- or overdosing of the neonate, and, because of the very low doses of drugs used in the premature neonate, small variations in infusion rate can have dramatic sequelae (11). Oral drug therapy is problematic due to unpredictable absorption from the developing gastrointestinal tract of the premature neonate (12). The high osmolarity of many medications may also irritate the immature gastrointestinal tract of the neonate. Transdermal drug delivery can thus provide a potentially noninvasive, continuous, and controlled delivery method for the neonate. Therapeutic concentrations of theophylline (13) and caffeine (14) have been achieved in premature neonates following transdermal delivery. A method to predict the permeability of premature skin would permit an appropriate delivery rate of drug to be matched to the gestational and postnatal age of the neonate (15).

Noninvasive transdermal monitoring of substances in the bloodstream would also be particularly beneficial for the neonatal population. Continuous, noninvasive monitoring of blood glucose levels has been a primary goal for the management of diabetes. A similar need exists for neonatal patients who are subjected to frequent, painful (16) heelsticks for blood glucose level monitoring. Such repeated heelsticks lead to extensive scarring, which may develop into lesions that persist for years (17,18). Recently, glucose monitoring in diabetic adults has been achieved by noninvasive transdermal extraction (19). Advantages of transdermally extracted glucose levels include greater accuracy, continuous monitoring, and significantly improved patient compliance.

In adults and full-term neonates, the outermost and least permeable layer of the skin, the stratum corneum (SC), forms the rate-limiting barrier to most topically contacting substances (20–22); premature neonates, however, have thinner SC with significantly fewer cell layers (23). The stratum corneum can be described simplistically as a brick wall: the fully differentiated, keratin-filled corneocytes form the bricks and the complex, intercellular, lamellar lipid domains serve as the mortar (21,24). The major penetration pathway is the intercellular lipid route (22), involving diffusion around the corneocytes of the SC (20). Conclusive evidence has been presented to show that the SC behaves as a lipoidal, rate-controlling membrane over a reasonable range of penetrant lipophilicity and structure (25). The lamellar packing of SC intercellular lipids is established (26), and several experiments have directly implicated these lipoidal domains as the integral components of the transport barrier (27,28).

As permeability is controlled by the biophysical, structural, and chemical characteristics of the SC, multiple techniques are necessary to fully characterize barrier function and its development in the premature neonate. Various strategies, which have been used to investigate adult SC, have also been used recently to probe the physical and biological characteristics of neonatal stratum corneum. In adult skin, Fourier-transform infrared spectroscopy (FTIR) and small-angle x-ray diffraction (SAXD) have provided insight into molecular structure and function of the SC. Morphological studies with transmission electron microscopy (TEM) have been instrumental in visualizing intercellular lipid lamellae in adult SC, and have provided a tool for investigating their mode of formation and role in barrier function. Analysis of the lipid composition of normal and diseased adult SC has implicated its importance in governing skin permeability. Measurements of transepidermal water loss (TEWL) across the skin of adults (and, indeed, neonates) have allowed indirect evaluation of skin barrier integrity.

Despite the known importance of the SC in barrier function, information about neonatal skin development is limited. It is therefore the goal of this chapter to provide an overview of past research pertinent to understanding the development of skin barrier function in the premature neonate (29–33) and to summarize the most recent developments in the field.

II. EMBRYOGENESIS OF HUMAN SKIN

In utero, the skin must continuously undergo two-dimensional growth to cover the surface area of the developing fetus, a complex and difficult process to investigate. To date, primarily morphological characterization has been accomplished (34–36). The skin of the human embryo begins to develop as a single-layered epithelium at 7–8 days postconception. Within the first 2 weeks of embryonic life, the epidermis and dermis are established. The dermis lags behind the epidermis in development, and, at full-term birth, it is about 60% of its adult thickness (35). Most dermal development takes place 3–5 months postnatally.

A second epidermal layer, called the periderm, forms toward the end of the first month of fetal life and is present only in embryonic and fetal skin. The periderm is a transient protective covering that is sloughed into the amniotic fluid when epidermal differentiation is complete. During the first 2 months of fetal life, the basal layer and periderm are filled primarily with glycogen, which may be important in energy metabolism, as a precursor molecule for lipid synthesis, or as an intermediary in the formation of the cell coat (37,38).

During the second trimester, two to three more intermediate cell layers, produced from the differentiating basal cells, are found between the basal layer and the periderm. In adults, infusion with [^{3}H]thymidine indicates that approximately 50% of the epidermal basal cells are not mitotically active and remain in a noncycling state (39,40). In the second-trimester embryo, basal cells have a 60% greater labeling index than that of the adult, correlating with the formation of the first intermediate layer and epidermal appendages, and the rapidly growing surface area of the fetus (41,42). The cytoplasm of the intermediate cells fills with glycogen and an increasing number of desmosomes and filaments. With the appearance of the intermediate cells, the basal cells show a reduction in glycogen and, at 4 months, the mitotic proliferation rate declines to adult levels (42). The periderm also undergoes differentiation, and is modified at the skin surface by microvilli and then by large blebs at the beginning of the second trimester (36). These modifications increase the surface area in contact with the amniotic fluid and suggest that the periderm cells may be involved in uptake from or secretion into the amniotic fluid (43).

In the fetal epidermis, intermediate cells bypass the granular stage temporarily and undergo a modified process of terminal differentiation, or "incomplete keratinization." The intermediate cells accumulate and release lamellar granules; they subsequently differentiate into cells with a cornified envelope and contain bundles of keratin filaments, glycogen, and organelles. Several layers of partially keratinized intermediate cells form beneath the periderm prior to differentiation of the granular and stratum corneum layers. A true granular layer develops at 20 weeks gestational age. In contrast to the intermediate cells, these cells contain keratohyalin granules and diminished amounts of glycogen. Below the granular layer, the spinous cells retain a significant amount of glycogen until the last weeks of gestation.

The stratum corneum is first visible at the end of the second trimester and remains a few cell layers thick until prior to birth, with the number of stratum corneum cell layers increasing during the third trimester, particularly near term (40 weeks gestation) (35). Morphologically, during the fifth and sixth months of gestation, the periderm thins. Studies with fetal skin in vitro suggest that percutaneous permeability decreases at 24 weeks gestation when keratinization begins (44). But, because it is difficult to study fetuses in the last trimester, little information is available on the time course of stratum corneum maturation in utero during this period.

The dermis is watery and cellular during the first 4 to 8 weeks of embryonic life. Collagen fibrils appear at about 7 weeks. During the second trimester, hair follicles develop in the dermis and subcutaneous fat cells become visible; collagen fibers also increase, the number of cells decrease, and the dermis grows in thickness. By the third trimester, elastic fibers appear to augment the collagen fibers in the dermis. But the dermis of the premature and full-term neonate has many fetal characteristics, and most dermal development takes place 3–5 months postnatally (35).

The dermal–epidermal junction (DEJ) is another important structure of the skin. In adult skin, the DEJ includes anchoring and attachment structures between the dermis and the epidermis. The DEJ is immature in the embryo and develops throughout the fetal period. During the third trimester, the attachment fibers of the DEJ are well established and the convolutions between the epidermis and dermis begin to form. However, the underlying dermis is edematous, with a loose organization of fibers, and is believed to provide less secure anchoring sites for the DEJ filaments (45). The immaturity of the dermis in the premature and full-term neonate may have implications for an increased risk of physical and mechanical injury.

III. CHARACTERIZATION OF STRATUM CORNEUM LIPID COMPOSITION IN NEONATAL SKIN

The unique mixture of lipids in adult human stratum corneum, composed primarily of ceramides, cholesterol, and fatty acids, prevents desiccation and serves as a barrier to the diffusion of substances across the skin. A knowledge of the composition of the lipids in the stratum corneum of the neonate is therefore critical to the understanding of barrier function development.

It has been suggested that variations in lipid concentration may explain the differences observed in permeation rates across different anatomic regions (46). For example, the fluxes of water and salicylic acid in vitro were inversely correlated to the lipid weight percent of body site (leg 3.0%; abdomen 6.8%) (46). It was also shown that the thickness and the number of cell layers in the stratum corneum were not correlated with permeability (46). Grubauer et al. have found that extraction of specific lipids from the stratum corneum results in increased water loss across the skin (47). Selective extraction using petroleum ether of nonpolar lipids (alkanes, sterol esters, triglycerides, free fatty acids) results in a small increase in TEWL, while acetone extraction of all lipids results in greater barrier disruption. Other investigators have concluded that a higher ratio of neutral lipids to sphingolipids is associated with lower permeability, suggesting that variations in neutral lipids are a potential cause of site-to-site differences in skin permeability (48).

Alterations in lipid content have also been found in skin diseases such as dermatitis and psoriasis (49,50), in which significantly increased transepidermal water loss is observed. Similar effects have been shown in essential fatty acid-deficient (EFAD) mice, whose diet is deprived of linoleic acid (24). The skin of these animals becomes hyperproliferative and exhibits decreased barrier function. Biochemical and structural analyses of the skin of EFAD mice indicate that oleic acid replaces linoleic acid in the acylglucosylceramide and acylceramide (ceramide 1) (51), and that both lamellar bodies and stratum corneum intercellular spaces lack an organized lamellar structure. In patients with atopic dermatitis, although no abnormalities were found in the relative amounts of the different stratum corneum lipid classes, a significant decrease was detected in the fraction of ceramide 1 (the least polar ceramide), as well as increased levels of ceramide 1 esterified with oleic acid, as compared to the skin of healthy controls (52). A 50% reduction in total lipid content has also been detected in these patients (53). A recent study of the stratum corneum lipids from psoriatic plaques also showed an abnormal distribution of ceramides; in particular, the amount of ceramide 1 was significantly decreased (50).

Moreover, abnormal lipid compositions have been observed in skin cell cultures and may account, at least in part, for the detected deficiencies in barrier function. Mak et al. measured a reduced free fatty acid content, especially of those with $C_{22}-C_{28}$ alkyl chains, an increased ratio of cholesterol to cholesterol ester, and an altered production of ceramides in cultured human keratinocytes (54). This model also displayed a threefold greater permeability to water transport than adult epidermis. Others have found significantly lower linoleic acid content and higher amounts of oleic acid, as well as elevated triglycerides, in cultured skin cells as compared to normal adult skin (55). Altered lipid composition in skin cell cultures has been associated with poorly organized intercellular stratum corneum lipids and reduced barrier function as measured by TEWL (54,55).

Neonatal skin lipid development has been studied in early human fetal skin and neonatal rat skin. From analyses of skin samples obtained from 50 to 140 days EGA (estimated gestational age) human fetuses, Williams and coworkers found increasing amounts of ceramides and glycosphingolipids in the epidermis after 110 days EGA, proceeding in the cephalocaudal direction (56). The appearance of the sphingolipids was believed to be correlated with the initial formation of keratinized stratum corneum (56), which is detectable toward the end of the second trimester (150–170 days EGA) (57). In the premature neonatal rat, ceramides 1 and 2 increased six- to sevenfold and cholesterol increased fourfold between 19 days EGA and term (22 days EGA), coinciding with the organization of the SC lipids into bilayers (29). Furthermore, there was a threefold increase in sterol esters, a twofold increase in triacylglycerols, and total SC lipid content increased approximately twofold between 19 and 21 days (29). In addition, the synthesis of cholesterol, ceramides, sphingomyelin, and fatty acids decreased from 17 to 21 days EGA (58). It was postulated that this decrease in synthesis coincided with the establishment of a competent barrier in rats.

The lipid composition of human neonatal stratum corneum has also been recently analyzed using high performance thin layer chromatography (59). Abdominal skin was obtained from the autopsy of neonates of varying gestational (24 weeks–40 weeks) and postnatal (1 day–6 weeks) ages. The concentration of all the lipid classes in the stratum corneum from a 40 week EGA, 1 month postnatal age (PNA) neonate was similar to that found in adult stratum corneum, a result consistent with the mature skin barrier function measured in full-term neonates (Fig. 1). The sample from a 37 weeks EGA, 7 days PNA neonate also contained adult levels of total ceramide and polar lipids, suggesting close to full-term maturity of the stratum corneum.

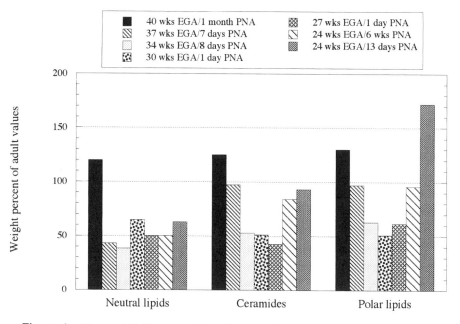

Figure 1 The total lipid composition of neonatal stratum corneum.

The total ceramide and polar lipid concentrations of the stratum corneum from a 24 weeks EGA, 6 weeks PNA neonate was similar to that found in the 37 weeks EGA, 7 days PNA neonate. In addition, the 24 weeks EGA, 6 weeks PNA neonatal sample contained greater concentrations of total ceramide and polar lipids than that found in the stratum corneum from the gestationally older (but postnatally younger) 34 weeks EGA, 8 days PNA, 30 weeks EGA, 1 day PNA, and 27 weeks EGA, 1 day PNA age neonates. These observations suggest that postnatal factors may play an important role in accelerating SC lipid development in the premature neonate.

The free fatty acid content of the samples obtained from neonates less than full-term (40 weeks EGA) was less than 40% of adult levels. In adult stratum corneum, free fatty acids comprise approximately 10–15% of the total lipid mass and consist predominantly of saturated 22- and 24-carbon chains (60). Free fatty acids and cholesterol sulfate are the only ionizable lipids in adult stratum corneum and may be required for bilayer formation in vivo (61). In addition, x-ray diffraction analysis of isolated skin lipids indicates that free fatty acids affect lipid phase behavior by increasing the solubility of cholesterol in the lamellar structures (62). Consequently, the decreased amounts of free fatty acids in the stratum corneum of the immature neonate may be correlated with decreased barrier function.

The greater sterol ester content in the SC of the 24 weeks EGA, 6 weeks PNA, and the 24 weeks EGA, 13 days PNA neonates, relative to the 30–37 weeks EGA neonates, may be associated with the development of sebaceous glands during the end of the second trimester (63). This time also marks the appearance of the granular layer and the beginning of keratinization of the interfollicular regions of fetal skin (63). Thus, cholesterol and cholesterol esters may be synthesized in greater concentrations within the granular cells in preparation for the formation of the initial stratum corneum layers within fetal skin, and may be extruded prematurely in the 24 weeks EGA neonate.

Among the neonatal SC samples, the amounts of the most polar ceramide fraction were highest and were most similar to adult "baseline" levels. The levels of the fifth most polar ceramide fraction, although less than adult values, were similar among the neonatal SC samples. Aszterbaum et al. also found that the more polar ceramide fractions remained relatively constant in fetal rats of varying gestational ages (29). However, the levels of the nonpolar ceramides were most decreased in the SC of neonates of younger postnatal age, regardless of EGA at birth. These SC samples were obtained from neonates (27–37 weeks EGA) whose postnatal ages ranged from 1 day to 8 days. In the fetal rat, the quantity of the two least polar ceramide fractions also increased most significantly with increasing gestational age and was coincident with the establishment of a functional barrier

(29). Decreased levels of the most nonpolar ceramide fraction have also been associated with atopic dermatitis (52,53) and psoriasis (50). Consequently, the levels of the most nonpolar ceramides in the human neonate may be dependent on postnatal factors and may be important determinants of barrier function maturation.

The stratum corneum from the 24 weeks EGA, 13 days PNA neonate contained the lowest concentration of cholesterol sulfate (~50% of adult levels). Kagehara and coworkers have shown that cholesterol sulfotransferase activity in fetal murine skin increases with the formation of a multilayered epidermis from 14 to 16 days EGA, and increasing concentrations of cholesterol sulfate are detectable from 15 to 17 days EGA (64). In the premature human neonate, reduced concentrations of cholesterol sulfate in the early postnatal period may also indicate the decreased activity of cholesterol sulfotransferase or other enzymes involved in cholesterol sulfate synthesis.

The stratum corneum from the 24 weeks EGA, 13 days PNA neonate also contained the greatest amounts of phospholipids, which have not been detected in adult stratum corneum (Fig. 2). Phospholipids, such as sphingomyelin, phosphatidyl-ethanolamine (PE), phosphatidylcholine (PC), phosphatidylinisitol (PI), and phosphatidylserine, constitute 53% of the total lipids of the viable epidermis in normal adult skin (65).

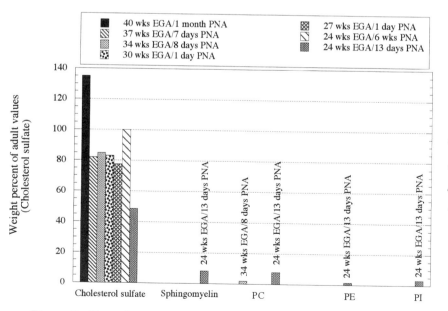

Figure 2   The polar lipid composition of neonatal stratum corneum.

The decreased amounts of phospholipids (21% of total lipids) in the granular layer compared to the basal and spinous cells (62%) of adult epidermis (65) is correlated with the catabolism of plasma membranes, mitochondria, Golgi apparatus, endoplasmic reticulum, and nuclei by phospholipases, sphingomyelinase, and phosphatases in the epidermis (66,67). Sphingomyelin accounts for about 21% of the phospholipids in adult human viable epidermis; this concentration is higher than that found in other mammalian tissues (65). It has been shown that sphingomyelinase in the adult human epidermis degrades sphingomyelin to ceramide (67), although it is not known whether the ceramides in the stratum corneum arise mainly from sphingomyelin metabolism or from glucosylceramide metabolism or a combination of the two pathways (68). Recently, Mao-Qiang and coworkers demonstrated that degradation of phospholipids to free fatty acids by an extracellular phospholipase is required for normal barrier function in murine skin (69). Furthermore, the topical application of palmitic acid, a product of phospholipid catabolism, reversed the barrier abnormalities (69). Thus, the isolated phospholipids from the stratum corneum of the 24 weeks EGA, 13 days PNA neonate may represent components of the viable epidermis that have not been catabolized at the stratum granulosum–stratum corneum interface, due to either an absence or an inactivity of phospholipases in premature neonatal skin.

The ceramides/cholesterol ratio does not appear to change appreciably as a function of gestational and postnatal age (Table 1). Similarly, Bouwstra and coworkers found that a change in the stratum corneum lipid structure of patients with lamellar ichthyosis was not related to a change in the ceramides/cholesterol ratio (70). They suggested that differences in the relative amounts of ceramide fractions, free fatty acids (FFA)/ceramide ratio, and FFA/cholesterol ratio could explain the abnormal lipid organization (as detected by x-ray diffraction) and reduced barrier function detected in these patients (70).

Decreasing gestational age corresponded with decreasing FFA/ceramide ratios, with the exception of the sample from the more postnatally mature neonate (24 weeks EGA, 6 weeks PNA). Thus, the differences in FFA/ceramide ratios were more indicative of the changes in the major lipid classes that occur with skin barrier development. This suggests that an optimal ratio of FFA/ceramides may be an important determinant of the intercellular lamellar structure of adult and neonatal stratum corneum. In addition, Bouwstra and coworkers have found that cholesterol/ceramide/FFA mixtures result in x-ray diffraction patterns similar to adult stratum corneum only when long-chain free fatty acids are used (62), indicating that the specific free fatty acid composition may play a role in proper lipid organization. Since the interactions of the aliphatic chains of the ceramides are believed

Table 1 Lipid Ratios of Major Lipid Classes in Adult and Neonatal Stratum Corneum

	Lipid ratio		
Age	FFA/ cholesterol	FFA/ ceramides	Ceramides/ cholesterol
Adult	0.87	0.27	3.4
40 weeks EGA, 1 month PNA	0.84	0.26	3.3
37 weeks EGA, 7 days PNA	0.77	0.22	3.0
34 weeks EGA, 8 days PNA	0.78	0.22	3.5
30 weeks EGA, 1 day PNA	0.64	0.20	3.2
27 weeks EGA, 1 day PNA	0.86	0.11	2.8
24 weeks EGA, 6 wks PNA	0.72	0.16	3.6
24 weeks EGA, 13 days PNA	0.52	0.13	3.8

to be crucial to the multibilayer structure in the stratum corneum, the altered distribution of ceramides in the stratum corneum of premature neonates may also be significant in barrier function development.

Thus, the ratio of FFA/ceramides may be an important marker of stratum corneum development, intercellular bilayer organization, and barrier function. The presence of phospholipids in the most gestationally and postnatally immature neonate suggests that improper lipid processing and/or distribution is indicative of barrier immaturity. In addition, nonpolar ceramide content appears to be related to postnatal age and may also be an important determinant of barrier development. Further work to elucidate the relationship between stratum corneum lipid composition, organization, and function will be crucial to improving the care of the premature neonate.

IV. CHARACTERIZATION OF THE MORPHOLOGICAL CHANGES IN NEONATAL SKIN

Biophysical, morphological, and biochemical evidence has indicated that the stratum corneum lipids are arranged into lamellar bilayers. These lipids originate from lamellar bodies within the stratum granulosum cells, and are transformed into bilayers at the stratum granulosum–stratum corneum interface. Because stratum corneum lipids are believed to constitute the major component of the permeability barrier, the spatial structuring of intercellular lipid lamellae is essential to the maturation of the permeability barrier in the neonate.

Alterations in the lamellar body extrusion process and in the lamellar pattern within the stratum corneum have been correlated with a number of skin diseases. In the SC of essential fatty acid-deficient (EFAD) mice, extensive membrane-depleted domains may account for this barrier's elevated permeability (71). In patients with chronic plaque psoriasis, shortened lamellar body disks persisted to higher layers within the stratum corneum and bilayers failed to form in the intercellular space (72). In another phenotype of psoriasis, lamellar bodies were found within the corneocytes and extracellular regions lacked lamellae (73). In both psoriatic phenotypes, abnormalities in lamellar body production and stratum corneum bilayer formation were associated with reduced barrier function (73). Similar disturbed mechanisms in lamellar differentiation have been found in atopic skin (74) and latex-occluded murine epidermis (75).

Abnormal ultrastructure has also been observed in human keratinocyte culture systems (55,76–78). For example, Ponec and coworkers found partially or completely empty lamellar bodies and altered stratum corneum lipid composition in air-exposed cultures (55). Kennedy et al. have shown that human neonatal foreskin keratinocyte cultures lack the basic lamellar unit and the covalently bound lipid envelope in the stratum corneum; furthermore, they found that the compromised barrier function of the cultured skin was also associated with altered lipid content (78). However, normal lamellar-body-derived membrane disks and a multilayered lipid structure have been observed in murine keratinocyte cultures, and indicate that the conditions required for differentiation vary among culture systems (79).

The overall structural organization of the skin also undergoes developmental changes with important physiologic implications for the neonate. In a histological study of human skin, Holbrook found a similar number of cell layers in the viable epidermis and the stratum corneum of adults and full-term neonates (80). In the skin of premature neonates (30–32 weeks EGA, within the first days of life), however, the epidermis was thinner, the epidermal cells were more compressed, and the stratum corneum consisted of only a few cell layers. Evans and Rutter detected a thin epidermis and poorly formed stratum corneum in neonates less than 30 weeks EGA, but a mature epidermal structure in neonates greater than 34 weeks EGA, suggesting that histological changes parallel the development of barrier properties (23). In the fetal rat, an increase in the number of stratum corneum cell layers has also been correlated with barrier function maturation (29). In addition, in the neonatal rat, maternal glucocorticoid treatment accelerates epidermal lamellar body secretion and the formation of the stratum corneum membrane ultrastructure, as well as barrier function as measured by transepidermal water loss (30). In adult skin, basal epidermal cells display numerous cytoplasmic projections into the dermis, resulting in a highly convoluted dermal–epidermal junction (DEJ). A flat DEJ has been observed in full-term and premature neonates (80); this flattened interface has also been observed in aged skin and is associated with a lower resistance to shearing forces (81).

Morphological changes in premature and full-term neonatal skin have also been studied with light microscopy and transmission electron microscopy (82). Skin samples were obtained from surgical procedures of neonates of varying gestational (23 to 40 weeks EGA) and postnatal age (7 days to 8 months).

Histological analysis of skin obtained from the 30–31 weeks EGA neonate and full-term neonate agrees with the previous findings (35,80) and further reveals the maturation process of the epidermis and dermis. In the skin from the 23–24 weeks EGA, 1 month PNA neonate, the epidermis was thin and the dermis contained a lower fiber content (Fig. 3). In addition,

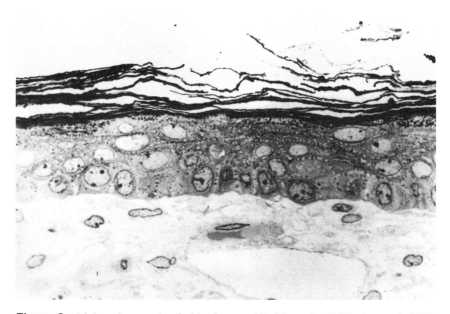

Figure 3 Light micrograph of skin from a 23–24 weeks EGA, 1 month PNA neonate. Toludine blue, 925×. (EGA, estimated gestational age; PNA, postnatal age).

although a thick SC was observed, the flat DEJ indicated that very premature but postnatally older neonates may be particularly susceptible to trauma. With increasing gestational age, the DEJ became more convoluted, the dermis less cellular, and the number of viable cell layers in the epidermis increased (Fig. 4). In the skin from neonates greater than 30 weeks EGA and greater than 8 weeks PNA, the viable epidermis appeared mature, with clearly visible basal, spinous, and granular cells. Although the dermis was more fibrous and less cellular in these neonates, an earlier study (35) had shown that the dermis of the premature and full-term neonate retains many fetal characteristics. The ratio of type I:type II collagen does not become equivalent to that in the adult until childhood, and fibroblastic cells are more abundant by comparison with those of the adult (35). Consequently, the dermis lags behind the epidermis in development.

The stratum corneum thickness of the neonatal skin samples also correlated well with previous results (23), which showed that the stratum corneum appears histologically normal within 2–3 weeks of postnatal age, regardless of gestational age, whereas neonates less than 30 weeks EGA and less than 11 days PNA have a thin stratum corneum. A better developed stratum corneum was observed in neonates of increasing gestational age. The skin from the 34 weeks EGA, 5 days PNA neonate exhibited a thick

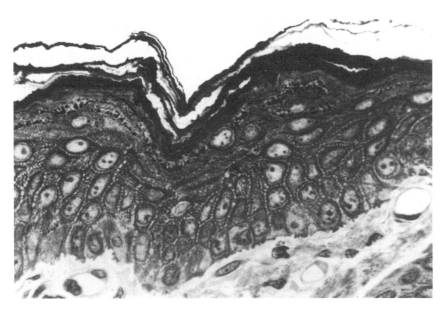

Figure 4   Light micrograph of skin from a 30–31 weeks EGA, 8 weeks PNA neonate. Toludine blue, 925×.

SC, indicating that stratum corneum thickness increases in utero between 30 and 34 weeks of gestation, or that the number of stratum corneum layers increases more rapidly immediately after birth in these neonates than in more premature neonates.

Micrographs of the skin from a 38 weeks EGA neonate showed that the lamellar body extrusion process and intercellular lipid organization is mature in full-term neonates and is similar to that found in adult skin. The stratum corneum from a 30–31 weeks EGA, 8 weeks PNA neonate, and from a 25 weeks EGA, 10 weeks PNA neonate also exhibited a normal SG–SC interface and normal stratum corneum lamellae. Thus, 8 weeks of postnatal life appears to be sufficient time for stratum corneum development in premature neonates born at less than 31 weeks gestational age.

At lesser gestational and postnatal ages, abnormalities were found in lamellar body processing and/or SC bilayer organization. Although a normal extrusion mechanism at the SG–SC interface and organized stratum corneum lamellar patterns were observed in the SC from a 34 weeks EGA, 5 days PNA neonate, regions where intercellular lamellae were discontinuous or absent were also visible (Fig. 5). Similarly, the skin from a 29 weeks

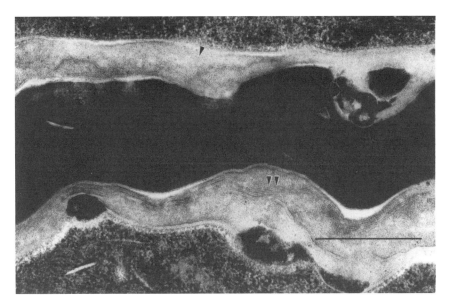

Figure 5 Transmission electron micrograph of the stratum corneum from a 34 weeks EGA, 5 days PNA neonate. Disorganized (∇∇, double arrow) and absent (∇, single arrow) lamellae are present throughout the SC. Reduced osmium stain (bar = 0.5 μm).

EGA, 11 days PNA neonate exhibited areas of normal and disorganized stratum corneum lipids. It has been suggested that extracellular or lamellar body enzymes are crucial to the fusion and proper reorganization of the lipids at the SG–SC interface (83). Consequently, immaturity in the enzyme systems and/or the structural and biochemical processing mechanism of the lamellar body lipids into bilayers may play a role in the abnormalities observed in the samples from the 34 weeks EGA and 29 weeks EGA neonates.

The stratum corneum from a 25 weeks EGA, 7 days PNA neonate displayed greater abnormalities in lipid ultrastructure, including circular and intracellular lamellae. Similar structures have been documented in cultured SC, and have been associated with abnormal stratum corneum lipid composition and infrared (IR)-detected conformational differences in lipid structure (78). In the stratum corneum from a 23–24 weeks EGA, 1 month PNA neonate, focal lamellar body extrusion was detected at higher layers within the stratum corneum, and has also been observed in atopic skin (74). In contrast to the stratum corneum from the 25 weeks EGA neonate, the sample from the 23–24 weeks EGA neonate contained lamellar regions with the normal alternating electron-lucent and electron-dense pattern. Consequently, postnatal factors would appear to accelerate stratum corneum maturation. Furthermore, the altered conversion of lamellar body lipids into stratum corneum lamellae in the 25 weeks EGA neonate may be due to an accelerated extrusion process for the rapid formation of a barrier, regardless of organization.

In conclusion, neonates less than 34 weeks EGA and less than 11 days PNA do not appear to have normal stratum corneum. However, well-organized SC lipid ultrastructure was observed in the samples from premature neonates greater than 4 weeks PNA. In addition, SC maturation appears to occur prior to completion of epidermal or dermal development, and is probably due to the more compelling requirement for a functional barrier.

V. BIOPHYSICAL CHARACTERIZATION OF NEONATAL SKIN

The biophysical properties of neonatal SC provide insight into the development of barrier function. For example, x-ray diffraction studies have proved useful in further understanding the lipid organization in mammalian stratum corneum (84–89). White and coworkers were the first to demonstrate that high-quality diffraction patterns could be obtained from intact murine stratum corneum (87). At room temperature, a lamellar structure, with a large repeat distance of 13.1 nm, which was attributed to a complex bilayer configuration, was deduced (87). It was suggested that the pattern may be produced by two apposed asymmetric lipid bilayers, or by bilayers with nonuniformly distributed proteins. When the stratum corneum samples were heated, changes in the diffraction pattern were observed at 45°C and 70°C, corresponding to the lipid phase transition temperatures measured by differential thermal analysis (90), differential scanning calorimetry (91,92), and infrared spectroscopy (93,94). From small-angle studies, the dimensions of the unit cell (i.e., the smallest repeating unit from which one can generate the complete structure) of the lipid lamellae have also been calculated. Different laboratories have reported diffraction patterns of adult human SC, which include a repeat distance of 6.5 nm (86), repeat distances of 6.5 and 4.5 nm (88), and repeat distances of 6.5 and 13.4 nm (95). Model calculations based on a complicated multibilayer unit cell with a length of 13.4 nm have suggested that it may contain three alkyl chain regions, of which two are identical (95).

Fourier-transform infrared spectroscopy (FTIR) is another useful technique. The infrared spectrum of adult SC is typical of hydrated biologic materials with strong amide and water absorbances in the ranges of 1500–1700 and 3300–3600 cm^{-1} (94). It has been shown that the CH_2 stretching vibrations primarily reflect the properties of the lipids in mammalian stratum corneum (94,96), since lipid extraction results in a 90% reduction in the intensity of the methylene stretching bands (96).

Attention has concentrated upon the response of the intercellular lipids of the SC to thermal perturbation. In adult stratum corneum, the phase transitions of the lipids have been monitored via spectral changes as a function of temperature. A broadening and a shift to higher wavenumber of the CH_2 stretching bands occur with increasing temperature. When heated from 25 to 115°C, adult stratum corneum shows two large and abrupt increases in the frequency of the symmetric CH_2 absorbance between 55–65°C and 70–85°C (94), consistent with the lipid-associated thermal transitions detected by differential scanning calorimetry and correlated with gel-to-liquid crystalline phase transitions.

The conformation of stratum corneum lipids has also been inferred from the scissoring and rocking vibrations. Alkyl chains that are tightly packed in an orthorhombic lattice result in factor group splitting of both the CH_2 scissoring and rocking bands, while chains that are more loosely packed in a hexagonal lattice give rise to a single vibrational band (97). Analyses of porcine and adult SC lipids indicate highly ordered chain packing of the alkyl chains at physiological temperatures (96,98). Thus, unlike most biological membranes, which exist in the more fluid, liquid crystalline state at physiological temperature, most of the stratum corneum lipids are present in the gel state, a conformation consistent with the large ceramide component of adult SC (99).

Infrared spectroscopy has also detected low-temperature (<65°C) lipid transitions in adult SC (98). A 35°C phase transition measured by differential scanning calorimetry (DSC) may be due to a solid-to-fluid phase change involving a discrete subset of lipids (98). Another transition at 55°C, sometimes observed by IR and DSC, may represent a complex change in the lipid lateral packing (100). Below 45–65°C, IR spectroscopy indicates an orthorhombic crystalline structure, which may be localized to the covalently bound ceramides of the corneocyte envelope (100). It has been suggested that the occurrence of such a tightly arranged lipid structure could account for the relatively low permeability of the corneocyte (100).

X-ray diffraction and FTIR studies have subsequently been used to examine the conformational structure of the lipids in neonatal stratum corneum. Comparison of the x-ray diffraction repeat distances and IR lipid phase transitions between adult and neonatal SC has been useful in characterizing lipid packing, and localized and long-range lipid order, as a function of neonatal gestational and postnatal age.

In the x-ray diffraction patterns of the stratum corneum from full-term neonates less than 7 days PNA, and from premature neonates, peaks occur at higher Q values, corresponding to significantly smaller d spacings than those found in adult SC (101). Reduced d spacings have also been found in the stratum corneum of patients with lamellar ichthyosis (70,102), in reconstructed epidermis (103), and in model lipid systems (62,104).

In the SC from a 29 weeks EGA, 10 days PNA neonate, a peak at d = 3.34 nm (not detected in healthy adult SC) may indicate the presence of crystalline cholesterol. Lipid analysis of neonatal stratum corneum shows decreased levels of free fatty acids and cholesterol sulfate in the SC from more gestationally immature neonates (59). Similarly, a diffraction peak at d = 3.4 nm and decreased amounts of free fatty acids have been reported in the SC of patients with lamellar ichthyosis (70). Since free fatty acids and cholesterol sulfate are required for the solubilization of cholesterol, reduced levels of these lipids may be associated with the phase separation of cholesterol in this neonate. Furthermore, a negative correlation between the free fatty acids–ceramide ratio and barrier function as measured by TEWL was found, reinforcing the importance of lipid composition for optimum barrier function (70).

A peak at approximately d = 4 nm, detected in the stratum corneum from the 40 weeks EGA, 7 days PNA; 40 weeks EGA, 2 days PNA; and 34 weeks EGA, 8 days PNA neonates, may also be related to differences in lipid content from that of adult SC. X-ray diffraction from mixtures of cholesterol/ceramides/free fatty acids (isolated from pig SC) revealed an additional 4.2 nm d spacing when short-chain free fatty acids (C_{14}–C_{18}) were used (62). This was not the case when long-chain free fatty acids were present. Although the chain lengths of the free fatty acids in neonatal SC were not determined, the presence of shorter d spacings suggests a different population of free fatty acids in neonatal SC. Equally, lipid analyses of premature neonatal SC have revealed an altered ceramide distribution compared to that of adult SC (59). Differences in the relative amounts of ceramide fractions and abnormal x-ray diffraction patterns have also been found in patients with lamellar ichthyosis (70). Furthermore, x-ray diffraction analyses of mixtures of various ceramides, in the presence of cholesterol and palmitic acid, indicate that the type of ceramide affects d spacing and lipid organization (104,105).

The x-ray diffraction data from the SC of neonates show that the intercellular lipids undergo complex structural changes as a function of temperature (82). For three of the samples studied (40 weeks EGA, 7 days PNA; 40 weeks EGA, 2 days PNA; and 34 weeks EGA, 8 days PNA), the changes in lipid phase at physiological temperatures may be correlated with lipid composition differences, relative to that of adult SC, and this may in turn explain the impaired barrier function. Similarly, in cultured SC, lamellar disordering occurs between 25 and 45°C, perhaps related to the presence of fatty acids with shorter alkyl chains, low amounts of ceramides, and high amounts of cholesterol (103). It may be postulated that the phase transitions at physiological temperatures support the presence of essentially "fluid" subsets of lipid populations in neonatal SC and that, in turn, this may contribute to the reduced barrier function observed.

In FTIR, the asymmetric and symmetric components of the CH_2 stretching vibrations are sensitive to the conformational order of the lipid alkyl chain. Thus, the temperature-dependent behavior of the CH_2 stretching vibrations can be used to monitor the phase transitions of the lipids in the stratum corneum. The peak frequency of the symmetric CH_2 stretching vibration $[\nu_s(CH_2)]$ in adult stratum corneum showed an increase from 2849.5 to 2850.7 cm^{-1} and a 33% increase in the bandwidth from 65 to 90°C. Over the same temperature range, the asymmetric CH_2 stretching frequency $[\nu_a(CH_2)]$ increased from 2917.5 to 2921.1 cm^{-1}, with a 53% increase in the bandwidth, in agreement with previously published data (94). Similar temperature-dependent changes in the frequency of the CH_2 symmetric stretching vibrations occur at the gel to liquid-crystalline phase transition for DPPC, and are associated with a sharp increase in the gauche/trans conformer ratio in the alkyl chains (106). The increase in bandwidth for these stretching vibrations has also been observed in DPPC (106) and other bilayers (107), and is due to the increased rates of motion and the larger number of conformational states of the chains in the liquid-crystalline state.

The temperature-dependent frequency profiles of the CH_2 symmetric stretching vibrations from the stratum corneum of the 42 and 40 weeks EGA neonates are qualitatively similar (Fig. 6), with an increase in v_sCH_2 of 2.2 and 1.7 cm^{-1}, respectively, as the temperature was increased from 65 to 90°C. These frequency shifts were greater than those observed for adult SC (1.3 cm^{-1}). The transition for the SC from the 29 weeks EGA neonate began at a lower temperature, ~60°C, and exhibited an increase in $v_s(CH_2)$ of 2.6 cm^{-1} from 65 to 90°C. The $v_s(CH_2)$ profile for the SC from the gestationally youngest neonate (29 weeks EGA) showed a number of striking differences, compared to the other samples: a) The "baseline" $v_s(CH_2)$ was ~1 cm^{-1} higher than the more mature SC samples (i.e., the lipids were relatively more disordered at temperatures *below* any solid–solid or solid–liquid phase transitions), b) the solid-liquid crystalline phase change commenced at a lower temperature, with the transition midpoint being ~10°C lower than the adult SC, and c) the SC lacked the phase transition near 35°C, characteristic of mature SC (see later discussion).

The effect of cholesterol on DPPC bilayers also indicates the importance of lipid composition on the temperature dependence of $v_s(CH_2)$ and

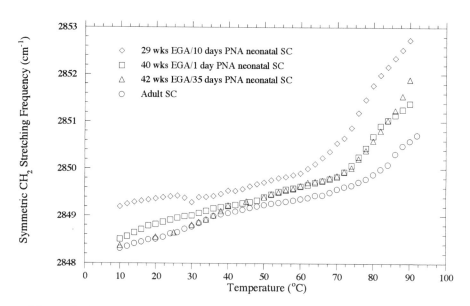

Figure 6 Temperature dependence of the symmetric CH_2 stretching frequency in adult and neonatal stratum corneum.

$\nu_a(CH_2)$. Cholesterol inhibits gauche rotamer formation above the phase transition temperature in DPPC and exhibits a strong ordering effect on the center of the bilayer (108). A similar effect in adult SC lipids may result in the smaller frequency change in the CH_2 stretching vibrations observed compared with those from the neonatal stratum corneum samples.

For all the neonatal samples studied, the thermotropic phase transitions were detected at lower temperatures ($5-10°C$) when assessed from the bandwidth of the CH_2 symmetric stretching vibration. Similar behavior has been seen in results from porcine (96) and mouse (109) stratum corneum, as well as membrane phospholipids (110). The lower transition temperature observed from the bandwidth, compared to the frequency, of the CH_2 stretching vibrations is due to the overlap of absorption bands resulting from the coexistence of at least two lipid phases (110).

The temperature-dependent frequency and bandwidth profiles of the CH_2 symmetric stretching vibration from adult and neonatal (40 and 42 weeks EGA) stratum corneum also show low-temperature inflections that are more readily observed by the bandwidth data. Low-temperature transitions have also been observed by IR in porcine SC (96) and by IR and differential scanning calorimetry in human SC (94,98). These low-temperature transitions, which correspond to an increase in $\nu_s CH_2$ and $\nu_a CH_2$, without an attendant change in the lateral packing of the hydrocarbon tails, have been attributed to a solid to fluid transition for a small fraction of SC lipids (96,98). The $\sim 35°C$ transition in adult SC reported here is consistent with these earlier observations, with respect to both the magnitude of the CH_2 symmetric stretching bandwidth and frequency changes. Similarly, a transition at $\sim 35°C$ is detectable by IR in the SC from the 42 weeks EGA neonate, while in the 40 weeks EGA neonate, this transition appears at a lower temperature ($\sim 25°C$) (101). Interestingly, in both the 40 and 42 weeks EGA neonatal samples, changes in the alkyl chain lateral packing also occur in this temperature range. Although concomitant alterations in lateral packing would not be anticipated in the event of a gel-to-liquid-crystalline phase change, this observation, nevertheless, needs further investigation. The absence of a low temperature inflection in the stratum corneum from a 29 weeks EGA neonate, together with an inherently higher value for the CH_2 symmetric frequency and bandwidth, suggests that this subset of SC lipids may already exist in a fluid state. Similarly, Mak et al. found that cultured stratum corneum did not exhibit a low temperature transition, and suggested that the cultured SC lipids may be physically organized and segregated differently to those in normal adult SC (54).

The frequency of the CH_2 rocking bands is a sensitive monitor of changes in the alkyl chain packing of the stratum corneum lipids. The existence of two peaks in the CH_2 rocking band from the adult stratum corneum in the temperature range 10 to 40°C (Fig. 7) is consistent with the previously reported factor group splitting of the scissoring mode (98), and is indicative of an orthorhombic lattice, but does not exclude the coexistence of a hexagonal network. At higher temperatures, however, the two peaks coalesce, reflecting the existence of a solid hexagonal phase below the gel-to-liquid-crystalline phase transition. Similar factor group splitting of the CH_2 rocking band was observed in the samples from 42 to 40 weeks EGA neonates. However, these neonatal samples demonstrated a solid–solid phase change at lower temperatures relative to adult SC. The CH_2 rocking region from the stratum corneum of a 29 weeks EGA neonate shows only one peak at the lower frequency at all temperatures above 10°C, demonstrating the existence of only one solid, loosely packed hexagonal phase below the gel to liquid crystalline phase transition. Similarly, IR spectra from cultured SC do not exhibit factor splitting of the CH_2 scissoring bands, an observation attributed to the presence of high cholesterol concentrations in these samples (78), on the basis that cholesterol (111) and other small molecules are reported to prevent the formation of the orthorhombic structure in lipid bilayers. The x-ray diffraction data indicate the presence of crystal-

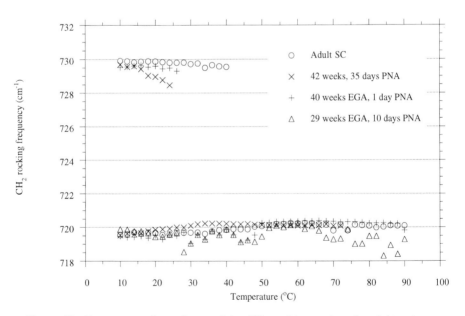

Figure 7 Temperature dependence of the CH_2 rocking regions in adult and neonatal stratum corneum.

line cholesterol in the stratum corneum from the 29 weeks EGA neonate, clearly suggesting that an abnormality in lipid composition may be responsible for the differences in alkyl chain packing and conformation in the stratum corneum of premature neonates.

Taken together, the x-ray diffraction and infrared spectroscopic results clearly demonstrate that the lipids of adult and neonatal SC are not organized in an identical manner. IR provides information about changes in alkyl chain organization which are complementary to the bilayer repeat distance detected by X-ray diffraction. For example, although the diffraction pattern of the stratum corneum from the 42 weeks EGA, 35 days PNA neonate was similar to that of adult at 25°C, IR results indicate that the SC lipids from the same neonate undergo gel-to-liquid-crystalline and solid–solid phase changes at slightly lower temperatures than adult SC lipids.

Both the x-ray diffraction data from the SC from a 40 weeks EGA, 2 days PNA neonate and the infrared results from a 40 weeks EGA, 1 day PNA neonate show a transition at 30–35°C. This suggests the existence, at physiological skin surface temperatures, of a subset of lipids in the liquid state in full-term neonates within the first days of life. In addition, the smaller d spacings found in the SC from the 40 weeks EGA, 2 days PNA neonate may be related to a different lipid composition or bilayer organization compared to that found in adult SC, and may further explain the lower solid–solid phase transition temperature detected by IR in this sample. Additionally, the disappearance of the 6.18 nm peak at 65°C in the scattering curve of the SC from the 40 weeks EGA, 2 days PNA neonate, agrees well with the gel-to-liquid-crystalline phase transition temperature observed by IR for the stratum corneum of the 40 weeks EGA, 1 day PNA neonate.

The x-ray diffraction pattern and IR spectra of the stratum corneum from a 29 weeks EGA, 10 days PNA neonate also provide complementary information. As mentioned previously, the presence of crystalline cholesterol, as detected by x-ray diffraction, may explain, in part, the lower lipid phase transition temperatures reported by IR spectroscopy. In addition, as with cultured SC, which is characterized by both hexagonal packing and an abnormal lipid composition (78), the decreased free fatty acid-to-ceramide ratio in the SC lipids from premature neonates (59) may be responsible for the existence of only the more loosely packed hexagonal phase and the absence of an IR-detectable solid–solid phase change at low temperatures.

VI. MACROSCOPIC CHARACTERIZATION OF NEONATAL BARRIER FUNCTION

Transepidermal water loss (TEWL) represents the passive diffusion of water through the epidermis and the secretion of sweat from the sweat glands.

TEWL measurements are noninvasive and provide information on the skin's function as a protective barrier. TEWL has been correlated with the percutaneous penetration of exogenous molecules (112,113) in adults and with the absorption of theophylline (114) in neonates.

In neonates with birth weights less than 1000 g, intensive monitoring is critical to minimize fluid loss due to evaporation. Early measurements of total insensible water loss (water loss from the skin and respiratory tract) indicate that preterm neonates may lose up to four times the volume lost by full-term newborns (115). TEWL may comprise about 75% of total water loss in neonates (31). It has been suggested that the skin may serve as the most important route of water depletion (both in total body water and extracellular volume) shortly after birth (116). TEWL measurements have also provided qualitative information about the maturation of skin barrier function in neonates. From TEWL measurements in 78 neonates, ranging from 26 to 41 weeks EGA, during the first 4 weeks of life, it was concluded that premature neonatal skin achieved maturation within 2 to 3 weeks of postnatal age, regardless of gestational age (4). All neonates maintained a constant skin surface temperature, although the environmental conditions (e.g., relative humidity, air temperature) were not reported. TEWL from 68 appropriate-for-gestational-age neonates (25–41 weeks EGA) was measured in environments of 50% relative humidity over the first 28 days of extrauterine existence (116). They found that TEWL was exponentially related to gestational age at all postnatal ages. Furthermore, TEWL in the very premature neonates remained higher than normal values after 4 weeks of postnatal age.

Recently, to investigate barrier function development in the premature neonate, TEWL and skin impedance were measured daily in neonates ranging from 23 to 32 weeks estimated gestational age (EGA) and from 1 to 64 days postnatal age (PNA) (117). Thus, longitudinal data were obtained to track the day-to-day changes in barrier function. In addition, the effects of environmental and individual-related variables on the changing barrier function of the neonate were studied. In neonates of less than 30 weeks EGA, it appeared that more than 4 weeks may be required for skin maturation, in agreement with earlier work (116), and challenging the widely accepted belief that 2 to 3 weeks of postnatal life are sufficient for skin barrier function development. In addition, the complementary impedance spectroscopy measurements validated the TEWL data and provided insight into the day-to-day changes in barrier function evolution. During the first weeks of life, abrupt changes in TEWL and impedance, followed by weeks of unchanging and approximately constant values, were observed. Mature levels of TEWL and impedance were reached in subsequent weeks.

In addition, important information about the skin developmental process in very premature neonates has been obtained. The dependence of skin impedance as a function of postconceptional age (PCA) in neonates born at less than 30 weeks EGA showed that acceleration in development of barrier function occurred at 200–220 days PCA. This suggests that 30 weeks PCA may represent a key milestone in the development process. It is likely that a combination of gestational and postnatal age, as well as environmental factors (medication, humidity, temperature), determines the state and rate of barrier function evolution. For example, a 31 weeks EGA (5–11 days PNA) neonate (Fig. 8) and a 32 weeks EGA (1–9 days PNA) neonate both displayed adult values of skin impedance and TEWL on each day of measurement.

This investigation also revealed that the effect of steroid administration on skin development requires further research. For example, the prenatal administration of betamethasone to a neonate born at 31 weeks EGA may explain the mature values of TEWL measured after only 1 week after birth. Although no previous studies have examined the effect of steroid administration on neonatal skin, maternal administration of betamethasone has been found to accelerate maturation of the stratum corneum lipid ultrastructure, lipid composition, and barrier function in the premature neonatal rat (30). Prenatal administration of glucocorticoids is also well established as a means of promoting fetal lung maturation (118,119), which occurs at a similar gestational age to stratum corneum lipid development in fetal rat skin (29).

Postnatal steroid administration may also influence skin development in the premature neonate. The administration of a tapered dosing regimen (120) of dexamethasone for 5–6 weeks in two neonates (23 weeks EGA, study period 1–64 days PNA, and 24 weeks EGA, study period 7–57 days PNA) appeared to improve barrier function shortly after the end of drug treatment. Among triplets who were treated with dexamethasone after 10 days PNA, the one triplet given the highest initial dose also required the longest time to exhibit mature skin impedance values (Fig. 9). Similarly, a twin who received more prolonged postnatal steroid therapy than its sibling required a longer time to achieve adult levels of TEWL and skin impedance. Glucocorticoid administration is used to treat chronic lung disease in premature neonates and has been shown to improve lung function (121). However, long-term use of topical glucocorticoids has been shown to cause thinning of the epidermis in adults (122,123). In addition, premature neonates receiving pulsed dexamethasone treatment have shown growth retardation, and reduced skinfold thickness during and after each steroid pulse. Nevertheless, following termination of steroid treatment, a period of "catch-up growth" ensued and, at full-term gestational age, the neonates had comparable body composition and size to control groups (124). Neonatal skin

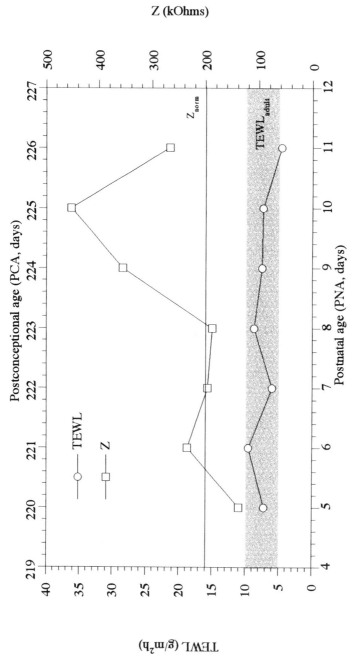

Figure 8   Changes in TEWL and low-frequency (1.6 Hz) impedance (Z) as a function of PNA and PCA for $30^{5/7}$ weeks EGA neonate.

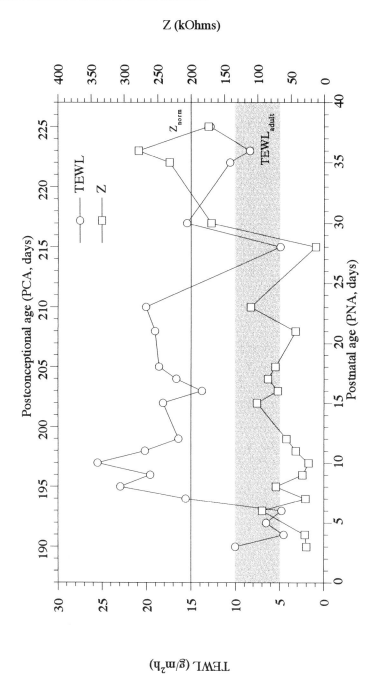

Figure 9 Changes in TEWL and low-frequency (1.6 Hz) impedance (Z) as a function of PNA and PCA for a triplet (26⁵/₇ weeks EGA).

development may also be retarded during the course of postnatal glucocorticoid treatment and consequently delay barrier maturation.

Humidity is also an important environmental factor affecting barrier function development (32). Although increased humidity (>80% relative humidity) reduces evaporative water loss, it is not used for extended periods of time in the neonate due to potential problems with infection and interference with nursing care practices. No long-term studies have been performed on the impact of humidity on skin development. Furthermore, the transition to a dry extrauterine environment may be involved in the biological signaling of stratum corneum maturation (21), and high-humidity environments may interfere with this process.

Although these recent results suggest that the pattern of changes in TEWL and impedance are discontinuous rather than steady, the absolute quantity of data and the limited number of neonates studied preclude remarks of any generality. It is evident that barrier evolution in these neonates is complex and sensitive to multiple factors, one or more of which in combination can provoke significant effects.

VII. CONCLUSION

Morphological, biophysical, and compositional analyses of mammalian stratum corneum attest to the complexity of this composite of multilamellar lipids and corneocytes. Although it has been established that the barrier resides within the intercellular lipids of the SC (100,125), the exact relationship between stratum corneum architecture and function continues to be an area of intensive research. Investigation of various diseased states, skin culture systems, and animal models indicates that environmental and biochemical signaling also play important roles in the barrier properties of the stratum corneum.

The depiction of the stratum corneum as a "brick and mortar" structure is a useful, albeit simplified, two-compartment model: a lipophilic/hydrophobic part (the intercellular lipids) and a more hydrophilic component (the corneocytes) (21). This basic concept of the stratum corneum has evolved to incorporate the concept of a dynamic skin barrier. Metabolic studies indicate that stratum corneum lipid synthesis is regulated by barrier requirements (126,127) and that extracellular enzymes are important in bilayer organization at the stratum granulosum–stratum corneum interface and throughout the layers of the stratum corneum (128). Recently, a domain mosaic model was developed to relate the heterogeneity and biophysical properties of the SC lipids to barrier function (129). This model depicts the bulk of the lipids as segregated into crystalline domains bordered by "grain borders" where lipids are in the fluid crystalline state. The fluid character of the "grain borders" represents areas where hydropyhobic molecules may diffuse through the SC.

Although a complete understanding of normal adult barrier function has not yet been achieved, studies in mammals and model lipid systems indicate that abnormalities in lipid composition are correlated to altered ultrastructure, which in turn may be responsible for reduced barrier function. Consequently, biochemical, morphological, biophysical, and functional in vivo analyses of neonatal skin are necessary to obtain a global perspective on skin maturation. Further studies of the maturation process may provide crucial insight into methods to ameliorate the impaired barrier function, which places the very premature neonate at risk due to problems with thermoregulation, fluid imbalance, and percutaneous toxicity. Future research into the biochemical and physical signals which control barrier function development will undoubtedly benefit the care of the premature neonate.

ACKNOWLEDGMENTS

Our work in this field has been supported by a grant from the National Institutes of Health (DA-09292). L. B. Nonato was supported by a National Science Foundation Fellowship and a University of California Berkeley Graduate Fellowship. We would like to thank Joanne Kuller, RN, MS, and the nurses of the Intensive Care Nursery, and Dr. Roger Williams and the Department of Pathology at Children's Hospital, Oakland, CA; Drs. Joke Bouwstra and Gert Gooris from the Leiden Amsterdam Center for Drug Research at Leiden University, the Netherlands; and Dr. Chris Cullander, Dr. Linda S. Franck, and the Department of Pathology at the University of California, San Francisco, CA.

REFERENCES

1. Klaus, MH, and Fanaroff, AA. Yearbook of Perinatal/Neonatal Medicine. Chicago: Year Book Publishers; 1987.
2. West, DP, Worobec, S, and Solomon, LM. Pharmacology and toxicology of infant skin. J Invest Dermatol 1981; 76:147–150.
3. Sinclair, J. Thermal control in premature infants. Annu Rev Med 1972; 23:129–132.
4. Rutter, N, and Hull, D. Water loss from the skin of term and preterm babies. Arch Dis Child 1979; 54:858–868.
5. Scholz, D. EKG electrodes and skin irritation. Neonatal Network 1984; 3:46–51.
6. Leyden, J. Bacteriology of newborn skin. In Neonatal Skin: Structure and Function, H Maibach and ER Boisits, editors. New York: Marcel Dekker; 1982:167–181.

7. Holbrook, KA, and Sybert, V. The structure and development of skin. In Dermatology in General Medicine, TB Fitzpatrick et al., editors. New York: McGraw-Hill; 1998:93–131.

8. Sammons, WAH, and Lewis, JM. Premature babies, a different beginning. St. Louis, MO: C. V. Mosby; 1985:3–7.

9. Lesko, SM, Epstein, MF, and Mitchell, AA. Recent patterns of drug use in newborn intensive care. J Pediatr 1990; 116:985–990.

10. Aranda, JV, Clarkson, S, and Collinge, JM. Changing pattern of drug utilization in a neonatal intensive care unit. Am J Perinatol 1983; 1:28–30.

11. Gould, T, and Roberts, RJ. Therapeutic problems resulting from the use of the intravenous route for drug administration. J Pediatr 1979; 95:465–469.

12. Hyman, PE, Clarke, DD, Everett, SL, et al. Gastric acid secretory function in preterm infants. J Pediatr 1985; 106:467–471.

13. Evans, NJ, Rutter, N, Hadgraft, J, et al. Percutaneous administration of theophylline in the preterm infant. J Pediatr 1985; 107:307–311.

14. Amato, M, Isenschmid, M, and Hippi, P. Percutaneous caffeine application in the treatment of neonatal apnoea. Eur J Pediatr 1991; 150:592–594.

15. Barrett, D, and Rutter, N. Transdermal delivery and the premature neonate. Crit Rev Ther Drug Carrier Syst 1994; 11:1–30.

16. McIntosh, N, Van Veen, L, and Brameyer, H. The pain of heel prick and its measurement in preterm infants. Pain 1993; 52:71–74.

17. Sell, EJ, Hansen, RC, and Struck-Pierce, S. Calcified nodules on the heel: A complication of neonatal intensive care. J Pediatr 1980; 96:473–475.

18. Fitzgerald, M, Millard, C, and McIntosh, N. Cutaneous hypersensitivity following peripheral tissue damage in newborn infants and its reversal with topical anaesthesia. Pain 1989; 39:31–36.

19. Tamada, JA, Bohannon, NJV, and Potts, RO. Measurement of glucose in diabetic patients using noninvasive transdermal extraction. Nat Med 1995; 1:1198–1201.

20. Michaels, AS, Chandrasekaran, SK, and Shaw, JE. Drug permeation through human skin: theory and in vitro experimental measurement. AIChE J 1975; 21:985–996.

21. Elias, PM. Plastic wrap revisited: The stratum corneum two-compartment model and its clinical implications. Arch Dermatol 1987; 123:1405–1406.

22. Flynn, GL. Mechanism of percutaneous absorption from physicochemical evidence. In Percutaneous Absorption, R Bronaugh and H Maibach, editors. New York: Marcel Dekker; 1989:27.

23. Evans, N, and Rutter, N. Development of the epidermis in the newborn. Biol Neonate 1986; 49:74–80.

24. Wertz, PM, Swartzendruber, DC, Abraham, W, et al. Essential fatty acids and epidermal integrity. Arch Dermatol 1987; 123:1381–1384.

25. Barry, B. Dermatological Formulations: Percutaneous Absorption. New York: Marcel Dekker; 1983:1–50.

26. Elias, PM. Epidermal lipids, barrier function, and desquamation. J Invest Dermatol 1983; 80:44–49.

27. Sweeney, TM, and Downing, DT. The role of lipids in the epidermal barrier to diffusion. J Invest Dermatol 1970; 55:135–140.

28. Nemanic, MK, and Elias, PM. A novel cytochemical technique for visualization of permeability pathways in mammalian stratum corneum. J Histochem Cytochem 1980; 28:573–578.

29. Aszterbaum, M, Menon, K, Feingold, KR, et al. Ontogeny of the epidermal barrier to water loss in the rat: Correlation of function with stratum corneum structure and lipid content. Pediatr Res 1992; 31:308–317.

30. Aszterbaum, M, Feingold, KR, Menon, GK, et al. Glucocorticoids accelerate fetal maturation of the epidermal permeability barrier in the rat. J Clin Invest 1993; 91:2703–2708.

31. Hey, E, and Katz, G. Evaporative water loss in the newborn baby. J Physiol 1969; 200:605–619.

32. Hammarlund, K, Nilsson, GE, Oberg, PA, et al. Transepidermal water loss in newborn infants. I. Relation to ambient humidity and site of measurement and estimation of total transepidermal water loss. Acta Paediatr Scand 1977; 66:553–562.

33. Harpin, V, and Rutter, N. Barrier properties of the newborn infant's skin. J Pediatr 1983; 102:419–425.

34. Holbrook, KA. Human epidermal embryogenesis. Int J Dermatol 1979; 18: 329–356.

35. Holbrook, KA, and Smith, LT, Ultrastructural aspects of human skin during the embryonic, fetal, neonatal and adult periods of life. In Morphogenesis and Malformations of the Skin, RJ Blandau, editor. New York: Alan R. Liss; 1981: 9–38.

36. Holbrook, KA. The fine structure of developing human epidermis: Light, scanning and transmission electron microscopy of the periderm. J Invest Dermatol 1975; 65:16.

37. Lev, R, and Weisberg, H. Human foetal glycogen: A histochemical and electron microscopic study. J Anat 1969; 105:337–349.

38. Sharp, F. A quantitative study of the glycogen content of human fetal skin in the first trimester. J Obstet Gynaecol 1971; 78:981–986.

39. Briggaman, RA, and Kelly, T. Continuous thymidine labeling studies of normal human skin growth on nude mice: Measurement of cycling basal cells. J Invest Dermatol 1982; 78:359.

40. Gelfant, S. "Of mice and men": The cell cycle in human epidermis in vivo. J Invest Dermatol 1982; 78:296–299.

41. Gerstein, W. Cell proliferation in human fetal epidermis. J Invest Dermatol 1971; 57:262–265.

42. Stern, IB. The uptake of tritiated thymidine by human fetal epidermis. J Invest Dermatol 1974; 63:268–272.

43. Parmley, TH, and Seeds, AE. Fetal skin permeability to isotopic water (THO) in early pregnancy. Am J Obstet Gynecol 1970; 108:128–131.

44. Lind, R, Parkin, FM, and Cheyne, GE. Biochemical and cytological changes in liquor amnii with advancing gestation. J Obstet Gynaecol Br Commonw 1969; 76:673–683.

45. Solomon, LM, and Esterly, NB. Neonatal dermatology. I. The newborn skin. J Pediatr 1970; 77:888–894.

46. Elias, PM, Cooper, ER, Korc, A, et al. Percutaneous transport in relation to stratum corneum structure and lipid composition. J Invest Dermatol 1981; 76: 297–301.

47. Grubauer, G, Feingold, KR, Harris, RM, et al. Lipid content and lipid type as determinants of the epidermal permeability barrier. J Lipid Res 1989; 30: 89–96.

48. Lampe, MA, Williams, ML, and Elias, PM. Human epidermal lipids: Characterization and modulations during differentiation. J Lipid Res 1983; 24:131–140.

49. Serup, J, and Blichmann, CW. Epidermal hydration of psoriasis plaques and the relation to scaling. Measurement of electrical conductance and transepidermal water loss. Acta Derm Venereol (Stockh) 1987; 67:357–359.

50. Motta, S, Monti, M, Sesana, S, et al. Abnormality of water barrier function in psoriasis. Role of ceramide fraction. Arch Dermatol 1994; 130:452–456.

51. Wertz, PW, Cho, ES, and Downing, DT. Effect of essential fatty acid deficiency on the epidermal sphingolipids of the rat. Biochim Biophys Acta 1983; 753:350–355.

52. Yamamoto, A, Serizawa, S, Ito, M, et al. Stratum corneum lipid abnormalities in atopic dermatitis. Arch Dermatol Res 1991; 283:219–223.

53. Imokawa, G, Abe, A, Jin, J, et al. Decreased level of ceramides in stratum corneum of atopic dermatitis: An etiologic factor in atopic dry skin? J Invest Dermatol 1991; 96:523–526.

54. Mak, VHW, Kennedy, AH, Golden, GM, et al. Are water permeability measurements sufficient to characterize in vitro cultured human skin surrogates? J Toxicol Cutan Ocul Toxicol 1993; 12:139–159.

55. Ponec, M, Weerheim, A, Kempenaar, J, et al. Lipid composition of cultured human keratinocytes in relation to their differentiation. J Lipid Res 1988; 29: 949–961.

56. Williams, ML, Hincenbergs, M, and Holbrook, KA. Skin lipid content during early fetal development. J Invest Dermatol 1988; 91:263–268.

57. Holbrook, KA. Structure and function of the developing human skin. In Physiology and Biochemistry of the Skin, LE Goldsmith, editor. Oxford: Oxford University Press; 1983:64–101.

58. Hurt, CM, Hanley, K, Williams, ML, et al. Cutaneous lipid synthesis during late fetal development in the rat. Arch Dermatol Res 1995; 287:754–760.

59. Nonato, L, Naik, A, Lund, C, et al. Neonatal stratum corneum lipids: Characterization and relationship to gestational and postnatal age. Submitted for publication.

60. Wertz, PW, Swartzendruber, DC, Madison, KC, et al. The composition and morphology of epidermal cyst lipids. J Invest Dermatol 1987; 89:419–425.

61. Wertz, PW, Abraham, W, Landmann, L, et al. Preparation of liposomes from stratum corneum lipids. J Invest Dermatol 1986; 87:582–584.

62. Bouwstra, JA, Gooris, GS, Cheng, K, et al. Phase behavior of isolated skin lipids. J Lipid Res 1996; 37:999–1011.

63. Holbrook, KA, and Hoff, MS. Structure of the developing human embryonic and fetal skin. Semin Dermatol 1984; 3:185–190.

64. Kagehara, M, Tachi, M, Harii, K, et al. Programmed expression of cholesterol sulfotransferase and transglutaminase during epidermal differentiation of murine skin development. Biochim Biophys Acta 1994; 1215:183–189.

65. Gray, GM, and Yardley, HJ. Lipid composition of cells isolated from pig, human, and rat epidermis. J Lipid Res 1975; 16:434–440.

66. Long, VJW. Phospholipase A and lysophospholipase activity of the epidermis. Br J Dermatol 1975; 92:603–610.

67. Bowser, PA, and Gray, GM. Sphingomyelinase in pig and human epidermis. J Invest Dermatol 1978; 70:331–335.

68. Gray, GM, and White, RJ. Glycosphingolipids and ceramides in human and pig epidermis. J Invest Dermatol 1978; 70:336–341.

69. Mao-Qiang, M, Jain, M, Feingold, KR, et al. Secretory phospholipase A2 activity is required for permeability barrier homeostasis. J Invest Dermatol 1996; 106:57–63.

70. Lavrijsen, APM, Bouwstra, JA, Gooris, GS, et al. Reduced skin barrier function parallels abnormal stratum corneum lipid organization in patients with lamellar ichthyosis. J Invest Dermatol 1995; 105:619–624.

71. Hou, SYE, Mitra, AK, White, SH, et al. Membrane structures in normal and essential fatty acid-deficient stratum corneum: Characterization by ruthenium tetroxide staining and X-ray diffraction. J Invest Dermatol 1991; 96:215–223.

72. Menon, GK, and Elias, PM. Ultrastructural localization of calcium in psoriatic and normal human epidermis. Arch Dermatol 1991; 127:57–63.

73. Ghadially, R, Reed, JT, and Elias, PM. Stratum corneum structure and function correlates with phenotype in psoriasis. J Invest Dermatol 1996; 107:558–564.

74. Fartasch, M, Diepgen, TL, and Bassukas, ID. Disturbed extruding mechanism of lamellar bodies in dry non-eczematous skin of atopics. Br J Dermatol 1992; 127:221–227.

75. Menon, GK, Feingold, KR, and Elias, PM. Lamellar body secretory response to barrier disruption. J Invest Dermatol 1992; 98:279–289.

76. Mak, VHW, Cumpstone, MB, Kennedy, AH, et al. Barrier function of human keratinocyte cultures grown at the air–liquid interface. J Invest Dermatol 1991; 96:323–327.

77. Fartasch, M, and Ponec, M. Improved barrier structure formation in air-exposed human keratinocyte culture systems. J Invest Dermatol 1994; 102:366–374.

78. Kennedy, AH, Golden, GM, Gay, CL, et al. Stratum corneum lipids of human epidermal keratinocyte air–liquid cultures: Implications for barrier function. Pharm Res 1996; 13:1162–1167.

79. Madison, KC, Swartzendruber, DC, Wertz, PW, et al. Lamellar granule extrusion and stratum corneum intercellular lamellae in murine keratinocyte cultures. J Invest Dermatol 1988; 90:110–116.

80. Holbrook, KA. A histological comparison of infant and adult skin. In Neonatal Skin: Structure and Function, H Maibach and ER Boisits, editors. New York: Marcel Dekker; 1982:3–31.

81. Lavker, RM, Zheng, P, and Dong, G. Aged skin: A study by light, transmission electron, and scanning electron microscopy. J Invest Dermatol 1987; 88:44s–51s.

82. Nonato, LB. Evolution of skin barrier function in premature neonates. Doctoral Dissertation. Bioengineering. Berkeley; University of California; 1997:177.

83. Elias, PM. Epidermal lipids, membranes, and keratinization. Int J Dermatol 1981; 2:1–18.

84. Wilkes, G, Nguyen, A, and Wilnauer, R. Thermal stability of the crystalline lipid structure as studied by x-ray diffraction and differential thermal analysis. Biochim Biophys Acta 1973; 304:267–275.

85. Elias, PM. X-ray diffraction analysis of stratum corneum membrane couplets. J Invest Dermatol 1983; 80:213–214.

86. Friberg, S, and Osborne, D. X-ray diffraction study of human stratum corneum. J Soc Cosmet Chem 1985; 36:349–354.

87. White, SH, Mirejovsky, D, and King, GI. Structure of lamellar lipid domains and corneocyte envelopes of murine stratum corneum. An x-ray diffraction study. Biochemistry 1988; 27:3725–3732.

88. Garson, J-C, Doucet, J, Leveque, J-L, et al. Oriented structure in human stratum corneum revealed by x-ray diffraction. J Invest Dermatol 1991; 96:43–49.

89. Bouwstra, JA, Gooris, GS, van der Spek, JA, et al. Structural investigations of human stratum corneum by small-angle x-ray scattering. J Invest Dermatol 1991; 97:1005–1012.

90. Bulgin, J, and Vinson, L. The use of differential thermal analysis to study the bound water in stratum corneum membranes. Biochim Biophys Acta 1972; 136:551–560.

91. Van Duzee, B. Thermal analysis of human stratum corneum. J Invest Dermatol 1975; 65:404–408.

92. Rehfeld, S, and Elias, PM. Mammalian stratum corneum contains physiological lipid thermal transitions. J Invest Dermatol 1982; 79:1–3.

93. Knutson, K, Potts, R, Guzek, D, et al. Macro- and molecular-physical considerations in understanding drug transport in the stratum corneum. J Control Rel 1985; 2:67–87.

94. Golden, G, Guzek, D, Harris, R, et al. Lipid thermotropic transitions in human stratum corneum. J Invest Dermatol 1986; 86:255–259.

95. Bouwstra, JA, Gooris, GS, Salomons-de Vries, MA, et al. Structure of human stratum corneum as a function of temperature and hydration: A wide-angle x-ray diffraction study. Int J Pharm 1992; 1992:205–216.

96. Ongpipattanakul, B, Francoeur, ML, and Potts, RO. Polymorphism in stratum corneum lipids. Biochim Biophys Acta 1994; 1190:115–122.

97. Cameron, DG, Casal, HL, and Mantsch, HH. Characterization of the pretransition in 1,2-dipalmitoyl-*sn*-glycero-3-phosphocholine by Fourier transform infrared spectroscopy. Biochemistry 1980; 19:3665–3672.

98. Gay, CL, Guy, RH, Golden, GM, et al. Characterization of low-temperature (i.e., <65°C) lipid transitions in human stratum corneum. J Invest Dermatol 1994; 103:233–239.

99. Potts, RO, Mak, VHW, Guy, RH, et al. Strategies to enhance permeability via stratum corneum lipid pathways. Adv Lipid Res 1991; 24:173–210.

100. Potts, RO, and Francoeur, ML. The influence of stratum corneum morphology on water permeability. J Invest Dermatol 1991; 96:495–499.

101. Nonato, LB, Bouwstra, J, Gooris, G, et al. Lipid biophysics of neonatal stratum corneum: Characterization by small-angle x-ray diffraction and fourier-transform infrared spectroscopy. Submitted for publication.

102. Ghadially, R, Williams, ML, Hou, SYE, et al. Membrane structural abnormalities in the stratum corneum of the autosomal recessive ichthyosis. J Invest Dermatol 1992; 99:755–763.

103. Bouwstra, JA, Gooris, GS, Weerheim, A, et al. Characterization of stratum corneum structure in reconstructed epidermis by x-ray diffraction. J Lipid Res 1995; 36:495–504.

104. Bouwstra, JA, Cheng, K, Gooris, GS, et al. The role of ceramides 1 and 2 in the stratum corneum lipid organisation. Biochim Biophys Acta 1996; 1300: 177–186.

105. McIntosh, TJ, Stewart, ME, and Downing, DT. X-ray diffraction analysis of isolated skin lipids: Reconstitution of intercellular lipid domains. Biochemistry 1996; 35:3650–3653.

106. Mantsch, HH, and McElhaney, RN. Phospholipid phase transitions in model and biological membranes as studied by infrared spectroscopy. Chem Phys Lipids 1991; 57:213–226.

107. Mantsch, HH, Martin, A, and Cameron, DG. Characterization by infrared spectroscopy of the bilayer to nonbilayer phase transition of phosphatidyl-ethanolamines. Biochemistry 1981; 20:3138–3145.

108. Davies, MA, Schuster, HF, Brauner, JW, et al. Effects of cholesterol on conformational order in dipalmitoylphosphatidylcholine bilayers. A quantitative IR study of the depth dependence. Biochemistry 1990; 29:4368–4373.

109. Knutson, K, Krill, SL, Lambert, WJ, et al. Physicochemical aspects of transdermal penetration. J Control Rel, 1987; 6:59–74.

110. Dluhy, RA, Chowdhry, BZ, and Cameron, DG. Infrared characterization of conformational differences in the lamellar phases of 1,3-dipalmitoyl-*sn*-glycero-2-phosphocholine. Biochim Biophys Acta 1985; 821:437–444.

111. Umemura, J, Cameron, D, and Mantsch, HH. A FTIR spectroscopic study of the molecular interaction of cholesterol with DPPC. Biochim Biophys Acta 1980; 602:32–44.

112. Dupuis, D, Rougier, A, Lotte, C, et al. In vivo relationship between percutaneous absorption and transepidermal water loss according to anatomical site in man. J Soc Cosmet Chem 1986; 37:351.

113. Oestmann, E, Lavrijsen, APM, Hermans, J, et al. Skin barrier function in healthy volunteers as assessed by transepidermal water loss and vascular response to hexyl nicotinate: Intra- and inter-individual variability. Br J Dermatol 1993; 128:130–136.

114. Cartwright, RG, Cartlidge, PHT, Rutter, N, et al. Transdermal delivery of theophylline to premature infants using a hydrogel disc system. Br J Clin Pharmacol 1990; 29:533–539.

115. Wu, PYK, and Hodgman, JE. Insensible water loss in preterm infants: Changes with postnatal development and non-ionizing radiant energy. Pediatrics 1974; 54:704–712.

116. Hammarlund, K, Sedin, G, and Stromberg, B. Transepidermal water loss in new born infants. VII. Relation to postnatal age in very pre-term and full-term appropriate for gestational age infants. Acta Paediatr Scand 1982; 71: 369–374.

117. Kalia, YN, Nonato, LB, Lund, CH, et al. Development of skin barrier function in premature infants. J Invest Dermatol 1998; 111:320–326.

118. Bourbon, JR, Farrell, PM, Doncet, E, et al. Biochemical maturation of fetal rat lung: A comprehensive study including surfactant maturation. Biol Neonate 1987; 52:48–60.

119. Kresch, MJ, and Gross, I. The biochemistry of fetal lung development. Clin Perinatol 1987; 14:481–506.

120. Avery, GB, Fletcher, AB, Kaplan, M, et al. Controlled trial of dexamethasone in respirator-dependent infants with bronchopulmonary dysplasia. Pediatrics 1985; 75:106–111.

121. Cummings, JJ, D'Eugenio, DB, and Gross, SJ. A controlled trial of dexamethasone in preterm infants at high risk for bronchopulmonary dysplasia. N Engl J Med 1989; 320:1505–1510.

122. Lehmann, P, Zheng, P, Lavker, RM, et al. Corticosteroid atrophy in human skin: A study by light, scanning and transmission electron microscopy. J Invest Dermatol 1983; 81:169–175.

123. Sheu, H, and Chang, C. Alterations in water content of the stratum corneum following long-term topical corticosteroids. J Formos Med Assoc 1991; 90: 664–669.

124. Gilmour, CH, Sentipal-Wilerius, JM, Jones, JM, et al. Pulse dexamethasone does not impair growth and body composition of very low birth weight infants. J Am Coll Nutr 1995; 14:455–462.

125. Elias, PM, and Friend, DS. The permeability barrier in mammalian epidermis. J Cell Biol 1975; 65:180–191.

126. Menon, GK, Feingold, KR, Moser, AH, et al. De novo sterolgenesis in the skin. II. Regulation by cutaneous barrier requirements. J Lipid Res 1985; 26: 418–427.

127. Grubauer, G, Elias, PM, and Feingold, KR. Relationship of epidermal lipogenesis to cutaneous barrier function. J Lipid Res 1987; 28:746–752.

128. Proksch, E, Elias, PM, and Feingold, KR. Regulation of 3-hydroxy-3-methylglutaryl-coenzyme A reductase activity in murine epidermis: Modulation of enzyme content and activation state by barrier requirements. J Clin Invest 1990; 85:874–882.

129. Forslind, B. A domain mosaic model of the skin barrier. Acta Derm Venereol (Stockh) 1994; 74:1–6.

50
Percutaneous Absorption of Sunscreens

Kenneth A. Walters
An-eX Analytical Services Ltd., Cardiff, Wales

Stephen D. Gettings
Avon Products, Inc., Suffern, New York

Michael S. Roberts
University of Queensland at Princess Alexandra Hospital, Brisbane, Queensland, Australia

I. INTRODUCTION

As a result of an increased awareness of the potential damaging effects of exposure of the skin to sunlight, the frequency of use of topical sunscreens has increased over the past several years. Furthermore, there is much evidence that the incidence of skin cancer is related to sun exposure (1), and cancer prevention organizations (including the American Academy of Dermatology) have undertaken major efforts to educate the public to avoid unnecessary exposure to solar radiation. For these reasons, it is now common practice for manufacturers to include sunscreen active ingredients in many traditional cosmetic products (e.g., lipsticks, foundations, moisturizers) as well as in so-called "beach sunscreens."

The U.S. Food and Drug Administration (FDA) has defined a sunscreen active ingredient as a chemical ingredient that absorbs at least 85% of the radiation in the ultraviolet (UV) range at wavelengths from 290 to 320 nm, but may or may not transmit radiation at wavelengths longer than 320 nm (2). However, FDA has since recognized that this definition is inadequate because it fails to include safe and effective ingredients whose absorption maxima are in the UVA wavelength range of 320 to 400 nm

Table 1 Some Examples of Commonly Used Sunscreen Actives

Chemical name	Other names
Homomenthyl salicylate	Homosalate
2-Ethylhexyl salicylate	Octyl salicylate
p-Aminobenzoic acid	PABA
Octyl dimethyl-p-aminobenzoate	Octyl dimethyl PABA, Padimate O
2-Hydroxy-4-methoxy benzophenone	Benzophenone-3, oxybenzone
2-Hydroxy-4-methoxy benzophenone-5- sulfonic acid	Benzophenone-4, sulisobenzone
2-Ethylhexyl-p-methoxycinnamate	Octyl methoxycinnamate
Butyl methoxydibenzoylmethane	Avobenzone

Source: Reference 27.

(e.g., avobenzone) (3). Most sunscreen actives (e.g., octyl salicylate, octyl methoxycinnamate) absorb in the UVB spectrum (290 to 320 nm). Some sunscreen actives absorb in both the UVA and UVB range (e.g., benzo-phenone-3, 200–350 nm). Some examples of commonly used sunscreen actives approved for products throughout the world are shown in Table 1.

An understanding of the potential for human systemic exposure is an integral part of the safety assessment of sunscreen actives used in consumer products (4). There is, however, little published data in the scientific litera-ture regarding the skin penetration of topically applied sunscreen actives. Although mathematical modeling from physicochemical data has suggested that the percutaneous absorption of certain sunscreen actives may be sig-nificant (5), it is important to appreciate that under conditions of actual exposure the rates of percutaneous absorption of such compounds will not be constrained solely by physicochemical properties but may also be strongly influenced by the nature and properties of the vehicle in which they are applied (4,6).

II. SKIN PERMEATION OF SUNSCREENS

A. Data from In Vivo Human Studies

The amount of published literature concerning the percutaneous absorption of sunscreens in human subjects in vivo from commercial products is lim-ited. Noting considerable variability in absorption between subjects (Fig. 1), Hayden et al. (7) found between 1 and 2% benzophenone-3 absorption 12 h after application to human skin. Benzophenone-3 has also been detected in breast milk of humans (8) following dermal application.

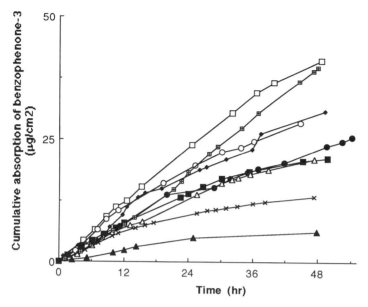

Figure 1   Skin absorption of benzophenone-3 following topical application of the sunscreen agent in a commercial formulation. Data calculated from urinary recovery of the sunscreen agent and its metabolites. From Reference 7.

Treffel and Gabard (9,10) examined the permeation of benzophenone-3, 2-ethylhexyl-4-methoxycinnamate, and 2-ethylhexyl salicylate (octyl salicylate) from a commercial vehicle (emulsion gel) and petroleum jelly, through human skin both in vivo and in vitro. Sunscreen formulations were applied to the skin surface in amounts reflective of consumer use ($\sim$2 mg/ cm^2), and samples were taken at intervals up to a maximum exposure of 6 h. For the in vivo study, samples of the stratum corneum of the treated area (back) were taken by tape-stripping at 0.5, 2, and 6 h following application of the formulations. Although there were no statistical differences in the amount of individual sunscreen agents found in the stratum corneum following 0.5 h of exposure, there was a significant vehicle effect in that the agents applied in the emulsion gel generated considerably higher stratum corneum levels than when applied in petroleum jelly (Fig. 2). These data were reflected in the in vitro study, where epidermal levels of the sunscreen agents were, for the most part, higher following application in the emulsion gel. In most cases, levels in the dermis and receptor fluid were below the level of quantification, except for benzophenone-3 (the least lipophilic sunscreen agent). Although higher levels of benzophenone-3 were found in the dermis

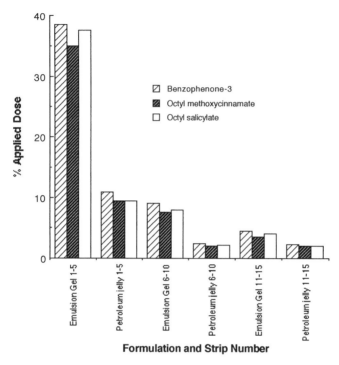

Figure 2   Effect of formulation on the amount of sunscreen agent recovered from stratum corneum tape strips following 30-min exposure in vivo. From Reference 10.

and receptor fluid following application in petroleum jelly, permeation to the receptor fluid also occurred from the emulsion gel. These data indicated that when applied in petroleum jelly approximately 5% of the applied dose of benzophenone-3 permeated through human skin within 6 h. The in vitro data generated in this study should, however, be viewed with caution since the skin was obtained from only two donors. Furthermore, all three sunscreen agents were detectable in samples of dermal tissue after only 2 min of exposure, which suggests possible contamination during the biopsy procedure.

Early work by Feldman and Maibach (11) indicated that a significant amount of p-aminobenzoic acid (PABA) applied to human skin (at a dose of 4 μg/cm^2) was excreted in the urine over 5 days (approximately 28% of the applied dose). It is important to appreciate, however, that these data were obtained under conditions in which the permeant was applied as a solvent-deposited solid. This application method can generate higher absorption values than those that would be obtained using more conventional application techniques. For example, Arancibia et al. (12) applied PABA in three different formulations (a hydroalcoholic gel, an oil-in-water emulsion at pH 4.2, and an oil-in-water emulsion at pH 6.5, each containing 5% PABA) to the face, neck, arms, and trunk of human males (20 g application) and subsequently measured urinary excretion over 48 h. There was no discernible difference in absorption between vehicles, although a large variation in the amount absorbed (1.6–9.6% of the applied dose) was observed.

Sunscreen absorption rates based on disappearance measurements after application to the skin may be used inappropriately to describe the extent of systemic absorption. Studies on commercial sunscreen products applied to human skin have shown that, while the more lipophilic sunscreen actives penetrate the stratum corneum, they do not tend to permeate through into the epidermis (13) and may be expected to be eventually lost by desquamation. In contrast, the more polar benzophenone-3 is less tightly bound by the stratum corneum, may permeate further, and thus may be available for systemic absorption (13). These observations are pertinent to the interpretation of the data of Hagerdorn-Leweke and Lippold (14), who determined the human skin permeation of several sunscreen actives in vivo. Each active (octyl dimethyl *p*-aminobenzoic acid, 4-isopropyl-dibenzoylmethane, 3-(4-methyl-benzylidine) camphor, isoamyl-4-methoxycinnamate, and benzophenone-3), was applied as a saturated solution in 30% propylene glycol–water to the skin of the upper arm using glass chambers. Reduction of the permeant in the donor vehicle was assessed hourly for 6 h. The calculated maximum flux ranged from 0.53 μg/cm^2/h for octyl dimethyl *p*-aminobenzoic acid (octanol–water log P = 5.75) to 4.93 μg/cm^2/h for isoamyl-4-methoxycinnamate (octanol–water log P = 4.83). There was a correlation between log P values and skin flux such that maximum flux decreased with increasing lipophilicity. These data were then used to estimate the amount of absorption of a saturated solution following total body exposure (surface area 1.8 m^2) for 1 h. The predicted values ranged from 10 mg for octyl dimethyl *p*-aminobenzoic acid to 89 mg for isoamyl-4-methoxycinnamate. It is important to point out, however, that such estimates can be very misleading and not particularly relevant for safety assessment. In this case, for example, the estimates were made using data obtained from repetitive infinite doses, of maximum thermodynamic activity, under fully occluded conditions. These conditions constitute an unlikely scenario during actual consumer use of sunscreen products.

B. Data from In Vitro Human Studies

Data from in vitro human studies should be interpreted carefully. Most such studies use isolated or dermatomed human epidermis and artificial receptor solutions. The choice of receptor conditions may greatly affect the observed extent of absorption and should be made with due consideration to the physicochemical nature of the permeant (6).

Similarly, the composition of the formulation vehicle may also significantly affect penetration. For example, the in vitro skin penetration of padimate O (octyl dimethyl *p*-aminobenzoate) from an alcoholic formulation was greater than from a lotion (15). It was postulated that evaporation of volatile vehicle contributed to the fourfold difference in penetration. Confirmation of the influence of the vehicle of application on the extent of penetration of sunscreen actives has been presented by Marginean Lazar et al. (16). These researchers evaluated the in vitro human skin permeation of octyl methoxycinnamate and butyl methoxydibenzoylmethane from a series of five emulsion vehicles (a conventional oil-in-water emulsion, an emulsifier-free oil-in-water preparation, a water-in-oil emulsion, a water-in-silicone emulsion, and an oil-in-water emulsion with lamellar liquid crystals). Although no sunscreen active was found in receptor solutions following 8 h of exposure to 2% butyl methoxydibenzoylmethane (irrespective of the vehicle), 8% octyl methoxycinnamate was detected at varying levels depending upon the vehicle. The greatest amount of octyl methoxycinnamate was found following application in an emulsifier-free oil-in-water preparation. The oil-in-water emulsion containing lamellar liquid crystals and the water-in-silicone formulation resulted in the slowest permeation rates of the sunscreen active.

Jiang et al. (17) have recently reported the in vitro human skin permeation of the commonly used sunscreen active octyl salicylate (2-ethylhexyl salicylate). Their data, obtained using infinite dose application of the sunscreen at high concentrations in liquid paraffin, indicated relatively low permeation. Furthermore, evidence of self-association of octyl salicylate, at high vehicle concentrations, was presented. More recently this group has evaluated the in vitro human skin penetration of sunscreen actives from commercial products intended for use by either adults and children (13). In this study, the penetration into and permeation across isolated epidermal membranes of sunscreen actives (benzophenone-3, octocrylene, octyl salicylate, octyl methoxycinnamate, and butyl methoxydibenzoylmethane) were determined following application of a finite dose (2.0–2.5 mg formulation/ cm^2). Only benzophenone-3 was detectable in the receptor phase (4% bovine serum albumin in phosphate-buffered saline), and the data indicated that approximately 10% of the applied dose had permeated over the 8-h exposure period. Significant amounts (5–25 μg/cm^2, representing 3–14% applied dose) of each of the sunscreen actives were recovered from the epidermal membranes following exposure (Fig. 3). It is interesting to note that penetration of benzophenone-3 through the skin was not significantly different

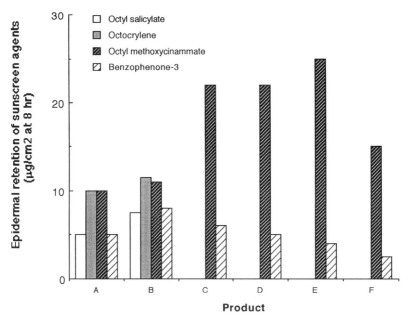

Figure 3   Amount of sunscreen agent recovered from human skin epidermal membranes following 8-h exposure to several commercial sunscreen products in vitro. Key: A, Product 1 (adults); B, Product 1 (children); C, Product 2 (adults); D, Product 2 (children); E, Product 3 (adults); F, Product 3 (children). From Reference 13.

for products intended either for adult use or for child use (Fig. 4). This latter observation is important since the skin surface–body weight ratio is higher in children than in adults and thus the amount available for absorption following an equivalent applied dose would be expected to be higher in children than in adults.

Finally, Walters et al. (18) have reported the in vitro human skin permeation of octyl salicylate from two vehicles (an oil/water emulsion and a hydroalcoholic formulation) that were representative of typical commercial sunscreen products. Human abdominal skin obtained at autopsy was heat separated to yield epidermal membranes (comprising stratum corneum and viable epidermis) and mounted in glass horizontal-type diffusion cells. Receptor phase solutions consisted of phosphate-buffered saline, pH 7.4, containing 6% Volpo N20. A finite dose of the oil/water emulsion formulation (5 mg/cm^2) and hydroalcoholic lotion (5 μl/cm^2) was applied to the skin surface. Permeation of ^{14}C-labeled octyl salicylate was determined by analysis of samples taken from the receptor phase at intervals over 48 h. The

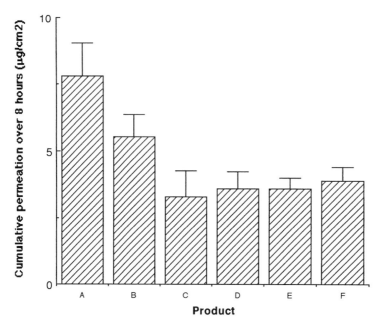

Figure 4 Amount of benzophenone-3 permeated through human skin epidermal membranes following 8-h exposure to several commercial sunscreen products in vitro. Key A, Product 1 (adults); B, Product 1 (children); C, Product 2 (adults); D, Product 2 (children) E, Product 3 (adults); F, Product 3 (children). From Reference 13.

data (Table 2) clearly show that the percutaneous penetration of octyl salicylate from typical sunscreen vehicles is low (<1% over 48 h), although higher than predicted by physicochemical data. The cumulative percutaneous penetration of ^{14}C-labeled material was very similar in each case (1.58 μg/cm^2 over 48 h), although the amount of applied material remaining in the epidermal membranes at 48 h was slightly higher for the hydroalcoholic solution (32.77%) than for the oil/water emulsion (17.18%) (Fig. 5).

Since it might reasonably be expected that vehicles of the type used by Walters et al. (18) would have a greater influence on the cumulative amount of permeant appearing in the receptor phase, these data reflect the importance of the use of final formulations (rather than simple solutions) in the risk assessment of substances intended for topical exposure. However, these data should also be interpreted cautiously. The amount of ^{14}C-labeled material recovered from the skin in vitro may not be truly predictive of the quantity of octyl salicylate remaining within the stratum corneum or epidermis under user-like conditions, since the surface rinsing procedure used

Table 2 Permeation and Recovery Data for Octyl Salicylate After Application as a Finite Dose In Vitro

Parameter	Hydroalcoholic lotion	O/W emulsion
Total permeated at 48 h ($\mu g/cm^2$)	1.58 ± 0.25	1.58 ± 0.36
Total absorption at 48 h (%)	0.59 ± 0.09	0.65 ± 0.16
Recovery in wash (%)	36.21 ± 5.98	36.66 ± 5.31
Recovery in skin (%)	32.77 ± 4.74	17.18 ± 1.28
Total recovery (%)	69.57 ± 6.84	54.50 ± 5.47

Note. Data are means $\pm$ SE ($n = 9$–11).
Source: Reference 18.

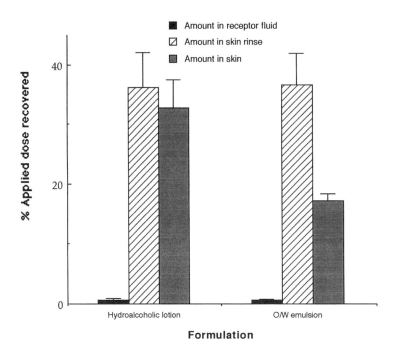

Figure 5 Recovery of octyl salicylate from various compartments following 48-h exposure to two different formulations in vitro. Human skin epidermal membranes. From Reference 18.

Table 3 Reported Human Skin Absorption of Sunscreen Actives

Sunscreen	Type of test	Percent of dose absorbed	Exposure time (h)	Flux ($\mu g/cm^2/h$)	Vehicle	Other	Ref.
p-Aminobenzoic acid	In vivo	1.6–9.6	48		Hydroalcoholic gel o/w emulsion		12
	In vivo	28.37 urine	120		Solvent-deposited solid radiolabel		11
Octyl dimethyl PABA	In vivo	Loss from donor	6	0.53	Propylene glycol	Permeability: 0.259 cm/h	14
Octyl salicylate	In vitro	1.9–7.7	6		o/w Emulsion gel and petrolatum	In skin only, none in receptor	10
	In vitro	0.59–0.65	48		Alcoholic and o/w emulsion	Absorption rate 1.58 $\mu g/cm^2$	18
	In vitro	5–7	8		Commercial formulations	In skin only, none in receptor	13
TEA salicylate	In vivo	0.8–26.8 urine	24		10% Cream		28
Homosalate	In vivo	1.08–4.43	24		8% Lotion		28

Compound	Method				Formulation		Ref.
Octyl methoxy-cinnamate	In vitro	10	16		o/w Emulsion		24
	In vitro	1.6–9.3	6		o/w Emulsion gel and petrolatum	In skin only, none in receptor	10
	In vitro	7–17	8		Commercial formulations	In skin only, none in receptor	13
Isoamyl-p-methoxy-cinnamate	In vivo	Loss from donor	6	3.16	Propylene glycol	Permeability: 0.105 cm/h	14
Benzophenone-3	In vivo	Loss from donor	6	4.44	Propylene glycol	Permeability: 0.053 cm/h	14
	In vitro	5.3–6.4	6		o/w Emulsion gel and petrolatum		10
	In vitro	5–10 in receptor 4–12 in skin	8		Commercial formulations		13
4-Isopropyl-dibenzoyl-methane	In vivo	Loss from donor	6	0.85	Propylene glycol	Permeability: 0.277 cm/h	14
3-(4-Methylbenzyl-idene)-camphor	In vivo	Loss from donor	6	2.11	Propylene glycol	Permeability: 0.091 cm/h	14
2-Hydroxy-4-methoxy-benzophenone-5-sulfonic acid	In vitro	6	16		o/w Emulsion, radiolabel		24

Note. Exposure periods and application conditions vary.

here was not particularly rigorous. Moreover, sunscreen products are typically formulated so as to provide a high degree of skin substantivity (i.e., the sunscreen active is delivered preferentially to the skin surface, where it remains in or on the upper layers of the stratum corneum). Table 3 gives a summary of some of the studies on human skin permeation of sunscreen actives.

C. Animal Models

In evaluating the significance of the results from studies utilizing animal models, it is very important to appreciate that, in the majority of cases, permeation data obtained using small-animal skin (e.g., rat, mouse, guinea pig) gives much higher absorption values than that obtained using human skin (19). Any data from studies using animal models should, therefore, be extrapolated to the human situation only with great caution.

Following dermal application of up to 800 $\mu g/cm^2$ of benzophenone-3 to rats, urinary and fecal excretion (collected over 72 h) amounted to 13.2–39.2% and 3.71–6.67% of the applied dose, respectively (20). More recently, benzophenone-3 was shown to readily cross rat skin (21). In contrast, quaternary ammonium benzophenones showed no detectable skin penetration over 45 h when applied to hairless mouse skin in vitro (22).

Brinon et al. (23) evaluated the effect of formulation vehicle on the permeation of benzophenone-4 across full-thickness hairless rat skin following application of an infinite dose. Using an aqueous solution of benzophenone-4 as the control, steady-state flux was determined over 48 h from a series of emulsions prepared using different surfactant systems. The results showed a variation in steady-state flux between 10 ± 6 $\mu g/cm^2/h$ and 86 ± 7 $\mu g/cm^2/h$ (aqueous solution control: 32 ± 24 $\mu g/cm^2/h$), depending on the surfactant system used. The data indicated that flux of the sunscreen active from simple emulsion systems was low compared to that from more complex liquid crystalline emulsions. It should be pointed out, however, that the concentration of benzophenone-4 in all systems was kept constant (2.5%), and it is possible that solubility of the sunscreen in the formulation may have varied, resulting in differences in thermodynamic activity of the permeant within the formulation. Nonetheless, these data suggest that manipulation of the formulation vehicle may be a means of reducing the percutaneous absorption of sunscreen actives.

More relevant results (i.e., pertinent to actual product use) were presented by Benech et al. (24), who applied a finite dose of an oil-soluble sunscreen active (octyl methoxycinnamate) and a water-soluble sunscreen active (benzophenone-4) in an oil-in-water emulsion formulation, to human and pig skin, in vitro. Approximately 10% of the applied dose of octyl methoxycinnamate and 6% of the applied dose of benzophenone-4 were recovered from the epidermis, dermis, and receptor solution following 16-h exposure in both species. However, data for the distribution of sunscreen active in the various skin compartments were not reported, and it is possible that the majority of the recovered dose in each case was in the outer layers of the stratum corneum.

III. PREDICTION OF THE SKIN PENETRATION OF SUNSCREENS

Watkinson et al. (5) have predicted the skin penetration of various sunscreen agents based on mathematical modelling. The model took into account the physicochemical characteristics of each permeant in approximating several parameters important in the prediction of percutaneous absorption: a) partition of the permeant from vehicle into stratum corneum; b) passive diffusion of the permeant across stratum corneum; c) partitioning of the permeant at the junction between stratum corneum and viable tissue; d) uptake into systemic blood circulation and elimination of the permeant, e) potential return of the permeant into vehicle from stratum corneum; f) potential return of the permeant to stratum corneum from viable tissue; and g) potential return of the permeant to viable tissue from blood.

As for any mathematical modeling of permeation, several assumptions were made in the generation of rate constants: a) the first-order rate constant described the partitioning from vehicle into stratum corneum (k_a, which is dependent on diffusion through the vehicle); b) partitioning into skin was a ratio of k_a and k_r (the rate constant for diffusion from stratum corneum back into vehicle); c) diffusion across stratum corneum was predicted from molecular weight (M) ($k_1 = D/h^2$, where D = diffusion coefficient, h = path length for diffusion); d) partitioning from the stratum corneum into the viable tissue was determined from the molecular weight and the octanol-water partition coefficient (calculated by the Hansch group contribution method) as the ratio k_2/k_3 (k_2 = rate constant from stratum corneum into viable tissue; k_3 = rate constant from viable tissue back into stratum corneum); e) permeants with melting points below ambient temperatures were assumed to possess very large stratum corneum solubilities; f) the applied dose was assumed as 40 μg/cm^2; g) exposure time was 12 h; and h) the area of application was 1.4 m^2, estimated as 75% of the total average skin surface area.

Thus, estimation of k_1 from molecular weight (M), for $M < 740$, and estimation of the diffusion across viable tissue, k_2, were made using:

$$k_1(h^{-1}) = 0.91M^{-1/3}$$
$$k_2(h^{-1}) = 14.4M^{-1/3}$$

and estimation of partitioning between the stratum corneum and viable tissue, where K_{oct} is the octanol–water partition coefficient, was made using:

$$k_3 = k_2 K_{oct}/5$$

Estimation of stratum corneum solubility was made using melting point (Mp):

$$\log[SC] = 1.911 \times 10^3/M_p - 2.956$$

Although, based on their predictions, the authors suggest that systemic absorption of some sunscreen actives across large areas of skin may occur at significant levels after long periods of exposure, only minor amounts of the most lipophilic compounds were expected to penetrate. For example, octyl salicylate had a calculated systemic availability over 12 h (from an application to 1.4 m^2) of 3.3 μg. Examples of predicted values for dermal absorption of typical sunscreen agents using this model are given in Table 4. It is important to appreciate, however, that predictive estimates of the extent of skin absorption are of limited value unless the estimates and assumptions within the model are rigorous. Models of this type often ignore the potential for biotransformation or degradation, which may modify the dermal penetration characteristics and toxicity profile of the original chemical. Furthermore, many variables associated with actual product use will alter the extent of skin penetration of sunscreen actives. For example, fre-

Table 4 Relationship Between Predicted Human Skin Absorption and Calculated Octanol–Water Partition Coefficients (log P) for Selected Sunscreen Agents

Compound	Absorption[a] (μg/1.4 m^2)	Calculated log P[b]
2-Ethylhexyl salicylate	0.50	6.19
p-Aminobenzoic acid	21,000	1.00
4-Isopropyl-dibenzoylmethane	61	4.06
2-Hydroxy-4-methoxybenzophenone	100	3.87
2-Ethylhexyl-p-methoxycinnamate	0.75	6.40

[a]Predicted total body absorption in 6 h.
[b]Calculated using Hansch group contribution method.
Source: Reference 5.

quent reapplication of product (as is recommended), vehicle release, and formulation excipients may affect skin permeability. It is essential, therefore, that for risk assessment purposes more accurate determinations of skin permeation are obtained (25).

IV. CONCLUSIONS

In most countries, the marketing of sunscreen products is subject to significant regulatory controls. Following extensive review of safety and efficacy, only certain sunscreen actives are permitted for use within allowable concentration ranges. In addition, combination of individual actives within the same product is also highly controlled. Because of the ingenuity of formulators and the careful use of combinations of sunscreen actives, today's sunscreen products contain lower concentrations of active ingredients than ever before. Despite the widespread use of sunscreens, only small numbers of adverse reactions, mainly consisting of contact or photocontact allergies, have been reported. Sunscreen products provide protection against skin cancer and guard against sunburn and many cumulative, suberythemal forms of skin damage. As such, sunscreen products may be assumed to be among the safest of consumer products. Nevertheless, as new information on the toxicological properties of individual sunscreen actives emerges, it may be necessary to conduct a risk assessment of the safety of sunscreen products. Under these circumstances, it may be useful to evaluate the percutaneous absorption of sunscreen actives (see e.g., References 4 and 26).

It may be concluded from the examples described here that there are two factors in particular that are important when considering the skin penetration and permeation of sunscreen actives: a) that the vehicle of application can markedly influence the rate, skin distribution, and extent of percutaneous absorption, and b) that those sunscreen actives with high lipophilicity will possess a greater affinity for the stratum corneum—which will tend to reduce diffusion to deeper layers of the skin. However, studies on the percutaneous penetration of octyl salicylate from representative sunscreen formulations through human skin in vitro (18) illustrate that although it is possible to predict the penetration of sunscreen agents and to speculate how penetration may be modulated by various types of vehicle, there is no substitute for conducting finite-dose experiments using human skin and actual formulations of interest, when the data are to be used for risk assessment of substances intended for topical exposure. The influence of formulation vehicle, quantification of penetration rates, and skin distribution using experiments designed to mimic actual product use, are essential prerequisites to the generation of realistic risk assessment values.

ACKNOWLEDGMENT

The authors are indebted to Cameron Hayden for his unreserved advice and assistance in the preparation of this chapter.

REFERENCES

1. Naylor MF, Farmer KC. (1997). The case for sunscreens. A review of their use in preventing actinic damage and neoplasia. Arch Dermatol 133:1146–1154.
2. Federal Register. (1993). Sunscreen drug products for over-the-counter human use; Tentative final monograph; Proposed rule. Fed Reg 58:28194 (May 12).
3. Federal Register. (1996). Sunscreen drug products for over-the-counter human use; Amendment to the tentative final monograph. Fed Reg 61:48645 (Sept 16).
4. Hayden CGJ, Benson HAE, Roberts MS. (1998). Sunscreens: Toxicological aspects. In Dermal Absorption and Toxicity Assessment, MS Roberts, KA Walters, eds. Marcel Dekker, New York, pp. 537–599.
5. Watkinson AC, Brain KR, Walters KA, Hadgraft J. (1992). Prediction of the percutaneous penetration of ultra-violet filters used in sunscreen formulations. Int J Cosmet Sci 14:265–275.
6. Gettings SD, Howes D, Walters KA. (1998). Experimental design considerations and use of in vitro skin penetration data in cosmetic risk assessment. In Dermal Absorption and Toxicity Assessment, MS Roberts, KA Walters, eds. Marcel Dekker, New York, pp. 459–487.
7. Hayden CGJ, Roberts MS, Benson HAE. (1997). Systemic absorption of sunscreen after topical application. Lancet 350:863–864.
8. Hany J, Nagel R. (1995). Detection of sunscreen agents in human breast milk. Dtsch Lebensm Rundsch 91:341–345.
9. Treffel P, Gabard B. (1996). Vehicle influence on the in vitro skin penetration of ultra-violet filters used in sunscreen formulations. In Prediction of Percutaneous Penetration, Vol 4b, KR Brain, VJ James, KA Walters, eds. STS, Cardiff, pp. 178–181.
10. Treffel P, Gabard B. (1996). Skin penetration and sun protection factor of ultra-violet filters from two vehicles. Pharm Res 13:770–774.
11. Feldman R, Maibach H. (1970). Absorption of some organic compounds through the skin in man. J Invest Dermatol 54:399–404.
12. Arancibia A, Borie G, Cornwell E, Medrano C. (1981). Pharmacokinetic study on the percutaneous absorption of p-amino-benzoic acid from 3 sunscreen preparations. Farmaco Ed Prat 36:357–365.
13. Jiang R, Roberts MS, Collins DM, Hoffmann NV, Benson HAE. (1998). Absorption of sunscreens into human skin: An evaluation of a number of commercial products. Br J Clin Pharmacol, in press.
14. Hagedorn-Leweke U, Lippold BC. (1995). Absorption of sunscreens and other compounds through human skin in vivo: Derivation of a method to predict maximum fluxes. Pharm Res 12:1354–1360.

15. Kenney G, Sakr A, Lichtin J, Chou H, Bronaugh R. (1995). In vitro skin absorption and metabolism of padimate ○ and a nitrosamine formed in padimate ○-containing cosmetic products. J Soc Cosmet Chem 46:117–127.

16. Marginean Lazar G, Baillet A, Fructus AE, Arnaud-Battandier J, Ferrier D, Marty JP. (1996). Evaluation of in vitro percutaneous absorption of UV filters used in sunscreen formulations. Drug Cosmet Ind 158:50–62.

17. Jiang R, Roberts MS, Prankerd RJ, Benson HAE. (1997). Percutaneous absorption of sunscreen agents from liquid paraffin: Self-association of octyl salicylate and effects on skin flux. J Pharm Sci 86:791–796.

18. Walters KA, Brain KR, Howes D, James VJ, Kraus AL, Teetsel NM, Toulon M, Watkinson AC, Gettings SD. (1997). Percutaneous penetration of octyl salicylate from representative sunscreen formulations through human skin in vitro. Food Chem Toxicol 35:1219–1225.

19. Walters KA, Roberts MS. (1993). Veterinary applications of skin penetration enhancers. In Pharmaceutical Skin Penetration Enhancement, KA Walters, J Hadgraft, eds. Marcel Dekker, New York, pp. 345–364.

20. El Dareer S, Kalin J, Tillery K, Hill D. (1986). Disposition of 2-hydroxy-4-methoxybenzophenone in rats dosed orally, intravenously and topically. J Toxicol Environ Health 19:491–502.

21. Okereke CS, Abdel-Rahman MS, Friedman MA. (1994). Disposition of benzophenone-3 after dermal administration in male rats. Toxicol Lett 73:113–122.

22. Monti D, Saettone MF, Centini M, Anselmi C. (1993). Substantivity of sunscreens—In vitro evaluation of the transdermal permeation characteristics of some benzophenone derivatives. Int J Cosmet Sci 15:45–52.

23. Brinon L, Geiger S, Alard V, Tranchant J-F, Pouget T, Couarraze G. (1998). Influence of lamellar liquid crystal structure on percutaneous diffusion of a hydrophilic tracer from emulsions. J Cosmet Sci 49:1–11.

24. Benech F, Berthelot B, Wegrich P, Schaefer H, Leclair J. (1995). A comparative study of in vitro skin models: percutaneous absorption of sunscreens on pig skin and human skin. In Prediction of Percutaneous Penetration, Vol 4a, KR Brain, VJ James, KA Walters, eds. STS, Cardiff, p. C68.

25. Gettings SD, Azri-Meehan S, Demetrulias JL, Dressler WE, Kasting GB, Kelling CK, Howes D. (1993). The use of in vitro skin penetration data in the safety assessment of cosmetic formulations. In Prediction of Percutaneous Penetration, Vol 3b, KR Brain, VJ James, KA Walters, eds. STS, Cardiff, pp. 621–637.

26. Agin P, Anthony FA, Hermensky S. (1998). Oxybenzone in sunscreen products. Lancet 351:525.

27. Klein K, Steinberg D. (1998). Encyclopedia of UV absorbers. In Sun Products, Protection and Tanning, Allured Publishing, Carol Stream, IL, pp. 11–65.

28. Federal Register. (1978). Sunscreen drug products for over the counter human drug use; Establishment of a monograph. Fed Reg 43:38206 (Aug 25).

51
Microemulsions

Reinhard H. H. Neubert
Martin Luther University, Halle/Saale, Germany

Ulrike Schmalfuß
permamed laboratories ltd., Therwil/Basel, Switzerland

I. INTRODUCTION

During recent years there has been interest in the development of new effective vehicle systems to modify drug penetration into the human skin. These studies started using liposomes. In recent years colloidal vehicle systems such as microemulsions (ME) have been included in the investigated spectrum of potential dermal therapeutics in order to obtain enhanced penetration.

ME are isotropic, transparent or slightly opalescent, low-viscosity, and thermodynamically stable systems usually consisting of water, oil, surfactant, and cosurfactant. The diameter of the disperse phase of MEs is in the range of 10 to 100 nm. These systems show structural similarity to micelles and inverse micelles, respectively. According to the content of water and oil, MEs can be classified into W/O (water-in-oil) and O/W (oil-in-water) types. They are highly dynamic systems showing fluctuating surfaces caused by forming and deforming processes.

MEs have been shown to exert a high capacity for incorporating both lipophilic or hydrophilic substances, depending on the composition of the formulation. Therefore, they have been considered as vehicle systems for drugs. In pharmaceutics in recent years, research work and commercial interest in MEs has increased.

Urogastron (human epidermal growth factor) was incorporated in the pseudomicellar phase of a W/O ME and showed high oral bioavailability (1). Sandimmune Neoral is a commercial available cyclosporine formulation containing the drug in an ME preconcentrate consisting of surfactant, ethanol, and a mixture of hydrophilic and lipophilic solvents (2). These waterless MEs containing, for example, lecithin, glycerol, and decane are estimated to be suitable vehicles for drugs, resistant against hydrolysis.

Furthermore, MEs have been used for ocular drug delivery and as model systems for studying lipid transfer into fibroblasts (3,4).

II. CHARACTERIZATION AND PREPARATION OF MICROEMULSIONS

For the characterization of microemulsions, various methods are available, whereby only the consideration of results from different experiments allows an elucidation of the structure.

> Microscopy: Polarization microscopy can be used to detect isotropy (ME) or anisotropy (mesogeneous phases). Electron microscopy combined with a freeze fracture technique is applied to determine the structure and the size of colloidal phases.
> Light-scattering experiments: Static laser light scattering and small-angle neutron scattering are used in order to measure the diameter of the pseudomicellar phase.
> Spectroscopic methods: Nuclear magnetic resonance (NMR) spectroscopy is capable of characterizing the colloidal phase of the ME.
> Other methods: The classical rheological methods are applied to determine the viscosity of the ME systems. The type of ME can be estimated by measuring the conductivity of the systems.

The formation of an ME system takes place spontaneously. The free total Gibbs enthalpy (ΔG^M) has a minimum when the ME is formed:

$$\Delta G^M = \sum n_i \mu_i + n_T (\pi r^2/3)[\gamma + 2K_b/d + 3kT/4\pi r^2 f(\Theta, r)] \qquad (1)$$

$$K_b = [\delta\gamma/\gamma \ (2/r)]_{T,p,\mu_i} \qquad (2)$$

where n_i is the amount of the applied compound, μ_i the chemical potential, n_T the number of droplets of pseudomicellar phase, r the radius of the droplets, T the temperature, γ the surfase tension, Θ the relation between molar volumes of pseudomicelar (colloidal) phase (droplets) and dispersion medium, and K_b the diffraction coefficient of the surfactant interphase.

The mechanism of the formation of ME is determined mainly by the tension of the surfactant interphase. Adsorption of the surfactants at the oil/water interface leads to a strong decrease of the surface tension. This results in an extension of the interface and therefore in the formation of highly dispersed droplets. However, prerequisite is a low solubility of the surfactants both in the oily and in the aqueous phase. The formation of a micro-emulsion may depend on the order of the mixing of the constituents.

The HLB scheme (hydrophilic–lipophilic–balance) developed by Griffin can be used to select surfactants and cosurfactants. The latter increase the entropy for the formation of the ME due to an increase of the fluidity of the interphase (5).

The formation of MEs has to be estimated using phase diagrams. For this purpose, Winsor developed the concept of the intermicellar phase equilibrium (6).

III. MICROEMULSIONS FOR DERMAL AND TRANSDERMAL DRUG DELIVERY

The penetration of drugs from MEs into human skin has been the objective of numerous studies. It was shown that these systems offer advantages concerning amounts of drug transported and the depth of penetration into skin. The focus of interest often has been substances that are considered to be problematic with respect to skin penetration.

Delgado-Charro et al. (7) compared the delivery of a model hydrophilic solute (sucrose) across hairless mouse skin in vitro from W/O and O/W MEs and an aqueous solution. While a constant flux value for the aqueous solution was reached after 5 h, the flux from the MEs continuously increased up to a maximum after 9 h, independent of the emulsion type. The transport rate from the microemulsion systems was about an order of magnitude greater than that from the reference. Generally, it was shown that the delivery of sucrose from microemulsions containing a higher percentage of aqueous phase was favored.

The enhancing effect for drugs contained in MEs in comparison to a creme and a gel formulation consisting of the same components was shown by Siegenmeyer and Führer (8). The in vitro permeation across skin membranes as well as the in vivo penetration of tetracycline hydrochloride was higher from an ME than from conventional systems. Thus it could be shown that in addition to the composition, the structure of each of the topically applied vehicle may play a dominant role in the process of drug penetration.

Systems for an administration to human skin have to be characterized according to their irritation and toxicity potential (9). It is known that MEs exert irritative effects, often by their high content of surfactants. It is possible to overcome this problem by the use of physiologically compatible nonionic and polymeric surfactants.

Furthermore, the irritation potential of the formulation depends strongly on its structure. Because of an equilibrium between MEs and liquid crystals, when brought into contact MEs may dissolve skin structures that are organized in liquid crystalline form. Thus an irritative effect is produced. Deduced from this, the nature of the system formed during the penetration process and the residue remaining on the skin surface are of importance in regard to skin irritation. Friberg (10) and Tadros (11) showed that the formation of a liquid crystalline structure during the penetration process will inhibit irritation of the applied ME system. This circumstance may be controlled using a composition that produces, due to evaporation, lamellar liquid crystalline structures.

IV. MECHANISM OF DRUG PENETRATION FROM MICROEMULSIONS

Drug transport from MEs is influenced by interactions within the vehicle as well as by interactions between vehicle and skin. In terms of the vehicle, physicochemical parameters of the drug influence the penetration process, such as the drug solubility in the formulation, the diffusion coefficient within the system, and partition coefficients (12–16).

Trotta et al. (12) investigated the dependence of drug liberation on the cosurfactant content using an in vitro membrane system. The diffusion of steroid hormones from O/W MEs containing varying amounts of butanol was influenced by the partition of the drug between the disperse, internal, and continuous phase, the droplet size, and the presence of aggregates in the external aqueous phase. It was established that an equilibrium between free diffusible drug, drug solubilized in aggregates, and that present in the disperse oil phase affected the diffusion rate. Thus, the composition and the ratio between the disperse and the continuous phase were considered to influence the release of drugs from MEs.

The relationship between drug transport and composition of the system as well as physicochemical properties of the active substances was studied by Pattarino et al (13). The focus of interest was the permeation of β-blockers of varying solubilities and partition coefficients in a O/W ME system through hairless mouse skin using Franz-type diffusion cells. A correlation between permeability coefficient, drug fluxes, and the partition coefficient of the drug, determined in the presence of cosurfactant, indicated the dependence of drug transport on the concentration in the external phase of the systems used.

Gallarate et al. (14) further studied the ability of the internal phase of a ME system to act as drug reservoir. It could be established that by varying the composition of the external phase and thus affecting the drug solubility of the drug in the ME "phases," a prolonged release can be maintained.

With respect to solubility parameters, Kemken et al. (15,16) investigated the pharmacological effects of selected β-blocking agents incorporated in water-free systems and applied to human skin under occlusion. By an uptake of water, evaporated from the skin surface, an in situ formation of MEs occured, whereby the solubility of the drug in the ME was decreased. Consequently, oversaturated systems were formed, which are thermodynamically favored for drug delivery.

Mechanistical investigations were also focused on the relationship between structural features of the system, such as droplet size and penetration properties. The transdermal absorption of nifedipine from ME with different contents of ethanol and different types of oils was studied by Thacharodi et al. (17). The results indicated that the diameter of the droplets of the colloidal phase of the ME depends strongly on the content of ethanol. The ME with smaller diameters of the colloidal phase showed higher penetration rates. A nonaqueous ME system consisting of 100% ethanol showed surprisingly not the highest enhancer activity concerning drug penetration. The results indicated that the action of both lipophilic and hydrophilic components of the ME enhances the drug penetration significantly.

Synergistic effects of the constituents were demonstrated also by other studies (18–20). Principally, it is assumed that hydrophilic as well as lipophilic constituents of the ME interact with those of the stratum corneum, leading to an enhancement of drug penetration. Lipophilic constituents of the ME may influence the structure of the stratum corneum lipid bilayers, thus changing their fluidity. Additionally, hydrophilic constituents of the ME entering the horny layer may exert hydrating effects, leading to an opening of the polar pathway even for hydrophilic drugs.

Osborne et al. (18) investigated the transport of glucose from MEs of varying water content across human cadaver skin. It was shown that the penetration of water from the vehicle was parallel to the transport of the hydrophilic drug.

The objective of our studies (19,20) was determining the penetration of the hydrophilic substance diphenhydramine hydrochloride from W/O ME systems. A standard system was modified by the incorporation of different additives, such as fatty acids, urea, cholesterol, or a glycolipid. The penetration of the drug as well as selected vehicle components was studied. Different effects in the presence of the modulators could be detected:

1. An increase or decrease of drug penetration.
2. A modified distribution of the drug between the skin layers.
3. An altered transport behavior of the continuous phase of the system.
4. An influence of the concentration and/or the combination of modulators on their penetration properties and distribution characteristics in the stratum corneum.

As an example (see Fig. 1), the incorporation of 5 or 10% of 10-methylpalmitic acid (5MPS and 10MPS) into a standard ME(ST) caused a decrease of the penetration of the hydrophilic drug with dependence on the fatty acid concentration. A considerable transport of the fatty acid could be shown, leading to the assumption that the penetrating additive caused an increase of the lipophilicity of the stratum corneum, whereby the entering of the hydrophilic drug was decreased.

The influence of the penetration of vehicle constituents on the penetration of drugs was also indicated by studies by Tacharodi et al. (17). An increased enhancing activity of MEs compared to solutions was due to the combined effect of lipophilic and hydrophilic components of the systems. The author concluded that lipophilic components of the vehicle are able to increase the permeability of the lipid pathway of the stratum corneum by inserting themselves between the chains of the lipid bilayers. Hydrophilic constituents may hydrate stratum corneum domains and proteins. Thereby a synergistic effect favors drug transport from MEs.

Additionally to these features, a reduction of the interfacial tension between vehicle and skin makes ME superior in comparison to other systems with respect to penetration properties (21). Linn et al. (21) studied the use of MEs as dermal pretreatment agents. The penetration of cetylalcohol from a creme and from a lotion could be increased by 50–250% after pretreatment

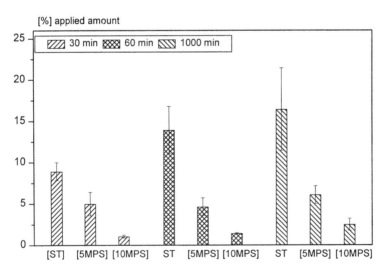

Figure 1 Comparison of the penetration of diphenhydramine from microemulsions: Standard ME [ST], 5% 10-methylpalmitic acid ME [5MPS], and ME 10% 10-methylpalmitic acid [10MPS] into human skin ex vivo (mean ± SD; $n = 3$).

with an ME system for 10 min in comparison to untreated skin. Cetyl alcohol and octyldimethyl-*p*-aminobenzoic contained in an ME penetrated into deeper skin layers compared with an incorporation into cremes and lotions. The authors discussed the ability of the system to lower the interfacial tension between the skin and the vehicle upon its intimate contact with skin polar and nonpolar constituents and thereby to facilitate the rate of transport as well as the depth of penetration.

It can be concluded that in general MEs undergo structural changes following an application to the skin because of the penetration and/or evaporation of constituents and under occlusion by the uptake of water from the skin surface (11,15,16,22,23). The formed structures and their penetration behavior finally influence the effectiveness of the systems for dermal drug transport.

V. CONCLUSIONS

In comparison to classical dermal formulations, MEs are very effective vehicle systems for dermal as well as for transdermal drug delivery, due to their high drug loading capacity caused by their colloidal structure. Furthermore, thermodynamic stability and a simple preparation process favor them to be considered as vehicles for skin application. These features could open for them a broad range of new approaches in the field of dermal use. Therefore, prospective studies should be aimed even more toward the clarification of transport mechanisms and toward an improvement of dermal tolerance.

REFERENCES

1. Luisi, P. L. Process for preparing a solution of inverted micellae. Patent application WO 86/02264 (1986).
2. Sarciax, J. M., Acar, L., and Sado, P. A. Using microemulsion formulations for oral drug delivery of therapeutic peptides. Int. J. Pharm. 120:127–136 (1995).
3. Keipert, S., and Schulz, G. Mikroemulsionen auf Saccharoseesterbasis, Teil 1: In vitro-Charakterisierung. Pharmazie 49:195–197 (1994.
4. Ekman, S. [^{3}H]Cholesterol transfer from microemulsions particles of different sizes to human fibroblasts. Lipids 22:657–663 (1987).
5. Dörfler, H.-D. *Grenzflächen- und Kolloidchemie*. Weinheim: VCH Verlagsgesellschaft (1994).
6. Mosinski, S. Microemulsions. Pestycydy (Warschau) 2:5–16 (1995).
7. Delgado-Charro, M. B., Iglesias-Vilas, G., Blanco-Mendez, J., Lopez-Quintela, M., Marty, J. P., and Guy, R. H. Delivery of a hydrophilic solute through the skin from novel microemulsion systems. Eur. J. Pharm. Biopharm. 43:37–42 (1997)

8. Ziegenmeyer, J., and Führer, C., Mikroemulsionen als topische Arzneiform. Acta Pharm. Technol. 26(4):273–275 (1980).

9. Dreher, F., Walde, P., Luisi, P. I., et al., Human skin irritation of a soybean lecithin microemulsion gel and of liposomes. Proc. 22nd Int. Symp. Controlled Release Bioact. Mater., 640 (1995).

10. Friberg, S. E. Micelles, microemulsions, liquid crystals, and the structure of stratum corneum lipids. J. Soc. Cosmet. Chem. 41:155–171 (1990).

11. Tadros, T. F. Future developments in cosmetic formulations. Int. J. Cosmet. Sci. 14:93–111 (1992).

12. Trotta, M., Gasco, M. R., and Pattarino, F. Diffusion of steroid hormones from O/W microemulsions: Influence of the cosurfactant. Acta Pharm. Technol. 36: 226–341 (1990).

13. Pattarino, F., Carlotti, M. E., Trotta, M., and Gasco, M. R. Release of β-blockers from O/W microeumulsions: Influence of vehicle composition on the permeation across hairless mouse skin. Acta Technol. Legis Med. VI(2) (1995).

14. Gallarate, M., Gasco, M. R., and Rua, G. In vitro release of azelaic acid from oil in water microemulsions, Acta Pharm. Jugosl. 40:533–538 (1990).

15. Kemken, J., Ziegler, A., and Müller, B. W. Investigations into pharmacodynamic effects of dermally administered microemulsions containing β-blockers. J. Pharm. Pharmacol. 43:679–684 (1991).

16. Kemken, J., Ziegler, A., and Müller, B. W. Influence of supersaturation on the pharmacodynamic effect of bupranolol after dermal administration using microemulsions as vehicle. Pharm. Res. 9:554–558 (1992).

17. Thachorodi, D., and Panduranga, R. Transdermal absorption of nifedipine from microemulsions of lipid skin enhancers. Int. J. Pharm. 111:235–240 (1994).

18. Osborne, D. W., Ward, A. J. I., and O'Neill, K. J. Microemulsions as topical drug delivery vehicles: in-vitro transdermal studies of a hydrophilic model drug. J. Pharm. Pharmacol. 43:451 (1991).

19. Schmalfuss, U. Investigations into the modulation of the penetration of a hydrophilic drug from microemulsion systems into human skin under ex vivo conditions. Doctoral thesis, Martin-Luther-Universität Halle-Wittenberg (1997).

20. Schmalfuss, U., Neubert, R., and Wohlrab, W. J. Control. Rel. 46:279–285 (1997).

21. Linn, E. E., Pohland, R. C., and Byrd, T. K. Microemulsions for intradermal delivery of cetylalcohol and octyl dimethyl PABA. Drug Dev. Ind. Pharm. 16: 899–920 (1990).

22. Trotta, M., Gallarate, J., and Gasco, M. R. Effect of the variation of microemulsion water content on the in vitro permeation of hydrocortisone. Acta Techn. Legis Medicamenti VI(2) (1995).

23. Carlfors, J., Blute, I., and Schmidt, V. Lidocaine in microemulsions—A dermal delivery system. J. Dispersion Sci. Technol. 12:467–482 (1991).

52
Bioavailability of [^{14}C]Hydroquinone in F344 Rat and Rhesus Monkey

Xiaoying Hui, Ronald C. Wester, Joseph L. Melendres, Steffany Serranzana,* and Howard I. Maibach
University of California School of Medicine, San Francisco, California

I. INTRODUCTION

Hydroquinone is produced in large amounts and used widely as an antioxidant and developing agent in industry. It is also used as a topical bleaching agent in dermatology and cosmetics. Studies found that hydroquinone was nonmutagenic by the Ames test but induced chromosome aberrations and karyotypic effects in eukaryotic cells. Carinogenic activity was found in some animal studies (NTP, 1989). Since hydroquinone can be considered an environment contaminant (Devillers et al., 1990), and potential dermal exposure is possible, knowledge of percutaneous absorption and topical bioavailability of hydroquinone in different species will help to assess its pharmacologic and toxicologic risks in humans. This percutaneous absorption study was done in the rat, a species in which hydroquinone toxicology has been studied, and the rhesus monkey, an animal model relevant to humans for percutaneous absorption (Wester and Maibach, 1993).

II. MATERIALS AND METHODS

Hydroquinone USF, a white crystalline solid with 99.6% purity, was obtained from Eastman Chemical Company (Kingsport, TN). [UL-^{14}C]-Hydroquinone (lot no. 930521), a colorless liquid, with specific activity of 20 mCi/mmol (MW 110) and 98.8% radiochemical purity measured by thin-layer

---

*Present affiliation: Penederm Inc., Foster City, California.

chromatography (TLC) and high-performance liquid chromatography (HPLC), was obtained from Wizard Laboratories (David, CA).

A. Dosage Formulation

Topical formulation was prepared following the manufacture's standard operating procedures (KIWI Brands, Inc., Douglassville, PA). The formulation consisted of Emersol, Drakeol, Adol, IPM, Brij, Amphisol, and hydroquinone to be a final 2% (w/w) hydroquinone cream. It contained 200 μCi/g cream of radioactivity for the F344 rat and 299 μCi/g for the rhesus monkey. The intravenous formulation was 2% (w/v) hydroquinone in 0.9% sodium chloride sterile solution, containing 20 μCi/ml solution of radioactivity for the F344 rat and 2 μCi/ml for the rhesus monkey. Oral formulation for the rhesus monkey was 2% (w/v) hydroquinone in water containing 6.9 μCi/ml radioactivity. The radioactivity of each dosing formulation was subsequently determined. The stability of each formulation was verified by HPLC.

B. Animals

Female adult Fischer 344 rats (F344 rats), weighing 151 ± 5.3 g, were obtained from Charles River Breeding Laboratories (Portage, MI). Upon arrival, the F344 rats were examined and kept for 2 weeks as a quarantine period before the experiment. Near the end of the quarantine period, the F344 rats were randomized to 14 treatment groups and 2 controls using a computer-generated random program. Each group consisted of four animals.

Four female rhesus monkeys, weighing 8.25 ± 0.95 kg, housed in the UCSF colony were used for this study. All animals were individually housed during the experiment. The rooms were environmentally controlled to maintain a room temperature of 23 ± 6°C, a relative humidity 50 ± 20% and a 12-h light/12-h dark cycle. The rats and monkeys were provided ad libitum access to drinking water and to Purina certified rodent chow and Purina monkey chow (Ralston Purina Company, St. Louis, MO), respectively.

C. Rat Study

Six groups of F344 rats were topically dosed on the back area (2×2.5 cm, near the interscapular) with 10 mg of 2% hydroquinone cream containing 20 μCi radioactivity. The hair of the area was clipped 1 day prior to dosing. At 2, 4, 8, 24, 48, and 72 h after dosing, the animals were sacrificed by CO_2 gas inhalation followed by exsanguination, and the dosed site skin and skin washing samples collected.

Seven groups were intravenously dosed in the lateral vein on the middle length with 0.5 ml of 2% hydroquinone solution containing 12.15 μCi radioactivity. The animals were sacrificed by CO_2 gas inhalation at 0.25, 1, 2, 4, 8, 24, and 48 h after dosing and blood samples were collected.

For urine collection, two additional groups of F344 rats were dosed topically or intravenously with the same doses as just described and individually housed in plastic metabolic cages (Maryland Plastics, Inc., Federalsburg, MD) to collect separated urine and feces. Urine sample was collected every 24-h period for 96 h. One topically treated group was used to collect blood, dosed site skin, and skin washing samples at 96 h.

D. Monkey Study

Four monkeys were dosed intravenously, topically, and orally with three different 2% hydroquinone formulations. After each dosing, the monkeys were placed individually into a metabolic chair for 24 h and then transferred to a metabolic cage until 96 h. At a set time, blood and urine were collected. Each treatment was conducted for 5 days, followed by a period of 3 weeks to depurate any remaining radioactivity in the body.

For intravenous administration, rhesus monkeys were injected with 0.5 ml of 2% hydroquinone solution containing 1 μCi radioactivity into the antecubital vein. After dosing, the animals were individually transferred to a metabolic chair for 24 h and then placed in metabolic cages.

For topical administration, 20 mg of 2% hydroquinone cream containing 5.8 μCi radioactivity was applied to a 10-cm^2 application site on the abdominal area of a rhesus monkey. The treated site skin was washed with 50% soap solution and distilled water five times at 24 h after dosing.

For oral administration, rhesus monkeys were placed in a metabolic chair. A gavage tube was inserted into the esophagus and 1 ml of 2% hydroquinone containing 6.9 μCi radioactivity in water was displaced into the tube, followed by 5 ml water.

E. Urine Collection and Analysis

The collection period was for 24-h intervals to 96 h postdosing. Urine aliquots were assayed to determine radioactivity level. Metabolic cages were washed with 50% Ivory soap solution and distilled deionized water five times at 96 h postdosing to determine residual radioactivity.

F. Blood Collection and Analysis

Blood was drawn from the posterior vena cava of F344 rats or the antecubital vein of rhesus monkeys. In F344 rats, blood samples were immediately centrifuged at $10,000 \times g$ for 3 min to separate the plasma and packed blood cells. Aliquots of 0.1 ml plasma were mixed with 10 ml of scintillation cocktail to determine radioactivity level. In rhesus monkeys, an aliquot of whole blood was precipitated in 10% trichloroacetic acid (1:1) to remove protein, followed by centrifuging at 10,000 $\times$ g for 5 min, and the aqueous phase was assayed for ^{14}C activity.

G. Treated Site Skin Washing and Analysis

Hydroquinone remaining on the treated site skin (not absorbed) was determined by measuring the radioactivity of skin washings. The washing procedure was as follows: a) 50% Ivory soap solution, b) distilled deionized water, c) 50% Ivory soap solution, d) distilled deionized water, and e) distilled deionized water.

H. Treated Site of Rat Skin Digestion

Hydroquinone remaining in the treated site skin of F344 rats was removed immediately after skin washing. Skin samples were placed in a borosilicate glass vial followed by adding 5 ml of Soluene 350 tissue solubilizer (Packard, Downers Grove, IL). The skin was solubilized after incubation at 37°C for 48 h. The sample was appropriately diluted with Hionic-Fluor (Packard, a Canberra Company, Meriden CT) and 5 ml of the mixture was counted.

I. Scintillation Counting

The scintillation cocktail used was Universal-ES (ICN Biomedicals, Costa Mesa, CA). Background control samples and test samples were counted in a Packard model 1500 counter (Packard Instruments, Downers Grove, IL). The counter was audited for accuracy using sealed samples of quenched and unquenched standards as detailed by the instrument manual. Control and test sample counts were transferred by hand to a computerized spreadsheet (Microsoft Excel 5.0 for Window) and the percentage of labeled material recovered from the applied dose was determined. From these values, the concentration of ^{14}C-labeled hydroquinone was calculated, based on the specific activity of unmetabolized material.

J. Data Analysis

The results were analyzed by Student's t-test or analysis of variance, followed by the Duncan multiple-range test with $p < .05$ as the level of significance.

Noncompartment pharmacokinetic parameters from plasma (F344 rats) or whole blood (rhesus monkeys) levels of each treatment group versus time data were determined by using the program PCNONLIN V4.2 (Scientific

Consulting, Inc., Cary, NC). C_{max} is the peak drug concentration; T_{max} is the time at which C_{max} occurred. The terminal rate constant (beta) was determined by linear regression analysis of the log-linear terminal phase of the plasma (or whole blood) concentration–time profile. Area under the curve (AUC) was determined as area under the plasma (or whole blood) concentration–time curve up to the last measured time point, calculated by the linear trapezoidal rule, plus any residual area, which was calculated as the concentration at the last time point, divided by the terminal rate constant. Total plasma (or whole blood) clearance (Cl) was calculated as the intravenous dose divided by the AUC. Volume of distribution (V) was calculated as the clearance divided by the terminal rate constant (beta).

Bioavailability was estimated by using cumulative urine data as follows (Rowland and Tozer, 1995):

$$F_{extravascular} = \left(\frac{Ae_{extravascular}}{Ae_{iv}}\right)\left(\frac{Dose_{iv}}{Dose_{extravascular}}\right) \times 100\%$$

$F_{extravascular}$ is either topical bioavailability or oral bioavailability. Ae is the total amount of ^{14}C content excreted from 0 h to the last following intravenous, topical, or oral administration. Dose is the total amount of ^{14}C content administered.

The flux rate for topically applied hydroquinone was calculated from the slop of plots of the amount of hydroquinone penetrated versus time:

$$Flux(\mu g/h/cm^2) = \frac{dc}{dt}\,F_{topical}$$

where dc/dt is the steady-state slope, which is obtained at 24 h after the beginning of the study, and $F_{topical}$ is the topical bioavailability calculated from total urinary ^{14}C excretion.

III. RESULTS

A. Percutaneous Absorption in F344 Rats

Following topical administration of [^{14}C]hydroquinone cream in F344 rats, the rate of percutaneous absorption appeared to be early and rapid. Peak plasma concentration occurred within 2 h (Fig. 1); radioactivity levels in the treated site skin reached steady state within 2 h (Fig. 2); and radioactivity levels in the skin washing samples significantly decreased within 8 h (Fig. 2).

Continuous topical exposure, however, did not increase [^{14}C]hydroquinone absorption through rat skin. Plasma concentration was 0.0104 ± 0.0024, 0.0097 ± 0.0035, 0.0060 ± 0.0024, and 0.0052 ± 0.0013 μg-Eq/

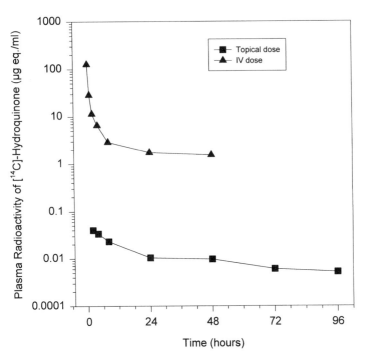

Figure 1   Plasma concentration of [^{14}C]hydroquinone versus time in F344 rats following topical or intravenous administration. Each point represents the mean of four
animals.

ml at 24, 48, 72, and 96 h, respectively (Fig. 1), and no statistical differences
were found. Radioactivity levels stayed at steady state up to 96 h (Fig. 2).
Radioactivity levels of skin washing samples were almost the same
(26.54–32.35%) from 24 to 96 h after dosing, and no statistical differences
were found.

The result of the rat study show that [^{14}C]hydroquinone was quickly
removed via urinary elimination. In intravenously dosed rats, plasma
[^{14}C]hydroquinone concentration was 126.35 $\pm$ 30.59 μg-Eq/ml at 15 min
and then decreased to 6.54 $\pm$ 2.04 μg-Eq/ml at 4 h. Most of dose (84.7%)
was excreted in urine within 24 h, and 96.7% by 96 h (Table 1). In topically
dosed rats, 17.61% of the dose was excreted in urine by 24 h and total
urinary recovery was 36.21% by 96 h.

Dose accountability was 96.7 $\pm$ 5.37% for the intravenously administered rats and 76.25 $\pm$ 5.67% for the topically applied rats (Table 2).
Topical bioavailability ($F_{topical/iv}$) therefore was calculated to be 37.45%.

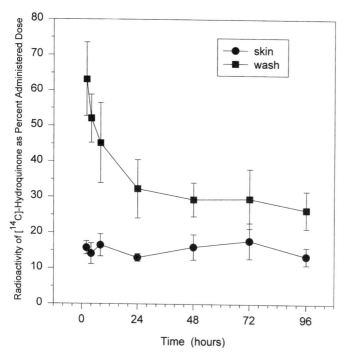

Figure 2 Treated skin site washing and treated skin radioactivities as percent dose in F344 rats following topical administration. Each point represents the mean ± SD ($n = 4$).

Pharmacokinetics parameters of intravenously and topically applied rats are shown in Table 3. In topically administered rats, T_{max} was reached at 2 h with [^{14}C]hydroquinone concentration at 0.04 μg-Eq/ml plasma. AUC was 1.35 μg-Eq·h/ml and $T_{1/2}$-beta was 45.07 h. In intravenously dosed rats, AUC was 211.75 μg-Eq·h/ml, $T_{1/2}$-beta was 14.28 h, Cl was 47.22 ml/h, and V was 944.47 ml.

B. Percutaneous Absorption in Rhesus Monkeys

In topically treated monkeys, blood concentration of [^{14}C]hydroquinone slowly increased until 24 h after dosing (Fig. 3). By 24 h, 11.92 ± 0.54% of the dose was excreted in urine (Table 1) and 65.76 ± 6.07% of the dose was recovered in the treated site skin washing samples. Total urinary recovery was 18.55 ± 0.78% by 96 h (Table 1). In intravenously dosed monkeys, blood concentration of [^{14}C]hydroquinone was quickly decreased from 0.81 ± 0.32 μg/ml at 15 min to 0.03 ± 0.03 μg/ml at 4 h (Figure 3). A

Table 1 Radioactivity as Percent Administered Dose Excreted in the Urine During the Indicated Intervals in Female F344 Rats and Rhesus Monkeys Following Differing Routes of Administration

Collection interval (h)	F344 rats[a]		Rhesus monkeys[a]		
	Intravenously	Topically	Intravenously	Topically dosed	Orally
0–24	84.70 ± 7.27	17.69 ± 5.17	53.21 ± 2.83	11.92 ± 0.54	52.34 ± 9.26
24–48	2.99 ± 0.83	7.69 ± 0.46	0.29 ± 0.20	5.40 ± 0.88	0.22 ± 0.07
48–72	1.95 ± 0.47	4.00 ± 0.43	0.19 ± 0.14	0.82 ± 0.44	0.07 ± 0.01
72–96	0.97 ± 0.28	2.55 ± 1.00	0.40 ± 0.22	0.45 ± 0.31	0.09 ± 0.02
Total	96.70 ± 5.37[b]	36.21 ± 5.38[b]	54.09 ± 2.58	18.55 ± 0.78	52.71 ± 9.39

[a]Values represent the mean ± SD for four animals per group.
[b]Values include radioactivity levels of cage washing samples.

Table 2 Dose Accountability and Bioavailability in Female F344 rats and Rhesus Monkeys Following Differing Routes of Administration of [^{14}C]Hydroquinone

Species	Route of administration	Dose accountability (%)	Bioavailability ($F_{extravascular}$, %)
F344 rats	iv	96.70 ± 5.37^a	—
	Topical	76.25 ± 5.67^b	37.45
Rhesus monkeys	iv	54.09 ± 2.58^a	—
	Topical	100.11 ± 6.13^c	34.34 ± 2.01
	Oral	52.71 ± 9.39^a	97.95 ± 20.70

[a]Dose accountability equals to radioactivity of total urinary recovery.
[b]Dose accountability equals to the sum of radioactivities of total urinary recovery, dosed site skin washing, and dosed site skin digestion.
[c]Dose accountability equals to bioavailability (as percent dose absorbed) plus radioactivity of dosed site skin washing.

total of $53.21 \pm 2.83\%$ of the dose was found in urine within 24 h after dosing and $54.09 \pm 2.58\%$ by 96 h (Table 1). In orally dosed monkeys, [^{14}C]hydroquinone urinary excretion as percent dose was similar to that of iv dosed monkeys: $52.34 \pm 9.46\%$ was excreted by 24 h and $52.71 \pm 9.39\%$ by 96 h (Table 1).

Dose accountability in intravenously administered monkeys was 54.09% (Table 2). The remainder is assumed to have been excreted in feces, although feces were not assayed. Dose accountability for topically treated rats was the sum of topical bioavailability and radioactive level of skin washing samples. Since topical bioavailability was 34.34%, dose accountability was 100.11% (Table 2). Oral bioavailability was calculated as $97.95 \pm 20.70\%$ dose.

Pharmacokinetic analyses show that in intravenously administered monkeys, AUC was 1.34 µg-Eq·h/ml, $T_{1/2}$-beta was 2 h, Cl was 7.46 L/h, and V was 10.97 L (Table 3). In topically dosed monkeys, T_{max} was 24 h with 0.0003 µg-Eq/ml(C_{max}). AUC was 0.01 µg-Eq·h/ml and $T_{1/2}$-beta was 34 h.

IV. DISCUSSION

These studies determined the topical bioavailability of [^{14}C]hydroquinone in female F344 rats and female rhesus monkeys using the analyses of radioactive levels in blood, urine, skin, and skin washing samples after topical, intravenous, and oral administration. The results of the F344 rat study found

Table 3 Pharmacokinetic Parameters of the Time Course of $[^{14}C]$Hydroquinone Equivalent in the Plasma (or Blood) of Female F344 Rats or Rhesus Monkeys Following Differing Routes of Administration

Parameters	Unit	F344 Rats		Rhesus Monkeys		
		Intravenously	Topically	Intravenously	Topically	Orally
T_{max}	h	—	2.00	—	24.00	2.00
C_{max}	µg-Eq/ml	—	0.04	—	0.0003	0.29
Beta	h^{-1}	0.05	0.02	0.68	0.02	0.24
$T_{1/2}$-beta	h	14.28	45.07	2.00	34.22	12.10
AUC, $0 \rightarrow \infty$	µg-Eq·h/ml	211.75	1.36	1.34	0.01	1.53
MRT, $0 \rightarrow \infty$	h	16.54	61.15	2.20	41.07	24.08
Cl	ml/h	47.22	—	7462.69	—	—
V	ml	944.47	—	10,974.54	—	—

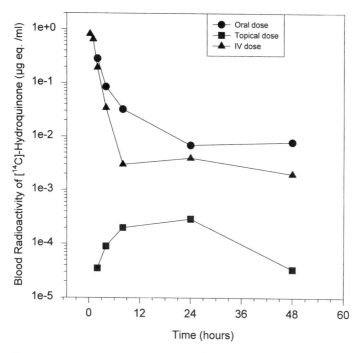

Figure 3 Plasma concentration of [¹⁴C]hydroquinone equivalent in rhesus monkeys following oral, topical, or intravenous administration. Each point represents the mean of four animals.

that intravenously dosed [¹⁴C]hydroquinone was completely eliminated by urinary excretion (96.7 ± 5.37%). In topically treated rats, total urinary recovery was 36.21 ± 5.38%, which was equal to 96% of absorbed dose. This suggested that for F344 rats, urinary excretion is the primary route of [¹⁴C]hydroquinone elimination following intravenous or topical administration. Marty et al. (1981) reported that in hairless rats, urinary excretion of [¹⁴C]hydroquinone was 89 ± 2% and 5.7% of dose 3 days after intravenous and topical administration, respectively. Urinary excretion also is the major elimination route after oral administration in rats. Marty et al. (1981) and Divincenzo et al. (1984) reported that 86 ± 3% or more than 90% of the dose was excreted in urine by 72 or 96 h following oral administration of [¹⁴C]hydroquinone in hairless rats or Sprague-Dawley rats, respectively. The results of these reports also indicated that total fecal excretion of [¹⁴C]hydroquinone, regardless of the route of administration, only accounted for a very small amount of the dose. In contrast, in rhesus monkeys urinary excretion of [¹⁴C]hydroquinone was about 50% of absorbed dose after topical, intravenous, and oral dosing. This suggested that urinary excretion is a major route of elimination but other routes are also important for rhesus monkeys.

The absorption pattern is different between F344 rats and rhesus monkeys. Topically applied [^{14}C]hydroquinone in F344 rats exhibited an early increased rate of percutaneous absorption with T_{max} at 2 h after dosing; then the process might have been limited by hydroquinone formulation. The rate of percutaneous absorption of [^{14}C]hydroquinone in rhesus monkeys was much slower and weaker when compared with that of rats, with T_{max} reached at 24 h. However, for both species topical bioavailabilities are similar, 37.45% for F344 rats and 34.34% for rhesus monkeys. The flux for topical hydroquinone is also the same, 0.6 μg/cm^2/h. Wester et al. (1998) found that following topical application of [^{14}C]hydroquinone (2% cream) in human healthy volunteers in vivo, flux rate was 1.9 μg/cm^2/h (forehead) and 1.0 μg/cm^2/h (forearm). Flux rate in vitro in human dematomed skin was 2.8 μg/cm^2/h. Comparing these data, flux rate for topical hydroquinone in humans is higher than in F344 rats and rhesus monkeys in vivo. Flux rates for rats and monkeys are also different from in vitro data. Barber et al. (1992) reported that absorption rate for hydroquinone through full-thickness F344 rat skin in vitro was 1.09 $\pm$ 0.65 μg/cm^2/h. Lehman and Franz (1992) reported that in vitro percutaneous absorption of hydroquinone was similar in Nemestrina monkeys and human skin, but Fasicularis monkeys absorbed less. These values suggested that different species, different regions of the same species, or different experimental conditions could cause different hydroquinone flux rates following topical application.

Results of this study indicated that in F344 rats the treated site skin quickly reached steady state after topical application. Extending topical exposure time did not increase percutaneous absorption and did not cause an accumulation of [^{14}C]hydroquinone levels at the application site. Marty et al. (1981) reported, that following topical application, radioactivity levels of [^{14}C]hydroquinone in muscle tissue of mice under and around the zone of application and found that [^{14}C]hydroquinone did not accumulate preferably at the site of application. These results suggested that percutanoeus absorption of hydroquinone was slow or weak.

In F344 rats and rhesus monkeys, 96% and 50%, respectively, of the topically absorbed dose was excreted in urine within 24 h. For human, Bucks et al. (1988) reported that a peak of urinary elimination was reached at 12 h after topical dose. Wester et al. (1998) also reported that 36.95 $\pm$ 9.84% (equal to 81.57% of total urinary recovery) was excreted in urine within 24 h after topical application of [^{14}C]hydroquinone on the forehead of human volunteers. This suggests that hydroquinone was rapidly removed by urinary elimination after being absorbed.

A result that should be mentioned is that oral bioavailability of hydroquinone in rhesus monkeys was 100%, which indicated that hydroquinone was completely absorbed after oral dosing. Similar results have been reported by Marty et al. (1981), in which oral bioavailability of [^{14}C]hydroquinone in hairless rats was 91%. Divincenzo et al. (1984) found that total recovery of [^{14}C]hydroquinone from the expired air, urine, feces, and carcasses of male Sprague-Dawley rats was 100% by 96 h after a single oral dose, in which urinary excretion covered 91% to 101% and fecal excretion covered 0.56% to 5.82%.

Hydroquinone risk assessment should include all sources of the chemical. This includes exposure to the skin via industrial chemical use, and exposure in pharmaceutical or cosmetic ingredients. Hydroquinone also is prevalent in many food sources (Deisinger et al., 1996), and since the oral bioavailability approaches 100%, all hydroquinone in food is completely absorbed. A worse-case situation would involve all hydroquinone sources at same time, which is entirely possible.

REFERENCES

Barber, E. D., Hill, T., and Schum, D. B. (1995). The percutaneous absorption of hydroquinone (HQ) through rat and human skin in vitro. Toxicol. Lett., 80: 167–172.

Bucks, D. A. W., McMaster, J. R., Guy, R. H. and Maibach, H. I. (1988). Percutaneous absorption of hydroquinone in humans: Effect of 1-dedecylazacycloheptan-2-one (azpne) and the 2-ethylhexyl ester of 4-(dimethylamino)benzoic acid (escalol 507). J. Toxicol. Environ. Health, 24:279–289.

Deisinger, P. J., Hill, T. S., and English, J. C. (1996). Human exposure to naturally occurring hydroquinone. J. Toxicol. Environ. Health, 47:31–46.

Devillers, J., Boule, P., Vasseur, P., Prevot, P., Steiman, R., Seigle-Murandi, E., Benoit-Guyod, J. L., Nendza, M., Grioni, D., Dive, D., and Chambon, P. (1990). Environmental and health risks of hydroquinone. Ecotoxicol. Environ. Safety, 19:327–354.

Divincenzo, G. D., Hamilton, M. L., Reynolds, R. C., and Ziegler, D. A. (1984). Metabolism and disposition of [^{14}C]-hydroquinone given orally to Sprague-Dawley rats. Toxicology, 33:9–18.

Lehman, P. A. and Franz, T. J. (1992). Percutaneous absorption of hydroquinone in human, pig, monkey, and baboon. Abstracts for the 1992 Annual Meeting of the Society for Investigative Dermatology, abstr. 589.

Marty, J. P., Trouvin, J. H., Jacquot, C., and Wepierre, J. (1981). Pharmacocinetique percutanée de l'hydroquinone ^{14}C. C. R. Congr. Eur. Biopharm. Pharmacokinet. Inst. 2:221–228.

National Toxicology Program. (1989). Toxicology and carcinogenicity studies of hydroquinone in F344/N rats and B6C3F1 mice (gavage). Technical Report Service report no. 366, NIH publication no. 90-2821.

Rowland, J., and Tozer, T. N. (1995). Absorption and disposition kinetics. In Clinical Pharmacokinetics, Concepts and Applications, eds. M. Rowland and T. N. Tozer, 3rd ed., William & Wilkins, Media, PA, pp. 11–52.

Wester, R. C. and Maibach, H. I. (1993). Animal models for percutaneous absorption in Topical Drug Bioavailability, Bioequivalence, and Penetration, eds. V. P. Shah and H. I. Maibach, Plenum Press, New York, pp. 333–349.

Wester, R. C., Melendres, J., Hui, X., Wester, R. M., Serranzana S., Zhai H., Quan, D., and Maibach, H. I. (1998). Human in vivo and in vitro hydroquinone topical bioavailability. J. Toxicol. Environ. Health (in press).

53
Dermatopharmacokinetics of Topical BCX-34 Cream in Human Skin

Xiaoying Hui, Ronald C. Wester, Steffany Serranzana,* and Howard I. Maibach
University of California School of Medicine, San Francisco, California

Tacey X. Viegas, Gerald M. Walsh, and George A. Omura
BioCryst Pharmaceuticals, Inc., Birmingham, Alabama

I. INTRODUCTION

BCX-34, 9-[3-pyridinylmethyl]-9-deazaguanine, or peldesine, is a small inhibitor that specifically acts on purine nucleoside phosphorylase (PNP) (Montgomery et al., 1993; Walsh et al., 1994). The enzyme activity of PNP is necessary for DNA replication and proliferation of human T cells, but apparently not for other cells (Bantia et al., 1996). Since mechanisms of several dermatologic diseases, such as psoriasis, cutaneous T-cell lymphoma, and allergic contact dermatitis, involve DNA replication and/or proliferation of human T-cells, control of PNP's activity may be an important aspect of treating these diseases. BCX-34 exhibits a complete inhibitory effect on the proliferation of T cells in an in vitro system, whereas no influence was observed in B-cell lines (Bantia et al., 1996). Previous clinical studies showed that, after topical application of BCX-34 dermal cream, little systemic absorption was observed in normal human subjects (unpublished observation, BioCryst, 1995). Preliminary clinical studies have suggested histologic and clinical improvement in psoriasis and cutaneous T-cell lymphoma (Walsh et al., in press).

**Present affiliation*: Penederm Inc., Foster City, California.

However, information about quantitative delivery of BCX-34 in dermal cream formulations to human skin and the retention time of BCX-34 in the skin has been lacking. This study defines the absorption and retention time from dermal formulations. The results should provide a rational dose (concentration) of BCX-34 cream and treatment interval for further clinical trials.

II. MATERIALS AND METHODS

A. Materials

^{14}C-Labeled BCX-34 dermal cream formulations were provided by BioCryst Pharmaceuticals, Inc. (Birmingham, AL) at concentrations of 0.1, 0.3, 1, 3, and 5% (w/w) doses. These creams are simple oil-in-water (o/w) emulsions containing the triethanolamine–stearate emulsifier with no added penetration enhancer; other notable excipients are cetyl alcohol, stearyl alcohol, and 10% glycerin. The purity of nonlabeled BCX-34 was 98.9%. The specific activity of ^{14}C-labeled BCX-34 was 55.9 mCi/235 mg and the radiochemical purity was 97.7%. The radioactivity of all concentrations of cream was 1 μCi/ 5 mg cream.

B. Human Subjects

Eight normal, healthy volunteers, male and females, ages 26 to 60 years, without preexisting dermatologic disease, were selected for study. Four human subjects were used for rational dose determination and the others for retention time determination. The study was approved by the Committee on Human Research at University of California, San Francisco, and each human subject provided written informed consent.

C. In Vivo Human Study

For rational dose determination, a single topical application of 0.1, 1, 3, and 5% creams was made unoccluded, at the same dose (5 mg cream containing 1 μCi [^{14}C]-BCX-34/cm^2 of surface area), on four 1.0-cm^2 skin sites on both forearms. All marked sites were located within the region on the ventral forearm that was 2–3 cm above the wrist and 2–3 cm below the antecubital fossa. Dosing sites were located on the skin surface such that marked sites were 3 cm (center to center) apart. Each marked site received a single application of cream with a single concentration of BCX-34 using a weighing spatula. The delivered dose was quantitated by weight difference of the weighing spatula before and after dosing. The subjects were requested not to touch or wash the dosed area until the planned skin washing and stripping. Six or 24 h after the topical application, the dosed sites on the left or right forearm, respectively, were washed followed by tape stripping.

For retention time determination, a single topical application of 5% BCX-34 cream was made unoccluded, at the same dose (5 mg cream containing 1 μCi [^{14}C]-BCX-34/cm^2 of surface area), on five 1.0-cm^2 skin sites on the left forearm. The location of the dosing sites, the distance of each site, and the method of dosing were the same as described above. Twenty-four hours after dosing, the dosed sites were washed. The tape stripping was conducted at set times after skin washing. Each site (dosed site number 1, 2, 3, 4, and 5) was stripped at 0, 1, 3, 6, and 24 h after washing, respectively, corresponding to 24, 25, 27, 30, and 48 h after dosing, respectively.

At the designated time after dosing, the dosed site was washed with a cotton application with 50% liquid soap (Ivory Liquid soap, Procter & Gamble, Cincinnati, OH) and water five times. Immediately after skin washing or at the designated time, the dosed skin site was stripped with cellophane tape (3M Commercial Office Supply Division, St. Paul, MN) 10 times. These samples were then individually placed in a borosilicate glass vial with 15 ml scintillation cocktail and assayed for radioactivity.

D. In Vitro Study

Dermatomed human skin $\sim$500 μm thick was obtained from the Northern California Transplant Bank. All skin sources were stored in Eagle's minimum essential medium (MEM) with Earle's BSS and used within 7 days postmortem. For each skin source, the test materials were added to one replicate of skin $\sim$1.0 cm^2, contained within a flow-through diffusion cell apparatus and run for 4 or 24 h. Receptor fluid, phosphate-buffered saline (PBS, Sigma Chemical Company, St. Louis, MO), was collected every 0.5 h for the 4-h study and every 2 h for the 24-h study. When the study was finished, the flow-through diffusion cell apparatus was disassembled and the surface of the skin was washed with 50% liquid soap and water five times to analyze any materials not absorbed or bound to the skin. After washing, the skin surface was tape-stripped 10 times. The residual skin sample was then digested using Soluene 350 tissue solubilizer (Parkard Instrument Company, Inc., Meriden, CT) at 50°C for 48 h.

E. Scintillation Counting

All nondigested samples used Universal-ES (ICN Biomedicals, Costa Mesa, CA). The digested skin samples were mixed with 15 ml of Hionic-Fluor cocktail (Parkard Instrument Company Inc., Meriden, CT) to measure radioactivity. Background control samples and test samples were counted in a

Packard model 1500 liquid scintillation counter (Packard Instruments). Radioactivity of each sample, percent of administered dose, and micrograms BCX-34 equivalent in each sample were determined.

F. Data Analysis

Statistical analyses of the data were performed with Sigma-Stat 1.0 for Windows using a t-test and one-way analysis of variance. Noncompartment pharmacokinetic parameters were determined using data from the receptor fluids of each in vitro skin source versus time by using the program PCNONLIN (Scientific Consulting, Inc., Cary, NC).

III. RESULTS

A. In Vivo Human Study

Table 1 gives the amount of residual BCX-34 cream removed from the dosed skin site by skin washing. After 6-h dermal exposure periods, radioactivity as percent dose remaining on the surface of the dosed site skin ranged from 53.41% to 90.32%. Increasing exposure time to 24 h did not significantly change the amount of removable material when compared with that of 6 h ($p > .05$).

Figure 1 shows the amount of BCX-34 in the stratum corneum measured by the dosed skin site tape stripping. After 6-h exposure periods, the average micrograms BCX-34 equivalent in the stratum corneum for 0.1% BCX-34 cream was 0.08 ± 0.05 μg/cm^2 (mean $\pm$ SD). Increasing the concentration of BCX-34 cream to 1% and 3% significantly increased the amount of BCX-34 to 0.42 ± 0.21 and 0.86 ± 0.13 μg/cm^2, respectively ($p < .05$). Further increasing the concentration of BCX-34 cream to 5% did not increase the amount in the stratum corneum. After application of 5%

Table 1 Radioactivity as Percent Dose of Skin Washing at 6 h and 24 h After Topical Application In Vivo

BCX-34 cream (%)	Radioactivity of skin washing sample	
	6 h	24 h
0.1	53.41 ± 12.33	49.88 ± 19.49
1	75.71 ± 11.52	70.16 ± 17.70
3	84.40 ± 9.39	64.78 ± 25.13
5	90.32 ± 11.62	64.43 ± 27.72

Note. Each value represents the mean $\pm$ SD ($n = 4$).

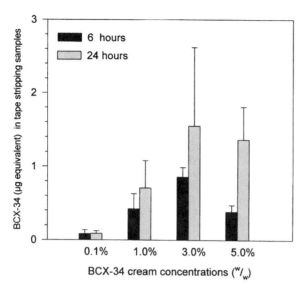

Figure 1 Comparison of amount of BCX-34 (μg equivalent) in the stratum corneum and dermal exposure periods following application of BCX-34 creams at 0.1%
to 5.0% dose levels in human skin in vivo. Each bar represents the mean ± SD (n
= 4).

BCX-34 cream, the microgram BCX-34 equivalent level in the stratum corneum was similar to those of 1% and 3% creams. After increasing dermal
exposure time to 24 h, the amount of BCX-34 in stratum corneum did not
significantly change at each dose level except for the 5% cream. The average
microgram BCX-34 equivalent level at 5% dose level after 24 h increased
significantly by 3.6-fold compared to the 6-h exposure period ($p < .05$), but
was similar to that of the 3% dose level at 24 h.

Table 2A shows that the range of average radioactivity of skin washing
from five dosed skin sites was 58.58% to 70.53% of administered dose at
24 h after topical application of 5% BCX-34 cream. The amount retained
in the stratum corneum was measured by tape stripping at 0, 1, 3, 6, and
24 h after skin washing (Table 2B). The highest BCX-34 as percent dose in
the stratum corneum was found in samples tape stripped immediately
following skin washing (0 h after skin washing or 24 h after dosing) as
1.31 ± 0.92% of dose. Intermediate levels were observed in those taken 1,
3, and 6 h after skin washing: 0.60 ± 0.30%, 0.64 ± 0.33%, and 0.56 ±
0.26% of dose, respectively. The lowest was observed in samples taken at
24 h after skin washing as 0.30 ± 0.13% of dose. Statistical analysis showed

Table 2 Radioactivity of Skin-Washing and Tape-Stripping Samples as Percent of Dose After Topically Dosing with 5% BCX-34 Cream In Vivo

A. Skin washing (24 h after dosing)

Skin site	Radioactivity (% dose)
1	58.58 ± 14.84
2	65.15 ± 9.06
3	68.94 ± 11.00
4	70.53 ± 16.13
5	66.12 ± 26.67

B. Tape stripping

Time after dosing (h)	Time after washing (h)	Radioactivity (% dose)
24	0	1.31 ± 0.92
25	1	0.60 ± 0.30
27	3	0.64 ± 0.33
30	6	0.56 ± 0.26
48	24	0.30 ± 0.13

Note. The skin site was treated with 5% BCX-34 cream. All dosed skin sites were washed at 24 h after topical dosing. Tape stripping was conducted at set time periods after skin washing. Each value is a mean ± SD ($n = 4$).

that those samples taken either at 0 h or at 24 h after washing were significantly different from the others ($p < .05$).

Figure 2 shows the distribution of BCX-34 (μg equivalent) in stratum corneum measured at different time after drug removed. The amount of μg BCX-34 equivalent in the first five-time stripped stratum corneum samples (representing the upper layer of the stratum corneum) was higher than that taken by the next five-time stripped samples (lower layer of the stratum corneum) ($p < .05$).

B. In Vitro Study

Table 3 gives the in vitro percutaneous absorption of BCX-34 in human skin following topical application of different concentrations of BCX-34 cream after 24-h exposure periods. The majority of the dose was removed by skin surface washing ($84.25 \pm 16.46\%$ to $93.89 \pm 4.96\%$). Recovery of drug in

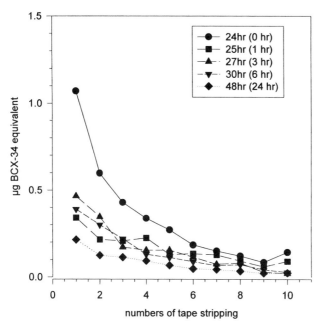

Figure 2 Average BCX-34 (μg equivalent) in the stratum corneum from tape stripping samples following topical application of 5% BCX-34 cream in human in vivo. The samples were taken at 24 to 48 h after dosing (0 to 24 h after skin washing). Each symbol represents the mean of four samples.

Table 3 In Vitro Percutaneous Absorption of BCX-34 in Human Skin After 24-h Exposure Periods

BCX-34 creams (%)	Percent dose absorbed			
	Receptor fluid	Skin	Surface Washing	Total
0.1	0.02 ± 0.02	1.31 ± 0.67	93.89 ± 4.96	95.22 ± 4.89
0.3	0.02 ± 0.01	0.56 ± 0.67	87.63 ± 12.69	88.37 ± 12.05
1	0.03 ± 0.03	0.45 ± 0.47	91.46 ± 6.59	91.94 ± 6.09
3	0.02 ± 0.01	0.33 ± 0.45	84.25 ± 16.46	84.60 ± 16.86
5	0.01 ± 0.00	0.14 ± 0.11	87.42 ± 3.14	87.57 ± 3.04

Note. Each value is a mean ± SD ($n = 3$). Chemical mass expressed as the mean value of micrograms (μg) BCX-34 equivalent associated with 1-cm^2 area of skin ($n = 3$).

the skin digests and receptor fluid was only $0.14 \pm 0.11\%$ to $1.31 \pm 0.67\%$ and $0.01 \pm 0.005\%$ to $0.02 \pm 0.02\%$ of the applied dose, respectively. Overall dose accountability was $84.60 \pm 16.86\%$ to $95.22 \pm 4.89\%$.

The lower dose concentration seemed the more efficient (higher percent dose absorbed) in delivering BCX-34 into human skin. However, percent dose absorbed only represents the efficiency of absorption. The mass of BCX-34 absorbed, expressed as μg BCX-34 equivalent/cm^2 skin area, gave the actual amounts of BCX-34 delivered into and through the skin. Figure 3 shows that 3% and 5% dose concentrations gave higher amounts of BCX-34 in the receptor fluid than did 0.1% and 0.3% concentrations ($p < .05$).

Table 4 gives the distribution of absorbed BCX-34 after 4-h or 24-h exposure periods. Whether concentrations are higher or low, BCX-34 (μg equivalent) was higher in the skin samples than that in receptor fluid samples ($p < .05$). Samples taken at 4 h were similar to those at 24 h for the corresponding dose level. Table 5 further shows that the concentration gradient of absorbed BCX-34 was from the upper layer of stratum corneum, to the lower layer of stratum corneum, to the dermis.

IV. DISCUSSION

An effective topical drug therapy for cutaneous disease requires drug uptake into a desired skin layer at sufficient concentrations over a particular period of time for maximal pharmacological activity. In this study, human skin absorption, distribution, and retention time of BCX-34 creams were examined with various concentrations and dermal exposure time under in vivo and/or in vitro conditions.

The skin washing method offers an effective evaluation of residual drug formulation on the surface of dosed site skin. The in vivo data show that without employing any protective method after topical application, 50% or more of the administered dose remained on the skin surface and there was no difference between 6-h or 24-h exposure periods ($p > .05$). This is closer to the in vitro data, in which 84.25% to 93.89% of dose was recovered with skin washing after a 24-h exposure. The results indicated the persistence of BCX-34 cream on human skin.

Quantitation of drug uptake into dosed human skin (stratum corneum) was measured by the tape stripping method. Results of in vivo data (Table 1) suggest that the concentration of BCX-34 cream is an important factor affecting the amount of BCX-34 penetrating into the stratum corneum. A dose of 0.1% probably would not be clinically relevant, but a 1% dose could be. An increase to 5% concentration would not increase absorption. The optimal concentration of BCX-34 cream for skin absorption in the pres-

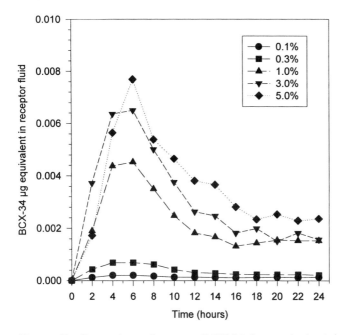

Figure 3 Comparison of average BCX-34 (μg equivalent) in the receptor fluid and incubation time following application of BCX-34 at 0.1% to 5.0% dose levels in human skin in vitro. Each symbol represents the mean of three samples.

ent study is in the 1–3% range. This is supported by the in vitro study showing that the mass of BCX-34 in the skin at the 0.1% and 0.3% dose levels was significantly lower than at the 3% and 5% dose levels ($p < .05$). For dose levels of 1%, 3%, and 5%, no statistical difference in BCX-34 skin content was observed ($p > .05$).

The dermal exposure period required for absorption was also investigated. Dermatopharmacokinetic analysis (Table 6) shows that the time for reaching the peak of skin absorption was about 6 h after topical application in vitro and did not show statistical differences among various BCX-34 dose levels ($p > .05$). In vivo data demonstrated that extending exposure periods from 6 h to 24 h in vivo did not increase the amount of drug absorption significantly ($p > .05$) at most dose levels except 5% BCX-34 (Table 1).

The distribution of BCX-34 after topical absorption was examined by tape stripping and/or skin digests. As shown in Table 3, only small amounts of BCX-34 were absorbed into and through the skin. Further analysis shows that BCX-34 was retained mainly in the upper layer of the stratum corneum when compared with that of dermis (Table 4). The in vivo data indicated

Table 4 BCX-34 as µg Equivalent Absorbed into and Through the Human Skin In Vitro

| BCX-34 creams (%) | BCX-34 (µg-Eq) absorbed | | | | | |
| | Skin | | | Receptor fluid | | |
	4 h	24 h		4 h	24 h	
0.1	0.1675 ± 0.0542	0.1100 ± 0.0600		0.0028 ± 0.0029	0.0020 ± 0.0018	
0.3	0.1730 ± 0.0563	0.2087 ± 0.1966		0.0060 ± 0.0034	0.0050 ± 0.0033	
1	1.1945 ± 1.2707	0.4881 ± 0.5370		0.0173 ± 0.0084	0.0309 ± 0.0313	
3	0.7177 ± 0.1495	0.9938 ± 1.3486		0.0349 ± 0.0262	0.0441 ± 0.0181	
5	1.3974 ± 1.0128	0.7900 ± 0.6618		0.0722 ± 0.0284	0.0490 ± 0.0097	

Note. Values are mean ± SD (n = 3).

Table 5 Distribution of BCX-34 as Percent Dose in the Human Stratum Corneum and Dermis In Vitro

Sample name	Exposure time (h)	Sample size (number)	Concentration of BCX-34 creams (%)				
			0.1	0.3	1	3	5
Stratum corneum	4	1	0.18	0.43	0.09	0.02	0.03
Dermis	4	1	0.44	0.12	0.04	0.02	0.02
Stratum corneum	24	1	1.13	0.13	0.20	0.06	0.08
Dermis	24	1	0.18	0.03	0.03	0.04	0.01
Stratum corneum	24	1	0.97	0.34	0.13	0.11	0.11
Dermis	24	1	0.56	0.14	0.05	0.15	0.06

Note. Dermatomed skin was dosed with BCX-34 creams. After 4-h or 24-h exposure periods, the skin was washed followed by tape stripping. The tape sample was counted to calculate radioactivity in the stratum corneum. The remaining skin was digested and counted to calculate radioactivity.

Table 6 Pharmacokinetic Parameters of BCX-34 Through Human Skin In Vitro

BCX-34 creams (%)	Pharmacokinetic parameters		
	T_{max} (h)	C_{max} (μg BCX-34 Eq)	AUC_{0-24} (μg-Eq $\cdot$ h)
0.1	6 ± 2	0.0002 ± 0.0002	0.0032 ± 0.0026
0.3	6 ± 2	0.0008 ± 0.0005	0.0090 ± 0.0054
1	6 ± 2	0.0047 ± 0.0049	0.0537 ± 0.0535
3	5 ± 1	0.0071 ± 0.0025	0.0767 ± 0.0306
5	7 ± 3	0.0085 ± 0.0008	0.0881 ± 0.0157

Note. Values were calculated using BCX-34 μg equivalent in receptor fluid after 24 h topical exposure of BCX-34 creams in vitro. Pharmacokinetic parameters were calculated by using the program PCNONLIN. T_{max} was the time to reach maximum amount of BCX-34 μg equivalent in the receptor fluid. C_{max} was the concentration of BCX-34 μg equivalent at steady state. AUC_{0-24} was area under the curve from zero time to 24 h. Values are mean ± SD ($n = 3$).

that the mass of BCX-34 in the first five strippings was higher than that of the next five (Fig. 3).

Tape stripping of dosed human stratum corneum at various intervals over time post drug removal offers a way to quantify drug elimination (retention time) from the skin. The in vivo data show that there were two phases of BCX-34 elimination from the stratum corneum (Fig. 4). First the amount of BCX-34 in the stratum corneum was quickly reduced. One hour after the drug removal, the amount in the stratum corneum was only one-half of the amount at 0 time after removal. At 3 h, the elimination rate was slower, and at 48 h after dosing (24 h after skin washing), it was one-fourth the amount of BCX-34 at 24 h ($p < .05$). The pattern of retention time of BCX-34 in the stratum corneum could be mathematically calculated as follows:

$$Y = 70.1 - 36.6X + 2.97 \exp(X) \qquad R^2 = .99$$

where Y is the calculated retention time of BCX-34 (µg equivalent) in the stratum corneum and X is dermal exposure time (h) after application of BCX-34 cream (5% w/w).

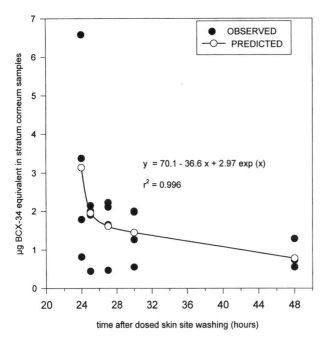

Figure 4 Amounts of µg BCX-34 equivalent in stratum corneum samples after topical application of 5% BCX cream in vivo versus time after skin washing. Each symbol represents the mean of µg BCX-34 equivalent at each time point.

In summary, following topical application of BCX-34 creams, an equilibrium was reached within 6-h exposure periods. Extending the dermal exposure time did not increase the drug penetration except at the 5% dose level. Concentrations of BCX-34 creams at 1%, 3%, and 5% had a similar skin absorption rate significantly higher than that of lower dose levels (0.1% and 0.3%) in vivo and in vitro. Only 1% or less of the applied dose was absorbed into and through the human skin. The highest amount of absorbed BCX-34 was found in the upper layer of stratum corneum, followed by the lower layer of stratum corneum and dermis. In vitro data show that in receptor fluid (equals systemic absorption), very little (less than 0.06% of dose) BCX-34 was detected. Low systemic absorption can equate with lower side effects; therefore, a topical BCX-34 dose of 1–3% for 6 h using the supplied o/w cream formulation should maximize skin concentration and minimize potential side effects.

REFERENCES

Bantia S, Montgomery JA, Johnson HG, Walsh GM. In vivo and in vitro pharmacologic activity of the purine nucleoside phosphorylase inhibitor BCX-34: The role of GTP and dGTP. Immunopharmacology 35(1996):53–63.

BioCryst Pharmaceuticals, Inc. Investigator's Brochure: BCX-34 (1995).

Montgomery JA, Synder HW Jr, Walsh DA, Walsh GM. BCX-34, purine nucleoside phosphorylase (PNP) inhibitor. Drugs of the Future 18(1993):887–890.

Walsh GM, Reddy NS, Bantia S, Babu YS, Montgomery JA. Development of inhibitors of purine nucleoside phosphorylase. Hematology Reviews 8(1994): 87–97.

Walsh GM, Omura GA, Montgomery JA, Franz TJ, Viegas TX, Cook WJ, Bantia S, Reddy N, Sams WM Jr, Bartolucci AA, Maibach HI. Purine nucleoside phosphorylase inhibition: a novel therapy for psoriasis. In: Roenigk H, Maibach H, editors. Psoriasis, 3rd ed. New York: Marcel Dekker, in press.

54

In Vitro Cutaneous Disposition of a Topical Diclofenac Lotion in Human Skin
Effect of a Multidose Regimen

Philip G. Hewitt
University of California School of Medicine,
San Francisco, California, and E. Merck, Darmstadt, Germany

Nicholas Poblete, Ronald C. Wester and Howard I. Maibach
University of California School of Medicine,
San Francisco, California

J. Zev Shainhouse
Dimethaid Research, Inc., Markham, Ontario, Canada

I. INTRODUCTION

Diclofenac, a nonsteroidal anti-inflammatory drug, has been widely used in the treatment of rheumatoid arthritis and osteoarthritis. However, oral delivery of this drug poses certain disadvantages, such as fast first-pass metabolism and adverse side effects (including gastrointestinal reactions and idiosynchratic drug reactions). Therefore, alternative routes of administration have been sought. The skin has become increasingly important to this effect, and many drugs have been formulated in transdermal delivery systems, including diclofenac itself (1–4). However, diclofenac sodium is not easily absorbed through the skin due to its hydrophilic nature (5). Much work has concentrated on using percutaneous absorption enhancers or cosolvents to increase penetration (6,7). A new diclofenac sodium lotion has been developed for topical application, which includes the absorption enhancer dimethyl sulfoxide (DMSO). It is expected that the addition of DMSO may increase the in vivo permeation rate of diclofenac through the skin into the deeper target tissues beneath the skin.

This study determines comparative bioavailability of diclofenac sodium lotion compared to an aqueous solution of diclofenac sodium after topical application to viable human skin in vitro. In addition, the difference between a single dose and multiple doses (8 times over 48 h) was also determined.

II. MATERIALS

[^{14}C]Diclofenac sodium was purchased from by Amersham Life Sciences and further purified by Wizard Laboratories (Sacramento, CA). Specific activity was 9.76 mCi/mmol and radiochemical purity was 99.0%. The Pennsaid lotion (lot number 977132), was provided by Dimethaid Research, Inc. (Markham, Ontario, Canada), with diclofenac sodium being removed. Lactate diagnostic kit was obtained from Sigma Chemical Co. (St. Louis, MO). All other chemicals were of analytical grade from standard suppliers.

III. METHODS

A. Dose Preparation

For the lotion, 6 mg of [^{14}C]diclofenac sodium was added directly to 400 μl lotion, giving a final concentration in the formulation of 1.5%.

To make the aqueous solution, 6 mg of [^{14}C]diclofenac sodium was added directly to 360 μl high-performance liquid chromatography (HPLC) grade water (40 μl ethanol was required to dissolve the solid diclofenac), giving a final concentration in the formulation of 1.5%.

B. Test System

Human cadaver skin was dermatomed prior to delivery, using a Padgett Electrodermatome, to a target thickness of 500 μm. Each skin sample was placed in Eagle's minimum essential media (MEM) and stored refrigerated at $1 \pm 4°C$, until use within 5 days after death. This preservation/use regimen follows that used by the human skin transplant bank (8) and the work of Bronaugh et al. (9). Wester et al. (10) has shown skin to be viable using this storage regimen.

C. Dose Application

The formulations are identified as A, B, C, D, E, F, G, H, and I. Distinctions among these are as follows:

1. 2 μl 1.5% (1000 ppm) [^{14}C]diclofenac sodium lotion. Single dose to 1 cm^2 skin area.
2. 5 μl 1.5% (1000 ppm) [^{14}C]diclofenac sodium lotion. Single dose to 1 cm^2 skin area.
3. 2 μl 1.5% (1000 ppm) [^{14}C]diclofenac sodium in HPLC-grade water (+10% ethanol). Single dose to 1 cm^2 skin area.
4. 5 μl 1.5% (1000 ppm) [^{14}C]diclofenac sodium in HPLC-grade water (+10% ethanol). Single dose to 1 cm^2 skin area.
5. 2 μl 1.5% (1000 ppm) [^{14}C]diclofenac sodium lotion. Multiple doses to 1 cm^2 skin area, at times 0, 4, 8, 12, 24, 28, 32, and 36 h.
6. 5 μl 1.5% (1000 ppm) [^{14}C]diclofenac sodium lotion. Multiple doses to 1 cm^2 skin area, at times 0, 4, 8, 12, 24, 28, 32, and 36 h.
7. 2 μl 1.5% (1000 ppm) [^{14}C]diclofenac sodium in HPLC-grade water (+10% ethanol). Multiple doses to 1 cm^2 skin area, at times 0, 4, 8, 12, 24, 28, 32, and 36 h.
8. 5 μl 1.5% (1000 ppm) [^{14}C]diclofenac sodium in HPLC-grade water (+10% ethanol). Multiple doses to 1 cm^2 skin area, at times 0, 4, 8, 12, 24, 28, 32, and 36 h.
9. Viability control (glucose metabolism to lactate).

The diclofenac sodium lotion (formulations A and B) contains 45% DMSO, 11% glycerin, 11% ethanol, 11% propylene glycol, and water, and this vehicle was the same as the formulated product, without diclofenac.

D. Diffusion Study

The human skin was clamped into a continuous-flow diffusion cell in a flow-through diffusion cell system (11). The glass cells had a 1-cm^2 surface area of exposed skin, which was maintained at approximately 32°C using a Lauda heating circulator (Lauda, West Germany). The receptor fluid, MEM, with 50 μg/ml gentamicin sulfate added to ensure sterility during the running of the experiment, was pumped at a rate of 3 ml/h. Receptor fluid samples were collected every 4 h up to 48 h.

Forty-eight hours after dose application, skin surface wash was performed on all skin samples (washed once with 50% Ivory Liquid soap/Nanopure water, and then rinsed twice with Nanopure water). Wash and rinse samples for individual skin surfaces were combined.

After skin surface washing, 10 consecutive cellophane tape strips were performed on all skin samples (except for the lactate viability control skin) and added to a separate vial. The remaining skin was placed in separate scintillation vials and digested by adding 4 ml of Packard Soluene 350 for 7–10 days at room temperature.

Duplicate aliquots (500–1000 μl) of the receptor fluid fractions were taken and assayed for radioactivity by liquid scintillation spectrometry (Packard 1500 counter), after addition of 15 ml scintillation cocktail (UniverSol, ICN). All other samples (tape strips, skin digest solutions, and skin surface washes) were also counted for radioactivity after addition of 10–15 ml UniverSol.

E. Lactate Assay

An assay for the energy metabolism, by the donor skin, of glucose to lactate in the receptor fluid perfusate (which contains glucose) was used to verify that individual skin sources were viable. One undosed, viability-control, penetration cell was run concurrently, specifically for this purpose. Receptor fluid samples from this cell were collected every 4 h up to 48 h and were frozen immediately to be analyzed. Samples were measured for lactate using a Sigma Diagnostic kit. A lactate calibration standard curve was constructed using the kit and MEM, and measured at 340 nm using a Hitachi spectrophotometer.

F. Statistical Analysis

Statistical analysis was performed using Student's t-test (both unpaired and paired). The limited number of skin samples precluded the use of more sophisticated tests, such as covariance.

IV. RESULTS

A. Percutaneous Absorption of Diclofenac Sodium

The in vitro percutaneous absorption values (receptor fluid + stratum corneum + skin residue) of diclofenac as mass (μg) absorbed for single and multiple doses are given in Table 1 (lotion) and Table 2 (aqueous solution), as well as Figs. 1 and 2, and the statistical summary for total absorption is given in Table 3. For single doses (2 μl vs. 5 μl) there was no statistical difference for lotion (A vs. B) and aqueous solution (C vs. D), nor was there any statistical difference between Pennsaid lotion and aqueous solution for the single dose (A vs. C; B vs. D). Multiple dosing was superior to single dosing (A vs. E; B vs. F; D vs. H) except for 2 μl aqueous solution (C vs. G). The absorption from lotion was statistically greater than from aqueous solution ($p = .005$; $p = .04$) for 2 μl/cm^2 and 5 μl/cm^2 multiple dosing, respectively, showing that a constituent of the lotion enhanced human skin absorption of diclofenac when used in a multidose regimen.

Table 1 In Vitro Percutaneous Absorption of Diclofenac from Pennsaid Lotion: Multiple Doses versus Single Dose

Parameter measured	A (30 μg, single)	B (75 μg, single)	E (240 μg, multiple)	F (600 μg, multiple)
Receptor fluid	3.4 ± 3.5	9.8 ± 12.4	4.9 ± 3.4	11.5 ± 11.3
Skin residue	3.3 ± 2.3	7.6 ± 5.5	10.5 ± 6.2	38.0 ± 24.0
Stratum corneum	3.5 ± 2.8	8.9 ± 4.9	24.8 ± 12.0	36.0 ± 30.5
Total absorption	10.2 ± 6.7	26.2 ± 17.6	40.1 ± 17.6	85.6 ± 41.4

Note: Values are mass (μg) recovered, mean ± SD, $n = 5$. Doses shown for formulations are total dose applied to skin.

B. Dose Accountability

The human skin surface was washed at the end of the experiment (48 h) to remove any residual radioactivity that did not penetrate the stratum corneum. Tables 4 and 5 give the data as percent of applied dose, including skin residue, skin surface wash, and dose accountability. The recovery from the skin surface ranged from 57.8 ± 26.3% to 94.2 ± 19.1%. Total recovery of radioactivity was good for the single-dose study, ranging from 92.8 ± 18.6% to 110.6 ± 15.1%. For the multidose study, recovery from lotion was also good (100.7 ± 7.1% and 90.6 ± 3.0%); however, only 82.0 ± 9.7% and 76.8 ± 8.2% was recovered when aqueous diclofenac was applied.

Table 2 In Vitro Percutaneous Absorption of Diclofenac from an Aqueous Solution: Multiple Doses versus Single Dose

Parameter measured	C (30 μg, single)	D (75 μg, single)	G (240 μg, multiple)	H (600 μg, multiple)
Receptor fluid	2.3 ± 2.4	3.0 ± 2.4	0.8 ± 0.7	5.8 ± 4.1
Skin residue	4.3 ± 0.8	3.8 ± 3.0	2.6 ± 1.5	8.5 ± 5.5
Stratum corneum	1.8 ± 0.9	5.5 ± 2.4	6.1 ± 4.3	21.4 ± 18.7
Total absorption	8.3 ± 1.5	12.3 ± 5.658	9.4 ± 2.9	35.7 ± 19.0

Note: Values are mass (μg) recovered, mean ± SD, $n = 4$ or 5. Doses shown for formulations are total dose applied to skin.

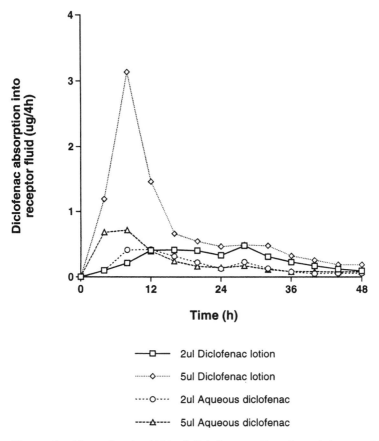

Figure 1   Absorption (μg/4 h) of diclofenac sodium through human skin in vitro after single dose of diclofenac lotion and aqueous diclofenac. Mean, $n = 4$ or 5.

C.   Steady-State Flux and K_p Values

Flux values are calculated during steady-state absorption into the receptor fluid, and are expressed a mass transfer/unit area of skin/unit time (namely, μg/cm^2/h). The apparent permeability coefficient (K_p), of particular value since it is independent of the applied dose, is calculated from the following equation: K_p = absorption (μg)/[surface area $\times$ concentration (μg/cm^3) $\times$ time). Table 6 shows the data calculated for all formulations. As observed previously, applying a single dose of diclofenac (in lotion or as an aqueous solution) had not effect on either the flux or partition coefficient. However, multiple doses of lotion significantly increased the penetration of diclofenac compared to application in an aqueous vehicle.

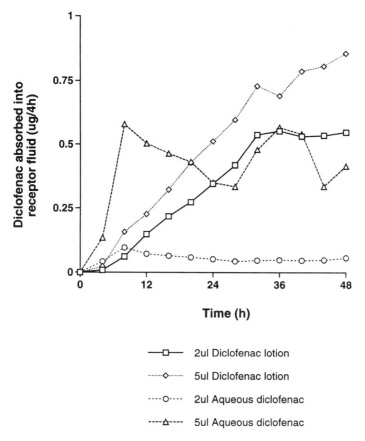

Figure 2 Absorption (μg/4 h) of diclofenac sodium through human skin in vitro after multiple doses of diclofenac lotion and aqueous diclofenac. Mean, n = 4 or 5.

D. Viability Control/Lactate Production

The calibration curve obtained for lactate showed a correlation coefficient of 1.00, and data for each skin source are represented in Fig. 3. Lactate was produced for each skin source (and hence it was viable) until approximately 32 h.

V. DISCUSSION

The in vitro percutaneous absorption of diclofenac sodium, applied as two different formulations (i.e., lotion and aqueous) and as a single- or multidose regimen, has been studied in viable human skin. Penetration of diclofenac

Table 3 Statistical Summary

Treatment	Statistic (p)[a]
Diclofenac lotion	
A vs. B, single dose	.09[b]
E vs. F, multiple doses	.05[c]
A vs. E, single vs. multiple	.007[c]
B vs. F, single vs. multiple	.02[c]
Diclofenac aqueous solution	
C vs. D, single dose	.16[b]
G vs. H, multiple doses	.02[c]
C vs. G, single vs. multiple	.42[b]
D vs. H, single vs. multiple	.03[c]
Diclofenac lotion vs. Aqueous	
A vs. C, single dose	.57[b]
B vs. D, single dose	.13[b]
E vs. G, multiple doses	.005[c]
F vs. H, multiple doses	.04[c]

[a]Student's t-test.
[b]Nonsignificant.
[c]Statistically significant.

sodium from the lotion vehicle through and into human skin was extensive after a single dose, reaching approximately 35% of the applied dose (either 2 μl or 5 μl). A high proportion of this actually penetrated through the skin, suggesting good delivery of diclofenac to the deeper layers of skin and underlying tissues in vivo. After a single dose of the lotion, absorption ki-

Table 4 In Vitro Percutaneous Absorption from Diclofenac Lotion: Multiple Doses versus Single Dose

Parameter measured	A (2 μl, single)	B (5 μl, single)	E (2 μl, multiple)	F (5 μl, multiple)
Receptor fluid	11.4 ± 11.6	13.1 ± 16.6	2.0 ± 1.4	1.9 ± 1.9
Skin residue	11.0 ± 7.8	10.1 ± 7.4	4.4 ± 2.6	6.3 ± 4.0
Stratum corneum	11.6 ± 9.3	11.8 ± 6.5	10.3 ± 5.0	6.0 ± 5.1
Total absorption	33.9 ± 22.3	35.0 ± 23.3	16.7 ± 7.3	14.0 ± 6.9
Skin surface recovery	62.5 ± 28.1	57.8 ± 26.3	84.0 ± 8.5	76.3 ± 9.7
Total recovery	96.4 ± 6.7	92.8 ± 18.6	100.7 ± 7.1	90.6 ± 3.0

Note. Values are percent of the applied dose recovered, mean ± SD, $n = 5$.

Table 5 In Vitro Percutaneous Absorption from an Aqueous Solution: Multiple Doses versus Single Dose

Parameter measured	C (2 μl, single)	D (5 μl, single)	G (2 μl, multiple)	H (5 μl, multiple)
Receptor fluid	7.6 ± 7.9	4.0 ± 3.3	0.3 ± 0.3	1.0 ± 0.7
Skin residue	14.3 ± 2.8	5.0 ± 4.0	1.1 ± 0.6	1.4 ± 0.9
Stratum corneum	5.8 ± 2.8	7.3 ± 3.2	2.5 ± 1.8	3.6 ± 3.1
Total absorption	27.8 ± 5.0	16.3 ± 7.5	3.9 ± 1.2	5.9 ± 3.2
Skin surface recovery	79.5 ± 19.9	94.2 ± 19.1	78.1 ± 9.9	70.8 ± 8.2
Total recovery	107.2 ± 22.7	110.6 ± 15.1	82.0 ± 9.7	76.8 ± 8.2

Note: Values are percent of the applied dose recovered, mean ± SD, $n = 4$ or 5.

netics were rapid over the first 12 h, after which time the rate slowed up to 48 h. This was in contrast to the absorption profile observed when the lotion was applied a multiple number of times. In this case the percentage dose absorbed was lower, but appeared to be linear over the time course of the experiment, and may even have still been increasing at 48 h. The linear kinetics after multiple dosing may be due to the presence of absorption enhancers in the lotion, namely, DMSO. Repeated application of DMSO may keep increasing the transfer of doclofenac through the skin in to the receptor fluid. It may also aid in "dragging" through diclofenac already within the skin from previous applications. Therefore, a multiple-dose regimen appears to produce a more sustained delivery of diclofenac, over 48 h, than that of a single dose. This increasing penetration rate of diclofenac

Table 6 Steady-State Flux and Apparent K_p Values

Formulation dosed	Flux (ng/cm²/h)	K_p (cm/h)
A	71.2 ± 72.7	$4.7 \times 10^{-6} \pm 4.8 \times 10^{-6}$
B	252.2 ± 271.8	$16.8 \times 10^{-6} \pm 18.1 \times 10^{-6}$
C	47.7 ± 49.0	$3.1 \times 10^{-6} \pm 3.2 \times 10^{-6}$
D	63.0 ± 50.7	$4.1 \times 10^{-6} \pm 3.3 \times 10^{-6}$
E	101.8 ± 70.0	$6.7 \times 10^{-6} \pm 4.6 \times 10^{-6}$
F	240.9 ± 235.6	$16.0 \times 10^{-6} \pm 15.7 \times 10^{-6}$
G	16.3 ± 17.7	$1.0 \times 10^{-6} \pm 1.1 \times 10^{-6}$
H	130.9 ± 94.7	$8.7 \times 10^{-6} \pm 6.3 \times 10^{-6}$

Note. Values are mean ± SD, $n = 4$ or 5.

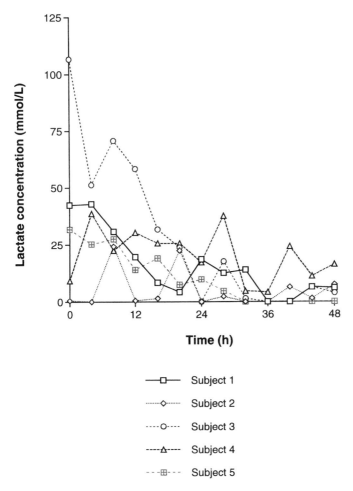

Figure 3 Lactate production and release into the receptor fluid during absorption through human skin in vitro.

sodium has been observed previously in vivo after multiple dosing (3). After multiple applications of diclofenac in a gel to human volunteers, Sioufi et al. (3) showed a prolonged absorption, with an increasing rate up to 24 h in 7 out of 10 subjects. They also reported great variability in plasma concentration levels, which is what was observed in our studies. This could be due to many factors of the individuals skin, namely, skin permeability, which will be dependent upon stratum corneum thickness, age, skin hydration, cutaneous metabolism during penetration, and the general condition of the donor skin (12). In addition, Wester et al. (13) reported a similar affect of azone, which could enhance its own absorption after repeated application to human volunteers, and could potentially accumulate within the skin.

Few reports have been published relating to multidose topical drug administration, despite the clinical and occupational relevance. Several workers showed no significant differences between a single dose and multiple doses in humans for compounds such as malathion, hydrocortisone, testosterone, and estradiol (14,15). Wester et al. (16) showed an increased absorption after long-term administration of hydrocortisone to the rhesus monkey. This has been suggested to be due to the chemical nature of hydrocortisone, which may have caused thinning of the skin during the 14 days of the experiment. Muller et al. (17) recently reported human in vivo studies with topical diclofenac. They concluded that transdermal penetration of diclofenac after single doses is unpredictable, as well as being strongly influenced by individual skin properties.

When diclofenac sodium was applied in an aqueous vehicle, a similar picture was observed as for the lotion used here. The percent dose absorbed after mutiple doses was lower than after a single dose. Again, after a single dose the absorption kinetics showed a slowing in the penetration rate between 12 and 24 h. When applied multiple times the rate of absorption was linear over 48 h. However, there was a much lower absorption into the receptor fluid, skin, and stratum corneum after multidosing compared to that from the lotion. This suggests that the increased absorption of diclofenac in the lotion multidose study was due to a constituent of the formulation (presumably DMSO), and not entirely the dosing regimen. DMSO is an absorption enhancer for compounds such as hydrocortisone (18) and salicylic acid (19) and is thought to function by solvation of the polar head groups of the stratum corneum lipids.

These in vitro data suggest a sustained level of diclofenac penetrating the skin. However, the biological efficacy of this lotion must be determined clinically. If the absorbed diclofenac is taken into the systemic circulation too quickly, then concentrations in the deeper target tissues (i.e., muscle and joints) may be inefficient.

In conclusion, the lotion exhibited enhanced diclofenac percutaneous absorption through human skin when used in a multidose regimen. This higher penetration was also constant over the time period studied, suggesting that under normal use conditions therapeutic levels may be maintained. It should be noted that the clinical regimen for the lotion is the effective multidose regimen shown in this study.

ACKNOWLEDGMENT

This project was funded by Dimethaid Research, Inc., Toronto, Canada.

REFERENCES

1. B. L. Seth. Comparative pharmacokinetics and bioavailability study of percutaneous absorption of diclofenac from two topical formulations containing drug as a solution gel and as an emulsion gel. Arzneim-Forsch. 42:120–122 (1992).

2. H.-O. Ho, F.-C. Huang, T. D. Sokoloski, and M.-T. Shue. The influence of cosolvents on the *in vitro* percutaneous penetration of diclofenac sodium from a gel system. J. Pharm. Pharmacol. 46:636–642 (1994).

3. A. Sioufi, F. Pommier, F. Boschet, J. Godbillion, D. Lauoignat, and D. Salliere. Percutaneous absorption of diclofenac in healthy volunteers after single and repeated topical application of diclofenac emulgel. Biopharm. Drug Dispos. 15:441–449 (1992).

4. S. P. Vyas, R. Singh, and R. K. Asati. Liposomally encapsulated diclofenac for sonophoresis induced systemic delivery. J. Microencapsulation 12:149–154 (1995).

5. T. Nishihata, K. Kotera, Y. Nakano, and M. Yamamazaki. Rat percutaneous transport of diclofenac and influence of hydrogenated soya lectin. Chem. Pharm. Bull. 35:3807–3812 (1987).

6. Y. Obata, K. Takayama, H. Okabe, T. Nagai. Effect of cyclic monoterpenes on percutaneous absorption in the case of a water-soluble drug (diclofenac sodium). Drug Des. Del. 6:319–328 (1991).

7. T. Nishihata, A. Kamada, K. Takahashi, K. Matsumoto, K. Shinozaki, Y. Tabata, M. Keigami, T. Miyagi, and N. Tastumi. Percutaneous absorption of diclofenac in rats and humans: Aqueous gel formulation. Int. J. Pharm. 46: 1–7 (1988).

8. L. N. Hurst, D. H. Brown, and K. A. Murray. Prolonged life and improved quality for stored skin grafts. Plas. Reconstr. Surg., 73:105–109 (1984).

9. R. L. Bronaugh, R. F. Stewart, and J. E. Storm. Extent of cutaneous matabolism during percutaneous absorption of xenobiotics. Toxicol. Appl. Pharmacol. 99: 534–543 (1989).

10. R. C. Wester, J. Christoffel, T. Hartway, N. Poblete, and H. I. Maibach. Cadaver human skin viability for *in vitro* percutaneous absorption: Storage and detrimental effects of heat-separation and freezing. Pharm. Res. 15:82–84 (1998).

11. R. C. Wester, P. K. Noonan, H. I. Maibach. Variations in percutaneous absorption of testosterone in the rhesus monkey due to anatomic site of appliation and frequency of application. Arch. Dermatol. Res. 116:186–188 (1980).

12. R. C. Wester, H. I. Maibach. Cutaneous pharmacokinetics: 10 Steps to percutaneous absorption. Drug Metab. Rev. 14:169–205 (1983).

13. R. C. Wester, J. Melendres, L. Sedik, H. I. Maibach. Percutaneous absorption of azone following single and multiple doses to human volunteers. *J. Pharm. Sci.* 83:124–125 (1994).

14. R. C. Wester, H. I. Maibach, D. A. W. Bucks, R. H. Guy. Malathion percutaneous absorption after repeated administration to man. Toxicol. Appl. Pharmacol. 68:116–119 (1983).

15. D. A. W. Bucks, H. I. Maibach, R. H. Guy. Percutaneous absorption of steroids: Effect of repeated application. J. Pharm. Sci. 74:1337–1339 (1985).

16. R. C. Wester, P. K. Noonan, H. I. Maibach. Percutaneous absorption of hydro-cortisone increases with long-term administration: In vivo studies in the rhesus monkey. Arch. Dermatol. 116:186–188 (1980).

17. M. Muller, H. Mascher, C. Kikuta, S. Schafer, M. Brunner, G. Dorner, H. G. Eichler. Diclofenac Concentrations in defined tissue layers after topical admin-istration. *Clin. Pharmacol. Ther.* 62:293–299 (1997).

18. R. B. Stoughton. Dimethylsulfoxide induction of a steroid reservoir in human skin. Arch. Dermatol. 91:657–660 (1965).

19. W. W. Shen, A. G. Danti, F. N. Bruscato. Effect of nonionic surfactants on percutaneous absorption of salicylic acid and sodium salicylate in the presence of dimethylsulfoxide. J. Pharm. Sci. 65:1780–1783 (1976).

55

An Innovative Method to Determine Percutaneous Absorption
Real-Time Breath Analysis and Physiologically Based Pharmacokinetic Modeling

Karla D. Thrall, Torka S. Poet, and Richard A. Corley
Pacific Northwest National Laboratory, Richland, Washington

From the 1950s through the 1970s, the rate of uptake of a chemical through the skin was generally estimated from studies of humans using radiolabeled compounds (1). More recently, estimates of dermal absorption have been made using animal (in vivo or in vitro) or human (in vitro) skin. In vivo percutaneous absorption in animals is usually determined by measuring radioactivity in blood and excreta following a topical application of the labeled compound. This indirect method of determining percutaneous absorption provides an estimate of the total absorbed dose, but often fails to reveal information on absorption kinetics. To date, the only way to determine the absolute bioavailability of a topically applied compound is to compare the concentrations in blood or urine following topical and intravenous administration (2). Since blood levels may be very low in these situations, this practice is often restricted by the sensitivity limits of the assay or analysis.

As an alternative to conducting radiotracer studies, a new, real-time breath analysis system coupled with physiologically based pharmacokinetic (PBPK) modeling has been developed to determine percutaneous absorption in rats and humans in vivo (3). With this system, a topical dose of a compound can be applied to a specified skin surface area, and the breath analysis system can be used to monitor the concentration of that compound and/or its metabolites in the exhaled breath. A PBPK model, modified to describe the dermal exposure, can than be used to determine the skin permeability constant under various exposure situations.

I. REAL-TIME BREATH ANALYSIS

The analysis of exhaled breath has long been recognized as an ideal non-invasive method for measuring recent as well as past exposures to chemicals (4). Traditional breath analysis techniques have been to collect breath samples in Tedlar bags or stainless steel canisters, then conduct the analysis— generally separation by gas chromatography followed by mass spectrometry (GC/MS). Disadvantages of these techniques include the possibility that the collection device (Tedlar bag or stainless steel canister) may alter the integrity of the sample, the time-consuming and costly analysis subsequent to collection of the sample, and the delay between sample collection and analysis.

To overcome these disadvantages, a breath-inlet device was developed that connects a human subject directly with an ion-trap mass spectrometer system (MS/MS) to provide a continuous, real-time analysis of the individual chemical components in the expired air (3). In operation, a human volunteer exhales through a one-way breathing valve via a large-diameter tube into a heated glass mixing chamber (3 L volume). The MS/MS, equipped with an atmospheric sampling glow discharge ion source (ASGDI), continually draws samples from the mixing chamber for analysis as often as every 1–2 s. Excess exhaled air is vented from the mixing chamber through a large borehole with negligible flow restriction. Since the subject is directly coupled with the analytical system, there is no sample collection device to alter the integrity of the sample. The human breath-inlet device has been modified to allow for the analysis of exhaled air from experimental animals (5).

The ASGDI MS/MS technique can simultaneously monitor for a number of volatile organic compounds directly from the exhaled air matrix by performing chemical ionization in air, thus eliminating the preconcentration step that normally precedes exhaled air analysis by GC/MS (6–8). This analytical method of detection yields sensitivities on the order of 1–10 parts per billion (ppb) for many of the commonly encountered volatile compounds. Gordon et al. (9) give an example of occupational monitoring using the breath-inlet device.

II. PHYSIOLOGICALLY BASED PHARMACOKINETIC MODELING

Physiologically based pharmacokinetic (PBPK) modeling is a toxicological tool that can be used to simulate a chemical exposure (regardless of route) and estimate the amount of the toxic moiety (parent compound, toxic metabolite, etc.) at a particular internal target. These models incorporate physiological parameters (i.e., tissue volumes, breathing rates, blood flow rates, etc.), biochemical parameters (i.e., metabolism), and chemical-specific parameters (i.e., solubility, partition coefficients, etc.) to describe the uptake, disposition, and elimination of a chemical and its metabolites in different tissues. A PBPK model, such as illustrated in Fig. 1, describes the body as a series of tissue compartments representing the probable route(s) of exposure, the metabolically active tissues, the target organs, and the excretion pathways. These models are typically developed and experimentally validated using common laboratory animals, then extrapolated to represent humans.

PBPK models have been developed to describe the dermal absorption of chemicals or chemical vapors (10–12). A PBPK model to describe dermal exposure can be used to determine the skin permeability constant (K_p) under various exposure situations. For vapors, the rate of change of compound in the skin, which is related to the rate of penetration through the skin (flux) and the rate of delivery due to blood flow and arterial concentration

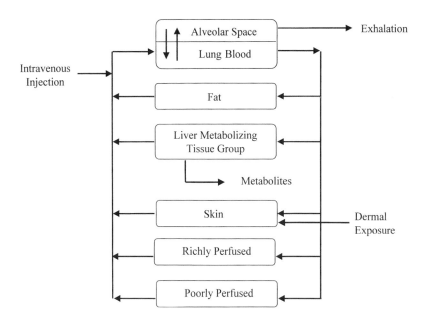

Figure 1 Generic PBPK model used to describe intravenous injection and dermal exposure.

(perfusion), can be described mathematically by the following mass balance equation:

$$V_{sk} \frac{dC_{sk}}{dt} = K_p A \left(C_{air} - \frac{C_{sk}}{P_{sk/air}} \right) + Q_{sk} \left(C_a - \frac{C_{sk}}{P_{sk/b}} \right)$$

where K_p is the permeability constant (cm/h), A is the exposed surface area (cm^2), Q_{sk} is the blood flow rate to the skin (ml/min), C_a is the arterial concentration (μg/ml), C_{sk} is the skin concentration (μg/ml), C_{air} is the ambient air concentration (μg/ml), and $P_{sk/b}$ is the skin:blood partition coefficient (dimensionless). A similar structure is often used to describe aqueous exposures.

III. APPLICATION OF THE SYSTEM FOR ANIMAL AND HUMAN STUDIES

A number of experiments in rats have been conducted using the breath analysis system to monitor the rapid exponential clearance of volatile compounds following dermal exposure to the aqueous compound or the compound in a soil matrix. For example, Fig. 2 demonstrates the real-time evaluation of methyl chloroform (MC) in a rat exposed dermally to 0.3% solvent in a dirt matrix. In this instance, a patch system was developed to maintain the soil in contact with the skin while simultaneously allowing volatilization of compound away from the skin. The day prior to exposure, male F344 rats were anesthetized with a ketamine/xylazine mixture and the hair on the lower back clipper shaved. A ring of DuoDerm (Bristol-Myers Squibb, Princeton, NJ) was attached to the clipped area using a cyanoacrylate adhesive. The topical dose of solvent-spiked soil was applied to the 8-cm^2 area bounded by the DuoDerm, and covered with Bioclusive transparent dressing (Johnson & Johnson, Arlington, TX). The transparent dressing allows free passage of moisture, thereby mimicking the real exposure scenario. To prevent the rats from breathing any compound volatilized from the patch, a weighboat with holes drilled in the center was placed over the transparent dressing and covered with a muslin patch containing activated charcoal (Fig. 3).

Immediately following dermal application of 600 mg soil containing 3 g MC/kg dirt (0.3% MC), the rats were individually placed in off-gassing chambers as described by Gargas et al. (13). Breathing air was continually supplied to the rat through the lid of the off-gassing chamber at a measured rate (200 ml/min). The MS/MS system draws a sample from the off-gassing chamber approximately every 5 s. Intensity data from the MS/MS system were converted to concentration (ppb) through the use of external standards and a calibration curve that was generated each day of experimentation.

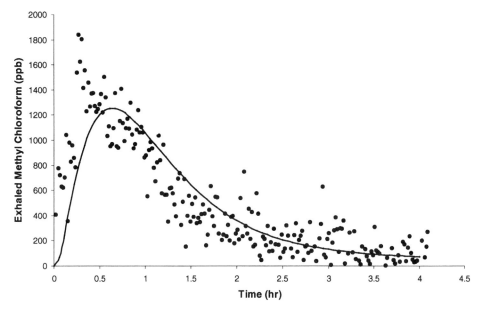

Figure 2 Data (symbols) and simulation (line) of the concentration of methyl chloroform in the exhaled breath of a rat dermally exposed to 600 mg soil containing 3 g methyl chloroform/kg dirt (0.3%). In this case, the PBPK model of Reitz et al. (17) was modified to describe percutaneous absorption.

Postexposure, each part of the patch system (i.e., charcoal, permeable barrier, DuoDerm, remaining dirt solution, etc.) was analyzed using a GC headspace method to quantitate the amount of solvent in each component. The concentration of the chemical in the soil changes over time due to the amount absorbed by the skin, the amount lost to the patch components, and the amount volatilized and trapped on the charcoal patch. Thus, the percutaneous absorption of the chemical can be calculated using Fick's Law with a changing concentration at the skin surface according to the following mass balance equations (14):

$$\frac{d\text{ABSORB}_{\text{skin}}}{dt} = \left(K_{\text{p}} \times \frac{\text{SA}_{\text{skin}}}{1000} \right) \times \left(\text{CONC}_{\text{surf}} - \frac{\text{CONC}_{\text{skin}}}{P_{\text{sk/b}}} \right)$$

$$\text{CONC}_{\text{surf}} = \text{CONC}_{\text{initial}} - \frac{(\text{ABSORB}_{\text{skin}} + \text{ABSORB}_{\text{patch}} + \text{ABSORB}_{\text{trap}})}{V_{\text{dirt}}}$$

Thus, a dermal PBPK model can be used to estimate percutaneous absorption when loss of the exposure compound via volatilization or to the patch system is adequately described. As shown in Fig. 2, the concentration

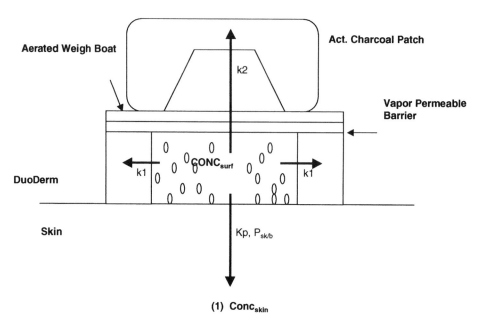

Figure 3 Diagram of the nonocclusive patch system developed for use for rat and human dermal exposures to volatile organic compounds in a soil matrix.

of methyl chloroform in the exhaled breath of a rat dermally exposed to 600 mg soil containing a 0.3% methyl chloroform is well described by a methyl chloroform PBPK model (17) modified to describe percutaneous absorption and the patch exposure system. Furthermore, this figure illustrates that methyl chloroform is rapidly absorbed through the skin of rats, with a peak concentration occurring within 20–50 min of exposure (Fig. 2).

The utility of the real-time breath analysis system to evaluate human exposures coupled with PBPK modeling is demonstrated in Fig. 4. In this study, a human volunteer was immersed in a hot bath filled with house tap water (assayed to contain 70 ppb chloroform) and was supplied with breathing air through a face mask (15). As the figure illustrates, a chloroform PBPK model (16) can be used to simulate the exposure and provide a prediction of the exhaled breath levels that agree well with the actual measured data.

IV. SUMMARY AND CONCLUSIONS

The sensitivity and specificity afforded by the ASGDI—MS/MS breath analysis methodology creates new opportunities for research into the dermal

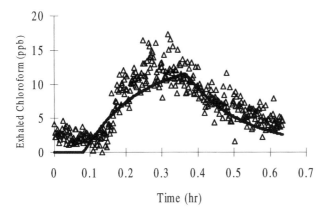

Figure 4 Data (symbols) and simulation (line) of the concentration of chloroform in the exhaled breath of a human volunteer before, during, and after a 20-min skin-only exposure to 70 ppb chloroform via bathwater. Simulations were conducted using an increased blood flow to the skin as a result of the increased water temperature (40°C).

bioavailability of volatile organic compounds at relevant environmental exposure conditions in both experimental animals and humans. Furthermore, the technique eliminates the need for application of radiolabeled tracer chemicals. The ability to measure multiple compounds at low levels in exhaled breath in real-time also provides unique opportunities to evaluate the kinetics of absorption and metabolism, as well as the potential for mixture interactions. In addition to tracking the exhalation of a parent compound or metabolite following exposure, volatile biomarkers of response can be simultaneously analyzed. For example, Fig. 5 illustrates the real-time analysis of carbon tetrachloride, its metabolite (chloroform), and ethane and acetone as biomarkers of lipid peroxidation (biological response) in the exhaled breath collected from rats receiving a 25-mg/kg body weight oral gavage dose of carbon tetrachloride.

PBPK modeling is ideally suited to provide the linkage between the kinetics of chemicals appearing in exhaled breath and the process of percutaneous absorption. Exhaled breath concentration data can be analyzed with a PBPK model to provide a realistic estimate of skin permeability constants. Once the dermal PBPK models are adequately validated, target tissue dose estimates of a chemical or its metabolite can be simulated for relevant dermal exposure scenarios. Thus, the combination of advancements made in real-time breath analysis, along with PBPK modeling, provides a unique opportunity to evaluate the importance of percutaneous absorption

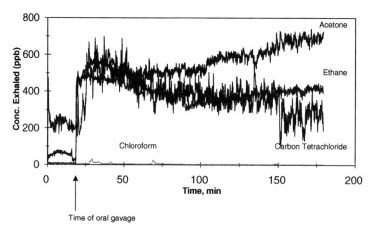

Figure 5   Real-time exhaled breath concentrations of carbon tetrachloride, chloroform, acetone, and ethane after an oral gavage dose of carbon tetrachloride (25 mg/kg in corn oil) to rats.

of chemicals and ultimately improve the scientific basis of human health risk assessments.

REFERENCES

1. Paustenbach, D. J., and Leung, H.-W. (1993). Techniques for assessing the health risks of dermal contact with chemicals in the environment. In: Health Risk Assessment. Dermal and Inhalation Exposure and Absorption of Toxicants, eds. R. G. M. Wang, J. B. Knaak, and H. I. Maibach, pp. 343–386. Boca Raton, FL: CRC Press.
2. Wester, R. C., and Maibach, H. I. (1993). Percutaneous absorption. In: Health Risk Assessment. Dermal and Inhalation Exposure and Absorption of Toxicants, eds. R. G. M. Wang, J. B. Knaak, and H. I. Maibach, pp. 63–74. Boca Raton, FL: CRC Press.
3. Kenny, D. V., and Kelly, T. J. (August, 1991). Apparatus and method for analysis of exhaled breath. U. S. Patent 5,042,501.
4. Fiserova-Bergerova, V., Pierce, J. T., and Droz, P. O. (1990). Dermal absorption potential of industrial chemicals: Criteria for skin notation. Am. J. Ind. Med. 17:617–635.
5. Thrall, K. D., and Kenny, D. V. (1996). Evaluation of a carbon tetrachloride physiologically based pharmacokinetic model using real-time breath-analysis monitoring of the rat. Inhal. Toxicol. 8:251–261.
6. Gordon, S. M., Wallace, L. A., Pellizzari, E. D., and O'Neill, H. J. (1988). Human breath measurements in a clean-air chamber to determine half-lives for volatile organic compounds. Atmos. Environ. 22:2165–2170.

7. Raymer, J. H., Pellizzari, E. D., Thomas, K. W., and Cooper, S. D. (1991). Elimination of volatile organic compounds in breath after exposure to occupational and environmental microenvironments. J. Exp. Anal. Environ. Epidemiol. 1:439–451.

8. Thomas, K. W., Pellizzari, E. D., and Cooper, S. D. (1991). A canister-based method for collection and GC/MS analysis of volatile organic compounds in human breath. J. Anal. Toxicol. 15:54–59.

9. Gordon, S. M., Kenny, D. V., and Kelly, T. J. (1992). Continuous real-time breath analysis for the measurement of half-lives of expired volatile organic compounds. J. Exp. Anal. Environ. Epidem. *Suppl.* 1:41–54.

10. McDougal, J. N., Jepson, G. W., Clewell, H. J. III, MacNaughton, M. G., and Andersen, M. E. (1986). A physiological pharmacokinetic model for dermal absorption of vapors in the rat. Toxicol. Appl. Pharmacol. 84:286–294.

11. Knaak, J. B., Al-Bayati, M. A., and Raabe, O. G. Physiologically based pharmacokinetic modeling to predict tissue dose and cholinesterase inhibition in workers exposed to organophosphorus and carbamate pesticides. In: Health Risk Assessment. Dermal and Inhalation Exposure and Absorption of Toxicants, eds. R. G. M. Wang, J. B. Knaak, and H. I. Maibach, pp. 3–29. Boca Raton, FL: CRC Press.

12. McKone, T. E. (1993). Linking a PBPK model for chloroform with measured breath concentrations in showers: Implications for dermal exposure models. J. Exp. Anal. Environ. Epidemiol. 3(3):339–365.

13. Gargas, M. L. (1990). An exhaled breath chamber system for assessing rates of metabolism and rates of gastrointestinal absorption with volatile compounds. J. Am. Coll. Toxicol. 9(4):447–453.

14. Jepson, G. W., and McDougal, J. N. (1997). Physiologically based modeling of nonsteady state dermal absorption of halogenated methanes from an aqueous solution. Toxicol. Appl. Pharmacol. 144:315–324.

15. Gordon, S. M., Callahan, P. J., Kenny, D. V., Brinkman, M. C., and Wallace, L. A. (1996). U.S. EPA Internal Grant 0608 (L. Wallace, Principal Investigator). Unpublished data.

16. Corley, R. A., Mendrala, A. L., Smith, F. A., Staats, D. A., Gargas, M. L., Conolly, R. B., Andersen, M. E., and Reitz, R. H. (1990). Development of a physiologically based pharmacokinetic model for chloroform. Toxicol. Appl. Pharmacol. 103:512–527.

17. Reitz, R. H., McDougal, J. N., Himmelstein, M. W., Nolan, R. J., and Schumann, A. M. (1988). Physiologically based pharmacokinetic modeling with methyl chloroform: Implications for interspecies, high dose/low dose, and dose-route extrapolations. Toxicol. Appl. Pharmacol. 95:185–199.

56
Role of Calcium Ions in Relation to Skin Barrier Function

Hanafi Tanojo and Howard I. Maibach
*University of California School of Medicine,
San Francisco, California*

I. INTRODUCTION

Recently many publications have shown the importance of calcium in the barrier regulation of skin. As the skin barrier properties become an essential subject for the study of the percutaneous absorption, it would be useful to understand the mechanism of calcium regulation in skin. Apart from the skin, this ion plays a crucial role in various processes in the body, including the growth, death, differentiation, and function of immune cells. The role of calcium in skin is incompletely understood, yet there are indications that calcium in skin may play a major role.

II. CALCIUM SIGNALING

On both an atom and weight basis, calcium, in the form of the ion Ca^{2+}, is the most abundant metal ion and fifth (after H, O, C, and N) most abundant element in the body. Over 90% of body calcium resides in bones and tooth enamel. The rest, described as mobile Ca^{2+}, is found throughout body fluids and takes part in various processes, including muscle contraction, blood clotting, nerve excitability, intercellular communication, membrane transport of molecules, hormonal responses, exocytosis, and cell fusion, adhesion, and growth (Sigel, 1984).

Calcium is a universal messenger, even in simple organisms and plants. The combination of its ionic radius and double charge may allow it tighter binding to receptors to the exclusion of other ions such as magnesium, leading to strong, specific binding (Carafoli and Penniston, 1985). The specificity enables cells to form special receptors to assess signals from calcium. For many parts of the body, Ca^{2+} often acts as a second messenger in a manner similar to cAMP. Transient increases in cytosolic Ca^{2+} concentration trigger numerous cellular responses including muscle contraction, release of neurotransmitters, and glycogen breakdown (glycogenolysis), also as an important activator of oxidative metabolism (Voet and Voet, 1990 pp. 496–498). Ca^{2+} does not need to be synthesized and degraded with each message transmission, so it is an energy-efficient signal for the cell (Fairley, 1991).

In skin, calcium can provide signals for the cells, either extracellular or intracellular (in the cytosol). The extra- and intracellular signaling may be connected to each other, but may also act separately. In cultured keratinocytes, extracellular calcium levels influence growth and differentiation (Hennings et al., 1980; Pillai et al., 1988). Low extracellular calcium levels (<0.1 mM) induce the growth of keratinocytes as a monolayer with a high proliferation rate, rapidly becoming confluent. In this condition keratinocytes never stratify, but possess many of the characteristics of basal cells; the cells synthesize keratin proteins and are connected by occasional gap junctions but not by desmosomes. High extracellular calcium levels (≥ 1 mM) induce differentiation of keratinocytes. Keratinocytes rapidly flatten, form desmosomes, and differentiate with stratification. Moreover, cornified envelopes form in cells of the uppermost layers (Hennings et al., 1980; Pillai et al., 1988).

The response to signaling is also shown in a progressive way. Keratinocytes grown in a low-calcium media proliferate. Increased extracellular Ca^{2+} inhibits proliferation, while it induces differentiation (Hennings et al., 1983). On the other hand, differentiation of keratinocytes caused a decrease in responsiveness to extracellular calcium, which may facilitate the maintenance of the high level of intracellular calcium required for differentiation (Bikle et al., 1996).

Intracellular Ca^{2+} increases with raised extracellular Ca^{2+} (Pillai and Bikle, 1989; Kruszewski et al., 1991; Reiss et al., 1992). This implies that increased intracellular Ca^{2+} is the actual signal to trigger keratinocyte differentiation. Intracellular Ca^{2+} signals are assessed through calcium-binding proteins to induce responses. The major calcium-binding protein in skin is calmodulin. Calmodulin regulates target protein by modulating protein–protein interactions in a calcium-dependent way. Calmodulin regulates many enzymes, for example, adenyl and guanyl cyclase, phosphodiesterase, ornithine decarboxylase, calcium-calmodulin-dependent protein kinase, transglutaminase, and phospholipase, which are also found in skin (Fairley, 1991).

III. REGULATION OF CALCIUM

The regulation of calcium in skin shows an ingenious adaptation of living organisms to the presence of ions. As Ca^{2+} cannot be metabolized like other second-messenger molecules, cells tightly regulate intracellular levels through numerous binding and specialized extrusion proteins (Clapham, 1995). The concentration of calcium in extracellular spaces (generally ~ 1.5 mM) is four orders of magnitude higher than in the cytosol (~ 0.1 μM). In excitable cells, for example, muscle cells, the extracellular concentration of calcium must be closely regulated to keep it at its normal level of ~ 1.5 mM, so that it cannot accidentally trigger the muscle contraction, the transmission of nerve impulses, and blood clotting (Voet and Voet, 1990, p. 1144). In other cells, including keratinocytes, the extracellular level is maintained in a specific equilibrium with the intracellular concentration.

Why does intracellular calcium level have to be maintained low? A low calcium concentration makes the use of the ion as an intracellular messenger energetically inexpensive. The movement of calcium ions across membranes requires energy, usually supplied by ATP. If the resting level of calcium in the cell were high, a large number of ions would need to be transported into the cytoplasm to raise the concentration by the factor of 10 that is ordinarily needed to activate an enzyme; afterward the excess calcium would have to be expelled from the cell. The normally low calcium level means that relatively few ions need to be moved, with a relatively small expenditure of energy, to regulate an enzyme. In contrast, energetic cost of regulation by the other important intracellular messenger, cyclic adenosine monophosphate (cyclic AMP), is high; it must be synthesized and broken down each time it carries a message, and both steps requires a significant investment of energy (Carafoli and Penniston, 1985). Furthermore, low intracellular calcium is a necessary condition for the phosphate-driven metabolism characteristic of higher organisms. The energy-rich fuel for most cellular processes is adenosine triphosphate (ATP). Its breakdown releases inorganic phosphate. If the intracellular concentration of calcium were high, the phosphate and the calcium would combine to form a precipitate of hydroxyapatite crystals—the same stony substance found in bone. Ultimately calcification would doom the cell (Carafoli and Penniston, 1985). This is likely the case with long-term occupational exposure to high levels of dissolved calcium, for example, in miners (Sneddon and Archibald, 1958), agricultural laborers (Christensen, 1978), and oil field workers (Wheeland and Roundtree, 1985), which can result in calcinosis cutis, a benign and reversible hardening of the exposed skin.

The large concentration gradient between extracellular spaces and cytosol is maintained by the active transport of Ca^{2+} across the plasma membrane, the endoplasmic reticulum (or the sarcoplasmic reticulum in muscle), and the mitochondrial inner membrane. Generally, plasma membrane and endoplasmic reticulum each contain a Ca^{2+}-ATPase that actively pumps Ca^{2+} out of the cytosol at the expense of ATP hydrolysis (Voet and Voet, 1990, pp. 496–498). Mitochondria act as a "buffer" for cytosolic Ca^{2+}: If cytosolic concentration of calcium rises, the rate of mitochondrial Ca^{2+} influx increases while that of Ca^{2+} efflux remains constant, causing the mitochondrial concentration of calcium to increase, while the cytosolic concentration of calcium decreases to its original level (its setpoint). Conversely, a decrease in cytosolic concentration of calcium reduces the influx rate, causing net efflux of concentration of calcium and an increase of cytosolic concentration of calcium back to the setpoint (Voet and Voet, 1990, p. 531).

Besides the already mentioned Ca^{2+}-ATPase, the transport of Ca^{2+} is regulated by a series of calcium pumps, transport systems and ion channels. The availability of certain regulatory systems is dependent on the activity of the cells. In excitable cells such as cardiac muscle, the influx of Ca^{2+} to cytosol is regulated by voltage- (or potential-) dependent channels, while the efflux (out of cytosol) is regulated by cation exchanger, such as Na^+–Ca^{2+} exchanger (Fairley, 1991). Undifferentiated keratinocytes in the basal layer have different sets of Ca^{2+} transport system than differentiated cells in the upper layers. In basal layer, the system consists of 14-pS non-specific cation channels (NSCC) (Mauro et al., 1993) and does not possess functional voltage-sensitive Ca^{2+} channels (Reiss et al., 1992). Differentiated keratinocytes are likely to possess at least two and possibly three pathways of Ca^{2+} influx: a) nicotinic channel (nAChR); b) voltage-sensitive Ca^{2+} channels (VSCC, which can be blocked by nifedipine or verapamil); and c) NSCC, which is not activated by nicotine (Grando et al., 1996).

IV. CALCIUM GRADIENT

As mentioned earlier, there is a high calcium gradient between extra- and intracellular domains of keratinocytes, which requires tight regulation. Moreover, a calcium gradient is present within the epidermis, with higher quantities of Ca^{2+} in the upper than in the lower epidermis (as the cell moves from the basal layer to the stratum granulosum) (Menon et al., 1985). Ca^{2+} concentration increases steadily from the basal region to stratum corneum, while this is not the case with other ions (Forslind et al., 1995). Figure 1 illustrates the calcium gradient in human skin in comparison with an actual literature data (Malmqvist et al., 1987). Such a gradient is not observed in skin abnormalities related to the formation of abnormal barrier function, such as psoriasis (Menon and Elias, 1991). Studies in mice and rats showed that this gradient exists at the same time as the formation of a maturing skin

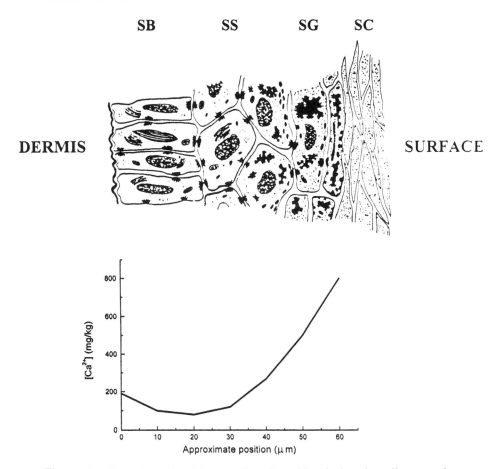

Figure 1 Illustration of calcium gradient in epidermis based on literature data (proton induced x-ray emission analysis of calcium in sectioned human skin) (Malmqvist et al., 1987). SB, stratum basale/basal layer; SS; stratum spinosum; SG, stratum granulosum; SC, stratum corneum.

barrier at the end of gestation. The gradient is then maintained from the newborn throughout the adult life (Elias et al., 1998).

It is not yet clear whether the calcium gradient leads to the formation of a mature barrier or the barrier caused the gradient. It may even be both, if the regulation uses a feedback mechanism, as the differentiation will eventually form a barrier leading to the accumulation of calcium ions in the upper epidermis. This high level of calcium will, in turn, guarantee the ongoing process of differentiation toward the formation of corneocytes (horny cells in the stratum corneum). The mechanism is thus almost completely autonomous, perpetual, and, if it runs smoothly, requires little correction from the body.

V. CALCIUM AND BARRIER REPAIR MECHANISM

Disruption of the barrier with acetone treatment or tape stripping depletes Ca^{2+} from the upper epidermis, resulting in the loss of the Ca^{2+} gradient (Menon et al., 1992; Mauro et al., 1996; Man et al., 1997). This is due to accelerated water transit that leads to the increased passive loss of Ca^{2+} into and through the stratum corneum (Menon et al., 1992; Man et al., 1997). One in vitro study showed that the permeability of human stratum corneum to Ca^{2+} dramatically increased after the stratum corneum was pretreated with acetone or sodium lauryl sulfate solution (Tanojo et al., in press). The decrease in Ca^{2+} levels in the outer epidermis is associated with enhanced lamellar body secretion and lipid synthesis (important components in repair responses) (Lee et al., 1992; Menon et al., 1992). However, if Ca^{2+} gradient is preserved by the addition of Ca^{2+} into the media, lamellar body secretion, lipid synthesis, and barrier recovery are inhibited (Lee et al., 1992). The inhibition raised by high extracellular concentration of calcium is potentiated by high extracellular K^+ (Lee et al., 1994). Another study confirmed that barrier recovery is accelerated by the low concentrations of calcium and also potassium during an increased water loss, since water loss may induce a decrease in the concentration of Ca^{2+} in the upper epidermis, which, in turn, may stimulate lamellar body secretion and barrier repair (Grubauer et al., 1987). Furthermore, the inhibition raised by high extracellular concentration of calcium is reversed by nifedipine or verapamil, specific calcium channel blockers (Lee et al., 1994). In another study, administration of Ca^{2+} free solutions by sonophoresis resulted in a marked decrease in Ca^{2+} content in the upper epidermis, and subsequently the loss of the Ca^{2+} gradient was accompanied by accelerated lamellar body secretion (a sign of skin barrier repair) (Menon et al., 1994). The process of barrier repair in connection with transepidermal water loss and calcium gradient is illustrated in Figure 2. Other aspects of the skin barrier homeostasis have been extensively reviewed in Feingold (1997).

VI. MECHANISM OF CELL SIGNALING

Both intracellular release and transmembrane flux contribute to the rise in intracellular Ca^{2+} (Kruszewski et al., 1991; Reiss et al., 1992). The rise in keratinocyte intracellular Ca^{2+} in response to raised extracellular Ca^{2+} has two phases: a) an initial peak, not dependent on extracellular Ca^{2+}, and

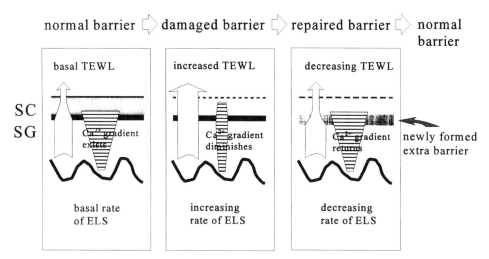

Figure 2 Illustration of skin barrier repair in epidermis. SC, stratum corneum; SG, stratum granulosum; TEWL, transepidermal water loss; ELS, epidermal lipid synthesis.

b) a later phase that requires extracellular Ca^{2+} (Kruszewski et al., 1991). An early response of human keratinocytes to increases in extracellular Ca^{2+} is an acute increase in intracellular Ca^{2+}. Stepwise addition of extracellular Ca^{2+} to neonatal human keratinocytes is followed by a progressive increase in intracellular Ca^{2+}, where the initial spike of increased intracellular Ca^{2+} is followed by a prolonged plateau of higher intracellular Ca^{2+} (Pillai and Bikle, 1991). The response of intracellular Ca^{2+} to increased extracellular Ca^{2+} in keratinocytes is saturated at 2.0 mM extracellular Ca^{2+} (Sharpe et al., 1989; Pillai and Bikle, 1991). The response of intracellular Ca^{2+} to increased extracellular Ca^{2+} in keratinocytes resembles the response in parathyroid cells, in that a rapid and transient increase in intracellular Ca^{2+} is followed by a sustained increase in intracellular Ca^{2+} above basal level. This multiphasic response is attributed to an initial release of Ca^{2+} from intracellular stores followed by an increased influx of Ca^{2+} through voltage-independent cation channels. The keratinocyte and parathyroid cell contains a similar cell membrane calcium receptor thought to mediate this response to extracellular Ca^{2+}. This receptor can activate the phospholipase-C pathway, leading to an increase in the levels of inositol 1,4,5-triphosphate (IP_3) and sn-1,2-diacylglycerol (DAG)—both of which are important messengers—as well as stimulating Ca^{2+} influx and chloride currents (Shoback et al., 1988; Brown et al., 1990). IP_3 causes release of Ca^{2+} from internal stores, such as endoplasmic reticulum, further increasing intracellular level to precede a number of calcium-stimulated cellular events (Berridge and Irvine, 1984). DAG forms a quatery complex with phosphatidylserine, calcium, and protein kinase C to activate the kinase. This will accelerate terminal differentiation (Hennings et al., 1983). The signal tranduction mediated through calmodulin induces other proteins, for example, desmocalmin, which is associated with the formation of desmosomes (Tsukita and Tsukita, 1985).

Keratinocytes grown in low-calcium medium (0.02 mM) maintained intracellular calcium levels adequate for arachidonic acid metabolism and actually showed increased prostaglandin (PGE$_2$ and PGF$_{2\alpha}$) production up to 4.5 times compared to cells grown at normal calcium level (1.2 mM) (Fairley et al., 1988). If this is true for the in vivo condition, a low level of extracellular calcium—for instance, due to a defective skin barrier—may cause an increase in prostaglandin synthesis, leading to hyperproliferative epidermal disorders, such as psoriasis, which are often associated with abnormalities in prostaglandin production (Hammarström et al., 1979).

VII. TRANSPORT OF CALCIUM IN THE SKIN

The permeability of skin to Ca^{2+} ions has been known from some dermatoses, such as calcinosis cutis (Sneddon and Archibald, 1958; Christensen, 1978; Wheeland and Roundtree, 1985) and perforating verruciform collagenoma (Moulin et al., 1995). In a shorter term, calcinosis cutis developed after a 24 h (at least) topical application of an electrode paste containing saturated calcium chloride solution, bentonite, and glycerin, used for examination by electroencephalography or electromyography (Mancuso et al., 1990; Johnson et al., 1993). The permeability of human skin to Ca^{2+} ions in vitro shows a marked dependence upon anatomic site. In agreement with the data observed for nonelectrolytes, permeation decreased in the following order: foreskin > mammary > scalp > thigh. Mouse and guinea pig skin show comparable permeability to that of human scalp. Ca^{2+} transport from dermis across epidermis is higher than that from epidermis to dermis (Stüttgen and Betzler, 1956, 1957).

Recently, a novel technique was developed to continuously monitor the low level of Ca^{2+} flux across human stratum corneum in vitro (Tanojo et al., in press). The technique utilizes an ion-selective glass microelectrode containing a highly selective Ca^{2+} ionophore (type 1 cocktail A, Fluka). The microelectrode was positioned 50 μm above the surface of the stratum corneum and scanned over a specific area to measure the Ca^{2+} current. Using the technique, Ca^{2+} diffusion was quantified as a function of time for up to 3 h, while measurements could be made every second, the spatial resolution was 2–4 μm, and the minimum concentration reliably detectable was 10^{-5}

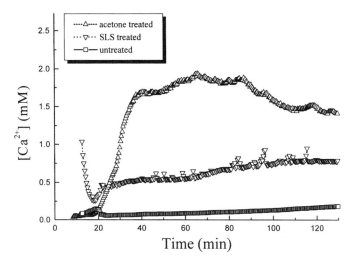

Figure 3 In vitro calcium flux across human stratum corneum, untreated and pre-treated with acetone or 10% sodium lauryl sulfate.

M calcium ion. The study showed that the flux through untreated human stratum corneum was sigmoidal. The steady-state flux had an average of 7 $\times$ 10^{-12} mol/cm^2/s. After the stratum corneum was pretreated with acetone or sodium lauryl sulfate, the shape of the curve was similar but the Ca^{2+} flux was significantly higher (Figure 3). This method may be useful in examining the changes in Ca^{2+} flux that take place during barrier formation and repair (Tanojo et al., in press).

VIII. CONCLUSION

Calcium ions play an important role in the homeostasis of skin barrier. A change in the barrier will change the calcium ion gradient in skin and lead to barrier repair process. A severe change might lead into a high degree of calcium signaling, which may induce the activation of various processes, from increased synthesis of skin components or messengers to the inflammatory reactions. Hence, the homeostasis of calcium in the skin should be taken into consideration while administering drugs into the skin.

ACKNOWLEDGMENT

We thank Dr. Theodora Mauro for tutorial assistance.

REFERENCES

Berridge, M. J., and Irvine, R. F. (1984). Inositol triphosphate, a novel second messenger in cellular signal transduction. Nature (London) 312:315–321.

Bikle, D. D., Ratnam, A., Mauro, T. M., Harris, J., and Pillai, S. (1996). Changes in calcium responsiveness and handling during keratinocyte differentiation. J. Clin. Invest. 97:1085–1093.

Brown, E. M., Chen, C. J., Kifor, O., Leboff, M. S., El-Hajj, G., Fajtova, V., and Rubin, L. T. (1990). Ca^{2+}-sensing, second messengers, and the control of parathyroid hormone secretion. Cell Calcium 11:333–337.

Carafoli, E., and Penniston, J. T. (1985). The calcium signal. Sci. Am. 253:70–78.

Christensen, O. B. (1978). An exogenous variety of pseudoxanthoma elasticum in old farmers. Acta Derm. Venereol. (Stockh.) 58:319–321.

Clapham, D. E. (1995). Calcium signaling. Cell 80:259–268.

Elias, P. M., Nau, P., Hanley, K., Cullander, C., Crumrine, D., Bench, G., Sideras-Haddad, E., Mauro, T. M., Williams, M. L., and Feingold, K. R. (1998). Formation of the epidermal calcium gradient coincides with key milestones of barrier ontogenesis in the rodent. J. Invest. Dermatol. 110:399–404.

Fairley, J. A. (1991). Calcium: A second messenger: In: Goldsmith, L. A., Ed., Physiology, Biochemistry, and Molecular Biology of the Skin, 2nd ed., Oxford University Press, New York, pp. 314–328.

Fairley, J. A., Weiss, J., and Marcelo, C. L. (1988). Increased prostaglandin synthesis by low calcium-regulated keratinocytes. J. Invest. Dermatol. 86:173–176.

Feingold, K. R. (1997). Permeability barrier homeostasis: its biochemical basis and regulation. Cosmet. Toilet. 112:49–59.

Forslind, B., Lindberg, M., Malmqvist, K. G., Pallon, J., Roomans, G. M., and Werner-Linde, Y. (1995). Human skin physiology studied by particle probe microanalysis. Scanning Microsc. 9:1011–1026.

Grando, S. A., Horton, R. M., Mauro, T. M., Kist, D. A., Lee, T. X., and Dahl, M. V. (1996). Activation of keratinocyte nicotinic cholinergic receptors stimulates calcium influx and enhances cell differentiation. J. Invest. Dermatol. 107:412–418.

Grubauer, G., Feingold, K. R., and Elias, P. M. (1987). Relationship of epidermal lipogenesis to cutaneous barrier function. J. Lipid Res. 28:746–752.

Hammarström, S., Lindgren, J. A., Marcelo, C. L., Duell, E. A., Anderson, T. F., and Voorhees, J. J. (1979). Arachidonic acid transformations in normal and psoriatic skin. J. Invest. Dermatol. 73:180–183.

Hennings, H., Michael, D., Cheng, C., Steinert, P., Holbrook, K. A., and Yuspa, S. H. (1980). Calcium regulation of growth and differentiation of mouse epidermal cells in culture. Cell 19:245–254.

Hennings, H., Holbrook, K. A., and Yuspa, S. H. (1983). Factors influencing calcium-induced terminal differentiation in cultured mouse epidermal cells. J. Cell. Physiol. 116:265–281.

Johnson, R. C., Fitzpatrick, J. E., and Hahn, D. E. (1993). Calcinosis cutis following electromyographic examination. Cutis 52:161–164.

Kruszewski, F. H., Hennings, H., Yuspa, S. H., and Tucker, R. W. (1991). Regulation of intracellular free calcium in normal murine keratinocytes. Am. J. Physiol. 261:C767–C773.

Lee, S. H., Elias, P. M., Proksch, E., Menon, G. K., Man, M.-Q., and Feingold, K. R. (1992). Calcium and potassium are important regulators of barrier homeostatis in murine epidermis. J. Clin. Invest. 89:530–538.

Lee, S. H., Elias, P. M., Feingold, K. R., and Mauro, T. M. (1994). A role for ions in barrier recovery after acute perturbation. J. Invest. Dermatol. 102:976–979.

Malmqvist, K. G., Forslind, B., Themner, K., Hyltén, G., Grundin, T., and Roomans, G. M. (1987). The use of PIXE in experimental studies of the physiology of human skin epidermis. Biol. Trace Elem. Res. 12:297–308.

Man, M.-Q., Mauro, T. M., Bench, G., Warren, R., Elias, P. M., and Feingold, K. R. (1997). Calcium and potassium inhibit barrier recovery after disruption, independent of the type of insult in hairless mice. Exp. Dermatol. 6:36–40.

Mancuso, G., Tosti, A., Fanti, P. A., Berdondini, R. M., Mongiorgi, R., and Morandi, A. (1990). Cutaneous necrosis and calcinosis following electroencephalography. Dermatologica 181:324–326.

Mauro, T. M., Isseroff, R. R., Lasarow, R., and Pappone, P. A. (1993). Ion channels are linked to differentiation in keratinocytes. J. Membr. Biol. 132:201–209.

Mauro, T. M., Rassner, U., Bench, G., Feingold, K. R., Elias, P. M., and Cullander, C. (1996). Acute barrier disruption causes quantitative changes in the calcium gradient. J. Invest. Dermatol. 106:919.

Menon, G. K., and Elias, P. M. (1991). Ultrastructural localization of calcium in psoriatic and normal human epidermis. Arch. Dermatol. 127:57–63.

Menon, G. K., Grayson, S., and Elias, P. M. (1985). Ionic calcium reservoirs in mammalian epidermis: Ultrastructural localization by ion-capture cytochemistry. J. Invest. Dermatol. 84:508–512.

Menon, G. K., Elias, P. M., Lee, S. H., and Feingold, K. R. (1992). Localization of calcium in murine epidermis following disruption and repair of the permeability barrier. Cell Tissue Res. 270:503–512.

Menon, G. K., Price, L. F., Bommannan, B., Elias, P. M., and Feingold, K. R. (1994). Selective obliteration of the epidermal calcium gradient leads to enhanced lamellar body secretion. J. Invest. Dermatol. 102:789–795.

Moulin, G., Balme, B., Musso, J., and Thomas, L. (1995). Perforating verruciform collagenoma, an exogenous inclusion-linked dermatosis? Report of one case induced by calcium chloride. Ann. Dermatol. Venereol. 122:591–594.

Pillai, S., and Bikle, D. D. (1989). A differentiation-dependent, calcium-sensing mechanism in normal human keratinocytes. J. Invest. Dermatol. 92:500.

Pillai, S., and Bikle, D. D. (1991). Role of intracellular-free calcium in the cornified envelope formation of keratinocytes: Differences in the mode of action of extracellular calcium and 1,25-dihydroxyvitamin D. J. Cell. Physiol. 146: 94–100.

Pillai, S., Bikle, D. D., Hincenbergs, M., and Elias, P. M. (1988). Biochemical and morphological characterization of growth and differentiation of normal human neonatal keratinocytes in a serum-free medium. J. Cell. Physiol. 134: 229–237.

Reiss, M., Lipsey, K. R., and Zhou, Z. L. (1992). Extracellular calcium-dependent regulation of transmembrane calcium fluxes in murine keratinocytes. J. Cell. Physiol. 147:281–291.

Sharpe, G. R., Gillespie, J. I., and Greenwell, J. R. (1989). An increase in intracellular free calcium is an early event during differentiation of cultured keratinocytes. FEBS Lett. 254:25–28.

Shoback, D. M., Membreno, L. A., and McGhee, J. G. (1988). High calcium and other divalent cations increase inositol trisphosphate in bovine parathyroid cells. Endocrinology 123:382–389.

Sigel, H., Ed. (1984). Metal Ions in Biological Systems: Calcium and Its Role in Biology, vol. 17. Marcel Dekker, New York.

Sneddon, I. B., and Archibald, R. M. (1958). Traumatic calcinosis of the skin. Br. J. Dermatol. 70:211–214.

Stüttgen, G., and Betzler, H. (1956). Zur Frage der Permeation von Elektrolyten durch die Haut. I. Mitteilung: Vitroversuche mit radioaktivmarkierten Ca^{++}, SO_4^{--}, und $PO_4^{(3-)}$-Ionen an Meerschweinchen- und Mäusehaut. Arch. Klin. Exp. Derm. 203:472–482.

Stüttgen, G., and Betzler, H. (1957). Zur Frage der Permeation von Elektrolyten durch die Haut. II. Mitteilung: In vitro- und vivo-Versuche an menschlicher Haut mit $^{45}Ca^{++}$. Arch. Klin. Exp. Derm. 204:165–174.

Tanojo, H., Cullander, C., and Maibach, H. I. (in press). Monitoring the permeation of calcium ion across human stratum corneum using an ion-selective microelectrode with high spatial resolution. In: Brain, K. R., and Walters, K. A., Eds., Perspectives in Percutaneous Penetration, vol. 6b. STS Publishing, Cardiff.

Tsukita, S., and Tsukita, S. (1985). Desmocalmin: A calmodulin-binding high molecular weight protein isolated from desmosomes. J. Cell Biol. 101:2070–2080.

Voet, D., and Voet, J. G. (1990). Biochemistry. John Wiley & Sons, New York.

Wheeland, R. G., and Roundtree, J. M. (1985). Calcinosis cutis resulting from percutaneous penetration and deposition of calcium. J. Am. Acad. Dermatol. 12:172–175.

Index